T0258474

# THE ORIGIN AND NATURE OF LIFE ON EARTH

## The Emergence of the Fourth Geosphere

Uniting the conceptual foundations of the physical sciences and biology, this groundbreaking multi-disciplinary book explores the origin of life as a planetary process.

Combining geology, geochemistry, biochemistry, microbiology, evolution, and statistical physics to create an inclusive picture of the living state, the authors develop the argument that the emergence of life was a necessary cascade of non-equilibrium phase transitions that opened new channels for chemical energy flow on Earth. This full-color and logically structured book introduces the main areas of significance and provides a well-ordered and accessible introduction to multiple literatures outside the confines of disciplinary specializations, as well as including an extensive bibliography to provide context and further reading.

For researchers, professionals entering the field, or specialists looking for a coherent overview, this text brings together diverse perspectives to form a unified picture of the origin of life and the ongoing organization of the biosphere.

ERIC SMITH is External Professor at the Santa Fe Institute, Research Professor at George Mason University, and Principal Investigator at the Earth-Life Science Institute, Tokyo Institute of Technology. He is a physicist specializing in the origin of life, non-equilibrium systems, economics, and the evolution of human languages.

HAROLD J. MOROWITZ is Clarence J. Robinson Professor of Biology and Natural Philosophy at George Mason University. He was the founding director of the Krasnow Institute for Advanced Study at George Mason University, and is Chairman Emeritus of the Science Board at the Santa Fe Institute.

# THE ORIGIN AND NATURE OF LIFE ON EARTH

## The Emergence of the Fourth Geosphere

ERIC SMITH

*Santa Fe Institute*
*George Mason University*
*Tokyo Institute of Technology*

HAROLD J. MOROWITZ

*George Mason University*

CAMBRIDGE
UNIVERSITY PRESS

Shaftesbury Road, Cambridge CB2 8EA, United Kingdom

One Liberty Plaza, 20th Floor, New York, NY 10006, USA

477 Williamstown Road, Port Melbourne, VIC 3207, Australia

314–321, 3rd Floor, Plot 3, Splendor Forum, Jasola District Centre, New Delhi – 110025, India

103 Penang Road, #05–06/07, Visioncrest Commercial, Singapore 238467

Cambridge University Press is part of Cambridge University Press & Assessment,
a department of the University of Cambridge.

We share the University's mission to contribute to society through the pursuit of
education, learning and research at the highest international levels of excellence.

www.cambridge.org
Information on this title: www.cambridge.org/9781107121881

© Cambridge University Press & Assessment 2016

First published 2016
Reprinted 2018 (version 3, September 2022)

Printed in the United Kingdom by TJ Books Limited, Padstow Cornwall

*A catalogue record for this publication is available from the British Library*

*Library of Congress Cataloging-in-Publication data*
Names: Smith, Eric, 1965– author. | Morowitz, Harold J., author.
Title: The origin and nature of life on Earth : the emergence of the fourth
geosphere / Eric Smith (Santa Fe Institute, George Mason University, Tokyo
Institute of Technology), Harold J. Morowitz (George Mason
University, Virginia).
Description: New York, NY : Cambridge University Press, [2016] | ?2016
Identifiers: LCCN 2015041999 | ISBN 9781107121881 | ISBN 1107121884
Subjects: LCSH: Life–Origin. | Earth (Planet)–Origin. | Biosphere. | Earth
sciences.
Classification: LCC QH325 .S56 2016 | DDC 570–dc23
LC record available at http://lccn.loc.gov/2015041999

ISBN 978-1-107-12188-1 Hardback

# Contents

# Preface

It has been almost a century since the inquiry into life's origin has been reinvigorated following several decades of quiescence influenced by Louis Pasteur's elegant experimental demonstration that microbes were not spontaneously generated in flasks of nutrient broth appropriately aerated to prevent the entry of airborne particles. It has been a century in which biochemistry and biophysics have completely altered our core understanding of the living world and consequently opened up a series of powerful experimental, theoretical, and computational approaches. In this new light, two of us independently thinking of first life were some fifteen years ago introduced by colleagues at the Santa Fe Institute who had detected some common points of interest in what we were investigating, one from the top down and the other from the bottom up – from the phenomenology of the living world with an emphasis on biochemistry and biophysics and from the underlying physics and chemistry and statistical theory that impose a necessary order.

A common theme was that life on Earth was not the outcome of an isolated event as suggested by the *Chance and Necessity* school but a planetary property that appeared early in the history of the planet and spread in a spontaneous way. What we designate life or proto-life has existed over most of the lifetime of planet Earth. The universality of the phenomenon and the massive flux of matter and energy due to the huge interaction of organisms and their products with the rest of planetary matter led us to return to the perceptive book *Geochemistry* by Kalervo Rankama and Th. G. Sahama [664], a work highly praised to one of us by the polymath G. Evelyn Hutchinson, dean of American ecologists. The two Finnish geochemists, acting as geological generalists, divided the planet into four *geospheres*: the lithosphere, the hydrosphere, the atmosphere, and the biosphere. The term "biosphere" had been introduced in 1875 by Eduard Suess and used in its modern sense some years later by the perceptive geochemist Vladimir Vernadsky.

What we wish to understand from a scientific point of view is how the newly formed planet, condensing approximately 4.6 billion years ago, was transformed over time into the present verdant world, home to millions of species and the abode of *Homo sapiens*, a taxon including individuals like ourselves, who are somehow impelled to ask the foundational questions that this work tries to answer. What was clear from the outset was that such a study must be embedded in the domains of physicists, chemists, geologists, and biologists aided by the analytical tools of mathematics and reified by database mining and computation. Using the baseball phraseology of the double play (Tinker to Evers to Chance), we are in the world of Pauli to Pauling to Woese. Wolfgang Pauli provided a formalism to generate the periodic table of elements within the domain of quantum mechanics and particle physics. Linus Pauling has provided the context to see in a systematic way how these atomic elements give rise to a huge number of structures, large and small, that, comprehensibly, are able to perform exquisitely complex chemical functions. Carl Woese showed why in the whole world of biota, phylogeny in its most general sense must be understood from its biochemical foundations as well as seen from its more classical perspectives. He worked at incorporating biochemistry and molecular biology into a more global biological perspective. The whole and its parts must be studied jointly. Several names of scientists have been and will be mentioned in this preface and throughout the book. We tend to think of the discipline in terms of those predecessors upon whose shoulders we stand. Science at its best is a community of scholars, and we have been privileged to have some of these as colleagues and friends.

The universe of entities we deal with has something on the order of twenty million species, yet the metabolic charts of all of these species map in major parts onto a single chart of intermediary metabolism, whose representation was developed and refined by the lifetime labors of scientist Donald Eliot Nicholson and his colleagues. This continuity of metabolism over four billion years and countless present and extinct species argues against the view of life as the outcome of an improbable event. We reject the idea developed in *Chance and Necessity* by Jacques Monod [561] that life is a result of chance whose origin is unfathomable. This view is inconsistent with the world that we know and with the way the living geosphere functions as part of it. We have thus rejected all reliance on "frozen accidents" as a dead end to our investigation of the earliest biosphere. We have also minimized appeals to panspermia, because there is often too much of a "biogenesis-of-the-gaps" character to arguments: we reason at the level of components where we have been unable to form a systems understanding of chemistry on Earth. When some components seem difficult to account for on this planet, we appeal to planets whose chemistry is less well understood than our own as a source for them, generating a host of new system-assembly and causation problems seemingly more intransigent than those with which we began.

We have tried to be guided by an "Earth Scientists Oath" or paraphrasing the prophet Micah we have tried "To achieve a scientist's love of honesty, to seek our limits of understanding, and to walk humbly in our universe." Where we have tried to say things that are genuinely new, we have kept our approach embedded in well-understood science.

The wide range of opinions expressed in the scientific community makes it clear that we do not yet have a paradigm, even in its broadest outlines, to explain in any detail the origin of life. The number of fundamental transitions it entailed has been large and some have probably been overwritten by subsequent innovations. The problem will be with us for a long time to come. Within these limitations we hope that this book will aid the community in framing the best and most productive questions within reach of current understanding. With this goal we include a more eclectic body of material than that usually found in writings on this subject. For example in Chapter 4 and Chapter 6 we review both experimental results on mineral organosynthesis and extensive analysis of variations within the post-LUCA diversification of metabolism, despite the fact the former are too simple and the latter too complex to serve as close models for the transition from geochemistry to the first biochemistry that is our main interest. We have attempted to make our choices so as not to bury the reader in details[1] but to illustrate the extent of repetition or convergence of key motifs, with citations to the primary literature that the interested reader might pursue. From each body of evidence we call out the patterns or mechanisms that we believe impose significant constraints on theories of origins at the difficult and obscure era when the metabolic substrate and its control systems were emerging.

From a somewhat different perspective, in Chapter 7 we provide a basic but wide-ranging review of the theory of stability from statistical mechanics and related disciplines. In part this is necessary because the key mathematical ideas underlying the theory of robust order, which have been one of the main achievements of mathematical physics within the last century, are poorly communicated outside the specific fields where they were derived. The application of formal principles of stability from non-equilibrium statistical mechanics to systems chemistry is a field currently in its infancy, bringing together complex suites of concepts from two already sophisticated fields. Yet it is an inevitable merger, because the emergence of the biosphere was not a compounding of misadventures but a restructuring of systems.

We believe that one of the greater services we can offer may be to lower the costs of entry for readers into each other's literature and main ideas. For human minds in society, professional disciplines have been the portals to expertise, but the emergence of the biosphere was not a respecter of human silos. Few readers who have spent a lifetime becoming experts in geochemistry, biochemistry, or microbiology will have happened across the fact that the theory of robustness in non-equilibrium systems continues seamlessly to the mathematics of asymptotically optimal error correction. Yet this continuity must have been fundamental to the emergence of hierarchical architecture capable of memory and control over metabolism on the route from minerals to cells. Conversely, for readers to whom the "thermodynamics" of error correction is every day's bread and butter (its carbohydrate and triglyceride), there are no easy paths into the enormous literatures of geochemistry and biochemistry that review, at affordable cost, the basic principles of organization in those disciplines. Yet it can only be in context of these principles that any theory of error correction might contribute to an understanding of biogenesis. Our choice of materials and approach is aimed at providing such bridges, so that as a community we might be less outsiders to each other's sensibilities to order in the world.

---

[1] Christian de Duve, in the fine book *Blueprint for a Cell* [183] deliberately avoided such byways in order to maintain focus on a central narrative and increase reader accessibility. His book, which rewards re-reading even after many years, deserves if anything more than the considerable influence it has already had.

Starting with the belief that conceptual frameworks as well as acquired empirical knowledge, technical methods, and examples from physics, chemistry, biology, and geology are necessary components of the study of biogenesis, we have tried to present material at levels satisfying to domain specialists as well as penetrable from a more general perspective. This sometimes involves saying things more than once at varying levels of detail. A series of boxes set off sub-topics or derivations that will be of interest to readers who want more detail, or glossaries of domain-specific terms to which the text refers repeatedly. Terms of art that might be mistaken for common-language expressions (a frequent problem in the shorthand of technical fields) are typeset in *italic* where they are first used in a chapter. A small subset of terms are highlighted with **boldface** where they are first discussed: these are topics that play a special role in the structure of arguments or signal key ideas. We have tried to strike a balance between inclusiveness and economy to engage the broad range of colleagues whose insights are essential to this story. Of course, our own limitations must be acknowledged.

In a more philosophical mode, we envision the goal of this branch of science as showing how the core of intermediary metabolism is a necessary consequence of galactic processes giving rise to some distribution of the elements of the periodic table, which under the right geochemical boundary conditions generate an autocatalytic network of chemicals capable of emerging subsequently through a series of transitions into more complex molecules and structures: the biosphere. This series of transitions follows matter and its internal states of organization from simplicity to complexity. At the lowest level, non-equilibrium thermal physics appears to be the analytical method of choice. At each transition, new paradigms become necessary as complexification leads to a broader range of possibilities and structures, and opportunities for path dependence. This we have envisioned as biology standing between physics and history, deterministic at the simplicity roots and rife with possibilities at its complexity shoots.

With the above in mind we turn to the organization of this monograph. The first five chapters deal with the phenomenological and its generalizations. What is life on the planet Earth like and how did it get that way? In choosing which aspects to review, we draw on what the core scientific disciplines have made it possible to understand, and within this we select the patterns that we believe best assemble into a system that is coherent within the more inclusive body of scientific laws. Chapter 6 sketches the synthesis that we believe the foregoing facts support.

Facts, however, are not self-interpreting, and proposed scenarios are not theories. Everything we review in the first five chapters is known within the scientific community, but the system we propose in Chapter 6 has not seemed an inevitable or even compelling interpretation to all readers, for reasons we understand. Chapters 7 and 8 introduce other parts of the reasoning that we believe are needed for inference from facts to theories. Chapter 7, like the first five chapters, does not speculate and sticks to what we know, but in this case we consider what is known about stability and its relation to information. To gently contradict Wittgenstein, the world is a totality of facts, *and* of things, and we believe the introductions to both work best when made in each other's company. Chapter 8 brings these two streams together in what we hope will provide a foundation for a proper theory. It explains some of the priorities we have expressed in the first five chapters, from the perspective in Chapter 7 that a formal theory of stability and robustness exists and that it has implications. We note there, too, that one measure of the plausibility of a theory of biogenesis that begins in geochemistry is how naturally it connects to the concepts and not only the metaphor of evolution as biology understands them today.

The opening chapter attempts to explain and justify treating the planet's biota as a geosphere. This perspective, which makes life a planetary property or perhaps a galactic property, opens the way to a more general treatment of biology at its core as a system science rather then a strictly historical discipline as the *Chance and Necessity* adherents or some evolutionary theorists would urge us to believe. It makes the case that the biological generalist must take geochemistry and geophysics seriously. From the other side, recent studies in geochemistry show why mineralogy must recognize major influences from the biosphere.

Chapter 2 is an overview of life as we know it, and our first attempt to come to grips with its extraordinary unity-in-diversity. The biosphere is a hierarchical complex system that has created novel order at scales ranging from electrons to ecosystems. Patterns at many levels can inform us about earlier ages, and more informative than any one pattern is the way they fit together as a system. Life on the planet Earth probably began 3.8 to 4.0 billion years ago. At first it seems extraordinary that present-day organisms and ecosystems could tell us much about the earliest ancestors, but when we consider that all are linked by laws as well as by descent, the diversity of modern life offers windows on the laws that created pasts older than the deepest we can reach with historical reconstruction. This prospect first came into focus in 1955 when it was noted that there was a single metabolic chart onto which all species mapped fully or in part. Thus metabolism is a living fossil possibly older than any extant rock. This amazing persistence becomes more compelling and more comprehensible as a feature of metabolism in physiological and ecological context, where metabolic order is recapitulated and given its functional semantics. We might note parenthetically that a 1999 paper of ours on the reductive citric acid cycle brought a letter of interest from the then 90 year old still hard working map maker of metabolism, Dr. Nicholson.

Our ability to comprehend life as a meaningful system, not only recording the past in extant diversity but also expressing elements of timeless order, has been aided by Carl Woese's revolutionary phylogeny and the growth of vast databases such as KEGG, Meta-Cyc, and Reaxys to mention just a few. At many points in Chapter 2 we are compelled to recognize ecosystems as forms of organization in their own right, not merely as communities of taxa but as the carriers of patterns that remain necessary even if all the taxa are changed. From the chemical propensities of metabolism to the invariants of ecology we are encouraged to look for empirical generalizations and theoretical principles applicable to non-equilibrium systems.

Chapter 3 considers energy sources for the kinetic processes of the maturing planet and energy exchanges among the geospheres. Sources include fusion energy coming as radiation from the Sun, fission energy from radioactive elements, and above all the disequilibrium between our planet's formative condition, still trapped in its interior, and an outer shell driven relentlessly away from that condition. The hero in this story is the active mantle, whether through volcanism, melt-separation of elemental states, or long-range subduction and upwelling in tectonics. These are the processes by which heat – alone too diffuse an energy source to drive chemistry far from equilibrium – is nonetheless the cause for a gateway through which the Earth's chemical disequilibrium is focused at narrow points in covalent bond dynamics. One of our most central messages will be that the biosphere, first and foremost, is a chemical conduit for the flow of electrons from higher to lower potentials. The thermodynamic "problem" of a lifeless Earth, which on our planet was "solved" by the emergence of a biosphere, was to facilitate the descent of electrons. In Earth's first life, and in some life today, part of this electron flux is not only coupled to, but occurs by means of, carbon fixation, the reduction of $CO_2$ by the formation of multi-carbon molecules. That conduit with accompanying chemistry self organized, we argue, a chemical phenomenon for which we find precedent in physical processes such as weather or lightning, the far from equilibrium yet inevitable ordered states of the atmosphere. This chapter focuses on the planetary restlessness that may have made it impossible to remain lifeless, and the ways the same processes sustain parts of the biosphere directly today.

Chapter 4 offers a detailed view of metabolism where the top down and bottom up approaches must make contact. There is now an enormous amount of data on genome sequence, chemical mechanism, macromolecular structure, and enzymology. By combining these, we may understand what functions molecules perform, what variations have been open to them, and how their present diversity came into existence historically. To understand metabolism we must recognize at least four levels of organization: the small-molecule synthetic networks, crucial intermediate cofactors (vitamins) that transfer components and complete networks, macromolecule synthesis and catalysis, and in some cases the organizational functions of cellular compartments. We argue that the historical sequence followed the molecular sequence from simple to complex, from small molecules to large, and that the layers of order we find in metabolism today likely reflect historical accretions that were watersheds, but not history-erasing floods. The elaboration of metabolism was not only hierarchical but also modular, indicating parallel streams of innovation, which could have occurred in somewhat independent places and times. The multilevel patterns of life suggest that constraints on what would ever be possible originated in metabolism and then propagated to higher levels.

On any account it is impressive that a small, network-autocatalytic cycle (the rTCA cycle) should be found running through the termini of the synthesis pathways of all essential monomers. Also remarkable is the almost mineral-like simplicity of direct reduction of one-carbon units used ubiquitously in extant life, sometimes furnishing alternative starting points to synthesize a few amino acids. The significance of these patterns compounds as we find that the same networks have anchored the very little innovation that ever occurred in carbon fixation, and that their catalysts are among the most distinctive and conserved in the biosphere. The bottom up focus on reaction mechanisms and considerations of control and stability privileges the same pathways that historically have been unchanging points of reference to which all innovations were anchored. Metabolism is not only a list of components but a logic of relations, and it is in that logic that we find non-arbitrary constraints strong enough to have entrained later accretions of complexity.

Metabolism by itself does not a geosphere make, so Chapter 5 turns to the hierarchy of biology, the series of transitions that moves to macromolecules, coding and memory molecules, control, containment, and the integration of bioenergetics. These higher-level systems define the context for the chemistry of life as we know it, and the question central to biogenesis is whether they inherited or dictated that chemistry's logic. Whether we can infer the path from geochemistry to the first cells by extrapolating modern biochemistry backward in time turns on the answer to this question.

The chapter considers three distinct higher forms of organization: the emergence of ribosomal translation of peptides and with it a genetic code, the integration of redox and phosphate energy systems, and the diverse forms and functions of compartmentalization realized in modern cells. We find, in all three cases, either common modes of constraint expressed in biochemistry and in its higher-level supports, or more remarkably, the imprint of metabolic logic in systems that have evolved to distance and insulate their internal functions from the substructure of metabolism. The modular unfolding of metabolism seen in Chapter 4 must have been interwoven with innovations at higher levels that increased the interdependence of prebiotic systems on each other and led to their progressive separation from lithosphere/hydrosphere chemistry to become a distinct geosphere. The layers in this interweaving have led to the dense interdependence of forms of order at many levels in modern life, which appears as a chicken-egg problem if viewed as a whole. The remarkable observation is that enough independent identity or partial autonomy has been preserved in different subsystems to suggest at coarse scale the alternation of moves through which the unification of cellular life and its distinct planetary role were attained. Cellular life is both unified and diverse because it is a confederacy, a gathering together of diverse opportunities for order from the abiotic geospheres to create a new identity that remembers its roots but is distinct from them all.

The first five chapters describe the world as it functions today, or as we can reconstruct its past from directly available evidence. We interpret some strong and invariant patterns as necessary features or as remnants of stages through which life must have passed, when the historical reconstruction and mechanism jointly support such interpretations. In Chapter 6 we weave these interpretations into a proposed web of stages in biogenesis. Our proposed sequence is not a detailed scenario – many important transitions occurred about which we can offer little or no detail. Our sequence is meant as a kind of skeleton provided by the strong constraints – those stages that seem to us sufficiently necessary that they should both constrain scenarios and guide the questions we ask about how to bridge the gaps between them.

Chapter 7 introduces the idea that gives precision to an intuition of many investigators that the emergence of life may have been stochastic, but it was not accidental. We introduce the ideas of both equilibrium and non-equilibrium thermodynamic states, and the robust transitions between them known as phase transitions. Our most important problem issuing from the first six chapters is how to describe and account for strong regularities in dynamical systems that also exhibit constant variation. Thermal phases provide the needed concept and a developed mathematics to study its ramifications. These phases are the carriers of physical laws even as the entities and events that make them up undergo constant change. This chapter provides a means to state, in language that is not metaphor or analogy, what we believe must be the foundation for a theory of the origin of life: via whatever chemical mechanisms are experimentally plausible, the emergence of the biosphere, particularly in its early stages, can only make sense as a cascade of non-equilibrium phase transitions.

Non-equilibrium states and their transitions have in recent years been extensively studied in certain areas of physics and chemistry. A gulf exists, however, between the relatively simple phenomena for which their potential is routinely explored, and the structurally rich phenomena that are the normal interests of most geochemists, organic chemists, biochemists, microbiologists, and specialists in other domains where they are almost unknown. We believe that any effort to separate what is chance in life from what is necessity, and to distinguish the patterns that reflect each, must eventually face questions of robustness and the relation between material and information, which will bring non-equilibrium statistical mechanics and structurally rich systems together to create a new frontier.

We provide systematic introductions to the most central concepts, which may be followed using elementary methods with some time and work. The reader who is unfamiliar with, or does not wish to push through, the various formalisms need not be put off. Our concern is with concepts, and although we will not traffic in metaphor, we describe these so that their applications to emergence can often be understood without heavy formalism.

One of the most important insights from the theory of phase transitions (equilibrium or otherwise) is that reductionist science is an enterprise of floors and ceilings. Each phase transition creates the robust forms of order that will participate in dynamics at or below its immediate scale. These ordered states, in chemistry, have been the building blocks from which life and evolution design. They remain in effect until the next phase transition constrains some of them, creating new levels of organization and typically greater complexity. Reductionist science, in its best sense, works because we can make falsifiable predictions at a given level without knowing what is above the ceiling or below the floor. In other words, the strength of reductionism is due to the simplifying effects of emergence. For the same mathematical reason, the emergence of a biosphere could only have been possible through a cascade of phases that buffered innovations at different levels from each other. If we can identify the transitions, we recognize the stages as lawful rather than accidental.

The earliest ordered phases in geochemistry were organizationally simple, though the search for them in the wide parameter range of the lithosphere/hydrosphere interface may be very difficult. As metabolic order developed, forms of individuality emerged, and evolutionary dynamics entered, they became increasingly complex.

An important consequence of the phase transition paradigm is that the origin of life and the organization of the biosphere cannot be understood as two separate topics. The emergence of the biosphere was the Earth's departure from a metastable lifeless state and its subsequent collapse into the more stable condition that includes life. Chapter 8 develops the ramifications of these conclusions, and argues that from them we must reconceptualize the nature of the living state.

This chapter introduces three new topics, which have been present in the context of all the earlier chapters, and which must now be addressed in their own right. The first is the modular and hierarchical architecture of life. Fully understanding hierarchy invokes all dimensions of the phase transition paradigm: material stability and layers, information requirements and buffering, the needs for control in hierarchical systems, and the way life meets these by following paths of least resistance. The second is the emergence of individuality and with it the Darwinian world. We invert the conventional view of life as a property inherent in individuals, and recognize individuality as one mode of organization – albeit an important one – among many that enable life. From this perspective for the first time, we find it most natural to expect what we observed empirically in Chapter 2: that metabolism and the ecosystem define two of life's most fundamental universals, to which individuals may form many different relations.

Finally we offer systematic arguments for a view of what is essential to the living state that could not have stood on its own from intuition and scenarios, but from the full development of the phase transition paradigm seems to us inevitable. The biosphere is, most fundamentally, the geosphere on Earth that opens otherwise unreachable domains of organic chemical states and processes. It is maintained for the same reason as it emerged: the chemistry of life and all the hierarchy of structures that maintain it constitute a channel for relaxation of redox and other energetic stresses. This view reaches much further than is at first evident. Chemistry unifies the extraordinary diversity of living order – its complexity, its stability, the particularity of its conserved features, and its tight integration with the other geospheres – to a degree that no other starting point can.

With these eight chapters, then, we will hope to have left the reader with the following main contributions.

1. The major transitions that constituted the emergence of the biosphere were not defined by single and rare microscopic events, but by regime shifts in the behavior that was typical in dynamical, stochastic ensembles. Some of these are the familiar phase transitions of equilibrium condensed matter, but the distinctively biotic ones are inherently dynamical.
2. One of us wrote in 1968 that "The energy that flows through a system acts to organize that system." Forty five years later two of us reiterate that point, but today a much more detailed picture can be drawn for the descent of electrons. A few channels for flow, used repeatedly, have given rise to the particular structures of biochemistry and life. Both from history and from mechanism, they are remarkably well positioned for the task. With a more advanced empirical and computational chemistry, it may not be beyond the ability of near generations to prove they are unique, at least in this planetary context.

3. A vision of the lawful nature of life might be expressed as "deriving intermediary metabolism from the periodic table of elements." The larger question of whether bio-genesis was a chance event or a necessary event is closely related. If it was necessary, to what extent are major features predictable from first principles? We argue for a large role for predictability in the earliest stages, with contingency and complexity enter-ing through later transitions. The periodic table is not all the law needed to derive metabolism, but significant structure in biochemistry does derive all the way from prop-erties of the elements. In some other cases we argue that it follows from higher-level but still essentially physical and energetic regularities such as orbital structure, network topology. and kinetics.

4. The phase transition paradigm, which accepts heavy dependence on complex and diverse boundary conditions (such as the Earth has likely always provided), but rejects a large early role for miracles even in catalysis, supports a view that the progression in molecular size and complexity defined the progression of emergence as well. Early geo-organic chemistry cast the die for metabolism, and metabolism then cast it for the rest of biochemistry. The earliest important regime change in catalysis was not through macro-molecules but through cofactors; the progression to macromolecules was a complicated transition, made possible by a biosynthetic foundation that was stable and already partly selective.

5. Historical reconstruction can sometimes help us see beyond the history we can directly reconstruct. With appropriate functional context, it can identify the relative ages and directions of change even of very ancient features. In some cases we can argue not only that universal features are primordial, but that they reflect causes that were at work before genetic history originated and perhaps before cells. These are windows on some of the earliest steps from prebiotic to nascent biotic chemistry.

6. Our understanding of the biosphere must be chemical. Organic chemistry is not an acci-dental stage on which abstract principles of life perform a play that could be performed elsewhere. Chemistry matters in detail because it matters in principle. Some of the most important sources of stability and complexity in life would not be expressible in any other system. The ecosystem is the bridge from geochemistry to life, and carries much of what is deterministic and necessary in metabolic order. Species emerge further into the domain of chance and are secondary. Thus we never need to call the ecosystem a "super-organism" to acknowledge its integrity, because we recognize this as in some ways prior to organisms.

7. Higher levels of living organization seem to require not single transitions, but many transitions for each form. Among these the most enigmatic is the emergence of an oligomer world in which mRNA, tRNA, rRNA, and proteins are unified by ribosomal translation. Yet, as complex as this sequence of transitions must have been, and with large gaps remaining in our understanding of its intermediate stages, it emerged carry-ing striking signatures of metabolic order that are not parts of its modern function. The long exploration that led to cells, at the end of its wandering, arrived still at the order of metabolism, and enables us to know some things about it for the first time.

# Acknowledgments

In the ten years prior to the authors of this book's beginning to interact, scientists and database builders were laying the foundations that we build on and informing us of their results. Jack Corliss told us of the theory of life's forming at undersea spreading centers and Larry Hochstein told us of the discovery of the reductive tricarboxylic acid cycle. The KEGG and Ecocyc databases were being assembled. Ribosomal sequencing was changing the world of taxonomy. Steen Rasmussen was developing artificial life and biogenesis studies at Los Alamos National Laboratories. Robert Hazen, George Cody, and Hatten Yoder were developing high-temperature, high-pressure organic chemistry at Carnegie Geophysics. Consultation with Walter Fontana and Leo Buss led to a study of the rTCA compounds in the Beilstein compendium with Jennifer Kostelnik and Jeremy Yang.

When we began to regularly interact it was during summers at the Santa Fe Institute, frequently joined by Shelley Copley. A phone call from an old colleague, Carl Woese, informed us of the opportunity of the FIBR grants: multi-year multi-institutional grants from the National Science Foundation. By this time we had been joined by Vijayasarathy Srinivasan. We were awarded a five-year, five-institution FIBR grant "From Geochemistry to the Genetic Code" centered at the Santa Fe Institute and Krasnow Institute of George Mason, with nodes also at University of Colorado, University of Illinois, and Carnegie Institution Geophysical Laboratory. This added to the senior staff Nigel Goldenfeld and Zaida Luthey-Schulten. Toward the end of the FIBR collaboration we were joined by Rogier Braakman. The insights, ideas, and interpretations of Shelley, George, Carl, Nigel, Rogier, and Vijay figure prominently in the following chapters as they do in our understanding of the subject.

The FIBR collaboration was made into an innovative and substantial educational effort through the gifted and committed work of Ginger Richardson, Mimi Roberts, Jim Trefil, Paige Prescott, Lokesh Joshi, Paul Cammer, and Carlos Castillo-Chavez.

Several colleagues have stood out as towering authorities or unique sources of vision and perspective as we have sought to find the correct frame through which to understand the living state. We must especially thank John Baross, Christian de Duve, Albert Eschenmoser, Georg Fuchs, Everett Shock, and Carl Woese.

The ways one depends on colleagues throughout the scientific network in coming to grips with a complex topic – for guidance and serendipity, for knowledge, for conceptual growth, and for shared work – are too numerous to list. For these and much more we are grateful to Ariel Anbar, Jakob Andersen, Gil Benkö, Aviv Bergman, Carl Bergstrom, Tanmoy Bhattacharya, Elbert Branscomb, James Brown, Leo Buss, Jim Cleaves, Ken De Jong, David Deamer, Jennifer Dunne, Brian Enquist, Doug Erwin, Jim Ferris, Jessica Flack, Christoph Flamm, Walter Fontana, Steve Frank, Murray Gell-Mann, Chris Glein, John Hernlund, Kei Hirose, Ivo Hofacker, Peter Hraber, Harald Huber, Nicholas Hud, Piet Hut, Sanjay Jain, Bill Jones Jr., Lokesh Joshi, Masafumi Kameya, Stu Kauffman, David Krakauer, Supriya Krishnamurthy, Irwin Kurland, Michael Lachmann, Manfred Laubichler, Pier-Luigi Luisi, Pablo Marquet, William Martin, Cormac McCarthy, Daniel Merkle, Stuart Newman, Harry Noller, Martin Nowak, Yoshi Oono, Stephen Ragsdale, Michael Russell, Van Savage, Peter Schuster, Cosma Shalizi, Peter Stadler, Yuichiro Ueno, Geoffrey West, George Whitesides, Loren Williams, Jon Wilkins, David Wolpert, William Woodruff, and Michael Yarus.

The manuscript has benefited greatly from review and recommendations by Elbert Branscomb, Nigel Goldenfeld, and Piet Hut, and from a gifted and expert statistics reviewer who has elected to remain anonymous. Our enduring thanks go to our publisher, Simon Capelin, for creating a beautiful volume and making it available to a wide readership, and to the excellent editors and designers at Cambridge University Press.

Our research over the years has been supported through the generosity of both institutions and individuals. Supporting institutions have been the Santa Fe Institute and the Krasnow Institute, the National Science Foundation, and the John Templeton Foundation. Individuals who have been staunch supporters as well as long-term friends are Jerry Murdock, William Melton, Jim and Celia Rutt (through the Proteus Foundation), and Bill Miller. Their belief in our work has not only been essential to its completion, but has been a source of deep personal gratification to us both.

The Santa Fe Institute has, uniquely, been home, and for its ongoing vision and able operation we are particularly grateful for Ellen Goldberg's tenure as president during the years that brought us together, and the skill and wisdom of Elisabeth Johnson as our grants manager. Wayne Coté and Bruce Bertram create a meeting environment that is second to none. Everything we have depended on in the Krasnow Institute has been possible only through the professionalism, energy, and goodness of Jane Wendelin and Jennifer Sturgis. We owe special thanks to Sara Bradley for obtaining permissions for graphics used in this volume.

Krastan Blagoev at NSF has been a long-term supporter and collaborator in a series of special workshops dedicated to recognizing the next paradigm. Jim Olds has been colleague and supporter for us both as director of the Krasnow Institute. DES has recently joined the Earth-Life Science Institute at Tokyo Tech where the manuscript was completed, and is a research scholar at the Ronin Institute in Montclair, New Jersey.

This book is dedicated to the memory of George Cowan, a brilliant scientist and institution builder and an exceptionally fine human being.

# 1

# The planetary scope of biogenesis: the biosphere is the fourth geosphere

The origin of life was a planetary process, in which a departure from non-living states led to a new kind of order for matter and energy on this planet. To capture the role of life as a planetary subsystem we draw on the concept of geospheres from geology. Three traditional geospheres – the atmosphere, hydrosphere, and lithosphere – partition terrestrial matter into three physical states, each associated with a characteristic energetics and chemistry. The emergence of life brought the biosphere into existence as a fourth geosphere. The biosphere is an inherently dynamical state of order, which produces unique channels for energy flow through processes in carbon-based chemistry. The many similarities, and the interdependence, of biochemistry with organometallic chemistry of the lithosphere/hydrosphere interface, suggests a continuity of geochemistry with the earliest biochemistry. We will argue that dynamical phase transitions provide the appropriate conceptual frame to unify chance and necessity in the origin of life, and to express the lawfulness in the organization of the biosphere. The origin of life was a cascade of non-equilibrium phase transitions, and biochemistry at the ecosystem level was the bridge from geochemistry to cellular life and evolution. The universal core of metabolism provides a frame of reference that stabilizes higher levels of biotic organization, and makes possible the complexity and open-ended exploration of evolutionary dynamics.

## 1.1 A new way of being organized

The emergence of life on Earth brought with it, for the first time on this planet, a new way for matter and energy to be organized. Our goal is to understand this transition, how it happened and what it means. The question how life emerged – what sequence of stages actually occurred historically – can at present be answered only at the level of sketches and suggestions, though for some stages we believe good enough arguments can be made to guide experiments. To arrive at a sketch, however, we cannot escape making many choices of interpretation, of things known about life and its planetary context.

Life emerged in an era not accessible to us through historical reconstruction. Our claims about what happened in this era will depend on the principles we use to generalize, simplify, and extrapolate from knowledge of modern life and a few fossilized signatures that become increasingly fragmentary and difficult to interpret on the approach to the beginning that we wish to understand. More important than any claim we can make about past events will be whether we frame the problem of emergence in terms that capture its most important abstractions.

Life appeared on Earth in a period known as the Hadean, the Earth's oldest eon, a reference to the etymology of Hades as "the unseen" [196]. Other than gross features of planetary composition, it has left no detailed signatures in the present because it is a time from which later eras preserved no memory.[1] The Hadean was, however, a time we think of as governed by laws of geophysics and geochemistry, and therefore open to understanding. (Indeed, the absence of memory makes the Hadean, more than later periods with accreted history, a simpler period to study with general laws.) The earliest stages of life were scaffolded by these geological laws, and in some respects may even have been continuous with them. Thus in geology the complement to historicity is lawfulness. What we cannot infer from preserved memories we seek to deduce by understanding the action of laws.

We argue in this book that the same is true for life. The complement to historicity in its earliest periods was not chaos, but lawfulness, albeit perhaps lawfulness of a statistical nature. Life inherited the laws of geochemistry, and grew out of geochemical precursors because some of those laws required the formation of a new state of order qualitatively unlike any of the lifeless states of matter. Life is still in large part lawful, if looked at in the right way, and some of the laws that govern the living world today are good candidates for laws that were at work during its emergence. However, modern life is also historical, to a much greater degree than modern geology, so one of the challenges to making the correct abstractions about its origin is to recognize and separate the contributions of law and of history.

The origin of life was a process of departure and a process of arrival: a departure from non-living states that we understand with natural laws, and an arrival at a new living mode of organization that is robust, persistent, and in its own respects law-like. To understand why a departure was prompted, and why the arrival has been stable, we must begin by recognizing that life is a planetary process, and that its emergence was a passage between two planetary stages.

### *1.1.1 Life is a planetary process*

The emergence of life was a major transition in our planet's formative history, alongside the accretion of its rocky core, the deposition and eventual persistence of its oceans, and the accumulation of its atmosphere. Several facts about the timing, the planetary impacts, and the organizational nature of living systems establish the context within which any theory of origins must make sense.

---

[1] As is fitting, one of the five rivers in Hades from Greek mythology is Lethe, the river of forgetfulness.

**Life apparently emerged early** Conditions on Earth earlier than about 4 billion years ago appear to have been too hot and desiccated from asteroid infall to permit even the chemical constituents that we now associate with life to exist, much less to permit its processes to take place.[2] Yet evidence, either from explicit microfossils or from reworking of element compositions and isotope ratios that we associate with life, suggests that as early as 3.8 billion years ago, and with much greater confidence by 3.5 billion years ago, cells existed that must have possessed much of the metabolic and structural complexity that is common to all life today [444]. Given the extremely fragmentary nature of the rock record from this ancient time, the existence of the signatures we know suggests that life first became established on Earth in a geological interval that was shorter than 200 million years – possibly much shorter – an interval that was also a period of geological transition on a still young planet.

**Living systems have (in some cases radically) altered planetary chemistry** Living systems have always altered the chemistry of their local environments (these changes are the most reliable ancient biosignatures), and they have gone on to change global planetary chemistry. The most striking change was the filling of the atmosphere and oceans with oxygen, which changed the profiles of elements in solution in the oceans, altered continental weathering, and increased as much as threefold the diversity of minerals formed on Earth [348]. By capturing trace elements in biomass, effectively creating microenvironments for them vastly different from the surrounding physical environments, organisms also govern the concentration and distribution of metals, phosphate, and sulfur, and influence the great cycles of carbon, nitrogen, and water [236].

**Living systems are ordered in many ways at many levels** Whether in terms of composition, spatial configuration, or dynamics, living systems are ordered in many different ways at many scales. The chemical composition of biomass is distinguished from non-living matter by at least three major classes of synthetic innovation, in small molecules (metabolites), mesoscale molecules (cofactors), and macromolecules (lipids, polynucleotides, polypeptides, and polysaccharides). These components are organized in physical-chemical assemblies, including phase separations and gels, non-covalently bonded and geometrically interlinked molecular complexes, various compartments, cells, colonies, organisms, and ecosystems of a bewildering array of kinds. The essential heterogeneity of the many different kinds of order, and the diversity of processes that have been harnessed to generate them, is a fundamental and not merely incidental aspect of life's complexity.

**Although diverse and heterogeneous, living order is also highly selected** At the same time as the diversity and heterogeneity of living order creates a complex challenge of

---

[2] Whether asteroid infall underwent a late pulse, known as the "late heavy bombardment," sufficient to melt and desiccate the entire Earth's surface concurrently, or tapered more gradually so that sub-crustal water was only locally and intermittently removed, is a point of uncertainty, and this creates uncertainty of as much as 200–300 million years in estimates of the earliest time the Earth could have sustained organic compounds.

explanation, it is important to recognize that, within each "kind" of order, the observed ordered forms comprise a vanishingly small set within the possible arrangements of similar kind.[3] We may characterize the sparseness of observed kinds of order by saying that each ordered form is "selected" – for stability, for functionality, or by some other criteria – but in first surveying the qualitative character of life, we wish to suspend theoretical assumptions about how the selection is carried out, because (we will argue) this turns out to be a complicated question to frame properly. Whether observed forms of living order are sparse because they are uniquely specified by first principles, or because the unfolding of a historically contingent evolutionary process has only sparsely sampled its possibilities, will be fundamental to our understanding of the role of laws in biology, including life's origin.

**An invariant simple foundation underlies unlimited complexity at higher levels**  At the core of life lies a network for the synthesis of the small organic molecules from which all biomass is derived. Remarkably, this core network of molecules and pathways is small (containing about 125 basic molecular building blocks) and very highly conserved. If viewed at the ecosystem level – meaning that, for each compound, one asks what pathways must have been traversed in the course of its synthesis, disregarding which species may have performed the reaction or what trophic exchanges may have befallen pathway intermediates along the way – the core network is also essentially universal. Some gateway reactions are strictly conserved. In areas where synthetic pathways do show variation, the network tends to be highly modular, and variations take the form of modest innovations constrained by key molecules that serve as branching points.[4] This universality made possible S. Dagley's and Donald Nicholson's assembly of a chart of intermediary metabolism [173] that generalized across organisms. Other universal features of life include its use of several essential cofactors and RNA, and some chemical aspects of bioenergetics and cellular compartmentalization. A more complex and enigmatic, but also nearly universal, feature of life is the genetic code for ribosomal protein synthesis. These higher-level universals are discussed in Chapter 5.

This small and universal foundation of life is a platform for the generation of apparently unlimited variation and complexity in higher-level forms. These range from cell architectures to species identities and capabilities, and ecological community assemblies and their coevolutionary dynamics. The contrast that is so striking is that in its invariant core elements, life is universal down to much more particular components and rules of assembly even than other broad classes of matter such as crystalline solids. Yet in its higher levels of aggregation, it appears to have open-ended scope for variation that has no counterpart in non-living states of matter.

---

[3] We return in Chapter 7 to a more systematic discussion of possible versus realized forms of order, and the significance of the fact that realized order has vanishingly small measure relative to possible forms.

[4] We provide a much more detailed discussion of metabolic modularity, conservation, and variations, to explain these claims, in Chapter 4.

**Life as a whole has been a durable feature of Earth** The presence of living systems has apparently been a constant and continuous feature of the Earth since their first appearance at least 3.8 billion years ago. The persistence or tenacity of life bears a resemblance to features that arise in the geological progression of a maturing planet, and we will see that this resemblance extends to the incremental elaboration of complexity in life's universal core as well. The perplexity in this observation is that other stable, invariant geological features result from physical processes under conditions that, at least in principle, we know how to produce in the laboratory or in computer simulations. The states of matter to which they correspond also tend to be reproducible under broadly similar conditions, so they tend to recur in broadly similar environments. In contrast, we currently have essentially no understanding of what laboratory conditions would reproduce the emergence of life. Current observation of non-Earth systems, including meteorites and other planets or moons, is also consistent with the absence of signatures we would characterize as unequivocally biotic. These observations have been interpreted by Francis Crick [168] and Jacques Monod [561] (among others) as circumstantial evidence that life is improbable or accidental.[5] While we believe this interpretation is unjustified, the observations do imply at least that the conditions for life to emerge and persist are much more *particular* than we have come to associate with robust and persistent physical states of matter.

The persistent presence of living states on Earth, including persistence of the universal core, is more striking because higher-level systems such as cell form and catalytic capability have undergone major episodes of innovation, while still higher levels such as species identities or ecological community structures exist in a state of almost constant flux or turnover. These higher-level systems appear, at any time, to be essential to carrying out core processes, yet they have been much more variable than either the core that depends on them, or the broad characteristics of the living state by which all life shares a family resemblance. Reconciling such signatures of accident and fragility, with other signatures of robustness that we normally associate with inevitability, is one of the longstanding puzzles in understanding the nature of the living state.

### *1.1.2 Drawing from many streams of science*

Natural languages for the origins of order can be drawn from many areas in biology, including functional and comparative studies of metabolism and cell physiology, molecular and cellular architecture, the nature of catalysis, genomic mechanisms of hereditary memory and regulation, the complex and multilevel character of individuality, the ways selection and regulation interact to produce developmental programs broadly construed, the reconstructed evolutionary history of many of these systems and their functions, and a host of regularities in ecological productivity, community assembly, and dynamics in both extant

---

[5]  Crick's characterization was "a happy accident, indeed nearly a miracle."

and reconstructed ecosystems. We will summarize some of these, and pursue others in greater depth, in Chapters 2 through 5.

In addition to the biological sciences, we have paradigms of architecture and control from engineering, and important theories of stability from physics and (closely related) of optimal error correction from information theory. These provide potentially useful abstractions of functions performed in living systems, and in some cases strong theorems about the limits of possibility. They will be developed in Chapters 7 and 8.

All of these provide windows on the nature of life. They capture patterns in the living world that do not exist without life, and in a piecemeal fashion, they often partially characterize mechanisms by which those patterns are created and maintained. We would like them to define the problem of departure from a non-living planet that must be understood.

## 1.2  The organizing concept of geospheres

The problem of unifying diverse phenomena is not new with biology. Similar problems have arisen in planetary science, involving as it does a variety of chemistries, physical phases of matter, and classes and timescales of dynamics. Here a traditional coarse-grained partition of planetary systems into "geospheres" has remained useful into the modern era. The 1949 text *Geochemistry* by Kalervo Rankama and Thure G. Sahama [664] partitioned the non-living matter of Earth into three "geospheres": the atmosphere, hydrosphere, and lithosphere.[6] The concern of the authors was to provide an overview of the chemical partitioning and physical states of all matter on the planet.

The "geosphere" designations are very coarse, and to understand their use it is helpful to keep in mind the kinds of distinctions they are *not* meant to make. The geosphere partition, for the most part, does not separate regions with sharply defined boundaries; their components often interpenetrate and interact. The geospheres also do not aim at strict chemical partitions. For example, water, the primary constituent of the hydrosphere and source of its name, is present in the atmosphere, and in the lithosphere both as hydrate of minerals and as a component in trapped fluids.

Despite (and in some ways, because of) its qualitative and approximating nature, the language of geospheres is useful because it groups *multiple classes of distinctions* that are inter-related and that share the same domains of space and often similar states of matter. A coarse partition into regions and aggregate states and dynamics, by pre-empting other classifications according to specific chemical identity or sharp spatial boundaries, emphasizes

---

[6] These three coarsely defined geospheres are all subject to much more refined description. In modern usage, the atmosphere subdivides into an ionosphere, mesosphere, stratosphere, and troposphere, and the hydrosphere layers into epipelagic, mesopelagic, bathypelagic, abyssalpelagic, and hadalpelagic zones plus crustal and sub-crustal water. The lithosphere, used by Rankama and Sahama to refer to the totality of rocky and metallic zones on Earth, is now refined into zones that are in some respects almost as different from one another as they are from the hydrosphere. The term "lithosphere" is now used specifically to refer to the crust and rigid layer of the upper mantle, followed in depth by the plastic asthenosphere, the stiff (though still plastic) lower mantle, the liquid iron/nickel metal outer core, and the solid (also Fe/Ni metal) inner core. The refinement, however, only changes in degree but not in spirit the original function of qualitatively partitioning fine-scale structures and dynamics into useful aggregate domains.

that the system-level relations and interactions are the unifying concept for each geosphere, rather than an exclusive list of material components.

### *1.2.1 The three traditional geospheres*

Each of the three traditional geospheres is associated with one or a few primary groups of chemical constituents, a primary phase of matter, and a characteristic class of chemical reactions.

**Atmosphere** Gas phase, composed of small molecules made principally from non-metals and noble gases, which exist as gases over large ranges of temperature from $\sim -40$ to $\sim 100\,°C$. Primary chemistry is photolytically excited gas-phase free-radical chemistry, with some ionization chemistry in the upper layers. High activation energies of excited states produce reactive compounds, which persist only at low density.

**Hydrosphere** Liquid phase, water solutions. Oxides of nitrogen and sulfur may be present as solutes in relatively high concentrations; the concentration of metals (particularly transition metals) depends sensitively on oxidation/reduction (or *redox*[7]) state through reactions to form insoluble compounds with non-metals. Primary chemistry is oxidation/reduction, acid/base, and hydration/dehydration chemistry. Radical intermediates have high energies and are not produced except very near the surface due to screening of light by liquid-water scattering and absorption, and they quench rapidly when formed. Acid/base and oxidation/reduction reactions may be coupled due to the high solubility of protons in water, in contrast to extremely low solubility of electrons.

**Lithosphere** Solid phase, dominated (outside the core) by the crystallography of silicate and sulfide minerals, with carbonates, hydroxides, and other metal oxides as lesser constituents. Much of the chemistry of the lithosphere is physical chemistry of phase transitions including melt-fractionation, dissolution, precipitation, and stoichiometric rearrangement in solid solutions. Many phase transitions involve changes in oxidation states of metals, driven by the crystallography of silicates as a function of temperature and pressure. Changes in compatibility of minor elements with temperature and pressure can be a large determinant of pH for included fluids. Although redox changes for transition metals often result from coordination changes in crystallographic contexts, they are of major importance to the chemical activity of the Earth as a whole. Mantle convection can convert heat energy, through long-range transport across temperature and pressure zones, into redox disequilibria that are too energetic to be created by thermal excitations in near-equilibrium conditions.

   We return to give a more detailed characterization of some of these properties of the Earth in Chapter 3.

---

[7] Oxidation and reduction are introduced in Chapter 2.

### *1.2.2 The interfaces between geospheres*

#### *1.2.2.1 Complexity often arises at interfaces where matter is exchanged*

Because the same chemicals can pass between geospheres, the interfaces between them can be concentrating centers for thermodynamic disequilibrium and the emergence of complexity. For example, volcanic outgassing, believed to be the main source of the present atmosphere, can release both methane and carbon dioxide from carbon trapped in the mantle when the Earth cooled. It also supplies hydrogen, ammonia, and hydrogen sulfide. Continental weathering is a process at the interface of the lithosphere and atmosphere also involving water, which alters minerals, replenishes trace elements (particularly $Ca^{2+}$) in the ocean, and plays a major role in sedimentation of carbonates and regulation of both the $CO_2$ partial pressure of the atmosphere and the atmospheric greenhouse. The ocean/atmosphere interface, where the cross section for absorption of sunlight energy changes drastically between two matter phases, is a primary generator of surface heat that powers evaporation and drives the global weather system. On the early Earth, it was also a boundary across which $N_1$ compounds could diffuse, between a region where only nitrogen oxides could survive and one in which only ammonia could survive. In the present Earth with its marine biota, the surface (photic) zone is the major zone of primary productivity. Organisms actively regulate not only light absorption and scattering, but also the viscosity of the air/water interface, controlling rates of evaporation, droplet formation, entrainment of bubbles, and thus gas exchange between the atmosphere and oceans. Finally, the lithosphere/hydrosphere interface is an extraordinarily rich zone of disequilibria in temperature, chemical potentials, geometries, and physical properties of matter, which we consider next.

#### *1.2.2.2 The lithosphere/hydrosphere interface is particularly important to life*

Of great interest to us, as we try to situate the materials and processes of life in their planetary context, will be convective currents of sub-crustal water near spreading centers and volcanos. This water is part of the lithosphere/hydrosphere interface, and is one of the most chemically active zones of terrestrial matter. Whereas local regions within the mantle, crust, or oceans generally exist very near chemical equilibrium, the interface between the hot, convected rock and surface water is constantly pushed far from equilibrium by the mismatch between the primordial reducing character of the bulk Earth and an atmosphere driven to be more oxidizing through escape processes. The mismatch at the interface is constantly replenished as a secondary effect of the dissipation of heat from fission of radioactive elements present when the planet formed. Sub-crustal convected water systems are a particularly interesting feature of the Earth, because they depend on its composition and its internal heating, and their chemical activity depends on its internal convection as well.[8] The chemical activity at the rock/water interface is closely connected to the chemical

---

[8] Whether tectonics in the current sense of oceanic basin subduction was a feature of the Hadean Earth is currently debated. We return to this question in Chapter 3.

activity *within* living systems, and modern-day hydrothermal vent systems host rich and ancient biota capable of exploiting this overlap.

Chemical systems tend to equilibrate to within the scale of thermal fluctuations ($k_B T \sim$ 0.026 eV at room temperature) if they are not continually re-energized. The thermal activation energy of typical covalent bond modifying reactions ($\gtrsim$0.5 eV) is at least 20 times the available thermal excitation energy under conditions where liquid water exists at surface pressures. Therefore covalent bond modifying chemical activity is seen only at extremely low rates in systems that are only activated thermally. The most important physics question for a chemical origin of life within geophysics is where on Earth chemical potentials can be sufficiently insulated from one another to form large differences, but then brought together rapidly enough to drive chemical reactions rather than simply dissipating as heat.

The key to this creation of sharp chemical disequilibria is **mantle convection**. The tendency of hydrogen to escape from planetary atmospheres, leaving complementary oxidants behind, is a ready source of disequilibrium between the interior and surface of the planet. The insulating layer of the crust provides a strong barrier between these systems so that their redox potentials can move far apart. Mantle convection, resulting in volcanism and under some conditions in plate tectonics, is the force that breaks through this insulating barrier to create local disequilibria. The surface phenomenon of water circulation through heated, cracked rock – a process that is particularly efficient and active at spreading centers and faulting systems on tectonically active planets – then leads to mixing zones where disequilibria that accumulate over millions of years are brought into contact on the molecular scale.

Before the 1970s, it was believed that all life on Earth ultimately owed its existence to energy captured photosynthetically from sunlight. The discovery of hydrothermal vent systems by John Corliss and collaborators [162] using the deep submersible Alvin first revealed a diverse and thriving biota existing out of contact from sunlight, and apparently fed by minerals dissolved in vent fluids and not by detrital carbon. This life was effectively decoupled from the solar energy system except by the existence of liquid water (and, though this is now known not to be limiting, the presence of oxygen produced by photosynthesizers).

Four decades of study of microbial metabolisms and energy sources [481] have gone on to show that an enormous diversity of bacteria and archaea obtain energy from geologically produced electron donors and acceptors in both surficial and subsurface environments,[9] and that this energy is sufficient to maintain self-sufficient life and growth from one-carbon inputs, molecular nitrogen, a few inorganic salts, and trace metals. Hydrothermal systems are profuse sources of these inputs, and support life in anoxic environments that provide better models for the early oceans than oxygenated environments such as surficial hot springs. Vents were quickly proposed [161] as plausible geochemical environments for the origin of life, and since phylogenetic reconstructions increasingly suggest reductive,

---

[9] Some vent environments support not only microbial assemblies, but complex ecosystems of worms, mollusks, and crustaceans supported by these microbes.

thermophilic metabolisms occupied all the deepest branches of the tree of life, this proposal seems historically plausible as well as energetically feasible.

Within the abiotic matter on Earth, the chemistry at the lithosphere/hydrosphere inter-face most closely resembles the chemistry of life both in its general character and even in detail. Biochemistry takes place in condensed phases, meaning either aqueous solution or microenvironments created by enzymes or membranes. Gas-phase chemistry is essentially impossible, and photoionization in the strict sense (such as occurs in space) is not used. Much of the bulk of biochemistry consists of reactions that are facile in water and involve either full bonding-pair exchange (oxidations and reductions), proton exchange, or group transfers. When radicals are used, they are formed at metal centers and either hosted on metal centers (as in ferredoxins) or transferred to a limited inventory of highly evolved cofactors. Several investigators (to whose ideas we return in Chapter 6) have emphasized the similarity of biological metal centers to metal sulfide minerals that would have been present in the surface and near-surface on the Hadean Earth. Even the temperatures at which biochemistry is carried out fall within the range found in hydrothermal systems.

### 1.2.3 The biosphere is the fourth geosphere

The three geospheres of Section 1.2.1 subsume, though only in general terms, the domains of scientific knowledge that would apply to matter and events on a lifeless planet. The part of Earth as we know it that is not even qualitatively accounted for within the three abiotic geospheres naturally defines a fourth geosphere. This is the **biosphere**, a term coined by Vladimir Vernadsky [832] to refer to the totality of living systems and their interconnec-tions, and approached by us as a component of Earth's matter and dynamics. The phase of matter in the biosphere is defined not only by its physical state but even more fundamen-tally by its necessarily non-equilibrium condition. Its chemical constitution draws from a sector of covalently bonded organometallic compounds, which are not produced by abiotic processes.[10] Its chemical process comprises the reactions that these compounds mediate and by means of which they are also produced and maintained. Its characteristic activating energy scales are the barrier- and reaction-free energies typical in reactions that make and break covalent bonds among C, H, O, N, and S atoms and phosphate groups, and dative bonds of O, N, and S to metals. Its characteristic temperature covers the range for liq-uid water in near-surface terrestrial (including submarine and sub-crustal) environments, $\sim 0-120\,°C$.

In attempting to characterize what the biosphere "is," it is important to us to recog-nize commonalities with the abiotic geospheres, along with all the levels of organization described within biology, but at the same time to recognize that the biosphere is more than any one of these alone. At the outermost level of abstraction, we emphasize that the order

---

[10] Speaking more carefully: some of the compounds are not produced at all, and others, which are produced at small rates in abiotic processes, are not produced with the selectivity, yields, or functions that they take in the biosphere, by many orders of magnitude of difference.

of the biosphere is fundamentally an order of processes, that the "internal" organization of the biosphere consists of flows anchored to the exchange at boundaries with the other geospheres, and that the biosphere as a whole, rather than any organism or ecosystem within it, is the level of aggregation in which to recognize the nature of the living state.

The biosphere is a set of patterns maintained by processes, and patterns *of* processes, and not merely a collection of "living things". Living matter, on one hand, is a subset of terrestrial matter organized into all the levels and patterns mentioned in Section 1.1.1, and maintained in this order by living processes. More fundamentally, it is the processes themselves that are maintained within a state of coordination and pattern. If we adopt the view that both the relations and the ordered state of processes are fundamental to the nature of life, it becomes clear how inadequate it would be to characterize the biosphere as merely a collection of "living things." For this reason, while our emphasis on systems and relations is not fundamentally different from Vernadsky's encompassing view of the biosphere, we approach it as a planetary subsystem to more strongly shift our emphasis away from entities and toward a focus on relations and processes. We will return in Chapter 8 to argue that even the appellation "living things" assumes a category error: life is not a property inherent in things so much as things are instantiations of organizational states that arise within a larger context of life.

In Section 1.2.2 we noted that the organometallic chemistry at the lithosphere/ hydrosphere interface resembles in character the chemistry of life, and also that the energy sources at the lithosphere/hydrosphere interface sustain an ancient and autonomous biota today. In both respects, the biosphere's order is anchored in boundary exchanges with abiotic geospheres. However, an important distinction is that the reactions characteristic of *interface* chemistry in the abiotic realm become the *constitutive* chemistry of the biosphere. This is another sense in which the ordering of processes pervades the nature of life in a way that it does not pervade the nature of non-living states. It has the consequence that the organization of living matter is anchored in the conditions for matter and energy exchange with its non-living context, at a finer level of detail than is true for physical phases of matter.

From these and a host of other related observations, we will arrive by the end of this book at the assertion that *the biosphere as a whole is the correct level of aggregation from which to define the nature of the living state.* It is a system of processes, anchored historically and causally in physical laws and geochemical circumstances at many places. Accounting for the robustness of life means accounting for the long-term persistence and stability of this integrated system, across all its levels.[11] It is necessary to understand why multiple kinds of order are possible, and the roles played by different levels of organization in creating a domain of stability for the living state and ensuring that the biosphere remains within

---

[11] Sometimes, as in the case of autotrophic bacterial or archaeal species, multiple levels coalesce into a single locus: the cell is both an organism and a biosynthetically complete ecosystem unto itself; the cell coordinates metabolism, compartmentalization, energetics, and molecular replication within the same aligning framework of individuals and generations. Both from history and from comparative and functional analysis of biodiversity, however, we have grounds to distinguish these kinds of organization, and to recognize that each may exist in more general contexts and systems of coordination than only those they possess in autotrophic cells. Therefore we believe it is correct to refer to the joint preservation of all of these kinds of organization, with their distinctions acknowledged, as the essential phenomenon underlying biological robustness.

that domain of viability. The aliveness of things is not defined as a property of structure or function inherent in the things themselves, but rather by their participation within the web of processes by which the systemic integrity of the biosphere is maintained.

## 1.3 Summary of main arguments of the book

In the next seven chapters we attempt to bring together empirical generalizations and functional knowledge about properties of life and its planetary context, with general mathematical principles about the nature of stability and robustness, to frame the problem of biological emergence and persistence, and to sketch major stages that we think can be proposed with some specificity. Our approach to interpreting regularities of life, and to adopting theoretical frames more generally, will be gradual and will proceed along several threads in parallel.

We will begin by characterizing the biosphere at a very aggregate phenomenological level. What makes life, in its planetary role, and viewed at a system level, unlike the union of the other three geospheres? We recognize the role that evolution plays as a mechanism for imparting and maintaining living order, but we also recognize that evolution as a distinct process depends on the prior organization of living matter into modes of individuality, which is a complex problem. We argue that evolution belongs within a wider class of order-forming processes, some of which have less complicated preconditions and play different roles in the maintenance of living order.

We believe that the problem of maintaining the biosphere within an asymptotically stable operating range, when faced in the full enormity of error, displacement, or degradation that can enter every atom, bond, structure, and process of life, presents the largest conceptual challenge to a theory of the origin of the biosphere and the nature of the living state. The paradigm most likely to address this problem correctly comes from the mathematics of cooperative effects responsible for thermal phases and phase transitions, which is also the mathematics of asymptotically optimal error correction. We argue that cooperative effects, acting to produce dynamical phase transitions, provide error buffering that is essential to maintaining the hierarchical complex systems that constitute the biosphere. Some kinds of dynamical phase transition arise in population processes and thus characterize the aggregate dynamics of evolution, but the concept applies much more widely within the domain of processes that we argue contribute to biological order.

Ordered phases form in response to their boundary conditions, and the boundary conditions for life are the chemical disequilibria created by planetary geochemical activity and (secondarily, we claim) by the flux of visible light from the Sun. The aggregate function of the many ordered phases of life is to conduct energy from sources to sinks through cycles of chemical reactions. The technical question whether the free energy in the non-equilibrium boundary conditions is sufficient to *drive* a biosphere into existence as an energy channel defines an appropriate criterion of necessity for the origin of life.

Phase transitions act through self-reinforcement to introduce robust order into nature by coordinating random small-scale events. Because part of the order in thermal phases is law-like and non-arbitrary, the points at which phase transitions can arise to buffer errors provide a skeleton of lawfulness that anchors the open-ended variation and complexity of evolution of particular species or ecosystems. Because some of the continuity of lawfulness reflects explicit properties within chemistry, we argue that in some chemical properties the biosphere is continuous back to the earliest metabolic departure from geochemistry. In this way we attempt to go beyond mere empirical generalization and provide a sketch of a way to *use* principles – if not an adequate demonstration at the required level of chemical detail – to connect the problem of the origin of life to the understanding of the organization, variation, and persistence of the biosphere as we know it today.

### 1.3.1 An approach to theory that starts in the phenomenology of the biosphere

To explain what we mean by the gross phenomenology of the biosphere, we imagine confronting the Earth as it might be experienced by an alien visitor who came here expecting to find a world of rocks, oceans, and atmosphere, but no life, and who instead found the planet we know. How are the most basic functions and structures of the planet different because it harbors life than they would otherwise be? How is life responsible for these differences, and what are the essential characteristics of living structure that span its internal heterogeneity and are common across historical eras? Essential properties of life, beginning with the most general and becoming progressively more specific, include the following.

**Living systems are chemical** The only life we know is a chemical system. This means at least three fundamental things. First, living states are delimited at the microscale by the quantum mechanics of atomic and molecular orbitals. Their dynamics is governed by the orbital-scale dynamics of reactions and a few extended-electron states in organic molecules, and by the physical chemistry of molecular assemblies. It is not necessary to probe scales below the quantum mechanics of electronic states to capture all essential foundations of living structure. Second, the living world inherits the complexity of chemical systems as its microscopic foundation. It may build further complexity by selecting among chemicals in a variety of ways, but it does not need to create the complexity of the chemical state space itself. Third, the combination of a quantum mechanically defined discrete state space, together with discrete reactions divided by energy barriers, enables chemical systems to maintain large differences of free energy within small distances – on the order of atomic radii. Life differs from the phenomena of weather and climate in taking its most basic structures from molecules rather than from soft structures such as diffusive boundary layers.[12]

**Life is dynamical** Life is an ordered assembly of processes. Living systems operate out of thermodynamic (principally chemical) equilibrium. Energy flowing through them is partly captured to construct states that would be improbable in equilibrium systems. In turn

---

[12] Although it may use the latter opportunistically, they are not a foundation for the overwhelming majority of its structure, and perhaps for any essential structure.

these disequilibrium states carry living processes that otherwise would not occur, including those that conduct energy flows between non-equilibrium boundary conditions that furnish sources and sinks. It is conventional (and correct) to say that the order in living states cannot (in most cases) be understood except in the context of the processes that build and maintain them, but it is desirable to go beyond this to emphasize the symmetrically interdependent character of states and processes in the biosphere. The most important meaning of life's being "dynamical" is that it maintains order inherently in a system of processes, as we noted in Section 1.2.3. Non-equilibrium processes may depend on more or less complex and extended histories, ranging from near-instantaneously determined outcomes to outcomes that are highly historically contingent.[13]

**No one level or form of biotic order serves as a source for all the others** The diverse forms and levels of organization we find in the biosphere do not all seem to be accounted for by dynamics at any privileged level *within* living systems alone. No one kind of living order serves as a foundation from which all the order in life grows, and no distinctively biotic process appears as the source of maintenance at all levels. As we show in Chapter 5, where we review some of the hierarchical complexity in cells, bioenergetics, and molecular control, life consists of subsystems, which are partly integrated and partly autonomous, and are brought together in cooperative assemblies to form living wholes. We characterize life as a "confederacy" of different sources of order, many of which we argue have independent origins within different domains of chemical or physical processes, or planetary conditions. The robustness of the full suite of living regularities results from a parallel appeal at many levels to boundary conditions and constraints of physical laws. The function that distinctively biotic dynamics performs uniquely is to *interconnect* these members of the confederacy into webs of mutual support and interdependence. A view of life as an integration of multiple disparate sources of order may explain how the emergence of life on Earth could have been at the same time an outcome of quite ordinary events, yet one that depends in detail on its planetary context.

**Ecosystems are more invariant than organisms** We noted in Section 1.1.1 that the small core of metabolism is essentially universal[14] if we define it by asking which pathways must have been traversed in the course of synthesis of essential molecules, anywhere within an ecosystem, without regard to whether an essential pathway was carried out within a single cell or distributed across cells by means of trophic exchange of pathway intermediates. As we will show in Chapter 2 in more detail, the relation of core metabolism to bioenergetics is also simple at the ecosystem level, with the major distinction falling between ecosystems that, in aggregate, rely on geochemically provided donors of energetic electrons for biosynthesis, versus those that produce their own electron donors using energy from sunlight.

---

[13] In adopting the term "historical contingency" we follow the usage established by Stephen Jay Gould [312].

[14] We provide a more precise characterization in Chapter 4, which takes account of variations in core pathways by showing how they may reflect redundancy, and factors out the ecological complexity that some organisms may synthesize complex metabolites that other organisms then degrade to produce their simpler precursors.

Moreover, in ecosystems that generate electron donors from sunlight, the biosynthetic networks into which those electrons are fed are essentially the same as those in ecosystems driven directly by geochemical free energy sources.

Any comparable simplicity or universality is emphatically *not* a property of most species considered separately, which consume organic carbon within complex ecological contexts, to provide either energy or biosynthetic intermediates.[15] Thus ecosystems, which in aggregate must be biosynthetically and energetically self-sufficient, assemble a limited inventory of core processes in ways that are much more invariant than the phenotypes of organisms, and are partitioned according to whether they use geochemical or light energy to generate the electron donors required to synthesize organic carbon.

**Universal metabolism is an ecosystem property** If the simplicity and universality of core metabolism and bioenergetics are expressed in ecosystems whereas they are not generally expressed at the level of organisms, metabolism is in some respects more a property of ecosystems as units of organization than a property of organisms. The added complexity and diversity found among species largely arises in response to the problem of becoming a complementary specialist within a community. Specialization requires the evolution of mechanisms to acquire, transport, and break down organic compounds to deliver to an organism those metabolites that it does not synthesize. Arriving at a stable community dynamic requires balancing trophic fluxes, as well as internal pathway fluxes, through a combination of physiological regulation within member species and adjustments in species' relative population numbers. An ecological community in steady state should minimize waste if it is not to be easily displaced by more efficient alternative community structures. The resulting complex network of constraints, involving gene gain or loss, regulation, and population dynamics, ostensibly supports an enormous variety of possible but mutually exclusive solutions [677].

Thus, while organisms provide the platforms within which metabolic reactions take place, ecosystems carry the patterns of metabolic invariance and record episodes of innovation that move the boundaries of aggregate metabolic constraints, for instance enabling new geochemical environments to be colonized. The organism, as a carrier of a pattern, is an enabler in the short term, but in the long term gene transfer permits metabolic capabilities to assemble in combinations different from those in which they originated,[16] making the organism as a unit of aggregation less important. At all timescales, however, organisms remain important in aggregate, as the carriers of complex networks of constraint for the problem of complementary specialization. These constraint networks may determine regulatory or adaptive flexibility, and the tempo and mode of innovation, which are aspects of community assembly more than of the fundamental chemical constraints on metabolism.

---

[15] The distinct ways of life available to carbon consumers are so diverse that current attempts to sample them using computational models of metabolism cannot even provide reliable estimates of their number (Andreas Wagner, personal communication). For efforts to sample the structure of this diversity, see [677].

[16] We provide numerous examples even within the restricted domain of carbon fixation in Chapter 4.

### *1.3.2  Placing evolution in context*

Without doubt, since Charles Darwin's 1859 publication of *On the Origin of Species by means of Natural Selection* [177], evolution in one or another variant on Darwin's framing has become biology's unifying explanatory system [311, 493, 534]. It plays the role in biology that mechanistic notions of causality play in the physical sciences. To the extent that it differs from physical causation – accepting history dependence and relatedness as explicit alternatives to prediction from first principles, as a criterion of scientific explanation – evolution is viewed by many as defining what makes life different from non-life.

As a consequence of the central place evolution is given in biological thinking, many approaches to the origin of life include evolution as a defining characteristic of the system they seek to explain.[17] In many cases, the path of origin proposed is explicitly motivated by a goal of arriving as directly as possible at a chemical system that can be described in Darwinian terms.[18] Thus evolution becomes not only the criterion by which an origin of life is defined, but also the mechanism by which it is assumed to occur.

From our phenomenological approach to life as a planetary subsystem, a central emphasis on evolution poses a problem. Evolutionary processes, as a class, are widely applicable mechanisms that produce a tendency toward order, but their scope is limited and they rely on relatively complex preconditions to be realized. Evolution is neither an exclusive nor an all-encompassing framework for the formation of dynamical order, but only one domain within a larger class of processes that must be considered. Here we will summarize the central concepts that define evolutionary processes as a coherent category, note their limitations, and explain the role we believe they play within a larger framework that is needed to understand the full variety of order that the phenomenology of life includes. We return to a more detailed treatment in Chapter 8.

### *1.3.2.1  Darwinian evolution as a Kuhnian paradigm*

> In the beginner's mind there are many possibilities,
> in the expert's mind there are few.
>
> – Shunryu Suzuki, *Zen Mind, Beginner's Mind* [788]

The acceptance of Darwinian evolution as an explanation for order and function in living systems was, perhaps more than many scientific revolutions, a paradigm shift in Thomas Kuhn's sense of the term [459], with both good and bad consequences.[19] To the good,

---

[17] An example is a widely circulated definition reached by a NASA panel: "Life is a self-sustaining chemical system capable of Darwinian evolution" [403].

[18] This is one of the expressed motivations to look for directly self-replicating RNA catalysts as a foundation for the departure from non-life to life [122, 488]. Non-RNA-based approaches likewise invoke evolution, as in the compositional inheritance models of Daniel Segré and Doron Lancet [721, 722, 723].

[19] Kuhn discussed Darwin's formulation of natural selection as the paradigm that had struggled to gain acceptance against entrenched ideas of goal-directedness, which had framed all theories of change in living systems, including Lamarck's version of evolution [467]. The limitations of Darwinian evolution as a paradigm in its own right were yet to be become clearly visible. Some gaps in Darwin's knowledge, when filled, only simplify and reinforce his formulation. For instance, Darwin did not write

evolution by natural selection is a correct framework within which to explain an enormous range of adaptive functions and ordered population states. For the first century after Darwin's *On the Origin of Species*, these included only populations of organisms, but within the past 50 years the explanation has been extended to include many kinds of sub-organismal populations – cell populations governing tissue formation in embryogenesis [567] or antigen specificity in immunogenesis [148], cell processes and synapses in brain development [126], etc. – so that evolution is also recognized as part of the mechanism by which developmental programs are implemented. From the work of William Hamilton and successors [273, 332, 333, 334], the scope of evolution has also been extended outward to describe competition and selection among a potentially unlimited variety of kinds of groups, within the same algorithm that applies to organisms. Evolution by natural selection is thus a very flexible and general algorithm for producing order in populations at many levels.

A detrimental effect of raising evolution to the status of a paradigm is that it creates a default explanation for biological order, which is becoming increasingly exclusive. It is difficult to find serious biological writing that does not suppose – does not feel *obliged* to suppose – that when a mechanism for producing order, function, or stability has been most fundamentally understood, that understanding will reduce to an explanation in terms of evolution. What is true in biology more generally is true for the origin of life in particular.

Imputing notions of cause or sufficient explanation is often one of the trickiest and most provisional efforts in science. The same empirical regularities, viewed through experience in different domains, can trigger very different default explanations, and each of these is a window on the phenomenon. For the origin of life, which is at the same time a phenomenon in geophysics and chemistry, and also the beginning of biology, it is perhaps easier to shift among paradigms than it is from the vantage point of any one discipline in isolation. We will argue, however, that what the origin of life pushes us to recognize, about sources of order and stability, should ultimately restructure our understanding of the living world including the role of evolution.

### 1.3.2.2 Three forms of evolutionary default interpretation to avoid

Three assumptions about the role of evolution either presume results that should be derived, or pre-emptively frame the problem of understanding the living state in terms that may not recognize all relevant mechanisms. We wish to avoid these assumptions.

---

about bacteria or other microbes, though they had been discovered by van Leeuwenhoek almost two centuries earlier, and were the subject of Louis Pasteur's experiments on sterilization, for which Pasteur won the Montyon prize [287] in the same year *On the Origin of Species* was published. Other omissions would require almost a century to gain sufficient coherence to enable a critical analysis of the Darwinian framework. Although Darwin was a consummate naturalist, aware at every turn of the complex dynamics of species interactions, the term *Ecology* would not be introduced to denote a scientific field until 1866 by Ernst Haeckel [326]. Regularities in macroecology [99, 521], the molecular biology of development and heredity [179, 295], and the diversity and complexity of lifecycles in many taxa of eukaryotic algae and small metazoans [354], which exemplify the complexity in formulating concepts of individuality, would not come to be understood even in outline until late in the twentieth century.

1. **Supposing that Darwinian selection has sufficient power and scope as an error-correction mechanism to explain all of living order** The potential error is one of false generalization: finding that selection is sufficient to trap errors in a subset of dimensions of variation, but then failing to quantify all dimensions of variation that produce error, and supposing that they are somehow trapped as well without requiring different mechanisms. Within the scope of population models, where the levels and units of selection are given as inputs, some criteria already exist showing the limits of selection's ability to maintain order even in the short term.[20] When the requirements are extended to indefinitely long-term maintenance of complex patterns, and the sources of error are recognized to include the full range of disruptions in both states and events reaching down to the chemical substrate, we anticipate that the problem of persistence will become more like the problem of forming long-range order in condensed matter physics [307, 885]. Here the difficulty of forming stable order has been found to be severe, despite the fact that the systems studied are much simpler than those studied in biology. We return in Section 1.3.2.5 to argue that the Darwinian framework for selection requires support from other error-correcting mechanisms that operate in simpler contexts, to arrive at a mechanism sufficient to explain the emergence, overall organization, and long-term persistence of life from non-living precursors.

2. **Supposing the distinctive character of life must be traceable to uniquely "biotic" order-forming processes** We wish to avoid supposing that because the living state is distinctive, that distinctiveness must have been produced by a process that is likewise distinct from processes at work in the non-living world. In particular, we will argue that Darwinian evolutionary dynamics arises as an emergent process *within the living context*, but that the reverse is not true: the distinctiveness of the living state cannot be accounted for solely in terms of the role evolution plays within it.

3. **Supposing the essential order-forming processes for life are of any single kind** Finally, while Darwinian evolutionary processes contribute to the dynamics of all living systems today, we believe it is an error of false conceptual reduction to suppose that competition and selection within Darwinian populations will thereby be the source of explanation for all relevant forms of order. The universality of metabolism offers a concrete case in point to illustrate that evolutionary mechanisms may be part of a system's dynamics but may not offer the level of description needed to understand its order. To be sure, selection has acted on genes, on chromosomes, on cells, and likely at many other levels, throughout the history of life. At the same time, the modes of evolution have changed significantly through major transitions in genome, cell, and organism organization [106, 180, 227, 790]. Simply knowing that competition and selection have occurred leads to no specific predictions for why metabolism is an ecosystem property, why the universal form we see exists, and why its conservation has apparently been unaffected by major changes in the evolving systems that carry it. More generally, the *co-evolutionary* dynamics among heterogeneous populations in ecosystems may show

---

[20] The best known are Muller's ratchet [581] and the Eigen error threshold [213, 214].

long-term constraints and convergences, which are not themselves traits evolved under competition and selection. Again, invoking the selection of the member species does little to elucidate the jointly formed pattern, though the pattern may be clearly expressed in other terms. We will list examples from macroecology in Chapter 8.

### *1.3.2.3 An alternative breakdown of biodynamics into three layers*

We think a more useful approach to the emergence of evolution is to recognize the full transformation as an accretion of three distinct layers of function having different levels of complexity.

1. **The ability to preserve a dynamical pattern essentially indefinitely** The universal feature of life to be explained is its capacity to preserve a distinctive, dynamical, chemical pattern in a planetary context, apparently indefinitely, and under the full range of planetary perturbations from microscopic fluctuations to astrophysical disturbances. Some details of the dynamical pattern, such as species identities and ecological community structures, change through time apparently without end, others change within limited ranges, and still others, such as chemical motifs in core metabolism, may not change at all. The existence and degree of change is *secondary* to the existence of a stable dynamical state of chemical order.

2. **The emergence of forms of organization that bring the Darwinian abstractions of replication, competition, and selection into existence** The second problem in the emergence of living dynamics to be explained is the emergence of organizational forms that can live and reproduce autonomously, and can therefore undergo competition and Darwinian selection. Note that the emergence of a Darwinian process does not by itself imply that the process supports unlimited variation. For example, within core metabolism, enzymes for specific reactions confer the capability of autonomous carbon fixation under wide but still finite limits of pH or oxygen fugacity (we will review these in Chapter 4). More generally, systematic adaptations in protein composition may shift optimal growth over finite ranges of temperature or salinity. For these features, Darwinian adaptation is a source of robustness and environmental flexibility, which does not require (and has not received) a wide range of innovation.

3. **The capacity to support sufficiently complex states that essentially "open-ended" variation becomes possible** Within the systems that undergo Darwinian evolution, we must then understand how it becomes possible to maintain such complex states that at some levels the evolving entities become capable of essentially open-ended variation. The most obvious horizon for the generation of a state space too large to be sampled was the production of oligomers of RNA, amino acids, and later DNA. The chemical underpinnings for such a transition, and especially the integration of RNA and peptide systems into the process of ribosomal translation, pose problems of enormous difficulty. For them to have arisen in an environment already possessing considerable chemical and energetic order is already difficult to understand; for them to have been a precondition for the creation of lower-level order seems to us impossible.

While all three of these capabilities are hallmarks of life, they are conceptually independent. Only the last two are evolutionary, and only in the last does adaptation become exploratory as opposed to merely responsive.

### 1.3.2.4 The universe of order-forming, Markovian stochastic processes

We wish to understand the special place of evolution within a context provided by the wider class of processes that share a concept of emergent order relevant to the structures of both matter and life. They have in common that they are all *stochastic*: the events of interest, at a microscopic scale, can all be treated as random. The emergence of order is defined by a law-like reduction in the range of this microscale randomness, but the fundamentally random nature at the small scale remains, and it is what makes the formation of order difficult. The processes, as a class, are also *Markovian* [232]: the present state of the world, described in sufficiently fine detail, contains all effects from the past that affect the trajectory of the future. This set includes the full range of phase transitions in equilibrium and non-equilibrium bulk processes [307, 506], it includes models of reliable error correction in information theory [732], and it includes Darwinian evolution. As we will explain in Chapter 7, a shared mathematics[21] associated with robustness lies behind all of these phenomena. We expect that, as a fuller understanding of development, physiology, ecological dynamics, and population processes is formed, many more classes of robust dynamics from these fields will be added as new distinct examples to the list above.

### 1.3.2.5 The framework of Darwinian evolution is predicated on the emergence of individuality

As Stephen Jay Gould argues in *The Structure of Evolutionary Theory* [312], the essential framework of evolution laid down by Darwin contains all of its major distinguishing assumptions, though details changed in the ensuing 100 years leading to the modern synthesis of Fisher, Wright, and Haldane [652], and even in some emphases made by Gould himself. A widely used concise statement of the key abstractions that define an evolutionary dynamic was given by Richard Lewontin in 1970 [482]. Paraphrased, they are the following.

Evolution is necessarily a *population process*. Members of the population must be sufficiently similar to be regarded as parallel copies of some common template, and to compete for the same niche. The population must persist via reproduction of its members, the members must be capable of some degree of variation, and variations (along with the common template) must be passed down more or less faithfully under reproduction.[22] The change

---

[21] This mathematics grows out of the combinatorial properties of large numbers, and it goes under the heading *large-deviations theory* [224, 811].

[22] Here we have deliberately used common-language terms, such as "members" of a population, and "reproduction," for many of the same reasons that we used descriptions in terms of gross phenomenology in Section 1.3.1. Technical terms of art, as a price of being more formal and explicit, often involve many theoretical premises, which we do not wish to take for granted and in some cases wish to modify.

A very common term of art that we will usually avoid is "replicator" [181]. This term presumes the existence of entities that are literally copied during reproduction, whereas most aspects of reproduction involve some degree of assembly as well

in the composition of the population over time results from random sampling in the events of reproduction and death, and from non-random selection by the environment of who reproduces (and how prolifically) and who dies. It is by means of the non-randomness in selection that information about the environment comes to be reflected in the composition of the population, a condition that is referred to as the population's becoming adapted to its environment.

The most important assumption that sets evolutionary processes apart, within the larger class of Markov processes, is what evolution assumes it means to be a "member" of a "population." The operative concept is one we will call **individuality**. We will characterize it informally, but ultimately it is a statistical concept extracted from properties of interdependence and autonomy among components within the physiology or reproduction of a living system.

We refer to the two concepts that set individual-based dynamics apart from continuum dynamics as "granularity" and "shared fate" of characters. Individuals in a population process are collections of parts, which are interdependent within an individual, and independent between individuals. The interdependence and independence may be matters of degree and need not be absolute, but in practice living systems often produce large, qualitative changes of dependence between intra-individual components, the individual and its environment, and between individuals. The step-like character of the degree of interdependence at the boundary of an individual identity gives the dynamics in individual-based systems a character we call **granularity** to contrast it with the behavior of continuous systems, much as the rigid interdependence of constraints in granular flows lead to dynamics very different from those in fluid flows [19, 46, 47, 489].

The granularity of individual states also leads to reproduction that is discrete in time, and the second characteristic essential to individuality is that the components within an individual tend to be lost or to be reproduced jointly, and thus to have **shared fates.**[23] Shared fate distinguishes individuals from randomly formed coalescences of components, making components that are reproduced together predictable by each other. It is the intergenerational counterpart to the intra-generational functional interdependence characterized by granularity.

Any order-forming process that qualifies as evolution is predicated on the existence of a corresponding form of individuality in terms of which competition and selection are defined. The emergence of forms of individuality is a process that we expect to be dynamically or algorithmically complex. Dynamical coordination of components is a problem of maintaining long-range order, and our experience with long-range order in equilibrium systems has suggested that this is possible only in limited circumstances. We also observe

---

as copying. Insistence on a materialist reification of a replicating entity has led to often unproductive debates on the validity of "genic" versus multilevel selection [181, 880], which we mention in Chapter 8. These obscure the more important point that incomplete or probabilistic transmission of patterns is the central process of interest, to which the construction of formal models must adapt.

We will use common-language terms as category terms, introducing technical terms where we can make them operational.

[23] Fate may be shared only probabilistically, at many levels in a hierarchical system. Therefore many nested notions of individuality may be appropriate to characterize a complex living system, such as gene, chromosome, or organism.

that only some of the robust patterns in the living world appear to have an individual-based organization. A case in point, as we noted above, is that although metabolism seems to require a cellular milieu to exist under the competitive conditions of a world with evolved organisms, the integration of biochemistry into a self-sufficient system does not usually depend on maintenance of genes for a complete biochemistry within a single genome. In most cases, it is maintained through feedback in the more fluid architecture of ecosystems.

The problem of stabilizing a form of individuality depends on a complicated process of selecting more robust individuals within a population, and creating environments in which the selective forces on components that make up individuals limit the forms of variation they can generate. Some of the process of stabilization may be mediated by transmitting selection criteria up or down within a hierarchy of nested levels of Darwinian dynamics [106]. However it is accomplished, the essential requirement is for sufficient system-level feedback to compensate for destabilization at all levels. This feedback may be carried by either individual-based or more continuous degrees of freedom. Stabilization becomes an easier problem for systems that are inherently capable of less open-ended variation, so we expect that these play an essential role as *reference states* for more variable forms. We emphasize the importance of distinctive but unchanging forms of biological order, such as metabolism at the ecosystem level, because we believe this order reflects the template that stabilizes the entire hierarchy of forms of individuality and their associated levels of Darwinian dynamics.

Understanding closure of error correction in hierarchical dynamical systems promises to be a complex and technically difficult problem even when the questions are properly framed. To understand why that problem has been solvable by living systems, we look for structure within the order-forming process that simplifies problems of error correction and stabilization. Although randomness, stochasticity, and error occur in all microscopic events in living and non-living matter, they are more contained and easier to correct in some domains than in others. This difference of containment in the universe of random events is what we refer to as structure within the order-forming process. Our argument will be that affordances for less costly and more reliable error correction determine to a considerable extent the organization of life today, and there is good reason both empirically and theoretically to believe they also dictated some stages in its emergence.

### 1.3.3 Chance and necessity understood within the larger framework of phase transitions

If life is a planetary phenomenon, then the emergence of life was a conversion of the state of the Earth. The question what "kind" of conversion this was includes the questions whether it was an unlikely or likely event sequence, whether the persistence of life indicates that its continuing existence is in some way favored over its spontaneous disintegration, and whether the life we know is somehow uniquely required by natural laws (at a coarse level if not in all details) or whether a starkly different alternative could have emerged and persisted in its place.

### 1.3.3.1 System rearrangement: collective and cooperative effects create global order from locally random events

> aye, chance, free will, and necessity – nowise incompatible – all inter-
> weavingly working together. The straight warp of necessity, not to be
> swerved from its ultimate course – its every alternating vibration, indeed,
> only tending to that; free will still free to ply her shuttle between given
> threads; and chance, though restrained in its play within the right lines
> of necessity and sidewise in its motions directed by free will, though
> thus prescribed to by both, chance by turns rules either, and has the last
> featuring blow at events.
>
> – Herman Melville, *Moby-Dick*, Chapter 47, The mat-maker [553]

Questions of chance and necessity, of predictability versus historical contingency, will not be answered by any simple appeal to empirical generalizations in extant life, or by merely listing facts about chemical synthesis from laboratory systems. They must be framed within a larger context of principles, to enable us to judge which facts are relevant and why, and to enable us to abstract from empirical generalizations to causes.

The framework that we propose should capture the roles of chance and necessity is one that originates in the theory of **phase transitions.** In the thermodynamics of ordered phases and the transitions between them, all events are random and unpredictable at the microscale in space and in time. Some chance events can propagate that randomness up to large-scale historical contingency, but for many others the only lasting consequence is a joint participation in a kind of "system rearrangement." The boundary conditions on a macroscopic system can act to filter collections of microscopic events, favoring configurations that respect certain forms of long-range order throughout the system. While the small-scale events are unpredictable, the favored states of order can be predictable and can depend in specific ways on the boundary conditions.

The abstract question, which an understanding of the detailed chemical mechanisms of life must teach us to pose in the correct way, is whether a plausible emergence of life could have occurred as a consequence of a unique and rare event sequence, or whether it must have resulted from a system-level re-arrangement, away from a less favored to a more favored organizational state of the Earth's matter and energy flows. The theory of phase transitions encompasses both the robust order within stable phases, and the amplifying effect of instability, at the cusp of a transition, on those few random events that are most likely to seed the transition. Even when a system changes phase – when a long-range internal rearrangement occurs – the change occurs along limited channels, and its likelihood or its uncertainty are governed by boundary conditions much as the ordered phases are.

The theory built up to explain long-range order in random systems also explains an empirical observation: that order forms under restricted circumstances. Attempting to apply the same filters to the rearrangement of terrestrial matter into a biosphere should distinguish plausible from implausible paths of emergence.

The mechanism that underlies the formation of ordered phases is mutually reinforcing interaction among many small-scale, individually stochastic degrees of freedom known as **collective or cooperative effects** [307, 308, 506]. Ordered phases form where the redundancy of cooperative effects creates a sufficiently strong tendency toward order to overcome a tendency toward disorder that is essentially combinatorial: the condition of being disordered is less restrictive and therefore can be met in more ways. Whereas disorder is generic, sufficiently coherent interactions to produce order are rare, and for this reason the plausible mechanisms to produce any form of order that we observe as robust and stable are limited.

Much of what is understood about phase transitions has accumulated over more than a century of experience with equilibrium systems, including the fundamental particles and forces and thermal states of matter. However, as the essentially mathematical nature of order created through cooperative effects has come to be better understood and communicated across scientific domains, it has become clear that phase transition is a mathematical concept, applicable in the domain of processes or information systems, as much as in the domain of classical theories of matter.[24]

### 1.3.3.2  The emergence of life was a cascade of phase transitions

Our thesis in this book is that the emergence of life should be understood as a cascade of dynamical phase transitions, as matter in an energetically stressed young planet was rearranged into conduits for energy flow. The function of these conduits, which comprise the ordered states and events of living matter, in a planetary context is to mitigate the accumulation of chemical potential stresses.

Before the period from roughly 1955 to 1975, phase transitions were seen as the objects of a domain-specific theory: a description of a restricted class of phenomena like many other descriptions in physics. The change that occurred over this period, through a coalescence of ideas in several domains, was that the theory of ordered phases came to be understood as an overarching framework for understanding robustness and stability. Phase transitions are not merely isolated events; they form hierarchies where they bring into existence the modes of order that we recognize as elementary entities and interactions. For the same reasons as phase transitions produce the stable states of matter, they also describe the limits of reliability in information systems, and form the basis of a very large part of our modern understanding of error correction and reliable inference.

Cascades of phase transitions organize states and dynamics in natural systems into layers or levels that, though inter-related, have internally consistent and somewhat independent characterizations. Even when we know that an ordered phase exists as one level within a cascade of stages of emergent order, we can characterize the level of interest without a complete knowledge of the hierarchy in which it is embedded, a phenomenon known as **universality**. The implication of a phase transition paradigm for life – that its stages of

---

[24] Manfred Eigen develops this perspective on overcoming the threshold for reliable replication in information systems, in [211, 212].

emergence and its internal modules are subject to some degree of universality – will allow us in Chapter 6 to sketch a sequence of stages of emergence. More importantly, it is the feature that has enabled the biosphere's own dynamics to assemble hierarchical complexity without drifting into intractable problems of error propagation. The biosphere is *reducibly* complex. In the phase transition framework, the source of biological reducibility is ultimately the same as the source that makes reductionism successful in the rest of natural science.

### 1.3.3.3 A theory of ceilings and floors

The picture of a cascade of transitions as the path to complex order has precedent in equilibrium systems, because it is the basis for our current hierarchical theory of matter. A cascade of nested "freezing" transitions at successively lower temperatures (reviewed in Chapter 7) creates the inventory of elementary particles and then the states of cold condensed matter.

We will argue for a similar cascade of transitions that produced living matter from non-living precursors on a prebiotic Earth. The difference between the phase transition cascade of matter, and the cascade to life, is that order in matter results from the constraints of limited energy, whereas the cascade to life results from the constraints of the need to support energy flows through chemical pathways. The difference between temperature and stress as sources of order is fundamental to much of what makes living matter different from the merely "physical" phases in non-living matter. Temperature as a boundary condition makes **energy** the constraining factor leading to equilibrium order. Stress as a boundary condition makes **transport currents** the constraining factor leading to biological order.[25] Transport currents are inherently dynamical properties; hence, the order of life can only be understood in dynamical terms.

A principle of fundamental importance, learned through experience with equilibrium phase transitions but applicable to phase transitions more generally, is that each transition is a kind of qualitative boundary that separates the descriptions required "above" and "below" it. In energetic hierarchies, a melted phase lies above any transition, and a frozen phase lies below it. In a cascade of transitions, every ordered phase has two boundaries: one above, which brought properties of that phase into existence through a freezing transition, and one below, which will go on to freeze out some of the current system properties and create an even more ordered (more intricately frozen) phase. The two boundaries make a kind of "ceiling" and "floor" for the scientific description of the ordered phase that falls between them. Most details of the fine structure that lies above the ceiling do not need to be understood to describe the dynamics in the phase below the ceiling, because the dynamics in the fine-scale details have been frozen out and are inaccessible. Likewise, any accidental properties of frozen order that may arise in the phase below the floor do not need to be predicted to know what constraints the dynamics above the floor place on all possible

---

[25] For a worked example in an extremely simple system, see [763].

ways of freezing.[26] A valid scientific description of any ordered state produced by phase transition, between its ceiling and its floor, can be largely self-contained.[27]

### 1.3.3.4 Emergence makes reductionism possible

One sometimes sees emergence put forward as an alternative to, or even a refutation to the validity of, reductionist science, but this is a misunderstanding of reductionism and ultimately a mis-appropriation of the term. Properly understood, reductionism consists of two observations. First, properties of components place limits on the kinds of assemblies that can be made from them. Second, although the possible assemblies of a collection of building blocks are typically much more numerous and diverse than the building blocks themselves, if the goal is to characterize the building blocks, the number of well-chosen experiments that must be performed is comparable to the diversity of the building blocks alone, not of the much more numerous assemblies that could be made from them. This remains true even if the building blocks are not directly accessible, and their properties must be inferred indirectly from experiments carried out on assemblies of them.

Ceilings and floors cut off a potential infinite regress in the reductionist program of inferring properties of building blocks. This cutting off has been essential to the formulation of a consistent theory of matter [861]. One only needs to look through one ceiling at a time, to characterize the relations of parts to wholes; one does not need to jump immediately to a theory of everything. In other words, emergence is the phenomenon that makes reductionist science possible in practice. We will be careful in extending lessons learned from equilibrium into the dynamical domain of life, where feedbacks across levels can become complicated. However, the reasons these relations are true in equilibrium are ultimately mathematical. While care may be required to understand where they are realized in the constraints on living systems, we believe the same ideas will inevitably apply.

### 1.3.3.5 The "collapse" into the order of life

One of the most longstanding questions physical scientists have posed about the biosphere is: Why doesn't it all collapse to disorder?[28] More formally: Why don't the arrangements of living matter degrade to distributions that more closely resemble those of equilibrium systems? In equilibrium systems we have come to associate disorder with collapse, because in closed systems (and even in many open systems), maximum entropy reflects both the greatest disorder and the largest likelihood.[29] Collapse occurs when a system wanders away from an orderly, and therefore improbable, initial condition, into more disordered conditions from which it never returns.

---

[26] Expressed in very informal terms, if bricks can be used to build houses, there are many things about the assembly of any particular house that do not need to be anticipated to understand the capabilities and limitations that the bricks place on all possible houses that could be built from them.

[27] An immediately familiar example is chemistry, which is a self-contained theory of dynamics that takes orbitals as its building blocks. Only a few parameters from the underlying quantum mechanics that derives those orbitals are required in order to entail all of their molecular consequences.

[28] See Section 7.6.1.1 for some historical examples of this question.

[29] For open systems in equilibrium, it is understood that the appropriate measure of entropy includes terms from both the system and its environment, so generally these are various *free energy* functions [441].

However, life is not merely (and not even principally) a collection of things, and the equilibrium entropy captures only part of the regularity we seek to explain in the biosphere. We will argue that life emerged early and has persisted robustly because the origin of life was actually a transition away from a less stable planetary condition devoid of life, and into a more stable condition that includes a biosphere. In entropic terms, this transition was still a "collapse" from an improbable to a more probable phase, but the stable phase in this case was the dynamically ordered living state.

Even at equilibrium, the idea of a collapse into order is not new or radical in open systems that can undergo phase transitions: it happens every time rapidly cooled water vapor nucleates its preferred state of frost, or whenever a supersaturated cloud condenses into a downpour. The frost and the liquid raindrop are both, in entropic terms, more ordered than the phases from which they formed, though the energy loss that makes them more ordered accounts for an even larger amount of entropy as heat in the environment. Systems that undergo non-equilibrium phase transitions can collapse into order in ways even more intuitively like the emergence of life: this happens whenever a fracture suddenly forms and propagates in a stressed elastic solid, or a lightning strike forms across a gap in the atmosphere between a charged cloud and the ground.

The last half of Chapter 7 and the synthesis in Chapter 8 explain how our picture of a "collapse of the Earth into life" is a natural conclusion from the principle of maximum entropy – the same principle that accounts for the tendency toward disorder in closed equilibrium systems – applied in an appropriate dynamical context. The main conceptual shifts are these.

To apply the concepts of cooperative effects and phase transition to a dynamical system like the living state, our ways of thinking about entropy must change from habits that have become long ingrained from experience with equilibrium systems. In equilibrium systems, entropy counts degeneracies of **states**. Information, or a reduction in entropy that often defines a relevant concept of order, measures the reduction in the number of states of *being* required to satisfy whatever constraints the environment imposes. Life, as we have emphasized, is a jointly ordered system of both **processes and states**. The information relevant to life must also measure the reduction in the number of ways of *doing something* required to satisfy the non-equilibrium constraints the environment imposes. Some processes play out over an extended interval of time, in the course of which they pass through series of states. In such cases, the relevant information must measure the reduction in the range of *histories* that perform a function, where each history is an integrated series of states and transformations. For each of these generalizations, worked examples of the principle of maximum entropy are understood,[30] though so far the examples are much simpler than realistic contexts for the emergence of life.

The required shift in our point of view does not entail a change in the meaning of entropy or its relation to information. It requires, rather, that we recognize different spaces of possibility as the domains within which different forms of order emerge. The same entropy

---

[30] For pedagogical expositions and a few simple worked examples, see [301, 649, 767, 768, 902].

concept then leads to different entropy functions for different classes of distributions. For applications to life, most of these differ from the equilibrium entropy.

Entropies of processes may be maximized along histories that do not necessarily pass through states that would be most probable in equilibrium. For some functions, no states near equilibrium may perform them robustly or at all, so they can be carried out only far from equilibrium. The entropic question then becomes: in the world of processes, why is a function possible at all that can only be carried on non-equilibrium states? Which functions does life uniquely perform that cannot be performed by simpler systems, and why, in a random world, are there more ways to perform those processes than to remain near equilibrium?

### 1.3.4 The emergence of the fourth geosphere and the opening of organic chemistry on Earth

Our contention is that, despite the remarkable complexity of living order, the aggregate function of the biosphere is a simple one: *it opens a channel for energy flow through a domain of organic chemistry that would otherwise be inaccessible to planetary processes.* It is analogous to a lightning strike through the graph of chemical possibilities, producing a channel that is stable at the system level but heterogeneous and far from equilibrium when viewed locally.

A planet with only three geospheres can still be a conduit for energy flow. The two primary long-term sources of free energy – disequilibrium between the bulk Earth and atmosphere, liberated by the release of radioactively generated heat, and direct coupling to the high-energy photon flux from stellar burning – are widely present in the universe. Three-geosphere systems can also host chemical interconversion that is limited in either form or extent. The volume of redox transformation in the Earth's mantle is large, though the forms that occur at large scale are limited. A greater complexity of organosynthesis is possible, even in planetesimals such as the parent bodies of carbonaceous meteorites, as attested in the organic contents of the Murchison meteorite [156, 704], but their concentration is more limited and it remains an open question whether a principle can be recognized in these systems that is chemically selective.

The emergence of a fourth geosphere introduces new channels for high-volume, steady energy flux through covalent bond chemistry, which may operate in parallel to, or may subsume, chemical interconversions within the other geospheres. The main network of these pathways on Earth today is **metabolism**. Chapter 4 is devoted to the metabolite inventory, network topology, functions, and historical diversification within this network, and summarizes our reasons for interpreting extant metabolism as a continuous outgrowth from prebiotic geochemistry. The causal link of metabolism to geochemistry in the first life was, we argue, the geoenergetics of electron flow from low-potential donors to high-potential acceptors.[31] The first life gave high-energy geochemically produced electrons paths for

---

[31] Due to the convention that the electron has a "negative" electric charge, and voltage is measured so that charge times voltage equals potential energy, a low-potential electron donor is a high-energy donor, while a high-potential acceptor is a low-energy acceptor.

relaxation through covalent bond organic and organometallic chemistry where no other geosphere did. Energetically, life facilitated the descent of electrons. Today geochemical redox energy remains essential to some ecosystems, but in terms of known primary productivity, the more complex but higher yielding harvest of light energy supersedes geochemical redox relaxation.

The opening and maintenance of protometabolic and eventually metabolic channels for energy flow is the aggregate property on which all living processes depend, and which in turn they all impact, whether constructively or parasitically. We believe, and will attempt to show in this monograph, that the establishment of these channels and of the energy flows through them is in a certain technical sense the *central* function of the system rearrangement that was the emergence of life.

Central properties of this kind are known in the theory of phase transitions as the **order parameters** of ordered phases: they are the statistically and causally primary aspects of novel order created in a phase transition. This means that all other forms of order can be explained in reference to them and to the pre-existing framework in which the new phase forms. On Earth the pre-existing framework came from the laws of physics and chemistry and the planetary composition and energy sources, and the new phase is living matter. All other order in the biosphere ultimately appeals to this energy flow through organic chemistry as a source of stability, whether directly or indirectly.

Those aspects of an order parameter that are determined fully by the boundary conditions are the *necessary* properties of the ordered phase. If it is correct to regard the aggregate energy flows through organic chemistry as the order parameter that defined the emergence of life, the relation between these flows and the free energy sources present on the early or contemporary Earth also defines the sense in which energy stresses can be said to have *caused* the emergence of life, and a biosphere can be said to be a necessary part of an energetically active planet like the Earth.

### 1.3.4.1 Not one phase transition, but many

The phase transition paradigm for emergence is a general claim that error buffering through cooperative effects is needed to permit the formation of hierarchical complex systems, especially those employing relations of control between levels in the hierarchy. While the formation of a chemical channel for energy flow may be the most fundamental dynamical-phase property in the biosphere and its ultimate reason for existence, the origin of life should *not* be understood as a single phase transition that created a single form of order. Requirements for buffering by ordered phases are found repeatedly in level after level of living order. This is why we have emphasized from the start the heterogeneous and multilevel character of the regularities of life that a theory of origin must explain. Our premise that major transitions in the origin and early evolution of life must correspond to emergences of new incrementally stable ordered phases will allow us to propose a sequence of steps in the origin of life that we believe has some theoretical justification, despite the fact that many links between these transitions are missing. A more detailed sketch of the hierarchy of transitions that we think provided a scaffold for the emergence of the biosphere is presented in Chapter 6. Here, to give a flavor of the kinds of differences that can be important, we compare three transitions in the sequence.

**The self-maintenance of metabolism** The difference between the complex but limited organosynthesis attested in meteorites, and the more concentrated synthesis of a few compounds that we believe must have preceded any accretion of higher-order structure in life, is one of *yield* and *selectivity*. In cellular life, yield and selectivity result from positive feedbacks that concentrate reaction flux in synthetic networks. Mechanisms of positive feedback include enhancement of specific reaction rates (often by large factors) by molecular catalysts selected under evolution, and also self-amplification from pathway loops that feed metabolites back as precursors to their own synthesis, a process collectively termed *network autocatalysis*.

The universal core metabolic network is remarkable for the presence of autocatalytic feedback in extremely short pathways that also constitute the center and the most invariant domain in the network [769]. The reactions in these short, central loops also exhibit other simplicities, redundancies, and analogies to reactions at mineral/water interfaces. These and other observations, reviewed in Chapter 4, lead us to argue (as many others before us have argued) that metabolism is continuous with geochemistry, and its first departure toward being an independent system was a transition in these autocatalytic core pathways through a threshold of enhanced selectivity and ultimately self-maintenance. Many of the steps in our proposed phase transition to self-maintenance remain undemonstrated – these are areas of ongoing work and incremental progress – and it is a matter of disagreement within the origins community whether such a phenomenon is plausible in geochemistry [614, 733]. However, the argument that the error-buffering character of ordered phases was required for the emergence of life is most important at this first transition, where we assert metabolism was selected for kinetic and topological properties, which continue to make it the anchor for the stability of higher-level structures in life.

**The rise of an oligomer world** If a phase transition to autocatalytic self-maintenance was the first selector of metabolic pathways, these probably have a minimal component of accident. The small-molecule world has little capacity for long-term memory: whatever is most facile and robust becomes most likely, moment-by-moment independently, and everywhere the boundary conditions provide similar energy supplies. The simplicity of extant metabolism suggests that the orderly core of a protometabolism would have been comparably simple. The main reactions and the primary fluctuations about them would have been fully sampled by the chemicals and reactions that led to the earliest cells.

At some stage, life began to make use of oligomers of large size, and from then onward the combinatorial possibilities for useful functions and structures became much too numerous to be sampled exhaustively by genomes, cells, or whatever were the relevant replicating units. The transition between a (putatively) unique metabolism and an undersampled world of oligomers marks a qualitative change in the problem of maintaining life on Earth. In the former case, self-reinforcement maintains a system around a unique solution. In the latter, selection (probably, we argue, scaffolded by the presence and uniqueness of the underlying metabolism) maintains a system despite the fact that its instantaneous states are not unique.

**Emergences of individualities** A different kind of character change occurred between chemistry in bulk phases (in volumes or on surfaces), and chemical reactions performed in cells or catalyzed by enzymes encoded in replicating macromolecules. In the bulk phase, selection takes place by means of reaction and diffusion kinetics. In the dynamics of either compartments or genomes, selection makes use of the duplication of many parallel platforms, performing nearly identical copies of the same function, which can be replicated or eliminated independently [184]. Compartments and genomes mark two forms of emergence of individuality, perhaps one of the most important characteristics of the dynamical living phase that has no counterpart in equilibrium phases. The existence of parallel units sensibly regarded as individual is the precondition for Darwinian evolution. Understanding why and how individuality emerges, how many forms it can take and how these interact or are related, will be key to understanding the relation between the more "thermodynamic" and more "Darwinian" aspects of the dynamics that contribute to the stability of life.

## 1.4 The origin of life and the organization of the biosphere

In Chapter 8, bringing together the empirical facts from Chapters 2 through 5 with the discussion of cooperative effects from Chapter 7, we argue that the origin and subsequent evolution of life have relied throughout on the stable forms of order created by phase transitions as the "building blocks" of emergence and adaptive design. The starting observation is that life is not a naked channel for energy flow through geochemistry, but a complex architecture of structures and functions maintained indirectly to support an energy-flow channel. To capture the problem in erecting and maintaining such an architecture, and to explain why it may be solvable but only in limited circumstances, we must understand not only isolated or abstract phase transitions, but the essential role played by cooperative effects in modular systems that make use of many kinds of ordered phases.

The problem of maintaining long-term stability in hierarchical complex systems may be understood with concepts from classical optimal control theory [604]. We argue that the pervasive role played by ordered phases is **error buffering**: in systems subject to errors in very many dimensions, cooperative effects can provide regression toward sufficiently low-dimensional spaces of variation that the residual errors can be managed within the limits of complexity of controllers and control signals. This buffering is only available, however, where cooperative effects are strong enough to cross thresholds to form ordered states. The capacity to self-buffer many dimensions of internal error is a form of *autonomy*, and we argue that this connection to error buffering and order through cooperative effects is the appropriate interpretation of the quasi-independent character of many modules we exhibit in the earlier chapters.

The deterministic character of phase transition provides a framework of lawful action and a notion of cause that can connect living processes across time and across scales: from the modern era where evolutionary variation provides much of our interpretive frame for comparative analysis, back to pre-cellular geochemistry, and from the aggregate dynamics that stabilizes complex communities of complex organisms, down to their aggregate effect of organosynthesis and energy flow. The unified picture of the emergence, organization, and persistence of life, and of the embedding of life in chemistry, consists of four main premises.

**Origin, ongoing organization, and persistence are not separate** We argue, from the existence of regularities in chemistry and ecosystem structure that are more universal than the individuals and species that carry evolutionary memory, that processes of biosynthesis and repair are at least partly a reflection of time-invariant *laws of organization*. The biosphere is self-renewing in a literal sense: biosynthesis directs matter into certain modes of organization now for the same reasons biogenesis first directed matter out of non-living states and into these modes of organization which were then novel. To the extent that biological evolution is constrained by time-invariant laws, the origin, organization, and persistence of the biosphere cannot be understood as separate problems. Each provides a view of the underlying constraints, though they act in very different contexts, ranging from pre-cellular geochemistry to genomically dictated physiology and complex population dynamics. We look for the action of laws in absences of evolutionary innovation, in long-range feedbacks that may give slight fitness advantages to organisms participating in favored networks and lead to long-term evolutionary and ecological convergences, and in regularities that are not governed by any one level of selection yet persist as features of coevolution.

**Common laws make present and past mutually informative** A framework of common laws is the only starting point from which we can reconstruct the origin of life with any specificity or confidence. It is only to the extent that current living processes and pre-cellular geochemistry reflect the same constraints, that we are justified in expecting continuity between prebiotic and biotic patterns, or in extrapolating existing patterns in evolution to an age before a record of evolutionary history was preserved in surviving diversity.

At the same time, a serious consideration of the difficulty of maintaining a complex, multilevel state of dynamical order suggests that it is implausible that order could be maintained without reference to supports that come from outside the biosphere itself, which would play the role of invariant laws with respect to the coevolutionary dynamics of member species.

**Evolution builds using the order parameters of phase transitions** Evolution is ultimately a commitment to the problem of *induction*. Among variations inherited from the past, a population is filtered according to advantages under present circumstances, and the filtering is more beneficial than chance only if the future reprises the present at least to some degree. It is difficult to select among unreliable components or components that respond to their environments with complex, multivariate dependencies; the future may be likely to resemble the past in coarse features, but is unlikely to repeat it in full detail.

We expect the successful outcomes of evolution to be concentrated among components and functions that best support induction: those that vary sufficiently to distinguish among environmental conditions but that are stable enough internally to permit selection in a few dimensions of variation. These are the kinds of systems, we argue, produced as ordered phases through cooperative effects. The building blocks of evolutionary "design" should be to a large extent the order parameters made available through phase transitions. The limitations in the availability of robust order then dictate limits on evolutionary innovation and maintenance.

**An invariant core is the reference enabling evolutionary variation** The problem of stabilizing a hierarchical complex system is mathematically equivalent to many problems of preserving messages sent through noisy transmission lines. Our theory of optimal information transmission, like our theory of physical stability, is a theory based on cooperative effects and phase transition.[32] The problem of maintaining an evolutionary system capable of open-ended variation is equivalent to the problem of maintaining an information system capable of preserving an unlimited variety of messages (though no one among the endlessly variable messages needs to be preserved forever). The problem of preserving information in messages leads to a problem of regress of stability. The transmission system protects the messages, but what preserves the integrity of the transmission system? If the system itself is a message, in what medium is it preserved, and how is that medium protected. *Quis custodiet ipsos custodes?*[33]

Considering the problem of regress in a system with even finite but large capacity for variation, we conclude that the system must ultimately have a reference outside itself from which to preserve its ordered state, and that reference must be invariant. For the biosphere, the natural candidate for this reference, from many considerations, is the universal core of small-molecule metabolism as it exists at the ecosystem level. The circumstantial evidence that it is a reference is its apparent universality and its existence as a property spanning all levels from cells to the biosphere as a whole, and all life as far back as we can see with evolutionary reconstruction. Causational arguments that the small-molecule core is a likely source of stability include its digital character, which facilitates error correction, the many reaction and network properties that support self-amplification in a compact system, making it robust, and its function as a biosynthetic gateway through which all matter passes in the course of biosynthesis.

Our claim that integrity and throughput in the universal metabolic network are the continuous and ongoing source of stability for the biosphere anchors the living state directly

---

[32] This equivalence is developed in Chapter 7. Although mathematically it is straightforward, we have not seen it emphasized as widely as we would have expected, given its interest and importance.
[33] Who will guard the guardians themselves?

in the laws of chemistry and in the composition and energetic circumstances of this planet. Because life depends in parallel on so many properties of its substrate that are uniquely provided by chemistry, we argue that the living state is fundamentally chemical before it is anything else, and that life will be the premier subject for the study of cooperative effects acting in the structured domain of chemistry.

# 2

# The organization of life on Earth today

The biosphere that exists today is complex and heterogeneous, not only with recent, history-dependent order, but with diverse ancient forms of order that all appear to be fundamental to the nature of the living state. Rather than propose an origin of life that projects away this diversity by seeking the emergence of a single kind of entity or process, we ask what essential functions and contributions to persistence of the biosphere come from forms or order at different levels. Fundamental constraints on the possible ways to assemble living systems are captured by grouping organism phenotypes according to their chemical energy sources for electron transfers (donors or acceptors), and according to whether they are metabolically self-sufficient or can only live using resources extracted from larger ecosystems. The same bioenergetic distinctions exist for ecosystems as for organisms, but as ecosystem boundaries can often be constructed to be metabolically closed, ecosystems are in a sense simpler and more universal than organisms. The universal aspects of metabolism are ecosystem properties, and core biosynthesis powered by electron donors is more fundamental, universal, and ancient than degradative pathways that contribute much of ecological complexity. Whether the emergence of our biosphere was surprising or inevitable is not well posed as a single question; some low-level features of biochemistry seem to have inherited nearly the inevitability of geochemistry, while the character of any particular species is an epitome of chance. Key patterns, many of chemical origin, are nonetheless recapitulated across the spectrum from necessity to chance, and in some cases we can make explicit arguments that constraint flowed upward in scale from metabolic foundations because chemistry dictates paths of least resistance for evolution across a wide range of scales.

## 2.1  Many forms of order are fundamental in the biosphere

In attempting to understand the origin of life, we begin with what is understood about life on Earth today. Rather than seeking to project the many things known about the nature of the living state among the different sciences onto one or a few abstractions, we think it is preferable to acknowledge that life employs a wide range and diversity of organizing motifs. A theory of biogenesis must ultimately account for why so many forms exist and why they are so qualitatively diverse, and what role each plays in the function and maintenance of the biosphere. Major motifs that we wish to acknowledge include the following.

- In the world of entities, we may cite: the biochemicals, of many sizes and kinds; cellular components and compartments, and whole integrated cells; biological energy systems; organisms that reproduce, compete, and die; ecological communities, and their constructed niches.
- In the world of processes, we note: metabolism and the cellular processes that regulate it; ecosystem cycles of elements and more complex compounds; replication and death of genomes, cells, and organisms; ecological fluctuations mediated by chemical or behavioral feedback; trophic shifts, short-term population adjustment by selection or drift, and long-term evolutionary convergence.

Even in these abbreviated lists, one is quickly overwhelmed with originalities. No one kind of entity or process above, when understood in detail, appears to be fully accounted for by the dynamics of the others. Conversely, no one level seems a sufficient starting point from which to derive the rest.

We believe this is important, not because it rules out a sequential and even a systematic or inevitable origin of life, but because it tells us that the sequence of dependencies may not have a simple packaging in terms of familiar entities and interactions. It is not a faithful response to the scientific facts to begin with one level, such as macromolecules or compartments as entities, and replication or competition as processes, and to expect from these to generate the correct abstractions of cause and constraint to explain why a biosphere exists, and why it has the organization we see. Instead we must arrive at the appropriate abstractions from evidence that may take the form of relations among patterns that exist in multiple levels of structure and process.

We also stress the diversity of life's patterns and the lack of any unambiguous hierarchy in their relations to one another because we believe that they partly arise in *parallel* from much lower-level contexts. After we have presented all of our facts and interpretations in the next five chapters, we return near the end of Chapter 8 to argue that the diversity of motifs in the above lists results from many distinct ways in which the chemical substrate of life supports spatial and temporal order. They cannot be understood only in reference to each other, outside of their actual and explicit context in chemistry. In a related, narrower argument for parallelism at the end of this chapter, we propose that the origin of life was not so much the creation of an autonomous system to which complexity was added, as it was the gathering of dispersed and heterogeneous mechanisms of order formation in geochemistry, into a system where their progressive integration and interdependence gradually made them autonomous.

To try to extract, from the above lists, which abstractions are most central to the nature of life and the constraints on its origin, we exploit the fact that the notion of what is essential (and how its signatures are recognized in data) varies across the sciences. It can turn out to be quite difficult to make arguments for cause that satisfy basic criteria of scientific common sense from many perspectives, so the intersection of many criteria helps us identify a few patterns that will be most central in our interpretations.

### 2.1.1 Three conceptions of essentiality

Every scientific domain has an order of priorities in the construction of theories. Certain principles are seen as first-order constraints, which must be satisfied before pursuing refinements and details. Central to this discussion are the following.

1. In statistical physics it is a premise that ordered states must satisfy criteria of robustness under perturbation, which turns out to be a severe filter for eligible forms of order. The robust forms of order tend to emerge progressively, they tend not to undergo large macroscopic fluctuations and rearrangements, and their particular forms tend to be dictated by the boundary conditions where they arise. These signatures will be important in distinguishing chance from necessity in a complex reconstruction of history from evolutionary clues.
2. In chemistry, some reaction classes and molecular configurations are potential generators of complexity while others lead to energetic or kinetic dead ends. Likewise, some are selective while others generate a kind of chemical chaos. The productive yet selective reaction classes define "paths of least resistance," which may shape what was likely to have been available in high concentrations prebiotically, and which afford the lowest selective costs if living systems, as they develop control over geochemical processes, can largely follow and remain within these paths. The relevance of any class of reactions is further conditioned on the energetic and chemical context of the planet.
3. In biology the dynamical (and in many cases, evolutionary) context of processes is expected to be central to understanding almost any kind of order produced in states. In order to understand what it is statistically easiest to *be*, we must understand what it is statistically easiest to *do* in a non-equilibrium context. The maintenance of order in processes is throughout a problem of error correction, and so we remain sensitive to how the problems of asymptotic stability are solved in the biosphere when both errors and error-correcting controls propagate across levels in complex hierarchical systems.

A plausible theory of how and why life emerged should meet the first-order constraints from all the disciplines that bear on it, before attempting to weave detailed scenarios from the *ad hoc* knowledge and conventions within any particular expert domain.

### 2.1.2 The major patterns that order the biosphere

This chapter lists properties of extant life that we think must guide the broad outlines of a theory of origins. They are summarized in the remainder of the introduction, and more detailed accounts of facts and arguments are given in the following sections. Many widely recognized facts, such as the ways in which macromolecules, cellular compartments, bioenergetic systems, and molecular regulatory processes support the execution of major living processes today, are well treated in standard texts on the molecular biology of the cell [15] and related topics. Therefore we direct attention here to system-level patterns that we believe have been under-emphasized relative to their importance.

### 2.1.2.1 Ecosystems and their affiliation with metabolism

We begin with the importance of the ecosystem as a level of organization, which we believe supersedes the importance of the organism for both the earliest stages of origin, and certain aspects of the long-term organization of the biosphere and constraints on evolutionary dynamics within it. We then consider the problem of how to classify kinds of biological order. A typological classification of metabolisms reflecting their energetics and synthetic chemistry (as opposed to a cladistic classification reflecting historical paths of descent) captures constraints from reactivity and network structure in both organisms and ecosystems, and we believe reflects laws of composition that were central before the advent of genetics and the historical contingencies to which genetic systems are subject.

We observe that the universal features of biochemistry at the ecosystem level are contemporaneous with, or antedate, the oldest mineral fossils on earth. Whereas the rock record effectively vanishes across the horizon to the Hadean eon, the profusion of life expressing the universals of metabolism provides a signature from antiquity that is the strongest it has ever been.

### 2.1.2.2 Multiplicity of scales and structures

We then turn to the segregation of essential functions in the biosphere across scales of aggregation and time. The living state draws on explicit properties of the chemical elements, and ramifies these in hierarchies of distinct chemical compounds and physical structures. Each level in each of these hierarchies has a different origin, composition, and class of functions. The control processes of life likewise exist in layers, and the effectiveness of control often depends on the separation between interacting with environments and maintaining internal states, which is served by the evolution of modular architecture and standardized interfaces.

We separate the abstract classes of living functions, and note their associations with structure, so that we can then consider the historical record of stability and universality associated with each. The empirical record is our best indicator of which levels of structure carry essential patterns of life with primordial origins, and thus which levels must have been realized in some forms from the prebiotic era into the present.

### 2.1.2.3 The great spread in degrees of chance and necessity

From an appreciation of the heterogeneity of the kinds of order that structure the biosphere, we revisit the question of chance and necessity, and argue that it cannot be understood as one question. Some characteristics of life, particularly metabolic universals at the ecosystem level, are nearly as steady and deterministic as geochemistry, and we believe, nearly as predictable. Others, particularly in complex organisms and ecological community relations, are fragile and ephemeral, and some even depend on programmed turnover to carry out their functions. All are aspects of life.

This is the first argument, from elementary observations, for a theme that we will develop in the remainder of the book. Life should not be viewed as one state of order. It must be recognized as a kind of unity-in-confederacy of many kinds of order. Different modes of organization are variously necessary or historically contingent (outcomes of chance). Such diversity is possible because, although processes at different levels mutually support each other (creating, on first exposure, networks of chicken-egg paradoxes for the investigator), those ties are of variable strength and some are substitutable. A view of life as a confederacy changes the questions we pose about its origin. Rather than seeking the emergence of a novel and unified form of order from the beginning, and asking about its progressive complexification, we seek the dispersed opportunities for chemical and physical order provided by the Earth's chemistry and dynamics, and ask how they were brought into interdependence and thus granted partial autonomy from the non-living geospheres.

### 2.1.2.4 Correlation and the propagation of constraint across hierarchical levels of order

When we examine the entities and processes in the lists at the beginning of this chapter, we recognize that they are not simple unitary wholes, but that each has a subdivision into modules that carry out internal functions quasi-autonomously, and which simplify the problems of control and stabilization by interacting through reduced and standardized interfaces. Remarkably, the module architecture in systems at multiple scales often shares the same boundaries across levels. This recapitulation of structure is one of our strongest pieces of evidence that the apparent chicken-egg paradoxes first encountered in a survey of life are resolvable, and that a progression of constraints from some "foundation" levels through the rest of the hierarchy will be identifiable. In particular, modules often first recognized in the architecture of metabolism are recapitulated at multiple levels from the structure and evolution of macromolecules, to cellular bioenergetics and even the genetic code, and to regularities in ecological dynamics.

Correlation, however, does not by itself imply causation or identify its direction. We will argue that, as a general pattern, constraints were laid down in geochemistry that became the earliest metabolism, and these constraints dictated at least partly the stable architectures for higher-level control systems as the latter emerged. To make this argument, we must furnish evidence for the direction in which constraint has flowed across levels. We list a few examples in which historical reconstruction indicates an explicit directionality in evolution in which the preservation of low-level metabolic chemistry served as a scaffold (the case of core pathways in *Aquifex aeolicus* [92]), and in which higher-level control systems (the translation system and genetic code [158]) record evidence of an earlier embedding within the structure of a simpler yet conserved metabolism.

## 2.2 Ecosystems must become first-class citizens in biology

Because the origin of life was a phenomenon in chemistry and physics before it was the horizon that brought biology into existence, it is not apparent to us that individual-based or organism-based dynamics – or even abstracted analogues to these – should have been more important in the earliest phases of biogenesis than dynamics in the chemical bulk or in some other aggregate phase. We therefore approach signatures of life as we would approach those of any other system in which the major sources of order are undiscovered: by asking which features seem oldest, most stable, most universal, or most unique. We try to form a coarse-grained sketch before asking how more variable features fill it in. From this direction of questioning, the ecosystem emerges as a level of organization more fundamental than the organism is, to many regularities that are our best candidates for constraints on the earliest stages of biogenesis.

We show in the next section the sense in which the universal and conserved network of core metabolic pathways comprising carbon fixation and intermediary metabolism is more closely associated with the ecosystem level of organization than with the dynamics of organisms. Metabolism at the ecosystem level is in its aggregate effect less variable and less arbitrarily complex than metabolisms of species, and its innovations are more simply tied to the abiotic context. Thus the ecosystem appears as the bridge between incipient order in geochemistry and the earliest biochemistry. Organisms are a derived level of organization that forms, in diverse ways, within a framework of constraints of biochemical and ecological order that are more conserved than the properties of the individuals that help to instantiate them.

Whether we try to understand the nature of metabolic universality, or the widely varying degree of chance and necessity among patterns that together produce the living state, we need a way to refer to ecosystems as primary entities in their own right and as carriers of patterns fundamental to life. Ecosystems must become "first-class citizens"[1] in biology, in some respects prior to and more fundamental than organisms.

### 2.2.1 No adequate concept of ecosystem identity in current biology

Biology in its descriptive tradition, as part of natural history, was sensitive to relations of many kinds, both in the growth and form of organisms and in their ecological relations. Darwin was a master of noticing and cataloging such relations. However, with the publication of *On the Origin of Species*, Darwin made a commitment to the individual as the sole sufficient level of selection to account for adaptation.[2] The effects of ecological

---

[1] We borrow the title of this section from a famous expression in computer science – *Functions as first-class citizens* – coined by Christopher Strachey in the mid-1960s [105]. First-class functions are not merely sequences of steps, but genuine *entities*, which can be passed as arguments to and from other functions in the same manner as other data types [10]. Languages that support this concept have a fundamentally greater expressive power than those that relegate functions to the status of "second-class citizens" relative to first-class "data" objects. Biology needs an analogous expressive power in order to refer properly to the role of ecosystems as carriers of fundamental patterns, and as entities parallel to and in some ways superseding organisms.

[2] See the argument that this was a strong commitment of Darwin's, and some of its consequences for the slow understanding of the relatively innocent concept of multilevel selection, in Gould [312].

context and all other forms of constraint were subsumed into the aggregate quantity "fitness,"[3] and since only fitness determined the change in population membership, this aggregate statistic was the gateway for entry of information into the biosphere. In the century following *On the Origin of Species*, the organism came to be viewed more and more as the carrier of biological information, which was imprinted at the population level through the selection of organism variations. The rise of gene selection [181, 880] shifted this emphasis somewhat away from organisms, but placed it onto an even finer-grained replicator: the (real or reified) gene.

Instead of developing recognized relations from the descriptive tradition, where appropriate, into diverse and heterogeneous concepts of identity at many levels, biology became increasingly focused on individual dynamics and particularly natural selection. At the same time as it progressively idealized the abstraction of the individual (in our view, to a point of harmfully oversimplifying this complex concept), it marginalized the concept of an ecosystem to that of merely an assembled community of member species. Macroecology would require another half-century [99] to form detailed and quantitative descriptions of ecological patterns, maintained by (among other factors) selection of organisms, but described inherently as aggregate phenomena. Thus biology failed to develop a natural way to refer to the role of ecosystems as carriers of essential aspects of the living state.

### 2.2.2 Ecosystems are not super-organisms

The lack of a concept of ecosystems as primitive entities becomes limiting when ecosystems are discovered to show homeostasis arising from coordinated evolution among many member species, or to possess patterns that are independent of the particular cohort of member species. Homeostasis has been regarded in biology as a regulatory capability of organisms, and so some treatments of ecological stability have sought to express the central concept by likening ecosystems to "super-organisms."[4]

While the adoption of an analogy in this situation is understandable, it is not desirable. Ecosystems are not organisms. They do not persist through replication or even through more general processes that could naturally be characterized as "reproduction." Their persistence is partly carried by reproduction of member individuals, partly through constructed niches [602], and partly through relations maintained dynamically both within and

---

[3] Although it was a central organizing concept for Darwin, and is invoked constantly in biological discourse today, the concept of fitness remains problematic and has an adequate formal foundation only in very restricted systems. Following Steven Frank's analysis [273], in our view the most important contribution by R. A. Fisher [256] was the recognition that fitness should be *defined* in terms of a summary statistic in a diachronic process: the relative numbers of offspring partitioned according to the parents' type. The existence of a well-defined summary statistic separated the definition of the concept from a secondary problem, which was finding good sample estimators for fitness which were functions of an organism's type in its population context. Fisher's definition, however, only applies to populations in which each offspring *has* a unique parent – in other words, to populations of strict replicators. Biology needs a more general concept of the individual and its reproduction than one that requires strict replicators. For these more general cases, however, Fisher's contribution must be carried out again starting from the phenomenology at each new level, by identifying summary statistics for the diachronic process that encode the structure of the population process. In general, this formalism has not yet been developed.

[4] The most widely known of these is the characterization of the whole-Earth ecosystem as a homeostatic organism-like system termed *Gaia* by James Lovelock [497].

across generations of member organisms. Variant ecological configurations, which may be explored as member species change their properties or relations, do not undergo Darwinian selection as populations of quasi-autonomous ecosystems compete. Rather, the state of the ecosystem as a more or less integrated entity fluctuates, and aggregate effects at multiple regulatory and evolutionary levels within it determine which forms prevail.

In other words, the abstractions of individuality and Darwinian dynamics that are the major hallmarks of organisms characterize ecosystems poorly or not at all. In our view, the correct response is not to shoe-horn ecosystems into a strained analogy with organisms, but to grant them their own status as entities that carry patterns essential to the living state, and to develop a language and theory of the ecological level of organization [521].

### 2.2.3 Ecological patterns can transcend the distinction between individual and community dynamics

Patterns carried by ecosystems often involve relations among member species, but they are not only constructs of relations, and this is why the ecosystem has a more funda-mental identity than that of a mere community. Persistent patterns may involve processes within organisms together with relations between them, and may be maintained by complex interactions of regulatory and selective dynamics. For example, the balance of stoichio-metric fluxes, which are constrained both by coupled rates within organisms and trophic exchanges between them, is a long-range pattern that may involve coordinated regulation and population dynamics of many member species [163, 779].

The boundaries between dynamics within organisms and dynamics between organisms may be fluid; the same balance that may be regulated metabolically within some species may be regulated through population balances in syntrophic consortia in other cases. Yet the patterns of metabolic completeness and optimization that are preserved may be broadly similar, transcending the distinction between physiological and macroecological properties.

### 2.3 Bioenergetic and trophic classification of organism-level and ecosystem-level metabolisms

Not all living systems are equally complex or complex in the same ways. The way we classify systems such as organisms or ecosystems reflects the sources of order that we believe are most central to the time period and generating process we wish to understand.

In this section we propose that the most informative questions, for very long-term organi-zation, that we can ask about organisms and ecosystems concern their biochemistry and the way its control is partitioned among one or more independently evolving genomic lineages that interact ecologically. These are fundamentally typological rather than phylogenetic classification criteria, so we first explain how the two differ and why typological classi-fication captures the distinctions most relevant to the first departures from geochemistry to life.

### 2.3.1 Divergence and convergence: phylogenetic and typological classification schemes

The dynamics of living systems derives from the interaction of history with laws or constraints on processes and the kinds of order they can create. Accordingly, living systems show relations of more than one kind. Some relations are due to the shared imprint of history, and are most readily seen from the idiosyncratic *divergences* of forms. Other relations reflect shared constraints or the structure of underlying laws, and are most readily seen from the absence of variation or from strong tendencies toward evolutionary convergence. Divergence and convergence (including constraint) are antitheses, and so we may partition classification schemes into those that are constructed to capture signatures of divergence and those that are constructed to capture signatures of convergence or constraint.[5] The former are the phylogenetic classification schemes, and the latter are various kinds of typological classifications.

The dominant classification scheme for species in biology is phylogenetic, because it is required to subsume the enormous diversity created by evolutionary dynamics in the era of organisms. For the investigation of origins, however, the role of laws may become more important relative to the roles of historical contingency, and typological classifications may capture more essential kinds of information. Moreover, organisms and species are not the only relevant levels of organization in the emergence of life and the organization of the biosphere, and so typological classifications are needed for systems that do not record their divergences in branching trees of descent. Here we argue that a typological classification based on the directions of electron flow in metabolism reveals the greatest simplicity in the grouping of organisms. More importantly, *the same* classification criteria apply at the ecosystem level, where they reveal greater universality than they do at the level of species. We arrive at a conclusion that constraints on ecological function are more central to the earliest stages of biogenesis than most properties of organisms, and that the ecosystem is the proper level of aggregation to recognize as the bridge from geochemistry to life.

#### 2.3.1.1 Why the primary biological classification of organisms is phylogenetic

> All animals have breasts that are internally and externally viviparous,
> as for instance all animals that have hair, as man and the horse; and the
> cetaceans, as the dolphin, the porpoise, and the whale – for these animals
> have breasts and are supplied with milk.
>
> – Aristotle, *Historia Animalium*, Book 3 [37]

> Be it known that, waiving all argument, I take the good old fashioned
> ground that the whale is a fish, and call upon holy Jonah to back me.
>
> – Herman Melville, *Moby-Dick*, Chapter 32 [552]

---

[5] Of course, complex forms of order such as species have many features, and so the same groups may sometimes be arrived at within both kinds of classification, as these apply to different signatures.

The appropriate way to group organisms in order to express patterns is a longstanding problem. The modern approach using nested hierarchies based on shared features is generally credited to Carl Linnaeus in the eighteenth century. Although most of Linnaeus' categories have not been retained, the combined use of shared features and nested hierarchies essentially define the Linnaean system as one suited to classification by divergences under descent.[6]

Even into modern times, arguments have been made in favor of classifications that emphasize aspects of form and function that are not based on lines of relatedness. Ernst Mayr, in a series of papers traded with Carl Woese [535, 536, 538, 888, 889, 896, 897], resisted molecular taxonomy in favor of a physiological grouping of "prokaryotes" versus eukaryotes. Lynn Margulis, while recognizing the fundamental signatures of relatedness in molecular phylogenetics, nonetheless argued for a classification of organisms into "five kingdoms" [518] which reflected distinctions of physiology and ecological context.[7] Taxonomic groupings are adopted to provide windows on the causes of regularity, which may be constraints of possibility, convergence on preferred forms, or as has been central to biological thinking since Darwin, limitations in the degree of differentiation among descendants of a common ancestor, which therefore show relatedness through their similarities.

Since grouping can reflect many kinds of relations, it is not clear that there should be one dominant way to group living systems. Ishmael's insistence that "the whale is a fish," perhaps meant to pick a fight with nineteenth century academics for whom the whale's mammalian status had been firmly established for almost a century, is argued on the basis of physiological convergences, which express the fact that life at high Reynolds number[8] limits the architectural freedoms open to evolution irrespective of ancestry.

Abstract criteria of function are usually much more difficult to consistently compare and classify, however, than explicit elements of structure. The need for consistent criteria of classification therefore drove biological taxonomics in the nineteenth and early twentieth centuries to depend more and more exclusively on structural homologies, particularly those that are discrete, such as topologies, counts, and discrete symmetries. Most of these, we now realize, reflect conserved aspects of developmental regulation preserved under descent [179]. Since Darwin, Linnaean taxonomy became not only implicitly but explicitly a scheme for expressing structures of relatedness.

---

[6] It is striking the role that discrete elements played in Linnaean classification, ranging from counts of stamens and pistils in plants, to teeth or teats in animals. These discrete features, to the extent that the are not accidental evolutionary convergences, presage the increasing reliance on topology and other discrete features of homology, as for bones, in nineteenth century taxonomy. This culminated in D'Arcy Thompson's insight that geometry could be understood as a developmentally malleable property of organisms under evolution, constrained by topology [802].

[7] Mayr's arguments against Woese were indefensible on the terms in which they were proposed, as Woese's own results lucidly and consistently demonstrated. However, if, as Nick Lane with several collaborators has argued [470, 472, 475], the architecture of the eukaryotic cell has an inevitable logic driven by the need to flexibly regulate metabolism in a large cell, particularly when oxygen is available as a terminal electron acceptor, then the grounds Mayr gave for emphasizing a eukaryote/prokaryote division may with time come to be understood as reflecting a natural typological classification.

[8] The reduction of so complex a phenomenon as evolutionary convergence to such a simple constraint as Reynolds number in fluid dynamics is an important function that a typological grouping can serve. Just as the macroscopic swimmers show strong convergences reflecting fluid dynamics at high Reynolds number, the diatoms and other microscopic swimmers, which could not differ much more than they do from large aquatic animals, nonetheless show strong structural and functional converges of their own as a consequence of life at low Reynolds numbers [413, 414, 653].

The degree to which developmental regulation is both preserved under descent and reflected in overt aspects of phenotype is remarkable, and accounts for the considerable success that classifications based on comparative anatomy and physiology had already achieved in the nineteenth century. However, development is plastic in some important respects, and not all signatures of development are transparently expressed in gross structure; hence, comparative classification is limited in the consistency it can achieve.

In the 1970s, Carl Woese and George Fox showed [893, 894] that the sequence fingerprints in genomes, despite their simple nature and lability at any single position, contained a *volume* of information so much greater than developmental signatures that they provide a more reliable, and much more detailed, basis for classification based on descent. Modern biological classification therefore aims to be wholly phylogenetic (limited only by the availability of data), and is exclusively derived from gene signatures where these are known.

### 2.3.1.2 Typological classification: expressing the structure in laws of assembly

The imprint of ancestry is strong on most aspects of biological function, and dominates most questions of detailed similarities among organisms, but historical features are not the only property of organisms. To express other patterns that are not due to historical contingency, other groupings can be useful. These are sometimes referred to as "phenetic" classification schemes [573], to emphasize that they concern aspects of phenotype that are not necessarily underpinned by genetic homology.

Evolutionary dynamics plays out under constraints on the kinds of organisms it is possible to assemble. One goal for phenetic groupings can therefore be to describe the way these constraints, acting together, assemble into a kind of typology of the possible.[9]

### 2.3.2 *The leading typological distinctions among organisms*

Finding a good phenetic classification of organisms is something like a game of 20-questions. Can we choose questions about whether an organism is in one class or another, so that the earliest questions in the list provide the most information and make the largest distinctions, and later questions capture distinctions of more limited consequence or scope? This is important because phenetic classification suffers all the problems of comparing and grouping functions, which at some stage become too complex to perform systematically.

The first two questions we would choose in such a game would group organisms in the manner shown in Figure 2.1. The first question concerns where the energy for metabolism, including biosynthesis, originates, which is a biochemical question. The second question concerns whether the metabolism encoded within a single organism's genome is

---

[9] It is important to remember that even phylogenetic classification usually serves a larger problem of jointly reconstructing both historical events and the non-contingent elements of a process model for evolutionary change. The typology of the possible is part of that process model.

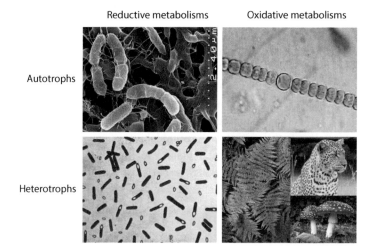

Figure 2.1 A two-by-two matrix, grouping organism phenotypes according to their primary chemical energy source, and their degree of ecological completeness. Organisms with reductive metabolisms obtain their energy from molecules that are electron donors relative to carbohydrate ($CH_2O$). Organisms with oxidative metabolisms obtain their energy from molecules that are electron acceptors relative to carbohydrate. Organisms that are self-sufficient with respect to biosynthesis, called *autotrophs*, can exist in a medium of only inorganic inputs. Organisms that depend on carbon or nitrogen compounds of biotic origin, called *heterotrophs*, can only exist through dependency on a larger ecosystem. Upper left: *Venenivibrio stagnispumantis* (Aquificales) [1]. Lower left: *Clostridium botulinum* (Courtesy CDC). Upper right: the cyanobacterium *Anabaena sphaerica* [2]. Lower right: the air-breathing, photosynthesizing world of plants [3], animals [4], and fungi [5].

self-sufficient, or whether the organism depends on an ecological embedding with other organisms in order to survive and reproduce.

### 2.3.2.1 The source of metabolic energy: oxidative and reductive metabolisms

Metabolism, the system that subsumes both biosynthesis and the foundation of bioenergetics, is a chemical process. Chemistry is in the end a dynamics of electrons transferred among chemical bonds, and the transfers and conversions of functional groups that result.

Major subgroups of metabolic functions, and in many cases metabolic networks as aggregate systems, can be characterized as **reductive** or **oxidative**, according to the direction in which the oxidation state of carbon is changed in the major energy-yielding processes. Biosynthesis starting from $CO_2$ is a process that in aggregate requires donors of electron pairs to produce the stoichiometry of living systems, which is roughly that of carbohydrate $CH_2O$. On purely net energetic grounds (that is, if there were no need to take into account the problem of constructing a reaction sequence to reach each compound), biosynthesis to carbohydrates could proceed simply from the donation of electron pairs to $CO_2$, under the scheme $4H_2 + 2CO_2 \rightleftharpoons 2CH_2O + 2H_2O$. Everett Shock with several collaborators has shown [21, 22, 545, 746, 749] that under conditions prevalent in hot springs, sufficient free energy is present to support the stoichiometric conversion of $CO_2$ and $H_2$ to biomass.[10]

Real systems are not governed only by net free energies of formation of biomolecules. Chemical synthesis requires compatible orbital structures, sufficient free energy to reach transition states for each reaction, and the assembly of reactions into networks that regenerate reactive functional groups to replace those that they consume, as part of the process of forming metabolic products. The resulting network of constraints to finding any biosynthetic pathway ensures that, even when catalysts are available to lower free energy barriers, no pathway is ideally efficient. Since any step-by-step energy mismatches must all be exergonic in order for a reaction network to execute in the forward direction, significant free energy is dissipated in all biosynthetic pathways as the cost of satisfying the constraint system.[11] Therefore, even in environments that could support stoichiometric formation of biomass, organisms require auxiliary energy systems, which directly transfer electrons between environmental donors and acceptors, harnessing the free energy of the transfer to augment internal biosynthetic reactions.

In the simplest case of reductive chemo-autotrophs such as methanogens or acetogens, and for the main electron-transfer reactions in ancient reductive citric acid cycle carbon fixers such as the Aquificales,[12] we may regard the organism metabolism both stoichiometrically and energetically as reductive, while at the same time it functions in the environment as a catalyst for electron transfer via the auxiliary energy system. The energy from the auxiliary electron transfers, along with part of the exergonic energy of formation of biomass, is dissipated as heat.

The reductive metabolisms stand in sharp contrast to organisms that use molecular oxygen or other strong oxidants to oxidize carbohydrates as their primary energy source. The oxygen may be provided (from the organism's perspective) environmentally, or it may be produced photosynthetically within the organism. In either case, the ultimate source of molecular oxygen as an electron acceptor in the modern biosphere is oxygenic photosynthesis. Derived oxidants, such as nitrites, nitrates or a variety of oxidized sulfur species, are produced in large volumes as well, by microbial metabolisms that oxidize ammonia, hydrogen sulfide, or elemental sulfur using molecular oxygen.

---

[10] See Section 6.3.1.3 later in the book for characteristic reaction free energies of this and related reactions. An important caveat is that the entry point to metabolism does not have merely the stoichiometry of sugars, but that of an *activated* compound – acetyl-CoA – from which the leaving-group energy of CoA enables subsequent biosynthesis. When the activating energy is taken into account, the conditions to support biomass directly from the free energy of $CO_2$ reduction become much more marginal. Thus originates the necessary role for elegant but complex couplings between carbon fixation and bioenergetic systems discussed further in Chapter 6.

[11] Extant organisms have evolved exquisite strategies to limit losses from mismatch, by "titrating" the potential of electron donation so that the simplicity of maintaining only a few populations of electron-donating cofactors can be combined with the flexibility of matching their potentials to a wider variety of electron acceptors in synthetic pathways. The mechanisms for this titration are discussed further in Section 4.5.5.1.

[12] Aquificales use elemental sulfur as the primary electron acceptor [92, 320, 321], which is only a mild oxidant, though they use molecular oxygen as a terminal electron acceptor for a few key reactions requiring high-potential acceptors.

### 2.3.2.2 *Fermenters: masters of efficiency and innovation*

A diverse but metabolically limited class of organisms cannot naturally be regarded as either oxidative or reductive. These are the fermenters, which use internal oxidation/ reduction reactions – inter-molecular or intra-molecular electron transfers in which both the donor and the acceptor are organic molecules – as energy sources. Fermenters exploit small "pools" of free energy that accumulate because the complex chemical constraints on biosynthesis create barriers to the full relaxation of non-equilibrium chemical potentials.[13] Because these small free energy stores are strewn throughout biosynthetic networks, and are often released into environments by organisms that have high-energy reductive or oxidative metabolisms,[14] the diversity of fermentative metabolisms is enormous. However, the energy budget of fermenters is small, so they typically minimize the organosynthetic component of their biosynthesis. The same organic carbon sources that provide their energy also provide starting materials for biosynthesis that require minimal electron transfer with the environment.

The fermenters, making a living from inefficiencies in higher-energy metabolisms, are secondary in either the energetics or the stoichiometry of metabolism relative to the abiotic environment. Capturing small free energy stores, however, requires innovations that more robust metabolisms forego, and the fermenters have been an important class of innovators in metabolic evolution. We cite just two examples: clostridia which ferment 4-hydroxybutyrate or 4-aminobutyrate [284, 720] appear to be the most likely innovators of pathway segments which were later co-opted to form the *autotrophic* 4-hydroxybutyrate pathway for carbon fixation, discussed in Section 4.3.6. Even more consequential, fermentative use of citric acid cycle reactions seems likely to have been the source of key enzymes that enabled the transition we call the "great reversal," from reductive to oxidative citric acid cycling, in response to the rise of oxygen, discussed in Section 4.4.6. Today, *Mycobacterium tuberculosis* and other mycobacteria retain enzyme systems from both reductive and oxidative pathways [56, 805, 852], which are used to switch between oxidative respiration and fermentation in which succinic acid is excreted as a waste product and electron sink.

### 2.3.2.3 *Autotrophic and heterotrophic ecological roles*

The second partition in Figure 2.1 distinguishes organisms that are metabolically self-sufficient, called *autotrophs* (for self-feeding), from those that are not, called *heterotrophs* (for different-feeding).[15] Although defined in terms of the input/output characteristics of individual organisms' biochemical networks, this criterion is ultimately a distinction among ecological roles that organisms take. Autotrophic organisms are ecosystems-unto-themselves, while heterotrophic organisms require an ecological embedding to define their

---

[13] Many of these reactions are *redox disproportionations*: reactions that begin with carbon at formal oxydation states near zero and convert it to mixtures of more oxidized and more reduced carbon species [856, 857].

[14] An example is the delivery of large stores of sugar in fruits by flowering plants.

[15] A broad miscellaneous set of organisms are categorized as "mixotrophs" [481], because (facultatively or obligately) they fix some carbon themselves and obtain some other carbon from the environment. We will argue (Chapter 4) that many autotrophs have a metabolic architecture more similar to mixotrophs than has been supposed.

biochemical function. For most heterotrophs, the environments in which they can survive and grow are restricted.

Another way to think about the distinction between autotrophy and heterotrophy is that each defines a particular relation between biochemistry and genomic control under the influence of natural selection. Autotrophs have the entire inventory of metabolic functions that they require under the control of a genome that undergoes selection for survival in each generation. Thus the components of the metabolic machinery are by and large[16] filtered for mutual compatibility in each generation. Ecosystems composed of heterotrophs that jointly account for the whole biosynthetic network partition the problem of metabolic control among multiple genomes, which undergo selection autonomously from one another in the generations of different species. Thus whereas the distinction between reductive or oxidative metabolic modes is fundamentally chemical, the distinction of autotrophy from heterotrophy concerns the assembly of the organism as a regulated entity.

The flexible relation of information and control to chemistry will be a recurring theme in this book. It will be one reason we view life as an assembly of qualitatively different modes of organization, so that at the same time as life has a unified character and integrated systems, it is also partly an assembly of autonomous elements.

### *2.3.2.4 Phenotypic modes or obligate traits?*

Both oxidation/reduction and heterotrophy/autotrophy are in the most general case modes of living among which organisms have to choose rather than fixed "traits" of organisms. Nonetheless many organisms predominantly or obligately use restricted modes, in which cases they may be thought of as traits. The phenetic classification into oxidative and reductive major modes of metabolism, and autotrophic or heterotrophic ecological roles, reflects basic architectural divisions in ways to compose metabolism or to assemble an organism, whether they are obligate traits or phenotypic modes among which organisms can switch more or less discretely. Although phylogenetic clades often have members that all occupy the same category, the distinctions in Figure 2.1 are not fundamentally phylogenetic in character or dependent on descent in their origin.

### *2.3.2.5 Three of the four groups are prokaryotic*

Note that three of the four groups in the classification of Figure 2.1 include only bacteria or archaea. All eukaryotes, and thus all multicellular organisms (in the strict sense) are heterotrophs supported by the world of oxygenic photosynthesis.[17] Even plants, the most nearly autotrophic eukaryotes, require reduced nitrogen from bacterial symbionts. The cyanobacteria, shown in the figure, contain members that have retained capacities for both oxygenic photosynthesis and nitrogen fixation.

---

[16] Thus we omit events of horizontal transfer in this characterization. Even when such events are comparatively common, they remain much less frequent for any gene than the events in which it is vertically transmitted as part of the parent's entire genome.

[17] Some eukaryotes, such as yeasts, are fermenters, to which we have not assigned a separate category in the figure. Since these are heterotrophs, they largely draw carbon from the pool supplied by oxygenic photosynthesis.

### 2.3.3  *Anabolism and catabolism: the fundamental dichotomy corresponding to the biochemical and ecological partitions*

The fundamental asymmetry between reductive and oxidative ecosystems – that the chemistry of reductive ecosystems is both simpler and more universal than the chemistry of oxidative ecosystems (of which the former is a core) – arises in part because the reactive functional groups that create the complexity of biomass are formed at carbon oxidation states which are reduced compared to the environmental carbon source $CO_2$. To synthesize biomass, oxidative metabolisms must shift the oxidation state of carbon in the direction opposite to the one from which they gain free energy. For reductive ecosystems the two processes operate in the same direction. This energetic asymmetry leads to an asymmetry also with respect to the building up or breaking down of molecular complexity.

Most metabolic pathways may be classified as either **anabolic** or **catabolic**. Anabolism is defined by the construction of larger molecules from smaller components, including one-carbon units. Catabolism is defined by the decomposition of larger molecules into smaller components. Like the notion of a geosphere, the anabolic/catabolic categorization is coarse. One can find reactions in which bond breakage is not degradation, but merely part of a larger synthetic system, as in certain cleavage reactions in the pentose phosphate reaction network (reviewed in Section 4.3) or salvage reactions that bridge or buffer pathways. By and large, though, pathways are either bond forming and perform functions of synthesis, or bond breaking and serve functions of degradation to provide either small-molecule building blocks or fuel for energy metabolism.

Anabolism begins from $CO_2$ and is in aggregate reducing toward the stoichiometry of biomass (approximately $CH_2O$). Catabolism can be more complex, with carbon exiting organisms as waste in a variety of oxidation states. Most carbon circulating in the surface biosphere is at some stage fully oxidized to $CO_2$ using molecular oxygen, making $CO_2$ the primary short-term abiotic reservoir in the global carbon cycle.[18]

Thus the anabolic pathways are bioenergetically associated with reductions, biochemically associated with the increase of complexity, and ecologically the sources of organic carbon. Catabolic pathways are generally either redox-neutral or oxidizing, they degrade complexity, and ecologically they either recycle or form net sinks for organic carbon.

#### 2.3.3.1  *Consequences for the "struggle for existence"*

The existence of reductive autotrophic metabolisms overturns a perception of the nature of the living state that has heretofore posed one of the greatest difficulties to seeing a route from geochemistry to life. This was the perception of an inherent contradiction between thermodynamic spontaneity and the generation of complexity.

---

[18] The long-term dynamics of global carbon is considerably more complex; part of this cycle is reviewed in Chapter 3. $CO_2$ incorporated into the mantle as mineral carbonates, either through hydrothermal precipitation or through sedimentation and subduction, has been a historically important reservoir affecting the atmospheric greenhouse. Some fraction of subducted carbon may be returned to the atmosphere on very long cycles through volcanism at convergent plate boundaries, creating a complex interplay of short-term and long-term cycles. Carbon may be sequestered in reduced molecules, eventually becoming coal or petroleum deposits, and the residual oxygen from $CO_2$ is the major source of biologically produced atmospheric oxygen. (An earlier, but smaller source, is left by atmospheric escape of hydrogen.)

Thomas Malthus famously characterized life as a "struggle for existence" [513]. The study of fermentative or oxidative organisms, through which biochemistry became a science, seemed to imprint this struggle on metabolism itself: it seemed that the generation of complexity must be an uphill climb against the direction of spontaneous reaction. The intuition, however, had largely been about the complexity of living *in an oxidizing world*. In reducing environments, the thermodynamically spontaneous direction of reactions is largely aligned with the direction that leads to biosynthesis – enough so that acetogens and methanogens can use the same pathway with only its terminal steps altered both to fix carbon and to recharge a bioenergetic system [523].

The struggle for existence is not entirely removed, nor are all the chicken-egg paradoxes that arise because life relies on machinery that it must itself synthesize in order to overcome intermediate energetic barriers and to integrate pathways. However, the broad compatibility between reducing metabolisms and biosynthesis diminishes the needed complexity and breaks the biosynthetic problem into smaller components, for which an emergence in stages is easier to conceive (see Section 3.6.4.1). More fundamentally, the formation of chemical complexity can be a means to the dissipation of free energy, not only indirectly through complex life but plausibly even in early geochemistry.

### 2.3.4 Ecosystems, in aggregate function, are simpler and more universal than organisms

It was not necessary, in the preceding section, that in seeking to characterize life we ask for a phenetic classification of *organisms*. The same problem could have been posed for ecosystems. One of the two informative questions from Figure 2.1 – whether the primary energy-yielding reactions reduce or oxidize $CO_2$ – remains informative for ecosystems. (Indeed it may be even more fundamental, to the extent that it reflects an environmental condition that is abiotic.) This distinction is shown in Figure 2.2, in the contrast between an iron-oxidizing microbial mat in a hydrothermal spring[19] and a photosynthesizing forest.[20]

For an ecosystem we may aggregate biochemistry across member organisms and their exchanges through trophic networks until only abiotic inputs or outputs are needed at the boundaries. For the biosphere as a whole this can always be done, with the result of a biosynthetically closed and self-sufficient system. For smaller ecosystems, aggregation can also yield a system that is closed in some approximation: generally if organic carbon or energy flows across the boundaries of a well-chosen ecological whole, the magnitude of

---

[19] The red color in the figure is rust, which is generated in large quantities and is insoluble in water, so it accumulates as an encrustation.

[20] Microbial mats at hot springs often include photosynthetic members, which are responsible for the pigmentation often seen in such mats. Important to the distinction we make here is that they also receive energy directly from geochemically produced electron-donating molecules. Everett Shock with numerous colleagues has compiled extensive lists [21, 23, 745, 747, 748] of the electron-transfer systems that sustain organisms in hot springs independent of photosynthetic energy, and compared the energy available from particular environmental electron-transfer processes to the energy requirements of elementary steps in biosynthesis.

Reductive ecosystems                      Oxidative ecosystems

Figure 2.2 Ecosystems are biochemically simpler to classify than organisms. To the extent that an ecosystem can be approximated as closed, it must be biosynthetically complete, so the organism distinction of autotrophy from heterotrophy disappears. Ecosystems remain distinguished according to whether they consume geophysically generated electron donors (reductants), as in the iron-oxidizing mat on the left (Courtesy of the National Oceanic and Atmospheric Administration Central Library Photo Collection), or whether they depend on oxidants, as in the temperate forest on the right [6]. On Earth today, oxygenic photosynthesis is the source of the primary oxidant $O_2$, and derived secondary oxidants such as nitrates and sulfates.

the flow and the complexity of the compounds that carry it will be much lower than the flux and complexity within the internal ecosystem dynamics.[21]

To the extent that ecosystems may be regarded as closed, the second distinction for organisms (between autotrophy and heterotrophy) disappears. All closed ecosystems are metabolically complete; the Earth today does not depend on abiotic sources of organic carbon or reduced nitrogen. One could say metaphorically that all ecosystems are "autotrophic," but we have argued that autotrophy in organisms properly refers to a relation among biochemistry, regulation, and ecological role, so we prefer not to borrow the term to refer merely to metabolic closure. The nature of the dynamics by which ecosystems regulate the biochemistry jointly performed by their member species is a separate problem that requires reference to a wider range of phenotypic regulatory and coevolutionary mechanisms.

The classification of ecosystems according to aggregate function is therefore simpler than the corresponding classification for organisms.

### 2.3.4.1 Reductive versus oxidative meta-metabolomes

The major chemical distinction among ecosystems that are closed with respect to biosynthesis concerns their use of abiotic reductants or oxidants, whether these are environmentally produced or self-generated and released as a form of environmental modification by biota.

---

[21] Coral reefs provide a good example. Forming in the photic zone in warm, generally nutrient-poor waters, they are islands of primary productivity. This is one reason they have been favored study systems for ecologists seeking closed-system models smaller than the whole biosphere.

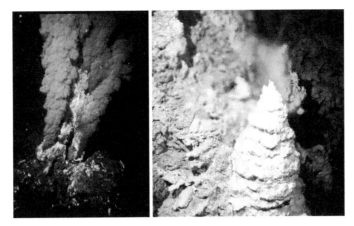

Figure 2.3 Hydrothermal vents. The left-hand panel is a magma-hosted "black smoker" from the Mid-Atlantic Ridge. These vents are very hot (as high as $350\,^\circ$C interior fluids), acidic (pH 3), and contain high concentrations of sulfur and reduced metals. The black "smoke" in this vent is a precipitate of iron sulfides that forms as the fluid enters cooler oxygenated seawater. Accumulation of the precipitate forms sulfide chimneys around the venting sites. The right-hand panel is a peridotite-hosted "white smoker" from the Lost City hydrothermal field on the Atlantis Massif near the Mid-Atlantic Ridge. White smokers are cooler ($50-90\,^\circ$C), alkaline (pH 9–11), and rich in dissolved hydrogen and methane but not in metals. Reductants at white smokers are created by oxidation of olivine to serpentine, brucite, and magnetite in the process known as *serpentinization*. Precipitation in white smokers forms large carbonate chimneys under the conditions of present ocean chemistry. Both types of hydrothermal fields host ecosystems that are supported by geochemical redox disequilibria. (Courtesy of the National Oceanic and Atmospheric Administration Central Library Photo Collection.)

Important and ancient kinds of ecosystems exist that depend on geochemical reductants independent of molecular oxygen [162, 427]. Deep-ocean hydrothermal systems, shown in Figure 2.3, are particularly good examples of systems hosting such ecologies. They are separated from sunlight by the depth and absorptivity of seawater, and to a large extent they are decoupled from the effects of molecular oxygen. Although seawater today contains oxygen of biotic origin, the water that emerges as vent effluent has passed through sub-crustal rock systems where nearly all oxygen is absorbed in conversion of reduced metals hosted in silicates to oxides such as magnetite.[22] Sub-oceanic vents, while not an ideal model for Hadean ecosystems because of changes in both rock and water chemistry, are likely to be among the most conservative environments relative to Hadean conditions available on the present surface of the Earth [668].

In contrast, ecosystems mediated by $O_2$ as electron acceptor are all ultimately photosynthetic. Therefore the $O_2$ is a recycled product of life rather than an input from "outside" the biosphere's processes.

---

[22] A particularly important process of this kind is the weathering of the depleted mantle mineral *olivine* to the more oxidized magnesium silicate–hydrate called *serpentine* [526, 717, 760]. We review the chemistry of the reaction and its consequences for hydrothermal fluids and life in more detail in Section 3.5.5.

### 2.3.4.2  Anabolism and energy systems: the reductive metabolism is a subset of other cases

All ecosystems require anabolism but only some make use of respiration. Therefore the two classes shown in Figure 2.2 are not actually parallel. The reductive metabolisms are both simpler than, and in a sense *contained within*, the more complex metabolisms driven by oxidants. Reductive metabolism consists primarily of anabolism. In chemoautotrophs, a comparatively simple external redox energy system converts geochemically provided electron donors such as $H_2$ into the higher-energy and standardized electron donors within the cell: various nicotinamide cofactors or flavins, small iron-sulfur proteins called ferredoxins, and a few other special cases [481].

Figure 2.4 shows the relation of anabolism to the energy systems required in non-reducing environments. The anabolic core is still required to synthesize organic compounds from $CO_2$ and differs little from the reductive case. In reductive organisms, the electron donors are externally provided. In more complex metabolisms depending on light, oxidants, or organic carbon, the anabolic core is surrounded by a much more complex layer of bioenergetic systems. This layer is reminiscent of a "space suit" that surrounds the core

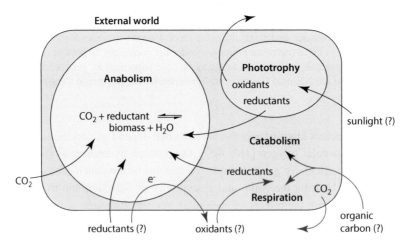

Figure 2.4  The anabolic core (blue), possibly including carbon fixation from $CO_2$, and responsible for the synthesis of the small organic metabolites, either consumes environmentally supplied reductant as in chemotrophic organisms, or is wrapped in an "energy shell" (red) that permits it to survive where environmental reductants are not provided. The energy shell feeds reductants to the core on the same cofactors that are charged directly from environmental reductants in chemotrophs. These may be generated from sunlight in phototrophy (green) with oxidants exhausted to the environment, or they may come from respiration of organic carbon with $CO_2$ exhausted. Both in respiration and in chemotrophy, the environment must provide terminal electron acceptors to form a redox-bioenergetic flow (red arrows). For respiration these must have higher midpoint potentials than biomass, but for chemotrophy they only need to have higher midpoint potentials than the environmentally supplied reductants. In fermentation (not shown) both electron-donating and electron-accepting potential are provided within the inventory of consumed organics.

and makes it viable in inhospitable geochemical environments that do not offer chemical electron donors directly.

The energy-supplying layer may consist of a photosynthetic apparatus that produces electron donors directly from a light-generated charge separation, generating electron acceptors such as elemental sulfur or $O_2$ as by-products. Alternatively, it may degrade organic compounds, transferring their electrons to oxidants provided by or stored in the environment, the process known as *respiration*. Respiration requires organic carbon as input, part of which can be used in biosynthesis without a large input of reductants (because it is already reduced to the stoichiometry of biomass), and part of which serves as net electron donor. The oxidation of the latter part of organic carbon through the sequence of membrane-bound proteins and cofactors known as the *respiratory chain* drives trans-membrane proton transfer, which is used to produce the other fundamental energy currency of life, the phosphoanhydride bonds of ATP and ADP, used in polymerization reactions. As a third alternative to photosynthesis or respiration, the energy shell may be driven by the much weaker intramolecular electron transfers of fermentation. Respiration and fermentation, along with the processes preceding them that break down large molecules to provide the small-molecule inputs to these low-level processes, are all forms of *catabolism*: the breaking down of biotically produced compounds for either energy metabolism or inputs to anabolism.

### *2.3.4.3 Ecosystem networks can be simplified by projection: metabolism is an ecosystem property*

The aggregate input/output behavior of an ecosystem-level meta-metabolome incorporates the input/output capabilities of the member species, but may be simpler than the species themselves are. In ecosystems that include heterotrophs, a large part of the aggregate pathway complexity comes from the biosynthesis of essential molecules in one species, which are then consumed and degraded by another. Often the biosynthetic and degradative pathways, viewed at the level of the small-molecule organic substrate, are reverses of one another.[23] Electron-transfer and group-transfer cofactors, and the use of auxiliary energy systems, will generally differ in biosynthesis and degradation, because the two pathways are exergonic when run in opposite directions, but the underlying organic chemical modifications remain the same.

Therefore, ecosystems contain large numbers of compound pathways that, were they to be carried out within single organisms, would be **futile cycles**: reaction sequences that produce no net change from the starting compounds, but dissipate free energy from bioenergetic subsystems.[24] Sometimes these compound pathways reflect one-way

---

[23] We argue in Chapter 4 that in the well-known cases where this is true, the degradative pathways evolved from their biosynthetic counterparts.

[24] We list several of the most important examples of such pairs in Chapter 4. They include gluconeogenesis/glycolysis, the malonate synthesis pathway/β-oxidation of fatty acids, oxidation/reduction of one-carbon groups on folate or pterin cofactors, and oxidative/reductive citric acid cycling. Many more examples could be provided. The extent to which, on average, degradative pathways retrace biosynthetic pathways in the opposite direction is roughly quantified by topological measures showing that metabolism in heterotrophic organisms is a "bowtie," with rays connecting both inputs and outputs to a common core [171, 672]. The interpretation of these topological measures is discussed in Section 4.3.1.

gain by parasites or predators, but in other cases they may reflect complementary specialization, as when plants trade organic carbon that they fix in a high-oxygen environment, for reduced nitrogen that their symbiotic bacteria fix in environments sheltered from molecular oxygen that is toxic to the nitrogen-fixation enzymes. At the ecosystem level, these "futile cycles" form loops, which may be contracted to leave a simpler network with the same aggregate biomass productivity as the ecosystem as a whole.

We may think of each ecosystem-level metabolome, then, as having two layers. The lower layer is the minimal network with the same biomass production as the actual meta-metabolome, but with all futile cycles contracted away that do not affect the stoichiometry captured in biomass. The upper layer contains the addition of the futile cycles resulting from parasitic, predatory, and symbiotic associations among organisms. The lower layers have roughly the complexity of, and overlap in large parts with, the metabolisms of minimal autotrophic organisms in similar abiotic contexts. Since minimal autotrophs may be found which have genomes considerably smaller than the genomes of typical heterotrophs – and much smaller than the aggregate metagenomes of complex ecological communities – most of the complexity of the meta-metabolome lies in the second layer reflecting community interactions.

The existence of a lower layer that separates ecological stoichiometry from questions of control through regulation and coevolutionary dynamics suggests the interpretation that *metabolism as a biosynthetic system is an ecosystem property*. The fact that each reaction in an ecosystem is carried out *within* some organism highlights the importance of the organism as a unit of organization for carrying out metabolic processes. However, the organism *as a locus of control* seems to add nothing to our understanding of metabolic closure beyond what is already explained from necessary flux balances at the ecosystem level.

Figuratively speaking, the partitioning of control among organisms feeds these universal networks into a "puzzle cutter," which divides their metabolic capabilities and partitions their fluxes among various member species. The complexity in the second layer of meta-metabolomes is not so much a reflection of properties of metabolism as it is of the complex problems of constraint satisfaction, leading to diverse possibilities for trophic ecology as a combination of species evolution and community assembly problems.

### 2.3.5 *Universality of the chart of intermediary metabolism*

When Donald Nicholson began to study metabolic pathways in 1946, about twenty were known, for the synthesis or degradation of a few major classes of compounds, and with a major emphasis on human metabolism. Nicholson's aim in his original, hand-drawn 1955 charts, was toward *integration* of that knowledge to obtain a system-level understanding of the way pathways assembled to form a system.

As the number of organisms studied has exploded in the succeeding sixty years, one of the remarkable observations from the modern version of Nicholson's chart is that the diversity in key core pathways has not likewise exploded. A survey of pathways in intermediary metabolism in KEGG [411] or MetaCyc [412], or of proteins by pathway annotation in UniProt [819], returns small numbers of pathway variants. We return in Chapter 4 to describe the small variation that is found in pathways for very fundamental functions such as carbon fixation.

While it must be borne in mind that only a fraction of microbial genes have been identified,[25] our current samples are consistent with a proposition that the common biosynthetic pathways in autotrophs and the aggregate networks of ecosystems are in fact a *universal* property of life on earth.

### 2.3.5.1 *Metabolic universality and trophic integration of ecosystems*

A shared inventory of small metabolites is a precondition to the integration of arbitrarily diverse organisms in trophic ecosystems. Organisms that do not initiate biosynthesis from building blocks that can be obtained from other organisms cannot benefit from predation, parasitism, or symbiosis.

We cannot rule out the possibility that ecological integration could in part be a *cause* for metabolic universality. If multiple, parallel "shadow biospheres" existed on Earth and competed for the same abiotic energy sources, organisms in the majority biosphere could benefit from resource sharing and – other factors being the same – come to constitute the sole biosphere through competitive exclusion.

From a variety of additional observations, we favor the interpretation that metabolic universality should be recognized as the *precondition*, and trophic integration as one of many consequences. Biosynthesis originates in a very small collection of compounds in or near the citric acid cycle, and respiration terminates in the same compounds. The densely cross-linked core of intermediary metabolism includes only on the order of 50 compounds – the entire inventory of monomers required to synthesize an autotroph numbers only about 125 [777] – and the rest of anabolism and catabolism is organized in rays from or to this core. We argue in Chapter 4 that the molecules and pathways in the core bring together many properties in organic chemistry, network topology, and overlaps with geochemistry that may make them uniquely likely to emerge and easy to maintain. The extent of universality in the deep core and the periphery seems to follow criteria of network centrality and evolutionary age. It is likely that these properties are favored under ecological fluctuations and coevolutionary dynamics, but it seems implausible to us that a modest selective advantage, such as being part of a majority trophic group, creates sufficient advantage to explain all of these biochemical properties together or their concordance in the layer structure of metabolism.

---

[25] Hugenholtz, Goebel, and Pace [380] estimate that fewer than 1% of microbes estimated to exist through culture-independent methods can be cultured and thus characterized biochemically. Even culture-independent methods based on polymerase chain reaction amplification are limited by the use of primers for conserved positions on molecules such as rRNA used as fingerprints [830], suggesting further systematic undersampling of extant species.

## 2.4 Biochemical pathways are among the oldest fossils on Earth

The existence of a common framework of core metabolism covering the diversity of known organisms is made more remarkable by the great antiquity from which this core has apparently been preserved. It is difficult to arrive at precise dates for the early transitions in cellular life, but a variety of evidence suggests that cells incorporating much of the molecular complexity of modern forms existed at or soon after the boundary between the Archean and the Hadean eons, conventionally placed at 4 billion years ago (4 Ga).

### 2.4.1 Evidence that currently bounds the oldest cellular life

William Schopf [708] identified microfossils from what were claimed to be 11 distinct taxa in the early Archean apex chert in Western Australia dated to 3.465 Ga. Although none of these is likely to be from cyanobacteria as the original publication suggested (see footnote 26), they do provide reliable evidence that filamentous bacteria of nearly modern morphology were not only present but differentiated by this date. Older chemical evidence of $^{13}C$ depletion which is generally believed to be of biological origin comes from carbonaceous inclusions in apatite, from the Isua supra-crustal belt (the oldest known preserved sedimentary formation) and Akilia Island in West Greenland, dated to 3.85 Ga [517, 560].

Calibrating a molecular clock for 16S rRNA phylogeny against a date of $2.65 \pm 0.03$ Ga for the origin of cyanobacteria from 2-methylhopanoids in late Archean shales from the Hamersley Basin of Western Australia,[26] Peter Sheridan and colleagues [736] published branching times of bacteria from archaea (corresponding putatively to the LUCA) of 4.29 Ga, and subsequent earliest branchings within both the archaea and the bacteria of 3.46 Ga. Many difficulties attend assumptions of constant mutation rates used in such reconstructions [464], and we expect that these will be systematically worse for the earliest divergences when cellular mechanisms of error correction were likely different and presumably less effective than those of modern organisms. However, acknowledging that such dates must be taken provisionally, these reconstructions are broadly consistent with life's having arrived at the integrated cell form characteristic of the LUCA at or within 100–200 million years after the beginning of the Archean eon, long enough ago to account for the isotope shifts in Isua sediments and the differentiated microfossils in Western Australia.

### 2.4.2 Disappearance of the rock record across the Hadean horizon

For comparison, few rocks exist that preserve sufficient structure to serve as fossils of *anything* at the Hadean/Archean boundary. Prior to 2008, the oldest known intact rock,

---

[26] The original report of hopanoids is due to Summons *et al.* [787]. The important difference between this signature and the microfossils analyzed by Schopf is that 2-methylhopanoids are considered a specific signature of cyanobacteria. The dating of the Hamersley Basin shales is broadly consistent with other evidence that places the origin of the cyanobacterial clade around this time [76].

which had not been metamorphosed by crustal recycling in the mantle, was a gneiss from the Acasta formation of the western Slave Province in Canada, dated by zircon inclusions to 4.03 Ga [86]. In 2008, Jonathan O'Neil and colleagues [606] reported a faux-amphibolite from the Nuvvuagittuq greenstone belt in Northern Quebec, which they dated at 4.28 Ga from neodymium isotope ratios. Beyond this horizon, only zircon crystals are known which have been preserved from dates as long ago as 4.4 Ga, but these are all inclusions in rocks that have been metamorphosed by heat and pressure in the mantle, and preserve no fossil record beyond possible isotopic inclusions within the zircons themselves.

To put this comparison into perspective, the rock record itself becomes vanishingly rare at the Hadean/Archean boundary. We have no claims of rock-preserved evidence of life earlier than 3.8 Ga, no morphological evidence older than 3.45 Ga, and only isolated instances of each of these. Yet the entire biota of Earth form a living fossil of core metabolic pathways that must have been in place by the era of the LUCA, and we conjecture well before the genetically recorded split of bacteria from archaea, which occurred around the same time. Whereas mineral fossils depend on the durability of crystals to preserve structure, biological fossils were carried for the first 2.3 billion years of life's history exclusively in the cells of archaea and bacteria, most of which could have had generation times on the order of hours.

### 2.4.3 Metabolism: fossil or Platonic form?

We will return in Chapter 8 to the interpretation of such dynamically preserved "fossils." It may be inappropriate to regard core metabolism as analogous to a fossil, because a fossil is something *arbitrary* with respect to the crystal matrix in which it is imprinted. The arbitrariness, the lack of any *necessary role* (for the fossil) to the crystal structure itself, is what makes fossils in all but the hardest encasing crystals fragile against erasure. A more apt comparison may be between core metabolism and the *crystal structure of the mineral itself*, a kind of inevitable form which is stable because it is a favored pattern.[27] An important limitation of this analogy is that, whereas crystalline unit cells re-form *de novo* under appropriate conditions, the preservation of core metabolism appears to have required continuity of the life forms that carried it, and in that sense it resembles true fossils.

The universality of core metabolism is an exemplar of the kind of regularity that is distinctive of life and that a theory of origin must account for. What kind of fossil is it that is carried by fragile and transitory individuals yet survives planetary conditions that destroy the hardest minerals? What kind of inevitable regularity might it be that is preserved like a crystal's Platonic form, yet gives no evidence of independent cases of emergence, so that all known instances have depended on continuity of the life forms that carried them?

---

[27] In making this argument for metabolism, we recall a similar argument originally made by Michael Denton [189] for protein folds: initially viewed as constructs governed entirely by arbitrary sequence evolution, folds are now understood to fall within a few families determined by the geometric constraints of spatial packing with a one-dimensional string. Sequence evolution, which may be very diverse, sorts proteins among as few as a thousand folding motifs, and then governs finer dynamical processes such as folding sequence and reliability.

## 2.5 The scales of living processes

Part of the challenge of understanding cause in the current biosphere, and making plausible conjectures for simpler stages that could have led to it, comes from the fact that living order involves interactions across a wide range of scales of aggregation and of time.

In this section we consider the multiple scales at work in biochemistry, in the physical organization of living systems, and in the network and catalytic interactions that make up internal information and control systems in organisms and ecosystems. We wish to understand what combinations of causes and constraints contribute to the formation of structure at each scale, and how this leads essential parts of living function, or contributions to the stability of the biosphere, to segregate among components at different scales.

As we see how different dimensions of living pattern are carried at different levels of organization, and how each level may work cooperatively with other levels to support stability of the living state, we will be discouraged from seeking explanations for living complexity in any one level of structure such as the cell or a single class of macromolecule such as RNA, or projecting the biological problem of error correction onto any single process such as replication with selection. In the next section we will consider the way different levels of organization are arrayed along an axis from chance to necessity, as an empirical guide to the levels of structure that are most likely to be primordial.

### 2.5.1  Scales of biochemistry

#### 2.5.1.1  Roles of the elements

The most basic determinant of the structure of biochemistry is invariant properties of the chemical elements [881, 882]. Distinctive roles associated with the most abundant and central elements include the following.

**Carbon, hydrogen, and oxygen for structure**  Stable bonds among carbon, oxygen, and hydrogen create the basic structural components of life. George Wald has elegantly written [848] on the reasons why the tetravalency of carbon, combined with its small size and relative hardness, make it unique among the elements as a carrier of the flexible structures in living systems. $CO_2$, unlike $SiO_2$, is a gas, carbon has a lower affinity for oxygen, and both pure carbon and carbon-oxygen systems form chains under some conditions, where silicon oxides tend much more strongly to form crystals. The same cooperative effects that make crystalline states highly ordered and durable also militate against internal variability, and in most cases associate a unique thermodynamic phase with each class of boundary conditions. We believe Wald has given the most thorough and sensible refutation to arguments that attempt to associate "life" as an abstraction with silicon on the basis of its limited electronic similarity to carbon.

**Nitrogen for catalysis and information**  The introduction of nitrogen to CHO backbones brings with it the first capabilities for electron-transfer catalysis. The similarity of the three unpaired valence electrons in nitrogen to the valence electrons of carbon permits nitrogen to enter stable backbone structures, while the transition between a non-bonding pair state for its two additional electrons, and ionized states (typically involving $\pi$ bonding in ring structures) enables electron transfers [587].

Since catalytic relations have the essential asymmetry that confers the first notion of "control" on relations among components, the incorporation of nitrogen also introduces the beginnings of hierarchical control into biochemistry. The diversity of properties available in amino acid chains, and the role of nitrogen heterocycles as the nucleobases, also make nitrogen-containing compounds the alphabets from which sequence-based information systems are made.

**Phosphorus and sulfur for energetic bonds** Phosphorus and sulfur create the principal energy-carrying bonds of organic chemistry [183]. Sulfur substitutes for oxygen in compounds such as thioacids and thioesters, but its larger size and chemical "softness" make these high-energy bonds, from which sulfur is a good leaving group. Phosphorus occurs in biology exclusively in the form of phosphate, and is responsible for dehydrations in 55 M water, a unique chemical property that enables polymerization to proceed as a non-equilibrium steady-state process. Phosphorus participates in reactions also involving sulfur because the thioester bond and phosphate ester bond are similar enough in energy that the two groups can be exchanged almost reversibly. Because high-energy thioester bonds can be formed using the energy from redox potential, the exchange of thioesters for phosphates in the process called *substrate-level phosphorylation* is a rare and important bridge afforded by the bond properties of organic chemicals, between the redox and phosphate energy systems of biochemistry (discussed further below). Frank Westheimer has summarized these and other essential roles of phosphorus and sulfur in a classic article [870].

**Transition metals for electron transfer** Metals – transition metals in particular – are the primary centers for many electron-pair transfers and most single-electron transfers that produce radical states on organic compounds. We consider the many roles of metals further in Chapter 4, so we limit the discussion here. One point of interest, however, is the relation between transition metals and the periodic table relations of nitrogen and sulfur to carbon. Nitrogen, as we have noted, structurally resembles carbon except that one valence electron is replaced with a non-bonding pair. Oxygen replaces two valence electrons with non-bonding pairs, as does sulfur, one level below oxygen in the periodic table. All transition metals in electron-transfer proteins or cofactors are bound with dative bonds to either nitrogen or sulfur, in which the non-bonding pairs on these atoms are donated[28] to fill orbitals on the metal.

This small inventory of distinctive and non-substitutable element properties is the foundation for all forms of hierarchy discussed in the remainder of the section and in later chapters. Their importance in creating both a rich and an orderly landscape of structure and

---

[28] The term "dative" comes from Latin *dare* "to give," also the root of "donate."

process, on which living systems depend in an incredible range of ways, cannot be over-emphasized. The actual structure of the life we know is deeply and pervasively anchored in the particularities of the periodic table. If one wishes to propose an abstraction of "life" away from its chemical substrate, one should first consider seriously the depth of this embedding, and should ask, if the chemical particulars were removed, how much of the structure we think of as living would remain to be abstracted. This embedding, of course, is also the basis for belief that the emergence of life was heavily scaffolded in the structure of physical laws, and thus not an arbitrary and vastly improbable discovery.

### *2.5.1.2 Biomolecules fall into three classes by size, composition, mechanism of biosynthesis, and function*

Both the functionality of biomolecules and the chemistry of their synthesis are segregated along similar boundaries of molecular scale [587]. These boundaries reflect shifts in the level at which information is both incorporated and used within living structures.

The **small-scale metabolites** include the ketoacids, amino acids, small hydrocarbons, simple sugars, a few kinds of carbon-nitrogen heterocycles and aromatics, and simple combinations of these. Some of these appear only as intermediates making up the biosynthetic network, while another subset also becomes the monomers from which larger structures are assembled. The small metabolites contain the main structural bonds among C, N, O, and H, and their size is limited to $\lesssim 20$ carbon atoms. This is roughly an upper size limit for molecules assembled by organic reactions.[29] Phosphates and thioesters may appear in intermediates of the biosynthetic pathways, but their role generally is to provide energy for leaving groups, and they are not retained as parts of the structure of the molecules. Most of the key chemical properties relevant to structure or catalysis are incorporated into biochemistry at this low level.

At intermediate scales are found a diverse collection of organic molecules called **cofactors**, which carry out electron transfers and several classes of essential group transfers [254]. Cofactors tend to have highly stereotypical functions defined at the level of the single molecule. As such they are central loci of control for the flow of material and energy through biosynthetic pathways, and key hubs for the integration of networks. Cofactors are the largest and most complex molecules in which the reactions of organic chemistry are the major biosynthetic process. Sulfur first enters several key groups as a structural element, and together with nitrogen can be key to tuning the properties of carbon atoms that serve as binding centers. Phosphorus enters structurally only as a linker, typically between a monomeric component that performs the essential molecular function, and a nucleotide component that appears to function mainly in manipulating the cofactor by larger molecules [132]. In this respect cofactors straddle a border between the small metabolites and the polymers.

---

[29] Even siderophores, among the most complex of widely used organic compounds, are often elaborations of functional centers that are small core metabolites, such as citrate [107, 841].

The large-molecule sector of biochemistry contains large lipids[30] and the three well-known classes of **oligomers**:[31] the polynucleotides, polypeptides, and polysaccharides. Use of phosphorus as a source of dehydrating potential and sometimes as a linker becomes the main chemistry in all groups except the lipids, and the specification of information shifts to the sequence (and for sugars, a few choices of topology) among the monomeric components. Unlike the functions carried by cofactors, the functions performed by large oligomers are not usually determined by chemical properties of individual monomeric components, and are only realized through the relations among multiple components [323].

### 2.5.1.3 Environmental heterogeneity and cellular homogeneity

Environments may provide many forms of chemical energy, most of this diversity taking the form of different combinations of molecules that can be paired as electron donors and acceptors to form redox couples.[32] Yet inside the cell, energy sources are universally converted to two energy "buses"[33] consisting of electron-transfer and phosphoryl group transfer compounds.[34] The electron-transfer compounds are principally nitrogen heterocycles, a small number of nitrogen-free aromatics (quinones), and a diverse collection of small metal-center proteins and cofactors. The principle phosphoryl-transfer cofactors are the nucleoside triphosphates (NTPs), with ATP serving as the major energy carrier and other NTPs performing special functions that include a signaling role because of their restricted uses. Inorganic polyphosphates have also been found to be present in vacuoles and other storage bodies in a very wide range of cells [100, 454, 462], likely functioning as long-term energy stores and perhaps participating directly in phosphorylation reactions.

The conversion from the diversity of environmental energy sources to a small number of more or less standardized cofactors decouples the problem of refining the many catalysts and pathways required in metabolism from the problem of responding to environmental variations. A theme that we will encounter repeatedly in this book is the importance of *standardization*, at all levels from the selection of a small-molecule inventory to the modularization of high-level architectures. Reducing the number of components responsible for some essential function, and ensuring that their forms are stable in time and reproducible over instances, provides a stronger and more consistent statistical signal for natural selection acting as an inference engine, in the evolution of higher-level architectures or hierarchical control structures. The question of how selective geochemistry might have

---

[30] We include the lipids in this group, although their maximal sizes are much smaller than those of characteristic polymers in the other three classes, because the lipids are synthesized by recursive attachment of either two-carbon or five-carbon units.

[31] Oligomers are polymers synthesized from single classes of monomers.

[32] Lengeler, Drews, and Schlegel [481] offer a representative summary in Chapter 10 (G. Keunen).

[33] Here we borrow a term from electrical engineering, to refer to a conduit for energy which is tapped by components at many points.

[34] Proton transfer is essential as an intermediate energy store that couples the redox and phosphate energy buses. In photosynthetic organisms the proton potential may be supplied directly by bacteriorhodopsin (in archaea) or photosystem II (in bacteria or chloroplasts) [358]. Some direct use of proton-motive force is made by machines such as flagellar motors. pH contributions to biosynthesis, however, come from the small-scale control of $pK_a$ values of organic monomers, and not from the contrasts of pH between bulk proton reservoirs.

been, over the early inventory of energy sources and organic inputs, will be fundamental as we try to infer from the present inventory of intracellular components what their geochemical precursors could most plausibly have been.

### 2.5.2 Scales of physical organization

Living systems likewise exploit order across a wide range of scales of physical organization. The smallest scales employ structural variations afforded by equilibrium thermodynamics, including contrasts between bulk phase and surface reactions, or between polar and non-polar solvent phases. Biochemical networks that are topologically modular may also be segregated in space, by cellular compartments or lipid/water phase distinctions, or in time by diurnal or other cycles. Cells or organisms furnish units of Darwinian selection, and the timescale for cell division or organism reproduction governs the rate at which selection can incorporate information into populations of genomes. The interactions that appear, in aggregate, as the "fitness" of individuals are mediated by long-range feedbacks and cycles of elements and compounds in ecosystems, and these interactions impose relations on the fitness functions of multiple member species. Weak fitness differences, which may be masked in the short term by the constant generation of novelty in complex organisms, may be expressed over the long term in evolutionary convergences.

We wish to keep multiple scales present in the analysis because errors from statistical fluctuations can propagate across scales, as do constraints from asymmetric interactions such as selective catalysis, actively driven transport, regulation of phenotypic states, or selection, which act as error-correction processes. If we think of error correction as a process that must not only be described qualitatively in terms of mechanism, but also accounted for quantitatively in terms of relative rates of entropy incursion and rejection, we realize that it is not known whether the signals within any one layer of biotic organization are sufficient to maintain stability within that layer or its environment. We will argue in later chapters that corrections originating from the small scales of reaction mechanisms and physical chemistry, and from the ecosystem scale of stoichiometry and aggregate growth, act synergistically as reference signals to support the stability of cells and organisms. This is the reason metabolism is found empirically to be stable and law-like at the ecosystem level, while permitting and even exploiting significant variation at the level of organisms.

### 2.5.3 Scales of information and control

Finally, the distinct functions of memory systems and the asymmetric interactions that implement control relations are also realized across a range of scales.

We will argue in Chapter 4 that the specific chemical functions offered by cofactors feed back to determine the subset of organic chemistry that makes up biochemistry. Thus the determination of the cofactor inventory is the first layer of control over metabolism – the evolution of oligomer catalysts and structures is more complicated, and we believe it arose later. How much of the selection of cofactor form and function is a reflection of chemical structure inherent in the early elaboration of organosynthesis, and how much is due to selection in the Darwinian era, must strongly affect our interpretation of the degree to which metabolism is entrained on organic chemistry or geochemistry. The premise that the origin of metabolism was in part a transfer of functions from mineral or bulk phase geochemical environments onto self-generated organics leads to our shift of emphasis, away from the invention of organosynthesis and toward the transfer of control and interdependence, in the earliest stages of biogenesis.

Within physiology organisms employ regulation from the short timescales of allosteric response in enzymes to slower active regulation of gene expression,[35] while on the timescale of generations selection can change population compositions. Ecological reconfiguration can take place on timescales shorter than the species persistence times, but can also be responsible for coordinated extinctions.

Longer terms reveal evolutionary convergences or the localization of classes of innovation within particular evolutionary contexts or intervals. The few variations known in carbon fixation all occurred before the rise of molecular oxygen. Following the rise of oxygen, pathway innovation became architecturally different and centered on mechanisms of organic carbon exchange (we provide a few examples in Section 4.4). Similarly, a shift appears to have occurred between innovation in organic chemistry in early life, and innovation in cell form (endosymbiosis), cell coordination (colonialism and multicellularity), and complex regulation (multicellular development) following the rise of oxygen.

Even at much larger and later levels of developmental complexity, evolutionary innovation appears to segregate into different stages in complexity and time. Early clades often develop metabolically costly but morphogenetically simple weapons of competition and conflict such as antibiotics or toxins (examples include cnidarians in the deep pre-Cambrian [229], and then arachnids as an early group within in the arthropods). Later clades appear to shift innovation away from exquisitely effective biochemistry and into the development of morphological features such as skeletons, musculature, armor, teeth, etc. (the Cambrian radiation) or refined mechanics and coordination (insects).

On the very longest terms the **major transitions** take place. While these may often be recognized by qualitative changes in the content and use of genomes [790], they are only understood when the genomic changes are seen as one component in larger system rearrangements that often involve ecological and even chemical or energetic shifts. Examples are too numerous to list, but include nitrogen fixation [87, 665], the

---

[35] An important question in microbiology is how little regulation an organism can employ and still survive and reproduce within a stable environment. A general expectation is that the number of regulatory elements within a genome must increase faster than the number of coding proteins. In bacteria and archaea, it appears that the scaling of regulatory/total genes is roughly quadratic (though estimated over only about 1.5 orders of magnitude in total genome size) [530]. Allosteric response of enzymes is a regulatory mechanism evolved into protein sequences, which acts in a predictive, "feed-forward" mode [604] on physiological timescales. In contrast, genome regulation is responsive to cell states and acts through feedback, a slower but more versatile mechanism. With both mechanisms at work, even "simple" prokaryotes with small genomes living in stable environments are potentially far from the kinds of passive chemical systems governed by environmental chemical potentials, which we expect the earliest geochemical networks must have been.

rise of oxygen [348], endosymbiosis [470, 475], multicellularity [354], or the Cambrian radiation [228].

In the next section we contrast the emergence of biochemical innovations – those that have been preserved – which seem stable and progressive, with the ongoing "Darwinian churn" that routinely extinguishes species and rearranges ecosystems. We suggest that biochemical innovations have a permanence but took time to discover or access, while the species level of organization is not similarly protected.

## 2.6  Diversity within the order that defines life: the spectrum from necessity to chance

> The Universe was not pregnant with life, nor the biosphere with man.
>
> – Jacques Monod, *Chance and Necessity* [561]

The question of "chance and necessity" [561] – whether the emergence of a biosphere on Earth was in some way an improbable or an inevitable occurrence – is part of a larger suite of questions about the role of historical contingency in the dynamics of living systems [311], and how this makes biology different as a science from the more deterministic sciences that have largely been the source of our concept of natural laws. Estimating the likelihood of biogenesis is complicated because it is not one question, but an amalgam of several different kinds of questions compounded by problems of ignorance.

The first group of questions concern what exactly are the transitions for which we wish to compute probabilities. How many must be understood, and are the likelihoods similar or very different for different transitions? Just as we have argued that "life" cannot be understood as a single kind of order, we argue that the emergence of the biosphere cannot have been a single transition or even a single kind of transition.

The second group of questions concern the relation between accident or historical contingency in the origin of life and their role in later evolutionary dynamics. To what extent do the early and the later eras even reflect the same constraints and processes?

In posing all of these questions, we are limited by a lack of compelling hypotheses for how most early order might have formed, and by a lack of comprehensive theory of biological causation that goes beyond the contributions of individual selection in evolutionary dynamics. We must not, however, conflate gaps in our own understanding with the absence of mechanisms in nature, and must try to identify what can be meaningfully said about likelihoods in biogenesis from a perspective severely limited by ignorance.

In this section we note that the essential forms of order in the biosphere, listed at the beginning of the chapter, vary enormously in their ages and persistence times, their degree of universality, and their characteristic dynamics. Sensible questions about the likelihood of biogenesis can only be framed by disaggregating this heterogeneity and evaluating the evidence for chance and necessity at each level.

### 2.6.1 How contingency has been extrapolated from modern evolution to origins

> a distinguished logician ... has well and quaintly said, that if a malign
> spirit sought to annihilate to whole fabric of useful knowledge with the
> least effort and change, it would by no means be necessary that he should
> abrogate the laws of nature. The links of the chain of causation need not
> be corroded. Like effects shall still follow like causes; only like causes
> shall no longer occur in collocation. Every case is to be singular; every
> species, like the fabled Phoenix, to be unique. Now most of our practical
> problems have this character of singularity; every burning question is a
> Phoenix in the sense of being sui generis.
>
> – F. Y. Edgeworth, Introductory lecture [208]

Biology requires some conceptual shifts that are new from a perspective on natural science born in physics and chemistry. The possibility for history-dependent dynamics is recognized in physics, but is largely understood in terms of sensitive dependence on initial conditions, created by mechanisms such as non-linear amplification.

In biology the capacity for history dependence becomes pervasive and takes on qualitatively new importance. Constellations of circumstance that may only form once can shape events of evolutionary innovation, and the consequences of innovation can alter the trajectory of living systems indefinitely in the future. Over a long and influential career of technical and popular writing [311, 312], Stephen Jay Gould fought against a kind of Laplacian determinist view of science inherited from physics, to assert that in biology the role of prediction as a criterion of understanding needed to be seriously reconsidered and for many cases either marginalized or abandoned.

Gould was only one of many influential writers to argue for the centrality of historical contingency in biology, and to insist on the importance of understanding that structures and functions of fundamental importance in life could be due in part to chance. This message, applicable to evolutionary dynamics throughout the history of life, became attached to theories of life's origin as these were framed in evolutionary terms. Jacques Monod's specialization was the genomic regulation of metabolism in modern and highly specialized bacteria [389, 562], and from this experience he argued in *Chance and Necessity* [561] in favor of an accidental origin of life. Francis Crick famously wrote in *Life Itself* [168] "the origin of life appears at the moment to be almost a miracle, so many are the conditions which would have had to have been satisfied to get it going."[36] While not making an explicit assertion about the origin of life, Gould wrote [311] with regard to the long-term trajectory of the evolution of life on Earth, that "Any replay of the tape would lead evolution down a pathway radically different from the road actually taken."

---

[36] Crick also concluded that the assignments of amino acids to trinucleotide codons in the genetic code was a "frozen accident" [166], a question we consider in Chapter 5. Writing with Leslie Orgel [169], and revisited in *Life Itself*, he would go so far as to propose "directed panspermia," a hypothesis not only that life originated somewhere besides Earth, but that its transport here was a deliberate event of seeding the Earth with life by a more advanced civilization.

What we must judge, when we return below to compare the long-term historical dynamics of living order at different levels of aggregation, is whether the model of integrated organisms or the varied character of subsystem histories carries more weight in our interpretation. Is there good reason to believe that the cross-level couplings in life have always been so tight and obligatory that all of life inherits the path dependence that Gould, Monod and others observed in the most complex developmental and regulatory systems? Or does the wide variation in stability, universality, and persistence of living order at different levels constitute evidence that their dynamics must be controlled partly by different factors, and that some of these are not particularly subject to historical contingency?

### 2.6.2  Natural selection for change and for conservation

One of the great conceptual challenges to understanding the nature of the living state is that evolution can produce open-ended variation, but it can also be the agent of absolute conservation of some characters of life. Foremost, in our view, among the conserved characters is the role and structure of metabolism, but several other elements of molecular biology and cell physiology can be cited as well. The puzzle in their conservation is that we see these structures and functions only within the domain of living systems, where they seem to rely inherently on complex and high-level entities tuned by natural selection. Yet the low-level patterns are much more invariant than the higher-level mechanisms on which they seem to "depend."

We think that this observation motivates a more quantitative awareness of the relative roles of adaptive versus stabilizing selection. Adaptation, and particularly "open-ended" variation, seems to be the major function of evolution stressed in most writing [828]. However, in biology one can find what one chooses to notice. Adaptations and arbitrary complexities abound for those who find them interesting. However, in purely quantitative terms, the "events" of stabilizing selection against deviations from a set template probably far outweigh events that favor innovation. Every moment is an opportunity to die if essential functions are not performed, whereas reproduction occurs only at the culmination of long intervals of survival.

We return in Chapter 8 to some theoretical considerations for why it might be impossible for evolution to retain exact characters indefinitely at a high level of complexity, but why at the same time, the preservation of a living state and the ability to preserve complex characters at all – even over finite durations – may depend on the existence of absolute invariants. Here we note some empirical patterns to furnish intuition.

### 2.6.3  Different degrees of necessity for different layers

In living systems, evidence for chance versus necessity is not distributed evenly across different scales of aggregation. Figure 2.5 arranges units of living organization according to criteria that we associate with seeming inevitablility or seeming accident. One end is anchored with phenomena in geochemistry, which are expected to be predictable in detail

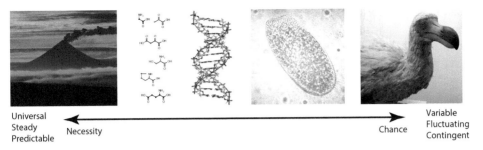

Universal
Steady        ← Necessity                                    Chance →        Variable
Predictable                                                                 Fluctuating
                                                                            Contingent

Figure 2.5 The living world contains organizational forms across a wide range from the law-like and seemingly necessary, to the deeply historically contingent. Biological order covers most of this range, and is by no means all clustered at the contingent end. Geophysical processes such as volcanism and its associated chemistry seem universal, predictable, and stable in the course of planetary maturation. By these criteria, the universal small metabolites are nearly in a category with geochemical phenomena, more universal than even the apportionment of tasks to RNA versus DNA and proteins, and in most cases much more universal than the homology classes of genes that encode their synthesis. More variable still is the inventory of cell forms, and most variable of all are the identities of particular species, which depend on accidents of circumstance to arise in their precise forms, and are inevitably lost through circumstances equally idiosyncratic. Image sources: volcano (Public domain); DNA [3]; paramecium [7]; dodo [6].

under natural law. The other end is represented by a developmentally and ecologically complex species such as the dodo, which evolved in idiosyncratic circumstances, persisted while those circumstances were preserved, and went extinct through a tiny number of even more arbitrary cross-species interactions.

### 2.6.3.1 Criteria of chance and of necessity

Phenomena that we regard as necessary tend to be universal in the way they respond to boundary conditions and steady if the conditions do not change. This regularity in many cases reflects a kind of "closure" in their causes, which over time we succeed in capturing in natural laws. At the other extreme, forms that we regard as due to chance show variability that does not follow in any simple way from boundary conditions; they are often fluctuating or ephemeral even when their circumstances are broadly stable, and thus their particular forms are often contingent in major ways on details of history. They are not repeatedly derived from evolutionary convergences, and are not re-discovered if they are lost. *By these criteria, different levels of order, each fundamental in its way to the living state, cover almost the entire range from chance to necessity.*

Geophysical phenomena, including their associated chemistry such as mantle convection, volcanism, or hydrothermal circulation, are examples of necessary processes. We noted in Section 1.2.2.2 that many of these are both similar, and coupled, to biochemical processes. We expect such geophysical processes to be predictable from first principles of natural law and a few aggregate parameters governing the composition and energetics of our particular planet. By the same criteria of both antiquity and universality, the small-molecule substrate of metabolism is hardly different from geophysical phenomena. The core metabolic inventory apparently came into existence in either the Hadean or early Archean, and has changed little in the history of life since.

As we move up in scale of aggregation and complexity, the transition from characteristics suggesting necessity to characteristics suggesting chance is never made in one large leap. The core metabolic substrate is somewhat more universal than the use of RNA and DNA, which shows small variability (particularly among viruses and between viruses and free-living cells) and is probably different today than in the earliest stages of RNA-mediated life. Reactions in the core metabolic network are, to different degrees, more conserved than the enzymes that catalyze them,[37] even when the latter are coarse grained to the level of homology classes.[38] Cell form is yet more variable, with notable groups including the archaea, Gram-positive and Gram-negative bacteria, and among eukaryotes, the protists, fungi, plants, and animals. Only at the highest levels of aggregation such as species do the roles of historical contingency and variation lead to an incessant churning that rearranges ecosystems and drives speciation and species extinction.

From these and similar observations we conclude that chance versus necessity is not a distinction that can be ascribed to life as a whole. The contributions of invariant laws and variable history must be assessed for each level of component and each class of relation. In this breakdown, the small core of metabolism appears closest in character to lawful geochemistry. Hallmarks of cellular life, including the use of RNA, DNA, and proteins, and cellular encapsulation itself, are all more variable and ostensibly further from the character of geochemistry. The species adapted by evolution, and the community-assembly aspect of ecosystems derived from species properties, are heavily and essentially contingent in the sense so persuasively described by authors such as Stephen Jay Gould, and summarized by Monod.

### 2.7 Common patterns recapitulated at many levels

The interdependency among levels in the hierarchy of living structure is reflected in the observation that, where living systems show a modular architecture, the module boundaries are often aligned across multiple levels and kinds of organization. At a minimum, this reaffirms the conclusion that to understand error propagation and error correction in living systems, we will need to trace the flow of control signals and constraints across layers. For some cases, however, a stronger interpretation can be given: the alignment of modularization is not merely a coordination, but a *recapitulation* of structure that originates in some levels, by structure that accretes at other levels. In most of these instances, the interpretation we give for the direction in which information has flowed will depend on

---

[37] Well-studied examples in core metabolism include the fructose-1,6-bisphosphatase [701] and bisphosphate aldolase [753], and the citryl-CoA synthase/lyase system for reductive citric acid cycling [35, 36].

[38] Some key enzymes, particularly for highly conserved reactions in core metabolism, may have only a single homologous class. Most show multiple homology classes. Among these, some of the greatest variability, suggesting repeated innovation, arises in the genes for related functions in DNA replication, repair, and error correction [45, 93, 681].

arguments about the structure of variation in a comparative analysis, but in a few cases we can trace directionality directly from evolutionary reconstructions.

The richest source of modularity and structure that we will argue is primordial, is the decomposition of the small-molecule metabolic network into the subsystems of carbon fixation and the anabolic pathways. It results from the complex interplay of chemical mechanisms, energetics, kinetics, and network topology, which we consider in Chapter 4. Higher-level systems, which we argue in Chapter 5 were entrained on this metabolic order and froze it into place, include the following.

**Catalysis of organic and phosphate-mediated reactions** We noted in Section 2.5.1.2 the abrupt transition from control over organic synthesis as the source of information in monomers and cofactors ($\lesssim C_{20}$), and phosphate-mediated polymerization and control over monomer sequencing, as the source of information in oligomers ($\gg C_{20}$). We believe the shift occurs because the combinatorial problems of controlling complex organosynthesis become prohibitive for molecules above a few tens of carbon atoms, whereas the sequence control problem remains tractable.[39] Directly reflecting this transition is a transition in the pattern of evolutionary conservation or convergence of catalysts. Catalysts for organosynthetic reactions require precisely tuned residues, and when these catalysts are conserved, the positions of the residues in the reaction site are targets for particularly strong stabilizing selection. The opposite behavior is shown among catalysts for DNA polymerization, in which the conserved motif, apparently a target for evolutionary convergence, is a general "right hand" geometry. The fingers and thumb are highly variable, while the palm region typically reflects one of two folds in the different replicase families [45, 681].

**Bioenergetic systems in relation to cell structure** The particular chemical requirements of organic synthesis and polymerization are supported by two largely decoupled free energy systems. One is the system of redox cofactors, and the other is the system of nucleoside phosphoanhydrides. These may be coupled through a few key organosulfur compounds, or through membrane-mediated proton transport. We will argue in Chapter 5 that the cell is not one fundamental kind of functional compartment, but several different kinds, and that the different functions of catalysis, homeostasis, and energy-system coupling are to some extent reflected in the subcompartmental structure of many kinds of cells.

**Amino acid biosynthesis and assignments in the genetic code** In Chapter 5 we show that the translation system from RNA to proteins – a system that has clearly been under evolutionary pressure to provide a buffer between the idiosyncratic structure of metabolism and the flatter opportunity space provided by sequence combinatorics – is organized in a way that heavily recapitulates the order of metabolism. Relations between the assignments

---

[39] A further shift in sequence control arises, following Denton [189], when the unit of coarse graining passes from the sequence to the fold, affording some robustness to the problem of selecting precise sequences.

of amino acids to nucleotide triplets in the *genetic code* have been recognized since the 1970s [901], but the information we assign to such associations becomes even greater when we recognize that it extends also to the selection of the biological amino acid inventory. Part of the reflection of biochemistry in the code can likely be accounted for by precisely the code's function of buffering [276, 327, 833]: to minimize the leakage of metabolic pattern through the coding process when errors occur, the assignments must make use of redundancies in the underlying chemistry. We will show, however, that the regularity of the code is even more fundamentally a biosynthetic pathway regularity, which can only be rationalized if it arose when the precursor to the modern translation system was an embedded system coevolving with metabolism.

**Metabolic structure and evolutionary history**  Finally, we will show that the patterns of innovation in metabolism trace many of the major divisions in evolutionary history of the bacteria and archaea, to a degree that is unexpected from the frequency of horizontal gene transfer in the deep tree of life. We noted in Section 2.5.3 that innovations in carbon fixation occurred early and seem to have saturated with a few major pathways. The character of pathway remodeling changed as innovation shifted away from carbon fixation and toward carbon exchange, following the rise of oxygen.

Working from this relatively rich set of abstractions about general patterns in living systems, we turn in the next three chapters to the facts in geochemistry, metabolism, and cell biology that in our view make this degree of conceptual breadth necessary. The reason the complexity of the biosphere cannot be projected down to a few privileged abstractions is that it grows out of complexity already inherent in the dynamics and interactions of the other geospheres. It is not any ability to bypass complexity, but rather the many threads of continuity in the complexity, that motivate us to argue for a necessary place for life on Earth.

# 3

# The geochemical context and embedding of the biosphere

The chemical non-equilibrium order of life grows out of a much larger context of non-equilibrium order, occurring at scales from the formation of the Sun and planetary system, to the complex chemical environments and dynamics within the Earth's core and mantle, oceans, and atmosphere. Many disequilibria on Earth result from slow relaxation timescales in late-stage planet formation; with respect to these the Earth is still a "young" planet. These include partitioning of redox states of transition metals in the core, mantle, crust, and oceans; the slow process of partitioning of volatiles among the lithosphere, hydrosphere, and atmosphere; the escape of hydrogen and atmospheric photochemistry that move the atmosphere far from redox equilibrium with the mantle; and tectonic circulation below the crust, ocean circulation through it, and weather above it. Many of these disequilibria focus energy to an extreme degree on the rock/water interface and in the mixing chemistry of fluids and volatiles in and near the crust. Complex processes of crust formation at spreading centers drive fluid/rock interactions that both depend on and modify the chemistry of the oceans and atmosphere. Ecosystems supported by rock/water chemical disequilibria are the phylogenetically most basal and biochemically simplest, and in some ways the most conservative living systems on Earth. They have been put forth as models for the first life, and may serve as useful models if we are careful to recognize several key respects in which the Archean Earth was different from the Earth today. Here we first ask whether the same chemical stresses that support life today could have driven the emergence of the biosphere as a necessary planetary subsystem.

## 3.1 Order in the abiotic context for life

The order of life is inherently dynamical and dependent on forces and flows that keep living systems driven away from thermodynamic equilibrium. This fact was appreciated by Boltzmann in 1886,[1] and has since been repeated sufficiently often [98, 570, 654, 711] to qualify as part of the common knowledge of the physics of living systems. However, to fully understand the non-equilibrium order of life it is necessary to recognize its context in a much larger field of non-equilibrium order that ranges in scope from the formation of the Sun and planetary systems, to the large-scale structure and energetics of the Earth, and down to complex webs of mechanism in planetary chemistry.

In this chapter we observe not only the truism that planetary and astrophysical processes must be supplying sources of free energy since we observe that life exists, but more importantly that the planetary and astrophysical processes are *themselves* richly structured in their non-equilibrium dynamics. The order of life is intricately coupled to the order in planetary energetics and chemistry, and if anything, was likely more intricately coupled to this order when life emerged than it is today.[2] We will argue in this and later chapters that in many respects the order of life grows out of the order of its planetary context, in some places enfolding planetary chemical mechanisms, in others extending them, and throughout coevolving with other planetary subsystems.

### 3.1.1 Many points of contact between living and non-living energetics and order

The points at which living order contacts the energetic order of the abiotic world are diverse and heterogeneous, so that today the biosphere can be seen as a kind of network or bridge, suspended across many independent energy-exchange mechanisms. The coherence and essential continuity among all living systems is not necessarily, in any direct way, a reflection of continuity or connection among the environmental sources to which life as a whole is anchored today. Moreover, the sequence in which we would naturally describe different energy sources from a point of view in astrophysics or planetary science is not the sequence in which their structure probably first dictated the order in the nascent biosphere.

The central topic in this chapter will be the geochemistry of rock/ocean interactions that is most closely associated with core biochemistry, because this is a kind of focal region for many planetary stresses where we believe the particular structure of chemical interactions determined the most details in the form of the emerging living state. However, we wish to stress that *we view the biosphere as an ordered system that has formed by a kind of accretion around a long arc of planetary disequilibrium.* Planetary dynamics in the interior, on the surface, and in the atmosphere and oceans may all have been essential links in this arc of disequilibrium. The rock/ocean interface is special because it is a

---

[1] The famous statement from [82] is:

> The general struggle for existence of animate beings is not a struggle for raw materials – these, for organisms, are air, water and soil, all abundantly available – nor for energy, which exists in plenty in any body in the form of heat Q, but a struggle for entropy, which becomes available through the transition of energy from the hot sun to the cold earth.

[2] In the process of gaining autonomy, living systems have formed numerous general-purpose interfaces, which weaken some of the links between details of the environment and details within organisms.

barrier to relaxation of a very general energy stress (the oxidation/reduction potential),[3] which despite its generality couples to a wide variety of quantum mechanically particular chemical transitions.

The disequilibrium within the Earth has its own context, though, within the larger disequilibrium of star and planetary system formation. We sketch this larger context both because it is interesting in its own right and provides an essential continuity to the cascade of emergences in which biogenesis was just one transition, and because it includes another particular energy source (solar radiation) that is a major anchor to the biosphere today and may have been important (though in different ways) for the biosphere's emergence.

### *3.1.2 Barriers, timescales, and structure*

We will present the hierarchy of disequilibria and structure starting from the more general and elementary and proceeding to the more particular and derived. This means beginning with the structure of chemical kinetics itself, then considering star and planet formation, and finally descending to the dynamics within the planet that produce the states of the mantle, crust, oceans, and atmosphere.

Everywhere a disequilibrium exists, and a free energy is available, some barrier exists against an energetic relaxation, and relative to the timescale to cross this barrier, some part of our universe is still young. The existence of these barriers, and the stress-driven systems that accumulate behind them and over them, is what creates structure as well as directionality in the aging universe.

### 3.2 Activation energy and relaxation temperature regimes in abiotic chemistry and metabolism

Part of the order of life can be traced to the kinetics and energetics of the rules of chemistry, prior even to the context provided by the Sun and Earth. At this very general level we find a similarity between geochemistry and biochemistry, which does not exist for either atmospheric or astrophysical processes.

In the most general terms, we may partition chemistry in the interstellar medium (ISM [352, 772]), in planetary atmospheres [557], and in subsurface oceans and crustal fluids [426, 427, 526, 837], into three major categories according to two key energy levels. Every reaction environment is characterized by a baseline energy (usually represented by a temperature) to which all species eventually relax or quench, and one or more activation

---

[3] The rock/ocean interface is also a barrier where proton-motive force, in the form of pH differences, can accumulate. Michael Russell and collaborators have argued for several years [473, 474, 523, 524, 525, 526, 687, 689, 690, 691, 692, 694] that this force alongside redox potential was important for the emergence of bioenergetics. There is a certain pleasing symmetry in this argument, as electron exchange among bonds is the universal currency of chemical reactions, while protons are the most soluble universal charge carrier in water, as well as the source of pH dependence of reaction kinetics. In our current evaluation, the importance of redox potential and its specific connection to rock/water interactions is well established and common to all views of early organosynthesis; whether proton transfer was also important early on is still a subject of ongoing experimentation and has a larger element of conjecture.

energies that permit reactions to occur. For each regime of temperature, a particular region of activation energies leads to the most interesting formation of chemical structure and complexity. Notably, as the background temperatures decrease, the most fertile activation energies increase.[4] The three classes of chemistry, in the ISM, in atmospheres, and in fluid phases, partition along this curve of temperature and activation energy into three qualitative clusters, along with other correlated changes in the density and other properties of the reaction medium.

**The interstellar medium** The most complex organosynthesis in the interstellar medium occurs outside the shock fronts of young stars, typically where organics condense on dust grains that are close to the 3.17 K temperature of the cosmic microwave background [878]. The primary activation process is ionization, by photons of energies ranging upward from ~10 eV. Quenching of the ions or recombination products followed by surface migration or concentration in molecular ices at very cold temperatures is required to produce complex organics from such highly excited intermediates before they relax to unreactive species.

**Planetary atmospheres** Temperatures in planetary atmospheres relevant to organosynthesis range from tens to one-hundred or two-hundred degrees Kelvin, in low-density gas phases. The primary activation process is photolysis by photons of energy ~1 eV, leading to free radicals[5] including OH, CO, and CN under Earth-like atmospheric compositions [432, 628, 703]. Some of these will be of importance in the discussion to follow here and in Chapter 6, because they are stable against photolysis by most of the photon flux that generates them.

**Solution-phase chemistry in fluids** Organosynthesis in aqueous solution occurs at temperatures greater than 0 °C and (at least near-surface) below the critical temperature of water at 417 °C. Much of it probably occurs below ~250 °C where at least some organics more complex than methane can be metastable against hydrolysis even in circulating fluids [737]. Activation energies are on the scale of covalent bond energies, typically ~0.1 eV, comparable to the free energies of oxidation/reduction (redox) reactions. Aqueous-phase reactions typically involve exchange of bonding pairs, because radicals are excluded at the activation energy of stable redox couples in water. Where radical chemistry occurs, it is primarily hosted by metal centers that can stably change the charge state by ±1 emu.

By these criteria, biochemistry resembles geochemistry much more than atmospheric chemistry[6] or the chemistry of the ISM.[7] The resemblance is, to some degree, an inevitable

---

[4] This observation was originally made by Rogier Braakman to one of the authors, for which we are grateful.

[5] Since gas phases have no process akin to solvation in polar liquids, ions are subject to rapid recombination. Radicals, as neutral species, are thus the longer lived excited intermediates in gas phases.

[6] The initial light-absorbing process in phototrophic systems involves the same energies ~1 eV as those responsible for photolysis in atmospheric chemistry, but they are (of necessity) very dissipative [77], leading to charge separations with free energies closer to the ~0.1 eV energy scale of covalent bonds. They thus have no close affinity with any of the three classes we have proposed. Perhaps some metal-center absorption processes in minerals are rough analogues [325].

[7] Here we are careful to distinguish the cold dust chemistry of the ISM from astrophysical chemistry more generally. Many astrophysical contexts host organosynthesis which has been proposed by various researchers to have been relevant to the

consequence of the fact that biochemistry takes place under warm, aqueous-phase conditions, and does not have access to the same regimes of density and temperature that exist in planetary atmospheres or in cold vacuum. The similarity therefore does not permit us to conclude that biochemistry *must* have emerged continuously from aqueous-phase geochemistry, but it does permit us to conclude that any such continuous transition would require the lowest level of innovation, allowing incipient life to "enfold" mechanisms already resident and active within the physical environment.

### 3.3 Stellar and planetary systems operate in a cascade of disequilibria

The two major classes of disequilibrium in star/planet systems are the rate-limiting processes for fusion of light elements, which lead to the regulated release of light energy from stars, and the variety of condensed matter rearrangements and radioactive decay of heavy elements in planets, which lead to their internal redox partitioning and dynamics. We consider first the disequilibria present over the course of stellar lifetimes, and the particular channels through which they deliver disequilibrating energy flows to planetary systems.

#### 3.3.1 The once young and now middle-aged Sun

The Sun is an average mass star that has lived through roughly one half of its life as a hydrogen-burning star on the *main sequence* of the Hertzsprung–Russell diagram.[8] The remarkable facts that the age at which the Sun entered the main sequence (about 4.6 billion years ago) is not much greater than the age of the Earth, and that the emergence of life on Earth appears to have been quite early, requires that life formed when the Sun's luminosity and color were both significantly different than they are today. The luminosity and color of the Sun today are well understood to have important consequences for the form of extant life, and the differences that have occurred between the Sun's youth and its current middle age place important constraints on the composition of the Earth's atmosphere during or soon after life's emergence.

##### 3.3.1.1 Gravitation-radiation balance and the fusion rates of nuclei create characteristic timescales for stages in stellar aging

Hydrogen-burning stars inherit the most fundamental initial-condition disequilibrium in the universe. The original episode of cosmic nucleosynthesis occurred at high temperature

---

origin of life. These include but are not limited to exoplanets (including moons) [417, 541, 550, 905], small rocky or metallic bodies which may be radioactively heated [156], and cometary ices [410, 554, 835]. The considerations of organosynthesis in these bodies depend in more detail on the particular, rather complex context provided by each. In that respect they are more like the discussion we provide below for the particular case of Earth's geochemistry, and we regard them as minor variants on condensed-phase planetary chemistry.

[8] The Hertzsprung–Russell (H-R) diagram [563] is an empirical plot of stellar absolute luminosities and spectral properties (which may be interpreted in terms of temperature), showing that all known stars fall along a few major paths in luminosity and temperature. Many arcs formed by stellar populations on this plot are now relatively well understood in terms of stellar models. The *main sequence* on the H-R diagram describes stars supported by hydrogen burning, which have not yet entered the late, brief stages of burning heavy elements that are their last phases before gravitational collapse.

in an entropy-dominated regime, producing almost exclusively hydrogen and helium. In contrast, heavy nuclei on the approach to $^{56}$Fe are the stable states for nuclear matter in the expanded, cool-to-cold, internal-energy dominated universe that (relative to nuclear energy scales) had evolved within only hours after the big bang [862]. Further condensation to form atoms would not occur for another $10^5$ years, but with respect to nuclear states, the universe had already frozen into a metastability, in which we still live today. The energy barriers to fusion of hydrogen and helium into heavier elements create special roles for stars as engines of nucleosynthesis reaching as high as iron. Supernovae are distinct engines of synthesis for nuclei heavier than iron.

The first energy barrier against fusion in low-density gases is simple Coulomb (electrostatic) repulsion. Overcoming this barrier requires the heat and density created by gravitational coalescence into stars. Within stars, a balance of radiation pressure against gravitational attraction segregates the burning of the elements into stages. Deuterium burning is a minor stage in early star formation (due to the small abundance of deuterium). Hydrogen burning is the first major stage in stars below about 8 solar masses, and the longest duration of any of the stages.

The long duration of hydrogen burning results from a second barrier inherent to the fusion reaction itself, which exists despite the fact that hydrogen burning is the most exergonic of the nuclear processes. This second barrier is the dependence of hydrogen burning on the initial formation of deuterons from pairs of colliding protons, a transformation mediated by the *weak interaction*, illustrated in Figure 3.1. The extremely low rates of weak-interaction events are, like the low rates of chemical reactions with high transition states, the result of an energy barrier. For proton-proton fusion the barrier comes from the energy of the very massive intermediate W boson (mass ~80 GeV), relative to which even the ten million degree temperatures (equivalent to ~1000 eV) in stellar interiors is negligible.[9] This rate-limiting step is responsible for the long-term persistence of hydrogen in main-sequence stars (characteristic lifetimes ~$10^{10}$ yr), and for the uniform delivery of visible to ultraviolet radiation over most of this term [142].

### 3.3.1.2 Other timescales and stellar generations

Systems such as stars and planets form only intermittently from dispersed gas and dust, even relative to the timescale of the 13.6 billion year age of the universe, nucleating a continual sequence of "young" and non-equilibrium systems as previously formed systems age. Because much of star formation in an interstellar medium now at low density results from gravitational instabilities driven by shocks from the collapse of existing stars as their fuel is exhausted, star formation is conducted on a "generational" cadence, with a characteristic timescale set by stellar lifetimes.

---

[9] Although energy barriers exist in both chemistry and nuclear physics, they do not function in exactly the same way. The process of positron emission is mediated by quantum tunneling through a *virtual* W boson state, rather than by production of a real W boson by thermal fluctuation. The tunneling process has a finite amplitude even at zero temperature, which approximates the rate in the stellar interior. If positron emission depended on classical excitation through a thermal fluctuation, the process would be so much slower as to be effectively forbidden.

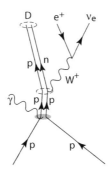

Figure 3.1 Diagram of the positron-emission process that permits two protons ($p$) to combine to form a deuteron ($D$). Two colliding protons first form an unstable state called a *diproton* through a collection of fast processes mediated by the strong and electromagnetic interactions (gray halo), culminating in emission of a high-energy photon ($\gamma$). Ordinarily the diproton would simply disintegrate but, rarely, one of the protons is converted into a neutron ($n$) with emission of a positron ($e^+$) and an electron neutrino ($\nu_e$). In isolation this process would be endergonic, but the binding energy in the deuteron (indicated by the dashed oval) stabilizes the neutron. The massive intermediate $W^+$ boson is responsible for the very low rate of this process.

The remarkable fact that the last phase of light-element burning in large stars – silicon burning to form the transition metals – lasts only one or two days, shorter than the 6-day half-life of $^{56}$Ni for decay to cobalt and then to iron, accounts for the presence of large quantities of $^{56}$Fe in the ISM following the collapse of first-generation stars larger than about $8-11\,M_\odot$.[10] These elements make up a large part of the inner planets in second-generation star systems.[11]

### 3.3.1.3 A faint young Sun emitted higher-energy light on average, but less of it

Stars on the main sequence of the Hertzsprung–Russell diagram begin fainter and bluer, and gradually become brighter and more red as they age toward the exhaustion of their hydrogen fuel. The Standard Solar Model gives an estimate that the luminosity of the Sun about 4 billion years ago was roughly 0.75 of its absolute luminosity today [670]. Brightening with age is a consequence of a combination of helium production and diffusive settling of helium into the core that contributes to gravitational pressure but not to radiation pressure. Increased gravitational attraction causes the hydrostatic equilibrium for hydrogen burning in the core to shift to higher densities and core temperatures, and higher overall luminosity.

---

10  $^{56}$Ni is the heaviest element that can be formed in the fast process of alpha-capture, but $^{58}$Fe and $^{62}$Ni have higher binding energies and would be the stable products of neutron capture, were there time for it to occur. However, stellar collapse and ejection of core material disperses much of the $^{56}$Ni to decay to iron in space, with only a fraction being converted to heavier nickel isotopes by fast neutron capture during the collapse and ensuing explosion.

11  Multiple mechanisms contribute to the further enrichment of inner planets in heavy elements, both during their initial condensation and in their early lifetimes as atmospheres are highly dynamical. We return to processes that differentially strip planetary atmospheres of lighter elements below in Section 3.3.2.15.

The continuum contribution to radiation in the far ultraviolet, as well as a spectrum of higher-energy emission lines, is often higher in absolute intensity in the young star, however, despite its lower overall luminosity. For instance, dwarf G-type stars in the Hyades cluster, with ages estimated at 625 million years, which are thought to be good proxies for the behavior of the Sun around the end of the Late Heavy Bombardment of the inner planets about 4 billion years ago, have intensities in the far UV ($\lambda \sim 2000$ Å) continuum about twice as large as those of the Sun [42].

Simplified models of the solar effective temperature at the top of the photosphere, which ignore magnetic activity, actually show a very slight *increase* with age between 0.6 Ga and today [670], as expansion of the photosphere partly compensates for increased total luminosity. The effective temperature at the photosphere in these models does initially change in the same direction as the core temperature, but by a much smaller degree. The higher UV and X-ray flux in young stars is instead dominated by changes in the delivery of energy to the solar atmosphere[12] by magnetic effects. We return to these below.

### 3.3.1.4 Lower solar flux requires a stronger greenhouse

Carl Sagan and George Mullen [697] first recognized that the lower luminosity of the young Sun would make it impossible, with early atmospheric compositions similar to those on Earth today, to account for abundant evidence that water on Earth's surface was liquid throughout almost all of the planet's history. The response to this difficulty (which is known as the "Faint young Sun paradox"), has been to propose a very large atmospheric greenhouse in the early Archean, which has diminished incrementally as the Sun's luminosity has also increased. The chemical composition for such a greenhouse remains a matter of uncertainty in atmosphere models, but most current scenarios require a large role for $CO_2$, with possible secondary roles for methane [251]. We return in Section 3.6.6.3, and later in Section 6.2 of Chapter 6, to other evidence suggesting elevated $CO_2$ levels in the early Archean, and to efforts to quantitatively model the greenhouse at this time.

### 3.3.1.5 Even stronger scaling for extreme UV and X-rays from the stellar atmosphere

An important feature of stellar structure that is strongly age dependent, and which affects chemistry in planetary atmospheres, is the typical presence of a magnetically activated *stellar* atmosphere that is much hotter than the radiatively heated photosphere – in other words, a very strong temperature inversion in the star's outer layers. While the black-body temperature in the photosphere of a G-type star on the main sequence is typically several thousand degrees (the Sun's photosphere is about 6000 K), the chromosphere may be at $\sim 10^4$ K and the corona may be heated as high as $\sim 10^6$–$10^7$ K. These regions produce a high-energy tail dominated by emission lines below about $\lambda < 1500$ Å that departs from

---

[12] The stellar *atmosphere* refers to the regions above the electromagnetically opaque photosphere. These consist of the chromosphere, transition region, and corona.

the black-body spectrum, and dominates radiation in the extreme UV and X-ray bands. The chromosphere is responsible for excess emission in the UV, the transition region in the far UV, and the corona in the extreme UV and X-ray.

Stellar atmospheric heating arises from magnetic fields produced in the dynamos of spinning stars, driven by the thermal convection of (electrically conductive) plasma and the Coriolis force. Stars generically form with non-zero angular momentum, which may vary over a large range within the first tens of millions of years after formation, making the historical reconstruction of atmospheric temperature in young stars a difficult problem. However, coronal mass loss is an efficient mechanism to transport angular momentum from spinning stars, leading to a tendency for dwarf stars[13] to converge toward a fairly uniform power-law decay of rotational speed with time at ages greater than $\sim0.5$ Ga. Therefore reconstruction of the Sun's history during the interval when life on Earth existed can be carried out with some confidence. From the Sun's current surface velocity of $\sim2$ km/s, this model suggests a rotational speed $\sim10$ km/s at age 0.5 Ga [670].

The dependence of magnetic field activity on rotational speed is also acceptably fit by a power law for stars older than a few tens of millions of years, and the resulting flux in the near UV is empirically modeled to decay with time $t$ approximately as $\sim t^{-1}$. For the young Sun at the dawn of the Archean, then, the ionizing UV flux in the neighborhood of the hydrogen Lyman-$\alpha$ line ($\lambda \sim 1215$ Å) is estimated to have been $\sim10\times$ the present level, a value considered typical for G-type stars.

While the total energy delivered to the Earth from high-energy photon emission (and particle emissions from a more vigorously driven solar wind) is only a small fraction of the total radiation budget determining average surface temperature, it may be important for processes such as light-element escape, because it is absorbed high in the atmosphere where the density is low (a region called the *exosphere*), and small energy changes may produce significant temperature changes in this outer envelope where escape occurs.

### 3.3.1.6 Greater luminosity today only avoids causing thermal run-away because greenhouse gases have declined since the Archean

The greenhouse that is empirically necessary to preserve liquid water in the early Archean would have led to a scorched Earth if it had persisted as the Sun's brightness increased through the middle and late Archean.[14] This state would have been an attractor. Since we are not in such a condition today, the greenhouse must have declined between the early and late Archean. An important set of questions concern the timing and mechanisms by which the decline occurred.

---

[13] With respect to the Hertzsprung–Russell diagram, *dwarf star* refers to all stars on the main sequence, which are supported by hydrogen burning in the core.

[14] Excessive greenhouse heating evaporates oceans, and the transition from a liquid to a vapor reservoir for water delivers much more water and, through photodissociation, much more $H_2$ to the top of the troposphere. The loss of *liquid* oceans therefore accelerates the loss of total planetary water through hydrogen escape, which in turn feeds back to impair removal mechanisms for $CO_2$. The feedback loop in a high-$CO_2$ atmosphere can spiral toward a steady atmospheric composition resembling the current atmosphere of Venus. We discuss these mechanisms in a more fully developed context of atmospheric escape in Section 3.3.2.16.

Indirect evidence for the evolution of atmospheric $CO_2$ levels can be obtained from records of the deposition of calcite ($CaCO_3$) in oceanic crust through hydrothermal alteration in mid-ocean spreading centers. We review evidence for the early Archean in Section 3.6.6.3 below, after we have explained mechanisms responsible for hydrothermal alteration. The general picture given by early Archean greenstones from the Cleaverville area of Western Australia [740] is that between about 3.2–2.8 Ga, very large carbonate volumes (ranging from 20–40% by volume) were emplaced to depths as great as 2 km into the oceanic crust by hydrothermal circulation. In the younger Fortescue group, aged 2.8–2.6 Ga, carbonatization continues but at reduced volumes, between 0–20%, while in the overlying Hamersley group, aged <2.6 Ga, carbonatization of basaltic components falls to single-percent levels or lower. The interpretation of such beds is complex because their composition includes lava flows interleaved with clastic (explosive surface eruption) and sedimentation layers, but Shibuya *et al.* [741] conclude that between 3.2 Ga and 2.6 Ga, crustal $CO_2$ content had decreased by at least an order of magnitude. By 2.6 Ga, carbonatization of oceanic crust appears to have reached levels comparable to those in the Phanerozoic (our current geological eon).

The decrease in oceanic crust carbonatization levels on Earth occurred at a time that was both geologically and biologically complex [741], with possible interactions between the two domains that make assignments of causation difficult. The first supercontinent is estimated to have formed at about 2.7 Ga, reflecting an increase in both continental crust volume and associated weathering by that time, and leading to the production of continental shelves and shallow marine platforms in convergence zones of continental plates. Efforts to date the cyanobacterial clade [76], believed to be the group in which oxygenic photosynthesis arose, place its origin around the same time (and perhaps mechanistically linked to the formation of supercontinent environs). Sediment accretion around continental margins, and $CaCO_3$ precipitation (as calcite or aragonite) which may have had a biological component, add other mechanisms besides hydrothermal alteration to remove $CO_2$ from a planetary atmosphere. All of these changes were therefore occurring during an interval between the early and middle Archean, over which the occurrence of sedimentary carbonates appears to increase at the same time as crust carbonatization through hydrothermal alteration declines.

Through whatever combination of circumstances, the global oceanic $CO_2$ concentration appears to have decreased, and the removal of $CO_2$ transitioned from mid-ocean hydrothermal systems to sedimentary deposition and subduction. The result has been a period of continuously present liquid oceans from the early Archean to the present.

### 3.3.2 Disequilibria in the Earth are gated by a hierarchy of phases and associated diffusion timescales

Stars, as we noted, remain "young," settling into states of long-lived dynamical order because nuclear disequilibria are bottled up behind the slow reaction rate of the weak interactions. Planets (especially those with high metal content) can also remain young – meaning that they retain significant memory of the conditions of their formation that are far from the equilibrium of the condensed state – and can set up even more heterogeneous hierarchies of structure than those in stars.

Many processes determine characteristic timescales for planetary relaxation, including dispersal of angular momentum, heat dissipation, and radioactive decay (processes also active in stars), but extending to diffusion in solid phases and convection in plastic phases as well as viscous fluids. Timescales for diffusion of matter in solids are much longer than the timescale for diffusion of heat that permits solid formation, so memory of initial conditions can be bottled up in planets by quenching, over longer timescales than stellar lifetimes.

Perturbations from faster processes that are energetically small on the planetary scale may be crucial in creating release channels for chemical free energy that could otherwise be kinetically inaccessible. These weak forces can overcome the much larger barriers imposed by cold condensed matter because they act over very large spatial scales, but as a reflection of their relative weakness, they can only permit energy release in small volumes. Such energy-release channels provide an energetic anchor for living systems on Earth today that is independent of direct delivery of energy from the Sun. They are also part of the larger arc of planetary disequilibria (which may also include links in the atmosphere and oceans driven by solar radiation) that we believe drove the emergence of life.

### 3.3.2.1 Radioactive decay in planets meters out quenched redox disequilibria

The continued release of redox disequilibria on Earth is due largely to a secondary phenomenon of radioactive heating, which in part determines its characteristic timescale. Radioactive decay is a lower-energy process than nuclear fusion, and the density of unstable elements in planets is a trace, in contrast to being the dominant component of main-sequence stars. Compared to the violence of stellar interiors, radioactivity in planetary interiors achieves only a "gentle warming" – although it must be remembered that radioactive decay is still a very high-energy process relative to chemical reaction scales.[15] It seems unlikely to us that chemical excitation by radiolysis was an essential source of energy in biogenesis derived directly from nuclear processes, though we cannot rule out some role for it.

Radioactive heating maintains warmth and mobility in planetary interiors, and depending on planet size and age, may enable the processes of mantle convection, volcanism, and in some cases plate tectonics to occur.[16] The characteristic timescales for decay of long-lived unstable nuclei, of order $\sim 10^9$–$10^{10}$ years, therefore permits planets such as the

---

[15] In total effect, the warming may for some purposes be far from "gentle." The temperature of Earth's core is estimated to be about 5700 K, only modestly cooler than the photosphere of the Sun. Still this is much cooler than the core temperature of 15.6 million Kelvin estimated for the Sun, which enables nuclear fusion.

[16] Radioactivity together with residual infall heat creates a liquid outer core on Earth, in which convection in a rotational setting creates a magnetic field. The Earth's magnetic field has a large influence on atmospheric radiation dynamics by deflecting the solar wind, which is important for life today, and could also have been important for organosynthesis around the time

Earth to meter out sources of free energy for redox reactions on timescales that are long compared to the planetary formation time, but still short compared to migration in a totally frozen planet, and comparable to the lifetime of the Sun.

### 3.3.2.2 Separation of the core from the mantle partitioned two major oxidation systems of iron

The average oxidation state of the Earth is near an equilibrium with metallic iron throughout most of the planet's interior. It rises to only 4–5 log units higher in the upper mantle, still far below the oxidation state in the atmosphere and oceans maintained by photolysis and escape processes (discussed below). Within this bulk average, however, the distribution of iron is complex. Its most basic structural feature is the separation of the core from the mantle.

Within 80 million years after the coalescence of the Earth's main mass, most of the present core of liquid iron and nickel had separated from a mantle of magnesium and iron silicates and oxides [29]. (Probably most of this separation was completed as early as 30 million years after planet formation.[17]) The outer core is liquid Fe/Ni alloy, and the inner core is a solid Fe/Ni alloy due to good heat conduction but greater pressure. The mantle consists of variously ductile silicate mineral assemblages. Both systems are convective, with the most active convection in the mantle occurring in the upper part known as the *asthenosphere*. However, the two convection systems are largely decoupled, separating the Earth's stock of iron into a metallic, electrically highly conductive, magnetically active inner ball, and a creeping, redox-active outer shell.

The transitions between ferrous and ferric states of iron together with the mobility of minerals in the mantle permits a variety of oxidation/reduction reactions with other elements or included fluids. Through such reactions, in a context of atmospheric processes that render the surface conditions much more oxidizing than the bulk-average planetary oxidation state, the mantle can buffer very long-lived non-equilibrium states in which reductants flow from the interior to the surface. These non-equilibrium states are of particular interest to us for the biosphere they sustain today, and the role we think they are likely to have played in biogenesis.

### 3.3.2.3 $f_{O_2}$ and $Fe^{3+}/\sum Fe$ characterize major oxidation properties of the mantle

Oxygen fugacity, denoted $f_{O_2}$, is a measure of oxygen activity expressed as the pressure of an ideal gas of $O_2$. It is used as a standard measure of redox potential throughout the

---

when life originated. A systematic exploration of consequences of a geodynamo, and the way this depends on the atmospheric structure and composition of the early Earth, is a complex problem that we do not pursue in this monograph.

[17] The formation of the Moon some time between 70 and 80 million years after the Earth's initial formation injected a very large amount of heat into the mantle, possibly creating a second major pulse of separation that complicates the analysis of signatures used to estimate the time of core/mantle separation.

mantle.[18] Where pure mineral phases coexist in equilibrium, they buffer $f_{O_2}$ as a function of temperature and pressure along a *coexistence curve* [441, 875]. (Specific mineral buffers are discussed in Section 3.3.2.9.) In the Earth, heterogeneities exist across scales from the level of crystals within a single rock, to the depth of the entire mantle, and multiple mineral phases are often co-present for which experimental calculations of buffered values of $f_{O_2}$ may be similar but not identical. These inconsistencies provide a way to quantify the degree of disequilibrium maintained within the mantle, as a function of spatial separation or age relative to transport and relaxation timescales.

Whereas $f_{O_2}$ measures a redox potential, the principal element that actually accepts or donates electrons in redox reactions in the mantle is iron. It is the fourth most abundant element in the Earth, and the most abundant element with a variable oxidation state. Because most iron in the mantle is included in silicate rocks in either $Fe^{2+}$ or $Fe^{3+}$ form, most redox reactions involve $Fe^{2+}/Fe^{3+}$ transformations. Therefore a useful measure of the capacity of mineral assemblages to change oxidation state is the ratio of ferric to total iron, denoted $Fe^{3+}/\sum Fe$.

$f_{O_2}$ and $Fe^{3+}/\sum Fe$ are independent measures: one is a potential and the other is a capacity. The two have a complicated relation within the mantle, as $f_{O_2}$ steadily decreases with depth in the mantle, while $Fe^{3+}/\sum Fe$ generally increases.[19] This situation results from the complex dependence of the activities of different ions in silicate minerals on pressure and crystallographic structure. It has important consequences for the partitioning of oxidation states of iron and other included elements with depth, and to the changes they undergo as mineral assemblages are transported vertically within the mantle.

### 3.3.2.4 Mineral phases share many aspects of complexity with life

Short of the diversity of life, the world of mineralogy probably contains the richest family of related yet distinct structures produced by a few underlying generating mechanisms. The mineral world is also an excellent example of the way complexity can be generated where a fine-grained structure of lawful constraints creates openings for chance to leave long-lasting signatures.

Mineral phases are, in principle, fully determined by atomic composition, temperature, and pressure. Yet the favored phase depends sensitively on small changes in these parameters. More than 4500 mineral phases are known on Earth today [345, 348], which combine in a bewildering variety of assemblages [347]. The *ab initio* prediction of the favored assemblage in many conditions is still an unsolved problem, so many conclusions about mineral phases in the deep Earth are arrived at semi-empirically using experimental models. Mineral phases also have long memory – in many cases comparable to or longer

---

[18] Just as electrons exchanged with a standard electrode may be used to decompose arbitrary electron-transfer reactions into half-cell reactions [330], compared according to voltage, $O_2$ exchange may be used to separate a wide variety of interactions of minerals with each other or with included volatiles.

[19] We explain below that the decrease in $f_{O_2}$ is *driven* by the increase in $Fe^{3+}/\sum Fe$ at depths below a few tens of kilometers.

than transport times in the mantle and crust – so that minerals may survive for extended periods outside the conditions where they are in equilibrium. As a result, many diverse phases can be brought together in a way that depends sensitively on accidents of history, despite the essentially deterministic physics of mineral equilibration.[20] The combination of determined elements, sensitive dependence on boundary conditions, and long memory also characterizes the biosphere. It is responsible for some forms of biological complexity in which the role of chance may be evident, but the role of necessity may be difficult to recognize.

The generating constraints behind the complexity of the mineral world are few: the charges and masses of nuclei in the periodic table, the representations of the rotation group in atomic orbitals and the energy spacings of these (which determine not only geometry but also electronegativity), the relative abundances of different elements in the volume of the Earth, and the problem of forming stable phases by packing atoms in three-dimensional space. Because three of these properties (charge, orbitals, and lattice packing) are *integer valued*, however, they admit an enormous number of similar but not identical solutions separated by barrier configurations, and from this diversity grows the complexity of mineralogy.

### 3.3.2.5 The stoichiometric oxide components of minerals change according to composition and pressure

Much of the regularity in mineral phases and their transformations may be understood in terms of the exchange or interconversion of a set of elementary oxides that define a basis set for mineral composition, among different crystallographic unit cell configurations. Some of the most abundant oxide components in the mantle, which are principally responsible for determining its redox state, are $(Mg^{2+}O^{2-})$, $(Fe^{2+}O^{2-})$, $(Fe_2^{3+}O_3^{2-})$, $(Al_2^{3+}O_3^{2-})$, and $(Si^{4+}O_2^{2-})$, as well as lower abundances of $(Ca^{2+}O^{2-})$ (in the form of carbonates), and $(C^{4+}O_2^{2-})$, and water near the surface, or $Fe^0$ and $Ni^0$ metal in the lower mantle on the approach to the core/mantle boundary.

Various proportions of these oxide constituents may be combined to produce the stoichiometry of more complex mineral phases. Because each represents a possible charge-balanced, filled orbital structure, it generally exists as a pure mineral phase in its own right under some conditions of temperature and pressure.

Two important properties that distinguish mineral groups are the ratio of heavy elements to silicon, written $(Mg, Fe)/Si$, and the proportion of ferric to total iron $(Fe^{3+}/\sum Fe)$ introduced above. An array showing several of the major constituents of the mantle is given in Table 3.1. Further mineral nomenclature is listed in Box 3.1.

---

[20] Consider that *gemstones* are economically valued not only because they are beautiful but because they are *scarce*. These are rocks composed of essentially homogeneous mineral phases, a condition that for most minerals is rarely attained at a scale larger than a few millimeters to centimeters. Geologists have, over many centuries, accreted a seemingly inexhaustible vocabulary to refer to rocks which are different juxtapositions of multiple mineral phases.

Table 3.1 *Pure phases of major mineral constituents of the crust and mantle, arranged in order of* (Mg, Fe) /Si *ratio and* $Fe^{3+}/\sum Fe$

| (Mg, Fe) /Si | Composition (phases) | | | |
|---|---|---|---|---|
| 0 | $SiO_2$ (quartz, stishovite) | | | |
| 1 | $(Mg, Fe)_2 Si_2 O_6$ (pyroxenes, silicate perovskite) | | | |
| 5/3 | | $(Mg, Fe)_3 (Fe, Al)_2 (SiO_4)_3$ (garnet, majorite) | | |
| 2 | $(Mg, Fe)_2 SiO_4$ (olivine, wadsleyite, ringwoodite) | | | |
| ∞ | $(Mg, Fe) O$ (ferropericlase) | | $Fe_3 O_4$ (magnetite) | $Al_2 O_3, Fe_2 O_3$ (corundum, hematite) |
| $Fe^{3+}/\sum Fe$ | 0 | 2/5 | 2/3 | 1 |

Parentheses such as (Mg, Fe) indicate free substitutions that result in solid solutions. Names of major crystalline phases associated with each composition are listed in order of increasing pressure at which they are stabilized. In both pyroxenes and perovskite, the representation $(Mg, Fe)_2$ may be replaced by $XY$, where $X$ may be any of a set of monovalent or divalent ions of larger size, and $Y$ is any of a set of divalent or trivalent ions of smaller size (see Box 3.1). The ambiguity of charge and valence in this shorthand reflects the possibility for much more complex modified mineral structures in which $Fe^{3+}/\sum Fe$ may be non-zero. A similar comment applies to the wadsleyite and ringwoodite phases of $(Mg, Fe)_2 SiO_4$. The trivalent metal oxides $Al_2 O_3$ and $Fe_2 O_3$, which include mineral forms corundum and hematite (respectively), are both listed because they become substitutable at high pressure in iron-aluminum perovskites (see text). However, they are listed separately as they do not form a solid solution under near-surface conditions where each exists in mineral form.

---

**Box 3.1   Brief glossary of mineral terms**

> The aim of the scientist is to say only one thing at a time, and
> to say it unambiguously and with the greatest possible clarity. To
> achieve this, he simplifies and jargonizes.
>
> – Aldous Huxley, *Literature and Science* [386]

The requirement that the elements that make up the bulk of the Earth must pack in space, and at the same time satisfy the orbital restrictions of quantum mechanics,

**Box 3.1    Brief glossary of mineral terms (cont.)**

imposes a complex web of constraint-satisfaction conditions on the formation of condensed matter. The solutions to these webs of constraint are the mineral phases.

Identifying the thermodynamically stable mineral states as a function of temperature, pressure, and element composition is an algorithmically hard problem, even though in principle equilibrium phase formation is an elementary and deterministic phenomenon. Much of mineralogy is therefore quasi-empirical. The result is a large almanac of terminology, both for pure mineral phases and for the combinations in which they are found to occur in natural rocks.

This box provides names and key properties for a few of the mineral phases that are most important in determining the composition of the bulk Earth, and the environments through which it interacts with the oceans or atmosphere. We group minerals into phases that share essential properties – usually stoichiometry, crystallographic form, or coexistence in solid solutions – and we order these roughly from low-pressure, more oxic and more silicic phases that occur near the top of the mantle or in the crust, descending to high-pressure, more reduced and less silicic phases.

**spinel** The general form $\left(A^{2+}O^{2-}\right)\left(B_2^{3+}O_3^{2-}\right)$, and the exemplar of the unit cell combining tetrahedral and octahedral metal centers in a cubic superlattice. The common inclusion in peridotites in the upper mantle is $(Mg, Fe)^{2+}(Al, Cr)_2^{3+}O_4$.

**magnetite** $\left(Fe^{2+}O^{2-}\right)\left(Fe_2^{3+}O_3^{2-}\right)$. A ferrous-ferric iron oxide in the spinel group.

**hematite** $Fe_2^{3+}O_3^{2-}$. A ferric oxide with stoichiometry and tetrahedral coordination of the Fe center the same as that of the ferric component in magnetite.

**greigite** $\left(Fe^{2+}S^{2-}\right)\left(Fe_2^{3+}S_3^{2-}\right)$. A sulfur analogue to magnetite, also called a *thiospinel*. An important ore component, probably precipitated in basalt-hosted vent settings, and interesting for a loose analogy to the active site structure in the CODH/ACS enzyme [688].

**pyroxene** A group of silicate minerals with the general formula $XYSi_2O_6$, defined crystallographically by having chains of $SiO_3$ tetrahedra. Pyroxenes crystallized in the monoclinic form are called clinopyroxenes, and those crystallized in the orthorhombic form are orthopyroxenes. X is monovalent or divalent, and may be any of $\left(Li, Na, Ca, Mg, Fe^{2+}, Zn, Mn\right)$. Y is generally a smaller ion, divalent or trivalent, and may be any of $\left(Al, Cr, Fe^{3+}, Mg, Mn, Sc, Ti, Vd\right)$. End members along the $(Mg, Fe)$ series in the monoclinic and orthorhombic groups are respectively termed (clino-)enstatite $(Mg_2Si_2O_6)$ and (clino-)ferrosilite $(Fe_2Si_2O_6)$.

**Box 3.1 Brief glossary of mineral terms (cont.)**

**olivine** $(Mg^{2+}, Fe^{2+})_2 SiO_4$. A solid-solution series with end members forsterite ($Mg_2SiO_4$) and fayalite ($Fe_2SiO_4$). Olivine in the upper mantle contains essentially no $Fe^{3+}$, but pressure-induced distortion in the lower mantle leads to increasing $Fe^{3+}$ incorporation (see main text), generally by joint substitution of $Si^{4+}$ with $Al^{3+}$.

**peridotite** An assemblage of minerals making up much of the upper mantle, consisting mostly of olivine and orthopyroxene and clinopyroxene. Minor inclusions of garnet and spinel are often parts of the assemblage. Peridotites are required to have less than 45% silica and are by this criterion ultra-mafic. The term comes from the name **peridot** for gem-quality olivine.

**wadsleyite and ringwoodite** High-pressure polymorphs of olivine. Wadsleyite is orthorhombic, and ringwoodite is a spinel. Both are characterized by a higher fraction of $Fe^{3+}/\sum Fe$ than olivine, and are believed to be major constituents of the transition zone in the mantle.

**garnet** A range of minerals with the formula $X_3^{2+}Y_2^{3+}\left(Si^{4+}O_4^{2-}\right)_3$. X may be (Ca, Mg, Fe, Mn). Y may be (Al, Fe, Cr). The iron-silicate end member $\left(Fe_3^{2+}Fe_2^{3+}(SiO_4)_3^{4-}\right)$ is named *skiagite*.

**majorite** A garnet mineral with formula $Mg_3^{2+}(Fe, Al)_2^{3+}\left(Si^{4+}O_4^{2-}\right)_3$. Majorite on the surface is associated with shock production, and it is also a component of the upper mantle. The term is sometimes applied to synthetic $MgSiO_3$, a polymorph of enstatite [539].

**silicate perovskite** (Mg, Fe) $SiO_3$. Stoichiometrically a polymorph of pyroxene. The crystallographic topology is that of the reference mineral *perovskite* with formula $CaTiO_3$ with oxygen in a face-centered cubic lattice. Silicate perovskites in the lower mantle are commonly distorted with $SiO_6$ octahedra tilted relative to the cubic superlattice.

**iron-aluminum perovskites** (Mg, Fe) (Si, Al) $O_3$. The fraction of $Fe^{2+}$ versus $Fe^{3+}$ can vary widely depending on the amount of $Al^{3+}$ substituted for $Si^{4+}$ in the second site. $Fe^{3+}$ may also substitute for $Si^{4+}$ in some cases, providing another way to incorporate $Fe^{3+}$ into the first site.

**wüstite** $Fe^{2+}O^{2-}$. Ferrous oxide. Because it is both reduced and silica free, it is a stable component in the lower mantle. Wüstite is one member in an important coexistence curve with metallic iron which governs oxygen fugacity.

**periclase** $Mg^{2+}O^{2-}$. Magnesium oxide. An end member in the more commonly encountered solid solution ferropericlase.

---

**Box 3.1   Brief glossary of mineral terms (cont.)**

**ferropericlase** $(Mg^{2+}, Fe^{2+}) O^{2-}$. A solid solution with end members periclase
and wüstite. Ferropericlase in the mantle may incorporate as much as 10%
$Fe^{3+}$ through joint substitution of other divalent cations with monovalent
ions such as $Na^+$ [539].

**stishovite** $SiO_2$. An extremely hard, dense silica phase with a tetragonal unit cell.
May be a significant component of the lower mantle.

**post-perovskite** High-pressure polymorph of perovskite, believed to constitute the
bottom $\sim$250 km of Earth's mantle. Increased $P$ and $S$ wave velocities, and
shear anisotropy of post-perovskite are believed to be due to its crystallographic lattice in which $SiO_6$ octahedra are arranged in sheets rather than
in a cubic lattice.

---

Most of the chemical formulae in the table can take on multiple crystallographic
phases, which are stable in different ranges of temperature and pressure. The alternative phases are referred to as *polymorphs*. For the formula $(Mg, Fe)_2Si_2O_6$ (sometimes
written $(Mg, Fe) SiO_3$), the phases in order of increasing pressure are *clinopyroxene* and
*orthopyroxene*, which occupy the upper mantle, and at much greater pressures *silicate
perovskite*, which is the dominant component of the lower mantle. For $(Mg, Fe)_2SiO_4$,
the low-pressure phase is *olivine*, the main constituent of peridotite in the upper mantle,[21] which is transformed into *wadsleyite* and then *ringwoodite* at increasing depth in
the middle layer of the mantle known as the *transition zone*. Multiple phases are grouped
together as *garnets*, with low-pressure phases (often called "garnet" in a narrow usage)
being minor constituents of peridotites and accumulating through melt separation in mid-
ocean ridge basalts (discussed below), and a separate phase called *majorite* becoming a
large component in the transition zone.

Since $(Mg, Fe) /Si$ ratios as well as metal/oxygen ratios differ in the different stoichiometries, when material is transported vertically in the mantle, the elementary
charge-balanced groups introduced at the beginning of the section must be partitioned or
interconverted so that the distribution of multiple phases in a mineral assemblage is consistent with conservation of element numbers and charge. The equilibria at which these
mixtures are balanced then influence the oxidation state of minor elements or included
volatiles as well.

### 3.3.2.6 Mineral phase and substitution effects determine redox states of included transition metals

Each major silicate phase has a default oxidation state for metal ions in the standard lattice,
in addition to which it may admit substitution effects or lattice defects that incorporate

---

[21] *Peridot* is a name given to gem-quality olivine.

minority oxidation states. The combination of major silicate phase assemblages and substitutions at each pressure and temperature determines the oxidation states of iron and other metals throughout the solid Earth.

The most important substitution effects on the redox state of the mantle are those that incorporate $Fe^{3+}$ ions in pyroxenes, garnet, and especially silicate perovskites with increasing pressure [279, 539]. Olivine in peridotites from the upper mantle contains almost no $Fe^{3+}$. Yet experimental syntheses of wadsleyite and ringwoodite under conditions characteristic of the transition zone show the equilibrium incorporation of $Fe^{3+}$ at a level of $Fe^{3+}/\sum Fe \sim 2\%$. Similarly, garnet in the deep upper mantle is characterized by $Fe^{3+}/\sum Fe \sim 2\%$. In contrast, majorite garnet in the transition zone has $Fe^{3+}/\sum Fe \sim 7\%$.

The most striking deviation is for pyroxenes and perovskite. The nominal stoichiometry in Table 3.1 indicates an oxidation state of $Fe^{2+}$ for both, which would be consistent with free substitution for $Mg^{2+}$ in a solid solution. However, orthopyroxene in the upper mantle may contain $Fe^{3+}/\sum Fe$ values ranging from 4–10%, while clinopyroxene may include as much as 10–40%. Silicate perovskites, which nominally are polymorphs of pyroxenes that are stable in the lower mantle, deviate significantly in composition from the formula in Table 3.1 (with substitution of aluminum, discussed below, being the most important variation). They may incorporate values $Fe^{3+}/\sum Fe \sim 50\%$, and experimental systems have been produced in which $Fe^{3+}/\sum Fe > 80\%$ [539]. This increase in $Fe^{3+}$ is the main factor driving $f_{O_2}$ in the lower mantle 4–5 log units *lower* than $f_{O_2}$ in the upper mantle.

### 3.3.2.7 Substitution effects in lower mantle perovskite

The case of lower mantle iron-aluminum perovskites illustrates the way substitution effects work. The trend for increasing $Fe^{3+}$ in high-pressure minerals is crudely a reflection of **ion size**.[22] $Fe^{3+}$ is a smaller ion than $Fe^{2+}$. At high pressure, silicate matrices, while preserving the overall topology that characterizes the crystalline phase, become distorted. In the case of perovskite, this distortion alters the crystal lattice from the cubic geometry of an ideal $(Mg, Fe)_2^{2+}SiO_4^{2-}$ perovskite to an orthorhombic geometry, in which the bi-capped trigonal prism sites normally containing $Mg^{2+}$ or $Fe^{2+}$ are compressed [305].

The compression can be accommodated by replacing $Fe^{2+}$ with $Fe^{3+}$, while balancing charge by jointly substituting $Si^{4+}$ in the octahedral sites of the lattice with $Al^{3+}$ or even a second $Fe^{3+}$. In terms of elementary oxides, the nominal structure of silicate perovskite is

$$Fe_2Si_2O_6 \leftrightharpoons 2\,FeO + 2\,SiO_2. \tag{3.1}$$

Jointly substituting one FeO with $\frac{1}{2}Fe_2O_3$, and one $SiO_2$ with $\frac{1}{2}Al_2O_3$ or $\frac{1}{2}Fe_2O_3$, on the right-hand side leads to the altered composition $Fe^{2+}Fe^{3+}Al^{3+}SiO_6$ or $Fe^{2+}Fe_2^{3+}SiO_6$ on the left-hand side.

---

[22] Here we omit a discussion of more complex spin effects, as $Fe^{3+}$ may transition between spin 1/2 and spin 5/2 states, which affect orbital geometry and participation in lattice spacing [279]. For details see [305].

The percentage of $Fe^{3+}$ is strongly correlated with the fraction of $Al^{3+}$ in experimental model perovskites, and enrichment in aluminum is believed to be the dominant mechanism by which lower-mantle perovskites incorporate ferric iron [539].[23] The resulting mineral, with partial substitutions, is referred to as iron-aluminum perovskite (or FeAlPv), and given the formula $(Mg, Fe) (Si, Al) O_3$.

### 3.3.2.8 Effects on mantle $Fe^{3+} / \sum Fe$ and $f_{O_2}$ with depth

Water, water, every where,
Nor any drop to drink.

> – Samuel Taylor Coleridge, *The Rime of the Ancient Mariner*

Through substitution effects, a mantle that is vertically mixed by convection can nonetheless maintain a systematic gradient of oxygen fugacity with depth. Olivine near the surface incorporates almost entirely $Fe^{2+}$, whereas clinopyroxene (the other major component of upper-mantle peridotite) incorporates only a few percent $Fe^{2+}$ on average. When these minerals are transported downward through the mantle, the steady lowering of the chemical potential for $Fe^{3+}$ relative to $Fe^{2+}$ drives a progressive oxidation of iron in mantle silicates, reducing included volatiles and lowering $f_{O_2}$ by 4–5 log units compared to the value with which upper-mantle silicates are in equilibrium. Throughout the bottom of the upper mantle and all of the transition zone and lower mantle, the potential for $Fe^{3+}$ is so low that it is controlled not by equilibria among silicate phases, but by disproportionation of $Fe^{2+}$ into $Fe^{3+}$ and $Fe^0$, which separates as metal inclusions alloyed with nickel. Depending on the nature of lower-mantle convection and the mobility of included metals at the core/mantle boundary, this may provide a mechanism to transport iron into the core, and slowly raise the average oxidation state of the mantle silicates. The result is a bottom 9/10 of the mantle's depth rich in ferric iron, yet highly reducing in character, with various iron-containing silicate phases in equilibrium with iron and nickel metal.

### 3.3.2.9 Mineral buffers and the mantle oxidation state

To understand quantitatively the dynamics of redox state in the mantle, it is necessary to list the mineral phases that mediate the dependence on temperature and pressure at each depth. Because multiple mineral phases coexist in each location, $f_{O_2}$ is *buffered* along the (approximate) coexistence curves of a collection of solid/solid equilibria. Within a given assemblage of phases, $f_{O_2}$ changes with depth through the dependence of the equation of state of each phase on pressure and temperature, and the slope of the curve shows discontinuities on the boundaries between the stability fields of different phases.

Here we consider mostly the behavior of the upper mantle (depth 0–410 km), since this is the source of magmas that interact directly with the hydrosphere and atmosphere, and the

---

[23] Aluminum incorporation may occur by the mechanism described above, which retains the oxygen content of the nominal mineral description, or alternatively by substituting two $Si^{4+}$ sites with an $Fe^{3+}$, $Al^{3+}$ pair and creating an oxygen vacancy [279].

buffer of the chemistry most likely to have affected the emergence of the biosphere. We provide only brief summary comments on the mineralogically complex transition zone (depth 410–660 km) and the lower mantle (depth 660–2891 km). It is possible that the detailed geochemistry in the lower regions of the mantle is important to the existence of tectonics on Earth,[24] but at present this chemistry seems well buffered, and we will suppose that these regions are mostly important as a large reservoir of heat and reducing equivalents. The discussion below follows reviews of Frost and McCammon [279] and McCammon [539], and references therein, to which the reader is referred for more complete explanations including reviews of the sources of evidence.

Two buffers, separated by 4–5 log units in $f_{O_2}$, provide the major references for redox potential respectively in the near-surface mantle, and at a depth $\sim$250 km where iron in silicates exists in equilibrium with phase-separated metal. Over most of this range, the variation in whole-rock $Fe^{3+}/\sum Fe$ is small, and $f_{O_2}$ responds principally to the partitioning of $Fe_2O_3$ among phases. Only near the surface (and in subducting slabs that partly retain compositions due to surface weathering) does increased $f_{O_2}$ reflect a significant increase in bulk-averaged $Fe^{3+}/\sum Fe$. These more oxidized mineral states result from interactions with oceans and the atmosphere which, even since the anoxic conditions of the Archean, have always been much less reducing than the bulk mantle.

The principal reference for oxidation states in the shallow and intermediate upper mantle comes from the oxidation of the FeO component of olivine to magnetite, with the separation of quartz into a distinct phase (which may vary depending on context). Where this equilibrium is computed for pure phases, using the iron end member fayalite of the olivine sequence (see Box 3.1), it is known as the **fayalite/magnetite/quartz**, or FMQ buffer. The equilibrium $f_{O_2}$ is determined from the activities of minerals along their coexistence curve by the equilibrium relation

$$\underset{\text{fayalite}}{3\,Fe_2SiO_4} + O_2 \leftrightharpoons \underset{\text{magnetite}}{2\,Fe_3O_4} + \underset{\text{quartz}}{3SiO_2}. \tag{3.2}$$

A second buffer, typically 4–5 log units below the FMQ buffer, results from the equilibrium

$$\underset{\text{iron}}{2\,Fe} + O_2 \leftrightharpoons \underset{\text{wüstite}}{2\,FeO} \tag{3.3}$$

between iron metal and FeO in the mineral phase wüstite. It is therefore known as the **iron/wüstite**, or IW buffer. This buffer provides a good approximation for a variety of coexistence curves where iron or nickel metal exists in equilibrium with silicates that may contain ferric as well as ferrous iron.

A sequence of approximate mineral equilibria cause $f_{O_2}$ to interpolate between the FMQ and IW buffers over the range $\sim$0–250 km depth. In the near-surface, partial melting at spreading centers separates the mixture of olivine, clinopyroxene, spinel, and garnet in the

---

[24] Seismic imaging suggests that subducted slabs carry ocean-bottom basalts into the lower mantle, requiring that some exchange of material between the lower mantle, transition zone, and upper mantle must be occurring, but the consequences of this transport for whole-mantle composition and oxidation state are not well understood.

bulk upper mantle into relatively (Mg, Fe)-depleted *magmas*, which cool to form **basalts**, and (Mg, Fe)-enriched **residual mantle peridotites**, which remain in solid phase below the surface. Mid-ocean ridge basalts (MORBs) show values of $Fe^{3+}/\sum Fe$ in the range $12\pm2\%$, and $f_{O_2}$ values roughly 0.4 log units below the FMQ buffer.[25] Residual peridotites tend to have median values of $f_{O_2}$ about one log unit below those of the MORBs that separate from them.

From 30–60 km depth **spinel** becomes a stable component of peridotites. Spinel may have ferric iron percentages in the range $Fe^{3+}/\sum Fe \sim 15-35\%$, making it the primary determinant of activity for $Fe_2O_3$. In the middle to lower depths of the upper mantle, a redox couple forms between FeO in olivine or orthopyroxene and $Fe_2O_3$ in **garnet**,

$$4\,\underset{\text{olivine}}{Fe_2SiO_4} + \ 2\,\underset{\text{orthopyroxene}}{FeSiO_3} \ + O_2 \leftrightharpoons 2\,\underset{\text{garnet}}{Fe_3^{2+}Fe_2^{3+}(SiO_4)_3}. \tag{3.4}$$

This buffer provides wide estimates of mantle $f_{O_2}$ bridging the spinel peridotites and extending down to depths where metal precipitation must first be considered, and is shown as the red curve [1] in Figure 3.2.

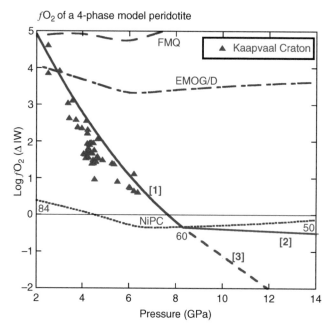

Figure 3.2 Mantle $f_{O_2}$, plotted relative to the iron/wüstite buffer (3.3), as a function of pressure. Curve [1] is the garnet buffer (3.4). NiPC is the nickel precipitation curve. From Frost and McCammon [279], compiled from the model of McDonough and Sun [547] (figure courtesy of the authors).

---

[25] It is estimated, from oxidation states of trace vanadium and scandium, that $f_{O_2}$ in Archean basalts has not differed by more than 0.3 log units over the past 3.5 Ga [486].

The garnet equilibrium may be used up to a pressure of $\sim 8\,\mathrm{GPa}$, corresponding to a depth of $\sim 250\,\mathrm{km}$. At this pressure the value of $f_{O_2}$ crosses the IW buffer, and further increase of $Fe^{3+}/\sum Fe$ must occur through disproportionation of $Fe^{2+}$ to $Fe^{3+}$ (in silicates) and $Fe^0$ as phase-separated metal. At this point, the properties of **nickel-iron alloys** become important in determining both $Fe^0$ and $Ni^0$ activities. The joint Fe/Ni precipitation curve differs by less than one log unit from the IW buffer across the entire range of pressures in the upper mantle (0–14 GPa). Disproportionation increases the average $Fe^{3+}/\sum Fe$ within the silicate assemblage, causing the mantle $f_{O_2}$ to depart from the garnet buffer (3.4) and remain near the nickel precipitation curve. A four-phase model of coexistence encompassing the FMQ, spinel, and garnet buffers, and showing the transition to the nickel precipitation curve, is shown in Figure 3.2.

In contrast to the large change in $f_{O_2}$ in the upper mantle, in the transition zone and lower mantle $f_{O_2}$ changes by $\lesssim 1.5$ log units. Conversely, whereas $Fe^{3+}/\sum Fe$ is relatively invariant (in the range 1–3%) in the upper mantle except in a small transition region near the surface, this quantity changes significantly in the transition zone and lower mantle, reaching values as high as 50% in the latter. This large variation in $Fe^{3+}/\sum Fe$ is made possible by disproportionation of iron into metal alloys with nickel.

### 3.3.2.10 Diffusion in mineral phases is too slow to support redox-driven dynamical order

The determinants of oxidation state in the mantle are important to life because the solid Earth can sustain persistent redox disequilibria – something that fluid phases, with their inability to hold persistent spatial structure, cannot do. These disequilibria may arise from initial conditions of planet formation frozen in by quenching, or they may be generated as fast processes drive the atmosphere and oceans far away from redox equilibrium with the planetary average composition.

However, the same stability that makes solids potential reservoirs of free energy may also cause them to deliver that energy too slowly to maintain non-equilibrium order. For example, partition coefficients of trace elements in basalts can be used to reconstruct $f_{O_2}$ levels from 3.5 billion years ago [486]. An even more striking example is the use of uranium/lead dating, which depends on the immobility over $\sim 4$ Ga of incompatible elements formed by radioactive decay, in zircon crystals that may be only a few hundred microns across.[26] The possibility to maintain redox-driven order on Earth therefore depends on the existence of timescales for transport phenomena that are slower than the rapid mixing times of fluids but faster than simple diffusion.

### 3.3.2.11 Convective transport of radiogenic and infall heat introduces new channels for redox as well as energy flow

The mechanism that turns quenched redox energy from the mantle/atmosphere system into an important driver of surface energy flows and chemical organization is the establishment

---

[26] See Knoll [444] or Erwin [226] for an introductory discussion of this method of dating and it uses.

of convective plumes and larger organized cells in the mantle. These cells are not driven to form as a relaxation channel for chemical energy, but rather for heat originally present from planetary accretion and maintained over the longer term by decay of radioactive elements. Heat diffusion in mantle minerals, though not nearly as slow as diffusion of matter, is sufficiently slow, relative to bulk mantle viscosity, that laterally heterogeneous accumulations of heat, and vertical transport driven by the resulting differences in buoyancy, can organize into self-perpetuating cyclic flows.

The classic example of a related (but much simpler) phenomenon, first studied experimentally in viscous fluids by Henri Claude Bénard [58], is known as **Bénard convection**. It occurs when the density difference between heated lower layers and cool upper layers crosses a critical threshold, where a local rotation can be driven sufficiently fast to maintain the temperature heterogeneity. The instability to growth of local rotations was first analyzed by Lord Rayleigh, so the phenomenon is also referred to as *Rayleigh–Bénard convection*.

Bénard convection is often put forth as a highly abstract model for the emergence of life as a form of dynamically maintained organization [303]. It is an example of the class of phenomena known as *non-equilibrium phase transitions*, which we will consider at length in Chapter 7. We will argue there that biogenesis was also the cumulative outcome of a cascade of non-equilibrium phase transitions in the geochemistry of early Earth. At this very broad category level, Bénard convection is a useful example to illustrate what it means for a phase transition to occur in a dynamical system. However, it is not otherwise an apt model for the kinds of chemical, physical, or informatic phase transitions that we will propose for biogenesis. Moreover, the kinds of convective structures that are important for redox dynamics in the Earth involve several additional levels of structural differentiation than those in simple viscous fluids, so we will not pursue Bénard convection further as a model in its own right.

### 3.3.2.12 Crustal cooling creates a redox-insulating layer between the mantle and the volatile geospheres

The existence of mantle convection alone may not be sufficient to create a one-way flow of reducing potential from the interior of the Earth to its surface, or a net change in internal oxidation state. Convection limited within the mantle, if it were sufficiently slow, could approximately follow mineral equilibria along pressure and temperature gradients, cycling iron between oxidation states among the silicate phases described in Section 3.3.2.9.

Such conditions can result if a thick rigid upper-mantle layer forms what is termed a **rigid lid** over more ductile layers corresponding to the asthenosphere. The lid limits the rate of mantle cooling at the surface, resulting in a hotter and also a more homogeneous upper mantle. Less temperature difference between the upper and lower layers in turn weakens the strength of the convective system. While an initial period of oxidation can still occur if a rigid lid forms beneath hydrospheric or atmospheric conditions that are made more oxidizing than the average mantle, this boundary layer becomes a diffusion barrier shutting off redox reactions.

### 3.3.2.13 Volcanism, possibly ordered in the long-range structures of plate tectonics, breaks through the crustal redox shield

It is very important to the geochemical context of life that the Earth is not a rigid-lid system, though the reasons it possesses the much more complex convection we observe are still incompletely understood.

The convection realized within the Earth is not limited to the interior of the mantle, but affects the surface as well. Transport of heat from deep regions to the near-surface permits the formation of melts which partition phases and can breach the surface in volcanic eruptions (in which we include basalt pillowing at spreading centers as well as more violent extrusions), creating direct contact between reduced minerals and the ocean. Transport of the rigid upper mantle and crust[27] in plate tectonic cycles moves hydrated basalts beneath continents (creating other forms of melts) and appears to preserve long-lived chemical disequilibria in diving slabs that may reach the lower mantle. The release of heat at spreading centers and transport of cold oceanic crust back into the mantle increases the temperature difference between lower and upper mantle layers, feeding back to strengthen convection. Finally, the flexing and cracking associated with oceanic crust formation creates other channels of contact between the hydrosphere and upper-mantle peridotites which are active zones of redox reaction.

We turn now to mechanisms that produce redox disequilibria particularly between the atmosphere and the mantle, as well as charging the atmosphere with other sources of chemical potential. These inputs to the planetary system enable the focusing of large chemical potential differences at the lithosphere boundary capable of activating covalent bond chemical rearrangements.

### 3.3.2.14 Outgassing plus late infall creates an atmosphere and oceans

Planetary atmospheres and oceans are supplied by an ongoing process of outgassing of volatiles from magmas, augmented by infall of ices from comets. These processes act continuously from the time of planet formation and core/mantle separation, though whether they continue depends on the kind and degree of long-term mantle convection formed. In the earliest stages of accretion of small, metal-rich planets, surface temperatures are hot enough to drive volatile gases to escape velocity, so the onset of atmosphere formation is more accurately an onset of atmosphere *retention*. Cooling at the planet's surface enables the retention first of heavier and then of lighter elements [121].

On a tectonically active planet, some part of the atmosphere is continually recycled through the mantle, with $CO_2$ being a particularly important component on Earth. Alkaline earth elements (particularly calcium) liberated through continental weathering precipitate $CO_2$ in carbonates, which become emplaced in altered lithosphere. Subduction of this lithosphere beneath continents liberates part of the $CO_2$, which is recycled in continental volcanism. The first formation and subsequent accretion of continents (discussed

---

[27] The rigid upper mantle and crust form the region termed the *lithosphere* in the precise terminology of modern Earth sciences.

below) is therefore a cause of large regime shifts in the dynamics of $CO_2$ over the first billion years of planetary evolution. Processes of atmospheric recycling are accompanied by the gradual release of new volatiles that are also mobilized in the course of convective cycling.

### 3.3.2.15 Hydrogen escape creates oxidation disequilibrium between the atmosphere/ocean and the mantle

The most important process driving Earth's atmosphere persistently out of redox equilibrium with the lithosphere is the differential loss of the atmosphere itself. Atmospheric molecular gases range in atomic mass from 2 ($H_2$) to 50 ($CO_2$) or 54 (Xe), and if one considers also the neutral atoms produced by photodissociation, the lower end of the mass range extends to 1 (H). The resulting factor of $\sim$50 in relative gravitational binding energies is so large that differential escape favoring loss of light elements and especially H, $H_2$, and He is driven by a wide range of mechanistically distinct processes, across a wide range of atmospheric structures and chemical reaction systems including those that may trap elements differentially. Because the strongest contrast of electronegativity among the abundant elements is between hydrogen (mass 1) and oxygen (mass 16), greater mass loss of hydrogen constantly drives the atmosphere toward higher $f_{O_2}$.

Because new atmosphere is constantly being produced by outgassing of volatiles from magmas, and (to a lesser extent over time) from cometary infall, the steady composition of the atmosphere at any time results from a balance of these supply processes against escape and against the chemical processes that recycle components of the atmosphere through minerals. Over geological timescales, the atmosphere is not generally stable either in composition or in mass. Factors such as continent formation and consequent weathering, increase in solar luminosity, and change in solar spectrum, qualitatively restructure the atmosphere on planets that are too small and too close to the central star to prevent escape of light elements.

Escape is a phenomenon in orbital mechanics at the molecular scale. Because both kinetic energy $mv^2/2$ and gravitational potential energy $GMm/r$ depend on the mass $m$ of the molecule of interest, the **escape velocity** $v_{esc}$ (at which kinetic energy equals gravitational binding energy) is independent of $m$. The escape velocity and associated escape energy are given by

$$\frac{E_{esc}}{m} = \frac{1}{2}v_{esc}^2 = \frac{GM_\oplus}{r}. \tag{3.5}$$

Here and below we use the mass of the Earth $M_\oplus$ for reference, and $G$ is Newton's gravitational constant. For all escape processes of interest on Earth, the atmospheric radii of interest are comparable (within factors of 2 or 3) to the radius of the Earth, denoted $R_\oplus$, so we may obtain order of magnitude estimates for escape velocities by using the escape velocity at the surface as a figure of merit. The surface escape velocity on Earth evaluates to $v_{esc} \approx 11$ km/s.

The gross physical structure of the atmosphere is determined by the interplay of three functions, which depend on independent parameters in planetary composition:

- the gravitational potential energy $GMm/r$, determined by planetary mass $M$;
- the thermal kinetic energy $k_B T$, determined by the balance of heat gain and loss; and
- the density $\rho$, set by the total atmospheric mass.

The *relative* values of density and pressure at different heights depend on the ratio of gravitational potential to thermal kinetic energies, known as the **Jeans parameter** and denoted $\lambda$ [836]. This parameter governs the log-decay rate of density and pressure in the atmosphere. Its value also determines whether regions of the atmosphere have sufficient kinetic energy to escape gravitational binding. $\lambda \gg 1$ corresponds to strongly bound atmosphere, and $\lambda \lesssim 2$ is generally interpreted as a threshold below which thermal excitation leads to escape.

A second relation, between the relative density profile, and the absolute scale of density, determines whether thermodynamic state variables such as pressure are well defined, and thus whether particle motion is governed by statistical mechanics or orbital mechanics. The absolute molar density $\rho$ matters because it is the mean free path for scattering, $l_{MFP} \equiv \rho^{-1/3}$ that determines whether local motion is thermalized. When the mean free path satisfies $l_{MFP} \geq r/\lambda$, a combination known as the **scale height** of the atmosphere (see Box 3.2), atmospheric density decays fast enough that upward-moving particles have a non-vanishing probability to avoid scattering until they either escape the gravitational potential or fall back to denser regions.

---

**Box 3.2    The Jeans parameter and atmospheric structure**

The Jeans parameter is a ratio of gravitational (potential) to thermal (kinetic) energy, which determines the log-rate of change of pressure and density in the atmosphere. To see how it arises, consider a small unit of solid angle $d\Omega$ in spherical coordinates, and consider the problem of force balance in the column of atmosphere that lies within that angle, around a planet of mass $M$. Let $m$ be a characteristic molecular mass of the atmospheric gas (the choice of which may depend on the mixing conditions as well as total atmospheric composition), and for simplicity suppose the atmosphere has a fixed temperature $T$.

Let $F_{d\Omega}(r)$ denote the force downward through the area $r^2 d\Omega$ at any radius, due to the weight of the overlying air. Let $\rho(r)$ denote the molar density of air molecules. Then $F$ and $\rho$ are related as

$$F_{d\Omega}(r) = \int_r^\infty dr' \frac{GMm}{r'^2} \rho(r') r'^2 d\Omega. \tag{3.6}$$

**Box 3.2    The Jeans parameter and atmospheric structure (cont.)**

The force also equals the pressure at $r$ times the area, $F = Pr^2 d\Omega$, and pressure (equal to energy density in the ideal gas approximation) is related to density by the thermal equation of state

$$P = k_B T \rho. \tag{3.7}$$

Thus we may recast Eq. (3.6) as

$$F_{d\Omega}(r) = \int_r^\infty dr' \frac{GMm}{k_B T r'^2} F_{d\Omega}(r'). \tag{3.8}$$

The differential equation satisfied by $F$,

$$\frac{d \log F_{d\Omega}}{dr} = -\frac{GMm}{k_B T r^2}, \tag{3.9}$$

has solution

$$F_{d\Omega}(r) = F_{d\Omega}(r_0) \exp\left\{ -\frac{GMm}{k_B T r_0} \left(1 - \frac{r_0}{r}\right) \right\}. \tag{3.10}$$

From Eq. (3.10) the solutions for $P$ and $\rho$ follow as

$$P(r) = P(r_0) \left(\frac{r_0}{r}\right)^2 \exp\left\{ -\frac{GMm}{k_B T r_0} \left(1 - \frac{r_0}{r}\right) \right\}$$

$$\rho(r) = \rho(r_0) \left(\frac{r_0}{r}\right)^2 \exp\left\{ -\frac{GMm}{k_B T r_0} \left(1 - \frac{r_0}{r}\right) \right\}. \tag{3.11}$$

The geometric prefactor does not alter the leading exponential dependence of the decay.

**The Jeans parameter**

$$\lambda(r) \equiv \frac{GMm}{k_B T r} \tag{3.12}$$

is called the *Jeans parameter*. In more realistic descriptions the molecular mass $m$ and the temperature $T$ are both functions of $r$. The logarithmic decay relation (3.9) may be written

$$\frac{d \log F_{d\Omega}}{dr} = -\frac{\lambda(r)}{r} \equiv -\frac{1}{h_{\text{scale}}}, \tag{3.13}$$

where $h_{\text{scale}}$ is called the *scale height* of the atmosphere at radius $r$.

Order of magnitude estimates for the Jeans parameter and scale height on Earth can be obtained by taking $M = M_\oplus$, $r_0 = R_\oplus$, and $T = 1000 \, \text{K}$, and referencing

---

**Box 3.2 The Jeans parameter and atmospheric structure (cont.)**

molecular mass to the mass of the hydrogen atom $m_H$. At these values the Jeans parameter evaluates to $\lambda(R_\oplus) \approx 7.57 \times (m/m_H)$.

Temperatures in the atmosphere between the Earth's surface and the base of the exosphere today range from about 200–1000 K, and relevant gases for both the contemporary and ancient atmospheres include $H_2$ ($m = 2m_H$), $H_2O$ ($m = 18m_H$), $N_2$ ($m = 28m_H$), CO ($m = 28m_H$), $H_2S$ ($m = 34m_H$), $CO_2$ ($m = 44m_H$), and a variety of atoms and radicals of intermediate molecular weight, as well as smaller amounts of ideal gases that do not influence atmospheric chemistry but are therefore useful as tracers of ancient atmospheric dynamics. $\lambda(r)$ on Earth therefore varies upward from 7.57 by factors ranging from 25–100 for most of the atmosphere over most of the age of the Earth.

**Non-finiteness of the integral** While $P(r)$ in Eq. (3.11) goes to zero geometrically, the volume integral over $\rho(r)$ diverges linearly in the upper limit of integration. This unphysical result is corrected by recognizing that either the *mean free path* $l_{MFP} = \rho^{-1/3}$ for scattering at some altitude becomes larger than the scale height

$$l_{MFP} \geq h_{scale}, \tag{3.14}$$

meaning that pressure $P$ is no longer defined as a local state variable, or else that forces *cannot* be balanced and the solution must be dynamical. The region where Eq. (3.14) is satisfied is called the *exosphere*, and its lower boundary is known as the *exobase*.

**Escape criteria** A rule of thumb, originally due to Jeans [395], is that when $\lambda(r)$ decreases through $\sim 2$, the atmospheric gas develops a significant escape current. If $\lambda \sim 2$ first occurs above the exobase then the escape has a single-particle character. If $\lambda \sim 2$ is reached below the exobase, then the solution is non-stationary and escape is hydrodynamic in character [836].

---

This region of low and decreasing density is known as the **exosphere**. Its lower boundary, called the **exobase**, is the threshold between thermal and orbital mechanics [121].

The way escape processes structure the atmosphere depends on the way the deposit of energy, and the exit mode of matter, act selectively on different chemical species. The following list notes several quantitatively relevant processes, and also whether they are classified as "thermal," driven by average energy, or non-thermal, resulting from discrete absorption or collision events.

**The Jeans mechanism (thermal)** Also known as "single-particle escape," the Jeans mechanism describes evaporation of particles in the high-energy tail of the thermal velocity distribution. A "rule of thumb" introduced by Sir James Jeans, the first to publish a systematic analysis of thermal escape [395], was that the escape flux becomes significant where the parameter $\lambda$ first decreases through $\sim 2$ (see Box 3.2). As long as this condition occurs above the exobase the underlying atmosphere is stable [836].

Heavy cold planets such as the gas giants around the Sun are too massive to experience significant Jeans escape of any elements, but Earth is well within the threshold for significant escape of H, $H_2$ and He, though not of water or heavier molecules. The temperature of the exobase on earth is about 1000 K, more than three times the temperature at the surface, and almost five times the temperature at the top of the troposphere where, importantly, most atmospheric water vapor condenses. Jeans escape is only responsible for about 30% of hydrogen escape on earth, because most hydrogen is bound up in water in the troposphere or oceans.

**Hydrodynamic escape (thermal)** Hydrodynamic escape [804, 836] occurs in atmospheres that are warm enough and dense enough that the condition $\lambda \lesssim 2$ is reached below the exobase, as first observed by Donald Hunten [384]. Under this condition stationary solutions with finite atmospheric mass cannot be maintained by gravitational binding energy, and the classical equations of fluid motion require a dynamical solution describing a *planetary wind*, so named by analogy to stellar winds.

Outside the *homopause* of the atmosphere – the radius where eddy mixing is no longer sufficiently strong to compensate for diffusive partitioning of particle species – different gases separate and the atmospheric pressure is no longer governed by a single average molecular mass. In this outer region lighter elements show the same increased susceptibility to hydrodynamic escape that they show in the Jeans mechanism. Therefore hydrogen escape will dominate the planetary wind in atmospheres that are not hydrogen depleted. Momentum transfer from the hydrogen wind dominates the escape of heavier atoms and molecules by hydrodynamic drag, an effect that is mass dependent and therefore partitions isotopes as well as elements. The Earth does not experience significant hydrodynamic escape today, but xenon isotope ratios on Earth have been interpreted as evidence of an earlier period of hydrodynamic escape [121]. While the phenomenon of planetary wind was expected for gas giant planets in close stellar orbits (so-called "hot Jupiters"), direct observational support has only recently been obtained for the planet HD 209458b. How much hydrogen the terrestrial atmosphere may originally have contained, or how long its blow-off took to occur, are still uncertain because of the combination of dynamic effects that act concurrently on early atmospheres.

**Charge-transfer mechanisms (non-thermal)** Characteristic ionization energies for atoms fall in the range $1-10\,\text{eV}$, whereas $k_B T \approx 0.086\,\text{eV}$ for $T = 1000\,\text{K}$.[28] Therefore

---

[28] This is also the reason mere thermal gradients do not offer the excitation energies required to drive non-equilibrium covalent bond chemistry.

the recombination event following any process of ionization provides a far higher potential source of escape energy than thermal excitation in planetary atmospheres. However, due to the very low photon momentum, the initial ionization energy is a center of mass energy, not coupled to the escape of a neutral species. A variety of single-collision or collective effects (mediated by electric field polarization) exist which can transfer momentum to ions which are subsequently neutralized at energies above $E_{esc}$.

If a sufficient population of ions is being maintained through photoionization, unscreened electrostatic accelerations among multiple (heavy) ions can transfer momentum as well as energy to the center of mass motion of subsets of these. Through subsequent collisions with neutral species, some of these energetic ions can exchange charge but not energy or momentum, so that the energetic species become neutral and can escape the atmosphere. It is estimated that such charge exchange processes account for 40% of current hydrogen escape from Earth [121].

**Photochemical escape (non-thermal)** An alternative to charge transfer among equivalent, light ions such as hydrogen, is for either direct photo-absorption, or recombination of molecular ions, to lead to photolysis, splitting the molecules into multiple, *heavy* neutral atoms or radicals. C, N, and O, as well as H, are susceptible to photochemical excitation to escape velocities. This is not an important process on Earth today, but on planets such as Mars with less dense atmospheres and thus lower exobase altitudes, it is an important loss mechanism.

**Polar wind (non-thermal)** A separate, cooperative effect arises from ionization because lighter electrons have a higher orbital radius on average than heavier nuclei. Ultraviolet and X-ray ionization which produce large charge separations can induce polarization of the ionized atmosphere, which can accelerate ions through the net electric field produced. Most ions accelerated in this way remain confined within trajectories that spiral around the closed field lines of the Earth's magnetic dipole. Interaction with the solar wind, however, causes field lines within narrow cones around the poles to close "at infinity," so that accelerated ions cycling around these field lines are "conducted" along escape trajectories. It is estimated that polar wind accounts for 15% of current hydrogen escape on Earth.

**Sputtering by stellar winds (non-thermal)** Sputtering affects planets that lack magnetic fields which can deflect the stellar wind from direct interaction with the atmosphere. Entrainment on magnetic field lines in the solar wind accelerates ions in the upper atmosphere, which through collisional charge exchange are then freed to escape.

Other processes also lead to atmosphere loss, with impact ejection by asteroids being quantitatively important for many planets in the early solar system, and with continuing effects for small planets into the present. Because this phenomenon – to the extent that it serves as more than a source of increased upper-atmosphere temperature – is a bulk-hydrodynamic effect, it does not lead to the same structured selection of different elements and isotopes as the foregoing processes, so we omit a discussion here. The interested reader is referred to discussions in [752, 834].

### 3.3.2.16 Sequestering of hydrogen in oceans and the troposphere

Escape tends to be an inherently unstable process, running to completion for those elements light enough to be affected (in the case of impact ejection, potentially including the whole atmosphere), and leaving a residual atmospheric composition that finds a new equilibrium with the remaining planetary composition. Therefore the long-term existence of non-equilibrium redox steady states depends on rate-limiting bottlenecks for hydrogen escape, just as we have seen for stellar evolution and mantle dynamics.

On Earth, which is not limited by UV energy delivery to the exobase, the rate-limiting factor for hydrogen escape is *sequestering* as water in the oceans and the troposphere. Almost all hydrogen outside the lithosphere today is bound in water.[29] To escape it must first be unbound by UV photolysis, which occurs in the stratosphere. The existence of a liquid reservoir of water in the ocean, and of condensation at the top of the troposphere (a boundary called the *tropopause*), keeps the stratosphere dry and limits the delivery of hydrogen to the exobase [432]. Condensation in turn depends on the Earth's being sufficiently cool.

The presence of oceans in turn regulates the concentration of other important atmospheric gases such as $CO_2$. Most of the Earth's $CO_2$ is sequestered as carbonates in sediments, or in the crust and hydrated depths of the upper mantle [69, 742]. Carbonate minerals require both a cation such as $Ca^{2+}$, and *hydrated* $CO_2$ in the form of carbonate $CO_3^{2-}$. The carbon cycle on Earth today is regulated significantly by the delivery of $Ca^{2+}$ from continental weathering and basalt interactions at spreading centers. In the absence of either water or exposed continents, $CO_2$ remains in the atmosphere as a greenhouse gas.

Venus and Mars furnish examples of planetary atmospheres not protected by sequestration of water. Both have very little hydrogen in the atmosphere and heavily oxidized surfaces. Venus has little water in its atmosphere and no oceans; its $CO_2$ is not sequestered in carbonates, but remains in the atmosphere at a surface pressure of 90 bars, and creates a greenhouse causing surface temperatures over 730 K. Models by James Kasting, James Pollack, and Kevin Zahnle [419, 907] suggest that Venus could have lost the hydrogen in a volume of water equal to that in Earth's oceans within only hundreds of millions of years by hydrodynamic escape, which would also have removed much of the oxygen liberated by UV photolysis of water. Mars was apparently oxidized early by hydrodynamic escape of hydrogen, and has since lost most of the remainder of its atmosphere through non-thermal escape of carbon, nitrogen, and oxygen as well.

If the presence of oceans is important to the persistence of plate tectonics on Earth, through its effects of lowering the melting temperature and strength of subducting oceanic crust, the feedbacks through the planetary water/mantle system may become very complex.

---

[29] It is an open modeling question whether methane and ammonia were important reservoirs in the Archean and before, but even if they were, these reservoirs were small compared to the ocean.

**Subduction** sequesters carbonates directly and, in the course of recycling them, produces the melt conditions that form continental crust (discussed in Section 3.4.2), the weathering of which seeds further carbonate precipitation in oceans. Amelioration of the $CO_2$ greenhouse in turn lowers the evaporation rate and keeps the stratosphere dry, reinforcing sequestration.

Subduction also increases heat transport out through the mantle, lowering the temperature at the core/mantle boundary and amplifying the disequilibrium that drives the **geodynamo** and the geomagnetic field. This field in turn shields the Earth from the solar wind and reduces photochemical and sputtering losses of atmosphere.

The foregoing discussion shows that the elementary fact of atmospheric redox disequilibrium has behind it a rather complex explanation. It results from a variety of mechanistically distinct processes, which have different sensitivities to initial conditions, and among which several may be at work at each stage in a planetary atmosphere's unstable history.

The story of the solar system, from stellar evolution, to planetary atmospheres, to the dynamics of crust, mantle, and core, is one of transience impeded and given structure by systems of barriers. Initial conditions that are no longer supported persist through these transients for long periods. From this perspective, the most natural interpretation to give to the biosphere is that it is one more tier in this cascade of transients. Continuing the way in which stellar and planetary system transients have interacted with each other in many details, so will the biosphere be anchored at many points to its sub-biotic context.

### 3.3.2.17 Gas-phase photochemistry and weather create two new channels for energy transport and transduction

Upon the foundation of redox disequilibrium created by hydrogen escape, multiple other disequilibria are created within planetary atmospheres, and some of these couple to the primary redox disequilibrium in ways that are important to life today and may have been important in biogenesis.

Photolysis of water and of $CO_2$ can act jointly to transfer reducing potential from hydrogen to carbon. A H radical created by photolysis of water can scavenge an O radical created by photolysis of $CO_2$, leading to accumulation of CO. The complementary OH, exchanged with seawater in peroxides, is recycled to water through oxidation of soluble metals or other reductants. We return to this transfer mechanism in Sections 3.6.6.5 and 6.3.1.

A secondary energy transduction mechanism within the atmosphere is electric discharge. Whereas photo-absorption decreases the concentration of energy from the photon to the dissociation or ionization products, discharge concentrates (a small fraction of) the otherwise diffuse energy of heat into accelerated collisions in a plasma energetic enough to drive chemical reactions. The electrolysis of $N_2$ to form either oxidized or reduced $N_1$ compounds overcomes one of the major barriers to incorporating nitrogen into organic compounds; subsequent interconversion of oxidation states is comparatively facile [784].

These and other mechanisms create numerous parallel channels for energy flow from sunlight to chemical transformations. The reader may find reviews of modern reaction-system models that attempt to produce calibrated, quantitative predictions of atmospheric chemistry in [139, 432, 628].

## 3.4 The restless chemical Earth

Above we have summarized the depth-dependent oxidation states and redox potentials in the solid Earth, and the surface disequilibria maintained by exposure to space and the solar environment. We now consider in more detail the processes that prevent these systems from settling into either an equilibrium or a long-lived quenched state. An equilibrium Earth would be a lifeless Earth. Even a geologically frozen Earth with an atmosphere driven to a shifted oxidation state (or stripped altogether) would be unable to support major subsystems of the Earth's observed biota. Because these particular subsystems are, we will argue, the most plausible models for the stages through which the biosphere emerged, we believe that an Earth that had never been geologically active, or that had quenched too soon or formed a rigid lid, would never have generated a biosphere at all – despite the existence of solar free energy sources that by appearances could sustain life that had achieved sufficient complexity.[30]

It is essential that our planet is a *restless* planet. The intricate dynamics within its fourth geosphere are a magnificently elaborated yet in some respects seamless continuation of the dynamics in the other three geospheres. As the extent, complexity, and structure of geodynamics become a habitual part of the frame through which we understand the living state, it becomes much more difficult to see life as a singularity, and more seemingly inevitable that it is an elaboration within an ongoing process of planetary maturation. Certainly, it contains rare events, and follows unpredictable trajectories, but even these occur within a framework of orderly and comprehensible probabilities.

As important as the maintenance of dynamics in the Earth, which drives energy flows through its subsystems, is the fact that Earth is a *chemical* entity, that its unrest is carried in the forces and flows of chemical states and processes, and that the continuity with the biosphere occurs in specific domains of organometallic chemistry. From a highly abstracted evolutionary perspective, one could mistake the chemical substrate of life as merely opportune, but we will argue throughout the rest of the book that the living state draws on chemistry for many essential forms of order and stability that are not available from any other source.[31]

---

[30] An important and interesting question, which we do not know how to answer, is whether the geological activity on Earth today, and the subsystem of the biosphere supported by means of it, continues to be essential to maintaining the biosphere as a whole, including its phototrophic constituencies.

[31] The role of heredity and selection in population processes is mathematically identical to the role of Bayesian updating in certain problems of optimal inference [729], a relation to which we return in Chapter 7. It is so general that at one level it describes everything, and for the same reason it inherently describes nothing in particular. The particularity of each inference problem must come from the structure of its state space, and for life, this is the space of states and processes in certain geologically important domains of organometallic chemistry.

In the remainder of the section we review the processes of melting that partition silicates and metal oxides, the complex rupture processes that form, age, subduct and reprocess the Earth's crust, and the profound influence of liquid water ranging from the differentiation of continental and oceanic crust to the formation of hydrothermal convection systems and the ecosystems they support.

### 3.4.1 Mafic and felsic: ocean basins and continental rafts

Iron and magnesium are the two most abundant metals in the Earth, roughly a factor of twenty (for iron) and ten (for magnesium) more abundant than the next three: nickel, calcium, and aluminum. The two most abundant non-metals, falling between iron and magnesium in abundance, are oxygen and silicon. This is why the main architecture and dynamical properties of the solid Earth are determined by the partitioning of Fe and Mg within, or apart from, matrices of Si and O.

Among silicate phases, those with lower (Mg, Fe) /Si ratios are less dense, and tend to partition toward the surface of the Earth. Geologists grade silicate phases by (Mg, Fe) /Si ratio, referring to those with more **Ma**gnesium and iron (**Fe**rrum) as *mafic*, and those with lower Mg and Fe as *felsic*. (The latter term refers to **Fel**dspar and **Si**lica, as felsic minerals are enriched both in $SiO_2$ and generally also in aluminum, potassium, and sodium, three principal constituents of feldspars.)

Minerals are categorized according to the weight-percent of $SiO_2$ that they contain, a quantity that can readily be compared among different mineral stoichiometries using the oxide decomposition introduced in Section 3.3.2.5. *Felsic* rocks are those containing >69% $SiO_2$, *intermediate-felsic* contain 63–69% $SiO_2$, *intermediate* contain 52–63% $SiO_2$, *mafic* contain 45–52% $SiO_2$, and *ultra-mafic* contain <45% $SiO_2$. Pristine mantle peridotites are ultra-mafic, basalts (created by melt separation from peridotites to form oceanic crust) are mafic, and granites are felsic.

The major chemical and structural distinctions in the Earth's crust are between continents and ocean floors. Continental crust forms thickened rafts of low-density felsic minerals (principally granites), which "float" in the heavier mantle peridotites. Only about 30% of the Earth's surface is covered by continents today, and the fraction covered was smaller in the past; continent formation is an ongoing process. The remainder of the surface consists of thinner, heavier basaltic (mafic) crust overlying depleted peridotites from which it was extracted. Because this crust is heavier it lies deeper, and so makes up the ocean basins.

The principal mechanism behind the partitioning that forms both continental and oceanic crust is melting. Melts below the surface are termed **magmas**, and those that breach the surface are termed **lavas**. Melts form in contexts created by mantle convection, which may also include migration or release of volatiles. Continental and oceanic crust are chemically very different because they are created by melts produced in different convective contexts, which we consider next. The contexts on the early Earth may have been less differentiated than they are today, and there is evidence that continental and oceanic crust were less differentiated in the early Archean than they have since become.

Melts are also regions that host outgassing, and they bring mineral materials into direct contact with the planetary surface. The conditions of melting are therefore important determinants of the mineral evolution and the redox activity of the planets.

### 3.4.2 Three origins of magmas

Although the layered structure of the mantle is complicated, especially in the transition through the asthenosphere to the rigid upper mantle and the crust, the sources of magma that breach the lithosphere may be coarsely grouped according to relatively simple topological distinctions. Two forms of melt generation follow the boundaries between large, topologically *cylindrical* transport cells, and therefore produce eruptions along *lines* in the crust. These are the **rifting lines** along spreading centers at mid-ocean ridges, and the **subduction zones** where oceanic crust is pulled beneath continental or other oceanic crust. The third form of melt generation occurs where *plumes* form in the mantle; where these breach the crust they appear as *point sources* rather than line sources of volcanism. These magma sources are termed **hot spots**, which are understood (although, with many complications) as the surface manifestations of plumes that originate as deep as the core/mantle boundary and are anomalously hot along the course of their vertical ascent.

Examples of spreading centers that form long lines are the Mid-Atlantic Ridge and the East Pacific Rise, seen clearly in the map of ocean-bottom crustal age in Figure 3.3. These are the locations along which ocean hydrothermal systems lie that will be of particular interest to us in the remaining sections of this chapter. Examples of magma sources along subduction zones are the arcs of island volcanism on the eastern Pacific extending from New Zealand through Polynesia, up the coast of Asia and through to the Aleutian Islands, which continue to coastal chains of volcanism along the west coast of the Americas from Alaska to Patagonia. This arc is the famous *ring of fire* outlining the Pacific Plate, seen in Figure 3.4. Two important hot spots with very different roles are the Hawaiian hot spot, near the center of the Pacific Plate, and the Iceland hot spot, which produces a region of much intensified volcanism along the Mid-Atlantic rift.

#### 3.4.2.1 Dipoles and monopoles: the incompatible topology of cylinders and spheres

The similarity of mantle flow to Bénard convection – however limited by the differences and added complexities of mantle and crust structure – suggests an interesting role of the very basic constraints of topology in creating both the regularity and the complexity of melt conditions on planets. Flow in the mantle is, to good approximation, incompressible. Therefore the velocity field $v$ can always be written as a curl of another vector field $A$ known as the *vector potential* of the flow: $v = \nabla \times A$.[32] The vector potential is a directional quantity, defining an inherently dipole structure within convective flow.

---

[32] The vector potential is related to the *vorticity* of the flow $\omega \equiv \nabla \times v$ by the identity for differential forms $\omega = \nabla \times (\nabla \times A) = \nabla(\nabla \cdot A) - \nabla^2 A$.

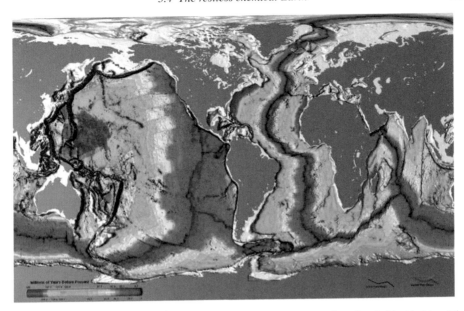

Figure 3.3 Satellite map of ocean-basin crust, showing youngest crust in red and oldest in blue. The Mid-Atlantic Ridge and East Pacific Rise are clearly visible as two lines of youngest crust. Also clearly visible is the basin of extremely old crust subducting beneath the Philippine Plate, reaching from the Japan trench through the Marianas to the Challenger Deep north of New Guinea. (Courtesy of the National Oceanic and Atmospheric Administration Central Library Photo Collection.)

Figure 3.4 World volcanos marked in red, with the "ring of fire" the most conspicuous feature in the figure. The complementary roles of subduction zones to mid-ocean rifting centers can be seen by comparing this figure to the East Pacific Rise shown in Figure 3.3. (Map courtesy of the Smithsonian Institution Global Volcanism Program [8].)

The problem of heat dissipation in a sphere, however, is topologically a monopolar problem. The impossibility of embedding a smooth dipolar field within a monopolar topology requires that the field have singularities at some locations, so even in the smoothest convective system, breaks or joins must occur.[33] Since, with respect to spherical topology, their location is inherently arbitrary, this topological mismatch creates a potential for complex and history-dependent dynamics.[34]

The reason convection arises may be expressed in terms of these elementary geometric concepts as follows. The fundamental elements of convective incompressible flow are rotation and shear. Shear is resisted by viscosity. Rotation transports matter by pivoting about a rotational center, enabling heat transport without internal shearing of the rotating mass. When a system of convective rolls forms, shear is pushed to the interaction zones with fixed boundaries or to the convergence and divergence zones of the rotating cells. Absolutely resistance minimizing configurations for Bénard convection would be those in which rotation organizes into cylinders, with the lines of vorticity $\omega \equiv \nabla \times v$ closing to form tori. The buoyancy forces of radial heat dissipation in a ball push the planes of maximal upwelling between rotating cells to migrate through great circles in spherical shells. However, no configurations of closed tori can be packed into a ball so that their meeting planes fall on such vertical cuts without singularities. Therefore the convective configurations that should be most stable are those with long organized vertical sheets of maximal upwelling and downwelling, which form a network of lines at the surface terminating in (typically three-way) point intersections. The surface manifestations of such sheets correspond to the networks of mid-ocean rifts and subduction zones. Such a topology can be re-enforced, but also rendered more complex, by the rigidity of plates in the lithosphere.

Plume convection, in contrast, is an inherently singular configuration in the vector potential for incompressible flow. The converging flow at the base and diverging flow at the top, as well as cylindrical shear along the ascent column, lead to a large ratio of viscous dissipation relative to advective transport, other factors assumed equal. Therefore we expect that plume flow will arise where other factors are not all equal, either forced as a topological singularity in larger sheet flows or enabled by a large reduction in viscosity through anomalous heating.

The following subsections summarize a few key properties of each melt source, and its role in planetary evolution.

---

[33] In the actual Earth not only topology (arbitrarily deformable) but also *geometry* matter. Where the cylindrical character of convection systems must be distorted by the spherical geometry of the globe, transverse fractures known as *transform faults* arise, to allow lateral displacement from the axis of the spreading center [477].

[34] Nowhere is the complexity of packing vector fields into spheres on more glorious display than in the turmoil of the Sun. The Sun and all other magnetohydrodynamic systems occupy a different order of topological complexity than convection in neutral fluids, because two vector fields – the velocity vector potential and the magnetic field – impose constraints on the flow. Convection cells in the Sun are ordered in space on much smaller angular scales than the long lines of rifting and subduction on the Earth, leading to the stippled pattern of solar luminosity, seen most dramatically in ultraviolet emission images. Yet even the Sun organizes into long-term regularities of formation and transport of large-scale magnetohydrodynamic structures, expressed in the 22-year sunspot cycle where they breach the surface of the solar atmosphere. Solar plasma convection creates singularities not only in space but also in time, as the "breaking" of macroscopically ordered magnetic flux tubes rearranges their spatial topology. This breaking is the source of the enormous energies to which ions are accelerated in ejecta from the solar corona.

### 3.4.2.2 Mid-ocean ridge spreading centers

Rifts at spreading centers form along lines of upwelling that are part of the cyclic flows largely within the asthenosphere (above depths ranging from 200–700 km), coupled to plate tectonics. Upwelling material replaces mantle displaced laterally as plates spread. The source regions are not regarded as being anomalously hot, which contrasts upwelling at spreading centers with "hot spots"; rather, a major driver of the tectonic cycle is the diving of *cold* lithosphere back into the mantle on far plate boundaries. The linear upwelling beneath rifts is topologically a meeting plane between two large counter-rotating cylindrical convection cells (though geometrically the rotations may be far from cylindrical).

The reason upwelling by mantle peridotites creates localized subsurface melts is illustrated in Figure 3.5. Melting temperature in minerals increases with pressure. The contour of first melting temperature as a function of depth, for pristine mantle peridotites, is called the *dry melting solidus*. The temperature profile referenced to stationary mantle, called the *geotherm*, decreases on approach to the surface due to simple heat conduction. The relation between the solidus and geotherm contours determines whether mantle is solid or at least partly melted. Under the stationary parts of oceanic lithosphere, conductive heat loss through the crust is sufficient to maintain peridotites in the solid phase throughout the upper mantle.

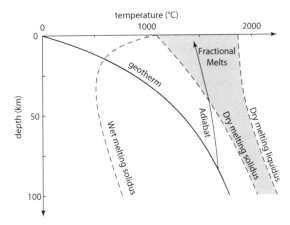

Figure 3.5 Decompression melting along the trajectory of upwelling mantle. The *geotherm* is the temperature profile of stationary mantle. The *dry melting solidus* is the curve of first melt separation for pristine mantle peridotites, and the *dry melting liquidus* is the curve of complete melting. The interval between the two is the *freezing range*, where a slurry of coexisting liquid and crystalline phases forms, with the liquid fraction increasing to the right. Perfect adiabatic upwelling from a point on the geotherm (thus, not taking into account heat dissipation to the surroundings) would transport mantle material on the adiabatic cooling curve. Where this curve crosses the solidus, decompression melting begins. In an alternative to decompression melting, introduction of hydrated minerals into the assemblage, which occurs in subduction zones, can produce a *wet melting solidus* with temperatures well below those of the geotherm, leading to forms of melt separation below island arcs and continental margins. (Material adapted from [555] and [669].)

Upwelling mantle transports heat from depth toward the surface by convection as well as conduction, remaining hotter at each depth along its trajectory than its surroundings (though not as hot as plumes are believed to be). Beneath spreading centers this heat transport is sufficiently greater than the combined effects of adiabatic cooling and dissipation that upwelling mantle crosses the solidus, resulting in **decompression melting**. Under current mantle conditions, decompression melting is partial, leading to magma chambers filled with slurries of melt and crystalline phases [477]. These slurries partition basaltic magma from depleted peridotite assemblages, and the magma, being fluid and buoyant, propagates to the top of the melt chamber, intrudes in fractures in the rigid lithosphere, and may breach the surface as lava.

The melts produced at spreading centers are rich in Mg and Fe, and they produce relatively low-viscosity magmas. Therefore it is possible for them to tumble out of ruptures and flow across the ocean floor, forming rocks known as *pillow basalts*, in contrast with the explosive eruptions that are common for some other kinds of volcanism.

### 3.4.2.3 *Subduction zones, volcanic arcs, and continent formation from hydration-induced melting*

An alternative mechanism of melt generation to transporting hot mantle upward through the solidus, is to change the chemical composition of emplaced mantle so that the solidus falls below the geotherm. This happens most ubiquitously on Earth today at subduction zones, through the introduction of water into the mantle from below. The curve labeled *wet melting solidus* in Figure 3.5 indicates the change that can be produced in melting temperatures with the introduction of even small quantities of water.[35]

Subduction of hydrated oceanic lithosphere beneath continental plates forms a region called the *mantle wedge* beneath the over-riding plate and the subducting slab. Dehydration of the subducting slab (which is cold) releases water which rises into the mantle wedge (which is still approximately at the temperature of the geotherm), weathering the mantle material and causing partial melting in the wedge known as **flux melting**.

Flux melting produces melts that are high in silica and also in incompatible elements including alkali metals (K, Na) and Ca, as well as aluminum. The resulting magmas are thus highly felsic in character, producing primarily granites as they cool. Flux melting in mantle wedges is the main process by which the lighter felsic minerals of continental crust are separated from mantle peridotites.

Subduction-induced melting is responsible for *arc volcanism*, which may form island arcs as occurs off the east Asian coast, or continental volcanic arcs as occurs on the west coast of the Americas. Continental arc volcanism delivers minerals directly to continental margins. Island arc volcanism occurs offshore, and the crust produced may later be accreted onto continents in the course of collisions at plate boundaries known as *convergent margins*. Only small quantities of light felsic crust are subducted in convergent margins;

---

[35] A figure of merit [421] is that the first melting point can be reduced by 150–250 °C per 0.1 weight percent water added to the bulk.

the majority separates from subducting lithospheric mantle to pile up on an accreting continental margin.

Felsic magmas are more viscous than the basaltic magmas formed beneath spreading centers. Therefore they lead more often to explosive eruptions with extensive degassing of volatiles, in contrast to the tumbling flows that produce pillow basalts at spreading centers.

Flux melting is associated with the production of highly silica-enriched magmas. One scenario for continent formation proposes an early stage with separation of small regions of modestly silica-enriched igneous rock from the shallow mantle, probably due to flux melting of a kind similar to that seen at convergent plate boundaries [564, 599], but occurring before the development of fully formed modern subduction. These early crusts would have been more basaltic in composition than modern arc magmas. As continental crust became more differentiated from oceanic crust, more ordered subduction could have led to increasing hydration, production of more felsic magmas, and increasing differentiation between continental and oceanic crust. Most of the land area of continents is believed to have accreted gradually at convergent boundaries with some volcanic contribution, over the interval from 4.0–3.0 Ga [117].

The buoyancy of continental rafts enabled parts of their interiors to escape recycling by mantle convection at least by the beginning of the Archean. As a result some of the same plates (the inner parts referred to as **shields**) have repeatedly collided, separated, and migrated to new configurations in the supercontinent cycle, preserving some of the oldest rock samples known, while the two thirds of the Earth's surface covered in oceanic crust was repeatedly recycled on timescales on the order of 250 Ma. In the course of shield motion, newly emplaced minerals, both igneous (for example, from island arc volcanism) and sedimentary, may be accreted onto shields in belts of new continent formation known as **orogens**. The world's crust, classified into geological provinces, is shown in Figure 3.6.

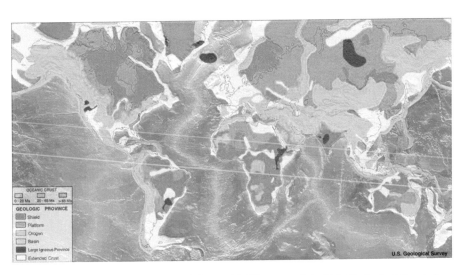

Figure 3.6 Geological provinces of the world. (Courtesy of the U.S. Geological Survey.)

From the oldest remnants, such as the 4 billion year old tonalites from the North American craton in Canada or the 3.8 billion year old Isua formation in western Greenland, it is believed that significant continent formation had taken place by the beginning of the Archean.

### 3.4.2.4 Active upwelling and hot spots

The empirical interpretation that the Earth possesses columnar sources of magma that are independent from spreading centers is almost as old as the acceptance of plate tectonics. The notion of a "hot spot" that delivered magma as plates moved over it was advanced by J. Tuzo Wilson in 1963 [884] to explain the existence of chains of volcanic islands and seamounts in the Pacific not directly associated with spreading centers. In 1971 the source of these hot spots was proposed by W. Jason Morgan [568] to be columnar plumes of heated mantle material, which also accounted for intracontinental volcanic chains (such as the Snake River Plain hot spot track terminating in Yellowstone), for a variety of (Mg, Fe)-rich continental flood basalts, and also for areas of anomalously high volcanism along spreading centers, such as Iceland.[36] Plumes, while not strictly stationary with respect to the bulk mantle, are less mobile than plates moving with the convecting asthenosphere, and therefore generate chains of eruption sites when they are in the interiors of plates moving over them. Hot spot eruption in the interiors of continents has also been proposed as an important mechanism of continental break-up.

A variety of compositional evidence about the sources of hot spot basalts, as well as trace isotope evidence such as $^3$He$/^4$He ratios, suggested a deep origin for the source material of hot spot magmas, which had preserved elements of the original planetary composition against contact with Earth's surface. The ability to image deep-mantle velocity structure with high resolution using seismic data is only about 15 years old at the time of writing [400], but several important hot spots, including Hawaii and Yellowstone [702], are now known to have large low-velocity columns beneath them reaching as far down as the D″ layer[37] which is a boundary layer between the core and the remainder of the lower mantle. These regions are interpreted as plumes of anomalously high temperature, and possibly also different chemistry from the surrounding lower mantle and perhaps melt inclusion.

The upwelling material in plumes is (mostly or entirely) solid in the deep mantle, but due to its anomalously higher temperature at the source, is sometimes described as "actively" convecting, to contrast it with "passive" upward transport in spreading centers [477]. Hot spot magmas form by decompression melting, like spreading center magmas, but they cross the solidus at greater pressures, generating larger magma volumes and also higher melt fractions, and thus different compositions from MORBs. While the concept and general character of plume convection is believed to be broadly understood, the interaction of plumes with the diversity of other structure in the mantle is a source of complexity

---

[36] Because the rate of magma delivery at spreading centers correlates with their elevation, Iceland may be viewed as a continuation of the Mid-Atlantic Ridge with such high rates of magma delivery that it is sub-aerial rather than submarine.

[37] The D″ layer is a ~200 km thick layer at the bottom of the mantle with a velocity discontinuity, believed to be due to a pressure-induced phase transition from perovskite to post-perovskite silicates.

and local idiosyncrasies, resulting in many variant plume models. The melts delivered by plume convection may be useful as an alternative magma model (to melting beneath modern spreading centers) for the Earth in the Hadean or very early Archean, where less fractionated melts were likely produced even from non-plume sources.

### 3.4.2.5 Differences in volcanism in the Archean

The Earth in the Archean was younger, hotter, and convectively probably more vigorous than it is today. Modern magma or peridotite systems may therefore be imperfect models for the kinds of rock/water interaction that existed especially in the Hadean, when the early stages of biogenesis probably occurred. In particular, the melting that produced many early erupting lavas was likely more nearly total than the partial melts at spreading centers today [760], and should thus have produced basalts and gabbros less differentiated from bulk mantle than those in the modern Earth's crust.

Preserved evidence of the difference in Archean magmatism comes from a class of ultra-mafic basalts limited to that eon known as *komatiites*, named for the Komati river in South Africa [117, 118]. These ultra-mafic flows are restricted to the shield provinces of continents preserved from the Archean. Their composition seems to require a melting temperature as much as 500 °C hotter than the melting temperatures of modern MORBs. The combination of low silica content (and particularly of very high MgO content), and very high temperature gave these magmas very low viscosity. Komatiites are plausible candidates for magmas formed in very hot plumes in the Archean mantle [118], with very high melt fractions. The weathering of komatiites would therefore have been more similar to metasomatism of depleted mantle peridotites such as olivine, than to the weathering of MORBs over magma-hosted hydrothermal systems today [690].

Here we have reviewed only the most general distinctions among processes of melting and magmatism. Many further details of melt composition, spatial extent, and migration; the way these affect the composition of crust and the near-surface rock/water environment; and the ways melt history is reconstructed through compositional and also seismic signatures, may be found in [117, 427, 477].

## 3.5 The dynamics of crust formation at submarine spreading centers

The complexity of mineral dynamics, and its consequences for chemistry, do not end within the mantle, but continue in increasingly intricate webs of processes as melts breach and form new crust. These include the dynamics of fracturing, flow of melts into and through fissures, the uplift, cracking, and stripping of crusts to expose underlying mantle, the generation by multiple heat sources of secondary ocean-water circulation systems through both crust and mantle, and processes of dissolution and precipitation carried out by these hydrothermal flows. These processes are jointly responsible for *weathering* of lithosphere, which in an aqueous context is referred to as *metasomatism*. Weathering accomplishes the oxidation of the lithosphere with transfer of reductants to the hydrosphere and ultimately the atmosphere, along with other alterations in mineral composition and phases.

We will focus in this section on submarine rather than sub-aerial volcanism because the kinds of steady-state disequilibria the submarine hydrothermal systems generate are coupled directly to the metabolic chemistry of biota today in a way that sub-aerial systems are not, as discussed in Section 3.6. For this reason as well as a number of others involving the essential continuity of the constraints from aqueous-phase geochemistry to biochemistry, we think the submarine systems are more directly relevant to the conditions leading to life.[38] Sub-aerial systems will continue to be indirectly relevant through the production of continental crust and particularly the delivery of alkaline earth cations such as $Ca^{2+}$ to seawater.

Each of the following subsections covers the stresses and structures that contribute to some stage in the formation and alteration of crust and upper mantle, and the attendant invasion and circulation of seawater in hydrothermal systems. Most of the phenomena occur far from equilibrium thermally as well as chemically, and this is a large part of the reason for their complexity. Dynamically complex processes such as fracture formation are pervasive and essential to both delivery and weathering of minerals, and complex geometries including fissures, pores, and granular media, together with the fluid flows they host, characterize most stages. The dynamical system includes feedbacks, as the cooling and hydration made possible by cracking amplify the thermal and mechanical stresses leading to further fractures.

### 3.5.1 Melt formation and delivery at mid-ocean ridges

The chemistry of the major channels through which reductants are delivered from the mantle to the oceans is governed on Earth today by the disproportionation of the metal/silicate stoichiometry of the pristine mantle between mid-ocean ridge basalts and residual depleted peridotites. Each of these hosts a hydrothermal system that directly drives biochemistry, but the two systems are markedly different in their energy sources, temperatures, pH ranges, and fluid chemistry including key energy-carrying species that can support redox reactions.

#### 3.5.1.1 Melt composition of mid-ocean ridge basalts

Melting at spreading centers, which produces mid-ocean ridge basalts, occurs at depths ranging from 90 km to 10 km subsurface. MORB is composed mostly of garnet and clinopyroxene, from magmas that usually involve melt fractions less than ~20% [427]. Low-fraction melts tend to concentrate incompatible elements, so MORBs tend to be relatively enriched in elements such as calcium relative to the source peridotites. The extraction of these basaltic melts leaves depleted upper-mantle peridotites enriched in olivine, which

---

[38] In abbreviating our treatment of sub-aerial volcanism, we also de-emphasize certain dehydrating environments, which are sometimes considered important particularly with respect to the production of anhydrous polyphosphates. Acknowledging that the existence of such mechanisms should be kept in mind, we reiterate the need to discuss mechanisms in a systemic context: it is not merely the *existence* of species such as polyphosphates that concerns us, but rather their availability out of equilibrium with their environments, whereby they can be sources of chemical work. In dry volcanic environments anhydrous polyphosphates are in equilibrium and have no work to do; to extract work from them involves spatial transport and transient timescales, and requires several further steps to propose a plausible system context.

means in particular that they contain very little $Fe^{3+}$. The $Fe^{2+}$ that is the near exclusive iron in olivine can later enter reactions where it provides a strong redox couple against near-surface levels of $f_{O_2}$, if depleted peridotites are exposed through transform faulting or uplift on continental margins.

### 3.5.1.2 Spatial geometry of melt zones

The displacement field beneath spreading centers, as upwelling mantle material must transition from vertical to lateral motion, together with the temperature profile along the decompression adiabat (see Figure 3.5), leads to a crudely triangular-prism shaped volume in which melting occurs [477] with the apex running beneath the spreading center. Much of the transport through this region is mediated by long arrays of fissures or conduits through which melts migrate toward the apex (see Figure 2a in [427]). Near the apex the fissures coalesce to form a magma chamber, consisting of the crystal mush produced by partial melting. Near the top of the magma chamber a pond consisting mostly of melt may form beneath a dome-shaped roof.

Divergent plate boundaries differ in their rates of spreading, and as a result they also differ in the delivery rates and volumes of magma. Faster spreading (which requires faster upwelling) correlates with deeper first melts and larger magma volumes, and frequently shallower ponding. Melt regions may extend for long distances beneath fast-spreading rifts. Slower spreading may lead to more intermittent and localized melt accumulation, deeper magma chambers, and larger fractions of unerupted melt that cools below the surface to form igneous inclusions known as *plutons*. Many further complexities of surface topography and subsurface structure also depend on spreading rate, for which we refer interested readers to [427, 477] for detail.

### 3.5.2 Tension, pressure, brittleness, and continual fracturing

As complex as the mineral phase diagrams already are that govern the distribution of silicates and metal oxides within the mantle, an increasingly *dynamically* complex and increasingly contingent class of processes delivers magmas to the surface to form crust. Their complexity and contingency stem from the pervasive role of fractures, fissuring, and fluid invasion in the multilayered upper mantle that encompasses the melt zone, the percolation zone of melt separation, and the basaltic layers made from separated magmas.

### 3.5.2.1 Tensile spreading and its associated stresses

Upwelling at spreading centers is termed "passive," as we noted in Section 3.4.2.4, because it is driven by traction on the mantle from overlying spreading plates. An important consequence of traction-driven upwelling is that plates are under tension at the spreading center, which governs the mechanisms of crust formation, and also away from it over distances of tens to hundreds of kilometers. Tensile stress, together with lateral stresses due to spherical curvature that lead to *transform faulting* [477], can result in crustal stripping and secondary weathering of depleted-peridotite mantle (described in Section 3.5.5). At the same time as the relatively thin, brittle, newly formed crust is under tension across its surface, combined effects of expansion due to melting, and possibly vaporization of volatiles, result in local topographic rise and bending stresses on the newly formed crust.

These long-range mechanical stresses are exacerbated by thermal stress from relatively rapid cooling across the steepest part of the depth gradient of the geotherm (see Figure 3.5), and also by mechanical stresses from hydration weathering in cases where ocean water leads to mineral alteration (see Section 3.5.5). Thermal and hydration-induced fracturing at the surface enables the intrusion of seawater, which greatly increases the convective transport of heat out of the upper layers of crust, increasing the rates of cooling, and steepening the geotherm gradient near the surface.

### 3.5.2.2 The upper lithosphere profile

The resulting lithospheric profile at spreading centers is a layered structure. Its basement is a generally ductile depleted peridotite. Overlying this is a layer of basalt cooled in-place, termed *gabbro*, which may be sheared and may include ultra-mafic minerals. This assemblage is called the *plutonic foundation*, produced by melt migration and complex processes of extraction, re-mineralization, and re-solidification of unerupted melts. Above the plutonic foundation is a layer with basaltic composition but considerable internal structure reflecting far from equilibrium physical processes associated with its emplacement.

### 3.5.2.3 Intrusion and extrusion: the record in a shattered crustal architecture

The surface layer of oceanic crust consists of rapidly erupted and quenched basalts termed *pillow lavas* or *pillow basalts*. Beneath these lies a region typically 1–4 km thick of more slowly cooled basalts pervasively fractured and filled in by melts and termed the *sheeted dike sequence. Dikes* are fissures, characteristically $\sim$1 m thick, spanning the entire 1–4 km depth of the basaltic layer, and tens to hundreds of kilometers in extent. Their orientation is mostly parallel to the spreading rift, along the cylindrical axis of uplift and doming of the brittle crust. These fissures form as fractures overlying magma chambers. Magma that fills dikes but cools before breaching the surface is termed *intrusive*, while magma that breaches to form pillow lavas is termed *extrusive*. The sheeted dike sequence is the region in and through which ongoing intrusion and extrusion from the magma chamber to the surface occur.

The whole of the oceanic crust, if it could be seen beneath its layers of pillow basalts and sediments, would be found to consist of vast systems of sheeted dikes, which record the singular and episodic events that form the crust. The events at which new dike systems form produce earthquake "swarms," associated with the propagation of the fracture front along the spreading center and the intrusion of magma into the fissure. Rates of fracture extension are on the order of several kilometers to tens of kilometers per day.

Episodic events associated with brittle failure, including fracture formation, melt intrusion, and degassing, may happen repeatedly through the same rock sequence. The recurring action of these spatiotemporally complex processes can produce patterns of mineral composition, volatile content, and isotope distributions that are challenging to interpret because the signatures of a variety of mechanisms are combined in event sequences that are idiosyncratic to each case and locale.

### 3.5.3 Fractures, water invasion, buoyancy, and the structure of hydrothermal circulation systems

Brittle fracture can be driven both by large-scale stresses, such as extension, uplift, or transform faulting, and also by local stresses including those from mineral expansion in the course of hydration. Diking over magma chambers permits invasion of melts from below, but fracture both at and away from spreading centers can also lead to invasion of seawater from above. In the presence of a heat source, seawater invasion may form circulation zones. Typically, water invades obliquely through networks of fine fissures or porous media. When it is heated, buoyancy drives exit typically through wider and more vertically oriented channels, resulting in net circulation [427].

In the process of mineral alteration by water, many chemical species are generally exchanged with the circulating fluid. These may include silica ($SiO_2$), transition metals, and dissolved carbon in a range of oxidation states. Elements termed "incompatible" with silicate matrices, including alkali metals (K, Rb, Li) and particularly calcium, are often rejected to the fluid, though very complex exchanges may take place in which the direction of uptake or release from the mineral phase depends on temperature and mineral environment.[39] Seawater composition, particularly the concentration of $CO_2$-derived carbonates that govern fluid alkalinity, may affect ion exchange by precipitating key ions such as $Ca^{2+}$ [739].

#### 3.5.3.1 Impulsive events versus steady-state systems

In the delivery of magma and the formation of crust, flow processes may act steadily over long periods, while fracture and intrusion/extrusion with subsequent quenching occur as discrete events. This division is reflected also in the interaction of water with the crust. Once faults, fractures, crack networks, or flow channels are in place, these may be extended or modified by dissolution and precipitation for long periods of quasi-steady flow, as seawater transports a current of heat away from magma sources or out of solidified crust or mantle rocks.

A class of impulsive events generally termed *diking/eruptive events* can produce large quantitative or even qualitative changes in the lithosphere/hydrosphere couple. Diking/eruptive events are often associated with delivery of pulses of volatiles from magma to water through rapid degassing, which may change concentrations of $H_2$, $H_2S$, $CH_4$, or other reduced species by an order of magnitude or more [427, 837]. Because dike intrusion, eruptions, or new fault generation alter flow channels, they may also punctuate long-term

---

[39] For a partial list of some of the exchange processes, see Section 3.5.4. A variety of zones of temperature and also mineral composition are distinguished, in which the directions of ion transfer may change sign in varying combinations. Much of the literature in this domain is based on calibrated computational solutions of complex matrices of reactions [17, 18, 268], and it can be difficult to make generalizations that are reliable beyond the level of coarse characterizations.

flows that deliver distilled low-salinity vapors with intervals that deliver the complementary brines that were sequestered near the bottom of the crust over an extended period of phase separation from vapors (see Section 3.5.4).

Finally, diking/eruptive events can disturb sub-seafloor biotic communities that would not normally be accessible to observation. Microbial mats may be torn apart and lifted into suspension in hydrothermal flows, which deliver them to the surface as debris where normally only cells in suspension would be recovered from samples. These events are informative about the presence of more extensive subterranean ecosystems than would be apparent from examination of vent effluents and structures under background conditions. Most diking/eruptive events take place within systems that already host steady-state flows, and much of what is observed in the impulsive event offers a dynamical probe of systems that have achieved a high level of chemical or biotic complexity.

### 3.5.3.2 Multiple venting geologies and edifices

Hydrothermal systems share the features of thermal disequilibrium (which drives convection), and chemical disequilibrium (between mantle minerals that originate from an $f_{O_2}$ close to the FMQ buffer, and more oxidizing surface conditions). They are, collectively, the major conduits for transfer of reducing potential from the very large reservoir of the bulk of the mantle to the hydrosphere and atmosphere. However, within this important but quite broad "family resemblance," venting systems also exhibit several qualitatively major differences [264, 427], which are relevant to their chemical effect on the oceans and to their connections to living systems.

Probably the two most fundamental axes of distinction concern the presence or absence of $H_2$ in vent fluids and the pH, which has consequences for a host of secondary properties such as the presence or absence of dissolved metals and sulfides. The presence or absence of these two components determines whether reducing potential is carried on $H_2$, on $Fe^{2+}$, or both.

The major determinant of the presence or absence of $H_2$ is whether hydrothermal fluids result from alteration of **basaltic crust** that formed by melt separation, and is typically higher in silica and aluminum, or from alteration of the **residual mantle peridotites** that are composed principally of olivine and clinopyroxene and orthopyroxene.[40] Basalt-hosted fluids typically contain small amounts of hydrogen from rock alteration (although some may be derived from magma degassing), while peridotite-hosted systems produce significant $H_2$ from oxidation of $Fe^{2+}$ to magnetite. Effectively, reducing potential is transferred from $Fe^{2+}$ which remains mineralized, to $H_2$ that escapes the system in peridotite-hosted fluids, whereas reducing potential must be carried out of basalt-hosted systems through mobilization of metals, sulfides, or some other reduced species.

Whether metals and sulfides are transferred to hydrothermal fluids by leaching is largely determined by the pH of the fluids. This variable is correlated with proximity to spreading

[40] Ultra-mafic systems in general predominate at slow and very slow spreading centers, where the volume of covering basalts is lower [760].

centers because it is largely controlled by the *temperature* of the venting system, and to some degree by the silica richness of the hosting rock. In basalt-hosted systems [837], high temperatures are necessary to permit fluid invasion and the establishment of circulation systems. The resulting alteration environment, characterized by relatively high $a$SiO$_2$, buffers pH at low levels, for reasons explained in Section 3.5.3.3 below.

A similar effect is produced in high-temperature peridotite-hosted systems [520, 633], though for reasons that are considerably more complicated to understand [17, 18].[41] Although they depend on kinetic effects that distinguish olivine and clinopyroxene across relatively narrow ranges of temperature, the outcome is qualitatively the same as in basalts: alteration is performed on mineral phases that produce a relatively high $a$SiO$_2$, and the resulting coexistence curves that buffer $a$Ca$^{2+}$ in relation to $a$H$^+$ produce low-pH fluids, with consequent leaching of metals and sulfides.

A starkly different behavior is produced in low-temperature peridotite-hosted hydrothermal systems [428, 429], where hydration of olivine leads to low $a$SiO$_2$ and buffering of $a$Ca$^{2+}$ relative to $a$H$^+$ along coexistence curves that produce much higher pH values. The resulting fluids are low in metals and sulfides, and although their precipitates depend on Ca$^{2+}$ as do those of most basalt-hosted systems, the counterion for low-temperature peridotite chimneys is carbonate rather than sulfate as for basalt-hosted chimneys [739].

The correlation of host rock temperature with pH results in overt similarities among vents that mask significant chemical and dynamical differences. Vents do not segregate into natural groups according to whether they are on or distant from spreading centers.

Both basalt-hosted and peridotite-hosted hydrothermal fields exist along spreading centers at mid-ocean ridges. Because these are all acidic and rich in metals and sulfides, they tend to produce at least some vents that at the surface are **black smokers**. Most, like the main vents at the *Trans-Atlantic Geotraverse (TAG)* or the *East Pacific Rise*, are basalt hosted and hydrogen poor, but a small number such as the *Rainbow* or *Logatchev* hydrothermal fields on the Mid-Atlantic Ridge are H$_2$ rich, because they result from very different alteration chemistry of ultra-mafic foundation rocks.[42]

Hydrothermal fields distant from heat sources in older rock are only known (at present) for peridotite-hosted systems, of which the *Lost City Hydrothermal Field* is the exemplar. Both chemically and phenomenologically these **white smoker** fields are very distinct, with clear vent fluids, carbonate chimneys, high pH, and little metal or sulfide [268, 428, 717]. Yet the host rock assemblages for white smokers bear more similarity with those at Rainbow than with those at the nearby TAG site. At the same time, minor vent fields such as the *Kremlin* field [803] on a flank of the TAG hydrothermal mound,[43] also vent clear fluids and precipitate Zn-rich chimneys. Yet their fluids are closely related to the basalt-hosted end-member fluids of the main TAG venting system, simply altered by seawater

---

[41] The cases of the Rainbow [520] and Logatchev [633] systems are further complicated by the possibility that the ultra-mafic host rocks may include gabbroic intrusions with elevated silica content [268].

[42] See Figure 3.8 for locations of different systems.

[43] See Section 3.5.4.3 below for more detail.

mixing and subsurface precipitation, and not similar in composition to peridotite-hosted white smokers.

### 3.5.3.3 Calcium, silica, and the control of pH

The pH of alteration fluids can vary over 9 log units, and is fundamental to all major chemical and morphological properties of hydrothermal systems. It is one of the more complex properties of vent chemistry to understand because it results from the interaction of several factors. pH in the reaction zone is determined by numerous ion exchanges, and deviations from neutrality are amplified at low temperature by changes in dissociation constants and solubilities of solutes. Efforts to quantitatively model fluid pH values are generally carried out using complex multi-species reaction network models solved computationally, and remain an ongoing area of work [268, 728].

Notwithstanding these complexities, it is possible to sketch a rough picture of the determination of pH by silica activity and calcium exchange, which in most systems are the dominant controlling factors. In this section we describe the calcium exchange buffers that control pH across a wide range of silica activities, following a model by Foustoukos *et al.* [268]. Later sections provide more detail about the hydrolysis reactions that determine silica activity in particular host rocks.

The relation between silica activity and pH is simplest for the dissolution of quartz, the reaction through which vent fluid pH affects the silica content of mound precipitates [739]:

$$
\begin{aligned}
SiO_{2\,(quartz)} &\rightleftharpoons SiO_{2\,(aq)} \\
SiO_{2\,(aq)} + H_2O &\rightleftharpoons \underset{silicate}{H\,SiO_3^-} + H^+.
\end{aligned}
\tag{3.15}
$$

Solubility of silica increases with increasing pH, as aqueous $SiO_2$ is drawn into the hydrate silicic acid which deprotonates to silicate and $H^+$. Conversely, high $aSiO_2$ will tend to acidify solutions, though silicic acid is a weak acid.[44]

Related exchanges of silica between solution and major-phase minerals generally act as a buffer for the relative activities of protons and exchanged ions (principally $Ca^{2+}$ and $Mg^{2+}$), in hydration reactions that are the proximal sources of protons in alteration fluids.

A tendency toward acidity in hydrothermal fluids comes from mechanisms that substitute $Mg^{2+}$ for $Ca^{2+}$ in the hydration of silicate phases, with accompanying generation of $H^+$. (An example of a specific exchange reaction, appropriate to the hydration of peridotite host rocks, is given in Section 3.5.5.2 below.) Whether Mg–Ca exchange actually makes fluids acidic, however, depends on the presence of other exchange reactions that may occur between calcic and non-calcic silicate phases. These reactions *trade* $2\,H^+$ for $Ca^{2+}$ through protonation of CaO components of host rocks, with release of water and also of $SiO_2$ in stoichiometries that depend on the phases involved. The coexistence curves of pairs of

---

[44] Often silica is not in equilibrium in venting systems, however, due to kinetic barriers to dissolution or precipitation.

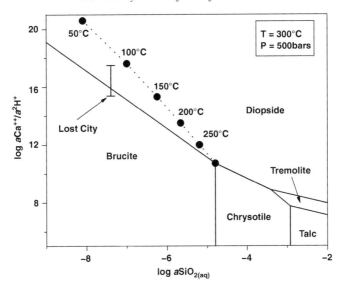

Figure 3.7 A mineral phase coexistence diagram for the MgO–CaO–SiO$_2$–HCl–H$_2$O system at 300 °C and 500 bars, after Foustoukos *et al.* [268]. Upper phases diopside and tremolite are calcic silicates, while lower phases brucite, chrysotile, and talc are non-calcic. Parallel sets of coexistence curves, displaced from the ones drawn, run through the dots at a series of temperatures between 50 °C and 300 °C (labeled). The dashed line affords an idea of the way $aCa^{2+}/a^2H^+$ would depend on $aSiO_{2(aq)}$ in a hydrothermal system where temperature also varied with depth due to conductive cooling.

calcic/non-calcic mineral phases, and the activities of SiO$_2$ in the fluid, create buffers for the ratio $aCa^{2+}/a^2H^+$.

A phase diagram for the exchange of calcium, protons, and aqueous silica among (Ca, Mg) silicate and hydroxide phases is shown in Figure 3.7, from [268]. The stable mineral phases are shown for 300 °C and 500 bars, spanning a range of activities of SiO$_{2(aq)}$ appropriate to peridotite-hosted (olivine/pyroxene-hosted) fluids.

The figure shows, for example, that at very low activity of aqueous silica (appropriate for fluids in hydration equilibrium with olivine, such as the Lost City hydrothermal field), brucite (Mg(OH)$_2$) and diopside (CaMgSi$_2$O$_6$) are stable phases. When both exist as solids, exchange of Ca$^{2+}$, H$^+$, and SiO$_{2(aq)}$ is governed by the equilibrium

$$\underset{\text{diopside}}{CaMgSi_2O_6} + 2\,H^+ \leftrightharpoons \underset{\text{brucite}}{Mg(OH)_2} + 2\,SiO_{2(aq)} + Ca^{2+}. \tag{3.16}$$

The coexistence curve implies a linear relation among chemical potentials

$$\mu Ca^{2+} - 2\mu H^+ = \left(\mu_{\text{diopside}} - \mu_{\text{brucite}}\right) - 2\mu SiO_2, \tag{3.17}$$

(so the activity ratio $aCa^{2+}/a^2H^+$ decreases as the second power of $aSiO_{2(aq)}$). At $aSiO_{2(aq)} \approx -7.5$ appropriate to Lost City, $aCa^{2+}/a^2H^+$ is buffered at a very high value of approximately 16. Therefore a low calcium activity, at or around the value maintained by exchange of Mg and Ca between olivine and seawater, maintains relatively high pH $\sim 6.5$ *in situ* (see discussion below in Section 3.5.5.3 for further detail).

This role of coexistence curves is general, and silica activity from major-phase hydrolysis then determines which coexistence curve governs $a\mathrm{Ca}^{2+}/a^2\mathrm{H}^+$. If $\mathrm{SiO}_{2\,(\mathrm{aq})}$ were progressively pumped into the mineral assemblage of Figure 3.7, as can occur through hydrolysis of certain major phases such as clinopyroxene, the raised $a\mathrm{SiO}_{2\,(\mathrm{aq})}$ would lead eventually to co-precipitation of chrysotile at $a\mathrm{SiO}_{2\,(\mathrm{aq})} \approx -4.9$, buffering $a\mathrm{Ca}^{2+}/a^2\mathrm{H}^+ \approx 10.5$ through the solid-phase equilibrium

$$\underset{\text{diopside}}{\mathrm{CaMgSi}_2\mathrm{O}_6} + 2\,\underset{\text{brucite}}{\mathrm{Mg(OH)}_2} + 2\,\mathrm{H}^+ \leftrightharpoons \underset{\text{chrysotile}}{\mathrm{Mg}_3\mathrm{Si}_2\mathrm{O}_5(\mathrm{OH})_4} + \mathrm{Ca}^{2+} + \mathrm{H}_2\mathrm{O}, \qquad (3.18)$$

until chrysotile had replaced brucite as the non-calcic phase. Under conditions of further $\mathrm{SiO}_{2\,(\mathrm{aq})}$ addition after brucite has been consumed, the diopside/chrysolite buffer would control $a\mathrm{Ca}^{2+}/a^2\mathrm{H}^+$. Yet further addition of $\mathrm{SiO}_{2\,(\mathrm{aq})}$ would cause tremolite to replace diopside as the calcic phase (at or below $a\mathrm{Ca}^{2+}/a^2\mathrm{H}^+ \approx 8.5$), and still further addition of $\mathrm{SiO}_{2\,(\mathrm{aq})}$ would lead to co-precipitation of talc at $a\mathrm{SiO}_{2\,(\mathrm{aq})} \approx -2.9$, buffering $a\mathrm{Ca}^{2+}/a^2\mathrm{H}^+ \approx 8$ through the equilibrium

$$2\,\underset{\text{chrysotile}}{\mathrm{Mg}_3\mathrm{Si}_2\mathrm{O}_5(\mathrm{OH})_4} + 3\,\underset{\text{tremolite}}{\mathrm{Ca}_2\mathrm{Mg}_5\mathrm{Si}_8\mathrm{O}_{22}(\mathrm{OH})_2} + 12\,\mathrm{H}^+ \leftrightharpoons$$

$$7\,\underset{\text{talc}}{\mathrm{Mg}_3\mathrm{Si}_4\mathrm{O}_{10}(\mathrm{OH})_2} + 6\,\mathrm{Ca}^{2+} + 6\,\mathrm{H}_2\mathrm{O}. \quad (3.19)$$

Above this range of high $a\mathrm{SiO}_{2\,(\mathrm{aq})}$, the ratio $a\mathrm{Ca}^{2+}/a^2\mathrm{H}^+$ may be driven as low as 6, so that if reaction temperatures and $a\mathrm{Ca}^{2+}$ were comparable to those at Lost City, the pH would drop by 5 log units.[45]

Different coexistence diagrams control calcium and proton exchange at higher aluminum concentrations in the feldspathic assemblages appropriate to basalt-hosted hydrothermal systems. A characteristic high-silicate, high-aluminum buffer appropriate to fluid interactions with basaltic rocks is the hydration of the end-member anorthite of the plagioclase feldspar series[46] to the end-member clinozoisite of the epidote series [728]:

$$2\,\underset{\text{clinozoisite}}{\mathrm{Ca}_2\mathrm{Al}_3\,(\mathrm{Si}_2\mathrm{O}_7)\,(\mathrm{SiO}_4)\,\mathrm{O}\,(\mathrm{OH})} + 2\,\mathrm{H}^+ \leftrightharpoons 3\,\underset{\text{anorthite}}{\mathrm{CaAl}_2\mathrm{Si}_2\mathrm{O}_8} + \mathrm{Ca}^{2+} + 2\,\mathrm{H}_2\mathrm{O}. \qquad (3.20)$$

In models of the interaction of basalts with Archean versus modern seawater profiles, Shibuya *et al.* [739] compute values $\log\left(a\mathrm{Ca}^{2+}/a^2\mathrm{H}^+\right) \lesssim 8$ for the buffer (3.20),

---

[45] Temperature and $a\mathrm{Ca}^{2+}$ relevant to actual peridotite-hosted systems are not comparable because other factors also change, but this comparison provides an illustration of the way silica controls calcium and pH buffers. See Section 3.5.5 for reports of calculations that control for multiple factors.

[46] The availability of anorthite in this mineral buffer may also be affected by exchange of $2\,\mathrm{Na}^+$ for $\mathrm{Ca}^{2+}$ in a process called *albitization*, but for this level of detail we must refer the reader to the specialist literature and computational models [727, 728].

comparable to the tremolite/chrysotile/talc buffer (3.19) appropriate to clinopyroxene hydrolysis in peridotites. This buffer is responsible for the neutral to slightly acidic *in situ* pH characteristic of basalt-hosted systems below the modern oceans [728]. The similarity of $\log\left(aCa^{2+}/a^2H^+\right)$ between basalt-hosted systems and high-silica peridotite-hosted systems explains the observation that the high-temperature peridotite-hosted Rainbow field has a low pH comparable to that of basaltic systems, rather than the high pH at Lost City.[47]

The *in situ* pH values of end-member fluids in calibrated models do not differ by 5 or more log units, but tend to vary within $\pm 2$ log units of neutral pH at the reaction-zone fluid temperatures, due to a variety of factors. Therefore, while major-phase hydrolysis and calcium exchange buffers ultimately *determine* the pH values of vent fluids, they do not directly *correspond* to the pH measurements reported from vent samples. In some systems where cooling and seawater mixing are sufficiently minor, *in situ* measurements are now possible [194], and agree well with model values [728]. Most pH values for hydrothermal systems around the world, however, are reported for reference temperatures of 25 °C in fluids that also have precipitated solid phases [194, 428, 808, 837].

The pH values derived from the solid-phase equilibria above are modified as vent fluids cool [268] (either conductively or through mixing with seawater) by increasing dissociation constants, and in the case of basalt-hosted systems, when they precipitate sulfide minerals [194]. The secondary alterations of fluid pH are responsible for the downward shifts to pH $\sim 2-5$ reported for black smokers, and the rise to pH $\sim 9-11$ for white smokers in the Lost City field.

### 3.5.3.4 Disequilibria of mixing as vent fluids re-enter the oceans

The seawater that enters hydrothermal systems is far from equilibrium with the rock it invades. The circulating fluid produced, while not necessarily brought to equilibrium with its host rock, is nonetheless pulled far from equilibrium from the seawater it re-enters as vent effluent, both thermally and chemically. The resulting mixed fluids contain unrelaxed redox couples that are far from chemical equilibrium, and the particulates and edifices they form through precipitation create ores that are highly concentrated compared with a uniform distribution. Some of the organometallic reactions involved are the closest geochemical analogues to biochemical reactions found among abiotic terrestrial processes.

Solvent species driven out of redox (and as a result, solubility) equilibrium include transition metals (Fe, Ni, Mn, Zn) leached from minerals. Non-metals in redox disequilibrium may include reduced states of sulfur and carbon such as $H_2S$, $CH_4$, and CO (as well as neutral carbon that may precipitate within convection channels), and molecular hydrogen. Some of these are by-products of mineral alteration processes, while others may be released as volatiles from magma. Aqueous silica ($SiO_2$) may be solubilized by high temperature and subsequently re-precipitated. Concentrated brines may form, which are important for solubilization of metals as complexes with $Cl^-$. Ions that are in solubility equilibrium with

---

[47] The reason high temperature causes Rainbow to be governed by the high-silica equilibrium of clinopyroxene, however, is a more complex kinetic effect to which we return in Section 3.5.5.6.

seawater may also be driven out of equilibrium by temperature change as hot vent effluent is intermixed. Disequilibria resulting in fast precipitation are responsible for the striking sulfide, sulfate, or carbonate chimneys that form at magma-hosted or peridotite-hosted systems in the present oceans. The co-present but unreacted oxidants and reductants from mixed fluids support a variety of prokaryotic metabolisms and the ecosystems that grow from them, which we consider in Section 3.6.

The many scales of planetary disequilibria that we have reviewed up to this point should serve to bring the following fact into sharp relief. *The co-presence of ions and molecules forming strong redox couples, which would never occur near equilibrium, is the joint product of all the above planetary forces: mantle stratification and convection, star-driven atmospheric oxidation, fracture-conducted melt concentration and hydrothermal alteration, and finally delivery by mixing into the hydrosphere. Each of these mechanisms has contributed in a particular way to the accumulation of a free energy behind a barrier. Jointly they culminate in the steady delivery of systems with moderate free energies of reaction $\sim 0.1\,eV$, comparable to the energy of covalent bonds.*

We next consider separately and in greater detail the particular properties of alteration fluids resulting from interaction with basalts or with peridotites.

### 3.5.4  Chemical changes of rock and water in basalt-hosted systems

Basalt-hosted hydrothermal systems are known along most spreading centers in the world's oceans, as shown in Figure 3.8. They occur at or near zones of new crust formation, where seawater invades mineral assemblages of garnet and clinopyroxene. Circulation is driven by buoyancy of heated water that has filtered in along faults or fractures, or through porous media, and which exits along a centralized upflow zone where flows coalesce. (Secondary convection and mixing may also occur through independent near-surface channels.)

Downward-flowing seawater in basalt-hosted systems may traverse several layers of chemically and thermally distinct rocks, and a complex collection of ion exchange and mineral dissolution and precipitation reactions alter the host rocks and the hydrothermal fluid. The fluid expelled from the central upflow zone, upon mixing with ambient seawater, generates extensive precipitates of metal sulfides and oxides, as well as inducing precipitation of sulfate salts from seawater, resulting in chimney and mound formation on the surface. Alternatively, mixing of separate sources of downwelling seawater with vent effluent may lead to precipitation *internal to* the mound, and venting of clear fluids with different pH and consequently different assemblies of dissolved metals [427, 803].[48]

A given venting system may be active for tens of thousands of years, though within this period particular flow channels may open, choke off, collapse, and be replaced on much shorter timescales [427]. Fluid circulation through basalt-hosted hydrothermal systems is the primary channel for the flow of reductant in the form of reduced metals from the lithosphere to the hydrosphere, and also a source of exchange of other metals,

---

[48] Compare with the discussion of the Kremlin site on the Trans-Atlantic Geotraverse in Section 3.5.4.3.

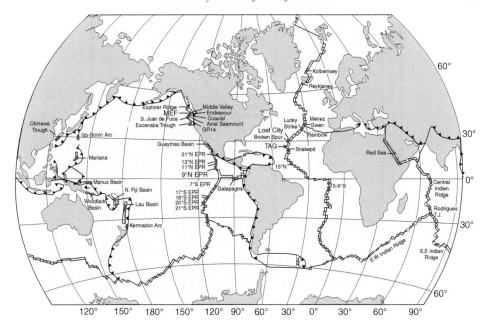

Figure 3.8 Map of known hydrothermal sites, after Tivey [808]. Red indicates known vents; yellow indicates areas of hydrothermal activity inferred from water-column chemical anomalies. Named systems in this chapter refer to this map.

oxides, and magma-derived volatiles, significantly affecting the chemistry of the world's oceans. Although the duration of any particular venting system is episodic on geological timescales, basalt-hosted hydrothermal systems have been continually present probably since the first existence of stable oceans.

The major feature distinguishing **non-magma-associated** from **magma-hosted hydrothermal systems** is the depth and steepness of the thermal gradient, and consequently the flow paths of circulating fluids. At slow spreading centers or in other regions (such as flanking ridges) remote from magma sources, basalts may be brittle and subject to fracture invasion through all three layers: surficial pillow lavas, sheeted dikes, and gabbros of the plutonic foundation [427]. The front of fracturing may extend to depths of 8 km. In contrast, the flow zones in the vicinity of magma chambers are governed by a very steep thermal transition between brittle and ductile phases, which defines a *cracking front* and the limit of downward-going fluid flow. Basalts immediately surrounding the magma chamber are too ductile to fracture, while the rapid export of heat by fluids slightly higher in the crust creates brittle rock and extensive fracturing. The limit of the brittle zone determines the cracking front. The ductile boundary layer between the magma chamber and the cracking front is estimated to be as thin as tens of meters [499], in order to account for the rates of heat transfer observed in magma-hosted systems.

The complexity of the downflow zone, or **recharge zone**, between the subsurface magma chamber and the surface venting site is considerable, in geometry and composition

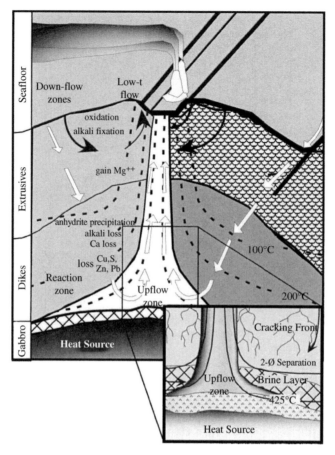

Figure 3.9 Geometry and composition of the subsurface flow zones for magma-hosted hydrothermal systems. (After Kelley *et al.* [427].)

of both minerals and volatiles, as suggested in Figure 3.9 from [427]. In early down-welling through pillow basalts, at temperatures below 150 °C, alkalis such as K, Rb, and Cs may be taken up in neoformed "replacement" mineral phases, while Mg, Si, S, and Ca are lost from the original phases. In deeper layers, Ca, Zn, Fe, S, and $SiO_2$ may be lost from mineral phases by leaching to the fluid. At temperatures above 150 °C, Mg is extracted from seawater to form clays (leaving vent fluids depleted in Mg relative to sea-water), while K, Rb, and Li in this region are transferred into the fluid. Together with Mg, Na and Ca may also be added to rock to form greenschist facies, or (NaSi) may replace (CaAl) in anorthite to form albite [728]. Reaction of water with minor compo-nents such as olivine (considered in more detail in Section 3.5.5) may also release $H_2$. Reduction of sulfate, as well as exchange from magmas, contributes $H_2S$ to the flu-ids. Fluids continuing to the reaction zone may undergo final alteration at temperatures as high as 500−700 °C, before entering the main zone up upwelling. In the reaction zone further leaching of S as well as metals from host rock contributes to final fluid composition.

The sequential effects of fluid/mineral exchange through all these zones determines the composition of vent fluids at the start of upwelling (known as *end-member* fluids) and the residual compositions of altered rock that become the stable oceanic crust.

### 3.5.4.1 Basalt-hosted vents are acidic, metal rich, and low to intermediate in $H_2$

Alteration fluids from basalt-hosted vents typically have pH values in the range 2–5 at a reference temperature of 25 °C [837].[49] Oxygen fugacity of vent fluids *in situ* is estimated to lie just above the mineral buffer pyrite/pyrrhotite/magnetite at both 350 °C and 400 °C [871], which is about 3 log units above the FMQ buffer (3.2) that serves as a reference for $f_{O_2}$ throughout most of the upper mantle.

Vent effluents produced by the sum of exchanges in the downwelling and reaction zones are enriched in alkalis (Li, K, Rb, Cs) and depleted in Mg. They typically include dissolved $SiO_2$ despite the low fluid pH because silica is inhibited from precipitating by kinetic barriers [728].

Volatiles include $CO_2$ (typical concentrations are 2–20 mmol/kg)[50] and $H_2S$ (0–20 mmol/kg), as well as He indicating that magma degassing is the source for at least part of the volatile content. $H_2$ may be present across a range from sub-$\mu$mol/kg to tens of mmol/kg concentrations, and $CH_4$ may be present in concentrations from 10 $\mu$mol/kg to 2.6 mmol/kg [808].

### 3.5.4.2 Fast-flowing, hot vents and sulfide chimneys

Vent effluent from magma-hosted systems is hot and fast flowing due to the high temperatures of the source rock maintained by heat from intruding magmas and the near-surface magma chamber. The combination of ferrous iron and $H_2S$ in the hydrothermal fluids, mixing with an ocean that today is oxidizing and cool, creates conditions for rapid precipitation of iron sulfides such as pyrite ($FeS_2$), chalcopyrite ($CuFeS_2$), and sphalerite ($(Zn, Fe) S$), as well as MgO and FeO. These precipitates are the black "smoke" in plumes that may rise 200 m above the seafloor before they reach neutral buoyancy. They also infuse and plate onto chimney structures, and remain as the main mineral deposits in orebodies formed from long-evolved precipitate deposits [683].

Metal sulfides, however, are not the first constituents of chimneys, which have been observed to grow in height at the astonishing rate of 30 cm per day [807] in a few cases, and may typically grow at rates of several meters per year [427]. The initially formed chimney – only centimeters thick and a mere scaffold for the edifice that will subsequently form – is anhydrite ($CaSO_4$) driven to precipitate by the high temperature of the venting fluid

---

[49] As noted in Section 3.5.3.3, source fluids at temperatures in the range 350–400 °C have near-neutral pH $\sim$ 5.5, a result predicted in models [871] and consistent with *in situ* sampling [194]. Lower pH at 25 ° C results from increased dissociation constants, and precipitation of Fe-bearing mineral phases.

[50] In diking/eruptive events where vent fluids directly receive volatiles from magma degassing, this concentration may transiently climb to 300 mmol/kg [427].

under conditions of turbulent mixing with seawater. The $SO_4^{2-}$ comes entirely from the seawater (due largely to oxidizing conditions on the surface today), and $Ca^{2+}$ comes partly from vent fluids (due to basalt alteration) and partly from seawater (due in significant part to continental weathering).[51] From an initially precipitated anhydrite chimney, the inner wall of a black smoker becomes plated in chalcopyrite. As more metal sulfides adsorb on, are transported through, or precipitate within the anhydrite matrix, the chimney becomes denser, less porous, and more metal rich.

Although the discovery of black smokers on the seafloor was a surprise for ocean geology, in retrospect it is clear that some unusual process must have been at work on the surface of the Earth to concentrate minor elements in **ores** that are commercially worth extracting. Few of these spontaneously form homogeneous phases, and most minor elements would be expected to be widely dispersed and dilute. The existence of orebodies requires the *work* of non-equilibrium processes. Interestingly, the low-temperature solubility of anhydrite, which is responsible for the availability of $Ca^{2+}$ and $SO_4^{2-}$ to initiate black smoker chimneys, also results in gradual dissolution of anhydrite from hydrothermal mounds as they age underwater, leaving orebodies dominated by sulfides and some oxides.

### 3.5.4.3 Warm venting fields from seawater mixing or secondary channels over magma-hosted fields

Not all vent effluent collects within focused, high-temperature flows. Some vent fluids mix locally with surrounding $2\,°C$ seawater to form diffuse, warm flows through and around the exteriors of chimneys or in small fissures around the main venting channels. These fluids have pH and temperature values, and compositions, reflecting the mixing ratios.

Extensive subsurface mixing between downwelling seawater and branches of the main focused flow up from the reaction zone may lead to fluids and precipitates with very different properties from those of end-member fluids reflected at the primary vent. An example is the set of vents producing the field known as "Kremlin" on the Trans-Atlantic Geotraverse [427, 803]. Here, subsurface mixing between end-member basalt-hosted fluids and seawater leads to extensive precipitation of anhydrite *within* the hydrothermal mound, so that the fluids emerging from the vent are depleted of $Ca^{2+}$. These fluids are clear, have temperatures of $265-300\,°C$, and precipitate spires composed of sphalerite (ZnS), amorphous silica, and $FeS_2$.

Other low-temperature flows may not involve direct mixing with fluids in the main upflow zone, but may come from heating in faults, crack systems, and porous media in the extrusive layer above the sheeted dike sequence (see Figure 3.9). In these shallow flow systems, alkali metals are lost from the seawater to form "replacement minerals" including clays, while the near-surface basalts lose Mg, Si, S, and possibly Ca to the convecting fluid [427].

---

[51] Note, however, that chimney properties commonly associated with basalt-hosted systems in the modern oceans may not have been present in the Hadean or early Archean, when seawater likely contained very little sulfate [463] and perhaps much less calcium.

### 3.5.5 Serpentinization in peridotite-hosted hydrothermal systems

The processes of crust formation typically leave depleted peridotites covered by basalts many kilometers thick, as indicated in Figure 3.9. Therefore direct hydrothermal alteration of these ultra-mafic minerals in the neighborhoods of active upwelling zones does not normally occur. However, in locations where tectonic disturbance acts on friable crust, uplift, fracture, and stripping of crustal basalts can expose depleted mantle peridotites to a variety of related hydrothermal alteration reactions collectively termed **serpentinization** [712]. This may occur near spreading centers, as in the case of the Rainbow [520], Saldanha [191], and Menez Hom [264] hydrothermal fields on the Mid-Atlantic Ridge, or several kilometers to tens of kilometers away from spreading centers, as in the case of the Logatchev [633] and Lost City [429] hydrothermal fields. They may be high-temperature ($\sim$350 °C), vigorously venting sites, such as Logatchev and Rainbow, intermediate temperature ($\sim$70 °C), such as Lost City, or low-temperature ($\sim$10 °C), such as Saldanha and Menez Hom.

The residual peridotites beneath oceanic crust, termed **abyssal peridotites**, contain more reduced iron than basalts, and are further from redox equilibrium with the oceans. Their alteration therefore delivers both reducing potential and some heat of hydrolysis. For the same reason, however – they are not naturally an exposed-surface phase – peridotites show more complex dependence on temperature in the relative rates of alteration of different components. Their phenomenology is much more qualitatively variable than the phenomenology of basalt-hosted systems, and more complicated to understand using equilibrium-phase alteration models.

Peridotites are composed of assemblages of olivine and clinopyroxene or orthopyroxene [17, 712]. Olivine is a high-$(Mg, Fe)/Si$ phase (see Section 3.3.2.5), and the dominant component of peridotites which accounts for their ultra-mafic composition. Pyroxenes are lower-$(Mg, Fe)/Si$ phases, typically accounting for a quarter or less by weight of abyssal peridotites. The relative stability of these phases changes with temperature at pressures relevant to serpentinization, with olivine hydrolysis favored at temperatures <350 °C and clinopyroxene hydrolysis favored at temperatures >350 °C [17]. The different $SiO_2$ stoichiometries of hydrolysis of olivines and pyroxenes, their different fractions in source rock, and the temperature dependence of their hydrolysis rates, leads to large qualitative differences among serpentinization zones resulting from relatively modest differences in temperature, water/rock ratio, and fluid circulation rates.

One might expect that the complementary composition of extracted melts and residual mantle would automatically set basalt-hosted and peridotite-hosted vents apart as natural categories. However, the variability of alteration conditions among peridotite-hosted systems militates against identifying any extensive suite of properties as "archetypal" of these systems as a class.

The first recognized peridotite-hosted hydrothermal system was the Logatchev hydrothermal field (1995), followed soon after by the Rainbow field (1997). Both are variant, $H_2$-rich "black smokers," with low pH values and high metal and sulfide concentrations resembling those of MORB-hosted vents, but orders of magnitude higher $H_2$, $CH_4$, and some organics [651]. When the first intermediate temperature system – the Lost City hydrothermal field – was discovered in 2003, with its clear venting fluids and towering carbonate chimneys,[52] these differences of chemistry and surface form seemed to set it apart more than its distinct host rocks as a different class of vent.

In the discussion that follows, we will characterize peridotite-hosted systems as more or less "simple" according to whether equilibrium phase relations are roughly adequate to understand their properties, or whether kinetics must be explicitly taken into account. The Lost City field is an example of serpentinization in old host rock, in which warm fluid is chemically consistent with alteration that involves significant hydrolysis of olivine, and which is well approximated with near-equilibrium models of the known mineral assemblages. The Logatchev and Rainbow hydrothermal fields are opposite: they are high-temperature systems in which fluid chemistry can only be accounted for in terms of predominantly pyroxene hydrolysis, with kinetic barriers limiting the degree of olivine alteration. Because this additional rate structure is important, Logatchev and Rainbow are in a sense a more "complex" hydrothermal system than Lost City.[53]

### 3.5.5.1 Hydration reactions for olivine and pyroxene phases

The shared property of peridotite-hosted vents – the production of $H_2$ as the primary emitted reductant – can be understood in terms of hydration reactions that have parallel forms for both their olivine and pyroxene components. The properties that make their temperature dependence different can also be understood in the contrast between the way hydration acts on the magnesium-bearing components of olivine and pyroxenes. The governing properties are the different behavior of iron and magnesium silicates under hydration, and the different silica affinities of the high-$(Mg, Fe)/Si$ olivine, and low-$(Mg, Fe)/Si$ pyroxene, phases [17, 18].

$Fe^{2+}$ can substitute for $Mg^{2+}$ in a variety of solid solutions in either olivines or pyroxenes, but $Fe^{2+}$ behaves very differently from $Mg^{2+}$ when these silicates are hydrated to clays. $Fe^{2+}$ silicates oxidize to magnetite, releasing $SiO_2$ to the solution, while $Mg^{2+}$ silicates may hydrolize at equal $SiO_2$ stoichiometry, or may even take up aqueous silica from fluids. The oxidation of $Fe^{2+}$ is responsible for the production of large quantities of $H_2$ that is a common characteristic of peridotite-hosted hydrothermal systems. The way in which this reaction combines with the more complex and variable hydration reactions for $Mg^{2+}$ silicates determines most other properties of these systems, particularly $SiO_{2(aq)}$ concentrations and pH, as explained in Section 3.5.3.3, and the variety of metal and sulfide fluid components that depend on pH.

**Olivine**, the principal constituent of peridotites, is a solid solution of iron and magnesium end-members fayalite and forsterite, respectively. In the presence of water, **fayalite** oxidizes to **magnetite**, in the reaction [17]

---

[52] The existence of peridotite-hosted vents with properties very different from those of black smokers was anticipated by Michael Russell in 1978 [683].
[53] We will refer mainly to Rainbow in the characterizations to follow, as most published quantitative models are calibrated to properties obtained from coring at this site.

$$3\,Fe_2SiO_4 + 2H_2O \rightleftharpoons 2\,Fe_3O_4 + 3\,SiO_2 + 2H_2. \tag{3.21}$$
$$\underset{\text{fayalite}}{\phantom{3\,Fe_2SiO_4}} \quad \underset{\text{magnetite}}{\phantom{2\,Fe_3O_4}} \underset{\text{silica (aq)}}{\phantom{3\,SiO_2}}$$

Oxidation of $4\,Fe^{2+} \rightarrow 4\,Fe^{3+}$ liberates hydrogen, yielding aqueous silica as a by-product.

The silica liberated by fayalite hydrolysis can be consumed in a compensating reaction that converts **forsterite** to **serpentine**:[54]

$$3\,Mg_2SiO_4 + SiO_2 + 4H_2O \rightleftharpoons 2\,Mg_3Si_2O_5(OH)_4. \tag{3.22}$$
$$\underset{\text{forsterite}}{\phantom{3\,Mg_2SiO_4}} \underset{\text{silica (aq)}}{\phantom{SiO_2}} \qquad \underset{\text{serpentine}}{\phantom{2\,Mg_3Si_2O_5(OH)_4}}$$

A combined, balanced reaction releasing no aqueous silica, may therefore be written

$$3\left(Fe_2SiO_4 + 3\,Mg_2SiO_4\right) + 14\,H_2O \rightleftharpoons 6\,Mg_3Si_2O_5(OH)_4 + 2\,Fe_3O_4 + 2H_2. \tag{3.23}$$
$$\underset{\text{olivine}}{\phantom{3\left(Fe_2SiO_4 + 3\,Mg_2SiO_4\right)}} \qquad \underset{\text{serpentine}}{\phantom{6\,Mg_3Si_2O_5(OH)_4}} \underset{\text{magnetite}}{\phantom{2\,Fe_3O_4}}$$

The silicate-consuming reaction (3.22) is normally available in excess because the FeO/MgO ratio in peridotites is typically below about 20% [17], so olivine hydrolysis ordinarily results in very low $a\mathrm{SiO}_{2\,(aq)}$ in hydrothermal fluids when it can run to an approximate equilibrium. Excess forsterite may, alternatively, be hydrated directly to form **brucite** without silica exchange [17]:

$$2\,Mg_2SiO_4 + 3\,H_2O \rightleftharpoons Mg_3Si_2O_5(OH)_4 + Mg(OH)_2. \tag{3.24}$$
$$\underset{\text{forsterite}}{\phantom{2\,Mg_2SiO_4}} \qquad \underset{\text{serpentine}}{\phantom{Mg_3Si_2O_5(OH)_4}} \underset{\text{brucite}}{\phantom{Mg(OH)_2}}$$

Mixtures of the reactions (3.23) and (3.24) are the dominant contributions to serpentinization in low-temperature systems where fluid properties are controlled principally by olivine hydrolysis.

A corresponding set of equations exists for the hydrolysis of **pyroxenes**, which are solid solutions of iron and magnesium end-members ferrosilite and enstatite. **Ferrosilite**, like fayalite, oxidizes to **magnetite** though with a larger transfer of $SiO_2$ to the fluid, in the reaction

$$3\,FeSiO_3 + H_2O \rightleftharpoons Fe_3O_4 + 3\,SiO_2 + H_2. \tag{3.25}$$
$$\underset{\text{ferrosilite}}{\phantom{3\,FeSiO_3}} \qquad \underset{\text{magnetite}}{\phantom{Fe_3O_4}} \underset{\text{silica (aq)}}{\phantom{3\,SiO_2}}$$

In place of the silica-consuming reaction (3.22) from forsterite, **enstatite** is hydrated to a mixture of **serpentine** and **talc**, in the reaction[55]

$$6\,MgSiO_3 + 3\,H_2O \rightleftharpoons Mg_3Si_2O_5(OH)_4 + Mg_3Si_4O_{10}(OH)_2. \tag{3.26}$$
$$\underset{\text{enstatite}}{\phantom{6\,MgSiO_3}} \qquad \underset{\text{serpentine}}{\phantom{Mg_3Si_2O_5(OH)_4}} \underset{\text{talc}}{\phantom{Mg_3Si_4O_{10}(OH)_2}}$$

---

[54] More precisely, the hydration product is antigorite, a mineral in the serpentine group.

[55] A corresponding reaction for admixture with the calcium-bearing pyroxene diopside produces the clay tremolite rather than talc, as:

$$2\,CaMgSi_2O_6 + 6\,MgSiO_3 + 3\,H_2O \rightleftharpoons Mg_3Si_2O_5(OH)_4 + Ca_2Mg_5Si_8O_{22}(OH)_2.$$
$$\underset{\text{diopside}}{\phantom{2\,CaMgSi_2O_6}} \underset{\text{enstatite}}{\phantom{6\,MgSiO_3}} \qquad \underset{\text{serpentine}}{\phantom{Mg_3Si_2O_5(OH)_4}} \underset{\text{tremolite}}{\phantom{Ca_2Mg_5Si_8O_{22}(OH)_2}}$$

Note from Figure 3.7 that talc/tremolite/diopside is not a buffer for $a\mathrm{Ca}^{2+}/a^2\mathrm{H}^+$ and $a\mathrm{SiO}_{2\,(aq)}$, though it is a stoichiometric conversion, because the three phases have no intersecting coexistence curves.

The lack of a silica-consuming reaction in the pyroxene hydrolysis system leads to much higher fluid $a\mathrm{SiO}_{2\,(\mathrm{aq})}$ values. In high-temperature systems where pyroxene rather than olivine governs fluid composition, greater $a\mathrm{SiO}_{2\,(\mathrm{aq})}$ results in the selection of different calcium-phase coexistence curves and much lower values of $a\mathrm{Ca}^{2+}/a^2\mathrm{H}^+$, and thus lower pH, as explained in Section 3.5.3.3.

### 3.5.5.2 Magnesium exchange for calcium

An important reaction occurring in hydrolysis of peridotites with seawater is replacement of $\mathrm{Ca}^{2+}$ in host rocks by $\mathrm{Mg}^{2+}$ from seawater, in reactions of the form

$$6\,\underset{\text{diopside}}{\mathrm{CaMgSi_2O_6}} + 6\,\underset{\text{enstatite}}{\mathrm{MgSiO_3}} + 5\,\mathrm{Mg}^{2+} + 9\,\mathrm{H_2O} \rightharpoonup$$

$$\underset{\text{tremolite}}{\mathrm{Ca_2Mg_5Si_8O_{22}(OH)_2}} + 3\,\underset{\text{serpentine}}{\mathrm{Mg_3Si_2O_5(OH)_4}} + \underset{\text{talc}}{\mathrm{Mg_3Si_4O_{10}(OH)_2}} + 4\,\mathrm{Ca}^{2+} + 2\,\mathrm{H}^+.$$

$$(3.27)$$

The equilibrium of reaction (3.27) is sufficiently far to the right that, in characterization of vent fluids, it is typical to plot concentrations of other components versus concentration of Mg, and to use the Mg $\rightarrow 0$ intercepts as proxies for properties of the end-member fluids originating from the reaction zone, with all Mg coming from admixed seawater [837]. Calcium-magnesium exchange is a source of protons driving peridotite hydrolysis systems toward low pH if $\mathrm{H}^+$ is not buffered by the clay-forming reactions of Section 3.5.3.3, for example, when $a\mathrm{SiO}_{2\,(\mathrm{aq})}$ is low.

### 3.5.5.3 Low-temperature serpentinization: the Lost City hydrothermal field understood in terms of near-equilibrium hydrolysis of olivine

The **Lost City** hydrothermal field [426, 428] is formed on an outcrop of the Atlantis Massif, located roughly 15 km from the Mid-Atlantic Ridge, where the spreading center intersects the Atlantis Fracture Zone. The age of the host rock is estimated to be $\sim$1.5 Ma. Although known systems of this kind are rare in comparison with basalt-hosted hydrothermal fields, peridotite outcrops are not uncommon along extensional faulting systems, and anomalies in fluid chemistry sampled over the rift mountains of the Mid-Atlantic Ridge may suggest that similar systems are widespread. Lost City is currently the only reported intermediate temperature peridotite-hosted undersea hydrothermal system, and the only vent system with temperatures far below those typical of basalt-hosted systems that shows chimney development. It is also the only currently reported actively venting hydrothermal system with high pH end-member fluids and little dissolved metal or sulfide.[56]

### 3.5.5.4 Lost City fluids are alkaline, metal poor, and hydrogen rich

Fluids emerge from vents in the Lost City field at temperatures ranging from $40-90\,^\circ\mathrm{C}$. Fluid pH is in the range 9–11 when measured for vent samples at a reference temperature of

---

[56] Several ophiolite systems are known, however, which produce similar alteration fluids.

25 °C. $H_2$ was measured in samples at concentrations of 249–428 mmol/kg, and $CH_4$ was measured in concentrations of 136–285 mmol/kg [428]. High-pH fluids do not leach metals from silicate minerals, so Lost City fluids are low in Fe and Ni, and also in S. Unlike basalt-hosted vents, in which significant reducing potential is carried in the vent fluid in dissolved $Fe^{2+}$ ions, the principal reductant carried in fluid from peridotite-hosted vents is $H_2$.

Most properties of Lost City vent fluids are well predicted by models that assume alteration of all mineral phases in the peridotite host rocks. One early combined computational/experimental model by Laura Wetzel and Everett Shock [871] predicted $aCa^{2+}$ roughly three times higher in peridotite-hosted fluids than in basalt-hosted fluids, and pH $\sim$ 6.5 *in situ* (about 1 log unit higher than the neutral pH of 5.5), which are in good agreement with Lost City observations.

The same model estimated $f_{O_2}$ to be about 1 log unit below the FMQ buffer at 350 °C, and about 1 log unit above FMQ at 400 °C (a range of alteration temperature they assumed in order to directly compare peridotite and basalt alteration).[57] This prediction is about 2 log units below the $f_{O_2}$ values for basalt-hosted systems at 400 °C, and nearly 5 log units below $f_{O_2}$ for basalt-hosted systems at 350 °C.[58] These very low $f_{O_2}$ values, together with the large reducing capacity of $H_2$, explain the remarkable observation that small amounts of Fe and Ni are reduced along the serpentinization front to *metal*, yielding alloys such as **awaruite** ($Ni_3Fe$) [760]. Such alloys are natural redox catalysts, and may also act as reductants in Fisher–Tropsch-type reductions of $CO_2$ in the vent fluid to $CH_4$. Dissolved sulfate, if present, should also be reduced to $H_2S$. However, whereas Lost City may be nearly in equilibrium with respect to olivine hydrolysis [17], the persistence of low levels of sulfate concurrently with $\sim$10 mmol/kg levels of $H_2$ indicate that the fluids are not near equilibrium with respect to conversion of sulfate to sulfide.

### 3.5.5.5 Carbonate/brucite chimney precipitation at Lost City

The fluids that exit chimneys and flanges at Lost City are mixtures of end-member hydrothermal fluids from the deep subsurface, with less altered or unaltered seawater than mixes in the near subsurface or in the vent-generated deposits themselves. Concentrations of both major and minor elements as well as temperature may be used to gauge the extent of mixing. K concentrations are within 3% of seawater values, and both Na and Cl concentrations are within 1% of seawater values, consistent with the low alkali content of the depleted peridotite host rocks, absence of anorthite phases subject to conversion to albite, and also the absence of phase separation between distilled vapors and brines, as occurs at the higher temperatures of basalt-hosted systems.

The main properties of Lost City fluids relevant to precipitating surface chimneys are high pH and $Ca^{2+}$ elevated to about twice the concentration of the input seawater. High pH shifts the equilibrium of dissolved $CO_2$ toward carbonate, which combined with elevated $Ca^{2+}$ induces precipitation of $CaCO_3$ when vent fluids mix with cooler seawater.

---

[57] This alteration temperature is actually too high, an error that leads to minor adjustments in the predictions made by Wetzel and Shock, but when used correctly requires a completely different interpretation of their model. We return to the revised interpretation in Section 3.5.5.6.

[58] See [871], Figure 8.

End-member high-temperature fluids are predicted to be very depleted in Mg relative to the source seawater, and samples taken at the venting field show nearly linear relations between Mg concentration and concentrations of other major elements, suggesting that most of the Mg in vent effluent comes from subsequent mixing. Hydration reactions lead to precipitation of some of this $Mg^{2+}$ as brucite ($Mg(OH)_2$). Kelley *et al.* [428] suggest that the presence of $Mg^{2+}$ aids precipitation of $CaCO_3$ as aragonite, though external aragonite precipitation also occurs along the strands of filamentous bacteria that live on Lost City structures. The result of these reactions is the precipitation of large chimneys – the largest precipitate structures known on the ocean floor – consisting principally of $CaCO_3$ as mixtures of calcite and aragonite, with minor components of brucite.

Reactive silica in samples of Lost City fluids is at levels comparable to or lower than in seawater, and silica is only a trace component of chimney precipitates. Dissolved concentrations of $SiO_{2(aq)}$ have not been reported for fluids from this site, but Foustoukos *et al.* [268] estimate that $aSiO_{2(aq)} \approx -7.5$ would be consistent with buffering of *in situ* pH in the observed range along the coexistence curve of diopside and brucite (see Figure 3.7).

### 3.5.5.6 High-temperature serpentinization far from equilibrium: the Rainbow hydrothermal field

> But this long run is a misleading guide to current affairs. In the long run we are all dead.
>
> – John Maynard Keynes, *A Tract on Monetary Reform*,
> Chapter 3 [431]

The model of Wetzel and Shock introduced in the last section was actually produced before the Lost City field was discovered. Its purpose was to compare fluid properties from alteration of representative suites of basalts and peridotites, so that deviations from basalt-hosted fluid compositions (which, at that time, were assumed to be the norm) could be interpreted in terms of influence from other mineral phases. It was for this reason that Wetzel and Shock chose a common range of alteration temperatures (350–400 °C) characteristic of basaltic systems for both rock types. Later analyses by Douglas Allen and William Seyfried [17, 18] concluded that the peridotite alteration in Lost City does not occur in the temperature range 350–400 °C, but more likely in the range 150–250 °C. A combination of subsequent conductive cooling and seawater mixing determines the final $\sim$75 °C temperature of the effluent. The nature of the change produced by shifting the alteration temperature is not large in equilibrium models, and so the model of Wetzel and Shock showed good agreement with Lost City observations with minor adjustments.

The striking observation from the fact that the model in [871] agrees well with Lost City, with minor temperature corrections, is that it is thus *far from agreement* with the observed properties of the Rainbow and Logatchev hydrothermal fields, both high-temperature peridotite-hosted systems for which the original model's alteration temperatures were valid, and both (unlike Lost City) known at the time. The resolution of his paradox goes beyond a minor adjustment, to force a reinterpretation of the high-temperature model.

Keys to the resolution are the activity of aqueous silica, which at Rainbow is low but not vanishingly small as in the models, and low pH resulting in large metal and sulfide concentrations. These disparities appear to have been successfully explained by Allen and Seyfried [17], who argue that *rate limitation* of olivine hydrolysis at high temperature causes the fluid properties of Rainbow to be governed mostly by its pyroxene component, despite the latter's being a minor fraction of the source peridotite host rocks.[59] The result is a partitioning by *kinetic factors*, between active alteration phases that can continue to be understood approximately with equilibrium phase relations, and effectively inert phases which (to a similar level of approximation) must be excluded by hand from the equilibrium calculations.

### *Rainbow hydrothermal field: vent properties*

The Rainbow hydrothermal field [520] is located on the Mid-Atlantic Ridge (MAR), south-west of the Azores, where faulting has created peridotite outcrops. This is a region of slow spreading, in which faulting and crustal separation enable invasion of seawater into peridotite host rocks, as at Lost City, though thermally mediated fracture invasion probably plays a larger role at Rainbow than at Lost City.

Fluids emerge from vents at the Rainbow field at $\sim 360\,°C$, and have values of pH $\approx 2.8$. Dissolved $SiO_2$ has concentrations $\sim 7$ mmol/kg. This is lower than the 15–20 mmol/kg characteristic of MORB-hosted vents, but much higher than the models or experiments of [871] for olivine-dominated alteration. It is also much too high to be compatible with the observed low pH, as observed by Foustoukos *et al.* [268] and explained in Section 3.5.3.3. Dissolved $H_2$ has concentrations $\sim 13$ mmol/kg, nearly two orders of magnitude greater than those at basalt-hosted black smokers nearby on the MAR [17], and some of the highest levels of methane, manganese, and helium reported for vents along the MAR.

The composition of precipitated chimneys at Rainbow is broadly similar to that of other Fe-rich and S-rich vents, but higher in Cu, Zn, Co, and Ni. It is also unusually high in chloride, which is consistent with the role of complexation with $Cl^-$ as a mechanism of leaching of these metals in low-pH fluids.

The interpretations by Allen and Seyfried [17] of how such properties arise were drawn from a set of hydrolysis experiments performed at 400 °C and 500 bars, which first isolated olivine and orthopyroxene as pure starting phases, and then studied assemblages of one or both with clinopyroxene. They found that *the reactions that did occur in olivine were slower and much less complete than those occurring in pyroxenes.* Specific aspects of this difference include the following.

---

[59] Rainbow could yet hold further complications, also connected to its location near a spreading center. Careful analysis of isotope chemistry by Marques *et al.* [520] provides evidence that not only heat but also some chemical properties at Rainbow may come from gabbroic host rock, making this venting system a mix of opposites.

- Olivine hydrolysis consumed only about 20% of initial fluid Mg by reaction (3.27), whereas orthopyroxene consumed 90% and clinopyroxene consumed 98%. At the compositions used, orthopyroxene released 20 times as much $Ca^{2+}$ to fluids as did olivine alone, and clinopyroxene-bearing phases released an additional factor of six times as much.
- Fayalite (by reaction (3.21)) and ferrosilite (by reaction (3.25)) both delivered $H_2$ to fluids, but due to the absence of reaction (3.22) to buffer $aSiO_{2(aq)}$, $H_2$ production was at least an order of magnitude lower for pure olivine than for either pyroxene phase, and even lower compared to olivine hydrolysis at low temperature, which proceeds rapidly and buffers $aSiO_{2(aq)}$.
- Hydrolysis of pure olivine generated very low levels of $SiO_2$, while orthopyroxene generated about 20 times more, and experiments including clinopyroxene generated about 40 times more, reaching respectively $\sim 2$ and $\sim 4$ mmol/kg, of the same order of magnitude as observed values at Rainbow. A significant observation is that reactions of assemblages bearing olivine and pyroxenes were not very different in $SiO_2$ levels from the pyroxene-only conditions. Thus the reaction (3.22) was not consuming $SiO_2$. The result is that the activity ratio $aCa^{2+}/a^2H^+$ in the phase diagram of Figure 3.7 was around 6, and the *in situ* fluids produced had pH values in the range 4.9–5.3, lower then MORB-hosted fluids at comparable temperatures.

The overall conclusion, then, is that whereas equilibrium hydrolysis of olivine produces fluids similar to Lost City but far from Rainbow, as demonstrated by Wetzel and Shock, for pyroxene hydrolysis the reverse is true. Because, at 400 °C, hydrolysis of forsterite olivine is slower by orders of magnitude than hydrolysis of either fayalite or either of the components of orthopyroxene or clinopyroxene, the main alterations of fluid chemistry under the conditions of the Rainbow field are governed largely by hydrolysis of clinopyroxene.

Thus, *equilibrium phase relations can be used to understand the properties of both Rainbow and Lost City, though kinetics is implicitly present in determining which phases are reactive and which are effectively inert on the relevant timescales.* At Rainbow, only the minor component is reactive, and thus governs fluid properties; at Lost City, the larger fraction of olivine determines fluid properties closer to a true equilibrium.

### 3.5.6 Principles and parameters of hydrothermal alteration: a summary

In the chemical diversity and complexity of hydrothermal systems, we see the beginning of a transition between relative simplicity in the determination of planetary redox state, and the complex realizations of redox flows across the Earth's rock/water interface that continue into biochemistry. Venting environments combine lawful necessity with significant contingency of context. The rules of chemistry, with their complex networks of non-linear constraints, interweave these two in ways that require a reduction to mechanisms to interpret. While a few properties, such as the correlation of low pH with incorporation of metals and sulfides into vent fluids, may admit simple empirical generalizations, most other major properties are not simply understood by empirical generalizations from surface manifestations. Moreover, the suites of major properties that may co-occur in a hydrothermal system do not cluster in simple ways into a few natural groups, as different properties are controlled by different parameters of environmental context.

The same diversity that creates a need for a certain amount of reductive theory, however, also provides a fairly rich array of contexts both for present life and for mechanisms that may have brought it into existence. Reducing power is carried on a variety of ions and volatiles, which may occur in many different combinations. Multiple catalytic phases may be produced, from metal clusters to oxides such as magnetite. A wide range of temperature, salinity, pH, and dielectric constant of water may be found, with strong changes in some of these parameters occurring within quite local regions. Thus whatever mechanisms brought life into existence need not have been as robust across environments as evolved cellular life is today, and need not even have all required identical conditions. They need only have been possible within the range of parameters present *in ensemble*, as long as suitable communication existed among these.

At a coarse level, the complexity of hydrothermal alteration systems is made comprehensible by a few rules of thumb.

- Hydrolysis of major Mg silicate phases governs $aSiO_{2(aq)}$.
- Exchange of $Mg^{2+}$ for $Ca^{2+}$ serves as a source of both $Ca^{2+}$ and $H^+$.
- Protonation of CaO components in both Ca-Mg and Ca-Al silicates exchanges $2H^+$ for $Ca^{2+}$ and various numbers of $SiO_2$. Because $Ca^{2+}$ and $SiO_2$ are co-produced in these reactions, $aSiO_{2(aq)}$ acts as a "control parameter" governing the ratio $aCa^{2+}/a^2H^+$; together with the $Ca^{2+}$ and $H^+$ generated by Mg-Ca exchange, these hydrations determine pH. Three-phase coexistence points at which clays buffer $aSiO_{2(aq)}$ may stabilize pH within high or low ranges.
- Some degree of quantitative correspondence in the relation of $aCa^{2+}/a^2H^+$ to $aSiO_{2(aq)}$ across Ca-Mg and Ca-Al silicates is responsible for the similar *in situ* pH values in basalt-hosted and pyroxene-hosted vent fluids.
- Hydration of Fe silicate phases to magnetite, also regulated by $aSiO_{2(aq)}$, governs the extent to which reducing potential is transferred from $Fe^{2+}$ to $H_2$ as the mobile species.

However, the details of these exchanges will depend on the particular stability diagram determined by the type of host rock, and may involve a half-dozen or more possible phases in each context. The mineral phases in a given environment depend, in turn, on factors ranging from upwelling and spreading rates via melt separation, to the complex interaction of diking/eruptive events interspersed with hydrothermal alterations, and tectonic disturbances such as deep faulting, outcrop, or stripping of crust. The alteration products also depend on composition of the input seawater and the water/rock ratio. Because the timescales for flow through hydrothermal systems can be comparable to hydrolysis rates for some minerals, kinetics may be fundamental in determining system behavior, and relations of flow rate to alteration temperature may partition the host rock into active and effectively inert components.

It appears, however, that although much more needs to be learned to provide a thorough account of circulation and alteration systems, enough is understood quantitatively already that the major variations in hydrothermal environments beneath the modern oceans can be consistently accounted for with both controlled experimental and computational models. Therefore it is not unreasonable to attempt to estimate properties of hydrothermal redox exchange in the lithosphere and hydrosphere of the Archean or perhaps earlier. Some properties will be common to those in modern oceans, but many will not. Before discussing this problem, we consider the last step in the sequence from the solar system to core biochemistry, which is the bridge between hydrothermal redox chemistry and microbial ecosystems.

## 3.6 The parallel biosphere of chemotrophy on Earth

Within only the past 30 years, an understanding of the biosphere has become possible that completely inverts the view that had been accreting for more than a century. A pivotal event in this reframing was the discovery of living systems that do not depend on sunlight.[60] These are the denizens of **chemosynthetic ecosystems** hosted in the Earth's dark, anoxic hydrothermal circulation zones. The discovery of a parallel chemosynthetic biosphere – around the same time as Carl Woese's discovery of Archaea [894] as the third domain of life, the invention of molecular phylogeny, and the discovery of ancient but previously unsuspected carbon fixation pathways including the reductive citric acid cycle [234] – fundamentally reorganized our understanding of metabolic architecture to one centered in microbiology. From this perspective, core metabolism is a natural outgrowth of carbon fixation, and direct connections between geochemical redox reactions and biosynthesis are reconstructed at the base of all major lineages of life.

### 3.6.1 The discovery of ecosystems on Earth that do not depend on photosynthetically fixed carbon

Prior to the late 1970s, it was a pillar of biological thought that all life on Earth depended ultimately on the fixation of carbon into organic molecules by phototrophic organisms. Organisms that are not themselves photosynthesizers were believed to depend on organic carbon fixed by photosynthetic species, often incorporated with the aid of energy either from fermentation or respiration of organic carbon. The required oxidants were also due directly or indirectly to the activity of photosynthesizers.

It thus came as shock when John Corliss and collaborators published a report in 1979 [162] of a deep-sea submersible discovery of lush ecosystems surrounding hydrothermal vents at mid-ocean ridges. The primary producers are archaea and bacteria, nourished by volatiles released through magma degassing and dissolved nutrients from water/rock

---

[60] They do not depend on sunlight directly at all. They may have some indirect dependence through limited use of biologically produced oxygen or detrital carbon, but this is generally regarded as a derived and complicating factor and not a fundamental dependence.

interactions that alter crust or mantle minerals. *The whole-Earth redox disequilibrium, which is focused in the chemistry of vent environments through the mechanisms described in the preceding sections, delivers chemical energy sources to these organisms which constitute points of direct continuation from geochemistry to electron-transfer reactions in metabolism.*

As of 2013, more than 150 active hydrothermal systems are known, in diverse tectonic settings, along the 60,000 km of ocean ridges and back-arc basins [22]. Ecosystems at these sites are colonized not only by rich prokaryotic colonies, but in most cases also by a variety of annelid worms, mollusks, crustaceans, and other multicellular animals (some of them quite large), which may feed from microbial mats or host symbiotic microbes. As the microbiology and ecology of vent environments has become better understood in the ensuing three decades, some of the more productive vent communities have been discovered to have among the *highest rates* of carbon fixation known on earth [504, 545], even though their contribution to total biotic carbon is small ($\sim$0.02%) because of their restricted environments. This is an important observation though it has taken several years to become fully understood: it is not incidental that vent environments support extreme rates of primary productivity; they do so because in key respects biosynthesis is *easier* under the conditions created by hydrothermal circulation.

### 3.6.1.1 Three stages in a dawning realization

The remainder of the section lays out three stages in the discovery of natural systems, and their progressive integration with experiments and computational models, which cumulatively change our understanding of the place of the biosphere on earth. They are presented in roughly their historical sequence, because they could only have been realized incrementally. The stages are as follows:

1. The discovery of life in venting systems, and by inference, in the much larger subsurface of fluid circulation zones, enlarged the class of environments in which it was understood that life could exist. The new domains are distinguished by their embedding in geoenergetics rather than sunlight energetics.
2. The realization that the newly recognized chemosynthetic organisms were not merely offshoot organisms or even specialized branches from the known tree of life, but rather a coherent biosphere unto themselves, which occupied a more basal position from which the previously known tree of life sprouted.
3. The appreciation that the biochemistry of the chemosynthetic biosphere is much less distinct from the geochemistry of its environment than the paradigm that had become established for the phototrophic biosphere. The requirement for complex evolved machinery to perform work against the chemical paths of least resistance is more limited, and in general the metaphor of a "struggle for existence" extends much less deeply into biochemistry than had come to be expected.

It is these interpretations that have led to proposals that the subsurface is the longest inhabited zone of life on Earth, that the chemotrophic biosphere today is the nearest model for the earliest stages of cellular life, and that the continuity of geochemistry with chemotrophic biochemistry holds the clues to life's emergence.

### 3.6.2 The evidence for a deep (or at least subsurface), hot (or at least warm) biosphere

The proposition that the deep history and biochemical diversity of life are carried by biota that grow from geochemistry rather than from surface exposure to sunlight is supported by a wide variety of chemical, biomolecular, cellular, and ecological observations, most of which have been systematically accumulated only within the past 40 years. They extend the notion of a planetary habitable zone from the near surface to the subsurface at least to a depth of several kilometers, and they extend the temperature range for autonomous life to above 110 °C, leading some investigators to suggest that upper limits as high as 120–150 °C are not out of the question [359].

With regard to questions about the origin of life, most investigators at present would probably assert that moderately thermophilic groups (with growth temperatures in the range 50–90 °C) are the most plausible models for early cellular life, balancing the facility of organosynthesis at higher temperatures against problems of maintaining stable macromolecules and secondary structure. We therefore expect that the most information about the origin of life will be gained from microbes living in the near sub-seafloor at warm temperatures, and that groups living at extremes of depth or temperature are specialists that have evolved to fill out the frontiers of viability.

#### 3.6.2.1 Many subsurface environments exist which host chemotrophic communities

Most of our discussion below concerns hydrothermal systems induced by crust formation or peridotite exposure associated with spreading centers. These are a large class of related systems, continually present at a set of constantly changing locations on the seafloor, probably since the formation of the oceans. They are connected to each other through mid-ocean ridge continuity and ocean circulation, and to the ongoing disequilibria in the atmosphere and oceans. Therefore they seem to us the systems from which we are most likely to learn about the large-scale and long-term structure of redox flows relevant to the origin of life.

However, it should be appreciated that Earth's subsurface is extensive and diversified, with a long history, and that many kinds of environments are known to host chemotrophic life. Examples include deep hard-rock mines [136, 487, 636, 795] and their associated waters [21], underground aquifers, and *ophiolites* [717] – peridotite outcrops on continents created when mantle material was driven onto surface rocks at convergent margins rather than being subducted – which may undergo weathering from newly supplied surface waters, or may release trapped fluids from periods as long ago as the Archean. Any of these is eligible as a source of information about the connection of geochemistry to life.

Diverse biotic communities are hosted across environments of different kinds. Methanogens, iron and sulfate reducers, and a variety of heterotrophs have been identified in basaltic and granitic aquifers at depths of one to several kilometers [21]. Bacteria believed to be formic acid metabolizers have been cultivated from fluids in gold mines at similar depths [136, 487]. Still, the most extensive systems by far, and probably the ones hosting the oldest communities and greatest diversity, are those in circulation contact with the world's oceans.

### 3.6.2.2 Direct examination of vent communities

The most direct information about chemotrophic communities comes from those observed and sampled on and in the chimneys of venting systems, and in the tissues and on surfaces of animals inhabiting the venting environments. Here they can be directly observed *in situ*.

Numerous researchers have observed that vent chimneys, with their diversity of mineral phases and their rapid transitions from hot, extreme pH, reducing fluids to cold, oxidized seawater across distances of centimeters to decimeters, form a microcosm of the less easily accessed subsurface from which the vent fluids issue. The subsurface is also a domain of mixing with seawater; it is also mineralogically diverse, and often involves the same kinds of precipitation reactions as occur in chimneys [786]. Young shallow crust can have porosities as high as 12–36% [427], so reaction volumes are also large. The subsurface, however, is much more extensive than most surface venting environments. Fluid alteration leading to chemistry that supports life is known to occur in lithosphere that is as old as 65 Ma. The total area covered by lithosphere of this age or less is comparable to the surface area of the present continents.

Individual vent fields are transient on the timescales expected for macroevolution or possibly for the full transition from geochemistry to cellular life. Individual hydrothermal circulation systems are typically active on timescales $\lesssim 10^5$ years. Within a single circulation system, however, particular flow channels may open or close, through fracture, faulting, or precipitation, on much shorter timescales. As a result the upper mass of a venting field will often be a mound of collapsed and cemented structures, and the subsurface may be a complex maze of filled or active flow channels collectively termed "stockwork," which record a history of repeated episodic flows. It is important, therefore, that circulation systems, although discrete in space and episodic in time, are connected by the extended subsurface and by ocean circulation.

**Vent chimneys** typically start out as highly porous structures formed by precipitation of sulfates or carbonates, which are progressively plated over and filled in by secondary precipitates and crystallites. These structures provide numerous surface and cavity microenvironments to be colonized, which are infused by fluids that vary widely in temperature, composition, and flow rate. Studies of chimney cross sections using both *in situ* DNA sequencing methods and culture-based extraction methods [427, 710, 794], show that walls are colonized by diverse clades of archaea and bacteria, within as well as on the surfaces. Mixing generally produces co-varying gradients of temperature (decreasing) and $O_2$ content (increasing), from the interiors to the exteriors of chimneys. Moderate temperature outer environments in the range $50-90\,^\circ$C are dominated by mesophilic micro-aerophiles, most of which are bacteria,[61] while inner environments at temperatures above 90 °C are dominated by hyperthermophilic archaea, which are generally strict anaerobes.

An example of cell densities and compositions is provided by samples taken from the Mothra hydrothermal field on the Juan de Fuca ridge [710] (see Figure 3.8 for locations). The numbers reported in this study are broadly similar to others from sulfide chimneys at sites around the world [427]. Sections of this chimney produced cell counts ranging from $10^5 - 10^9$ per gram, with the highest counts at a distance of 3–5 cm inward from the outer chimney surface. In these samples, 40% of microorganisms near the surface were archaea, and the percentage increased to 90% in the interior, near the 300 °C venting fluid. On the surfaces of sulfide chimneys, Fe- and Mn-oxidizing bacteria are often found. Turbulent mixing zones, which may contain both high concentrations of $H_2$ and significant molecular oxygen, are dominated by clades such as the Aquificales, which are distinctive in being strict chemoautotrophs for which $O_2$ is an obligate terminal electron acceptor.[62]

Other evidence of a **subsurface biosphere** comes from observations of cells in environments that, for one reason or another, cannot have hosted them long term [188, 360]. These may be vent plumes which are inherently transient convective formations, or they may be vent fluids or ambient seawater, from which cell populations are routinely cultured that cannot grow under the fluid conditions of the samples from which they were extracted. The signatures may be cell counts or DNA, or they may be chemical alterations in which isotopic signatures attest to the action of microbial metabolisms. Examples include microbial communities living within vent plumes, which harvest energy from sulfide mineral precipitates, or bacterial methane oxidation that is observed in plume samples, for which the species responsible have not yet been cultured.

Some genera found in diffuse-flow fluids include *Thiobacillus*, *Ferrobacillus*, and *Thiomicrospira*.[63] The latter are especially interesting as they appear to be the closest sequenced relatives to a diversified collection of endosymbiotic bacteria of vent animals, which we consider next. *Thermococcales* are particularly useful indicators of a subsurface biosphere, because they are hyperthermophilic and are strict anaerobes, but have been cultured from vent fluids sampled worldwide which are too cool and too oxidizing to support their growth.

Some of the cells obtained from plume samples occupy biofilms on particles, which are likely to reflect samples of microbial assemblages from the subsurface. In support of this interpretation, it is often found that bacteria and archaea attached to particles show a greater diversity of taxa than those cultured from free cells in solution.

### *Mats, metazoan ecologies, and endosymbionts*

In terrestrial ecosystems supported by phototrophy, the complex roles and interdependencies among members can obscure the dominant place held by prokaryotes. In vent

---

[61] Frequently $\epsilon$-proteobacteria are the prevalent clades in these environments.

[62] The history of gene transfer in this clade [84], and its position within a group of otherwise strict anaerobes, lead us to believe that the *Aquificales* are a branch from an ancient group that became specialized to such mixing environments when oxygen became a significant constituent of the seawater component near the beginning of the proterozoic.

[63] For details on microbial species and hosts, see [427] and references therein.

environments the lines of dependency are simpler and the foundation role of prokaryotes is evident. There are no species holding roles analogous to the quasi-autotrophic primary-producer roles held by plants[64] in terrestrial ecosystems. Trophic webs have few levels and few non-prokaryotic members. All dependencies for energy, fixed carbon, and fixed nitrogen terminate quickly in prokaryotic ecosystems,[65] either through grazing or through symbiotic associations. Rather than exhibiting complex vertical hierarchies of predation, vent ecosystems exhibit a striking array of parallel direct dependencies of metazoans on bacteria and archaea.

Some vent-hosted organisms are known that can fix $N_2$, but much of the bio-available nitrogen (in the form of $NH_4^+$) is believed to be produced by reduction of nitrate from biotic sources. A dependence on nitrate is a link of vent ecosystems to the modern oxidizing conditions of the oceans. This is a complicating factor in the interpretation of vent biochemistry in the modern oceans, but as we observe below in Section 3.6.6.2, it is not an ancient dependency as the earliest oceans contained little nitrate.[66]

Primary production for vent ecosystems comes from two sources that are phylogenetically distinct. One comprises assemblages of bacteria that form **mats** over chimney surfaces and surrounding precipitates, which can serve as food. The other includes species of **endosymbiotic bacteria** which are often specific to their metazoan host, and distinct from bacteria known from other environments.

Bacterial mats are grazed upon by amphipods and copepods (two groups of small crustaceans), by polychaetes such as *Paralvinella sulfincola* (sulfide worms), and by limpets (marine snails), clams, and even mussels [427, 825]. A few predators occupying higher trophic levels include galatheid crabs (squat lobsters), brachyuran crabs ("true" crabs), and anemones, which are relatively late colonizers in the ecological succession that builds a vent ecosystem. Predation is not considered to be a significant factor in vent ecology.

Many vent animals, including some from clades whose members normally subsist by grazing, can directly capture energy from endosymbionts or surface-hosted symbionts (called *epibionts*). The polychaete *Alvinella pompejana* (Pompeii worm) hosts a variety of bacteria that are believed to be symbiotic. They may provide insulation against the hot surfaces to which the worm attaches, or detoxify it of heavy metals [14],[67] or they may be chemolithotrophic sources of nutrients [116]. The shrimp *Rimicaris* aff. *exoculata* hosts

---

[64] Plants are quasi-autotrophs because, although they can fix carbon, they cannot fix nitrogen from $N_2$, which they must obtain from bacteria through a variety of strategies. Most cofactor biosynthesis is also restricted to bacteria.

[65] Dependencies on detrital carbon, to the extent that it comes from ocean-surface life and not from the vent organisms themselves, along with micro-aerophilic dependencies on $O_2$ as a terminal electron acceptor, create a kind of hybrid ecology between chemotrophy and phototrophy. Although it would not be correct to characterize such dependencies as merely "outposts" of the phototrophic biosphere in vent environments – the heterotrophs and micro-aerophiles are true vent denizens – the complex metabolic strategies made possible by $O_2$ and detrital surface carbon are elaborations on a foundation of chemotrophy that would likely be sparser and simpler without them. We will not qualify each statement in the following discussion to take account of the ways it may be modified for $O_2$ or detrital carbon, acknowledging its presence as a general complicating factor.

[66] The capacity to fix $N_2$ is a more recent innovation than the common ancestor of cellular life [87, 665], and is unlikely to have had an analogue in the early Archean and before. Therefore some coupling to a source of $N_1$ compounds, such as atmospheric electrolysis of $N_2$ [784] or possibly direct reduction in the course of mineral alteration [94, 761], is likely to have been needed for ammonium to have been present in significant concentration in early vent fluids.

[67] An important balance in the concentration of heavy metals must be reached in vent communities. Metals are required in low concentrations for enzymes, but bacteria can more readily evolve protections against them than many multicellular animals can. Therefore vent community structure can depend sensitively on the concentration of heavy metals in vent fluids.

colonies of epibiont $\epsilon$-proteobacteria on the walls of their branchial chambers, which may be a more important source of organic carbon than grazing for these animals [673].

The most charismatic denizens at sulfide-rich vents are the **tube worms**. Although related to annelid worms that inhabit many other ocean environments, the vent-associated tubeworms are remarkable in physiology, size, and growth rate. The largest, *Riftia pachyptila*, can grow to 2.5 m in length and 4 cm diameter, and can tolerate high $H_2S$ levels. These worms can grow in excess of 85 cm/year (the fastest known growth rate of a marine invertebrate) [504], reflecting the very high organosynthesis rates of microbial communities in vent environments.

The coelomic cavity of *R. pachyptila* contains an organ called a *trophosome* [519] which hosts symbiotic sulfide-oxidizing bacteria using $O_2$ as terminal electron acceptor. The bacteria produce organic molecules on which the heterotrophic worm can feed. Although the tube worms depend on $O_2$, and hence on the modern ocean conditions, they still live in environments with low oxygen activity, and have evolved special high-affinity hemoglobins to transport oxygen to their own tissues and to their endosymbionts. The chemosynthetic symbiont bacteria also reduce $NO_3^-$ to $NH_4^+$ from which they synthesize amino acids used by the worms.

Remarkably, the vent symbionts have not been cultivated outside the host, and all sequenced *R. pachyptila* specimens are bacteria from the same species. Moreover, this species clusters with other uncultured symbionts from vents, and distant from cultured bacteria [209]. The closest known sequenced species, as noted above, are *Thiomicrospira* sp. strain L-12 [195].

### 3.6.2.3 *Diking/eruptive events deliver hidden life to the surface*

The foregoing observations concern essentially stable vent populations, or cells delivered into solution under steady venting conditions. A distinct body of evidence for a subsurface biosphere not directly accessible from surface samples comes from **diking/eruptive events** that occur beneath established vent fields. Because the processes that form crust are largely discrete failure events of fracture, faulting, intrusion, and eruption, vent communities are frequently perturbed by hard shocks that disrupt and displace their constituents. Observations of such transient events have only become available within about the past 15–20 years [188, 360, 427]. Investigators measure seawater and surface properties starting days to weeks after the diking event, and then monitor the return to steady venting conditions over the span of a year or more.

A variety of physical changes are associated with the delivery of new lava to the seafloor environment. The melt volumes delivered to the surface during an eruptive event are typically $10^6-10^7 m^3$ in all, which are emplaced over a span on the order of hours. In addition to direct contact of lavas with seawater or other vent fluids, the collapse of chambers or tubes that have released lavas creates sudden opportunities for entry of seawater into unweathered mineral assemblages, resulting in rapid release of heat and alteration chemicals. In the immediate aftermath of an eruption, very wide zones over the collapse sites may vent very hot water for extended periods. It is also possible for subsurface-hosted fluids which are brought into direct contact with melts to be expelled, providing information on matured *in situ* fluids. Megaplumes, also called "event plumes" [427, 503], form in the water column over the course of a few days following submarine releases of lava, which may contain $10^{16}-10^{17}$ J of excess heat energy. They may reach 800 m above the seafloor, and are enriched in volatiles of magmatic origin. A sequence of such releases can proceed at the same site over hundreds to thousands of years.

The salinity of vent fluids can also change as flow reservoirs are shifted or flow paths are altered. The occurrence of an eruptive event at an established venting site can provide information both on the processes of boiling that separate low-salinity vapors from concentrated brines, and on the distribution of brines that had accumulated in the subsurface. It is common that in the immediate aftermath of an eruption, venting fluids decrease in $Cl^-$ from values that may already be from 10–20% of the values of ambient seawater. Over the following several years, these concentrations may increase tenfold or more, as cooling of the host rock pulls fluid circulation to deeper levels and mixes established brines into the vent fluids [427].

Diking/eruptive events may lead to the delivery of more or different life to the surface, relative to the *in situ* surface communities. An important signature of the presence of a subsurface biosphere that precedes and extends beyond the surface venting structures comes from the presence of hyperthermophiles in fluids that vent at low temperatures ($10-25\,^{\circ}$C) in regions of new eruption where surface chimneys were not previously present. It is common to culture thermophilic organisms from fluid samples with temperatures more than $50\,^{\circ}$C lower than the minimal growth temperature for the cultured cells. It is necessary in interpreting such observations to ascertain that the cells cultured did not come from the reservoir of dormant cells in seawater that may provide part of the connection among different venting systems,[68] but numerous observations exist for which this is considered well established [188, 360, 785].

In addition to the sudden appearance of novel organisms unfit for life in seawater, the large increase of cell counts in megaplumes associated with increased volatile concentration suggests that *blooms* form from reservoir populations of dormant cells in response to increased supplies of limiting nutrients. Sea-surface films formed from megaplumes may contain viable hyperthermophiles at cell counts of $10^9$ per liter [427]. The immediate aftermath of eruptions (on timescales of weeks), in which diffuse venting fluids are high in $H_2S$ and Fe, may also show blooms of sulfur-oxidizing bacteria believed to be derived from communities resident in the sub-seafloor.

Finally, during disruptions, microbial mats and other biotically produced matter may be ejected at the surface. Just as microbial mats are part of the foundation of ecosystems on vent chimneys or on the surfaces of newly emplaced lavas – indeed, these generally show the highest cell counts per volume of all locations where cells are found – it is expected

---

[68] While dispersal worldwide in seawater is suggested to be more rapid and effective than might have been expected for microbes outside their growth conditions [236], the constant presence of selective forces makes it possible to distinguish the slow emergence of dormant species under new conditions from rapid dispersal of *in situ* communities recently disturbed [786].

that these "constructed niches" would be an essential component of the larger subsurface biosphere. Eruptive events can make it possible to study the establishment of mats on newly formed and chemically reactive lava surface, and can also deliver intact samples of mats and other material from subsurface microbial assemblages.

A characteristic by-product of mesophilic metabolisms of communities that live in high-porosity layers, and bloom when the resident vent fluids rich in reductants or sulfides are mixed with infusion of seawater enriched in $O_2$, nitrate, or sulfate, is a white, clumped material known as "floc"[427, 730]. Floc is high in sulfur and low in carbon, and filamentous in structure. It is often ejected in large volumes with vent effluent through cracks and pits in newly created venting zones, creating the impression of snowstorms. The sulfate fibers that make up floc are known to be produced by both surface-associated microbes that live on sulfide precipitates, and by oxidizers of $H_2S$.

The same species that produce the sulfate fibers in floc, along with iron oxidizers, make up microbial mats of filamentous bacteria *as much as 1–10 m thick* that cover fresh lava flows. An important species in mat formation is the bacterium *Beggiatoa* spp. In addition to the by-products of sulfide oxidation, flocculated mat fragments may be lifted into suspension during subsurface disturbances, and ejected in vent fluids as much as 200 m above the seafloor.

The aggregation of observations of these kinds constitute our knowledge about the existence and composition of sub-seafloor microbial communities associated with neo-formation of crust. We learn more about the connections of biota to their geochemical environments by studying the variation among communities hosted in different settings.

### 3.6.3 The complex associations of temperature, chemistry, and microbial metabolisms

Bacteria and archaea supported by vent fluids or precipitates require three kinds of inputs: redox couples to furnish sources of energy, sources of C, N, and P that can be incorporated through chemotrophic pathways, and transition metals to provide catalytic reaction centers in enzymes and key metal-center cofactors. These must be provided in some suitable combination from the composition of end-member fluids generated by hydrothermal alteration and leaching of minerals, in combination with admixed seawater. Within each category, however, significant diversity of sources is possible. For example, $H_2$, $CO_2$, and $CH_4$ concentrations can readily span more than two orders of magnitude among vent fluid samples, and $H_2S$ concentrations easily span factors of more than 30. Dissolved iron may vary from tens of mmol/kg in acidic vents to only $\sim$10 $\mu$mol/kg at Lost City.

Summaries by Kelley *et al.* [427] (Table 4) provide useful reminders of the diversity of macroenvironments and microenvironments that host microbes, including mounds, smokers, sediments, mats, specialized tissues and both inner and outer surfaces of animals, pore spaces and surfaces in precipitates and rocks, plumes, diffuse-flow vents, and diverse rock settings in the sub-seafloor. The differences of kind and of relative supply of the three essential groups of inputs in each environment determine the classes and complexities of ecosystems they can support.

As the previous discussion has shown, the literature on redox couples, pathway inputs, and their phylogenetic distribution is now extensive and detailed. The specificity with which organism distinctions are coupled to details of environment are important to appreciate, but are beyond the scope that we can treat here. We will mention a few high-level trends as examples.

**Temperature and anoxia** We noted above that, due to the correlation of temperature and anoxia, strict anaerobes dominate environments hotter than $\sim 50\,^\circ$C. Most of these are archaea and in many samples they are found to be exclusively either methanogens or anaerobic heterotrophs. Flanges on sulfide structures are particularly favorable environments, because they are regions of rapid transition. Hyperthermophiles at cell densities of $10^8$/g-sulfide were found in discrete mineral layers in flange samples from the Endeavor Black Smokers [427].

Calculations of the available free energy and equilibrium molecular distributions from various mixtures of end-member vent fluids and seawater [545] suggest that methanogenesis should be supported in very reducing fluids, while both methane oxidation and $H_2$ oxidation, and also sulfate reduction, should be supported in mixing zones where $O_2$ or $SO_4^{2-}$ from seawater become available as terminal electron acceptors.

**Metals as catalysts and reductants or oxidants** The principal metal requirement for all organisms is iron, which is the main structural element of Fe-S clusters and also frequently the catalytically active ion, and which is also found in tetrapyrrole cofactors. The most widely used substituent for Fe in Fe-S clusters is nickel. Since these elements both have low solubility in today's cool, oxic oceans, the co-occurrence of ferrous or metallic Fe and Ni with volatile reductants in vent fluids is an important factor in their ability to support productive and diversified ecosystems.

$Fe^{2+}$ is also oxidized by a variety of bacteria on chimney surfaces (see Figure 2.2 for a mat encrusted with oxidized iron). Bacteria may also oxidize Mn in early event plumes, along with $H_2$ and sulfur.

A less common metal, but one that is essential in some gas-handling enzymes such as nitrogenase, is molybdenum. In some hyperthermophilic organisms, tungsten substitutes for molybdenum. An interesting question is whether use of tungsten is ancestral or is an adaptation to colonize extreme environments.

**Differences between high-temperature sulfide-precipitating systems and low-temperature carbonate-precipitating systems** A striking difference exists between the complex, diversified ecosystems hosted at high-temperature vents with metal-rich fluids, and a comparative lack of animal life at the Lost City carbonate field with its abundant $H_2$ but low concentration of metals [426]. The Lost City field precipitates, which are both porous and in initial stages fibrous, host both archaea and bacteria, with cell counts as high as $\sim 10^7$/g-carbonate. Scaffolding of deposits of aragonite on filamentous bacteria may be responsible for part of this architecture. Cultures from chimney samples, in media designed to support methanogens, methane oxidizers, and heterotrophs, yield extensive colonies, and it is believed that most or all of the $CH_4$ sampled from vent effluent is produced by methanogens.

### 3.6.4 Major classes of redox couples that power chemotrophic ecosystems today

#### 3.6.4.1 Not as much of a struggle for existence

Chemotrophic life provides an alternative paradigm to the concept of a *struggle for existence* that has dominated biological thought since Malthus and Darwin. Not only competitive replication, but also biosynthesis came to be perceived as a struggle at every step. Energy captured from light was needed to do the Sisyphean work of organosynthesis against a tendency to degrade toward one-carbon compounds. Such a view drives a wedge between the delicate, complex, and interdependent mechanisms of life and the robust and spontaneous processes we expect from the non-living world. It can hardly come as a surprise, within this view, that Francis Crick could characterize the emergence of life as "a happy accident," "an infinitely rare event," or "almost a miracle" [168].

Chemotrophic life offers a different paradigm: that life's robustness comes from following *paths of least resistance*. What chemotrophy shows is that in appropriate environments, such paths of least resistance may include a large part of biosynthesis and bioenergetics. Struggles still occur, but they involve fewer elements and they are supported by a foundation that uses many spontaneous steps.

The replacement of the paradigm of struggle by the paradigm of least resistance occurs repeatedly, as we first shift away from an emphasis on phototrophy and toward chemotrophy, and then again as we compare chemical environments at different temperatures and chemical compositions. Whereas the synthesis of almost all biotic carbon compounds from $CO_2$ is energetically uphill in oxidizing environments, it is now understood that synthesis of many of them becomes exergonic in reducing environments, even at millimolar concentrations of key oxidants and reductants appropriate to hydrothermal mixing systems. In particular high-temperature and highly reducing environments on Earth today, and more widely in the early Archean or the putative Hadean, an increasing fraction of biosynthetic reactions become exergonic [20, 21, 22, 545]. In these environments, the concept of a separate step to "harness" free energy becomes unnecessary for larger pathway segments or sectors of metabolism. This simplifies the problem of being cellular to a collection of less interconnected problems, in which more of the requirements for organosynthesis can be studied at the level of the pathway.

The progression of understanding is one of **localization** and **modularization**. In the view of life that comes from phototrophy, free energy dissipation occurs within an aggregate system of both Earth-bound chemicals and sunlight. Complex coupling mechanisms are needed to force dissipation in the electromagnetic spectrum to do work *against the chemical subsystem*. In passing to chemosynthesis, we eliminate the electromagnetic system from consideration, and the net process is exergonic within the chemical system alone. More limited coupling is still needed to force exergonic transitions in some subsystems to do work against other subsystems. The favorable environments or eras for chemosynthesis are those where the scope is narrowed still further, and more and more of the cross-coupling among pathways can be removed.

We think the problem of biogenesis is how to arrive at a total decomposition: how to find a connected set of environments that either provide, or remove the need for, all forms of cross-coupling so that the organosynthesis responsible for a biosphere is reduced entirely to a parallel collection of spontaneous processes. This does *not* mean eliminating the chemical diversity of life; it means eliminating a reliance on its vertical organization.

### 3.6.4.2 *Vent environments and organosynthesis*

Jan Amend and Everett Shock [21] emphasize two points that organize the study of geo-chemical free energy sources. First, a reaction must be *thermodynamically favorable but kinetically inhibited* to serve as an energy source. For example, under mixing conditions that would yield fluids of $100\,°C$ at solvent compositions found on the East Pacific Rise, reduction of $O_2$ with $H_2$ (known as the *Knallgas reaction*) would yield 100 kJ/mol of electrons transferred. This is sufficient energy per molecule to fix $CO_2$ or to generate both phosphoanhydride bonds of ATP. Yet the Knallgas reaction, as well as $H_2$ reduction of $CO_2$ (the reaction of methanogenesis) and even elemental sulfur are all sluggish reactions under abiotic vent conditions despite having large negative free energies.[69] Living systems function as catalysts for particular exergonic reactions, and by controlling the flows through these channels they can direct the resulting free energy toward other, non-spontaneous reactions in the form of chemical work.

Second, at sufficiently high temperature, dissipative reactions are fast enough that introducing catalysis confers reduced advantage. Amend and Shock argue that the disappearance of a sufficient *separation of timescales* between spontaneous and catalyzed reactions, and not biomolecule stability, determines the upper temperature limit for life. We believe that quantitatively, reduced advantage and thermostability may be inherently confounded, as the maintenance of life is a *competition of rates* between building up and falling apart. Therefore the threshold at which catalysis no longer provides a sufficient energy supply to maintain integration and function is likely to depend on the thermostability of assembled components, outside the reaction network under consideration even if not within it. However, their emphasis on the importance of a separation of timescales is one that we agree with, and they show that thermostability is not inherently a problem by showing that complex organics can be equilibrium species in geologically relevant ensembles.

The chemical species that persist for long times in redox disequilibrium due to kinetic inhibition include several carbon and sulfur volatiles and a suite of metal ions. Among the most abundant inputs and outputs in hydrothermal systems are $H_2$, $H_2S$, $CO_2$, and $CH_4$, any of which may be of magmatic or mineral-alteration origin. Ferrous iron is the most common and highest concentration reduced metal species. Its concentration in vent fluids ranges from $10\,\mu\text{mol/kg}$ to 20 mmol/kg.

---

[69] $H_2$ reduction of $CO_2$ is sluggish even up to $500\,°C$, compared to biological methanogenesis.

Most primary production in vent environments is carried out by bacteria that oxidize sulfur, methane, hydrogen, and iron. The primary $CO_2$-incorporating organisms are sulfur-oxidizing bacteria, and these are also the microbial partners in most symbioses with animals. Methane oxidation coupled to sulfate reduction (known as anaerobic methane oxidation) is performed by acetoclastic methanogens, and by *Methanosarcina*.

Sulfur plays a special role in microbiology because of the flexibility it affords in redox reactions. Partially oxidized sulfur may occur as thiosulfate ($S_2O_3^{2-}$), tetrathionate ($S_4O_6^{2-}$), or a variety of longer more sulfur-rich chains called *polythionates*. Any of these may serve as electron donors or acceptors depending on the chemical context. The majority of thermophiles and hyperthermophiles in culture use electron transfers involving species in the sulfur redox system [481].

Against these reducing species, seawater in the modern oceans provides $O_2$, nitrate, nitrite, and sulfate, which are the principal electron acceptors. The compendium of Lengeler, Drews, and Schlegel [481] (Chapter 10), and [427] (Table 3) provide more extended lists of redox couples used by bacteria and archaea, many of these brought together in hydrothermal systems. McCollom and Shock [545] have estimated that the chemical potential flux available for chemosynthetic primary production from these sources at deep-sea hydrothermal vents is about $10^{13}$ g biomass per year.

Some vent-microbial pathways involve oxidation of organic carbon, and thus are not autotrophic at least at the organism scale. The degree of coupling they reflect between vent and surface life depends on the ratio of detrital carbon from photosynthetic systems, to *in situ* fixed carbon that is simply recycled by these respiratory pathways. For example de Angelis *et al.* [182] have estimated that carbon fixation in plumes over the Endeavor field on the Juan de Fuca ridge exceeds deposition of organic carbon from the surface over the venting area, making the coupling to the surface secondary. The many animals supported by symbiotic chemotrophs are also part of the source of detrital carbon used by ubiquitous and diversified clades of Crenarchaeota in vent environments.

**Phosphorus** in hydrothermal systems is a potentially limiting nutrient. Its concentration in modern seawater is $\sim 2.5$ $\mu$mol/kg, and that may diminish by factors $\sim 5$ in vent fluids, through scavenging to apatites (highly insoluble calcium phosphate minerals). The central role and high concentration of phosphorus in biochemistry is disproportionate to its availability in most geochemical settings on Earth today. There are good reasons that no other chemical group may be able to do what phosphorus does in biochemistry [870], but its inevitability alone does not address the problem of either obtaining it or using it as a vehicle of disequilibrium. We return to the puzzles surrounding phosphates, and some suggested solutions, in Chapter 6.

### 3.6.4.3 Equilibrium models of reactions in mixed vent fluids and seawater: abiotic organosynthesis and constraints on biological pathway energetics

Constraints on the possibilities for organosynthesis can be computed from the *equilibrium* states reachable from mixtures of end-member vent fluids and seawater. To carry out such calculations, numerical models of the two fluids are calibrated to empirical samples from representative vents, and the mixing ratio is used to compute both temperature and composition of the resulting far from equilibrium solution. Minimization of total free energy yields a model of the equilibrium to which such a mixture could relax. Equilibrium models distinguish *possible* from *impossible* spontaneous processes, and thus provide constraints on pathways in microbial metabolism. However, most current calculations exist only for free energies of formation of final compounds, so they omit some endergonic intermediate steps which in microbial metabolisms require coupling to cellular energy systems.

Numerous equilibrium models have now been computed in the work of Everett Shock, Thomas McCollom, Jan Amend, and others, for carbon speciation, or for the synthesis of major classes of biomolecules. About 400 inorganic, organic, and organometallic molecular species, ions, mineral phases, and gases have been taken into account in such models, for mixing temperatures ranging from $2-200\,°C$. Reviews of models and available thermodynamically consistent databases may be found in [21, 22].

Two results from this large body of work will illustrate the changes in our view of the requirements for life brought about by the study of vent environments.

**Dependence of speciation on $f_{O_2}$** An analysis by Shock and Schulte [746] considered the energetics of $CO_2$ reduction by $H_2$ in basaltic versus ultra-mafic alteration fluids. Two different fluid oxidation states were referenced to the two primary buffers that control $f_{O_2}$ in the crust and upper mantle. The fayalite/magnetite/quartz (FMQ) buffer of Eq. (3.2), was used to characterize upper-mantle peridotites, and the pyrite/pyrrhotite/magnetite (PPM) buffer roughly 2 log units higher (over the temperature range of interest) was used to represent more oxidized crustal rocks on Earth today. They studied temperatures ranging from $0-350\,°C$, and found that within this range large fractions of carbon in the equilibrium mixtures were incorporated into carboxylic acids, alcohols, ketones, and in some parameter ranges alkenes. Percentages of carboxylic acids ranged from 45% for fluids in equilibrium with the PPM buffer to 90% for fluids in equilibrium with the FMQ buffer.

The speciation patterns in some reactions were very sensitive to variations in $f_{O_2}$ of less than 0.5 log units over some temperature ranges. Under very reducing conditions that are plausible in models of the Hadean ocean, the calculations showed that effectively 100% of the carbon could be converted to organics at equilibrium. A summary of these and other results [22], interpreted in terms of vent conditions found on earth today, suggests that near the PPM buffer, the temperature range $100-150\,°C$ is the most favorable for formation of stable organics, and that acetic acid (including the anion acetate) is the dominant compound formed.[70]

**Energetics of synthesis of the biological amino acids** In another study, Amend and Shock [20] modeled $350\,°C$ end-member vent fluids from $21\,°N$ on the East Pacific Rise, mixing with $2\,°C$ seawater, to assess the whole-pathway free energy of formation of the biological amino acids. They showed that for mixtures at $100\,°C$, biosynthesis of 11 of the

---

[70] We will see, in a survey of metabolism in Chapter 4, that acetic acid is by any biosynthetic or evolutionary measure the most central molecule in biochemistry.

20 amino acids is exergonic [20] (Table 3). In contrast, the biosynthesis of all 20 amino acids is strongly endergonic in mixtures that produce 18 °C oxic seawater.

Building on this work, Ref. [22] shows that the net free energy of biosynthesis of the complex of amino acids in typical proteins is exergonic by up to 8 kJ/mol protein. That is, biosynthesis could provide part of the free energy of polymerization if it could be captured with low dissipation. The fact that primary biosynthesis is exergonic in mixed vent fluids may provide part of the explanation for the extraordinarily high primary productivity of vent environments [504] (see Section 3.6.1). A similar calculation, adjusted to the much lower $f_{O_2}$ in plausible models of a Hadean ocean component in mixed vent fluids, suggests that instead of only 11/20 amino acid pathways being exergonic, synthesis of most of the amino acids may have been exergonic on the early Earth.

The shift in thinking that these results require may be summarized as follows. First, *organosynthesis under geochemical conditions is often a mechanism of free energy dissipation.* The production of molecular complexity is not necessarily antithetical to entropy maximization even in equilibrium, and many core biomolecules can be formed without external inputs of chemical work. Our thinking about the origin of life must shift from treating synthesis as the key problem, to accounting for the complexity and selectivity of the synthesized compounds. Second, *many biosynthetic pathways are exergonic under anoxic and disequilibrium environmental conditions;* they only survived the rise of oxygen, which has made them endergonic, through the support of highly evolved, integrated cellular energy systems which were already in place by the time of that transition.

### 3.6.4.4 A thermodynamic account of the partitioning of vent environments

The same calculations provide an explanation for the segregation of microbial metabolisms found in samples of vent chimneys with rapid gradients of temperature and oxygen fugacity. Reference [545] showed that mixing environments at temperatures below ∼38 °C, appropriate to mesophiles, also have fluid compositions which at these temperatures favor oxidation of $H_2S$, $CH_4$, $Fe^{2+}$, and $Mn^{2+}$. In contrast, methanogenesis and reduction of sulfate or $S^0$ are favored at higher temperatures. Thus they explain quantitatively the observation that thermophilic methanogens and sulfate/$S^0$ reducers are the primary inhabitants of the hotter parts of vent chimneys, with mesophiles occupying regions that are both cooler and enriched in $O_2$.

Another study [192] considered the energetics of biosynthesis for entire proteins along the cooling and oxidizing gradients in hot springs. It was shown that protein composition changed with oxidation state and temperature, and that the energy requirements for making proteins at each stage along the flow path were tuned to the activities and temperatures of the solutes in the vent fluids. Such surface fluids are complex, hybrid environments involving oxygen, chemotrophy, and often phototrophy as well, and thus more complex to analyze than those we believe were relevant to the earliest forms of life. Nonetheless, they serve to demonstrate the very fine sensitivity with which organism biochemistry can be entrained by parameters in the geochemical environment.

### 3.6.5 Why the chemotrophic biosphere has been proposed as a model for early life

Soon after vent ecosystems were discovered, sub-seafloor hydrothermal circulation zones were proposed to be the longest continuously inhabited class of environments on Earth [161], and vent microbes were put forth as the nearest models for the last common ancestors of cellular life. In other chapters we take up the problem of framing the role that hydrothermal systems may have played in the emergence and structuring of the biosphere; it is important throughout to remember that this role exists only in the larger context of planetary disequilibria that we have sketched in this chapter. We will review, here, the arguments from planetary science, biochemistry, and phylogenetics that are the basis for existing views of the importance of vents and the status of modern chemotrophs as models for early life.

#### 3.6.5.1 Vents are probably the most nearly preserved early environments on the modern Earth

Magmatism has probably been an active process on Earth through the planet's entire history. At least since the middle Archean, convection in the upper mantle has not only delivered magmas to the surface, but has probably been organized into tectonic cycles, though how much longer this may have been the case is currently an area of uncertainty (see the discussion in Section 3.6.6.1 below). Oceans have also been continuously present, apparently since the end of the late heavy bombardment, which would have spanned the boundary between the Hadean and Archean eons. Thus it seems among the safer assumptions about planetary history to suppose that hydrothermal intrusion into newly formed crust and perhaps upper mantle, with its associated alteration chemistry and mixing, has been continuously present on Earth during the entire history of life.

The character of magmas and the composition of the uppermost layers of the mantle itself have probably changed somewhat over this time, and the chemistry of the oceans has changed significantly. Still, the major composition of upper mantle has been stable since cooling from the Moon-forming event. The volume of rock in the upper lithosphere is enormous, and the buffering capacity of crustal rocks for the chemistry of alteration fluids is considerable. Therefore vent end-member fluids have probably retained many of the same solutes (with possible exceptions mentioned in Section 3.6.6.3 below) through most of Earth's history. With a few changes in mixing conditions with seawater, which can be accounted for in calculations and laboratory models, hydrothermal fluids and the structures they precipitate are probably the most conserved chemically active subsystems on Earth [668]. This is especially true for hotter fluids that even on the present Earth have low $f_{O_2}$.

The existence of conserved environments is important because all life on Earth has been subjected to evolution over the same 4+ billion years. We cannot expect that the modern world contains *ancient* organisms. We can, however, recognize that much of the stress driving evolutionary change comes from environmental variability, and also that simpler organisms with higher levels of gene transfer can be less able to maintain standing variation in populations that share an environment, so they are more likely to be closely entrained by their environmental boundary conditions. In highly conserved environments, we can expect more *conservative* organisms. The particularities of vent conservation are therefore key to interpreting the metabolic capabilities of their biota.

### 3.6.5.2 Organosynthesis is common to vent environments and metabolic chemistry

Section 3.6.4 reviewed thermodynamic analyses showing that organosynthesis is energetically favored in far from equilibrium mixed vent fluids. Numerous experiments have now exhibited mechanisms enabling this organosynthesis to proceed, in the presence of catalysts that are plausible or are known to be produced in mineral environments. Therefore, a hallmark of biological activity – the synthesis of organic compounds – is not a process that distinguishes life from non-life, as was believed for centuries. Rather, it is a process shared at some level by biochemistry and geochemistry. In Chapter 4 we review pathways of carbon fixation and core metabolism in more detail, and argue that the directions of synthesis and even some of the mechanisms are likely to be continuous from the mineral world to biological catalysts and cofactors.

### 3.6.5.3 Chemotrophic pathways correspond to the core metabolic pathways

We also show in Chapter 4 that metabolism has a strikingly modular and hierarchical network topology. In the core of this network are pathways carrying out the exchange of one-carbon compounds between organic and inorganic species, which include as intermediates the starting compounds for synthesis of all major classes of biomolecules. A variety of arguments from biochemistry and evolutionary reconstruction all indicate that the network of these core pathways is the most ancient and conserved layer within metabolism. In a large and ancient class of chemotrophic organisms, the core pathways operate in the direction of *incorporating* one-carbon and one-nitrogen compounds. Thus carbon fixation and *de novo* biosynthesis are integral to the deepest and oldest layers in the metabolic network.

### 3.6.5.4 Phylogenies reconstruct heat-loving, and sometimes acid-loving, reducing chemotrophs across the whole deep tree

It has been recognized since the mid-1990s that deep branches in the phylogenetic tree of life are disproportionally populated by heat-loving, chemotrophic, and anaerobic or micro-aerophilic organisms. Some of these are also acid loving or salt tolerant. As methods of phylogenetic reconstruction have become more sophisticated, and able to take into account gene transfer and to recognize likely directions of evolutionary change, it has become possible to reconstruct ancestral states with some confidence. We believe it has become consensus that chemotrophy and some degree of thermophily were the *ancestral* traits in most of these deep branches, and thus of the Last Universal Common Ancestor (LUCA) of life.[71]

---

[71] Anna-Louise Reysenbach and Everett Shock [668] make an important observation that the ability to *form* a deep tree of thermophilic lineages attests not only to the existence of early environments that could host these organisms, but to the continuous inhabitation of such environments.

### *3.6.5.5 Brines and the odd presence of halophiles in many deep branches*

Phylogenetic reconstructions also show, interspersed within deeply rooted clades, a sur-prising frequency of halophile lineages [612].[72] Though not as universal and not as deeply branching as thermophiles, reductive chemoautotrophs, or even acidophiles, the frequency of halophiles has still seemed a curiosity. These are organisms that live not only at the salin-ity of seawater, but at much higher salinities reaching as far as the precipitation equilibrium with solid salt. Their environments seem marginal for life, and we might have expected that such specialists would only arise infrequently as a late evolutionary optimization. They are not common enough in the deep tree to support a persuasive argument that life emerged in concentrated salt conditions.

The suggestion that the vent biota we see are only an echo of a much more extensive subsurface biosphere, and the evidence that a majority of both basalt-hosted and peridotite-hosted vents with reaction-zone temperatures above 400 °C partition distilled vapors and deep brines through either supercritical or subcritical phase separation (boiling), may pro-vide part of the answer [424]. Large quantities of brines may accumulate at the base of the sheeted dike sequence (see Figure 3.9). Intermittent disturbances such as diking/eruptive events that alter flow paths may lead to intervals in which these brines rather than low-salinity phases dominate vent effluents. The *generic* presence of brines in environments where the first life diversified would make these among the earliest environments to be colonized by variant physiologies. It would make halophilicity one of the competences present in many ancient microbial communities, and thus likely to be preserved in the phylogenetic record of the earliest branchings.

### *3.6.6 Differences between hydrothermal systems today and those in the Archean*

Before detailed knowledge about present hydrothermal systems can be used to understand the emergence and early evolution of life, we must take into account any major differences in composition and dynamics of the atmosphere, oceans, or mantle on the early Earth that could have led to different vent conditions.[73] The most important differences that must be understood in order to make quantitative models of vent environments concern the presence or absence of tectonics, the gas profile of the atmosphere, and the progress of continent formation.

---

[72] Clades that comprise entirely or almost entirely halophiles include the *Halobacteriales* (Euryarchaeota), *Halomonadaceae* (gamma-proteo), *Halanaerobiales* (Firmicutes). Bacterioidetes also have some members. The Firmicutes are the deepest branching of these major clades, but for purpose of this discussion the existence of a group such as the *Halanaerobiales* is still a relatively early origin to be preserved as a clade.

[73] Very early in Earth's history, changes could have been so large and occurred so quickly that they confront us almost with a problem of modeling from first principles. It has been argued that in the aftermath of the Moon-forming event, much of the ferric iron that is now present in the mantle (in silicate phases), was produced by the reduction of $4.5 \times 10^{22}$ mol of water to $H_2$ [760]. A further mantle component of siderophile elements (principally platinum group elements often termed a "veneer" because they were not partitioned into the core to the extent expected), require another $7.5 \times 10^{22}$ mol of water. The combined $H_2$ released from these reductions could have contributed 50 bars of hydrogen to the atmosphere, in the early Hadean. The competition between production and escape of such a large hydrogen atmosphere would have led to a very dynamic era in surface chemistry, especially if the deposition of the oceans was occurring around the same time. These changes are so extreme and require so many assumptions to model that we are not qualified to address them. We will suppose, reassured by the affinity between vent chemistry and biochemistry that has persisted even into the present, that life emerged in a later, more stable era that can be extrapolated backward from known conditions with some confidence.

### 3.6.6.1 Uncertainty about when plate tectonics began and the forms it may have had historically

Given the centrality of the processes that form new crust, to redox exchange between the lithosphere and atmosphere, an important area of uncertainty about the early Earth is when plate tectonics began, and the range of forms it may have taken over time. Although this is a subject of intense interest in geology and geophysics, and has received significant effort in the interpretation of historical signatures and also through modeling, it turns out to be a very hard problem and most of the central questions are still areas of significant imprecision and often significant debate even regarding major qualitative alternatives.

The absence of plate tectonics does not imply absence of magmatism or absence of delivery of separated mantle material to the surface to form new crust. Therefore it does not imply absence of redox connections to the hydrosphere and atmosphere. It would, however, entail the loss of spreading centers and subduction zones with their resulting arc volcanism, reducing both the volume and the diversity of magmatic systems and rock/water interactions. It would thereby diminish the contact of reduced metal species with seawater, and the accessibility of peridotites to water incursion and the resulting alterations of ultramafic minerals.

Most interpretations of geological signatures place the origination of plate tectonics some time in the early to middle Archean, though the geochemistry of ancient zircons, as well as modeling considerations, do not reject the possibility that the Earth has had organized tectonics since the Hadean [365]. The distinction is of major importance, since life had probably reached a quite complex cellular stage of organization by the middle Archean, and the question whether the major transitions in organosynthesis and energetics leading to cells had occurred in tectonic or non-tectonic environments affects the processes and model systems we investigate to try to reproduce these transitions.

Some conclusions that currently seem safe to draw are the following [453]:

- the mantle in the early Archean was likely $\sim$200 K hotter than at present and has been cooling steadily since then;[74]
- mid-ocean ridge spreading rates, and plate tectonic motion in general, were slower on the early Earth than they are today, and the speed of plate motion has been increasing over time [9, 453]; through the same anticorrelation between spreading rate and crust thickness that distinguishes spreading centers today, the early Earth's crust was thicker; common estimates are 15–20 km thickness (in comparison to the 7–8 km typical of modern ocean-bottom crust);
- at least by the late Archean, the Earth is believed to have hosted more but smaller continental cratons, and cumulative plate boundary length is believed to have been longer;
- the Earth has had a liquid ocean since some time in the Hadean. The oldest currently known zircons, indicating the presence of liquid water, date to 4.4 Ga [365].

---

[74] The total heat flux through the Earth's surface may or may not have been roughly constant over that interval.

Hard to constrain factors that make it difficult to deduce from these general trends whether mantle convection was more or less vigorous in the past (despite its higher temperature), and for how long it may have been organized by tectonic motion, include the volume of oceans, the degree of hydration of the upper mantle, the existence or extent of continental craton formation, and the relative buoyancy of oceanic lithosphere and continental lithosphere (whenever the latter came into existence). It is consistent with current knowledge that, although the Hadean and early Archean crust were thicker and thus stiffer, the mantle was much drier, providing the needed mantle viscosity to drive tectonics over the Earth's entire history.[75]

### 3.6.6.2 Stable sulfur isotope signatures: lack of $O_2$ in the early Archean and consequences for nitrogen oxides

The composition of the Archean atmosphere is difficult to constrain directly, and all interpretations rely to some degree on chemical modeling. However, a set of extremely useful signatures, which place multiple constraints on atmospheric composition and lead to several conclusions that are qualitatively insensitive to most aspects of modeling ambiguity, come from the stable sulfur isotope chemistry in the rock record. The richness of the sulfur system, both chemically and isotopically, provides fortuitous direct insight into properties of atmospheric structure and dynamics that would otherwise be very difficult to reconstruct from other evidence.

Sulfur occupies a kind of Goldilocks zone in the periodic table for preserving traces of atmospheric chemistry through their effects on isotope separation, because it combines some properties of heavy elements with other properties of lighter elements. The key characteristics that contribute to a multidimensional isotopic signature are the following.

1. Sulfur, positioned late in the third period of the periodic table, has a large enough atomic number to possess a range of oxidation states, giving it a rich chemistry of redox reactions, particularly involving bonds with oxygen and hydrogen, and to a lesser extent with carbon. This is the same richness of oxidation states and small-molecular compounds that makes sulfur a key intermediate in many microbial redox systems.
2. Sulfur is light enough that many of these compounds are atmospheric gases, including $SO_2$, $SO_3$, $H_2SO_4$, OCS, and $H_2S$. Several others are aerosols, such as $S_8$ and some organosulfur compounds. The networks of gas-phase chemical reactions in which these compounds are interconverted are sensitive to the intensity and spectrum of ultraviolet radiation, the presence of other oxidants (particularly free radicals such as OH or O), and more general properties such as pressure, that mediate relaxation of excited states via collisions.

---

[75] Jun Korenaga has argued [453] that the subsequent recycling of oceanic crust has been a process of progressive hydration of the asthenosphere and lowering of its viscosity, which has compensated for the cooling of the mantle to keep the Earth within a parameter range that enables tectonics into the present, but we understand this topic to be an area of active debate. The interested reader is referred to the specialist literature for much more extensive discussion of these questions.

3. Sulfur is a heavy enough nucleus to possess four stable isotopes: $^{32}$S, $^{33}$S, $^{34}$S, and $^{36}$S (in fact it is the lightest element to possess four or more; the next is calcium with five). The most abundant sulfur isotope is $^{32}$S. The existence of one odd-numbered and two even-numbered minor isotopes provides both a range of mass-shift effects and also the ability to control for spin-dependent properties. Differences in nuclear mass and spin lead to small but important variations in the radiative absorption and chemical reaction properties of compounds with different isotopes, known as *isotopologues*.

4. Many of the gases or aerosols that participate in atmospheric chemistry are also soluble in rain water and seawater, and can be mineralized, providing several parallel **exit channels** for isotopically fractionated sulfur from the atmosphere to the mineral systems that make up the rock record. Key compounds include mineral sulfates (particularly with Ca and Ba) and sulfides such as pyrite (FeS$_2$). Sulfates can be captured by sedimentation or hydrothermally driven precipitation, as described above in Section 3.5.4.2, and pyrites can be formed *de novo* from hydrothermal alteration of Fe-S minerals with H$_2$S.

5. Sulfur in multiple oxidation states, dissolved in seawater or seawater-derived hydrothermal fluids, can also provide sources of redox energy for microbial metabolisms, including both sulfate reduction and sulfide oxidation. (The former is by far the more likely to be important for questions about the early Archean or Hadean eons.) Microbial metabolisms can mix sulfur oxidation states across reservoirs, as well as adding their own fractionation signatures. Whether these are important biosignatures from the early Earth, or additional sources of model complication and ambiguity, depends on the question of interest. All sulfur isotope samples from the Archean were produced during an era influenced by microbial metabolism. Therefore, as a practical matter, controlling for metabolic signatures is an essential part of calibrating atmospheric and hydrothermal models, before these can be used in extrapolations to earlier, possibly prebiotic eras.

Elements which possess three or more stable isotopes enable a distinction among mechanisms that partition the isotopes, which cannot be made from the distribution of any pair of isotopes alone.

Both equilibrium distributions and reaction rates are potentially sensitive to the atomic masses of isotopes, the former through the dependence of zero-point energies on atomic mass and the latter through the relation of zero-point to transition-state energies. The most basic mass-dependencies, which are common to both abiotic and biological reactions, lead to isotope selectivities known as **mass-dependent (MD)** fractionation effects. The signature of MD fractionation is that the degree of fractionation is (to leading order in small quantities) proportional to the mass difference between the isotopes being compared [383]. Thus the enrichment or depletion of $^{33}$S relative to $^{32}$S is roughly half the comparable shift between $^{34}$S and $^{32}$S, while the shift between $^{36}$S and $^{32}$S is roughly double that between $^{34}$S and $^{32}$S (see Box 3.3).

**Box 3.3    Sulfur isotope fractionation**

The enrichment or depletion of a minor isotope (relative to the major isotope, which furnishes a background) is denoted with small-$\delta$ notation. Isotope shifts are often on the order of several parts per thousand, and are therefore multiplied by 1000 to yield a number in permil (‰).

For sulfur, the major isotope is $^{32}S$, and any of $^nS$, $n \in \{33, 34, 36\}$ may be the minor isotope. To obtain $\delta^nS$, the ratio of the mole fractions of minor and major isotopes in a sample is compared to their ratio in a reference mixture. For geological sulfur samples the reference used is often the troilite phase of the Canyon Diablo meteorite (abbreviated CDT) [383], a proxy for the sulfur composition of both the modern and the primitive mantle, which has been the source of atmospheric $SO_2$ and $H_2S$ through volcanic outgassing throughout the Earth's history.

Letting $\mu^nS$ denote the mole fraction of $^nS$ in either the sample or the reference, the enrichment in permil is given by

$$\delta^nS \equiv \left[ \left( \mu^nS/\mu^{32}S \right)_{\text{sample}} / \left( \mu^nS/\mu^{32}S \right)_{\text{ref}} - 1 \right] 1000 \ (\text{‰}) . \tag{3.28}$$

In samples that show **mass-dependent fractionation**, the isotope shifts fall very close to the linear relations[a]

$$\delta^{33}S \underset{\text{MD}}{\approx} 0.515 \, (\pm 0.0025) \times \delta^{34}S$$
$$\delta^{36}S \underset{\text{MD}}{\approx} 1.91 \, (\pm 0.01) \times \delta^{34}S. \tag{3.29}$$

Residual isotopic enrichments, relative to the MD reference, are denoted with capital-$\Delta$ (and also measured permil). Residual enrichment may be referenced to the MD line in a variety of ways that are equivalent for small $\delta^nS$ values; a standard form of the projection used in many papers is [239]

$$\Delta^{33}S \equiv \delta^{33}S - 1000 \left[ \left( 1 + \delta^{34}S/1000 \right)^{0.515} - 1 \right]$$
$$\Delta^{36}S \equiv \delta^{36}S - 1000 \left[ \left( 1 + \delta^{34}S/1000 \right)^{1.91} - 1 \right]. \tag{3.30}$$

Several strong and consistent patterns of correlation among $\delta^nS$ and $\Delta^nS$ values serve to separate MD and MIF signatures in samples, and to give specific characterizations of the MIF signatures. For example, plots of $\Delta^{33}S$ versus $\delta^{34}S$ for the pre- and post-2.45 Ga samples from Figure 3.10 show nearly orthogonal patterns

---

[a] These results were first reported for organic matter from a variety of carbonaceous meteorites [383], but are now recognized to characterize a much wider range of abiotic and also biological processes.

---

**Box 3.3  Sulfur isotope fractionation (cont.)**

of variation (see [238], Figure 2). They also delimit the range of ambiguity in samples with small variation to $\Delta^{33}S \in [-0.2, -0.4]\,‰ \times \delta^{34}S \in [-8, 10]\,‰$.[b] A second, important correlation found in MIF sulfur samples worldwide is that $\Delta^{33}S \approx -0.9\,\Delta^{36}S$, indicating that the source of $\Delta^{33}S$ is not a hyperfine interaction (which would produce $\Delta^{36}S = 0$ [237]).

---

[b] A somewhat different pattern of variation, still independent but no longer orthogonal, was reported subsequently for samples from the Dresser formation in Northwestern Australia [815], showing much larger values of $\Delta^{33}S \in [-1, 4]\,‰$ and a correlation $\Delta^{33}S \approx 0.56\,(\pm 0.37)\,\Delta^{36}S$.

---

A class of independent and more idiosyncratic isotope selectivies arise from differences in level spacing, spin-orbit coupling, and level crossing, as these affect photolysis, photoexcitation, and the subsequent relaxation dynamics of products of radiative interactions. These, although they are also related to nuclear mass as well as spin, do not produce simple proportionalities with isotope mass differences, and so are termed "non-mass-dependent" or more commonly (and more confusingly) **mass-independent fractionation (MIF)** effects [239].

Neither for MD nor MIF isotope shifts do the relative dependencies on atomic mass difference stipulate the absolute magnitude of isotope selectivity, which depend on molecular and reaction context. Absolute magnitudes of fractionation can span a wide range, and the mass shifts produced by many mechanisms overlap. The sulfur in a given pool may have drawn from several different reaction channels in parallel, or may have been cycled through the same reactions an unknown number of times. (The latter becomes important particularly in the case of microbially mediated redox conversions.) Therefore the absolute magnitude of enrichment or depletion of any one minor isotope in a sample is not likely to be discriminating of its sources.

However, the proportionalities of MD shifts cause all of the sulfur pools that have been influenced primarily by these to lie along a *ray* in the multidimensional plot of isotope fractionations. Such linear relations may be projected out of mass shift data. The residual deviations due to MIF effects, while more complex, are still sufficiently distinct that many may be interpreted as signatures of particular classes of mechanisms. The more dimensions are provided by multiple isotope fractionation signatures, the more distinct mechanisms may in principle be resolved.

Box 3.3 summarizes basic notation and relations that are fundamental to the interpretation of the sulfur isotope signature from the early Earth. We return in Chapter 6 to consider some of the detailed constraints these signatures place on the Archean atmosphere, and the problems of model calibration and extrapolation needed to apply them to the earliest organosynthesis relevant to the origin of life. The remainder of this section concerns the first and still the most robust conclusion drawn from sulfur isotope fractionation, which

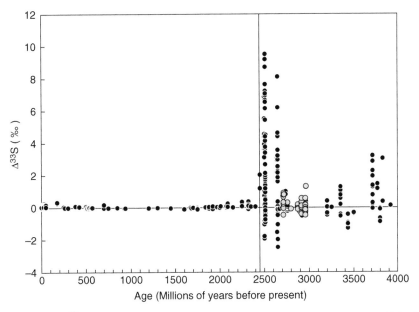

Figure 3.10 The $\Delta^{33}$S signature of mass-independent fractionation of sulfates and sulfides in the Archean (from [238]). The abundance and strength of the MIF sulfur signal before 2.45 Ga stands in sharp contrast to its almost entire elimination since that time.

was a bound on $f_{O_2}$ in Earth's atmosphere before 2.45 Ga, far below present levels, and its consequences for other gases.

### *MIF sulfur signatures in Archean sulfates and sulfides*

James Farquhar *et al.* [237] first discovered the existence of large and consistent signals of MIF sulfur partitioning in the residual mass shift denoted $\Delta^{33}$S from Archean sulfates and sulfides (shown in Figure 3.10), and interpreted these as strong evidence of anoxia in the Archean atmosphere before about 2.45 Ga. The general interpretation of this signature today remains the one put forth in [237], though many details of the mechanism are uncertain and appear much more complex to interpret than the original analysis suggested. Dissociation of atmospheric $SO_2$ due to UV absorption[76] results in a partition between an oxidized sulfur pool derived from $SO_3$, which combines with water to exit the atmosphere as $H_2SO_4$, and a complementary reduced sulfur pool derived from SO, which probably exits the atmosphere as sulfide aerosols such as $S_8$.

In a reducing atmosphere, broadband UV radiation propagates into the troposphere where most atmospheric $SO_2$ is found. Both photolysis and photoexcitation of $SO_2$ occur,

---

[76] Dissociation of $SO_2$ may result directly from photolysis, following absorption of UV photons in the band 180–220 nm, or indirectly following photoexcitation by UV photons in the band 250–350 nm [876]. Whether photolysis or photoexcitation is the primary or exclusive source of $SO_2$ dissociation in planetary atmospheres remains an important open question. Its answer depends on modeling of atmospheric chemistry and the many possible factors influencing the solar atmospheric radiation spectrum, termed the *actinic flux*.

and the oxidized and reduced sulfur pools remain segregated until each is sequestered in mineral form through independent processes.[77] The Archean rock record suggests that the oxidized atmospheric sulfur pool was depleted in $^{33}$S, while the reduced sulfur pool was $^{33}$S enriched.[78]

With the **rise of oxygen** in the atmosphere, the formation of an ozone layer shielded the troposphere and hence most atmospheric $SO_2$ from UV irradiation. Moreover, reactions with oxygen species re-oxidized any products of radiative fractionation back to $SO_2$, homogenizing the atmospheric sulfate pool before it could be sequestered and erasing the fractionation signature. Thus, an era from almost 3.9–2.45 Ga, from which isotopically fractionated sulfur is continuously observed in the rock record (though with varying absolute magnitudes of fractionation), abruptly ended with the rise of atmospheric oxygen.[79] Quantitative bounds on solar flux and re-oxidation rate currently place upper limits on $f_{O_2}$ in the Archean at $10^{-5}$ of present atmospheric levels (PAL) [627], a much stronger constraint than could be placed by other geochemical proxies [139]. The region of uncertainty, however, is large: the plausible range is currently estimated to be anywhere between $10^{-12}$ and $10^{-5}$ PAL.

It is generally believed that any mechanisms capable of accounting for the MIF sulfur signatures in Archean rocks will originate principally in $SO_2$ photolysis or photoexcitation, and thus that the mere existence of large $\Delta^{33}$S excursions is sufficient to imply anoxia in the Archean before 2.45 Ga. Other MIF signatures, involving correlation of $\Delta^{36}$S and $\Delta^{33}$S, as well as correlations of $\Delta^{33}$S with $\delta^{34}$S in some deposits, potentially place further informative constraints on the early Earth's atmospheric composition. We return to discuss these in Section 6.2.2.

Low $f_{O_2}$ in the Archean has several consequences for atmosphere and hydrosphere redox dynamics, with ramifications for the viable forms of life. A low-oxygen atmosphere, transparent to UV radiation well into the troposphere, is favorable for the reduction of $CO_2$ to CO or even to organic carbon species, as noted in Section 3.6.4.3. At the same time, however, low $f_{O_2}$ depletes mixed vent/seawater fluids of terminal electron acceptors that are an important source of energy today. Not only molecular oxygen, but also nitrate and nitrite,

---

[77] Evidence of mixing of these pools by microbial sulfate reduction appears in some samples as early as 3.4 Ga, particularly pyrite samples included in barite veins [237], but its overall importance as a confound in the interpretation of isotope patterns is suggested to be isolated and sporadic.

[78] The original photochemistry experiments of [237] yielded mass shifts with the same signs as those in the Archean rock record, using excitation by a narrow-band ArF excimer laser at 193 nm, and this was taken as evidence for a mechanistic connection. Subsequent work [814] has indicated that products from a narrow-band source are not representative of products from more realistic spectra, and that identification of a full mechanistic connection and account of the sign of isotope shifts is likely to be a considerably more complex problem.

[79] Today stable isotope signatures primarily reside in fractionation of $^{16}$O, $^{17}$O, and $^{18}$O, with small residual MIF sulfur signatures found in glacial sulfate deposits associated with pyroclastic volcanic eruptions which deliver $SO_2$ to the stratosphere [339, 629]. These fractionated sulfates share signatures associated with $SO_2$ photodissociation in Archean sediments and precipitate minerals, including correlation of $\Delta^{33}$S with $\delta^{34}$S, and correlation of $\Delta^{36}$S with $\Delta^{33}$S. Elaborate models, representing the altitude structure of actinic spectra as well as absence of water and associated OH chemistry in the stratosphere, purport to correctly account for the signs, correlations, and also time course of these signatures from photodissociation initiated by photoexcitation of $SO_2$ by UV wavelengths in the band 250–320 nm. Although such intermittent and small modern signatures carry little information about the structure of the atmosphere below the stratosphere, the ability to calibrate models against glacial sulfur deposits has been important in efforts to identify primary phenomena responsible for MIF sulfur in known atmospheric conditions.

were probably present in seawater only in very low concentrations. Some NO is formed in atmospheric processes from $N_2$ and $CO_2$ (see Section 6.3.3), but the yields are low, and in seawater with reduced iron it is expected to be rapidly converted to ammonia [784].[80]

### 3.6.6.3 Potentially much more carbon dioxide in the atmosphere and oceans

Section 3.3.1.4 noted that the faint young Sun requires a more intense atmospheric greenhouse in the early Archean than the present day, to account for the existence of liquid water throughout the Earth's history. The main contributions to this greenhouse are still debated, but independent evidence from the rock record suggests a much larger partial pressure of $CO_2$ in the early atmosphere than exists today. A large increase in atmospheric $CO_2$ would have changed ocean pH and alkalinity, and perhaps the chemistry of alteration fluids in systems with high water/rock ratios. It could also have affected the concentrations of reactive carbon species, which would be of interest for early organosynthesis.

Evidence about $CO_2$ content in the early atmosphere comes from the observation of *extensive and deep carbonatization of Archean basalts* [740]. The Pilbara Granite-Greenstone Terrane in Western Australia, a region of exposed middle-Archean basalts, shows carbonatization (mineral volume fractions) between 20 and 40% to depths as great as 2 km, and samples in the range 10–20% to depths slightly greater than 3 km. This extensive carbonatization is interpreted to indicate very large-scale precipitation of $CaCO_3$ from hydrothermal fluids in the Archean. Estimates of total carbonate in the oceanic crust, together with assumptions about rates of crust production, lead to estimates that carbonate deposition in the early Archean occurred at rates about two orders of magnitude higher than present deposition rates. Analysis of the sedimentary record suggests concentrations of $CO_2$ well into the early Archean that were 100–1000 times present levels, and possibly as high as several bars [498], consistent with the estimates from mantle carbonatization.

The drawing down of atmospheric and oceanic $CO_2$ is associated with the increase of continental surface area, and a displacement of hydrothermal subsurface precipitation by ocean-bottom sedimentary precipitation as the main removal mechanism. Continental weathering delivers $Ca^{2+}$ to seawater which leads to calcite precipitation in sediments (and under biotic activity, to synthesis of aragonite as well as calcite biominerals). Evidence from $NaHCO_3$ evaporites [498] has been interpreted as suggesting that surface temperatures $\sim 70\,^{\circ}C$ existed in the interval from 3.5–3.2 Ga,[81] accompanied by very aggressive weathering of land surfaces, increasing the sensitivity of ocean chemistry to land surface area beyond what would be true under present climate conditions in the absence of biologically mediated weathering. Sediment precipitation is believed to have depleted atmospheric $CO_2$ and with it the deposition in mid-ocean ridge hydrothermal systems.

---

[80] The oxidation of ammonia to nitrite and nitrate in the oceans today is performed by aerobic nitrifying bacteria. Paul Falkowski [235, 236] argues on this basis that the anaerobic denitrifying bacteria, which use oxidized nitrogen species as terminal electron acceptors, could only have evolved after the advent of oxygen from oxygenic photosynthesis.

[81] See cautions about this conclusion in [418].

The most developed current atmosphere models for the Archean [418, 432] indicate a composition of predominantly $CO_2$ and $N_2$, with minor $H_2$, and reduced C, N, and S gases. The total $CO_2$ in the early Archean is difficult to constrain, and geochemical $CO_2$ barometers give estimates that, while generally high, easily vary over factors of 100–1000. Greenhouse constraints do not help much because the early-Earth greenhouse is likely to have been contributed to significantly by methane as well as $CO_2$ [251, 627, 697]. Models have therefore considered a range of $CO_2$ concentration from 2500 ppmv $CO_2$ (about six times present long-term averages) to as high as 10 bars [416], with 2500 ppmv a common calibration point for the late Archean (2.8 Ga).[82]

It is likely that high atmospheric $CO_2$ would have led to moderately acidic pH in the oceans due to dissociation in the $H_2CO_3/HCO_3^-$ system. This is a contrast from the mildly basic pH of 8.2 in the Phanerozoic, our current eon.[83] Although carbonic acid would have lowered pH, the presence of the bicarbonate/carbonate buffer would have contributed high alkalinity to this system.

### 3.6.6.4 Changes in mineral alteration and vent chemistry of fluids that start with high $CO_2$ concentrations

It has recently been argued [739] that high $CO_2$ concentration in the Archean oceans could have qualitatively changed the alteration chemistry of hydrothermal systems, causing the end-member fluids in basalt-hosted systems to more nearly resemble the high-pH fluids of the present-day Lost City field, and leading to precipitation of quite different suites of minerals. The model of Shibuya *et al.* [739] predicts an *in situ* pH $> 10$ for alteration of Archean basalts by high-$CO_2$ seawater (to be compared with pH $\sim 5$ for modern basalt-hosted vents), because precipitation of $CaCO_3$ keeps $aCa^{2+}$ low. These model predictions are higher than experimental values obtained by the authors, but the experimental pH values are still high enough to drive down dissolved $Fe^{2+}$ concentrations (by precipitation of iron carbonates), and to increase solubility of silica in vent fluids, leading to silicate rather than sulfide chimney formation. Whether actual Archean vent fluids would have had such high pH depends on water/rock ratios and residence times. The reaction path for water on Earth today may reach to depths of 8 km, and on an early Earth with crustal thickness of 15–20 km this path could have been longer and residence times greater.

### 3.6.6.5 CO, HCN, and $H_2S$ were reservoirs of reducing power and reactivity

One of the intriguing possibilities created by an atmosphere with large concentrations of $CO_2$ and possibly methane, no $O_2$ and moderate $H_2$, in the context of a UV-active Sun, is the accumulation of CO [432] and HCN in the atmosphere [906]. These compounds are significant sources of free energy, and are reactive in many environments. CO readily undergoes addition reactions under reducing conditions at mineral surfaces [31], and

---

[82] Since the major decline in $CO_2$ is believed to have occurred around the interval 3.2–2.7 Ga [498], this calibration for the late Archean is consistent with much higher levels in the interval 3.8–3.5 Ga.

[83] Acidification in the modern era, sometimes termed the *Anthropocene*, has led to a reduction to pH $\approx 8.1$ in the modern oceans.

HCN spontaneously polymerizes [617], with subsequent hydrolysis of the polymer to produce a range of complex partially oxidized carbon species. Both compounds have therefore been of great interest as possible inputs to the organosynthesis leading to the emergence of life.

CO and HCN can accumulate in atmospheres because their triple bonds are too strong to be photolyzed by the solar UV spectrum. When oxygen radical, $O_2$, or $H_2O_2$ are absent, or when sufficient $H_2$ is present to scavenge O or OH radicals, CO and HCN can be produced faster than they recombine. Recent atmosphere/ocean/ecosystem models by Kharecha *et al.* [432], with $CO_2$ concentration fixed at 2500 ppmv (based on volcanic rates of delivery), show large sensitivity to $H_2$ concentration. At total $f_{H_2} = 200$ ppmv, CO concentration in the upper atmosphere is computed to be ~one fifth that of $CO_2$, while at $f_{H_2} = 800$ ppmv, the concentration of CO becomes nearly two orders of magnitude *larger* than that of $CO_2$. With insufficient biotic sinks for CO, or higher $CO_2$ concentration, the models may even show CO runaway. HCN concentrations in models (which are regulated by the concentration of $CH_4$) are several orders of magnitude lower, but are still large enough to make this a possibly important species for early surface organic chemistry. We return in Section 6.2 to possible roles for CO and HCN as important mediators of both free energy and reactivity between the atmosphere and the subsurface.

### 3.6.7 Feedback from the biosphere to surface mineralogy

In this chapter we have emphasized a one-way flow of influence, from stars and mantle convection to atmospheres, oceans, and magmas, and from these to the chemical raw materials that support life. However, as life became an autonomous planetary subsystem, it increasingly fed back to influence composition and dynamics in the abiotic geospheres. Just as the evolution of much of the biosphere's complexity has been *coevolution* among species in ecological contexts, the mineralogy of the lithosphere has been a component of this context and a participant in its evolution.

Minerals have figured centrally throughout our discussion of the energetics and catalysis of the Earth's major chemical energy flows. Thus it is important to appreciate that the mineral diversity present on Earth today has accumulated over the planet's history, and several stages in that diversification are dependent on conditions particular to the Earth, including the presence of life. Robert Hazen and colleagues [345, 348] have estimated that, of the ~4500 known minerals on Earth, perhaps 3000 depend in some way on the prior existence of a biosphere. For many of these the dependence is through the action of molecular oxygen, so they are contingent on oxygenic photosynthesis, a late and complex evolutionary innovation.

The remaining ~1500 minerals, which created the conditions for the emergence of life, may be grouped into stages of mineral diversification, occurring progressively in the formation of the stellar disk and the planet. The first ~250 are associated with the formation of the stellar disk and the Sun, and are common among meteorites, small asteroids, and moons. The formation and differentiation of larger rocky planets expands this number to ~350 in volatile-poor conditions, or ~500 for planets with water. Hazen [343] lists roughly 420 rock-forming minerals that would have been volumetrically important and widely distributed on the Hadean Earth.[84]

Remelting of initial basaltic crusts on large planets with significant sustained internal heating concentrates rare elements into distinct phases, and may produce a further ~500 minerals, but only the establishment of fully developed plate tectonics with subduction of water and the resulting changes in melting and volcanism, as well as emplacement of salts and sulfides, leads to the inventory of ~1500 minerals that were probably present on Earth by the early Archean. In this respect, even without life, the Earth would have been distinctive among the inner planets of the solar system.

### 3.7  Expectations about the nature of life

The theme of this chapter has been the way barriers structure the chemical dynamics of the abiotic geospheres. Barriers to energy flow and equilibration within a continually aging universe lead to concentration of stresses. The paths of least resistance over those barriers become the scaffolding on which new levels of organization form. Stress paths and focusing centers, however – whether in physical space or in a space of chemicals and processes – are the epitome of complex patterns. The details of the state spaces that host them and the processes that carry them matter. From knowledge of those details, much about the stress paths can be predicted from conventional reductionist science, showing how although they are complex, they are neither accidental nor surprising.

Some level of abstraction away from the details is also possible. With incomplete knowledge of states and mechanims, we may be unable to predict the particular paths that dominate redox flow on a young planet, but if it is driven sufficiently far from equilibrium chemically and thermally, we may anticipate that *some* concentrating relaxation path will form.[85]

We think this characterization of the order in geochemistry is also the correct one for the nature of the living state. The beginnings of the continuation can already be seen in the relation of the chemotrophic biosphere to its planetary context. Life *conforms* to the points of highest stress and paths of least resistance, within the chemical and geological scaffold of redox potentials and flows. It is sensitively dependent on the particularity and locality of its boundary conditions, yet those dependencies are no more unusual, and no less robust, than the dependency of other forms of order on the focusing centers that create them.

---

[84] The minerals in Box 3.1 are all in this set, as are most of the clays invoked as buffers of $aSiO_2$ and pH in Sections 3.5.3–3.5.5 (with tremolite as a possible exception [343]).

[85] As the discussion of tectonics shows, however, the range of activation energies in a planetary context can be enormous. Without knowledge of particular mechanisms, we may be very limited in predicting quantitatively how much stress is "sufficiently far" from equilibrium to create a self-reinforcing path.

The order begun in geochemistry continues in the deep history and universal elements of core metabolism. To understand the system of rules that erect biochemistry as an invariant scaffolding for life, from a few anchors in organometallic oxidation/reduction chemistry, and the way these have been expressed in evolution, we must study the network topology, elementary functions, variability, and history of metabolism in the microbial world. These topics are taken up in Chapter 4.

# 4

# The architecture and evolution of the metabolic substrate

Life at the ecosystem level is characterized by a nearly universal chart of core metabolism, comprising a few alternative paths for carbon fixation together with anabolic pathways for a standard set of small metabolites including cofactors and the three major classes of monomers (amino acids, nucleotides, and sugars). The universal chart is small, but serves as a foundation for all higher-level biochemical diversity. The patterns and functions in metabolism express a chemical logic of constraints and rules of assembly that make it in many ways a simple system, relying on surprisingly few fundamental mechanisms. The network architecture of metabolic reactions decomposes into modules and layers. Important motifs include autocatalytic loops either within or across hierarchical layers, repeated sequences of functional-group rearrangements, and distinctive and conserved dependence on certain catalysts – especially those involving metal reaction centers. By bringing together an analysis of functional dependencies, with comparative (phylogenetic) analysis where pathway variants exist, we argue that the layers and modules correspond to a historical sequence of accretions. For the known carbon fixation pathways we can propose a tree-like sequence of elaborations from an explicit root phenotype. Although the comparative analysis is carried out within the era of genomic evolution, many of the central organizing motifs originate in low-level chemistry, topology, or feedback dynamics, and would be expected to have constrained geochemical organization before the first cells. Chapter 5 will show that many motifs of apparently chemical origin in metabolism are recapitulated in higher-level systems such as the genetic code.

## 4.1 Metabolism between geochemistry and history

The layer of life that converts material and energy from the abiotic geospheres into the biomass that carries out all living processes is metabolism. It is both the interface layer that anchors life within planetary processes, and the first level in the synthetic hierarchy of living matter. Although metabolism has been the foundation that enables the variability of the rest of life, much of its own architecture is essentially constant across the tree of known organisms. Among the deepest core pathways, even quite detailed features such as the roles of specific molecules or reaction sequences show only modest and tightly structured variation. In this sense metabolism seems to straddle the transition from the necessity of geochemistry to the chance of cellular evolution.

Chapter 3 described the long arc of disequilibrium on Earth running through the lithosphere/hydrosphere/atmosphere system, the way it is carried on a subset of molecular species and focused in particular locations such as hydrothermal systems, and the way these focusing centers sustain an ancient and embedded biosphere. Chapter 6 to Chapter 8 will build the argument that *a biosphere on Earth is necessary, that it emerged along the arc of disequilibrium because it provides an independent channel from geochemistry to relax redox stresses, and that the universal core metabolic pathways we find today were the favored and perhaps unique solutions to this relaxation problem in an abiotic Earth.* The ground for that argument will come from the facts about the structure and evolution of metabolism that we review in this chapter.

The chapter presents the architecture of intermediary metabolism in relation to its context in geochemistry, microbiology, and deep evolutionary history. We review what is currently known about the functional properties, distribution, and variations in core biochemical pathways, emphasizing in particular the process of carbon fixation and the provision of a small set of organic backbone molecules which serve as precursors to all other biosynthesis. We highlight the many forms of modularity and redundancy in core metabolism, and the way these have governed, and have changed within, the course of its evolution. These facts serve as the specific basis for claims about the constraints and causes under which metabolism emerged out of geochemistry. We describe a logic in the structure of metabolism that we believe can be interpreted in terms of function, but we are equally concerned with the scope of variations, which enable a comparative analysis and provide independent evidence for the relation between structure and cause.

In some approaches to the origin of life, metabolism is viewed merely as the source of material for the biosphere. We wish to emphasize its role as a key source of *information* for the structure of the living state. As we develop in detail in Chapter 7, information is a property of systems that are subject to limits on their variation. So far as the constituents of metabolism and their network relations are dictated by the composition and energetics of the Earth, the production of selected metabolites is the first and deepest source of limits on the forms the biosphere can take, or could ever have taken.

Information in evolution is usually associated with processes of adaptive change, because that is how we recognize that variation is possible *ex ante* but restricted *ex post.* The information in a unique and universal metabolism would be of a different kind, because it would preclude variation in the processes that permit structure to form at all. We will argue that this is the information carried in *paths of least resistance.* It is essential to the organization of the biosphere because only along these paths are the residual problems of robustness in a hierarchical system simple enough to have evolutionary solutions.

Section 2.5.1.2 explained that biochemistry comprises three qualitatively distinct classes of molecules: the small elementary substrate molecules, the cofactors, and the oligomeric macromolecules. They are distinguished in their chemistry of formation, their structure, and their functional and network roles. Most of this chapter will concern the order in the network of small metabolites, with a discussion of cofactors in Section 4.5. We will make observations about macromolecules mostly in regard to key catalytic functions, or important patterns of conservation in the evolutionary record. We return to the emergence of the oligomer world as a subsystem, and to aspects of dynamics within that system, in Chapter 5. The relations and interplay we most want to emphasize in the description of the small-molecule substrate and the cofactor layer are the way the substrate reflects patterns in elementary organic chemistry, and connections to geochemistry, and the way the structures within the cofactor layer dictate the way that layer feeds back to influence the architecture of metabolic flows, and dictates the evolutionary paths of metabolism as a system.

The goal in a structural and comparative analysis is to propose **laws or causes** for the emergence of metabolism and life, on the basis of the particularities that define metabolism on Earth.

In a structural analysis we will emphasize the many ways in which metabolism, although complex as a whole, decomposes into simpler subsystems. In some cases we will argue that these subsystems are dictated by the limited ways certain functions can be performed in organic chemistry, or by the need, primordially, to have been embedded within geochemical settings seen in Chapter 3. In other cases we will argue that the requirement for simplicity was itself the reason a metabolism could only have arisen if it did so by assembly from modular, initially self-maintaining subsystems. On the basis of the asymmetric dependency some metabolic subsystems have on others, in Chapter 6 we will suggest a sequence in which they came to be assembled into the multifunctional system that cellular metabolism has become.

Interpretations of structure in terms of cause are always confounded in biology with the possibility that universality could simply be a reflection of descent from a common ancestor. When variations occur that permit a comparative analysis, they can provide ways to work at least partially around this confound. Comparison carried out in the form of evolutionary reconstruction provides a way to correct for correlated observables, and may generate better models of ancestral forms than extant organisms provide.

Indirect inferences about causation can sometimes be drawn from the *absence of variations* in a comparative analysis. The lack of variation in certain features against a background of system-level change can suggest that these features are constrained or subject to strong evolutionary pressure against deviation. These are evolution's "dogs that didn't bark," immortalized in Sir Arthur Conan Doyle's Sherlock Holmes memoir *Silver Blaze* [230]:

> Gregory: Is there any other point to which you would wish to draw my attention?
>
> Holmes: To the curious incident of the dog in the night-time.
>
> Gregory: The dog did nothing in the night-time.
>
> Holmes: That was the curious incident.

The metabolic substrate shows many invariant patterns in small-molecule chemistry against a background of change in enzymes, cofactors, or higher-level cellular systems. The argument that these reflect laws is most compelling where conserved features follow subsystem boundaries that we would argue independently are unique or optimal solutions to problems of function or robustness.

## 4.2 Modularity in metabolism, and implications for the origin of life

The most diverse and most informative evidence that metabolism has a comprehensible and even rule-like order comes from the many ways it decomposes into quasi-independent subsystems and redundant elements. The general importance of modularity, for the robustness that makes the emergence and evolvability of hierarchical complex systems possible, will be the subject of Chapter 8. In this chapter we focus on a descriptive account of the kinds of subsystem decomposition within metabolism, and the different ways they suggest origins in small-molecule geochemistry or early sources of robustness and selectivity during the emergence of metabolism and life.

### *4.2.1 Modules and layers in metabolic architecture and function*

The subsystems that give order to metabolism are of two kinds, which we will call "layers" and "modules." Modularization exists with respect to many criteria: reaction mechanisms, network topology, energetics, and correlations between these and patterns of historical diversification. In many instances the boundaries drawn by different criteria coincide.

#### *4.2.1.1 Modules are informative about emergence because they suggest partial autonomy of subsystems*

By "module" we mean a structural or functional subsystem that is internally integrated but less tightly coupled to the rest of the system. We are particularly interested in kinds of integration and isolation that suggest a module could have arisen or could function in a less developed or supportive context than its context in modern life. A module might pose a simpler problem of innovation and emergence than the system as a whole, if it employs a few related kinds of reactions repeatedly,[1] or it might suggest autonomy from the surrounding system because it depends on support from catalysts that would serve even if they were non-specific,[2] or that might once have had mineral analogues.[3] Module boundaries may separate different domains of variation, as when highly stereotypical group transfers from a few cofactors serve the biosynthesis of a variety of small-molecule reaction networks, or a sequence of enzymes is reused (the identical enzymes or homologues with high sequence and structural similarity) to catalyze parallel sequences of reactions on several substrates.

---

[1] Examples include aldol reactions in networks of sugar phosphates, or a very heavily repeated sequence of reductions from carbonyls and hydroxyl isomerizations reviewed in Section 4.3.7.
[2] Examples from core carbon fixation and from amino acid biosynthesis are reviewed in Section 4.4.5.
[3] Most cofactor functions are candidates for this interpretation. For cofactors involved in one-carbon reduction, reviewed in Section 4.3.4 and Section 4.5, plausible mineral analogues have been demonstrated.

Modular decompositions are associated with interfaces, which are the boundaries through which different modules interact, but which serve to some degree to insulate their internal workings from each other. The encapsulating effect of interfaces can simplify problems of control for modern living systems, and many have likely arisen as products of evolutionary refinement.[4] However, in deep metabolism some module identities are recapitulated by many criteria, and are more robust than the control relations beneath which they are assembled. These, we argue, had pre-Darwinian and at least partly independent origins. In most cases a correspondence of these divisions with very low-level aspects of chemistry or geochemistry is apparent.

### 4.2.1.2  Layers suggest dependencies in sequence

Layers refer to subsystems whose interdependencies place them in a natural order. For instance, many of the simple amino acids are synthesized by homologous reductive transaminations of $\alpha$-ketone groups on core metabolites containing only carbon, hydrogen, and oxygen. The amino groups are transferred (in modern organisms) from glutamine which acts as an activated amino group transfer cofactor. The reductive transaminations act as a layer over CHO core metabolism, which initiates amino acid biosynthesis. As another example, cofactors as a class have characteristic sizes, stereotyped functions, and roles as hubs in biosynthetic networks, which make them a natural control layer over the small-molecule metabolic substrates as a group. Accretion of layers is the process by which the emerging biosphere gained catalytic control over organic chemistry, sometimes recapitulating or incorporating geochemical mechanisms.

### 4.2.2  Reading through the evolutionary palimpsest

Metabolism has rightly been called a palimpsest [60],[5] but we will argue that its first layer of writing remains visible, and that subsequent layers have created mostly local disturbances in the original outline. Through evolutionary tinkering and opportunistic modification, module boundaries, layers, or other basic architectural motifs that may have been regular in early organisms, can have their edges blotted out or obscured by idiosyncratic innovations in modern organisms. Redundancies that may once have acted simply and independently, to dictate the contents of the small-molecule substrate or its reaction network, may have since become complicated by exceptions and special cases.

Evolutionary (phylogenetic, functional, and causal) reconstruction can sometimes identify a direction to these changes. The nature of change is not necessarily a reduction or an increase in the degree of modularity, but rather a change in the criteria for modularization, according to the technologies of the era. Early eras of few and simple enzymes

---

[4] Module boundaries provide interfaces where change for the sake of adaptive variation may occur readily, while leaving dependencies within individual modules in place [296, 440]. They thus permit adaptation while also serving robustness, and so we recognize the boundaries as places where change has concentrated, when we catalogue extant diversity of metabolic phenotypes.

[5] The emphasis in [60] was overwriting by an RNA World. We refer to the same kind of overwriting, which occurs constantly in the course of evolution in any era.

favor redundancy and non-specific use. They exercise control at the level of functional groups, and induce modularity by selecting for networks that reuse those functional-group transformations in the smallest and most robust pathways available. Later eras capable of providing a more diversified inventory of more specific catalysts permit integration of and optimization of pathways as a whole. In Section 4.4.5, we will argue that the bacterium *Aquifex aeolicus* has preserved enzyme redundancies and a modularization based on the homology of small-molecule functional groups that most other bacteria have replaced. In this instance a direction of evolutionary change can be constructed from the molecular biology and its evolutionary context.

### 4.2.3  Support for a progressive emergence of metabolism

The kinds of modularization that we identify as early are those that provide isolation and stability within subsystems with a minimum of dependency on hierarchical control or intersystem coupling. For instance, the short, redundant small-molecule pathways that would be stable given few and non-specific catalysts are of a type that creates the least interdependence between organosynthesis and reliable genome replication and translation. Weakening and ultimately eliminating a dependence on macromolecular memory systems, and shifting the dependence for stability onto chemical network effects, is an essential step in arguing for an emergence of life through metabolism. Therefore it is important that we can see evidence of a change in this direction with distance into the evolutionary past. Many of the subsystems are also clearly associated with particular organometallic chemical reaction mechanisms that suggest either limitations in available geochemistry or limiting innovations in the first departure from it. These features suggest that low-level constraints from chemistry and energetics are defining the interface positions in a modular architecture and the opportunities for evolutionary variation. A geochemically provided stability from simple mechanisms would not only have been necessary for the emergence of metabolism; it would have provided an essential scaffold for the earliest cells, enabling the gradual emergence of more reliable macromolecular systems of memory and control.

### 4.2.4  Feedbacks, and bringing geochemistry under organic control

An important generalization we will draw from the evolutionary transitions in metabolism is that major transitions are driven by feedback from higher levels in the metabolic hierarchy onto processes at lower levels. The first of these, we propose, were successive waves of cofactors, and (perhaps interleaved) later layers were macromolecular catalysts. On the basis of this generalization, we conceive of biogenesis in its early stages not as a replacement of geochemistry, but rather as an *enfolding* of geochemical mechanisms under increasing degrees of control by organic systems. Through the emergence of control, low-level organosynthesis was gradually rendered autonomous from the supports of geochemistry, at the price of greater interdependence among molecular systems. The emergence of the biosphere was as much a process of the achievement of autonomy as it was a process of the innovation of new mechanisms and new functions.

This set of premises about the origin of the biosphere is often called a hypothesis of "autotrophic origins," to contrast it to the Oparin–Haldane conjecture [329, 611] of an early organic-rich medium[6] from which emerging "heterotrophic" life obtained materials and energy catabolically. We will avoid the use of terms "autotrophy" and "heterotrophy" because they analogize the biosphere's relation to abiotic processes, to an organism's relation to its ecological environment. This analogy conjures a false divide between living processes and non-living processes, as if life were somehow "other" than physics and chemistry, on which it could rely as a kind of "environment."[7] We emphasize instead the difference between enfolding geochemistry through the innovation of feedbacks, and replacing geochemical (or astrochemical) synthetic patterns while preserving some of their intermediates.

### 4.2.5 The direction of propagation of constraints: upward from metabolism to higher-level aggregate structures

Modules that we will introduce in a metabolic context in this chapter will reappear in Chapter 5 in higher-level molecular and cellular systems. We will argue that the structures in these cases *originate* in metabolism and are propagated to higher levels. This is an argument that the paths of least resistance into which protometabolism settled were preconditions for the later architecture and roles of macromolecules and cells. Not only did hierarchical life build from what metabolism made available, but the influence of that constraint is recorded as signatures with the structure of metabolic pathways in systems that naively should have evolved independently of metabolism.

Propagation of a set of constraints upward in a hierarchical system is also propagation of information "from the bottom up." This runs counter to the assumption of many approaches to the origin of life such as the "RNA-first" premise [589], in which information flows only "from the top down": from catalytic and competitive interactions among RNA eventually down to the selection of a metabolism capable of resupplying RNA. A top down information flow is intuitive because information flow follows known mechanisms for the propagation of constraint: selection imposes information from the environment "inward" on frequencies of genomes, and catalysis then propagates constraints from genomes onto the formation of phenotypes. The notion of "upward" flow of constraint and information is

---

[6] The difference between the Oparin–Haldane conjecture and a conjecture of self-provisioning concerns mostly the richness and diversity of the soup, and the continuity of mechanisms for its generation. Oparin and Haldane imagined a medium stocked with sufficient complex molecules to initiate in parallel many functions of biochemistry. The self-provisioning hypothesis envisions a relatively narrow range of early organics with sufficient concentration, and proposes that the enhancement of that concentration and control over its outputs were the most probable pathways for the synthesis of more complex organics. Consequently the low levels of metabolism determined in large part which complex molecules could be added to augment them.

[7] We return in Chapter 8 to a careful discussion of the emergence of individuality and the consequent creation of ecological community structures, and emphasize that these are innovations of organizational frameworks within life rather than determinants of the nature of the living state.

more subtle because it must either result from limiting the variations that can be produced, or else it must be carried on the coevolution of relations among multiple components.

### *4.2.5.1 Mechanisms of pathway evolution*

The mechanisms by which metabolic pathways evolve as a result of enzyme evolution tend to propagate the patterns of chemical relations that make early pathways robust upward into the more refined pathways that replace them over time. They thus provide a mechanism for the upward flow of information, within the overall control system of evolution. The upward flow of information happens indirectly: innovation mechanisms are not biased to preserve pathway relations among whole molecules (though requirements of viability and stabilizing selection may tend to lock in pathways this way). Rather, enzymatic innovation tends to preserve the co-presence of reaction mechanisms. However, since the molecular pathways that were robust initially in geochemistry would have been largely determined by redundancy of low-level reaction mechanisms at the functional-group level, they would also have been preferentially carried through time as long as the lower-level relations are preserved.

We briefly review the mechanisms believed to dominate pathway evolution in the molecular age, as part of the general knowledge about metabolism that underlies our interpretations in later sections. In the particular case of the reconstruction of *Aquifex* in Section 4.4.5, we can see direct evidence for evolution along the trajectory from generality to specificity.

Innovation always requires some degree of **enzymatic promiscuity**, which may be the ability to catalyze more than one reaction (catalytic promiscuity) or to admit more than one substrate (substrate ambiguity) [434]. Pathway innovation also requires **serendipity** [438], which refers to the coincidence of new enzymatic function with some avenue for pathway completion that generates an advantageous phenotype from the new reaction. Although most modern enzymes are highly specific, broad substrate affinity is no longer considered rare, and is even explained as an expected outcome in cases where costs of refinement are higher than can be supported by natural selection, and in other cases by positive selection for phenotypic plasticity [796].

When enzymes are specific they must also be diversified in order to cover the broad range of metabolic reactions used in the modern biosphere. Serendipitous pathways assembled from a diversified inventory of specific enzymes will in most cases be strongly historically contingent as they depend on either overlap of narrow affinity domains or on "accidental" enzyme features not under selection from pre-existing functions. Such pathways therefore seem unpredictable from first principles; whether they are rare will depend on the degree to which the diversity of enzyme substrate affinities compensates for their specificity.

The opposite pattern characterizes systems in which enzyme diversity is low and therefore enzymes must have broad affinity and must be redundantly used. The patterns that pathway innovation governed by such non-specific catalysts should most readily preserve are those already present in the inventory of metabolites that determine which parallel paths are populated and which are not.

### *4.2.5.2 Reaction mechanisms are generally more conserved than substrate specificities*

Modern enzymes both create reaction mechanisms and restrict substrates, but the two functions can evolve to a considerable degree independently.[8] Roy Jensen argued in the 1970s [399] that early catalysts should be expected for a variety of reasons to be *non-specific*. The first forms were determined by the constraints from a pre-hierarchical world, not by optimization for the functions they would later take, and not yet refined by the billions of years of evolution that have subsequently transpired. They arose in a context of unreliable molecular replication in which only short sequences could evade an Eigen error catastrophe. Both the specificity and the diversity of catalysts available under such conditions are expected to be low. The only way such catalysts can support reaction networks of growing complexity is through redundant use along parallel pathways.[9] Modern reviews of the mechanisms underlying functional diversity, promiscuity, and serendipity [157, 433, 434, 601] confirm that substrate ambiguity is the primary source of promiscuity that has led to the diversification of enzyme families.

### *4.2.5.3 Specificity and diversity in the past versus the present*

We will show in Section 4.3.7 that the use of homologous enzyme sequences (and sometimes even the same enzyme sequence) to catalyze chemically similar parallel pathways is common in core metabolism even in modern organisms, and will present evidence in Section 4.4.5 that it was even more extensive in the LUCA. We conclude that serendipitous pathway formation was facile for the key innovations that led to variants in carbon fixation and biosynthesis of core metabolites, and that it was structured according to the same local-group chemistry around which the substrate network is organized.

## 4.3 The core network of small metabolites

Because biochemistry partitions organosynthetic reactions into three size classes, introduced in Section 2.5.1.2, we can refer with little ambiguity to a **core** of metabolism. The core is the network of pathways that include essentially all *organic* reactions: those that alter covalent bonds among C, O, N, H, and S atoms, augmented by a modest number of dative bonds of N or S to transition metals. Molecules in the core include a diverse set of small metabolites, a subset of which are *monomers* that are building blocks of macromolecules. Molecules in the core range in size from 2 (acetate) to about 20 carbon

---

[8] An extreme example of the potential for separability between reaction mechanism and substrate selection is found in the polymerases. A stereotypical reaction mechanism of attack on activating phosphoryl groups requires little more than correct positioning of the substrates. In the case of DNA polymerases, at least six known categories (A, B, C, D, X, and Y) with apparently independent sequence origin have converged on a geometry likened to a "right hand" which provides the required orientation [681].

[9] This argument was largely a rebuttal of an earlier proposal by Horowitz [370] for "retrograde evolution" of enzyme functions. The 1940s witnessed the rise of an overly narrow interpretation of "one gene, one enzyme, one substrate, one reaction" (a rigid codification of what would become Crick's *Central Dogma* [167]), which in the context of complex pathway evolution appeared to be incompatible with natural selection for function of intermediate states. The Horowitz solution was to depend on an all-inclusive "primordial soup," in which pathways could grow backward from their final products, propagating selection step-wise downward in the pathway until a pre-existing metabolite or inorganic input was found as a pathway origin.

atoms (tetrahydrofolate).[10] Larger molecules are made almost exclusively by assembly of metabolites from the core [572, 776]. Nearly all biomass in macromolecules consists of polymers formed by phosphate-driven dehydrations, with minimal further modification by organic reactions. Some molecules such as fatty acids and isoprene alcohols, although comparable in size to cofactors, have a structure of repeated two-carbon or five-carbon units, and thus something of the character of polymers. A few molecules of intermediate size, most of them cofactors, are formed from small numbers (two to five) of heterogeneous monomer components, such as amino acids, fatty acids, or RNAs.

Including the monomers or elementary repeated units, the essential core metabolism required for autotrophic life contains only about 125 distinct compounds [776, 777].[11] There are many ways to classify core metabolites because a few biosynthetic pathways are reused to produce a few catagories of active functional groups. Several of these will be reviewed in the following sections. One classification that is important to the scale dependence of metabolism is that the monomers are of only three types: amino acids (20 standard acids and a few idiosyncratically used variants), nucleic acids (four RNA and four DNA of which three share the same nucleobases), and sugars (a diversified group of which most simple members have 3–6 carbon atoms). All essential macromolecules larger than cofactors are *oligomers*, meaning that each polymer is made from only one kind of monomer.

**Intermediary metabolism** is the standard term used to refer to the metabolism that is within cells or organisms, as distinct from metabolic reactions that may require transfers between organisms through trophic links in ecosystems (such as predation, parasitism, etc.). Since autotrophic organisms are biosynthetically complete, their intermediary metabolism is also an ecologically self-sufficient metabolism. Within their intermediary metabolism, by separating the macroscopic but chemically simple and redundant oligomer sectors from the core, we may obtain a model for the organic chemistry of minimal autotrophy. In [777], a model of minimal intermediary metabolism was made by beginning with the annotated genome of the hyperthermophilic chemoautotroph *Aquifex aeolicus*, and comparing its gene assignments to comprehensive databases of known reactions, to propose complete reaction networks that are also plausible as models of early autotrophy. In [92], the genome of *Aquifex* was used as the starting point for construction of a *stoichiometric flux balance model* [621], the first such computer-verified model of autotrophy constructed. An extensive comparative genomic analysis and phylogenetic reconstruction was performed to infer corrections and completions of the annotated genome to produce a valid model capable of synthesizing biomass from $CO_2$, reductant, and other inorganic inputs consistent with reported growth media for *Aquifex*. A part of that reconstruction is described in Section 4.4.

---

[10]  A few cofactors, such as heme, cobalamin, or coenzyme $Q_{10}$, may have as many as 40–60 carbon atoms, but these already have a structure of repeated units such as pyrroles or isoprenes in the 4–5 carbon range, or are assembled from pre-existing building blocks such as RNA.
[11]  This number may increase by a few dozen if distinct membrane lipids are counted, rather than treated as repeats of the basic hydrocarbon units.

An important observation about relations between heterotrophic and autotrophic organisms is that *many catabolic pathways used to break down either stored or consumed organics as sources of material and energy are approximate reverses of anabolic pathways used to synthesize those organics.* This suggests that, if the "meta-metabolome" of a trophically complete ecosystem can be constructed as gene inventories become more complete and annotations more reliable, it will be possible to collapse pairs of anabolic/catabolic pathways in different organisms to "futile cycles," which at the ecosystem level are equivalent to pure dissipation reactions. We expect, from experience with current pathway databases, that most and perhaps all catabolism will be removable under such projections, and the net metabolism of any complete ecosystem will be equivalent to some combination of autotrophic networks found in individual organisms. This is the microscopic description (at the network level) of the observation made in Chapter 2 that the deepest universal features of metabolism are reliably seen not at the individual, but at the ecosystem level. The ecologically conservative core is the part of the network that remains after futile cycles have been projected away, which are created in various forms by trophic links and catabolic pathways within member species.

A reason to expect that projection of ecosystems may have any resemblance to autotrophic species at the network level – rather than generating completely unrelated network completions – is that results of phylogenetic reconstruction of extant autotrophs reveals a highly modular architecture within metabolism. Variations in subnetworks such as carbon fixation are essentially independent of variations in other networks such as amino acid biosynthesis, because in all organisms the two are connected through a conserved set of small-metabolite precursors which are more highly conserved than networks in modules on either side of the interface. These universal precursors are introduced in Section 4.3.2 below.

### 4.3.1 The core in relation to anabolism and catabolism

> All happy families resemble one another; every unhappy family is
> unhappy after its own fashion.
>
> – Leo Tolstoy, *Anna Karenina* [129]

The core may also be recognized topologically and functionally. All heterotrophs, and most modern organisms with some phenotypic plasticity, employ some catabolic pathways to break down stored or environmentally consumed organic molecules for energy or inputs to biosynthesis. All organisms, of necessity, also employ suites of anabolic pathways to synthesize all molecules that they cannot consume in exactly the forms they require.

Many different characterizations of metabolic network topology, function, or control now characterize heterotrophic or plastic metabolism as a "bowtie" [171, 909]. Catabolic pathways are rays inward toward a "knot"; anabolic pathways are rays outward from it. Topologically, the inward and outward rays are distinguished from the knot because the knot is extensively cross-linked, and the rays much less so [672]. The lack of cross-linking among rays makes the shortest paths between most pairs of metabolites paths that pass through the knot. In comparative analysis across organisms or across modes of metabolism used facultatively by an organism, the knot is roughly invariant, while the input and output rays change roles from anabolism to catabolism as they change direction. The most widely seen examples are the switch between gluconeogenesis (anabolic) and glycolysis (catabolic) for sugars, and between fatty acid biosynthesis (anabolic) and $\beta$-oxidation.

The bowtie motif has become a frequent theme in literature on metabolic network topology, because most study organisms are heterotrophic or metabolically plastic, and so have extensive anabolic and catabolic sectors. Such motifs are also easy to identify with statistical measures against a null model of a random network that treats all metabolites equally. Many of these analyses attempt to explain the bowtie as an optimal solution to throughput, control, or adaptation problems, taking the inventory of metabolites as given parameters of that problem. We argue that, while each of these does capture some relation between a network and its inputs and outputs, the real explanation for both the bowtie and the metabolites it connects cannot be found in heterotrophy. The bowtie is a derived consequence of commitments to metabolism made in an autotrophic world.

Autotrophic organisms can, in principle and under benign conditions of minimal fluctuation, function without catabolic pathways. *They can synthesize all required biomass from $CO_2$, ammonia, orthophosphate, and a few other inorganic inputs.* Chemoautotrophs, by using environmentally provided oxidants and reductants, can do this most simply, not requiring light energy or the additional layer of biosynthetically complex molecular assemblies needed to capture it. Such organisms take no organic inputs from their environments, and in benign stable conditions have no need to store and later recycle organic stocks internally. The thermophilic, micro-aerophilic, chemolithoautotrophic bacterium *Aquifex aeolicus* [186] provides a reasonable model of such minimally complex autotrophic metabolism.[12]

In a minimal, chemoautotrophic metabolism, the core becomes the nucleus of a network, from which anabolic pathways radiate like rays of a fan. Catabolic pathways, in organisms that use them, act on the carbon skeletons of their intermediates in most cases as the reverses of the anabolic pathways that produce those intermediates.[13] The ability to reverse many pathways within the core permits the reversal of rays in the fan, from anabolism to catabolism, as a way to create phenotypic plasticity, to adapt to reducing, fermentative, or oxidizing redox conditions, and to become heterotrophic and ecologically specialized as ecosystems become complex.

---

[12] We note as a caution, however, that even *Aquifex* is a modern organism with significant internal regulatory complexity in some features. The most prominent of these govern its response to the oxidation state of the sulfur that it uses as a terminal electron acceptor [92, 320, 321] *Aquifex* may form either a sticky, sessile, non-flagellated calix and undergo steady growth if elemental sulfur is available, or a smooth, flagellated mobile form if thiosulfate or tetrathionate is the only environmental sulfur present. In the sessile form, *Aquifex* accumulates black, shiny internal nodules of organic sulfur compounds. Their function is not known, but they may be associated with survival for limited periods in the motile form, in thiosulfate environments where *Aquifex* cannot continue to grow indefinitely and ultimately dies.

[13] The energetic requirements of running metabolic pathways usually involve dissipation of heat in many reactions, both because most reaction free energies are not exactly zero and because reaction speed can be increased with non-zero dissipation. Therefore reversing a pathway often involves changing oxidants, reductants, or functional-group donors or acceptors, to change the free energies of electron pairs or leaving groups with respect to the backbone of C or CN that undergoes modification along the pathway. In most cases, the overall structure of the pathway remains clearly invariant due to commitments to pass through a sequence of essential intermediates. A modest number of the so-called "salvage pathways," which accumulate especially in complex and later branching organisms, are significant evolutionary innovations, and these provide many of the cross-links between catabolic and anabolic pathways outside the core that we describe, or that is recognized as the core in bowties.

We will argue that the correct understanding of the core and anabolic pathways, both in the fan and in bowties, comes from the simpler realization in minimal anabolism of the kind found in *Aquifex*. The core contains carbon fixation and the biosynthesis of the small number of universally essential metabolites. Anabolic pathways reflect opportunities to assemble those building blocks that the core has made available. They are topologically simpler than the core pathways because they consist in large part of repeated reactions acting recursively to incorporate core building blocks into larger structures. They are much less cross-linked because they involve few or no novel organic modifications beyond those used to build the core components. The exhaustion of innovations in reaction sequences outside the core is the original source of selection of the molecules present in biochemistry: *life is made of those compounds that can be synthesized with few and simple reactions from components in the core.*

We may now understand why, in heterotrophs, if an input/output pair is taken as given, the shortest path between them typically consists of a pair of rays connecting both input and output to the core, and this shortest path may often be characterized in terms of the number theory of small integers from a string chemistry representation of molecules [672].[14] Shortest paths typically modify an input to a compound with a multiple of a small integer, such as 2, 3, or 5, of carbon atoms. The catabolic path cleaves its intermediates by reducing the multiple of these small-integer factors, until a metabolite in the core is formed. The core metabolites may be modified in various complex ways that depend on the input/output pair considered, but which are limited by the small size of the core network itself. The anabolic ray to the product then does the reverse of what was done in the catabolic ray, building its intermediate up as an increasing multiple, of the same or another small-integer factor. The number theory of small prime divisors reflects the repeated-unit assembly that characterizes many anabolic pathways, and the small integers reflect sizes of simple metabolites in the core.

We lose most of the explanatory power of this analysis, however, by taking the input and output pairs of metabolites as boundary conditions to the optimization problem, given by some "other" criterion.[15] A much more comprehensive explanation, for both pathway topology and the inventory of universal metabolites, may be sought if we take the *prima facie* limits to complexity in the core and anabolism as a source of *constraint* on what is biologically possible, and then identify the reasons for this constraint.

### 4.3.2 The core of the core

The structure of the biological molecules is determined largely by their carbon backbones. Reactivity is conferred by the addition of oxygen-containing groups. Activation energy is carried in bonds to sulfur and phosphate. The capacity for catalysis (beyond the most general acid/base catalysis) is first introduced with the incorporation of nitrogen

---

[14] A *string chemistry* is the "chemical formula" of a molecule which counts atoms but does not represent bonding configurations.

[15] Implicitly, they are often supposed to be selected for function in a Darwinian competition for fitness. While fitness criteria may provide a partial explanation, we think that for core metabolites, the notion of *ex post* function and fitness provide less of the explanation than *ex ante* constraints on the opportunities for biosynthesis.

in heterocycles.[16] Reaction mechanisms involving radical intermediates are first made available with the chelation of transition metals. Reflecting this partition of functions, the metabolism of even the essential 125 core metabolites has a substructure of layers and modules.

At the core of the core is a network of CHO molecules and reactions, which in important ways lays the foundation for the rest of the structure of biochemistry, and hence for the biochemically determined structures of life. This subnetwork contains the crucial functions of carbon fixation (in which environmental carbon ultimately originating in $CO_2$ is first incorporated into organic compounds), the biosynthetic pathways for the two major CHO metabolite classes (sugars and lipids), and the precursors for *all* other classes of metabolites: the amino acids, nucleic acids, cofactors, and other essential functional groups such as aromatics and pyrroles. The CHO core also displays a crucial – although only partial – form of self-sufficiency, in that it is the smallest subnetwork within which the self-amplification of *network autocatalysis* can be seen.[17] This inner-core network also has a remarkably universal interface to the remainder of biochemistry, which we describe in detail below. Nitrogen is added at only a few points to produce all aminated compounds and heterocycles. Sulfur and phosphate are added in a very limited number of ways. Although the deep core, as we will present it, shows the possibility for some modest variation, the interface it presents to the remainder of biosynthesis is universal.

Within the CHO core, a moderate but important diversity of organic reaction mechanisms is found. Outside the core, the decrease in reaction diversity already begins to appear. Most carbon or nitrogen functional groups enter on a limited set of cofactor carriers, and are incorporated through a few kinds of reactions. The synthesis of larger core metabolites comes to rely increasingly on assembling, or restricting the opening and closing, of motifs such as heterocycles.

### 4.3.3 Precursors in the citric acid cycle and the primary biosynthetic pathways

Right as diverse pathes leden the folk the righte wey to Rome.

– Geoffrey Chaucer, *Treatise on the Astrolabe* [129]

Among the regularities in core metabolism, the one that stands out above all others is the universal centralizing role [769] of the *citric acid cycle*, also known as the **tricarboxylic acid** or TCA cycle, illustrated in Figure 4.1. The TCA cycle is a network of eleven reactions and eleven carboxylic acids, of which the smallest (acetate) has two carbons and the largest (citrate, isocitrate, *cis*-aconitate, and oxalosuccinate) have six.

---

[16] Catalysis in nitrogenous compounds often employs nitrogen's capacity to pass between states with a non-bonding pair and a double-bond/radical combination, while preserving the primary $\sigma$-bond skeleton of carbon backbones.

[17] The self-sufficiency is partial in the sense that these networks are only self-amplifying if their reactions can first be carried out, which presumes the existence of catalysts and either cofactors or plausible prebiotic surrogates that provide functions now provided by essential cofactors. Macromolecular catalysts and cofactors are not part of the minimal CHO network that we discuss here. We return to consider what this kind of partial independence means in the remainder of this chapter and in Chapter 8.

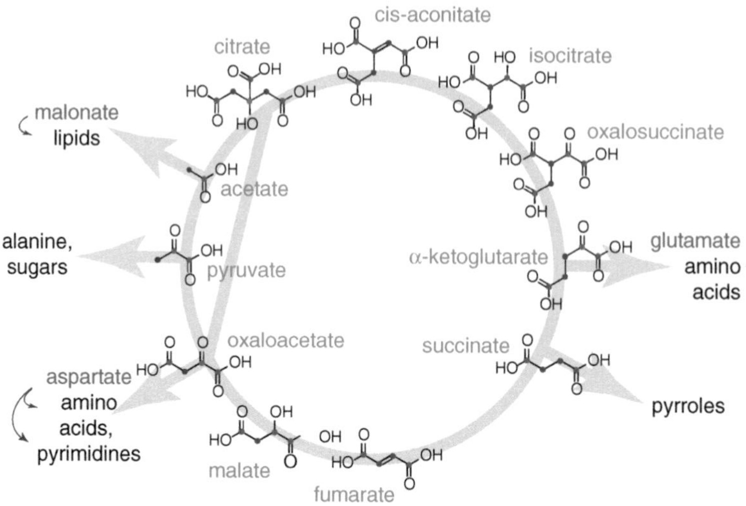

cis-aconitate
citrate
isocitrate
oxalosuccinate
malonate
lipids
acetate
alanine,
sugars
pyruvate
α-ketoglutarate
glutamate
amino
acids
oxaloacetate
succinate
aspartate
amino
acids,
pyrimidines
pyrroles
malate
fumarate

Figure 4.1 The five "pillars of anabolism" are intermediates of the citric acid cycle, and starting points of all major pathways of anabolism (arrows). The precursors and their downstream products are acetate (fatty acid and isoprene alcohol lipids), pyruvate (alanine and its amino acid derivatives, sugars), oxaloacetate (aspartate and derivatives, pyrimidines), α-ketoglutarate (glutamate and derivatives, also pyrroles), and succinate (pyrroles). Molecules with homologous local chemistry are at opposite positions on the circle. Oxidation states of internal carbon atoms are indicated by color (red oxidized, blue reduced). (After Braakman and Smith [91], Creative Commons [153].)

Four of the intermediates (pyruvate, oxaloacetate, α-ketoglutarate, and oxalosuccinate) are *α-ketoacids*. Three of these (excluding oxalosuccinate), plus the α-ketoacid glyoxylate (considered in Section 4.3.6 below) are precursors to all the amino acids. The amino acids aspartate (from oxaloacetate) and glycine (from glyoxylate), in turn, are precursors to the nucleobases. Succinate is precursor to pyrroles, which chelate transition metals in essential cofactors such as cobalamin and heme.[18] Pyruvate, via the activated form phosphoenolpyruvate (PEP), is the precursor to sugars and other carbohydrates derived from them, through the pathway known as *gluconeogenesis* from pyruvate to fructose (shown below in Figure 4.5). Acetate is the precursor to both fatty acids and isoprene alcohols, and as we show in the next subsection, may also be an entry point in some carbon fixation pathways to arcs of the TCA cycle that supply the other essential intermediates.

The five compounds acetate, pyruvate, oxaloacetate, succinate, and α-ketoglutarate are the **standard universal precursors** to all of biosynthesis.[19] Because of the centrality and universality of these precursors, the TCA cycle is the central organizing topological

---

[18] Pyrroles also chelate the non-$d$-block ion $Mg^{2+}$ in chlorophyll, in a somewhat unusual pattern for this class. Since chemoautotrophs can function without chlorophyll, we have not listed it as "essential" along with the transition metal binding tetrapyrroles.

[19] In some organisms, α-ketoglutarate rather than succinate is used as the precursor to pyrroles [839], and phylogenetic reconstructions indicate that this is more plausible as the ancestral form. Therefore as few as four TCA intermediates may be sufficient precursors for autotrophic life.

network in any graph of not only universal metabolism, but even the aggregate union of known metabolic pathways. As we will see in Section 4.3.6, even when variation in the use or completeness of the pathway exists, the universal precursors and paths to reach them retain their central place.

The TCA cycle remains topologically central, even after aggregating the known variations in core metabolism, because it is also phylogenetically central. A subset of its intermediates, and some subset of its reactions to reach them, is found in the biosynthetic pathways required by any organism (either directly or through trophic links from heterotrophs to primary producers). The complete cycle is found in organisms in either of two opposite forms: the oxidative *Krebs cycle* (run clockwise in Figure 4.1), which was the first form discovered as a part of energy metabolism in heterotrophs [456], and the **reductive TCA (rTCA) cycle** (run counterclockwise in the figure), used as a carbon fixation pathway in several clades of bacteria [102]. Incomplete arcs of the cycle are present in numerous fermenters,[20] and they also provide what are called *anaplerotic* pathways for other carbon fixation networks [481]; these are pathways that produce the essential precursors which may not be intermediates in the fixation pathway itself.

The key reaction by which the rTCA pathway differs from a cycle is a retro-aldol cleavage of citrate to produce acetate and oxaloacetate, making this a branched pathway. The two branches coalesce, however, because acetate is converted to oxaloacetate through two carboxylations and a reduction. The result of branching is that the TCA cycle (traversed in either direction) is *network catalytic* through oxaloacetate. Its oxidative form, the Krebs cycle, is catabolic, respiring pyruvate or acetate to $CO_2$, which is a waste product, and delivering reducing equivalents. The rTCA cycle is *network autocatalytic*, taking in $CO_2$ and reducing equivalents to produce biosynthetic precursors, and even to replenish the network catalyst itself as it degrades or is consumed in anabolism. It hence has the capacity to be self-amplifying, the property we have noted is required both for self-repair and for chemical competitive exclusion: the concentration of matter flows and energy flows preferentially into reactions in the cycle. To be realized, of course, this capacity requires the coupling of particular energy sources (particularly phosphates and thioesters), as well as specific catalysts so that key reactions such as reductions affect only some molecules in the cycle (malate and oxalosuccinate) but not others (pyruvate and $\alpha$-ketoglutarate). To date, no experimentally successful reaction systems have produced compelling models for the biological cycle, and this fact remains one of the principal challenges in understanding what the biological form captures about lawfulness in metabolism.

The forms of the molecular intermediates in the TCA cycle are shown in Figure 4.1. Brief inspection shows that the chemistry of the cycle is redundant: molecules with homologous functional groups are found at diametrically opposite points on the cycle, and the reactions between them (where they connect homologous molecules) perform the same functional-group modifications. We will return to the interpretation of this redundancy as

---

[20] Representative examples are found among $\gamma$-proteobacteria [253] or mycobacteria [852].

a simplification of the cycle, in the context of a more complete analysis of variability in Section 4.3.7.

Some organisms using the rTCA pathway for carbon fixation, such as the Aquificales, branch very deeply in the bacterial phylogenetic tree. If these branches are taken together with those that use parts of the pathway as anaplerotic reactions, *the use of the pathway segments to arrive at cycle intermediates as biosynthetic precursors is universal.* From this fact and the network autocatalysis of the small-molecule reaction sequence, many observers including us [574, 769, 844] have proposed that the cycle contains the *functional* precursors to metabolism because it emerged as the *temporal* predecessor to the rest of biosynthesis. The rapid growth of knowledge within the past two decades, about biosynthetic mechanism, variability, and phylogenetic distribution, provides additional input from which to try to draw a more specific hypothesis. From considerations in Chapter 7 and Chapter 8 of what causation means in stochastic, error-prone, or loosely organized networks of the kind that must have characterized the earliest geochemistry, we may also refine the way we interpret evidence from alternative pathways that overlap more or less with rTCA, in terms of constraints on the emergence of metabolism that are now reflected in the roles of the citric acid cycle in modern organisms.

The main question that we believe is forced upon us by the universality of the TCA intermediates – even more forcibly because the intermediates are more universal than the complete pathway – is *why and by what agency these intermediates were selected, so strongly that they constrain all metabolism both below and above.*

### 4.3.3.1 Energetics of reductive carbon fixation

The presence of the universal precursors as a subset of cycle intermediates causes the TCA cycle in Figure 4.1 to reflect, in this deepest of core pathways, the relation between bowtie and fan topologies that characterizes core metabolism more generally. Oxidative TCA is a bowtie. It must be fed organic inputs via one intermediate in order to have precursors to biosynthesis at other intermediates, but in producing these it loses a fraction of its carbon. The status and centrality of such a cycle as the organizational center of metabolism should have seemed a puzzle (if it did not at the time) when only the oxidative Krebs cycle was known. Why should metabolism be so strongly organized around a cycle, if its primary functions were to convert some complex organics into others? The discovery of rTCA [234], even apart from evidence about its phylogenetic distribution, suggests the resolution of the puzzle. rTCA is the nucleus of a fan. It converts inorganic inputs to the precursors of biomass, in a one-way flow from $C_1$ simplicity to organic complexity.

Metabolism exists not only as a topological network but as a dynamical system driven by free energy. To understand the energetic content of the TCA cycle we show standard-state $\Delta G$ of formation of the cycle intermediates, from $CO_2$ and $H_2$ reducing equivalents, in Figure 4.2.[21] The abscissa plots the ratio $[H_2] / [CO_2]$ consumed to form each cycle

---

[21] Although organisms do not live under standard-state conditions, this is a useful starting point of reference to understand the reduction chemistry of carbon. Corresponding free energies have been computed under conditions calibrated to geochemical settings, which reflect the same central message. We return to these below and in Chapter 6.

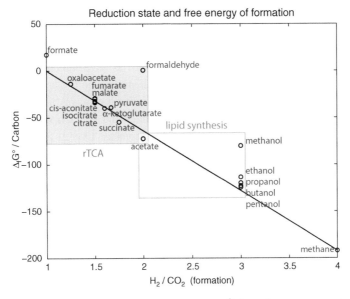

Figure 4.2 Free energies of formation per carbon atom $\Delta_f G^0 / [CO_2]$ in kJ/mol(C) for synthesis of TCA intermediates from $CO_2(aq)$ and $H_2(aq)$ versus reaction stoichiometry $[H_2] / [CO_2]$. Data for TCA intermediates and formate from Miller and Smith-Magowan [558]. Data for formaldehyde from Thauer *et al.* [800], and for other compounds from Plyasunov and Shock [639]. Long-chain alcohols suggest the limit point for aliphatic compounds approaches the linear relation passing through most TCA intermediates and methane.

intermediate, with excess O carried away as water, and the ordinate plots $\Delta_f G^0$ (free energy of formation).

The first important observation is that all free energies of formation are *negative. The standard-state formation of biomass from $CO_2$ in an environment of $H_2$ is net exergonic.* When the chemical potentials of the inputs are corrected to reflect the concentrations found in hydrothermal vents and hot springs fed with reductants from the oxidation of ferrous iron, the energetic picture is both more complicated and dependent on the particular environment studied. However, as noted in Chapter 3, Amend and Shock, in an extensive series of calculations [20, 21, 22] considering the CHO sector and also compounds including nitrogen and phosphorus, have shown that the free energy of synthesis of both large subsets of small metabolites, and whole organism biomass, can be negative under reducing conditions generated in known hydrothermal systems.

As we noted in Section 3.6.4.1, the study of organisms in the world of oxygen and sunlight has established a pattern of thinking about biochemistry as a struggle at all points against the tendencies of thermodynamics: free energy in the environment must be "captured" or "harnessed" to provide work, which then can be used to drive unfavorable reactions. The metaphor is "no such thing as a free lunch" – doing work against the forces of equilibrium is intrinsic to survival. However, conversion of free energy to chemical work is made possible only through the agency of machines, which are themselves constructs of life. This image suggests an intransigent circular dependency.

A different metaphor is suggested by the environments on Earth that generate reductants out of equilibrium with ambient carbon as well as environmental oxidants. It does not eliminate the role of machines, because even chemotrophic life uses chemical work to overcome locally endergonic steps in biosynthesis, but it removes the image of work as an all-encompassing barrier between thermodynamics and life. Chemotrophic organisms "capture" free energy *by the process of building biomass*. Everett Shock has characterized geochemical reductant consumed by chemotrophs as "a free lunch you are paid to eat" [749]. This point will be central in our attempt to understand *why* a biosphere exists, so we pause to list the interpretive steps in order, which are suggested by the energetics of reducing metabolism, particularly as it occurs in autocatalytic networks such as rTCA.

1. Free energy relaxation, or the conversion of internal energy to heat, is thermodynamically favored, if there are channels through which it can occur.
2. Synthesis of organic molecules from $CO_2$ in a reducing environment is exergonic, and hence provides such a channel. Therefore reducing anabolism is expected to happen spontaneously if it can.
3. The problem for understanding the existence of a biosphere is that in metabolism this flow is concentrated in only a few pathways, and concentration is not a phenomenon expected within equilibrium thermodynamics alone.
4. Therefore we must explain why the formation of relaxation channels via anabolism is *possible*, and what forms of *selectivity* have concentrated this flow in a few pathways. These are inevitably problems of kinetics at many levels.
5. The metaphor of "no free lunch" captures the intuition that in heterotrophic systems multiple coupled processes are needed to maintain energy flow, and their complexity seems a barrier to emergence. In the small molecules and short pathways of the rTCA cycle, the process seems much simpler, and therefore more plausibly one that could have emerged spontaneously.

Although it is important that the energetic "downhill run" of Figure 4.2 might allow a perfectly efficient, *aggregate* metabolism to operate without input of free energy, the component reactions that are aggregated to form the intermediary metabolisms of cells do not individually have the same property. Each reaction wastes some heat, and transferring chemical work from some reactions to others to minimize the aggregate heat generation depends on biological machinery. Imagining the same chemistry without the support of the coupling machinery that it is required to manufacture – which is the problem of extrapolating from extant life to origins – shifts the consideration to more disaggregated reactions, which interpose endergonic steps in most pathways.

Even the standard-state TCA cycle, operated either oxidatively or reductively, includes individual steps that are endergonic: although all the net free energies of formation fall below the source molecules $CO_2$ and $H_2$ in Figure 4.2, individual reactions oscillate up as well as down the ramp. Evolution has not succeeded in finding routes that are everywhere downhill within the carbon chemistry itself, because feasible reactions are not determined only by energy but by possible bond structures and available reaction mechanisms. All known chemotrophic organisms, in parallel to biosynthesis, must operate other channels for electron flow from environmental reductants to terminal electron acceptors, and harness a part of this energy to overcome unfavorable individual steps in biosynthesis [96]. The bioenergetic systems further increase the total dissipation of waste heat per carbon fixed.[22]

Initially, the compelling observation from reducing metabolism seems to be about energy, but upon reflection, it really is about *reduced complexity*. Ultimately, all of life at the system level must satisfy the same description: synthesis is net exergonic when the environment is taken into account. Ludwig Boltzmann more than a century ago [81] had fully understood this point, to which we return in Chapter 7. In oxidative metabolisms, the anabolic *subsystem* does not satisfy it individually (this is why biomass can be set on fire in an oxygen atmosphere), whereas in reductive metabolisms an ideally efficient anabolism alone could be net exergonic. In reductive chemotrophic metabolisms, electron transfer is both the source of free energy and the mechanism of chemical synthesis, so an emergent biosphere did not need to generate mechanisms to "couple" the two processes. In contrast, for metabolisms operating in environments that are more oxidizing than the average carbon in biomass, free energy originating in solar flux must be imported to biochemistry as a whole. It must first be converted to a charge separation (a difficult chemical problem), and the charge separation must then be stabilized either in an oxidant/reductant pair, or in a proton-motive potential [77]. Only through control of the synthesized reductant can biosynthesis through reduction of $CO_2$ proceed. Therefore the *control* requirements to carry out chemotrophic anabolism are much less stringent than those needed to regenerate both reductants and terminal electron acceptors through processes such as photosynthesis.

Shown together with the TCA intermediates in Figure 4.2 are a collection of long-chain alkane alcohols, for reference. These provide an idea of the range of $\Delta_f G^0$ for the stoichiometry $[H_2]/[CO_2] = 3$ that asymptotically characterizes long-chain lipids (and is exact for the alkane monols). Acetic acid, with the stoichiometry $(CH_2O)_2$ of sugars, is the most reduced compound in the TCA cycle, and the least reduced compound in the lipid synthesis networks. We will see in the larger network of Figure 4.5 below that *acetate is a topological "gateway" molecule*, between the universal precursors and the lipid synthesis pathways. Figure 4.2 shows that, in the energetic context of a network driven by reduction of carbon, it is a low-energy drain of the TCA intermediates, and a high-energy source for lipid synthesis.

Figure 4.3 shows the molecular conversions around the cycle of Figure 4.1, with only carbon atoms illustrated, and the oxidation state of each atom indicated by a color. Each bond to the electron acceptor O is counted as withdrawing one "unit" of (negative) charge

---

[22] Reference [686] compares ATP cost per carbon fixed for acetogenesis and rTCA carbon fixation. Reference [90] provides a comparison in terms of ATP and reductants ($H_2$ equivalents) for several pathways in the synthesis of serine and glycine.

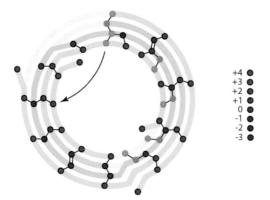

Figure 4.3 TCA cycle showing trajectories and oxidation states of individual carbon atoms. State +4 represents four bonds to oxygen (hence $CO_2$), and state $-4$ represents four bonds to hydrogen (methane $CH_4$). C–C bonds are treated as neutral. The reductive cycle (counterclockwise rotation) has the character of a vortex, incorporating fully oxidized carbon at each end of the molecules and reducing it as its position becomes more interior within the molecule.

from the bound carbon atom, and each bond to the electron donor H is counted as adding one unit.[23] Bonds between carbon atoms are counted as zero. Under this accounting, $CO_2$, with its four C–O bonds, contains a carbon atom with four "units" of excess (positive) charge. The carbon in methane would be the complement, reduced by four units, but the most reduced carbon among the TCA intermediates is a methyl group with only three C–H bonds. The characteristic median stoichiometry of biomass, $CH_2O$, is assigned a count of zero.

From the free energies of formation at the whole-molecule level in Figure 4.2, we may treat the redox index of each carbon as a crude proxy for its contribution to the free energy of formation of the corresponding molecule. Figure 4.3 shows that the turning of the rTCA cycle incorporates oxidized carbon at the molecule ends, and progressively moves it toward the molecule interior while simultaneously reducing it, until it reaches the most reduced form in the methyl group of acetate or the pair of aliphatic carbons of succinate. rTCA in a reducing environment may be likened metaphorically to a vortex, suggesting the one-way flow of carbon jointly in oxidation state and in biosynthetic complexity.

### 4.3.3.2 Carbon fixation between Scylla and Charybdis

The progression of carbon around the rTCA cycle, which is a progression up and down the ramp of average oxidation state seen in Figure 4.2, is a delicate balance to remain within the domain of reactivity between too oxidized and too reduced. The TCA intermediates (and many compounds adjacent to them in a network of common variations, which we discuss next) incorporate carboxyl, ketone, and alcohol groups juxtaposed in a variety of ways.

---

[23] These units are less than the charge on one electron, and are meant only as a coarse characterization of the asymmetry of orbitals between atoms of unequal electronegativity.

Electron-withdrawing groups such as ketones, adjacent to methyl groups, can lower the $pK_a$ of protons permitting facile ketone/enol interconversions that enable carboxylations and aldol additions.

To retain the capacity to generate complexity, the TCA intermediates must steer between two chemical dead ends, and various compounds show the proximity to those hazards.[24] Too many adjacent oxidized carbons lead to decarboxylations to bicarbonate or $CO_2$. Oxaloacetate is unstable to $\beta$-decarboxylation for this reason, and its homologue oxalo-succinate is so unstable that it is known as a free reagent in the cytoplasm only in a few organisms (in most lineages, it is only formed as an intermediate between $\alpha$-ketoglutarate and isocitrate, and remains bound to the fused isocitrate dehydrogenase that carries out both the $\beta$-carboxylation and the subsequent reduction). In respiratory metabolism, repeated oxidation from $NAD^+$ or other oxidants results in progressive elimination of $CO_2$.

The distinguishing step of the rTCA cycle – a reductive carboxyl insertion at a thioester formed between either acetate or succinate and CoA – *weakens* the oxidation of the terminal carbon by forming the thioester. A direct reductive carboxylation of a carboxylic acid group is never used.

On the other, reducing end, TCA intermediates approach the dead end of aliphatic C–C chains. Within the cycle, the aliphatic central pair of succinate provides a good example: in fermentative metabolisms of *Mycobacteria* succinate is a waste product excreted to obtain energy from glycolysis.[25] The almost fully reduced fatty acids are not targets for further carbon capture, and are used in membranes for the sake of their physical properties.

The TCA reactions must also strike a balance between enabling the formation of **complexity** and offering a certain inherent **selectivity** from the small-molecule network itself. In modern organisms, selection is provided by specificity of catalysts, but if the many compelling features of the rTCA cycle are a reflection of forces that have acted continuously since the pre-living geochemical era, we must understand what forms they would have taken in a world where less specific catalysts were available.

We will consider below some reaction networks that are both energetically and kinetically accessible, such as the very complex (indeed, quasi-intractable) network of aldol reactions from formaldehyde to form sugars. The reactive functional groups of rTCA are not as permissive of open-ended complexity, but if they can be formed, they do still open networks of possibility that risk becoming chaotic. We will see the first indications of proximal pathways that are known from actual organisms in Section 4.3.5, which appear to explore some of the easy forms of variation but stop short of molecular chaos. Perhaps the greatest challenge in thinking about the emergence of a selective but self-sustaining metabolism from geochemistry must come from understanding why two such balances –

---

[24] We owe this abstraction of the problem to colleagues George Cody and Chris Glein (personal communication).

[25] Tian *et al.* [805] and Baughn *et al.* [56] argue that this ketoglutarate dehydrogenase (KGD)-dependent pathway, which proceeds via succinate semialdehyde, is specifically regulated to operate with glycolysis, while a distinctive ketoglutarate oxidoreductase (KOR)-dependent pathway that proceeds directly to succinyl-CoA is regulated to operate with $\beta$-oxidation of fatty acids. The more remarkable feature of this KOR-dependent pathway is that the enzyme (ketoglutarate oxidoreductase) is the homologue of the enzyme commonly used in the *reductive* TCA pathway, but adapted to be oxygen tolerant, which the normal rTCA KOR enzyme is not.

between oxidation and reduction, and between productivity and selectivity – would be a *robust*, and in that sense "necessary," feature of planetary organic chemistry.

### 4.3.4  One-carbon metabolism in relation to TCA and anabolism

The core of the core has a second sector, different from TCA in the questions it raises about the emergence and organization of metabolism, but similarly essential and universally distributed. This sector consists of the direct reduction of one-carbon units. The intermediate oxidation states of carbon – formate, formaldehyde, and methanol – have much increased standard-state free energies of formation relative to the line passing through the rTCA intermediates in Figure 4.2. Therefore one-carbon reductions are not carried out on free molecules in solution, but rather on carbon atoms bound to a special class of cofactors known as *pterins*.[26] Pterins are unique in possessing a pair of subadjacent nitrogen atoms with finely tuned charge distributions, which may bind to carbon in different combinations through one, two, or three bonds. This allows pterins to bind highly oxidized formyl ($-CH=O$) groups, and continue to carry them through oxidation states of methenyl ($=CH-$), methylene ($-CH_2-$), and methyl ($-CH_3$). Three of these (other than methenyl) can be either used in direct exchange reactions or transferred to secondary specialized cofactors to support a diverse metabolism of $C_1$ exchanges.

#### 4.3.4.1  The hidden centrality of ancient one-carbon metabolism

At a first look, the $C_1$-pterin sector has the appearance of a biosynthetically complex "refinement" of metabolism, rather than an obviously central organizing motif like the TCA cycle. Its cofactors are larger than small-molecule metabolites and are by many forms of evidence (considered below) highly refined by evolution, while it contributes only singletons to biosynthesis. Even until very recently, arguments that this one-carbon sector was highly relevant to the emergence of metabolism [63, 282, 525, 689] turned mostly on its role as the reduction sequence in the *Wood–Ljungdahl* carbon fixation pathway to acetyl-CoA, which we consider in detail in Section 4.3.6.5 below. These arguments are bolstered by the great antiquity of the Wood–Ljungdahl pathway attested by its wide (though by no means universal) phylogenetic distribution, and are supported by important experimental results on one-carbon reduction and addition chemistry. However, they are also limited because they implicate a special role only for acetate (though in the very important activated thioester form acetyl-CoA). Acetate, recall, is the most reduced among the universal precursors and the one that, save for activation by CoA, would be a dead end and drain of redox energy.

Two more recent reviews of available evidence by Braakman and Smith [90, 91] support the conclusion of a very central role for direct $C_1$ reduction in early evolution but also complicate its interpretation. We will spend the remainder of this section, and Section 4.4,

---

[26] *Pterin* is a name referring to the class of cofactors, including folates and the methanopterins, which are both derived from a neopterin precursor [509]. We will return to say a great deal about them in Section 4.5 below.

reviewing the main points of that argument. Their multi-factor analysis of the requirements for early autotrophic metabolism suggests that $C_1$ reduction on pterin cofactors is much more widespread than has been recognized, and not necessarily associated with the complete Wood–Ljungdahl pathway. The presence of such a one-carbon sector makes sense of patterns of gene presence or absence in relation to the biosynthesis of glycine and serine and other amino acids derived from them,[27] and appears necessary in any consistent phylogenetic *and metabolic* reconstruction of the earliest bacterial and archaeal metabolisms. In suggesting a much more universal role for $C_1$ reduction than for the more complex Wood–Ljungdahl pathway, it emphasizes a geochemically more primitive process. At the same time, however, in the era of protein enzymes and resolvable phylogenetic lines of descent (the era following Woese's Darwinian Threshold), the only enzyme in the Wood–Ljungdahl pathway that is homologous across the tree of life is the last step – the step that is not universally distributed – the earlier one-carbon reduction, although ubiquitous in the small-molecule substrate, is supported by cofactors with many apparently independently derived biosynthetic steps [773].

In this situation two kinds of universality – gene presence/absence signatures and sequence similarity signatures – can be found in the same pathway, both suggestive of properties of the earliest metabolism, yet with quite different patterns of distribution. The difficulty of reconciling the two patterns suggests that the perceived similarity between Wood–Ljungdahl organisms and geochemical pathways may be premature, and we may require a better understanding of the evolutionary interval between the last common ancestor and modern clades, to recognize the correct relation.

### 4.3.4.2 The problem of understanding glycine synthesis in the early tree of life

Glycine is the simplest amino acid, and invariably appears as one of the most primordial in attempts (from almost any premises[28]) to assign a sequence to the inception of the amino acids within biochemistry. Yet the ketoacid precursor to glycine, glyoxylate, alone among the $\alpha$-ketoacid precursors of amino acids is not directly an intermediate in the TCA cycle.[29] It occupies an odd network position, connected to two side-pathways from TCA and also to the one-carbon reduction sequence. This position, which from a viewpoint that tries to emphasize any one network motif would make glycine seem more peripheral than we would expect, given its simplicity and antiquity, turns out to suggest a different kind of central organizing role in early metabolism.

We noted in the last section that all amino acids can be formed from precursors which are intermediates in the TCA cycle. Almost all of these are *only* found to be synthesized from these precursors in known organisms. The exceptions are glycine and serine, and the series of more complex amino acids synthesized from them. For these alone, a direct

---

[27] Like regularities in the rTCA cycle, regularities in amino acid biosynthesis from $C_1$ precursors are reflected in the genetic code, as we show in Chapter 5.

[28] For a review of the literature assigning an order to the use of amino acids as of 2004, see Trifonov [812].

[29] The only other $\alpha$-ketoacid which we treat as a direct precursor when describing some regularities of the genetic code in Chapter 5, and which is not a TCA intermediate, is 2-ketoisovalarate. However, this $\alpha$-ketoacid is produced by a branched-chain synthesis from pyruvate, from which it directly inherits the relevant $\alpha$-keto group.

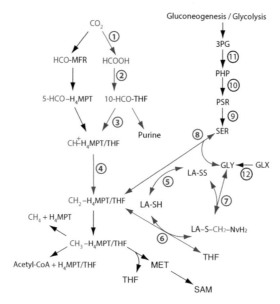

Figure 4.4 Schematic diagram of the reactions in direct $C_1$ metabolism. Reactions on the left-hand side involve $C_1$ bound to folate or pterin cofactors, called the "central superhighway." The cycle on the right involving lipoic acid (LA) interconverts, synthesizes or degrades glycine directly from ammonia, $CO_2$, and methylene groups from the superhighway. Other reactions on the right are alternative connections to glycine and serine either via phosphoglycerate or via glyoxylate. Reactions 1–8 are reconstructed in [90] as the ancestral route to glycine and serine in bacteria. The bottom reactions in the central superhighway include the synthesis of acetyl-CoA in the Wood–Ljungdahl pathway where the direct $C_1$ sequence was discovered, and also methane production in methanogens and a variety of methyl-group chemistry mediated by $S$-adenosyl-methionine. Abbreviations: MFR, methanofuran; $H_4$MPT, tetrahydromethanopterin; THF, tetrahydrofolate; MET, methionine; LA, lipoic acid; GLY, glycine; SER, serine; PSR, phosphoserine; PHP, 3-phosphohydroxypyruvate; 3PG, 3-phosphoglycerate; GLX, glyoxylate. (After Braakman and Smith [90], Creative Commons [153].)

biosynthetic pathway from $CO_2$ exists, by way of the $C_1$-pterin system. The direct synthesis, in bypassing the TCA intermediates, also bypasses *all* other known carbon fixation pathways (besides Wood–Ljungdahl), which we will review in Section 4.3.6.

Figure 4.4 shows the three pathways for glycine synthesis known in extant organisms, two by way of branches from TCA, and the third by $C_1$ reduction. The two versions of direct $C_1$ reduction in archaea and bacteria are shown on the left in the figure. Archaea, the left-most branch, carry $C_1$ on any of a diverse set of pterin cofactors, which we will argue in Section 4.4 diversified within the archaeal domain. The archaeal pterins are charged at the $N^5$ nitrogen atom, with an activated formyl group transferred from a methanofuran (MFR) cofactor. The bacterial path, the more central branch in the figure, carries $C_1$ on the simpler (and we will argue, older) pterin cofactor tetrahydrofolate (THF),[30] which is

30 In mammals, THF is derived from folic acid, also known as vitamin B9.

charged directly from formic acid, usually (though perhaps not always) on the $N^{10}$ nitrogen atom. The subsequent reductions, through a sequence of cyclic intermediates in which $C_1$ is bound to both $N^5$ and $N^{10}$ positions, have been termed the **"central superhighway" of THF metabolism** [509]. Formyl, methylene, and methyl groups may be donated to several other pathways, indicated schematically in Figure 4.4 and considered in more detail below.

The three routes to glycine, and their relation to the central superhighway, are the following.

**The oxidative pathway from serine** In eukaryotes and most late-branching bacteria, glycine is made *catabolically*, in the reaction sequence 9–11 in Figure 4.4 [507, 587]. The first step in the reaction is *oxidation* of 3-phosphoglycerate, an intermediate in the gluconeogenic pathway branching from the TCA cycle at pyruvate (shown below in Figure 4.5) to 3-phosphohydroxypyruvate (PHP). PHP is then reductively aminated to form phosphoserine. Finally, cleavage of phosphoserine, with transfer of a methylene group to THF, produces glycine. Glycine may be further broken down in what is known as the *glycine cleavage system* [435], to $CO_2$ and ammonia, donating a second methylene group to THF. In these organisms the folate pathway may be traversed in the oxidative direction, converting $C_1$ groups cleaved from serine and glycine to formyl groups needed in purine, thymidilate, and formyl-tRNA synthesis [509]. For reference, in analyses of phylogenetic distribution of pathways in Section 4.4, we will refer to this sequence as the *oxidative* route to glycine.

**Direct synthesis from the central superhighway** The glycine cleavage system is so named because it was discovered as a breakdown pathway in mammals, but it has since been understood to be a reversible reaction cycle [50, 54, 435, 842]. The $C_1$-THF system, with the energies of intermediate oxidation states of the bound carbon matched by evolution to the electron-accepting or electron-donating cofactors, can also be reversible [509] – this is one of the reasons for the appellation "central superhighway." All requirements for its use as a synthetic pathway are therefore met in the context of one-carbon reduction from formyl to methyl groups in folate-based carbon fixation via the Wood–Ljungdahl pathway. Numerous clades within bacteria and the Thermoplasmatales and Halobacteriales within archaea contain members with gene complements, indicating that these organisms directly synthesize glycine and serine through the reverse of the glycine cleavage system and serine hydroxymethyl transferase system.[31] We refer to this as the *reductive* pathway to glycine. The argument goes that, since the pathways are reversible, we should not exclude the possibility that the reductive synthesis is used much more widely than in the Wood–Ljungdahl context. Many pieces of evidence from gene distributions make it difficult to reach any other conclusion. A wide and diversified use of reductive glycine synthesis would also seem to provide a pathway-based rationale for certain regularities in the genetic code that are otherwise surprising. We review these in Section 5.3.5.

---

[31] For a cladogram showing their distribution see Figure 3 of [90].

Because direct reductive synthesis does not employ any precursors in the TCA cycle or its branches, it also does not require passage through acetyl-CoA, and hence can be considered independently of the presence or absence of the terminal enzyme synthesizing acetyl-CoA in the Wood–Ljungdahl pathway. Braakman and Smith [90] argued from a combination of comparative genomics, phylogenetic reconstruction, and network modeling in Section 4.4, that this route to glycine is both the ancestral route, and among extant clades is much more widely preserved than Wood–Ljungdahl carbon fixation.

**Reductive transamination from glyoxylate** Finally, glycine may be synthesized directly by reductive transamination of glyoxylate (GLX) [507], the same process used to produce the first amino acids from all $\alpha$-ketoacid precursors in the TCA cycle. We refer to this as the *glyoxylate* route. Glyoxylate itself is not a TCA intermediate, but is found in the network of aldol reactions that branches from the TCA cycle, shown in Figure 4.5 and discussed at length in Section 4.3.6. Glyoxylate may be present as an intermediate in organisms that use the glyoxylate shunt as a heterotrophic pathway, and it is also produced in a salvage pathway to remove the harmful metabolite 2-phosphoglycolate produced in a process called *photorespiration* [215, 216]. Photorespiration apparently occurs as a result of a harmful failure of specificity in the enzyme RubisCO which fixes carbon in many photosynthesizers, and glycine production from glyoxylate is biosynthetically important in many such clades, including cyanobacteria and plants.

Following glycine synthesis by transamination, serine may be formed by first degrading one molecule of glycine in the glycine cleavage system, and then using the produced methylene-THF with a second molecule of glycine to form serine, as in the direct reductive synthesis. In this synthesis, folates neither accept nor donate net carbon, and the oxidation state of $C_1$ bound to THF does not change, but one $CO_2$ and one ammonia are lost from glycine cleavage.

### 4.3.5 The universal covering network of autotrophic carbon fixation

The universal character of metabolism is, on the one hand, both a stronger and more striking fact of life from the vantage point of modern molecular biology than Donald Nicholson could have foreseen when he began his compilations more than a half-century ago, and on the other hand, it is a challenging regularity to express precisely. Many metabolites required by all organisms, if traced to their ultimate origin in $CO_2$ and other inorganic inputs, require some key intermediates or reaction sequences that show no variation among known organisms. In autotrophic organisms they may be synthesized directly along these essential pathways, while in heterotrophs the dependence may be very indirect, involving multiple species and trophic exchanges, salvage pathways, and other such twists and turns, making general statements difficult in biology. Universal constraints on synthesis exist, but they require the correct unit of analysis and aggregation scale to express in rules. For this purpose, biological units such as organisms are not always the appropriate ones.

For some essential functions such as carbon fixation, it would be incorrect to say that any single pathway, defined as a sufficiently complete reaction sequence to provide this function, is required by all living systems, even indirectly. *Six pathways are now known to support autotrophic carbon fixation* [381], which are in part genuinely different and independent. Yet the variations are mostly minor; all carbon fixation pathways show large overlaps of reaction sequences, and even larger homologies of reaction mechanisms and local-group chemistry. Where truly significant innovations have occurred, they appear to have been constrained to integrate within the same overall system, and therefore they feed their products back through a small standard set of precursors to anabolism and other key intermediates. For such cases, universality is not a property of any particular pathway, but must be recognized in the relations among pathways and the rules that have apparently governed the evolution of variants.

To describe metabolic universality, one needs a multilevel representation, in order to see that some regularities exist at the level of metabolites, reactions, or pathway segments, while others exist as patterns of local-group chemistry or limits on innovation imposed by a need to integrate within a self-maintaining system. We will introduce the idea of a **universal covering network** of autotrophic carbon fixation as the first level in such a description of regularities. The covering network, shown in Figure 4.5, is the smallest network that contains all known carbon fixation pathways, together with the universal anabolic precursors and the branch pathways to the major CHO compounds: sugars, fatty acids, and isoprene alcohols. (In Figure 4.5 and several related figures, in place of the traditional multi-arrow notation for reactions, we use a discrete bipartite-graph notation, explained in Appendix 4.8.)

The three most striking features of the covering network are (1) that it is *small*, (2) that it is easy to define a relatively unambiguous boundary, and (3) that it has significant internal structure.

**The covering network is small but inclusive**  The figure shows only 38 organic intermediates, yet includes the six currently known carbon fixation pathways, all five universal CHO precursors to anabolism,[32] the gluconeogenic pathway as far as fructose, and the entire surrounding pentose phosphate pathway including erythrose, ribose, and ribulose, and the key reactions in both fatty acid and isoprene synthesis. Any subsequent elaboration of hexose sugars is simple and repetitive: most sugars are formed from glucose which is a more stable isomer of fructose due to its planar arrangement of hydroxyl groups, and much of complex sugar chemistry consists of the assembly of polysaccharides. Fatty acid extension occurs by recursive application of the malonate pathway shown, and isoprene extension occurs by ligation of isopentenyl and its isomer dimethyl-allyl, formed by reduction and decarboxylation from hydroxymethylglutarate (shown).

**The boundaries of core CHO chemistry are largely unambiguous**  It has been possible to draw only the CHO skeleton of the covering network – omitting phosphorylated forms and thioesters, because the role of phosphate and sulfur in the core is very limited.

---

[32] Recall that these are acetyl-CoA, pyruvate or phosphoenolpyruvate, oxaloacetate, succinate or succinyl-CoA, and $\alpha$-ketoglutarate.

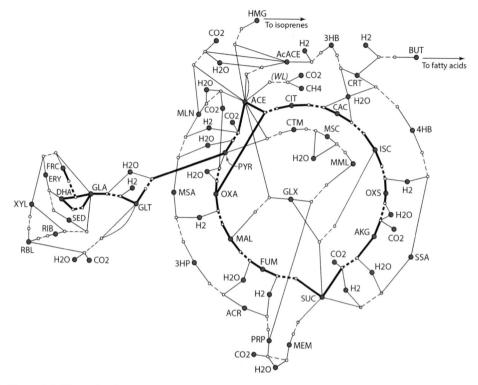

Figure 4.5 The projection of the universal covering network of core metabolism onto its CHO backbones. Phosphorylated intermediates and thioesters with coenzyme-A are not shown explicitly. Arcs of the reductive citric acid cycle, which pass through the universal biosynthetic precursors of Figure 4.1, and gluconeogenesis are marked in bold. The Wood–Ljungdahl pathway (labeled *WL*), without its cofactors and reductants shown, is represented by the last reaction of the acetyl-CoA synthase, which absent bonds to cofactors is the inverse of a disproportionation of $CO_2$ with $CH_4$. Abbreviations: acetate (ACE); pyruvate (PYR); oxaloacetate (OXA); malate (MAL); fumarate (FUM); succinate (SUC); $\alpha$-ketoglutarate (AKG); oxalosuccinate (OXS); isocitrate (ISC); *cis*-aconitate (CAC); citrate (CIT); malonate (MLN); malonate semialdehyde (MSA); 3-hydroxypropionate (3HP); acrolyate (ACR); propionate (PRP); methylmalonate (MEM); succinate semialdehyde (SSA); 4-hydroxybutyrate (4HB); crotonate (CRT); 3-hydroxybutyrate (3HB); acetoacetate (AcACE); butyrate (BUT); hydroxymethyl-glutarate (HMG); glyoxylate (GLX); methyl-malate (MML); mesaconate (MSC); citramalate (CTM); glycerate (GLT); glyceraldehyde (GLA); dihydroxyacetone (DHA); fructose (FRC); erythrose (ERY); sedoheptulose (SED); xylulose (XYL); ribulose (RBL); ribose (RIB). (After [91], Creative Commons [153].)

They provide leaving groups and stabilize certain intermediate compounds, but they are not incorporated as structural elements. The most extensive use of phosphates occurs in the sugar phosphate network, where they occupy alcohols that *do not* participate in the addition reactions shown.[33] The variant carbon fixation pathways shown do not introduce

---

[33] The sole use of phosphate as an energetic leaving group occurs in gluconeogenesis/glycolysis, where 3-phosphoglycerate is converted to 1,3-bisphosphoglycerate and dephosphorylated to form glyceraldehyde-3-phosphate. Here phosphate performs a dehydration jointly with oxidation of NADH to $NAD^+$, to reduce the terminal carboxyl group on phosphoglycerate. Source: MetaCyc [412].

ambiguity in the boundaries of the CHO network, because their products are always directed back to the precursors in the TCA cycle (shown in bold). The addition of ammonia defines the point of departure into the CHON network, which includes amino acid and nucleotide biosynthesis, and the formation of other heterocycles and amino sugars. Subsequent addition of sulfur as a structural element within these heterocycles first occurs at the level of the sulfur amino acids (cysteine and methionine) and more complex cofactors including thiamin and biotin. Phosphate enters as a permanent structural element primarily in ester bonds of oligomers and RNA-derived cofactors, and in phospholipids. All of these subsequent network elaborations contain some universal reactions and many variations, but those variations are essentially independent of variations in the CHO covering network, separated through the boundary of the TCA intermediates and gluconeogenesis/glycolysis.

**The network shows significant internal structure** Finally, it is apparent that the core network is a highly non-random graph. It is organized around the TCA cycle and the gluconeogenic pathway (also shown in bold). Not all reactions and not all metabolites in this pathway are present in all organisms, but subsets of these arcs are used to connect any variant carbon fixation pathways to those universal precursors that they do not include.

The primary connected component of the covering network is laid out in a series of concentric rings, with inorganic inputs and outputs appearing as rungs of a ladder between the rings. As we show in Section 4.3.7 below, this topological structure arises as a reflection of redundant chemistry at the local-group level in the pathways which are different rings. Two compounds clearly organize the entire network: acetyl-CoA which is *the most central compound in biochemistry*, and succinate or succinyl-CoA (which are not distinguished in the skeleton CHO graph). These are the two thioesters in the rTCA cycle, formed by substrate-level phosphorylation. Removal of these two compounds separates the concentric rings into two independent modules (one containing short metabolites and the other containing long metabolites which are the homologues to the short ones). Variation in carbon fixation paths has largely occurred by matching different segments from the short-molecule module with different segments from the long-molecule module (shown below in Figure 4.6).

A second major structural feature of the covering network is the strong modular separation of the integrated collection of core loops from the pentose-phosphate subnetwork, with only the gluconeogenic pathway through (phospho-glycerate) connecting the two. A less obvious module boundary separates the loops from the one-carbon reduction sequence on folates. In Figure 4.5, the latter is shown as a simple disproportionation (labeled *WL*) between $CO_2$ and $CH_4$, which is the CHO skeleton of the acetyl-CoA synthase reaction (discussed in Section 4.3.6.5). The cobweb-like cross-linkages both within the pentose phosphate pathway, and bridging the loops in the central network through glyoxylate, are formed by aldol reactions, and express the (sometimes chaotic) diversity created by this class of reactions. A topologically important connection that the covering network does not show, because it arises through one-carbon exchanges and amines, is the three-way linkage of glyoxylate, 3-phosphoglycerate, and acetyl-CoA introduced previously in Section 4.3.4.

### 4.3.6  Description of the six fixation pathways

We now provide brief descriptions of the six fixation pathways individually, plus the glyoxylate shunt which is not a fixation pathway but is related to the others in both its reaction network and many of its functional attributes. It will be possible to identify a basic template of reactions within which each pathway is distinguished by a few reaction mechanisms. From these variations, their consequences for the fitness of organisms in different contexts, and (considered below) their phylogenetic distribution, it will be possible to argue for a sequence in which the pathways came into use, and for the adaptations that drove each innovation.

As we review the pathways, we wish to keep always in mind that each of these is found in *modern organisms*: the pathway intermediates are few and their reactions are exclusive because they are catalyzed by protein enzymes that are both powerful and specific. To interpret the scope of pathway variability with regard to the emergence of biochemistry, we must always be prepared to think of these reactions in contexts of less powerful and less specific catalysts. In such a context many clusters of pathways are minor variations on a single chemical theme, and only distinctions among large modules in the covering network of Figure 4.5 reflect significantly different chemical domains. Within any cluster of similar pathways, the variants may be thought of as "fluctuations" within the central theme, depending only on a few reaction variations.

The co-presence of all of the pathways within a cluster should tell us something about the possibilities or limits on the emergence of metabolism, beyond what any one pathway tells on its own. They are all "present" in the chemistry of the easily accessible, even if most of them are not realized in a sample consisting of any particular clade or species. We are inclined to interpret the modern pathway variants as producing a distribution in clades, which roughly recapitulates the distribution in fluctuating paths for carbon-flux in the less tightly focused geochemical networks that preceded modern cells. In this sense, while cellular evolution strongly changes the mechanisms of variation on the short term as species partition niches into roles for which they compete to exclude each other, the full scope of variation on the evolutionary long term may continue to reflect lower-level kinetic and topological features of the underlying network.

We will return to this theme repeatedly throughout our argument. The powerful mechanisms of cellular evolution may succeed in forcing many variants into existence, simply as a consequence of random exploration of phenotypes in different lineages. It is also an observed property of many communities that species assemblies may partition quite close ecological roles among competitors, each becoming highly specialized and exclusive even though both perform overall similar functions, and differ only in a few fine details [286]. Ultimately, however, all strategies that survive in the long term must do so under competitive pressure. Strategies that deviate in core metabolism from the paths of least resistance, even by small degrees, take on fitness costs either in energy demand, overcoming of side-reactions, maintenance of enzyme specificity, or other basic physiological functions. Because core metabolism carries such high flux as the source of all biomass, even small costs per reaction are amplified when they occur in the core network. In similar fashion, even though short-term variation may be common simply because it is continually generated, variants that survive in the very long term must show consistent fitness advantages sufficient to overcome even small differences in physiological costs. Both network centrality and long-term selection are amplifiers of small distinctions. Therefore, while we expect that short-term Darwinian adaptations to local environmental details may mask small chemical differences in short-term evolution, we also expect that progressively finer distinctions in what constitutes "least resistance" will be exposed in progressively longer terms of evolution.

The first four pathways, shown in Figure 4.6, are all **autocatalytic loops** at the level of the small-molecule substrate, so the three non-TCA pathways are functionally similar to the rTCA loop. All four pathways may be understood as minor variants on the rTCA structure. Most of the complete pathways share reaction sequences with rTCA or with each other. Where they differ, they take the form of one or a few innovative reactions followed by similar sequences of downstream transformations. We show the explicit pathway overlaps in Figure 4.6, and return to the other similarities of reaction mechanism in Section 4.3.7.

### 4.3.6.1 Reductive citric acid (rTCA) cycle

The rTCA cycle is the form of the citric acid or tricarboxylic acid (TCA) cycle introduced in Section 4.3.3, traversed in the reductive direction. The existence of the cycle was first worked out by Evans, Buchanan and Arnon [234]. It was quickly recognized by Jack Corliss, John Baross, and Sarah Hoffman [161] as a plausible candidate for the earliest carbon fixation cycle and even the origin of metabolism. This proposal or variations on it were subsequently developed in more detail by Günter Wächtershäuser [843, 844, 845], Harold Morowitz [574], and others. rTCA is unique among the carbon fixation pathways in that its intermediates are *direct* precursors to all of biomass. Therefore, even as other carbon fixation pathways have been discovered, this one remains special as a candidate for the emergence of metabolism. Considerations of distribution and phylogenetic reconstruction below strengthen this interpretation, though they also suggest that consideration of rTCA alone is incomplete.

The rTCA loop is a sequence of eleven intermediates and eleven reactions, shown in black in Figure 4.6, which reduce two molecules of $CO_2$, and combine these through a substrate-level phosphorylation with CoA, to form one molecule of acetyl-CoA. In the cycle, one molecule of oxaloacetate grows by condensation with two $CO_2$ and is reduced and activated with CoA. The result, citryl-CoA, undergoes a retro-aldol cleavage to regenerate oxaloacetate and acetyl-CoA. Here we separate the formation of citryl-CoA from its subsequent retro-aldol cleavage, as this is argued to be the original reaction sequence [35, 36], and the one displaying the closest homology in the substrate-level

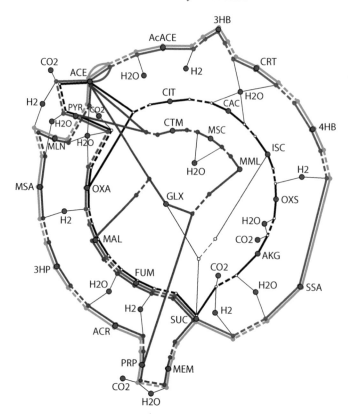

Figure 4.6 Four autocatalytic-loop carbon fixation pathways are contained in the covering graph of Figure 4.5, and pass through some or all of the precursors of anabolism. Color scheme is: rTCA cycle (black); dicarboxylate/4-hydroxybutyrate cycle (red); 3-hydroxypropionate bicycle (blue); 3-hydroxypropionate/4-hydroxybutyrate cycle (green). Acetate (ACE) and succinate (SUC) (both as coenzyme-A thioesters) are intermediates in all four pathways, and define the boundaries on which distinct pathway segments are combined to form different loops. Radially aligned reactions are homologous in local-group chemistry; deviations from strict homology in different pathways appear as excursions from concentric circles. (After [91], Creative Commons [153].)

phosphorylation with that of succinyl-CoA.[34] A second arc of reactions then condenses two further $CO_2$ with acetyl-CoA to produce a second molecule of oxaloacetate, completing the network-autocatalytic topology and making the cycle self-amplifying.

The distinctive reaction in the rTCA pathway is a carbonyl insertion at a thioester (acetyl-CoA or succinyl-CoA), performed by a family of conserved ferredoxin-dependent oxidoreductases which are triple-$Fe_4S_4$-cluster proteins [123]. The sulfur-containing

---

[34] Thioesterification of acetate and succinate is required to enable the reductive carboxylation reaction (termed "carbonyl insertion") that forms pyruvate and $\alpha$-ketoglutarate, respectively. The distinction between two variants of the oxidative reaction at succinate, and the fermentative pathways in which they are used, crucially involves the energy difference between succinate and succinyl-CoA. The direction of the pathway, and the requirements on secondary energy systems, depend on whether succinate is to be excreted as a low free energy waste product or reactivated as a cycle intermediate [56, 805].

cofactor thiamin pyrophosphate is an essential mediator of these reactions in all versions of the cycle [123], and reducing potential is derived from a class of low-potential ferredoxins [436]. Other essential reactions are $\beta$-carboxylations from carboxyl groups carried by biotin [34, 495] (another sulfur-containing cofactor), reductions (of two kinds, NADH- or NADPH-dependent reductions of alcohols, and an $FADH_2$-dependent reduction of fumarate), dehydrations and one rehydration. The aconitate intermediate in the dehydration/rehydration sequence isocitrate/*cis*-aconitate/citrate is also the site of aldol cleavage.

A reductive pathway, rTCA is the energetic reverse of the oxidative Krebs cycle, and is related to a variety of mixed oxidative-reductive split pathways that are roughly redox neutral and have been fully characterized as fermentative pathways in mycobacteria [56, 74, 338, 805, 852]. The oxidative and reductive reaction sequences are distinguished at some points by a few key enzymes and electron donors or acceptors, and extensive minor variations of these are known both in organisms that use the full loops and in fermenters. The cycle is found in many anaerobic and micro-aerophylic bacterial lineages, including Aquificales, Chlorobi (where it was discovered), and $\delta$- and $\epsilon$-proteobacteria. It was believed for a time that the rTCA cycle was present in both bacteria and archaea, but current evidence indicates that it is an exclusively bacterial pathway. The archaeal pathway that was mistaken for rTCA is the dicarboxylate-4-hydroxybutyrate cycle, considered next.

The distribution of full rTCA cycling correlates with clades whose origins are placed in a pre-oxygenic earth, while fermentative TCA arcs and the oxidative Krebs cycle are found in clades that by phylogenetic position (such as $\gamma$-, $\beta$- and $\alpha$-proteobacteria) and biochemical properties (membrane quinones and intracellular redox couples involving them [707]) arose during or after the rise of oxygen. The co-presence of enzymes descended from both the reductive and oxidative cycles in members of clades that at a coarser level are known to straddle the rise of oxygen, such as actinobacteria and proteobacteria [707], may provide detailed mechanistic evidence about the reversal of core metabolism from ancestral reductive modes to later derived oxidative modes. (We consider the rise of oxygen and the role of rTCA reactions and precursors in this transition in Section 4.4.6 below.)

### 4.3.6.2 Dicarboxylate/4-hydroxybutyrate cycle

The dicarboxylate/4-hydroxybutyrate (DC/4HB) cycle [63, 379], shown in red in Figure 4.6, is an archaeal cycle initially mistaken for rTCA because it shares the same reaction sequence from acetyl-CoA to succinate. The archaeal cycle differs by the 4-hydroxybutyrate (4HB) pathway emerging from succinate, which incorporates no new carbon, and self-amplifies by splitting its $C_4$ product acetoacetyl-CoA rather than the $C_6$ rTCA product citryl-CoA. Acetate is therefore the network catalyst in the DC/4HB cycle rather than oxaloacetate as in rTCA, and the cleavage reaction directly produces two copies of the network catalyst (although a second CoA thioester must be formed to allow both copies to be used, without further carbon-incorporation steps). Every turn of the complete DC/4HB cycle therefore incorporates only two $CO_2$ molecules, rather than the four incorporated to regenerate one oxaloacetate in a full turn of rTCA.

Because the first five reactions in the cycle are identical to those of rTCA, the DC/4HB pathway employs the same ferredoxin-dependent carbonyl-insertion reaction, though only at acetyl-CoA. The second arc (the 4HB arc) uses distinctive reactions associated with 4-hydroxybutyrate fermentation. In particular, the dehydration/isomerization sequence from 4-hydroxybutyryl-CoA to crotonyl-CoA is performed by a flavin-dependent protein containing an $Fe_4S_4$ cluster. This reaction involves an unusual ketyl-radical intermediate that is the distinguishing innovation of this pathway [528, 577]. Because $\alpha$-ketoglutarate is not in the primary pathway, an anaplerotic pathway must be used to produce $\alpha$-ketoglutarate as a precursor to amino acid biosyntheses that proceed via glutamic acid.

The DC/4HB cycle has so far been found only in anaerobic Crenarchaeota, but within this group it is believed to be widely distributed phylogenetically [63, 379]. The 4HB segment begins with reactions also found in 4-hydroxybutyrate and $\gamma$-aminobutyrate fermenters in the Clostridia (a subgroup of Firmicutes within the bacteria), and terminates in the reverse of reactions in the isoprene biosynthesis pathway, which is the source of membrane lipids in the archaea. The arc thus has a dual affinity to Firmicutes and archaea that, as we note in Section 4.4 below, characterizes many kinds of phylogenetic reconstruction that include these two groups. Both ends of the 4HB pathway have been found in Clostridia, though they are not known to form a continuous arc capable of fixing carbon, anywhere in this group. The most striking case is *Clostridium kluyveri*, which contains pathway segments to reach all DC/4HB intermediates, but (like fermentative TCA organisms) initiates the segments from both ends (succinate and acetoacetate) to perform fermentation rather than carbon fixation [720]. We return to implications for the origin of this pathway in Section 4.4.

### 4.3.6.3  3-Hydroxypropionate bicycle

The 3-hydroxypropionate (3HP) bicycle [908], shown in blue in Figure 4.6, has the most complex network topology of the fixation pathways, using two linked cycles to regenerate its network catalysts. The initiating 3-hydroxypropionate (3HP) pathway, first discovered in *Chloroflexus auranticus*, is an alternative departure from rTCA, this time along the small-molecule arc beginning from acetyl-CoA, rather than along the large-molecule arc from succinate as the 4HB pathway is. Although the presence of an innovative 3HP pathway in *Chloroflexus* was recognized as early as 1987 [363, 364], another 22 years was needed to solve for the completion of the bicycle, and thus to confirm that it is an autocatalytic carbon-fixation pathway.

The reactions in the bicycle begin with the biotin-dependent carboxylation of acetyl-CoA to form malonyl-CoA, from the fatty acid synthesis pathway, followed by a distinctive thioesterification of propionate and a second, homologous carboxylation of propionyl-CoA (to methylmalyl-CoA) followed by isomerization to form succinyl-CoA. This initiating pathway continues, to form a first cycle, by tracing the *oxidative* TCA arc to malate which undergoes a retro-aldol cleavage to regenerate acetate and produce a glyoxylate that must be recycled. A second cycle is initiated by an aldol condensation of a second molecule of propionyl-CoA with the glyoxylate produced in the first cycle, to yield

$\beta$-methylmalyl-CoA, which follows a sequence of reduction and isomerization steps through an enoyl intermediate (mesaconate) similar to the large-molecule arc of rTCA. Cleavage of the $C_5$ product (citramalate) of the second cycle yields acetate and pyruvate. Acetate is therefore the network catalyst in both parts of the bicycle, and pyruvate is the output from fixation of three carbon atoms. For the cycle to be self-amplifying, pyruvate must be recycled through some anaplerotic pathway to one of the cycle intermediates. As for the DC/4HB cycle, $\alpha$-ketoglutarate must also be produced anaplerotically to support amino acid biosynthesis.

The 3HP pathway innovation is believed to represent an adaptation to alkaline environments in which the $CO_2/HCO_3^-$ (bicarbonate) equilibrium strongly favors bicarbonate. All carbon is drawn from bicarbonate in the form of carboxy-biotin, thus bypassing the carbonyl insertion step required in the rTCA and DC/4HB pathways. While topologically complex, the bicycle makes extensive use of relatively simple aldol chemistry, in a network also used in the glyoxylate shunt (discussed in Section 4.3.6.7.) From the centrality and apparent facility of these reactions, we will argue in Section 4.4 that the evolutionary innovation of the bicycle was not as surprising as its complex topology might suggest.

### 4.3.6.4 3-Hydroxypropionate/4-hydroxybutyrate cycle

The 3-hydroxypropionate/4-hydroxybutyrate (3HP/4HB) cycle [799], shown in green in Figure 4.6, is an archaeal parallel to the bacterial 3HP bicycle. Unlike the bicycle, it is a single autocatalytic loop, in which the first arc is the 3HP pathway, and the second arc is the 4HB pathway. Like DC/4HB, 3HP/4HB uses acetyl-CoA as network catalyst and fixes two $CO_2$ to form acetoacetyl-CoA.

The pathway is found in the Sulfolobales (a group of Crenarchaeota), where it combines the crenarchaeal 4HB solution to autotrophic carbon fixation with the bicarbonate adaptation of the 3HP pathway. Like the 3HP bicycle, the 3HP/4HB pathway bypasses the carbonyl-insertion reactions of rTCA and DC/4HB, and is thought to be an adaptation to alkalinity. Because the 4HB arc does not fix additional carbon, this adaptation resulted in a simpler pathway structure than the bicycle. Since the 3HP/4HB loop fails to pass through *all three* of pyruvate, malate, and $\alpha$-ketoglutarate, several anaplerotic reactions are required to connect its intermediates to the precursors of anabolism.

The 3HP pathway uses two key biotin-dependent carboxylation reactions (to malonyl-CoA and methylmalonyl-CoA), suggesting a bacterial origin as this enzyme class features prominently in the biosynthesis of the fatty acids. Since both in bacteria and archaea it appears to be a derived pathway, its presence in both domains presents a phylogenetic puzzle. A comparison of the bacterial and archaeal enzymes for thioesterification of propionate to propionyl-CoA has been interpreted as implying convergent evolution for these essential steps in the pathway [799]. However, to the extent that specialization to alkalinity leads to both a restricted common environment, and a stressful environment possibly inducing increased lateral gene transfer [97], a lateral transfer followed by subsequent divergence remains a possible explanation.

### 4.3.6.5 The Wood–Ljungdahl pathway: one-carbon reduction and
### acetyl-CoA synthesis

The Wood–Ljungdahl (W-L) pathway [63, 491, 492, 821] is the only one among the known carbon fixation pathways that does not recycle a fraction of its small-molecule substrate as network catalysts. Instead, it employs cofactors in an expanded role as carriers of $C_1$ groups through most of the process of reduction from $CO_2$ or formate to bound methyl ($-CH_3$) groups, and only in its final reaction assembles the activated $C_2$ compound acetyl-CoA. When projected onto the small-molecule substrate, W-L is a "linear" pathway. Self-amplifying feedback loops are first found only in the much larger and more complex network that includes folate biosynthesis. The contrast between the existence of some positive feedback through simple molecules in short pathways, below the cofactor level, in the other fixation pathways, and feedback only through cofactors in W-L, requires us to think differently about how the two classes of organosynthetic reactions could first have transitioned from geochemistry and been brought under organic control.

Most of the reactions required for fixation of oxidized carbon take place along the "central superhighway" on folates or pterins, introduced in Section 4.3.4 and shown in Figure 4.4. As that introduction explained, the folate pathway has links to $C_1$ chemistry at many points, including transfers of one-carbon groups at several oxidation states, and amino acid metabolism involving glycine and serine. We will review evidence in Section 4.4 that these functions of the pathway are widely dispersed phylogenetically and define its essential roles for most organisms. Against this background of very widely distributed $C_1$ chemistry, the terminal step leading to acetyl-CoA synthesis, and rendering the complete W-L pathway sufficient for autotrophic carbon fixation, appears at the substrate level to be an isolated innovation, yet as we have noted, from the perspective of enzyme homology, exactly the opposite is the case.

The W-L pathway consists of a sequence of five reactions that first reduce one $CO_2$ to a methyl group, while in parallel a second $CO_2$ is reduced to CO, after which the methyl and CO groups are combined with each other and with a molecule of CoA to form the thioester acetyl-CoA. The reduction to CO, and the synthesis of acetyl-CoA, are both performed by the bi-functional CO-dehydrogenase/acetyl-CoA synthase (CODH/ACS), a highly conserved enzyme complex with Ni-$[Fe_4S_5]$ and Ni-Ni-$[Fe_4S_4]$ centers [176, 197, 726, 789]. Methyl-group transfer from the folate or pterin carrier to the ACS active site is performed by a corrinoid iron-sulfur protein in which the cobalt-tetrapyrrole cofactor cobalamin (vitamin $B_{12}$) is part of the active site [49, 59].

Acetyl-CoA synthesis in the CODH/ACS is similar to glycine synthesis in the reductive pathway from Figure 4.4. The CODH/ACS combines $-CH_3$ with CO and reduces the product (somewhat) with HS-CoA. The glycine decarboxylase complex (sometimes also called the glycine synthase) combines $-CH_2-$ with $CO_2$ and reductively aminates the resulting $\alpha$-ketone. We return in Section 4.5.2.2 below to the quite detailed similarity of roles for sulfur as carbon carrier and co-reductant in the cofactor systems for these two pathways, which apparently reflect convergent evolution in very early small-metabolite cofactors.

W-L itself produces only acetyl-CoA and no other anabolic precursors. To serve as an autotrophic fixation pathway it must be followed by any of several combinations of incomplete TCA arcs to form what are collectively known as the **reductive acetyl-CoA pathways**.

### A more evolved dependence on cofactors

The nature of cofactor support in W-L is subtly, but we believe importantly, different from cofactor support of the other autotrophic fixation pathways. It is certainly the case that *all* carbon fixation pathways in extant organisms employ distinctive and apparently unique enzymes and inventories of essential cofactors. For example, the $\beta$-carboxylations in both rTCA and the 3-hydroxypropionate pathway require biotin as a source of sufficiently energetic carboxyl, leaving groups to support the required group transfers. The rTCA cycle further relies on reduced ferredoxin, a simple iron-sulfur protein, and on thiamin in its reductive carbonyl-insertion reactions. In all the latter cases, however, the cofactors mediate individual group transfers in single contexts. The function provided by pterin cofactors in W-L is distinct and arguably more complex. Pterins undergo elaborate multistep cycles, mediating capture of formate, reduction of carbon bound to one or two nitrogen atoms, and transfer of formyl, methylene, or methyl groups, for which they provide a finely tuned free energy landscape across multiple oxidation states of the bound carbon atom [509]. In Section 4.5.2 we review evidence that the landscape is evolved as a whole in different pterin families to support different metabolic strategies that may involve quite long-range changes in the paths by which carbon flows to essential metabolites, most importantly the biosynthetic pathways for glycine and serine. The distinctive use of cofactors within W-L continues with the dependence of the acetyl-CoA synthesis on cobalamin, a biosynthetically elaborate and highly reduced tetrapyrrole capable of two-electron transfer.

W-L is phylogenetically widely distributed, and is the only one among the six autotrophic pathways found in both bacteria and archaea. It is found in acetogenic bacteria and methanogenic euryarchaeota (where in addition to fixation of some $CO_2$ the excretion of reduced carbon is used to drive an energy metabolism), as well as in sulfate reducers, and possibly anaerobic ammonium oxidizers.

Because of the wide distribution of the W-L pathway, the existence of one-carbon reduction in geochemical processes, and the analogy of the addition reaction performed by the CODH/ACS enzyme to Fischer–Tropsch-type synthesis [31] (discussed at length in Chapter 6), this pathway has frequently been promoted as a model for the earliest metabolism [63, 525, 689]. The different patterns of universality between the substrate chemistry and the cofactor biosynthetic enzymes and pathway enzymes, however, as well as striking differences between the energetic systems in acetogenic bacteria and methanogenic archaea (discussed in Section 6.3.1.3), for which pterin cofactors must be finely tuned to enable the required efficiency, lead us to believe that the interpretation of these patterns will not be a simple matter. The cofactor and bioenergetic systems that render carbon fixation through one-carbon units a self-sustaining system reflect an extensive history of the evolution of *control, integration, and autonomy* of what once must have been background geochemical processes. Much of the evolutionary history of cofactors and especially proteinaceous enzymes likely reflects incremental advances in the parallel solution of each of these problems.

### 4.3.6.6 Calvin–Benson–Bassham network

The Calvin–Benson–Bassham (CBB) cycle [55, 792] is responsible for most of known carbon fixation in the biosphere. Despite its enormous importance for current carbon cycling, and the apparently great phylogenetic depth of the common ancestor to the key enzymes [793], a variety of evidence places the use of the CBB cycle as an autotrophic carbon fixation pathway relatively late, and apparently distinctive to bacteria and eukaryotic endosymbionts of bacterial origin. We therefore regard the CBB cycle as a derived fixation pathway, dependent on earlier metabolic networks of considerable complexity and not directly informative about the earliest stages of metabolic emergence, and we give it a relatively brief account here.

In much the same way as W-L adds only the distinctive CODH/ACS reaction to an otherwise widely distributed folate pathway, or as the 4HB pathway adds a single key reaction to connect multistep fermentative and biosynthetic reaction sequences apparently evolved to support other functions, the CBB cycle adds only a single reaction to the pentose phosphate network. This otherwise universal network of aldol additions and cleavages among sugar phosphates, shown as the satellite network on the left in Figure 4.5, consists of the gluconeogenic pathway to fructose-1,6-bisphosphate and the network of side-reactions leading to ribose and ribulose-1,5-bisphosphate (as well as other biosynthetically central sugar phosphates).[35] The pentose phosphate pathway is needed to form all sugars from 3-phosphoglycerate and therefore would seem to be a precursor not only to cellular metabolism, but to the elaboration of any network capable of producing RNA, the large group of redox cofactors formed from phosphoribosyl-pyrophosphate, or the cyclic hydrocarbons formed from erythrose-4-phosphate.

The distinctive CBB reaction that extends reductive pentose phosphate synthesis to a carbon fixation cycle is a carboxylation performed by the ribulose-1,5-bisphosphate carboxylase/oxygenase (RubisCO), together with cleavage of the original ribulose moiety to produce two molecules of 3-phosphoglycerate. The Calvin cycle resembles the 4HB pathways in regenerating two copies of the network catalyst directly, not requiring separate anaplerotic reactions for autocatalysis. However, unlike the 4HB product acetyl-CoA, 3-phosphoglycerate is not a central biosynthetic precursor, and so it must be fed back into (either oxidative or fermentative) TCA reactions in order to support anabolism.

---

[35] The universality of this network requires some qualification. We show a canonical version of the network in Figure 4.5, and some variant on this network is present in every organism that synthesizes ribose. However, the $(CH_2O)_n$ stoichiometry of sugars, together with the wide diversity of possible aldol reactions among sugar phosphates, make sugar rearrangement a problem in the number theory of the small integers, with solutions that may depend sensitively on allowed inputs and outputs. Other pathways within the collection of attested pentose-phosphate networks are shown in Ref. [25].

In addition to carboxylation, RubisCO can react with oxygen in a process known as *photorespiration* to produce 2-phosphoglycolate (2PG), a precursor to glyoxylate that is independent of rTCA cycle reactions [215, 216, 272]. 2PG inhibits the CBB cycle, and must either be excreted or be recycled through other mechanisms (considered further in Section 4.4.3.2 below) if it is not to become a growth inhibitor. Either of these mechanisms is costly as it sacrifices from one, to all four, fixed carbons in each two molecules of 2PG. The susceptibility of RubisCO to photorespiration, despite its central place in biotic carbon cycling and so presumably its exposure to intense selection pressure, suggests that the ene-diol intermediate which is the key structure produced by RubisCO is not readily formed without an enzyme and difficult to control even with one. Between this inference about mechanism, and the dependence of the RubisCO reaction on an existing pentose phosphate pathway, it is difficult to suggest a pre-enzymatic analogue to the CBB cycle as an early carbon fixation pathway.

Bona fide RubisCOs, and hence functioning CBB pathways, are very widespread phy-logenetically, found in many euryarchaeota and in cyanobacteria and proteobacteria. In addition, several classes of RubisCO-like proteins with functions such as sulfate reduc-tion are present in organisms that do not use them in a Calvin cycle. However, the best current estimate of their function in euryarchaea such as Methanomicrobia – the putative common ancestor of the carbon-fixing RubisCOs – is as salvage pathways for $C_5$ sugar phosphates rather than as primary carbon fixation pathways. A single lateral transfer from a euryarchaeon to the common ancestor of cyanobacteria and proteobacteria is the likely origin of all bacterial RubisCOs [793]. Their role in autotrophic carbon fixation is closely linked to photosynthesis because CBB is an energy-intensive pathway (discussed further in Section 4.4), and is dominated today by cyanobacteria and their derivatives, a clade with estimated depth around 2.7 GY [76].

### *4.3.6.7 The glyoxylate shunt*

The glyoxylate shunt (or glyoxylate bypass) is not a carbon fixation pathway but rather a heterotrophic pathway fed by acetate. It pertains to the discussion of deep core metabolism because the shunt and the 3HP bicycle, taken together, employ the full network of aldol reactions closely associated to rTCA and the precursors to anabolism. Aldol reactions are potential sources of enormous complexity in CHO chemistry. They are kinetically facile enough to take place in simple sugar–water mixtures, and the $CH_2O$ stoichiometry of sugars provides a large combinatorial space of aldehyde and alcohol groups capable of undergoing additions or forming as a result of cleavages. The aldol reaction is also an internal redox reaction, requiring neither net oxidant nor net reductant, so it can proceed in isolated CHO mixtures as long as the compounds possess some undissipated internal free energy [858]. The sugar phosphate reactions of the pentose phosphate pathway offered a hint of the complexity of aldol networks,[36] and far greater diversity is generated in the

---

[36] Reference [25] suggests the much greater complexity that exists in variants on this network.

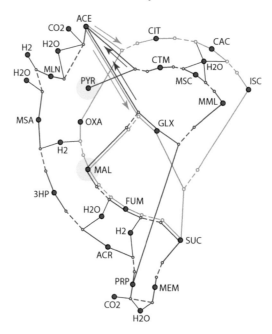

Figure 4.7 The carbon-fixing 3-hydroxypropionate bicycle (blue) and the acetotrophic glyoxylate shunt (orange) compared. Directions of flow are indicated by arrows on the links to acetate (ACE). The common core that enables flux recycling in both pathways is the aldol reaction between glyoxylate (GLX), acetate, and malate (MAL). The four other aldol reactions (labeled by their cleavage direction) are from isocitrate (ISC), methylmalate (MML), citrate (CIT), and citramalate (CTM). Malate is a recycled network catalyst in both pathways. Carbon is fixed in the 3HP-bicycle as pyruvate (PYR), so the cycle only becomes autocatalytic if pyruvate can be converted to malate through anaplerotic (rTCA) reactions. (After [91], Creative Commons [153].)

so-called *formose* reaction of formaldehyde and glycolaldehyde [109, 859].[37] Any discussion of the network of variant carbon fixation pathways should therefore include the subnetwork of cross-links from all possible aldol reactions. The shunt is also of biological interest in its own right because it possesses, in a heterotrophic context, some of the robustness derived from autocatalysis.

The glyoxylate shunt is shown in orange in Figure 4.7, against the 3HP bicycle for comparison. Reaction directions are indicated by arrows in this figure, to make it easier to see where the shunt and the bicycle use the same reactions in either the same or opposite directions. All aldol reactions that can be performed starting from rTCA intermediates appear in this pathway, either as cleavages or as condensations. In addition to condensation of acetate and oxaloacetate to form citrate (as in the oxidative Krebs cycle), these include cleavage of isocitrate to form glyoxylate and succinate, and condensation of glyoxylate and

---

[37] A discussion of the problem of selection that this diversity poses for the formose reaction as a prebiotic source of ribose, and an illustration of one mechanism that can prune the diversity, is given in [671].

acetate to form malate (reverse of the bicycle). The shunt is a weakly oxidative pathway (generating one $H_2$-equivalent from oxidizing succinate to fumarate), and is otherwise a network of internal redox reactions. Although it produces less energy than a respiratory pathway such as the Krebs cycle, it can deliver metabolic precursors and some energy without requiring regeneration of an oxidant such as $NAD^+$, and is therefore frequently used in organisms that cannot recycle NADH by membrane oxidation.[38]

The shunt contains two loops, much like the 3HP bicycle. It may be regarded as a network-autocatalytic pathway for intake of acetate, using malate as the network catalyst around an incomplete Krebs cycle, and regenerating a second molecule of malate from two acetate molecules. As in the 3HP bicycle, glyoxylate is the crucial intermediate coupling the two loops. It is also a precursor to glycine, and a molecule that may exchange $C_1$ groups with the folate pathway either as donor (in glycine cleavage) or acceptor (in serine synthesis). Hence the aldol reactions create a third independent connection between the TCA loops and their variants, and one-carbon reduction, besides the direct synthesis of acetyl-CoA in W-L, or the synthesis of serine from 3-phosphoglycerate.

Oxidative pathways such as the Krebs cycle are the opposite of self-maintaining. They catabolize their inputs (acetate or pyruvate) as long as the network catalyst is preserved, and if it is lost the oxidative cycle does not regenerate it. The shunt, in contrast, is autocatalytic, though from $C_2$ rather than $C_1$ inputs. The flexibility of redox-neutral pathways leads to many circumstances in which they may be up-regulated, but a particularly interesting case is up-regulation of the shunt in the Deinococcus-Thermus family of bacteria in response to radiation exposure [490]. We speculate that this may confer some additional robustness from network topology under conditions when metabolic control is compromised.

### 4.3.6.8 Seven building blocks of carbon fixation

In summary, we find that known diversity in carbon fixation comes from the assembly of pathways from a set containing only seven reaction sequences. These include the short-molecule and long-molecule arcs of rTCA; their respective alternatives, the 3HP and 4HB pathways; the pentose phosphate network made cyclic by RubisCO, the mesaconate arc together with a cross-linking of aldol reactions used to close the 3HP bicycle, and direct one-carbon reduction culminating in the CODH/ACS reaction. Not all possible combinations are used: the 3HP pathway could combine with the long-molecule loop of rTCA from succinate (including succinyl-CoA synthase) to form an autotrophic pathway, though it is difficult to imagine an environment to which it would be well adapted.

Other decompositions besides pathway segmentation can also be considered as measures of the redundancy of metabolic pathways. We turn now to consideration of reaction mechanisms and local-group chemistry.

---

[38] An example is its up-regulation in *E. coli* under iron-starved conditions, perhaps because depletion of cytochrome function impairs regeneration of $NAD^+$ (T. Hwa, personal communication).

### *4.3.6.9 Branched-chain amino acid biosynthesis*

Most of the discussion we will present concerning amino acid biosynthesis will be in relation to the assignment of amino acids in the genetic code, considered in Chapter 5. All except one of these $\alpha$-amino acids (the exception being histidine) are formed by **reductive transamination** of an $\alpha$-ketoacid using amino groups from glutamine, along with a collection of more heterogeneous side-chain modifications. Three groups of these have a particularly simple relation to the TCA intermediates: their standard and putatively ancestral biosynthetic pathways begin in either pyruvate, aspartate, or $\alpha$-ketoglutarate.

Another important set of amino acids is obtained by reductive transamination of more complex, **branched-chain $\alpha$-ketoacids** which are not intermediates in the TCA cycle, but for which the putative ancestral pathways still occur entirely within the sector of CHO chemistry that elaborates the universal covering network of Figure 4.5.[39] Here we briefly review the pathways reconstructed by Rogier Braakman [92] by joint flux balance and phylogenetic arguments for the deep-branching and conservative reductive chemoautotroph *Aquifex aeolicus*, which were argued to be the ancestral pathways for both bacteria and archaea, because they relate to the patterns observed for carbon fixation at many points.

The branched-chain amino acid biosynthetic pathways closely follow the modular structure of innovative initial reactions followed by a standard reduction sequence, which we will illustrate in Figure 4.9 in the next section for carbon fixation. In many reaction sequences in *Aquifex*, homology in the local-group chemistry is recapitulated much more extensively by either homology or even promiscuity of enzymes, than is the case for core carbon fixation pathways, which have undergone some evolutionary innovation. The branched-chain amino acid biosynthetic pathways also appear to be the first essential domains in which innovations of later importance to the core were established.

The isoleucine pathway partly overlaps (though in the oxidative rather than reductive direction) with the terminal reactions in the 3HP pathway that salvage methylmalate and render the bicycle autocatalytic. Interestingly, whereas the input $\alpha$-ketobutyrate is derived in many modern organisms by deamination of threonine [818] (hence ultimately by reduction of oxaloacetate), in *Aquifex* it follows an N-free pathway until the last step, the initial intermediate of which (citramalate) is also one of the decarboxylation products of citrate.[40] Although not a carbon fixation pathway, the pathway through 2-isopropyl malate (to leucine) follows the same sequence of aldol addition and oxidations, with decarboxylation to bicarbonate, which will be the most conserved sequence in carbon fixation, shown in Figure 4.9 below. The top and bottom rows of Figure 4.8 show these two homologous sequences, with the active carbon bonds highlighted in red. The use of the cofactor lipoic acid in both of these branched-chain pathways is the same as its later use in the reversal of rTCA to the oxidative Krebs cycle, and may be seen as one of the pre-adaptations that made the reversal of the cycle possible.

---

[39] The fascinating and enigmatic signatures of ancient glycine synthesis in the phylogenetic record and the genetic code are considered in Section 4.4 and again in Chapter 5.
[40] For a discussion of this pathway, see [128, 903].

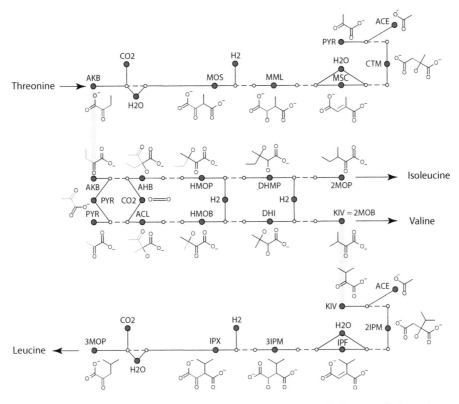

Figure 4.8 Branched-chain amino acid biosynthesis contains two distinct motifs in series, and each motif appears in two pathways homologous at the local-group level. The upper branch in isoleucine biosynthesis can be used to synthesize oxobutanoate, and is argued in [92] to be the ancestral form. This pathway reverses reactions found in the 3HP bicycle shown in Figure 4.6, and is mirrored in leucine synthesis from 2-ketoisovalarate. The middle two branches to isoleucine and to valine involve a unique TPP-dependent ketoacid reductioisomerase (cyan highlights). In *Aquifex aeolicus*, these reaction series are performed by a single sequence of non-substrate-specific enzymes. New abbreviations: (MOS) methyl-oxalosuccinate; (AKB) α-ketobutyrate; (AHB) 2-acetyl-2-hydroxybutyrate; (HMOP) 3-hydroxy-3-methyl-oxopentanoate; (DHMP) 2,3-dihydroxy-3-methyl-pentanoate; (2MOP) 2-methyl-oxopentanoate; (ACL) acetolactate; (HMOB) 3-hydroxy-3-methyl-oxobutanoate; (DHI) 2,3-dihydroxy-isovalarate; (KIV) keto-isovalarate, also 2-methyl-oxobutanoate (2MOB); (2IPM) 2-isopropylmalate; (IPF) isopropyl-fumarate; (3IPM) 3-isopropylmalate; (IPX) 3-isopropyl-oxalosuccinate; (3MOP) 3-methyl-oxopentanoate. (Figure modified from [92].)

The two central ranks of Figure 4.8 show the distinctive step in biosynthesis of all three branched-chain amino acids. A *decarboxylating condensation of pyruvate* (green in the figure) with either oxobutanoate or a second pyruvate yields (respectively) acetyl-hydroxybutyrate or acetolactate. These two are isomerized (blue in the figure) by a single distinctive enzyme [903] which uses thiamin pyrophosphate in the active site, to (respectively) hydroxymethyl-pentanoate or hydroxymethyl-butanoate. These are the two branched-chain backbones that lead to isoleucine and valine, respectively. An exchange of an acetylation for a $\beta$-decarboxylation (following the TCA pattern) then yields 3-methyl-oxopentanoate, the precursor to leucine.

### 4.3.7 Pathway alignments and redundant chemistry

A brief inspection of the TCA cycle in Figure 4.1 showed already that the complexity of the cycle is less than the count of its eleven intermediates and eleven reactions would suggest. Succinate is a symmetric copy of two acetyl moieties, and the downstream reactions from acetyl-CoA to fumarate, and from succinyl-CoA to *cis*-aconitate (shown diametrically opposed in the figure), are exactly homologous with respect to the acetyl moieties on which they act. They have comparable reaction free energies [558], and are catalyzed by homologous enzymes in organisms that preserve what are believed to be the ancestral forms [32, 33, 34, 35, 36]. The rTCA cycle is better understood as having six essential chemical forms (including citrate), and eight distinct reaction types (including reduction of fumarate, hydration of *cis*-aconitate, aldol cleavage of citryl-CoA, and the pair of thioesterifications not shown in Figure 4.1).

The reuse of local-group chemistry within rTCA extends to a significant reuse of the same chemistry *across* alternative carbon fixation pathways as well. Figure 4.9 shows the decomposition obtained by splitting the network from Figure 4.6 at acetate and succinate, and aligning the three groups of pathways that result so that similar functional-group transformations are arranged in columns. In this figure, as in Figure 4.1, molecular forms are shown, and the active acetyl moiety or semialdehyde in each reaction sequence is shown in red. The five arcs in Figure 4.6 are the short-molecule and long-molecule arcs of rTCA, their respectively parallel 3HB and 4HB pathways, the mesaconate arc used to complete the 3HP bicycle (or its reverse in isoleucine synthesis), and the homologous isopropyl-fumarate arc in leucine biosynthesis.

The reason for the *ladder-like topology* of reactions involving reductant ($H_2$), water, and some $CO_2$ incorporations, in the covering network now becomes clear. A reduction sequence from $\alpha$-ketones or semialdehydes, to alcohols, to isomerization through enoyl intermediates, is applied to the same bonds on the same carbon atoms from input acetyl moieties in rTCA, 3HP, and 4HB pathways, and to analogous functional groups in the bicycle. Furthermore, in the cleavage of both citryl-CoA and citramalyl-CoA, the bond that has been isomerized through the enoyl intermediate is the one cleaved to regenerate the network catalyst. Through this homology, *17 reactions that are distinct at the molecular level reduce to only four distinct local-group transformations.*[41]

Strict homology of reaction sequences leads to simple ladder topologies; deviations in one or another branch from the main template indicate innovations in the deviating pathway. These include the carbonyl insertions at thioesters in rTCA, the carboxylation

---

[41] The number of reactions included in this compressed description could be expanded slightly by considering homologous carboxylations from activated biotin, although for these some details of the substrate and reaction mechanism differ enough that we do not consider them strictly homologous.

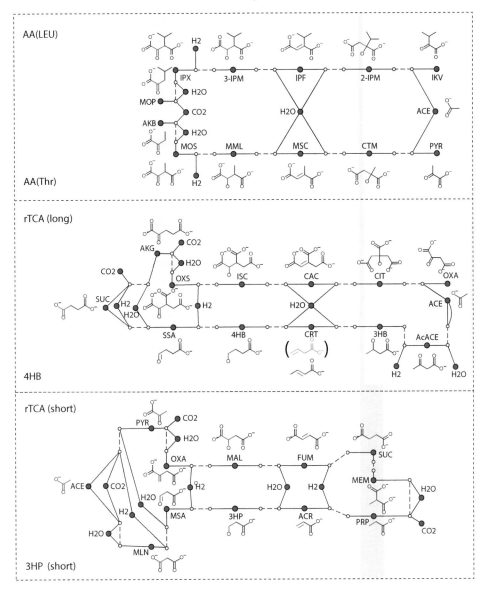

Figure 4.9 Comparison of redundant reactions in carbon fixation pathways and branched-chain amino acid biosynthesis. From Figure 4.6, rTCA and the 3HP/4HB loop pathways are split at acetate and succinate into "long-" and "short"-molecule segments. Molecule forms are shown next to the corresponding tags. Bonds drawn in red are the active acetyl or semialdehyde moieties in the respective segments. Vertical colored bars align homologous carbon states. The yellow block shows retro-aldol cleavages of citrate, citramalate, or isopropylmalate. Two molecules are shown beneath the tag CRT (crotonate): vinylacetate (greyed out, in parentheses) would be the homologue to the other aconitase-type reactions; actual crotonate (full saturation) displaces the double bond by one carbon, requiring the abstraction of the $\alpha$-proton in 4-hydroxybutyrate via the ketyl-radical mechanism that is distinctive of this pathway [577].

to form methylmalate in the 3HP pathway, and the oxidation of 4-hydroxybutyrate to acetoacetate in the 4HB pathway, which otherwise follows the template of either reductive or redox-neutral reactions.

The ketyl-radical reaction in the 4HB pathway does not appear as a topological deviation, but as a deviation in molecular form. The actual molecule (crotonate) has a double bond displaced from the homologous position for this class of dehydrations. Even so, the distinctive step to crotonyl-CoA in the 4HB pathway creates an aconitate-type intermediate, and the enzyme responsible has high sequence similarity to the acrolyl-CoA synthetase, whose output (acrolyl-CoA in the 3HB pathway) follows the standard pattern.

Duplication of local-group chemistry in diverse reactions appears to have resulted (at least in most cases) from retention of reaction mechanisms as enzyme families diverged. All enoyl intermediates are produced by a widely diversified family of aconitases [317], while biotin-dependent carboxylations are performed by homologous enzymes acting on pyruvate and $\alpha$-ketoglutarate [34].

This relation extends to the two thioesterifications in the rTCA cycle, if thioesterification of the acetyl moiety in citrate is compared before citrate cleavage to thioesterification of succinate. Aoshima *et al.* [35, 36] have shown that in the bacterium *Hydrogenobacter thermophilus*, a form of the citryl-CoA synthase independent of citryl-CoA lyase (the latter catalyzing the aldol cleavage), and succinyl-CoA synthase are homologous enzymes, and they have argued that these homologous enzymes conserve the ancestral form used in the reductive citric acid cycle. In a similar analysis, the same authors [32, 33, 34] have also argued that the carboxylation of $\alpha$-ketoglutarate and the subsequent reduction of oxalosuccinate to isocitrate, although performed by a single enzyme in most organisms, are catalyzed in *H. thermophilus* by independent enzymes homologous to those for pyruvate and oxaloacetate, and that these likewise conserve the ancestral forms.[42]

The functional-group homology shown in Figure 4.9 allows us to separate stereotypical sequences of reactions that recur in diverse pathways, from the key reactions that distinguish the pathways. The repeated homologous reactions catalyzed by related enzymes all represent relatively facile chemistry in water. They lie downstream, in most cases, from non-homologous reactions that initiate each leg and which are distinctive for each pathway. The widespread use of a few reaction types by a small number of enzyme classes/homologues may reflect their early establishment by promiscuous catalysts [157, 632], followed by evolution toward increasing specificity as intermediary metabolic networks expanded and metabolites capable of participating in carbon fixation diversified.

### 4.3.8 Distinctive initial reactions and conserved metal-center enzymes

The structure of pathway redundancy in Figure 4.9 – pathways are distinguished by an initial carboxylating reaction, and then proceed by common, repeated reactions with

---

[42] It currently appears that thioesterification of propionate in the 3HP pathway is performed by distinct enzymes in bacteria and archaea, and this observation has led Teufel *et al.* [799] to propose that these otherwise identical pathways have resulted from convergent evolution.

homologous or even shared non-specific catalysts – suggests that the initiating reactions may have been the crucial innovations for pathway emergence. They are in some way either chemically more difficult, or they require coupling to a source of free energy that is more complex, so that they were the bottlenecks to innovation of metabolic diversity. This interpretation would seem to be supported by the fact that many of these reactions are catalyzed by highly conserved enzymes (suggesting that innovation leading to evolution-ary convergence was difficult or that their mechanisms are unique for the reaction types). Most also use metal centers in the enzymes and sometimes in essential cofactors or aux-iliary peptides such as ferredoxins. The widespread use of metal centers is important both as a mechanism enabling radical intermediates, and for the possible association of modern catalysts with prebiotic mineral substrates.

Examples of key metal-center enzymes include the following.

1. **Pyruvate: ferredoxin oxidoreductase**, which catalyzes the reversible carboxyla-tion of acetyl-CoA to pyruvate, and contains three [$Fe_4S_4$] clusters and a thiamin pyrophosphate (TPP) cofactor [123].
2. The **carbon monoxide dehydrogenase/acetyl-CoA synthase** enzyme that catalyzes the final acetyl-CoA synthesis reaction in the W-L pathway, which uses [$Fe_4S_4$] clus-ters for electron transfer, employs a Ni-[$Fe_4S_5$] cluster in the CODH active site to reduce $CO_2$ to CO [197], and employs a Ni-Ni-[$Fe_4S_4$] cluster in the ACS active site to condense CO with –$CH_3$ from a folate donor [176, 726, 789]. (Recall also from Section 4.3.6.5 that cobalamin mediates the transfer of this methyl group.) For further details on this reaction see [91].
3. The enzyme **4-hydroxybutyryl-CoA dehydratase** responsible for the ketyl-radical conversion of 4-hydroxybutyryl-CoA to crotonyl-CoA is a flavoprotein that uses a sequence of single-electron transfers through unusual Fe-S clusters to accom-plish the cleavage of an *unactivated* $\beta$-C–H bond, a highly unusual reaction in biochemistry.
4. The widely diversified **aconitase** family of enzymes [317], which contains members in multiple carbon fixation pathways and also in branched-chain amino acid biosynthesis, is unusual in that it uses a non-redox-active iron-sulfur center. In this family, it is the *crystallographic* geometry of the cubic $Fe_4S_4$ center, together with dative bonding of substrate oxygens to iron atoms, that permits the crucial "flip" of the aconitate inter-mediate about its double bond, thus enabling stereospecific dehydration at one carbon atom and rehydration at the adjacent carbon [494].

Michael Russell and Alan Hall [688] have emphasized the similarity of several biological Fe-S motifs to subregions within the sulfospinel structure of the mineral greigite.

### 4.3.9 The striking lack of innovation in carbon fixation

This overview of carbon fixation pathways allows us to seek rules of composition that take into account the scope of natural variations. To conclude this section, we summarize the forms of redundancy, stressing the remarkable *lack of innovation* seen in carbon fixation across the biosphere.

- The extensive diversification of secondary and catabolic metabolism has grown out of six distinct autotrophic pathways, formed by combining subsets of only seven pathway segments.
- The innovation of a pathway segment does not require innovation of all of its reactions; most reactions in most pathways are redundant, and could plausibly have been provided by broad-affinity catalysts in the past. A number, approximately five, of these are elementary reactions involving reduction and hydration/dehydration of small water-soluble molecules; a comparable number involve carboxylation from $CO_2$ or carboxybiotin to moderately or significantly distinct substrates; aldol reactions account for most internal redox reactions and some cleavages and additions; and acetoacetate undergoes a distinct hydration/cleavage. Finally, the one-carbon system from Figure 4.4 adds a dozen reactions of which four are reductions and the remainder are (possibly aminated) $C_1$ group transfers.
- The "difficult" innovation, and the one that distinguishes a pathway, is often a single reaction. Usually this is the initiating carboxylation, though in some cases it is an unusual pathway completion (4HB) or intermediate reaction (3HB). Cofactors matter, as catalysts matter, in all pathways, but only for $C_1$ metabolism does the cofactor reflect a free energy landscape tuned to embed the entire reaction sequence in the redox exchange network of an integrated bioenergetic system. (We return to the problem of cofactor evolution in Section 4.5 below.)
- Most initiating carboxylations (and in some pathways, also certain key later steps) are associated with metal-center enzymes, and may reflect evolutionarily "difficult" innovations because they require radical intermediates in aqueous solution. At the same time, they may suggest prebiotic mineral affinities. In particular, the metal centers are Fe-S clusters, sometimes substituted with nickel or employing cobalt cofactors, suggesting rather narrow roles for this sequence of $d$-block elements.
- The lack of innovation around these metal centers further suggests that the conserved reaction mechanisms are not limited by evolution in the oligomer world but rather by the properties of the elements themselves.

The fact that carbon fixation is not only simple, but that it is simple *in this way* – with explicit links to minerals and the periodic table – leads us to propose that the same set of constraints can have been both *causes* of limited evolutionary innovation in the world of modern cells, and yet *common* to the cellular world and back through pre-cellular life and protometabolism to prebiotic geochemistry. The plausible continuity from minerals to modern enzymes leads us to characterize the emergence of metabolism as an "enfolding" of geochemical mechanisms in organic control systems, as much as the creation of a genuinely novel system.

Insofar as an extant pathway approximates the central tendency of the distribution of variants, it is clearly the rTCA cycle among the loop pathways. We will argue below that rTCA is not the *only* central tendency of carbon fixation, but it is the only one visible in a comparison of pathway structure. The other dominant theme is the role of direct one-carbon reductions, but to see this it is necessary to interpret extant variation in terms of an evolutionary sequence.

## 4.4 A reconstructed history of carbon fixation, and the role of innovation constraints in history

Theodosius Dobzhansky famously proclaimed [198] that "Nothing in biology makes sense except in the light of evolution."[43] Much of our argument in this book is for the proposition that a great deal in biology makes sense as time-invariant law or constraint, which is both prior to and outside of the Darwinian evolutionary dynamic. However, the presence of variation in core metabolism, the ambiguity of multiple non-aligned distributional signatures, and the great refinement of extant organisms, make clear that we are seeing these laws as they are instantiated in a process of historical development. Evolutionary reconstruction, so far as it is possible, is the principled way both to peel back the layers of accreted history, and more generally, to deconvolve the signatures of chance and relatedness in a comparative analysis, to more accurately represent the rules governing change.

This section reviews an evolutionary reconstruction of the history of core carbon fixation carried out by Braakman and Smith [90], jointly with construction of a computational model of the metabolism of *Aquifex aeolicus* [92] as a skeleton of the minimal requirements for chemoautotrophy. Key features of the analysis are that it is not a consensus study of reconstructed histories of genes, but rather a historical reconstruction of *functional systems*. Imposing a constraint of system-level function on past as well as present states enables a comparative analysis of gene presence/absence data independently of sequence homology or even exact biosynthetic analogy. A result will be that the problem of glycine synthesis introduced in Section 4.3.4.2 plays a central role in historical interpretation. We will arrive at the conclusion that direct one-carbon reduction, and a few crucial synthetic pathways that branch from it, have been as universal and consequential constraints on the major early branchings in the tree of life as the arcs of the rTCA cycle, and have been more fundamental than any of the 3HP, 4HB, or CBB pathways, which were later derivations.

### 4.4.1 Three reasons evolutionary reconstruction enters the problem of finding good models for early metabolism

An evolutionary reconstruction of carbon fixation leads to some rather different conclusions from those that generally arise when the origin of life is treated as a mere problem of finding synthetic pathways. In context-free organic chemistry, glycine seems among the least likely molecules to pose a "problem," whereas in the evolutionary reconstruction its synthesis seems to have been an important source of constraint. It is useful therefore

---

[43] This was the title of an education-and-advocacy article.

to explain why evolutionary reconstruction provides a relevant context for geochemical hypotheses.

Clearly, evolutionary reconstruction using gene signatures delivers no direct evidence about history before the last genomic ancestor. In three respects, however, it places important constraints on both structural models and process models for early geochemistry.

### 4.4.1.1  Reconstructed ancestral forms will be at least as good for models as extant organisms

First, we note that it is common to use extant organisms as models for the earliest cellular or pre-cellular metabolism, implicitly invoking a continuity hypothesis. Within the continuity hypothesis, the first reason to perform evolutionary reconstructions is that a well-reconstructed ancestral genotype and phenotype should be at least as good a model for earlier cells as any modern organism. This is well illustrated by the consistent reconstruction of thermophily in the deep-branching bacterial and archaeal clades [668], even though thermophiles are not a majority type among known modern organisms. The reason reconstruction accomplishes such corrections was first recognized by the statistician Francis Galton: the frequency of a type does not necessarily correlate with its informativeness, since the counts may not be independent. What one needs to obtain an accurate picture of the historical process is a reconstruction of *events* of change.

The problem of using modern organisms as models is most severe for the comparison of rTCA and W-L autotrophy: only W-L is found in bacteria and archaea, but the role of the rTCA precursors to anabolism is more universal than the distribution of any carbon fixation pathway. Within the W-L organisms, we have the problem noted above: that the CODH/ACS enzyme is homologous across the group, but the $C_1$ reduction pathway that feeds it is split between bacteria and archaea. We encounter the problem again in Section 6.3.1.3 where we consider the bioenergetic systems in which the acetogenic and methanogenic versions of the W-L pathway are embedded, and which split into two domains in their fundamental architecture. For this split the difficulty of interpretation is especially severe because all bioenergetic systems are highly refined, and it is difficult to see how any chain of viable organisms could interpolate between them. To the extent possible, phylogenetic reconstruction should provide a way to weigh such diverse signatures to arrive at consensus phenotypes in the deep past.

### 4.4.1.2  Outgroup comparison through phylogenies improves models of extant organisms

A second reason evolutionary reconstruction can provide better models for early life is that many extant organisms are incompletely or incorrectly characterized. This is a particular problem with gene annotation that is not backed by thorough biochemical characterization. Genes may be missed in sequence comparisons, and annotations made on the basis of reasonable criteria of sequence similarity may still be wrong. Genes may also have unidentified functions. The problem of misidentification was well illustrated in the work on rTCA genes in *Hydrogenobacter* [32, 33, 34, 35, 36].[44] The problem of unidentified genes or unidentified functions is proposed in [90] to be at the root of the glycine biosynthesis ambiguity.

Outgroup comparison can improve the modeling of extant organisms by pooling data that may cover different functions in different groups. The proper way to use such information is through phylogenetic reconstruction. In a reconstruction, the genotypes and phenotypes of ancestor nodes are *summary statistics*, given which the variations in descendant nodes should be conditionally independent.[45] This means that a reconstructed ancestor is the "best" average from information in the outgroup, to be informative about a descendant node that may be incompletely or incorrectly characterized. The reconstructions of [90, 92] make extensive use of outgroup comparison through phylogeny to provide gap-fills and to propose gene functions for reactions essential to autotrophy in *Aquifex*.

### 4.4.1.3 Reconstruction jointly infers the diachronic history of events and the invariant rules of dependency and causation

Finally, the existence of variations, which makes comparative analysis possible at all, provides a window on laws that would be absent if core metabolism had a single unambiguous universal signature. While a single universal metabolism might provide the surest model for the earliest cells, it might provide little insight into which constraints forced that universality. The presence of variations that show *structure* can suggest which constraints permitted variation and which precluded it. The absence of certain variations which might have been imagined, in the era with evolved genes and enzymes, is particularly important if it can be associated with chemical "paths of least resistance"; if any process should be capable of reinventing such paths it would be a genomic one. Chemical constraints that preclude variation in a genomic era are good candidates for constraints on pre-cellular geochemistry as well. In a historical reconstruction that shows directionality, the attribution of *causes* along the links of innovation may further illuminate the limitations on the ancestral forms.

### 4.4.2 Phylogenetic reconstruction of functional networks

The challenge in reconstructing the history of a functional system such as metabolism is to find a formal representation of the function that is applicable to all the members being compared, and which can be imposed as a constraint on all reconstructed ancestor states. In general, this is difficult for evolving systems, because many functions may be active in a system at the same time [313] and new functions can be introduced, or old functions lost, repeatedly during its history.[46] However, for autotrophic core metabolism a formal

---

[44] A particularly striking example occurs in the biochemistry of transamination in *Hydrogenobacter thermophilus* TK-6, worked out by Kameya *et al.* [409]. Three families of transaminases are found, which catalyze formation of aspartate (reversible), alanine, glycine (irreversible), and glutamate (from phosphoserine, reversibility not stated). For four of the six reported enzyme activities, the enzyme responsible in *Hydrogenobacter* clusters phylogenetically on the basis of sequence similarity most closely with enzyme families that in most or all other known bacteria have *different* aminotransferase activities.

[45] In an optimally reconstructed tree, for a process for which trees are an apt description, ancestral nodes should be *sufficient statistics*. See Box 7.2 of Chapter 7 for definitions and arguments.

[46] This is why most historical reconstruction is carried out for genetic sequences, with relatedness inferred from degrees of sequence similarity. Among Ernst Mayr's many colorful criticisms of molecular reductionism, which in his view omitted the

representation is readily available from **flux balance analysis** (FBA) [621]: any autotroph must be capable of providing all precursors to biosynthesis from an invariant input set comprising $CO_2$, one or more reductants and perhaps auxiliary oxidants, an inorganic source of nitrogen (in the era before nitrogen fixation this can be any of a range of $N_1$ compounds), and inorganic sources of sulfur, phosphorus, and metals in the form of minerals or salts.

If relatedness is to be expressed in terms of trees, a criterion must also be chosen for tree selection, and the meaning of the trees should be interpreted appropriately for that criterion. The criterion in [90] was **maximum parsimony.** We will sketch below the interpretation that this carries, in terms of serial dependency among innovations in carbon fixation.

### 4.4.2.1 Carrying out a phylogenetic reconstruction for a set of autotrophic flux balance networks

The reconstruction of core carbon fixation in [90] was carried out for what we have called the "core of the core": the fixation pathway itself and anaplerotic pathways to reach the universal precursors, the gluconeogenic pathway, and the initial steps in the biosynthetic pathways for amino acids. Gene presence or absence data were compiled for each extant species and used to produce a model of the species' metabolic capabilities. A requirement was imposed – both for extant organisms and for reconstructed internal nodes – that these produce an autotrophic network from $CO_2$ (or, where appropriate, formate) as the sole carbon input. The interpretation of gene presence/absence signatures was not separately modeled for all genomes surveyed, but made use of a much more extensive curated flux balance model for *Aquifex aeolicus* (described in Section 4.4.5) as a template. That model enforced consistency on the crucial pathways to amino acid precursors (particularly glycine and serine). Where alternative core carbon fixation pathways differed by novel reactions from the *Aquifex* model, the genes and functions in those pathways were all established biochemically in the primary literature.

For extant organisms the purpose of the FBA constraint was principally to overcome inconsistencies or gaps in gene annotations (since they are all known to be autotrophs). For internal reconstructed nodes, network models require a different interpretation. FBA in its most basic form[47] is sensitive only to stiochiometric input/output relations. An autotrophic flux network would take the same form whether it modeled the metabolism of a single species or a syntrophic microbial community. Therefore we cannot simply assume that all autotrophic species today descended from autotrophic ancestor species in the past, and that the tree reflects these lines of descent. It is consistent with the reconstruction that communities of species that were jointly autotrophic could have contributed genes in some cases to extant organisms in which autotrophy is an innovated trait.

---

richness of compositional and functional relations that were the substance of physiology, was that gene-counting methods were "bean-bag" genetics [533]. To move beyond "bean-bag phylogenetics," while retaining systematic and quantitative methods that are not always required in descriptive physiology, will be the challenge for the next generation of historical reconstruction.

[47] The most basic form includes only chemical input/output stoichiometry, and does not attempt to unify metabolic flux constraints with larger models of regulation and reproduction. Models attempting to integrate metabolic flux balance constraints with models of regulation are currently the leading edge of development [242].

Using either extant or proposed autotrophic graphs as nodes, a tree was formed by introducing links to indicate innovations connecting adjacent phenotypes. Because the relations among the core pathways described in Section 4.3.6 involve substitution of large pathway segments through a few central substrate molecules, the links could be taken to represent entire pathway replacement rather than single-gene replacement. If, as is suggested by the redundancies discussed in Section 4.3.7, much of pathway evolution turned on innovation of individual crucial reactions,[48] followed by routine sequences of redox transformations that could be catalyzed by non-specific enzymes, there might have been no difference between adding a pathway segment and adding its crucial reaction.

### 4.4.2.2 The meaning of a parsimonious tree of autotrophic metabolisms

The objective of the reconstruction was to identify a *maximum-parsimony tree*, in which the nodes are connected so that each innovation occurs independently on as few nodes as possible, preferably only once. In cases where complete parsimony was not possible – meaning that some innovation must be repeated on at least one pair of nodes – a judgment was made to duplicate whichever innovation is most plausible as a consequence of convergent evolution or gene transfer.

A phylogenetic tree of carbon fixation networks that has high parsimony is one in which all innovation is well explained by a few paths of serial dependency, which ramify independently of one another. In conventional trees, where the ancestral nodes are genes or organisms, the reason for parsimony is that the vertical process of descent has caused the ancestor to be the best predictor of the form of the offspring, independent of the states of any other nodes in the tree. For a reconstruction of autotrophic metabolisms, it is not appropriate to simply adopt the organism model as an analogy; we must limit our interpretations to the dependency relations that are actually demonstrated.

A literal description of a highly parsimonious tree is that *a single level of organization is sufficient to account for all dependencies.* That level could be the species, but it could also be a completely different factor such as a set of chemical environments, each admitting only one preferred metabolism, which are invaded in a particular order. Reticulation in the tree would indicate that more than one level of structure is required to explain the variation in the population. For instance, in organism phylogenies based on multiple genes, lateral transfer of a subset of genes which causes an ambiguous reconstruction reflects essential roles for both vertical descent within the gene-receiving species and the composition of the population of other gene-donating species in which it was a member. For metabolic networks, it would indicate that a network drew on innovations that had previously existed only in two or more different preceding networks. The cases in which violations of parsimony in the metabolic tree are inescapable provide evidence that multiple levels of organization were causally essential to the course of innovation.

---

[48] Usually this is the initiating carbon-intake reaction, though not always.

The sequence of pathway-segment replacements to form a tree of autocatalytic carbon fixation loops is unproblematic, and we return to that in the final presentation of the reconstruction in Section 4.4.4. The difficult and surprising part of the reconstruction is the assignment of $C_1$ reduction pathways and glycine/serine biosynthesis to the gene annotations of extant organisms, and the ways in which these two assignments depend on each other, so we consider that problem first.

### 4.4.3 Functional and comparative assignment of biosynthetic pathways in modern clades

The evolutionary diversification of carbon fixation indicates a much more central role for the reductive biosynthetic pathway to glycine than for either glyoxylate transamination or oxidative synthesis from 3-phosphoglycerate. Reductive glycine synthesis requires a complete $C_1$ reduction sequence *on THF in particular* (because of the energetics of methylene group transfer on this cofactor), and thus is unproblematic for acetogenic bacteria. However, to occur more widely, as the gene signatures indicate it does, requires $C_1$ reduction on THF in non-acetogenic organisms. Some steps required to create those pathways are novel, and would not be detected in Wood–Ljungdahl autotrophs.

A straightforward but somewhat intricate argument leads to the conclusion that those pathways must exist. The key observations behind the argument are: (1) genes indicating that glycine is synthesized from methylene-THF and not in any other way; (2) genes indicating that formate is used as a carbon source; and (3) the absence of genes for pathways indicating any other use for formate except as an input to $C_1$ reduction.

#### 4.4.3.1 The importance of alternations to the analysis

The alternation of gene signatures, rather than simply gaps in functional annotations, is important in arguments of this kind. Gaps that make it impossible to assign any pathway would suggest a lack of evidence from poor biochemical characterization of unusual autotrophic organisms. However, in the analysis of [90], only a small percentage of bacteria (sample size 359) and a somewhat larger percentage of archaea (sample size 92) show no gene assignments, indicating the correct reconstruction of biosynthetic pathways to glycine and serine. The presence of one of the three pathways of Figure 4.4 (or a close variant) in most members of the sample suggests that the known mechanisms account for much of actual variation, especially among the bacteria. Among archaea the somewhat larger fraction of undetected genes may be related to a greater diversification of archaeal pterins and enzymes involving them [218, 316].

Genes for glycine and serine biosynthesis also show a strong *anticorrelation* (they behave like "radio buttons") between presence and absence of genes for the oxidative pathway and the glycine cycle. This signature supports the interpretation that reductive and oxidative glycine synthesis are alternatives: in reductive organisms where the former is possible, it is energetically more efficient. Among these, only in methanogens, where the properties of $H_4MPT$ make reductive synthesis impossible, is oxidative synthesis forced.

### 4.4.3.2 Assignment of glycine biosynthetic pathways and reductive $C_1$ pathways in extant clades

Table 4.1 shows the distributions of genes from 359 bacteria and 92 archaea in the UNIPROT database [819], compiled by Braakman in [90], where pathways reference the scheme shown in Figure 4.4. The patterns of alternation require the following reconstructions of pathways at the bases of major branches.

**Glyoxylate transamination is chemically simple but does not reconstruct as basal**
First, the distribution of the glyoxylate transaminase is very limited. It is rare or absent in all bacterial clades except the Cyanobacteria (and present in about 10% of Clostridia), and rare (less than 25%) or absent in most archaeal clades, except the Halobacteria and one of two Marine I species samples. Even in the two groups where it is most common, the transaminase is present in only $\sim$50% of listed species. As noted in Section 4.3.4.2, it functions in cyanobacteria as a salvage pathway to recover part of the carbon from 2-phosphoglycolate produced in photorespiration.[49] Thus, although transamination is a simple reaction, it appears to have had limited impact on the evolution of carbon fixation, probably due to the limited context in which glyoxylate has been available.

**Wide distribution of the full reductive pathway beyond its use in Wood–Ljungdahl organisms** In many groups where the glyoxylate transaminase is absent, particularly deep-branching anaerobes or micro-aerophiles such as Thermotogae and Aquificales, two or all three of the genes for the oxidative pathway from 3-phosphoglycerate are also absent (as would be expected in groups with reducing metabolisms). In some of these, the full reductive pathway is annotated, including among bacteria the Thermotogae (100%), Bacillales (79%), Chloroflexi (in which rTCA was discovered, 57%), and Clostridia (48%). Among archaea it is also found in all listed Thermoplasmatales, in a quarter of the listed Halobacteria, and in *Isosphaera pallida* of the Planctomycetes.

*All eight* of the genes in this pathway must be identified for the pathway to be reported as present, so the likelihood of an error in this conclusion arising from gene misattribution seems remote. Many of the organisms in which it is found are *not* Wood–Ljungdahl organisms capable of synthesizing acetyl-CoA using a terminal CODH/ACS enzyme, so the flux of reduced $C_1$ compounds that we must conclude is being produced is serving other metabolic functions. Notably, the full reductive pathway is present in groups using other autotrophic loop pathways from Section 4.3.6, including the rTCA cycle (the Nitrospirae), the 3-hydroxypropionate (3HP) bicycle (the Chloroflexi), and the Calvin cycle (Cyanobacteria), along with a wide collection of anaerobic and aerobic heterotrophs (the *Thermotogae* and *Isosphaera pallida* of the Planctomycetes), and also several archaea.[50]

---

[49] Genes for the glycine cleavage system are present in almost all Cyanobacteria and Halobacteria. In organisms with the transaminase route to glycine, it may be functioning (in the cleavage direction) as a source of methylene groups used to synthesize serine from part of the glycine formed by transamination. Its presence is also consistent with a use in reductive synthesis in all these groups.

[50] For original data see Table S1 of the supplementary information of [92].

Table 4.1 *Counts of bacterial and archaeal species grouped by class or phylum, in which sequenced genomes indicate presence of variant pathways to glycine and serine*

| | Pathway | | | | | |
|---|---|---|---|---|---|---|
| | $Glx^a$ | $Ox^b$ | $Red^c$ | $Red(-2)^d$ | $GlyC^e$ | $FDH^f$ |
| **Bacteria** | | | | | | |
| Aquificales (7) | 1 | 0 | 0 | 3 | 3 | 5 |
| Thermotogales (11) | 0 | 0 | 11 | 0 | 11 | 1 |
| Firmicutes (257) | | | | | | |
|   *Bacilli, Bacillales* (90) | 1 | 13 | 71 | 18 | 89 | 66 |
|   *Bacilli, Lactobacillales* (90) | 0 | 16 | 0 | 0 | 0 | 5 |
|   *Clostridia* (77) | 6 | 6 | 37 | 1 | 39 | 27 |
| Deinococcus-Thermus (9) | 0 | 0 | 2 | 6 | 8 | 5 |
| Chloroflexi (14) | 0 | 2 | 8 | 0 | 8 | 9 |
| Chlorobi (11) | 0 | 0 | 0 | 9 | 11 | 2 |
| Planctomycetes (4) | 1 | 1 | 1 | 3 | 4 | 2 |
| Nitrospirae (2) | 0 | 0 | 1 | 0 | 1 | 2 |
| Verrucomicrobia (4) | 1 | 1 | 4 | 0 | 4 | 3 |
| Cyanobacteria (40) | 22 | 1 | 0 | 38 | 39 | 4 |
| All (359) | 31 | 40 | 135 | 78 | 217 | 131 |
| **Archaea** | | | | | | |
| Thermoproteales (8) | 2 | 1 | 0 | 0 | 0 | 7 |
| Desulfurococcales (8) | 2 | 0 | 0 | 0 | $5^g$ | 3 |
| Acidilobales (1) | 0 | 0 | 0 | $1^g$ | $1^g$ | 1 |
| Sulfolobales (12) | 1 | 0 | 0 | 0 | $12^g$ | 12 |
| Marine I (2) | 1 | 0 | 0 | 0 | 0 | 0 |
| Korarchaeota (1) | 0 | 0 | 0 | 0 | $1^g$ | 0 |
| Aciduliprofundum (1) | 0 | 0 | $1^g$ | 0 | $1^g$ | 1 |
| Thermococcales (7) | 0 | 0 | 0 | 0 | $7^g$ | 7 |
| Thermoplasmatales (3) | 0 | 0 | 3 | 0 | 3 | 2 |
| Halobacteria (13) | 6 | 0 | 3 | 7 | 11 | 8 |
| Methanococcales (12) | 3 | 1 | 0 | 0 | 0 | 10 |
| Methanobacteriales (6) | 0 | 1 | 0 | 0 | 0 | 6 |
| Methanopyrus (1) | 0 | 0 | 0 | 0 | 0 | 0 |
| Archaeoglobus (3) | 0 | 0 | 0 | 0 | 0 | 2 |
| Methanosarcinales (6) | 0 | 3 | 0 | 0 | 0 | 6 |
| Methanomicrobiales (8) | 1 | 2 | 0 | 0 | 0 | 8 |
| All (92) | 16 | 8 | 7 | 8 | 41 | 73 |

Table modified from [90], incorporating the glyoxylate transaminase reported for *H. thermophilus* in [409].

Notes: $^a$Glx Glyoxylate transaminase (reaction 12); $^b$Ox oxidative pathway (reactions 9–11), $^c$Red Full reductive pathway (reactions 1–8), $^d$Red(-2) reductive pathway missing only reaction 2 ($N^{10}$-formyl-THF or $N^5$-formyl pterin synthase), $^e$GlyC glycine cycle (reactions 5–8), $^f$FDH formate dehydrogenase (reaction 1), $^g$18 archaeal strains that lack only reaction 5 in the glycine cycle were included. For seven such Sulfolobales and the Acidilobales a BLASTp search finds a protein that closely matches this protein in strains lacking the annotated gene.

*The use of a folate pathway to directly reduce* $CO_2$ *therefore appears to be more widely distributed than any single primary carbon fixation pathway.*

**A seven-reaction reductive sequence deduced to be a carbon incorporation pathway**
In most of the groups where transamination and oxidative synthesis are not found, however, a full reductive pathway cannot be inferred from the gene annotations for some species. The same reaction is absent – the $N^{10}$-formyl-THF synthase that attaches formic acid to THF, labeled reaction 2 in Figure 4.4 – although *all seven other* reactions in the pathway are found. These seven-reaction sequences, labeled Red(-2) in Table 4.1, are more frequent and widespread than the full eight-reaction sequence: they are found in Cyanobacteria (95%), Chlorobi (82%), Planctomycetes (75%), Deinococcus-Thermus (67%), Aquificales (43%), and Bacillales (20%). A similar truncated cycle is found in about half of Halobacteria, and may be present in the Acidolobales (dicarboxylate/4-hydroxybutyrate carbon fixers), among the archaea.[51]

Multiple lines of evidence suggest that the Red(-2) pathway, like the full reductive sequence, is the source of carbon for glycine synthesis in the organisms in which it is found. First, the glycine cycle is almost universally present[52] with both reductive sequences. In most of these organisms (with the exception of the cases noted, containing glyoxylate transaminase, and the possible exception of heterotrophs), the glycine cycle cannot be operating as a cleavage system, because no glycine source exists. The glycine cycle in these organisms would also have no source of $C_1$ groups if the reductive pathway did not supply them, and thus no identifiable function to explain the presence of all three genes. Because the Red(-2) organisms also are not W-L autotrophs, the presence of the seven genes in the reductive pathway would have no explanation if they were not functioning as the source of $C_1$ to methylene- or methyl-group chemistry either via *S*-adenosyl-methionine or via the glycine cycle.

**An alternative mechanism to charge THF** Braakman and Smith [90] proposed that the Red(-2) pathway is a carbon-uptake pathway in the organisms where it is found, and that they are also using the glycine cycle to synthesize glycine, and subsequent methylene transfer to synthesize serine. The puzzle of the missing $N^{10}$-formyl-THF synthase is overcome, they argue, because these organisms possess an alternative mechanism for formyl groups to enter the superhighway. The interpretation is supported by the presence of a formate-dehydrogenase in all of the same groups.[53] In some of these, the FDH is used to return formic acid to $CO_2$, but in Thermotogales – the closest sister group to the Aquificales – in which a partial biochemical characterization is reported, the enzyme has been tested and

---

[51] A greater number of archaea are not annotated with genes for the glyoxylate transaminase or for the oxidative or reductive pathways, including many methanogens. This omission is a likely result of poor sequence overlap, as the archaea use a variety of pterin cofactors that are differentiated from one another and from THF, and likely employ different enzymes in the reduction sequence.

[52] The only exception in [92] is *Thermodesulfovibrio yellowstonii* among the Nitrospirae, where reactions 6 and 7 from Figure 4.4 are not annotated.

[53] Note that the nomenclature refers to the reaction in the oxidative direction, from formate to $CO_2$. The stoichiometry of the reaction, however, is independent of direction, and the same reaction in reverse is $CO_2$ reduction to formate.

found not to catalyze the oxidative reaction [57, 382].[54] This fact, although circumstantial based on an outgroup comparison, suggests that the full FBA reconstruction of *Aquifex* must produce formate, and must also include a reductive pathway not fed by any other source, unless an alternative formyl-THF synthesis reaction is present. To the extent that the conclusion seems unavoidable from the more extensive reconstruction and close outgroup comparison possible for *Aquifex*, we believe that it is a plausible conclusion for other instances of the Red(-2) pathway as well.

The proposal for an alternative formyl-THF synthase is compelling for another reason as well. The formation of $C_1$–$N^{10}$-bonded THF is always associated with hydrolysis of one ATP, as a consequence of the $pK_a$ and leaving-group energy of $N^{10}$-bound carbon [507, 509] which we argue in Section 4.5 is one of the central evolved features of THF. In organisms that synthesize $N^{10}$-formyl-THF, cyclization to $N^5$, $N^{10}$-methenyl-THF is spontaneous. Almost all of the organisms lacking reaction 2 possess an ATP-dependent $N^5$, $N^{10}$-methenyl-THF cycloligase [374, 780]. This enzyme, previously studied only as part of a human salvage pathway, has no plausible function in deep-branching bacteria unless an $N^5$-formyl-THF compound exists, on which it can act. The most likely solution appears to be either the presence of a previously unidentified $N^5$-formyl-THF synthase, coupled to the standard function of the cycloligase, or more simply, *that the cycloligase itself has a previously unidentified function as a formyl-THF synthase in these ancient clades.*

**Pathway completions possible through the glycine cycle**  To interpret certain alternations of oxidative or reductive pathways with the glycine cycle, recall from Section 4.3.4.2 the different pathways that can be completed using this cycle. Operating as the reverse of the glycine cleavage system, it is required to synthesize glycine from methylene-THF in the reductive $C_1$ pathway. Operating as the cleavage system, the cycle is required to produce *serine* if both the oxidative pathway from 3-phosphoglycerate and the reductive $C_1$ pathway are absent, leaving the transamination pathway is the only source of both the glycine precursor and methylene groups. The cycle is not required to produce either serine or glycine if the oxidative pathway is present. The reductive $C_1$-THF sequence also is not required as a source of methylene groups if serine cleavage to glycine is present, since this process produces methylene-THF directly.

**Alternation of the glycine cycle with the oxidative pathway from 3-phosphoglycerate**
The following alternations are seen in Table 4.1. In the Aquificales, Thermotogales, Deinococcus-Thermus, Chlorobi, and Nitrospirae, and in almost all non-methanogenic archaea, either two or all three genes in the oxidative pathway from 3PG are not found. In these groups the presence of the oxidative pathway is strongly correlated with the absence of both the glycine cycle and the reductive pathway. In archaea the same alternation is even stronger: the oxidative pathway occurs *only* where the cycle is absent, and

---

[54] Unfortunately, these experiments did not test whether it catalyzes the reductive reaction.

all organisms with this pattern in the sample except one member in the Thermoproteales are methanogens.

At the same time, the presence of the glycine cycle in bacteria predicts, almost without exceptions (213 out of 217 cases), the presence of either the 7-reaction (Red(-2)) or 8-reaction (Red) reductive pathway.[55] This implication strongly suggests that the absence of both the transaminase and the oxidative pathway is not a result of detection failure or gene misannotations. Both pathways are also found in most non-methanogenic archaea. In Wood–Ljungdahl acetogens, it is known to be responsible for reductive glycine synthesis, so its association with the other organisms possessing either the Red or Red(-2) pathways suggests a similar function.

### 4.4.3.3 Ramifications of widely distributed reductive synthesis of glycine

On the basis of these assignments of glycine biosynthesis pathways to members of extant clades, it appears necessary to reconstruct direct $C_1$ reduction, via either the Red or Red(-2) pathway, and glycine synthesis from methylene-THF, to the ancestors of both Crenarchaeota and Euryarchaeota, and within the bacteria to the ancestors of all four of the Firmicutes, Aquificales/Thermotogales, Chloroflexi, and Cyanobacteria, and possibly several other groups containing mostly heterotrophs today. The glyoxylate transamination pathway appears to be derived rather than basal within the Cyanobacteria (and still to be used only by a subset of them), and also within the Halobacteria, perhaps independently or perhaps by gene transfer. The oxidative pathway is also derived, and plausibly emerged independently in several groups of Firmicutes and other major groups such as proteobacteria (probably after the split from Aquificales/Thermotogales). Among archaea, the emergence of the oxidative pathway is notable among the methanogens, for reasons we will review below. By coalescence of the ancestors of all these groups, the combination of reductive $C_1$ on THF and reductive glycine synthesis is assigned to the common ancestors of both bacteria and archaea.

These assignments require several changes in our understanding of the major modes of autotrophic carbon fixation, and the ways in which they have governed the very earliest diversifications of bacteria and archaea.

First, the reductive synthetic pathway to glycine and serine through THF appears to be ancient and ubiquitous, making the direct $C_1$ reduction sector a second organizational center (together with the TCA intermediates and arcs) around which very long-term evolutionary diversification has been structured. To accommodate this fact in network models of autotrophy, several other reconceptualizations follow.

Second, the ancestors to most deep-branching clades with autotrophic members appear – from this enzymatic signature – to have fixed carbon via a pair of disjoint networks: one is the direct $C_1$ reduction pathway to glycine and serine; the other supplies carbon to the rest of metabolism from *any one* of the other five loop-fixation pathways, which feed carbon

---

[55] This kind of correlation, in which the presence of one trait almost perfectly predicts the presence or absence of other traits, is known as *implicational scaling*, also called *Guttman scaling* after its originator [324].

back into the four or five universal TCA intermediates. The only exception is the set of bacteria that fix carbon directly through the Wood–Ljungdahl pathway. This claim leads to a rather stark reconceptualization of the basic architecture of autotrophic carbon fixation, as having some of the pathway independence characteristic of **mixotrophy**.[56] Heretofore it has been verified in some organisms but assumed, in general, that autotrophs fix carbon through a single pathway, from which that carbon is directed to all of biosynthesis. That assumed generalization may be incorrect.

Third, for nearly 20 years now, it has been difficult to reconcile the distribution and functions of rTCA organisms and Wood–Ljungdahl organisms in any common terms. Both are widely but not universally distributed, and each possesses some features of a plausible prebiotic organosynthetic network that the other lacks. The paradox with respect to rTCA is that, where the entire loop is present, network autocatalysis could be invoked as the reason: the self-amplifying loop was the original mechanism that selected the universal precursors, yet the precursors are universal whereas the autocatalytic loop is not. A similar problem for W-L, as noted, is that its homologous enzyme is not universal, whereas the $C_1$ reduction sequence that is only one part of the pathway does not show homology suggesting a single origin.

The arguments put forth in this section appear to define a role for the folate pathway that is comparably universal to that for the universal TCA intermediates. It also makes the comparison of rTCA to W-L much more nearly one of complementarity: the bacterial pattern is reductive synthesis of glycine, possibly with carbon capture to other precursors through loop pathways. Loops and direct $C_1$ reduction are disjoint only because they lack the linking CODH/ACS enzyme. W-L shows the complementary pattern. Where the CODH/ACS is present, full TCA autocatalysis is lost, but the arcs remain as anaplerotic pathways.

### 4.4.4 A maximum-parsimony tree of autotrophic carbon-fixation networks

The three questions asked in constructing a maximum-parsimony tree of autotrophic carbon fixation are:

1. whether all known autotrophic carbon fixation phenotypes may be linked phylogenetically in a tree with one connected component;
2. which phylogeny requires the fewest or least improbable repeated innovations on distinct links;
3. which networks, unobserved in extant biology, must be posited to connect all autotrophic phenotypes that are observed.[57]

---

[56] *Mixotrophy* [481] refers to the derivation of carbon for anabolism from more than one source. So far it has been applied to organisms that fix the carbon they require for a subset of biosynthetic pathways, but obtain their required carbon for other pathways heterotrophically.

[57] The object of our concern is the innovation dependency among autotrophic metabolisms, and in general we expect that a similar reconstruction for heterotrophs would yield very low parsimony and highly ambiguous dependencies. However, we include two specific heterotrophic leaves in the reconstructed tree. One leaf covers a broad class of non-methanogenic Euryarchaeota, because this group contains no autotrophs. The second leaf represents a large group of heterotrophic Firmicutes including *Clostridium kluyveri*, which have close and remarkable affinities to autotrophs among the Crenarchaeota.

The manually constructed tree from [90] that proposes answers to these questions is shown in Figure 4.10. For loop-fixation pathways, the "subway map" of Figure 4.6 gives a natural suggested branching: 3HP results from a pathway substitution from rTCA, and CBB arises from the advent of RubisCO. The distribution of the archaeal 4HB pathways and their overlap with fermentative pathways in the Firmicutes requires the innovation of 4HB from W-L phenotypes, and the 3HP/4HB substitution is then a secondary refinement. Methanogens and acetogens form a continuum at the base of the bacterial tree and continuing into at least some branches in the Euryarchaeota.

Each internal node of the tree is labeled with an organism or clade possessing the indicated pathway, except for three. Between the acetogenic common ancestor of Euryarchaeota and the modern methanogens, we have proposed a W-L organism possessing the oxidative pathway but not yet having lost the glycine cycle. The oxidative pathway is a precondition to certain optimizations in the methanopterin cofactors of methanogens, as we explain below, and to maintain a tree with a single gain or loss on each link, we have separated these two events. A similar insertion of a common ancestor to Crenarchaeota and hydroxybutyrate/aminobutyrate-fermenting *Clostridia* proposes that fermentative arcs of the 4HB pathway must have evolved as facultative pathways before the loss of the CODH/ACS enzyme from acetogenic networks rendered them incapable of autotrophy. The presence of these arcs was a precondition to either the innovation of DC/4HB autotrophy or the heterotrophy of *Clostridia*. The third inserted node lies between the non-acetogenic bacterial clades and the acetogenic ancestor to Firmicutes and all archaea, and is discussed in detail in Section 4.4.4.3.

We may now consider what such a reconstruction suggests about the causes of evolutionary change along links,[58] and what differences the reconstructed root node has as a model for the LUCA or earlier metabolisms, relative to the pathways in modern organisms.

### *4.4.4.1 Causes along the arcs*

The tree in Figure 4.10, with the direction of time suggested by its overlay onto the major phylogenetic divisions of bacteria and archaea, suggests significant progressions of three kinds in going from the root to the leaves. One is from strictly anoxic conditions to conditions where $O_2$ is present in some environments either as toxin or energy source. A second is from a regime that we may think of as undiversified "first attempts" at metabolism into a more refined regime where branches become differentially optimized. The third is from an early regime which we argue favors robustness, to a later regime in which efficiency is selected. The reason these progressions can be overlaid on an undirected tree is that reasonable causes that act unidirectionally can be assigned to many branches in a mutually consistent way. The factors appear in the tree as follows.

**$O_2$ or other oxidants as poisons of the CODH/ACS** The major split between the non-acetogenic bacteria and all acetogens occurs by two different losses. From a proposed

---

[58] More detailed versions of these arguments are given in [90].

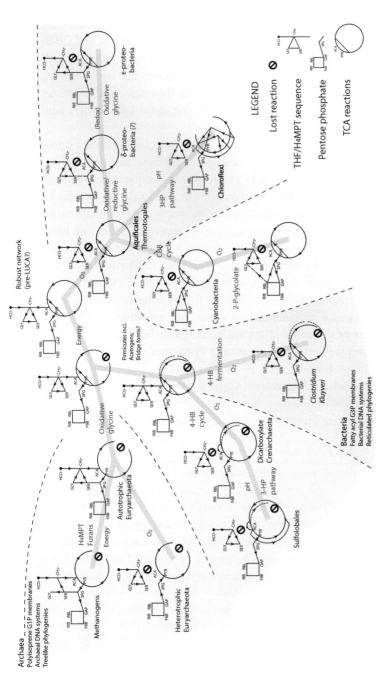

Figure 4.10 A maximum-parsimony tree (gray links) of autotrophic carbon fixation phenotypes. The three major modules, from the covering network of Figure 4.5, and the one-carbon subnetwork of Figure 4.4, shown schematically (legend) are pterin/folate-$C_1$ metabolism, the pentose phosphate pathway, and the TCA cycle reactions. Each link represents a loss or addition of a single pathway segment. Lost reactions (⊘) include the acetyl-CoA synthase (in $C_1$ metabolism), and ferredoxin-dependent succinyl-CoA synthase (in the TCA loop) or citryl-CoA synthase (not shown). Abbreviations: HCO−, formyl; −CH₂−, methylene; ACA, acetyl-CoA; PYR, pyruvate; GAP, glyceraldehyde-3-phosphate; F6B, fructose-1,6-bisphosphate; RIB, ribose phosphate; RBL, ribulose phosphate; ALK, alkalinity; others as in Figure 4.4. Arrows indicate reaction directions; dashed line connecting 3PG to SER indicates intermittent or bidirectional reaction. Highlighted in yellow are the innovations underlying divergences, while red labels on links indicate evolutionary forces associated with each innovation, explained in the text. Beneath each diagram is an extant species or clade name where the phenotype is found. Colored regions indicate domains within which all known instances of the indicated phenotypes are restricted. (After [90]. Creative Commons [153].)

root node with both autocatalytic rTCA cycling and a complete W-L pathway, the loss of the CODH/ACS enzyme requires all autotrophic bacteria along this branch to use one or another autocatalytic loop pathway to fix carbon. A similar (but later) loss in the Crenarchaeota, between a common-ancestor node[59] with the Firmicutes and the DC/4HB autotrophic Desulfurococcales and Acidolobales requires closure of fermentative arcs to form the 4HB pathway. The CODH/ACS enzyme has restricted distribution on Earth today because it is one of the most oxygen-sensitive enzymes known [662, 908], so we have labeled these links with "$O_2$." Molecular oxygen may not be a plausible toxin to label the loss leading to Aquificales/Thermotogales on the first branch. Although Aquificales are micro-aerophiles, the Thermotogales remain strict anaerobes, and this split from acetogens must have occurred very early, in or before the early Archean, when all current estimates of $f_{O_2}$ indicate very low activities. However, the CODH/ACS is also disabled by elemental sulfur[60] and perhaps other oxidants, and we cannot rule out the accumulation of low levels of $O_2$ or $H_2O_2$ in local environments due to processes such as radiolysis of water, so we tentatively retain a toxicity explanation for the loss of the CODH/ACS.

**$O_2$ as energy source** In later links, particularly in the proteobacteria, the accumulation of small but non-vanishing levels of $O_2$ may have favored the transition to an oxidative pathway for glycine synthesis, leading to the connected autotrophic networks in which rTCA is a sole carbon-uptake pathway, as is now found in $\varepsilon$-proteobacteria such as *Sulfurimonas denitrificans*.[61] The rise of significant levels of $O_2$, due to cyanobacterial oxygenic photosynthesis and saturation of oceanic iron buffers, coincides roughly with the perimeter of the reconstruction shown in Figure 4.10. We return to consider the gross qualitative changes in metabolic evolution at this horizon in Section 4.4.6.

**Shedding of ATP-dependent steps in Wood–Ljungdahl organisms** The alternative branch from the root in Figure 4.10 is characterized by the breaking of the rTCA loop to form the non-autocatalytic arcs of the acetyl-CoA pathways common to acetogens, methanogens, and heterotrophs using the Wood–Ljungdahl pathway. This link is labeled "energy" in the figure, because the loss always occurs at one of the two ATP-dependent thioesterification steps in rTCA: either formation of citryl-CoA prior to retro-aldol cleavage or formation of succinyl-CoA. Shedding of the ATP-dependent step results in a lower ATP cost per carbon fixed, which makes the acetyl-CoA pathway more energy efficient than rTCA [686] in environments where the CODH/ACS can function.

An extreme refinement occurs in methanogens, which have altered the free energy landscape of the pterin tetrahydromethanopterin ($H_4$MPT) relative to that of THF, so that a formyl group may be transferred to $N^5$ from formyl-methanofuran and then cyclized to $N^5$,

---

[59] We reiterate that nodes in the tree trace dependencies among innovations in carbon fixation, and need not correspond to species. The dependency enabling the evolution of the 4HB pathway could be on a microbial consortium of Crenarchaeota and Firmicutes, which follows a different branching pattern than 16S rRNA. Branching of Firmicutes with archaea in reconstructions of many gene families [138] provides independent support for some systematic association or common ancestry of some characters.

[60] S. Ragsdale (personal communication).

[61] R. Braakman (personal communication).

$N^{10}$-methenyl-H$_4$MPT without ATP hydrolysis.[62] The free energy required for this attachment is retained in a membrane ion potential, in an ATP-free Fe-S system [415, 509].[63] The price of this optimization is that H$_4$MPT can no longer serve as methylene donor to glycine or serine synthesis, so the emergence of the oxidative pathway to these amino acids is a precondition to the energy refinements of methanogenesis.

**"Irreducibility" at the energy frontier of methanogenisis** Methanogens approach an efficiency frontier that seems difficult to surpass, and which creates long-range constraints among pathways that make the phenotype seem "irreducible." The elimination of one ATP hydrolysis has led to the need for a large-scale rearrangement of metabolic fluxes; in particular, it demands the oxidative synthesis of serine and glycine in a reductive chemoautotroph. Methanogens are the only group in which the ATP cost to form a metabolite does not increase in at least linear proportion to the number of carbon atoms in a metabolite, because the Wood–Ljungdahl pathway operates within an iron-sulfur-based redox cycle (see Figure 6.2 for more detail). That saving permits the ATP cost for glycine synthesis via the *oxidative* pathway in methanogens to equal that via the reductive pathway in other organisms, and the ATP cost for serine synthesis to be one unit lower in methanogens than via the reductive pathway.

However, only an organism capable of the complete W-L synthesis can benefit from this refinement. For all other carbon fixation pathways, the ATP cost of glycine or serine synthesis (or both) is higher for the oxidative than for the reductive pathway. Notably, in heterotrophic archaea that have lost the CODH/ACS, there is no way to capture such an advantage, so the optimization of H$_4$MPT confers no added fitness. Therefore we can understand why the special changes to H$_4$MPT were most plausibly evolved in methanogens, and only later transferred to bacteria [83, 306, 524, 840] where they supported C$_1$ oxidation rather than C$_1$ reduction.

**Alkalinity, photorespiration, and other innovations** In assigning the 3HP pathways as alkaline adaptations in the Chloroflexi and the Sulfolobales, we follow standard interpretations. As suggested in Section 4.3.6, although RubisCO-like proteins are ancient and widely distributed, their role in primary carbon fixation pathways appears to be late and to have originated in bacteria. The joint direct-C$_1$/CBB pathway is most parsimoniously placed as a single innovation from the joint direct-C$_1$/rTCA pathway, which permits reversal of the direction of TCA cycling. Some Calvin cycle photosynthesizers occupy low-oxygen environments [285], and this presumably must have been the

---

[62] All the non-folate pterins eliminate the carboxyl group of para-aminobenzoate, replacing it with a ribitoyl moiety, before attaching it to the neopterin from GTP. The carboxyl of pABA in THF lowers the $N^{10}$ electron charge and hence the p$K_a$ is as much as 6.0 natural-log units below that of N$^5$ [407]. The resulting higher-energy C–N bond cannot be formed without hydrolysis of one ATP, either to bind formate to N$^{10}$ of THF, or to cyclize $N^5$-formyl-THF to form $N^5$, $N^{10}$-methenyl-THF (see Figure 4.15 below). In contrast, in H$_4$MPT the difference in p$K_a$ between N$^{10}$ and N$^5$ is only 2.4 natural-log units. The lower C–N$^{10}$ bond energy permits spontaneous cyclization of $N^5$-formyl-H$_4$MPT, following (also ATP-independent) transfer of formate from a formyl-methanofuran cofactor. H$_4$MPT goes further with the addition of methyl groups which are believed to further lower the bound-C$_1$ energy by changing $\Delta S$ of bond formation.

[63] See the diagrams of these systems in Figure 6.2 of Chapter 6 for more detail.

ancestral form before oxygen buffers had been saturated, so the emergence of the glyoxylate-transamination pathway as an adaptation to photorespiration should only have been adaptive after the accumulation of significant environmental $O_2$.

**A cryptic cyanobacterial phenotype for anoxic environments** The reconstruction of the cyanobacteria favors an ancestral form with hybrid CBB and THF-$C_1$ carbon fixation. (Recall from Table 4.1 that nearly all have the glycine cycle and the reductive pathway, while only about half have the transaminase.) Whereas direct reductive synthesis is among the most ATP economizing pathways to glycine and serine, synthesis (via any pathway) using carbon fixed by RubisCO is among the most ATP costly [63, 89]. Therefore, if RubisCO were reliable, we would not expect oxidative serine synthesis to have arisen in the Cyanobacteria. The use of the glyoxylate transaminase pathway, as we have noted, can only be energy efficient if it is the only alternative to excretion of carbon erroneously fixed in 2-phosphoglycolate. We have therefore proposed [90] that in Cyanobacteria which fix carbon in anoxic environments, where oxidation of phosphoglycolate to glyoxylate and subsequent transamination are not available, rather than substitute the oxidative pathway to serine, these organisms should revert to the direct reductive $C_1$ biosynthetic pathway.

### 4.4.4.2 The incomplete registry of the phylometabolic tree with other trees

The tree of Figure 4.10 violates two groupings that by other criteria are fundamental. First, Crenarchaeota branch with Firmicutes rather than with Euryarchaeota, in the metabolism of hydroxybutyrate or aminobutyrate. A possible interpretation is that large cassettes of metabolically related genes were first evolved in fermenters and exchanged between Firmicutes and Crenarchaeota, where they enabled the emergence of a new mode of autotrophy. Since innovations in heterotrophy occur by definition in ecosystem contexts, they are inherently ambiguous if projected onto a tree of individual lineages, which the autotrophy tree of Figure 4.10 imposes.

A more complicated question is raised by our reconstruction of W-L autotrophy including *folate cofactors* at a node ancestral to archaea as well as bacteria. Homology in the CODH/ACS enzyme is consistent with a common ancestry of the acetyl-CoA pathways in acetogens and methanogens, but this alone is not sufficient to imply common ancestry of the $C_1$-reduction branch of the Wood–Ljungdahl pathway, against the many signatures of independence in the archaea and bacteria. Filipa Sousa and William Martin have argued [773], based on a study of sequence homology of biosynthetic enzymes for THF, $H_4MPT$, and molybdopterin,[64] that the two $C_1$-reduction branches were independently derived in acetogens and methanogens.[65] Our reconstruction of a folate-dependent ancestor turns essentially on the problem of amino acid biosynthesis for glycine, serine, and

---

[64] In Section 4.5.2.1, where we describe the biosynthetic sequences, it will be clear that a much larger difference in the architecture of the biosynthetic pathways lies behind much of the sequence independence.

[65] Martin has argued for this interpretation for several years [523, 525]. For a variety of reasons, detailed in Section 4.5.2.1 and again in Section 6.3.1.3, we are unable to place methanogens as a basal clade among the archaea; they seem by too many criteria to be a highly refined and derived group. That is a finer-scale distinction than the broad grouping discussed in this section.

the sequence of acids derived from them, and on the genomic evidence for the reductive synthetic pathway from the $C_1$ superhighway. This is the function that $H_4$MPT is unable to perform. We return to the problem of interpreting organisms using the W-L pathway as models in Section 6.3.1.3.

### 4.4.4.3 Character of the root node

The most important proposed node in Figure 4.10, not yet known to be instantiated by a modern organism, is the node we have inserted between rTCA autotrophs and acetyl-CoA pathway autotrophs. As we noted in Section 4.3.6, both pathways feed carbon back network-autocatalytically, but the rTCA network does so in the small-molecule substrate, while the acetyl-CoA pathways do so by producing THF (or another pterin) which is a network catalyst. Either the breaking of the rTCA loop, or the loss of the CODH/ACS from the W-L pathway in acetogens, severs a link between $C_1$ input and completion of the catalytic network, destroying autotrophy. Since the two losses are distinct, no single innovation can connect rTCA carbon-fixers and acetyl-CoA pathway carbon-fixers. Any intermediate lacking either completion is not an autotroph. Therefore the only *topological* completion that contains rTCA and acetyl-CoA pathways in a tree with a single connected component must pass through a phenotype in which *both* forms of network autocatalysis exist.[66] The proposed root node is redundantly autocatalytic, a property that none of the nodes instantiated in presently known organisms possesses. A variety of considerations have led us to regard this inserted node as the most natural root of the tree. The properties that make our inserted root node different from the metabolisms of extant organisms come from the relation of network topology and redundancy to self-amplification.

**Network autocatalysis** is an essential feature for the emergence of kinetic *order*. It amplifies flow into the network in proportion to the instantaneous flux, leading to a form of chemical-kinetic *selection*. In reaction networks where the same inputs feed both auto-catalytic and non-autocatalytic networks, selection by differential amplification leads to chemical *competitive exclusion*. The presence of these properties in a bulk chemical process would provide a very powerful explanation of biochemical order without invoking the Darwinian dynamic of replication with heritable variation.

Network autocatalysis achieves these functions because the pathway flux around the cycle is *independent* from the chemical potentials of the inputs and outputs.[67] Increasing the concentration of pathway intermediates proportionally increases the rate constants of the input/output conversion relative to pathways that do not increase the concentration of their own catalysts, and if the accumulation of a network catalyst is not limited, the flux through the corresponding pathway can grow without limit to dominate other pathways.

---

[66] The *unavoidability* of such an inserted node reflects the deep differences that have made it difficult to compare W-L and rTCA models in the past. The identification of the Red(-2) pathway in Section 4.4.3 is the major revision that makes a connected tree possible with only *one* such insertion. Without independent evidence that this pathway is well supported, the reconstruction of any connected tree would have become much more speculative.

[67] The *proportional* rate of change in the flux is determined by the input and output chemical potentials, at any concentration of cycle intermediates.

However, independence of cycle flux also makes network-autocatalytic pathways *fragile*. In a linear pathway, depletion of the inputs results in a drop of chemical potential that eventually stops driving the reaction. Autocatalytic pathways possess no such inherent self-regulation. The pathway intermediates can pass smoothly from a concentration at which the pathway self-amplifies to a concentration at which it reproduces below replacement. Then the cycle spirals down to extinction. Because the chemical potential of the network catalyst appears in both the inputs and outputs to the network, it cannot exert feedback to stabilize the throughput reaction.

**Redundant autocatalysis** between W-L and rTCA can stabilize a network against the fragility of collapse in either of the two subsystems separately. To illustrate this effect, we show the input/output characteristics of a minimal kinetic model of the coupled networks in Figure 4.11. The eleven intermediates of rTCA are collapsed down to a single network catalyst, by a network-reduction procedure detailed in Appendix B of [91]. The catalyst in the resulting lumped-element reaction network (which we may take to be oxaloacetate) is both input and output of a reaction with the scheme

$$2CO_2 + 4H_2 + OXA \rightleftharpoons 2H_2O + CH_3COOH + OXA. \tag{4.1}$$

Fraction of equilibrium acetate from driving rTCA and WL in parallel

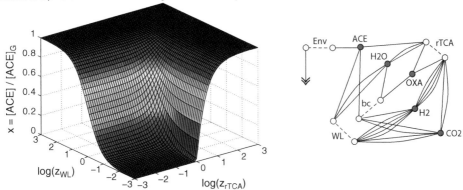

Figure 4.11 Network dynamics of connected W-L and rTCA pathways. Right-hand graph shows the stoichiometry of the rTCA network as a lumped-element graph (from [91]). W-L has the same input/output characteristics but is not autocatalytic in a presumed mineral-catalyzed environment. Left-hand graph shows conductance of the chemical network as a function of the driving chemical potential and the rate of the W-L pathway as a "feeder" stream for acetate. The quantity $x$ on the $z$-axis is the fraction of the acetate concentration [ACE] relative to the value it would take in an equilibrium ensemble with carbon dioxide, reductant, and water. The value $x = 1$ corresponds to an asymptotically zero impedance of the chemical network, compared to the rate of environmental drain. The parameter $z_{rTCA}$ is a monotone function of the non-equilibrium driving chemical potential to synthesize acetate, and $z_{WL}$ measures the conductance of the feeder W-L pathway. At $z_{WL} \rightarrow 0$, the W-L pathway contributes nothing, and the rTCA network has a sharp catalytic threshold at $z_{rTCA} = 1$. For non-zero $z_{WL}$, the transition is smoothed, so some excess population of rTCA intermediates occurs at any driving chemical potential.

An anaplerotic aggregate pathway interconverts acetate[68] and oxaloacetate with the scheme

$$CH_3COOH + 2CO_2 + H_2 \rightleftharpoons OXA + H_2O. \tag{4.2}$$

Acetate is drained by a linear reaction, representing simple diffusion. Finally, a pathway to acetate with the same conversion as Scheme (4.1) operates in parallel with fixed rate constants. Before cells this might be a model of the equivalent Fischer–Tropsch-type reaction [31] maintained on a geological substrate such as iron or nickel sulfide mineral surfaces. In early cells it could model actual W-L carbon fixation from a population of cofactors that fluctuates more slowly than the fluctuations in a poorly regulated rTCA cycle. The background reaction acts as a "feeder" of a small stream of acetate that does not depend on autocatalysis.

The important control parameters for the network are the rate constants of the cycle and the W-L feeder reaction, the rate constant for loss of acetate, and the chemical potential that drives the system, which is given by

$$2\mu_{CO_2} + 4\mu_{H_2} - 2\mu_{H_2O} - \mu_{CH_3COOH}. \tag{4.3}$$

The measure of network efficiency is the ratio of the concentration of acetate achieved in the reaction system, [ACE], to the value that would be in Gibbs equilibrium with the environmentally supplied concentrations of $CO_2$, $H_2$, and $H_2O$, which we denote $[ACE]_G \equiv K_{rTCA}[H_2]^4[CO_2]^2/[H_2O]^2$, in which $K_{rTCA}$ is the equilibrium constant determined by the reaction free energy. For a passive one-step reaction with fixed rate constants and without autocatalysis, the ratio $[ACE]/[ACE]_G$ would be a constant $<1$ at all values of $[ACE]_G$.[69] When the autocatalytic loop has such a large concentration of oxaloacetate serving as a network catalyst as to (asymptotically) create no resistance, the same ratio $[ACE]/[ACE]_G$ saturates at **unity**, which is equivalent to the condition that the chemical potential difference (4.3) vanishes.

Two lumped parameters governing the throughput characteristics are denoted by $z_{rTCA}$ and $z_{WL}$. Up to constant offsets, $\log(z_{rTCA})$ equals 3/2 times the log of the activity $[ACE]_G$ that drives Eq. (4.3).[70] $\log(z_{WL})$ is (again up to constant offsets) the negative of the transition-state chemical potential for the environmentally catalyzed W-L reaction.

---

[68] Here, as elsewhere in the figures of this chapter, acetate serves as a placeholder for the activated thioester, in a simplified model to quantify the properties of catalytic loops. Chemically realistic models would include both the activating paths and the free energy sources besides the $H_2/CO_2$ couple that drive them.

[69] For a multistep reduction sequence with carbon and reductants added sequentially, the rate will depend on both $[CO_2]$ and $[H_2]$ separately, but still saturates.

[70] Full expressions for the control parameters, given by the steady-state conditions of the network shown in Figure 4.11, are given by

$$z_{rTCA} = \frac{\sqrt{k_{rTCA}\bar{k}_{rTCA}}}{k_{Env}} \frac{K_{OXA}}{K_{rTCA}} \frac{[CO_2][H_2O]^2}{[H_2]} [ACE]_G^{3/2}$$

$$z_{WL} = \frac{\bar{k}_{WL}[H_2O]^2}{k_{Env}}.$$

Thus $\log(z_{WL}) \to -\infty$ describes an infinitely high reaction barrier that shuts off the W-L reaction, and $\log(z_{WL}) \to \infty$ describes no barrier, or perfect environmental catalysis.

The figure shows that when the rate of "feed" acetate through the linear pathway approaches zero ($\log z_{WL} \to -\infty$), rTCA has the fragility of a pure autocatalytic loop. Above a critical value (here $\log z_{rTCA} = 0$), the loop rapidly approaches saturation, but on the half-line $\log z_{rTCA} < 0$ it produces no output. The addition of a W-L "feeder" smoothes this transition. The cycle supports some flux for all values of $\log z_{rTCA}$, though its response is still sigmoidal. If the flux is perturbed below the autocatalytic threshold, the cycle can recover instead of crashing. *Even when it operates chronically below the threshold, an rTCA cycle with a feeder produces some concentration of its intermediates, selecting these intermediates in excess of the background from uncatalyzed reactions.*

The selecting effect of autocatalysis was the motivation for Günter Wächtershäuser's original use of rTCA cycling as the template for his "pyrite-pulled" theory of surface metabolism [843, 844]. It also provided a background context in the earlier experimental study by Claudia Huber and Günter Wächtershäuser [377] demonstrating thioester formation from $CH_3SH$ and CO as a mineral surrogate for the W-L pathway, in which they suggested that the resulting acetyl thioesters could be fed into such an autocatalytic cycle.

### 4.4.4.4 Robustness of flux and robustness of selection

We may summarize these points by saying that Wood–Ljungdahl confers robustness on pathway *fluxes* in autocatalytic loops, while self-amplification in loops confers robustness on the *selection* of the precursors of biosynthesis including cofactor biosynthesis. Synergy between these, conferring robustness, would have been important in at least two levels of organization.

If the metabolism-first hypothesis that biochemistry emerged from geochemistry is correct, functions now provided by cofactors must once have been provided by mineral surfaces, metal-ligand complexes, or some other geochemical substrate. Since such catalysts must have been non-specific against side-reactions, the problem of chemical order would have included a problem of protecting fluxes in catalytic networks against fluctuations and spiral down to collapse. In this context, a redundant network such as the root node in Figure 4.10 would have conferred an inherent robustness from its topology, which modern optimized and pruned networks do not.

The problem of protection of network catalysts from side-reactions arises at many levels. Chapter 2 noted that the biosphere as a whole is network-autocatalytic, and living matter is its network catalyst. Biosynthesis is another form of "parasitic side-reaction" from the

---

Each control parameter is a ratio of lumped half-reaction rates that feed [ACE], over the environment dilution constant $k_{Env}$ through which it drains. Here $k_{rTCA}$ and $\bar{k}_{rTCA}$ are the forward and reverse rate constants for the rTCA network determined from the graph reduction procedure in Appendix B of [91] assuming steady-state flow. (These effective lumped-parameter rate constants depend on the concentrations of inputs, but not on [ACE]. Their dependence is on linear combinations of chemical potentials independent of $\log([ACE]_G)$, the combination in Eq. (4.3), so contours exist along which $[ACE]_G$ can be varied while the lumped-parameter rate constants are held fixed.) $K_{rTCA} = k_{rTCA}/\bar{k}_{rTCA}$ and $K_{OXA}$ are the equilibrium constants for the reactions forming acetate and oxaloacetate, respectively. $k_{WL}$ is the forward rate constant for the W-L pathway, which we take as a parameter.

flux in a core autocatalytic network. If catalysts emerge which enable secondary reactions, their concentration or power must itself be sufficiently regulated that biosynthesis does not extinguish its own source of inputs. In a primitive cell with unreliable regulation occupying a fluctuating environment, W-L and rTCA would operate synergistically. W-L, stabilized by slowly varying cellular populations of THF, would make rTCA cycling robust against rapid fluctuations in the concentrations of small-molecule cycle intermediates. In turn, rTCA would contribute carbon fixation to the network synthesizing THF.

As cofactors, catalysts, and regulation of cellular homeostasis became more refined and reliable through the process of cellularization, and perhaps even into the early genomic era, the excess redundancy would have become less necessary, and other factors such as energy efficiency could have been selected. This is the argument for unidirectionality under which the inserted node in Figure 4.10 is interpreted as a root, and other nodes are interpreted as descendants from it.

### 4.4.4.5 Parsimony violations in autotrophy

The ability to arrange autotrophic metabolic networks on a tree in Figure 4.10 is not tautological. This can be seen from the two minor duplications that were unavoidable in the maximum-parsimony tree.

The first duplication is the convergent evolution or transfer (or combined transfer/convergence) of the 3HP pathway in Chloroflexi and in the Crenarchaeota, which entails duplication of an entire (and rather elaborate) pathway segment, and not merely a single key gene. It is favored in specialized environments which we would expect to create long-term association between inhabiting species.[71] These environments contain a stressor (alkalinity) which we expect to induce gene transfer [97]. Moreover, relative to the very ancient divergences in our tree, the innovations of 3HP occur late, at an era when we expect organism lineages to have evolved refractoriness to many forms of gene transfer, along with more integrated control of chromosomes. The second parsimony violation (noted above) is the likely convergent evolution of the oxidative serine pathway in (at least) proteobacteria and methanogenic archaea, using common reactions catalyzed by widely diversified enzymes. Much more striking violations of parsimony in core metabolism occur after the rise of oxygen, to which we return in Section 4.4.6.

### 4.4.4.6 Parsimony in an era of gene transfer: invasion of geochemically defined niches

The tree of Figure 4.10 mostly overlaps with major branchings of bacteria and archaea, with the important exceptions noted in Section 4.4.4.2. We cannot rule out the possibility that the descent and diversification of autotrophy followed the descent and diversification of *autotrophic species*, giving the tree the usual interpretation of a lineage phylogeny. However, this interpretation seems implausible at points such as the failure to produce bacteria and archaea as monophyletic groups. More generally, when many genes

---

[71] Similar long-term associations in anaerobic environments such as coastal muds are believed to have led to the (otherwise uncommon) aggregate transfer of a large complement of operational genes from $\epsilon$-proteobacteria to Aquificales [84].

are known to have been under frequent exchange among ancient groups, the absence of reticulation in a metabolic tree is a somewhat surprising result that deserves an explanation.

An explanation that becomes stronger in the presence of gene exchange than in its absence is *serial adaptive invasion* of geochemically defined niches. Starting from a crude ancestor that could survive only because it employed redundant and inherently robust networks, a variety of refinements is possible to resist oxygen, reduce energy consumption, compensate for $CO_2/HCO_3^-$ equilibria, etc. Each new environment may have been colonized from ancestors for which an adaptation required few steps. As a kind of allopatric speciation, structured by geochemical rather than merely geographic separation, different environmental conditions may have kept groups with different pre-adaptations out of mutual contact, as each left descendents that colonized new settings. The tree of Figure 4.10 and the inventory of known fixation pathways are not complicated, so this population separation could have required only very coarse environmental distinctions to explain the reconstructed dependencies.

Since the environment is *geochemically* defined rather than defined primarily in terms of an ecological community composition, different fixation networks are kept separate whether they are realized within individual species or at the consortium level. Thus the serial dependencies in innovations reflect *relations among geochemical environments* more than accidents in the order of gene acquisition. It may even be that species branchings should be understood as following, rather than leading, innovations in carbon fixation.

The presence of gene transfer should further inhibit reticulation, because gene transfer militates against the preservation of standing variation within single populations.[72] For reticulation to arise in a reconstruction, two or more lineages must have diversified and subsequently been brought into contact. In autotrophs whose phenotypes are closely entrained by their chemical environments, however, the communities cannot be brought into contact unless the environments are merged, tending to erase differences in their members which might leave phylogenetic signatures.

### 4.4.5 A reconstruction of **Aquifex aeolicus** *and evidence for broad patterns of evolutionary directionality*

The complete flux balance model of *Aquifex aeolicus* metabolism constructed by Rogier Braakman [92] permits us to see more general patterns of change between ancestral and derived organisms. The *Aquifex* model was not computed in isolation from other strains, but using joint phylogenetic and metabolic criteria to identify ancestral as well as derived forms for each collection of pathways. *Aquifex* is an outlier from the most widely attested

---

[72] An example of the difference between selection in the presence or in the absence of gene transfer is given by the model of optimization of the code due to Vetsigian *et al.* [833], to which we return in Box 8.3 (Chapter 8). When genes may be freely transferred they are optimized in a common context of the community gene pool rather than in subgroups partitioned against gene exchange. The population then shares innovations, and selection leads rapidly and robustly to optimal solutions among the set compatible with free exchange, but not to the preservation of standing variation.

forms for many pathways even in bacteria, yet it is possible to show that (for reasons that we can suggest only partially) the family Aquificaceae of which it is a member is a remarkably conservative clade of ancestral states at or very close to the root of the bacterial tree.

Three patterns of directional change that are clearly visible, in which *Aquifex* more closely approximates (but does not fully conserve) the form of the LUCA and most organisms show more divergence are: (1) the duplication and divergence of enzyme families for parallel pathways with the same local-group chemistry, enabling greater substrate specificity and improved kinetics; (2) optimization of established pathways by the fusion of enzymes or their organization into complexes; and (3) the reduction of ATP costs by coupling exergonic and endergonic reactions. (Recall our discussion of a similar energetic optimization in methanogens in Section 4.3.6.5 and Section 4.4.4.1.)

### 4.4.5.1 rTCA enzyme homology reflects substrate homology

In most modern organisms, the homology of local-group chemistry that we illustrated in Figure 4.9 is not exactly recapitulated in homology of their enzymes. We noted in Section 4.3.7 that a very common joint citryl-CoA synthase/citryl-CoA lyase alternates with a much rarer enzyme pair in which the citryl-CoA synthase is homologous to the succinyl-CoA synthase [36] and the citryl-CoA lyase is a separate aldolase [35]. Similarly, a two-step enzyme performs the $\beta$-carboxylations of $\alpha$-ketoglutarate to oxalosuccinate and subsequently reduces this to isocitrate, and never releases oxalosuccinate into solution. This mechanism contrasts with the $\beta$-carboxylation of pyruvate and subsequent reduction of oxaloacetate, for which the two steps are performed by distinct enzymes, and oxaloacetate must be released because it is a precursor to aspartate biosynthesis. Yet, as Miho Aoshima *et al.* showed in a remarkable sequence of papers [32, 33, 34], both of these common asymmetries are absent in *Hydrogenobacter thermophilus* (and more generally in the Aquificaceae), which uses pairs of enzymes whose homology follows that of the local-group chemistry. They go on to present evidence that the asymmetric forms are innovations, and that the *Hydrogenobacter* gene profile is the ancestral form in bacteria.

For the $\alpha$-ketoglutarate carboxylase/reductase,[73] the evolutionary advantage of an enzyme fusion in almost all organisms is easy to propose. $\beta$-carboxylation of pyruvate or $\alpha$-ketoglutarate is endergonic, and requires hydrolysis of one ATP. The subsequent reduction is highly exergonic, and if the energy from the reductant cannot be salvaged, performing the two reactions in sequence needlessly dissipates the chemical work performed by ATP hydrolysis as heat in the subsequent reduction. In the fused enzyme, the free energy of reduction can be coupled directly to carboxylation so that no ATP is hydrolyzed. The combined carboxylation/reduction is also a faster process, because the intermediate is not released to diffuse between two enzymes. (The latter advantage applies also to a fused citryl-CoA synthase/lyase.) A third possible, although likely minor, advantage to the

---

[73] This enzyme in databases is labeled the *isocitrate dehydrogenase*, or ICDH. Like most enzymes in databases, it is labeled with its function in oxidative organisms where it was first characterized, but the family is homologous and the pair of reactions, from the free energies shown in Figure 4.2, is nearly reversible.

two-step carboxylation/reduction arises because $\beta$-carboxylations are *kinetically* facile, making oxaloacetate unstable in water and oxalosuccinate even more so. Preventing the release of oxalosuccinate avoids events in which spontaneous decarboxylation results in an ATP-consuming futile cycle.[74] The sequestration of oxalosuccinate and also citryl-CoA is an evolutionarily accessible innovation because it requires no compensating innovation in other systems, as neither of these metabolites is a precursor to any other pathways.

### 4.4.5.2 Duplicated versus non-duplicated enzyme families

The pattern of enzyme homology in *Aquifex* reflects a more modest form of divergence – sequence divergence without discrete events of fusion or replacement – from an earlier stage with only one enzyme family for each sequence of local-group transformations. While one interpretation of such an earlier state is that only one part of a pathway was specifically catalyzed (so that enzyme innovation was required to enable the other part of the pathway), such an interpretation makes no sense to us, since carbon-fixing loop pathways have no function at all unless they close. Therefore it seems necessary to suppose that the ancestral form with a single enzyme sequence used non-specific enzymes to catalyze both pathway segments. Relative to such an ancestor, *Aquifex* and its relatives still show significant evolutionary refinement. Despite the high sequence similarity between the pyruvate and $\alpha$-ketoglutarate carboxylases in *H. thermophilus*, Aoshima *et al.* showed that the two enzymes possess no detectable cross-affinity for each other's substrates.[75] It is likely that the increase in specificity enabled an increase in catalytic rate, and that for core pathways which must supply almost all biosynthetic carbon, the rate advantage is a more important contributor to fitness than the cost of maintaining two gene families.

It is interesting that in *Aquifex* the dependence of this trade-off can be recognized because in branched-chain amino acid biosynthesis the balance appears to favor gene reduction. The condensation enzymes, which form acetolactate and 2-acetyl-2-hydroxybutyrate, shown in Figure 4.8, are the only two distinct enzymes in the homologous branches to form isoleucine and ketoisovalerate. The TPP-dependent ketoacid reductioisomerase and subsequent enzymes performing the C–C bond rearrangement and two reductions are shared by the two pathways [903]. An argument proposed in [92] is that the lower carbon flux to this subset of branched-chain amino acids (than the total flux in rTCA), together with the greater novelty and complexity of these reactions, has not been sufficient to select a duplication and divergence in this gene family.

### 4.4.5.3 Amino group metabolism and a caution

The flux balance model of *Aquifex* reconstructs reductive glycine synthesis from the $C_1$ superhighway, following the arguments given in Section 4.4.3.2. This is the part of the reconstruction about which we are most cautious. The study of nitrogen metabolism by

---

[74] The frequency of spontaneous decarboxylations will depend on the ratio between the lifetime of oxalosuccinate in the cytoplasm, and the typical diffusion time between enzymes in a bacterial cell.

[75] The statement in [34] is that "PVC [pyruvate carboxylase] did not function as a carboxylating factor for HtICDH [*Hydrogenobacter thermophilus* isocitrate dehydrogenase]."

Kameya *et al.* [409] reports a complex set of fully *five* different aminotransferase activities catalyzed by three proteins. Two of these form glycine from glyoxylate by transamination from glutamate, and a third forms glycine by transamination from alanine. All three reactions were unidirectional under the purified-enzyme conditions studied. However, no source of glyoxylate to serve as substrate for these enzymes has been identified biochemically or proposed based on sequence comparisons. A fourth transaminase activity involves phosphoserine (in the oxidative pathway), but the direction shown was *de-amination* of phosphoserine to form glutamate from $\alpha$-ketoglutarate. The sequence/function relations in these enzymes are strikingly different from those in most other bacterial groups (see footnote 44 above), so there is a possibility of systematic misassignment of sequences in other Aquificales that have not been characterized biochemically. If a more complete biochemical characterization of *Hydrogenobacter* demonstrates either the transaminase or the oxidative pathway as the route to glycine and serine, sequence fingerprints based on this metabolism could lead to different interpretations elsewhere among deep-branching clades as well.

### 4.4.6 The rise of oxygen and the attending change in metabolism and evolutionary dynamics

The leaves on the tree of Figure 4.10 fall within a horizon corresponding loosely to the rise of oxygen. The changes across that horizon in metabolism, ecological structure, and the evolutionary relations among these are drastic in almost all respects. While the rise of oxygen was a transition in the four geospheres near the end of the Archean [238] and long after the period of emergence that is our main interest, the changes it brought about throw the previous period of evolutionary history into often informative relief.

#### 4.4.6.1 The great reversal

Most fundamentally, oxygen brought about a reversal in the direction of the rTCA cycle to the oxidative Krebs cycle, while retaining the essential carbon backbones and conversions, and thus much of the core of intermediary metabolism. The availability of $O_2$ as an abundant terminal electron acceptor from carbohydrates and lipids enabled reversal of gluconeogenesis to glycolysis, and of fatty acid synthesis to $\beta$-oxidation, supplying carbon to oxidative TCA for respiration. Increased oxygen activity also energetically favored the adoption of the oxidative pathway to serine, and enabled reversal of the $C_1$ central superhighway in some clades (the example of methylotrophy is described below).

With increased catabolism, the diversity in the possible forms of heterotrophy and ecological trophic exchanges exploded. A suggestion of the complexity of the possibilities from known reactions, for a heterotroph taking a random walk in the space of flux balance models, has been published by João Rodrigues and Andreas Wagner [677]. At the same time as innovations in carbon transfer between species increased, innovations in autotrophic carbon fixation appear to have ended. All innovations shown in the tree of Figure 4.10 can be assigned to clades in a reducing world; the last innovation that suggests a response specifically to increasing $f_{O_2}$ is the phosphoglycolate salvage pathway in cyanobacteria.

### *4.4.6.2 Clades spanning the transition*

The great reversal was a complicated transition, and (like emergence) presumably could not have happened in one event, but must have occurred in stages. While near-equilibrium reactions can be reversed by changing the chemical potentials of inputs and outputs through concentration, most biological pathways include reactions sufficiently far from equilibrium that electron donors/acceptors with different midpoint potentials must be substituted to reverse pathway direction, along with the mediating enzymes. For the TCA cycle, this includes replacing reduced ferredoxin as the electron donor in the rTCA reductive carboxylations of thioesters, with a complicated cascade in the Krebs cycle, where lipoic acid is an initial reductant and carrier of acetyl groups to form acetyl-CoA, and is subsequently recycled via FAD and NAD. A substitution of membrane quinones (from menaquinone to the oxygen-insensitive ubiquinone with a higher midpoint potential [707]), coupled to the fumarate/succinate transition, also appears to have been part of cycle reversal.

Clades that underwent diversification during periods spanning the rise of oxygen likely faced variable environments and required parallel enzyme and cofactor systems for many reactions. Their descendants, some of which have retained this metabolic complexity and plasticity, are well suited to the complex lifecycles and resistances faced by pathogens. Some of these pathogens may serve as time capsules for stages of innovation in the great reversal.

Fermentative TCA pathways, containing both oxidative and reductive arcs, are an important component of resistance to anoxia in *Mycobacterium tuberculosis* when it is sequestered by macrophages in mammalian granulomas. *M. tuberculosis* and a group of related mycobacteria contain homologues to the rTCA succinyl-CoA carboxylase,[76] and to *two different* classes of oxidative TCA enzymes for the reverse $\alpha$-ketoglutarate decarboxylation [56]. Remarkably, in this group the reductive enzyme has evolved to be resistant to deactivation by $O_2$, while the two oxidative enzymes are both used in fermentative pathways to succinate, but are coupled alternatively to glycolysis or to $\beta$-oxidation as primary carbon sources.

The $\gamma$-proteobacterium *E. coli* offers another example of parallel systems in a clade that branches between the anoxic $\delta$- and $\varepsilon$-proteobacteria, and the clearly oxygenic $\alpha$- and $\beta$-proteobacteria. *E. coli* carries genes for three quinones: the oxygen-tolerant ubiquinone used as an electron acceptor from succinate in Krebs cycling, and two different oxygen-sensitive menaquinones typically used to reduce fumarate in rTCA [734].

### *4.4.6.3 Essential reticulation in the post-oxygenic tree: the example of methylotrophy*

The phylometabolic tree becomes essentially reticulated after the rise of oxygen, in sharp contrast to its simple ramification in the anoxic world. That this should be so at first

---

[76] This enzyme is annotated in Mtb as $\alpha$-ketoglutarate decarboxylase [805, 852].

seems surprising, given the generally increasing frequency of gene transfer with depth and approach to the LUCA in most phylogenetic reconstructions. For a tree not of lineages, but of *dependencies*, the direction of change becomes sensible. The increasingly chimeric character of metabolic networks in the post-oxygenic world, which makes reticulation unavoidable, can perhaps be understood in terms of qualitative changes in the nature of gene and pathway exchange with increasing integration and power density of organisms, and complexity of ecosystem interactions.

Nowhere is the change in character of core networks more strikingly visible than in the *serine cycle* of methylotrophy [133, 134, 840], shown in Figure 4.12. Methylotrophs are a diverse group of $\alpha$-, $\beta$-, and $\gamma$-proteobacteria which oxidize a variety of reduced one-carbon compounds as their sole sources of both carbon and energy.

The serine cycle, shown here from the $\alpha$-proteobacterium *Methylobacterium extorquens*, is a true "Frankenstein's monster" of metabolism. The core pathway of formaldehyde assimilation, overlaid in gold on the autotrophic pathways from Figure 4.6 is a bicycle formed from segments of *all four* other autotrophic pathways (besides the Calvin cycle), plus a segment from the glycolytic pathway. The segment from archaeal 4HB carbon fixation is reversed, and the 3HP pathway is converted from a branched bicycle at propionate to a unidirectional cycle. Glycine is formed via the transamination pathway and converted to serine as in the folate pathways, but serine is then fed through the reverse of the oxidative biosynthetic pathway to produce 3-phosphoglycerate and (via glycolytic reactions) pyruvate.

Methylotrophs likewise possess a chimeric central-$C_1$ superhighway, which contains the only known instance among bacteria of tetrahydromethanopterin [135, 840]. $H_4MPT$ is used exclusively for *oxidation* of $C_1$ from methyl groups (like 4HB, reversing the archaeal direction), in some organisms producing $CO_2$ which is then fixed by the Calvin cycle. The THF pathway, which is ancestral in bacteria, is retained and used for $C_1$ reduction.[77] Methylotrophy is thus both biochemically and ecologically a layered system, to which pathways from archaea were transferred and reversed, to enable respiration of reduced carbon exchanged trophically.

The preservation of such distinct pathway components among organisms in a common environment is the feature not seen in the pre-oxygenic tree. Perhaps it reflects simply the passage of time and evolutionary refinement of genome integration, and lower overall rates but more structured forms of gene and pathway exchange. Perhaps these refinements, like those leading to endosymbiosis or multicellularity, were in turn dependent on the greater power density of respiratory metabolisms. These questions may become answerable as mechanisms of genome regulation and gene exchange, and their connections to metabolism, become better understood for a larger range of organisms. At the same time, the diversification of methylotrophic and methanotrophic metabolisms reflects

---

[77] Recall that, among the archaeal pterins, $H_4MPT$ is distinctive in altering the free energy ramp from formyl to methyl groups [509], making the pathway easier to reverse to the oxidizing direction. In the context that the pterin pathway is used oxidatively, we believe this is why only this pterin was transferred and adopted by methylotrophs.

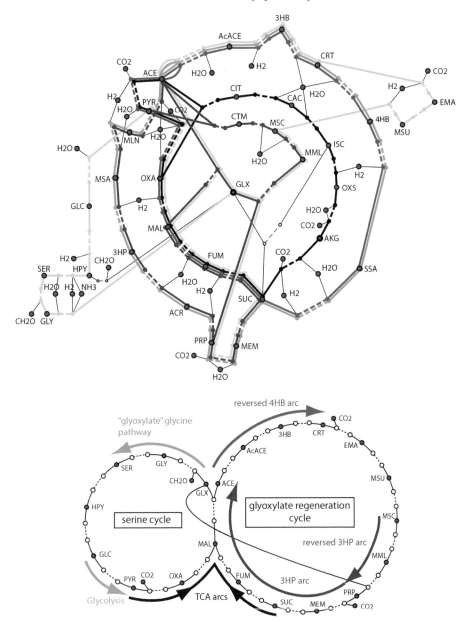

Figure 4.12 Two representations of the "serine cycle" of methylotrophic $\alpha$-proteobacteria. Like the 3-hydroxypropionate pathway and the glyoxylate shunt, it is a bicycle using aldol reactions involving glyoxylate to regenerate intermediates. Upper panel shows the stoichiometric pathway overlaid on the autotrophic loop pathways from Figure 4.6. Lower panel gives a projection of the serine cycle and glyoxylate regeneration cycle showing pathway directions and segments from all four fixation pathways that are used in this composite cycle. (After [91], Creative Commons [153].)

an exploration of complex, persistently differentiated ecosystem structures for microbial consortia. The rules of composition for such consortia will need to be understood as an additional evolutionary level to interpret the post-oxygenic tree.

### 4.4.7 *Chance and necessity for oxidative versus reductive TCA*

The oxidative and reductive TCA cycles may both be privileged as pathways in planetary organic chemistry, and in that sense predictable from some kind of laws or first principles, but the sense and context for their predictability is quite different. The functions that make rTCA distinctive are connected to $CO_2$ and reductant as the carbon and energy sources on early Earth, to metal-mediated reaction mechanisms, and to whatever limited spectrum may exist of pathway completions that are autocatalytic over short loops.

Oxidative TCA seems predictable only as a reflection of commitment to anabolic pathways established during the Archean. Its emergence may, further, have been possible only in the context of certain pre-adaptations, such as the existence of lipoic acid evolved to serve acyl and $C_1$ transfers, as in the acyl-carrier proteins and the glycine cycle, or of reductants and quinones with appropriate midpoint potentials (although the latter seem more malleable and subject to convergence rather than lock-in). Although oxidative TCA was in that sense "frozen,"[78] it was not a frozen accident in the sense of Crick [166]. It is indirectly connected to low-level organic chemistry and planetary composition and energetics. The oxidative cycle may be inevitable, but only through the route that the earlier reductive cycle was inevitable and then created a locked-in context.

The question of predictability is distinct from the question of certain desirable properties of cycles, which TCA retains in the oxidative as well as reductive direction. Cyclic pathways provide routes to convert some functional groups while conserving matter within the system, which in chemistry can require solution of complex networks of constraints. Similar conservation of captured carbon is seen in the complex but elegant network of aldol reactions in the pentose-phosphate pathway. Variant reaction sequences, all within a covering network of reactions from glyceraldehyde phosphate to ribulose phosphate and ribose phosphate, can generate a remarkable diversity of carbohydrate inputs to biosynthesis, while recycling all carbon that is not delivered to anabolism.

We return in Chapter 7 (Section 7.6.1.4) to discuss in general terms the status of such regularities in non-equilibrium systems, which may be predictable from first principles, but for which the prediction includes identification of dependency pathways that may be

---

[78] Such frozen constraints on evolution recur at many levels, suggesting that they are a general feature of non-equilibrium dynamics in complex configuration spaces. In the regulation of metazoan development, early differentiation networks may become locked in as increasingly complex downstream development becomes dependent on them in many steps. Eric Davidson and Douglas Erwin have come to refer to these as *kernels* [180, 227]: they are products of early evolution which come to act as invariable constraints on later evolution. The approximation of kernels across embryogenesis as a whole is seen in the reduced variability of a "phylotypic stage" in many clades [388]. The manner in which embryogenesis either before or after the phylotypic stage is variable, and the two variations are somewhat conditionally independent given the phylotypic stage, may resemble the independent variations in carbon fixation and in later intermediary metabolism, given the invariant universal precursors. If this resemblance is more than an accident, its explanation must be sought in problems of architecture and stabilization of non-equilibrium, order-forming stochastic processes.

complex and indirect, and tied to a historical sequence. It is not clear, for instance, whether oxidative TCA has any special chemical status in reaction conditions driven by oxidants, which have not been preceded by long periods of anoxia.

## 4.5 Cofactors and the first layer of molecular-organic control

Cofactors form a unique and essential class of components within biochemistry, both as individual molecules and as a distinctive level in the control over metabolism [254]. In synthesis and structure they form a subset of the metabolites, including many of the largest and most complex products of organosynthesis, but unlike amino acids, nucleotides, sugars and lipids, they are not primary structural elements of the macromolecular components of cells. Instead, cofactors provide a limited but essential inventory of functions, which are used widely and in a variety of macromolecular contexts. Nearly half of enzymes require cofactors as coenzymes. If we extend this grouping to include chelated metals and clusters, ranging from common iron-sulfur centers to the elaborate metal centers of gas-handling enzymes [261], more than half of enzymes require coenzymes or metals in the active site.

Whereas cofactors are "only" required in about half of all metabolic reactions, *for all the key carbon, nitrogen, and energy-incorporation steps in core metabolism they are essential.* The more structurally complex cofactors such as thiamin or biotin also tend to be associated with more catalytically complex functions within carbon fixation. Specialized cofactors are also required in many of the special reactions that enable complex and diverse biosynthesis, such as thiamin in the ketoacid isomerizations of branched-chain amino acid biosynthesis shown in Figure 4.8, or the carbonyl insertion reactions of rTCA.[79] Lipoic acid exchanges a variety of fatty acids in acyl-carrier proteins, and is particularly unique in providing reducing power with immediate transfer of acyl groups to one of the lipoate sulfur atoms, from which it can be transferred to CoA.

The combination of reaction mechanisms, enabled by enzymes, and group transfers or electron transfers mediated by cofactors *determines in large part which transformations from organic chemistry have become parts of biochemistry.* In any era or clade, the inventory of group transfers made available by existing cofactors may limit the pathways that can be formed. Correspondingly, innovations in cofactor synthesis can have major consequences for the large-scale structure of evolution.

### 4.5.1 The intermediate position of cofactors, feedback, and the emergence of metabolic control

Cofactors in extant metabolism are, in three respects, a class in transition between the core metabolites and the oligomers. We therefore propose that they were also temporally

---

[79] The special role of TPP in these reactions comes from the facile dissociation of the proton bound to carbon between S and N in the thiazole ring. The resulting carbanion readily attacks the keto carbon of $\alpha$-ketoacids, as an initial step in multiple reactions [254].

intermediate along the path of biogenesis, and are responsible for a large part of the specification of biochemistry.

### 4.5.1.1 Cofactors as a distinct class in transition between small metabolites and macromolecules

**As structures** The cofactors considered as structures include some of the largest directly assembled organic monomers (pterins, flavins, thiamin, tetrapyrroles), but many also show the beginnings of polymerization of standard amino acids, lipids, or ribonucleotides. These may be joined by the same phosphate ester bonds that link RNA oligomers or aminoacyl-tRNA, or they may use distinctive bonds (e.g. $5'$-$5'$ esters) found only in the cofactor class. The polymerization exhibited within cofactors is distinguished from that of oligomers by its heterogeneity. In contrast to oligomers, cofactors often include monomeric components from several molecule classes. Examples are coenzyme-A, which includes several peptide units and a $3'$-phosphorylated ADP; folates, which join a pterin moiety to para-aminobenzoic acid (pABA); quinones, which join a chorismate derivative to an iso-prene lipid tail; and a variety of cofactors assembled on phosphoribosyl-pyrophosphate (PRPP) to which RNA "handles" are esterified [132]. As noted in Section 2.5.1.2, the border between small and large molecules, where most cofactors are found, is more fundamentally a border between the use of heterogeneous organic chemistry to encode biological information in covalent structures, and the transition to homogeneous phosphate chemistry, with information carried in sequences or higher-order non-covalent structures.

**As autonomous carriers of functions** The relation between structure and function in the cofactor class is different from that of other monomers such as amino acids, RNA, or even monosaccharides in being determined mostly at the single-molecule scale. The monomer constituents of oligomer macromolecules often have rather general properties (e.g. charge, $pK_a$, hydrophobicity), and only take on more specific functional roles that depend on location and context in the assembled molecule [323].[80] In contrast, the functions of cofactors are specific, often finely tuned by evolution, and deployable in a wide range of macromolecular contexts. Usually they are carriers or transfer agents of functional groups or reductants, but this may extend to exchange roles as part of the active sites within enzymes. An example in which a cofactor creates a channel is the function of cobalamin as a $C_1$ transfer agent to the nickel reaction center in the acetyl-CoA synthase from a corrinoid iron-sulfur protein [659, 660, 661]. An example of cofactor incorporation in an active site is the role of TPP as the reaction center in the pyruvate-ferredoxin oxidoreductase, which lies at the end of a long electron-transport channel formed by Fe-S clusters [123].

The most complex and subtle aspects of cofactor function are those in which the type of function is dictated by the single-molecule chemistry of the cofactor itself, but the exact

---

[80] We will note a few interesting exceptions to this rule for complex amino acids in Section 4.5.3.

reaction properties are tuned by context. An example is flavin, the electron-transfer component of the redox cofactor FAD. The heterocycle structure of the molecule dictates its mechanism and role as an electron-transfer cofactor, but the actual midpoint potential for electron exchange may be set across a range of hundreds of millivolts by the context for FAD in its flavoprotein holoenzyme [718].

**As the sources of links that complete reaction networks** In consequence of their function as transfer agents, cofactors also occupy an intermediate place in network completion between the places of small metabolites and of enzymes. Small metabolites can act as network catalysts, as we have shown in Section 4.3.6, but while the molecular forms may be regenerated, individual atoms flow through such networks and change properties such as redox state, as illustrated in Figure 4.3. In a systematically opposite pattern, macromolecular catalysts mediate typically single reactions,[81] and are not altered as the substrate is transformed.

Cofactors, like small metabolites, can complete autocatalytic networks as group carriers, but more like enzymes, they conserve a large core of atoms that are not altered during the reaction. They may require only single intermediate states to perform transfers, as does glutamate/glutamine for activated amino groups, or biotin for activated carboxyl groups, or they may undergo cycles in which the cofactor is altered by the attachment of energetic leaving groups or reductants, as occurs for the (homocysteine/*S*-adenosyl-homocysteine/*S*-adenosyl-methionine/methionine) cycle, or the oxidation/reduction cycles of disulfides in lipoic acid. In the most elaborate cases, such as folates, the transported carbon atoms may undergo multiple chemical changes attached to the cofactor. Because cofactor functions are few in number compared to either substrate molecules or enzymes, and are relatively context-independent, the cofactor role of completing networks gives them the highest connectivities among the metabolites [335].

### 4.5.1.2 Cofactors as intermediates in the emergence of biochemistry

In this section we review selected aspects of cofactor chemistry that seem to us most essential to overall metabolic architecture and evolution, with the goal of framing questions as much as answering them. Much of origins research has focused either on properties of the small-molecule substrate [572] or on properties of oligomers (particularly RNA) as catalysts and, it is often proposed, as early replicators [589]. The possible role of cofactors in bringing into existence the organization of biochemistry and defining the chemical dimensions of the nature of the living state has in our view been underestimated. Cofactors have been suggested to be late additions to metabolism, descended from a world of hypothesized RNA replicators [872], but the view that they may have driven the rise of catalysts is receiving increasing (and we believe well-deserved) attention [159, 904].

---

[81] Multifunctional enzymes often arise as a result of fusion of previously disjoint, monofunctional enzymes. These are often identifiable as evolutionary refinements, against a background of modular architecture created by distinct catalysts for single steps in reaction sequences, as noted in Section 4.4.5.1.

If one accepts the premise that the early stages of biogenesis were most plausibly defined by feedbacks from organic chemistry to incrementally supplant a rock-hosted organosynthesis, then it is natural to try to situate the emergence of different cofactor groups in time, roughly as they fall in molecular size and complexity: as intermediates between the small-metabolite and oligomer levels [159]. A rock world capable of reaching intermediate levels of organic complexity should also have supported the early emergence of at least the major redox and C- and N-transfer cofactors. Conversely, the pervasive dependence of biosynthetic reactions on cofactor intermediates makes the expansion of protometabolic networks most plausible if it was supported by contemporaneous emergence and feedback from cofactor groups.

The emergence of the cofactor layer therefore would define a transitional phase when the reaction mechanisms of core metabolism came under selection and control of organic as opposed to mineral-based chemistry, and they provided the structured foundation from which the oligomer world grew.

### 4.5.1.3 Cofactors are targets of intensive natural selection

Because a single cofactor appears in a large number of reactions – typically much larger than the number in which any other metabolite appears as a reagent or any enzyme appears as a catalyst – it is to be expected that cofactors have been under intense pressure of natural selection. An example of the refinement that such selection can produce is given by the tuning of the free energy landscape of bound $C_1$ units at different occupation states on $H_4MPT$ in relation to THF, reviewed in Section 4.4.4.1. Possibly for this reason, the biosynthesis of cofactors involves some of the most elaborate and least understood organic chemistry used by organisms. The pathways leading to several major cofactors have only recently been elucidated, and others remain to be fully described. Their study continues to lead to the discovery of novel reaction mechanisms and enzymes that are unique to cofactor synthesis

As a consequence of the same centrality that focuses selection, when cofactor innovations do occur, they can result in large-scale repartitioning of flows in metabolism, or clade-level distinctions in evolution. The innovation of oxidative serine/glycine synthesis in methanogens as a non-local (in the network) response to a fine but difficult energy optimization is an example. Both $H_4MPT$ and deazaflavins (diverged from flavins by substitution of a benzene ring from chorismate for a second pterin found in other flavins) are cofactors specific to methanogens, and the methanogenic adaptation of the W-L pathway to an energy system is arguably the innovation that sets methanogens apart from the other euryarchaeota, which are heterotrophs. Another similar example comes from the quinones, a diverse family of cofactors mediating membrane electron transport [151]. The synthetic divergence of menaquinone from ubiquinone follows the pattern of phylogenetic diversification within proteobacteria. $\delta$- and $\epsilon$-proteobacteria use menaquinone, $\gamma$-proteobacteria use both menaquinone and ubiquinone, and $\alpha$- and $\beta$-proteobacteria use only ubiquinone. Because menaquinone and ubiquinone have different midpoint potentials, it was suggested that their distribution reflects changes in environmental redox state as the proteobacteria diversified during the rise of oxygen [594, 707].

### 4.5.2 Key cofactor classes for the earliest elaboration of metabolism

The universal reactions of intermediary metabolism depend on only about 30 cofactors [254]. Major functional roles include:

1. transition metal mediated redox reactions (heme, cobalamin, the nickel tetrapyrrole $F_{430}$, chlorophylls[82]),
2. transport of one-carbon groups that range in redox state from oxidized (biotin for carboxyl groups, methanofurans for formyl groups) to reduced (lipoic acid for methylene groups, *S*-adenosyl methionine, coenzyme-M and cobalamin for methyl groups), with some cofactors spanning this range and mediating interconversion of oxidation states (the folate family interconverting formyl to methyl groups),
3. transport of amino groups (pyridoxal phosphate, glutamate, glutamine),
4. reductants (nicotinamide cofactors, flavins, deazaflavins, lipoic acid, and coenzyme-B),
5. membrane electron transport and temporary storage (quinones),
6. transport of more complex units such as acyl and amino-acyl groups (pantetheine in CoA and in the acyl-carrier protein (ACP), lipoic acid, thiamin pyrophosphate),
7. transport of dehydration potential from phosphate esters (nucleoside diphosphates and triphosphates), and
8. sources of thioester bonds for substrate-level phosphorylation and other reactions (pantetheine in CoA).

Some features of the key classes used in core metabolism are listed in the following subsections. More detail on each class is provided in [91].

#### 4.5.2.1 The cofactors derived from purine RNA

An important set of related cofactors that use heterocycles for their primary functions have biosynthetic reactions closely related to those for purine RNA. These reactions are performed by a diverse class of cyclohydrolase enzymes given the 3.5.4 classification in the E.C. classification scheme, which are responsible for the key ring-formation and ring-rearrangement steps. The cyclohydrolases can split and reform the ribosyl ring in PRPP, jointly with the five-membered and six-membered rings of adjacent guanine and adenine. Five biosynthetically related cofactor groups are formed in this way. Three of these – the folates, flavins and deazaflavins – are formed from GTP, while one – thiamin – is formed from amino-imidazole-ribonucleotide (AIR), a precursor to GTP via the same cyclohydrolase reactions, as shown in Figure 4.13.

---

[82] It is natural in many respects to include ferredoxins (and related flavodoxins) in this list. Although not cofactors by the criteria of size and biosynthetic complexity, these small, widely diversified, ancient, and general-purpose $Fe_2S_2$, $Fe_3S_4$, and $Fe_4S_4$-binding polypeptides are unique low-potential (high-energy) electron donors. Reduced ferredoxins are often generated in reactions involving radical intermediates in iron-sulfur enzymes, described below in connection with electron bifurcation.

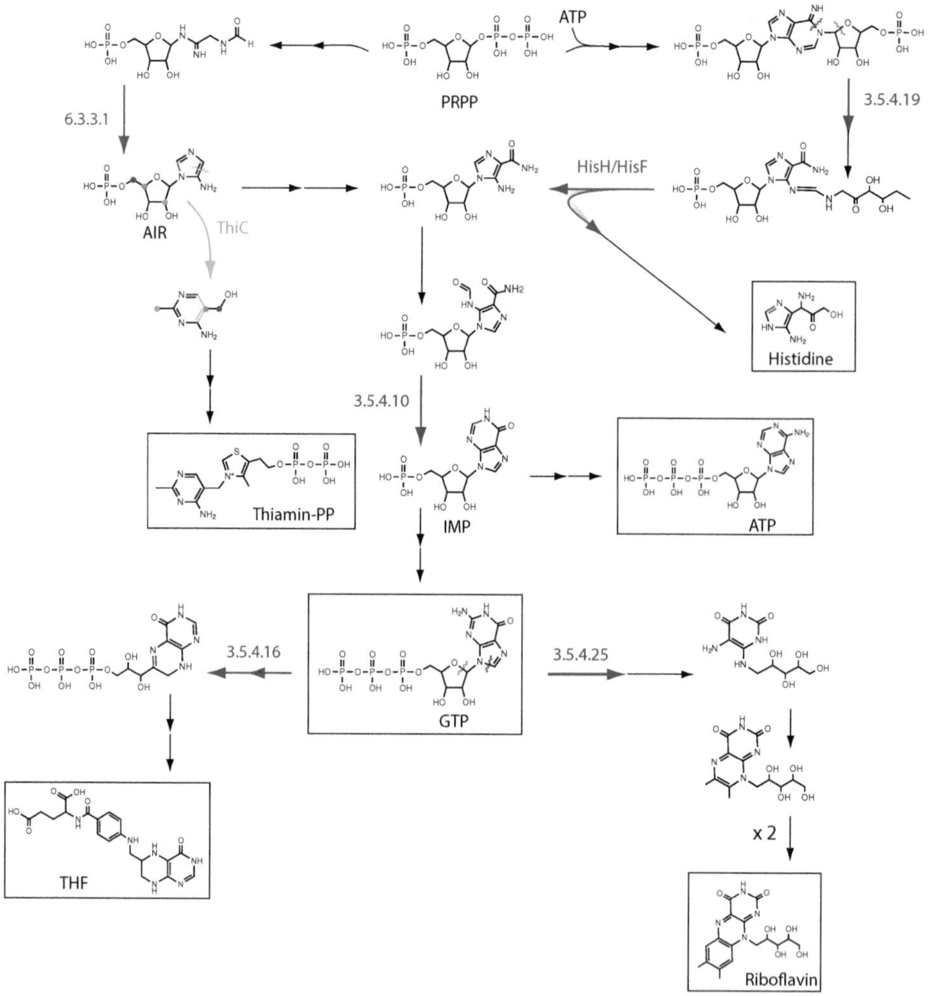

Figure 4.13 Key molecular rearrangements in the network leading from AIR to purines and the purine-derived cofactors. The 3.5.4 class of cyclohydrolases (red) convert FAICAR to IMP (precursor to purines), and subsequently convert GTP to folates and flavins by opening the imidazole ring. Acting on the six-member ring of ATP and on a second attached PRPP, the enzyme 3.5.4.19 initiates the pathway to histidinol. The thiamin pathway, which uses the unclassified enzyme ThiC to hydrolyze imidazole and ribosyl moieties, is the most complex, involving multiple group rearrangements (indicated by colored atoms). This complexity, together with the subsequent attachment of a thiazole group, lead us to place thiamin latest in evolutionary origin among these cofactors. (After [91], Creative Commons [153].)

**Folates and the central superhighway of $C_1$ metabolism** The folates are structurally most similar to GTP, but have undergone the widest range of secondary specializations, particularly in the archaea. Members of the folate family carry $C_1$ groups bound to either the $N^5$ nitrogen of a heterocycle derived from GTP, an exocyclic $N^{10}$ nitrogen derived from a para-aminobenzoic acid (pABA), or both, during reduction from formyl to methylene or methyl oxidation states. The two most common folates are tetrahydrofolate (THF), ubiquitous in bacteria and common in many archaeal groups, and tetrahydromethanopterin ($H_4$MPT), essential for methanogens and found in a small number of late-branching bacterial clades. Other members of this family are exclusive to the archaeal domain and are structural intermediates between THF and $H_4$MPT.

Two kinds of structural variation are found among folates, as shown in Figure 4.14, which tune the free energy landscape of the bound carbon across its oxidation states. First, only THF retains the carbonyl group of pABA, which shifts electron density away from $N^{10}$ via the benzene ring, and lowers its $pK_a$ relative to $N^5$ of the heterocycle. All other members of the family lack this carbonyl. Second, all folates besides THF incorporate one or two methyl groups that impede rotation between the pteridine and aryl-amine planes, changing the relative entropies of formation among different binding states for the attached $C_1$ [507, 509]. The tuning of free energies for an entire sequence of $C_1$ oxidation states, which determines at the same time their properties as leaving groups and also the free energies of oxidation/reduction transitions between states, is the remarkable feature that fits individual pterin cofactors into tightly optimized networks of redox exchanges, to which we return in Section 6.3.1.3.

The synthesis of THF from neopterin and pABA, as shown in Figure 4.14, is a straightforward sequence of group additions. THF is also distributed throughout bacteria and is found in some clades of archaea, where it remains an open question whether it is ancestral (as we have argued), or rather a result of very deep-rooted horizontal transfer. In contrast, the synthesis of the non-folate pterins, all of which besides $H_4$MPT are exclusive to archaea, is more elaborate and satisfies an implicational scaling with respect to the addition of methyl groups (magenta in the figure). The $-C^{13a}H_3$ methyl is present only if $-C^{12a}H_3$ methyl is present, and this is also the order in which these groups are added biosynthetically [91] (original biochemical characterization of these cofactors in [657, 658, 824, 873, 874, 910]).[83] $H_4$MPT is the end member of these modifications, consistent with a picture of exploration of folate modifications within the archaeal domain, which produced methanogenesis as a derived and highly refined phenotype. This end member is maximally different from THF, and was well suited to horizontal transfer back to bacteria which use it for $C_1$ oxidation.

Folates mediate a diverse array of $C_1$ chemistry, various parts of which are essential in the biosynthesis of all organisms. Figure 4.15 provides a more detailed description of the primary network of one-carbon reactions, which were sketched in Figure 4.4. Functional groups supplied by folate chemistry, the transfer of which depends on their binding energies to the $N^5$ or $N^{10}$ nitrogen atoms, include (1) formyl groups for synthesis of

---

[83] It would be desirable to phylogenetically reconstruct the history of introduction of these modifications, to determine whether their implicational and biosynthetic order is also their historical order, which we suspect it to be. A similar elegant relation – an identical implicational, biosynthetic, and evolutionary order for oxidized positions on cholesterol – was demonstrated by Konrad Bloch [79], and was further linked to a sequence increasing oxygen fugacities for the equilibria of each position. That relation provides a strong causal interpretation of the elaboration of cholesterols during the rise of environmental oxygen.

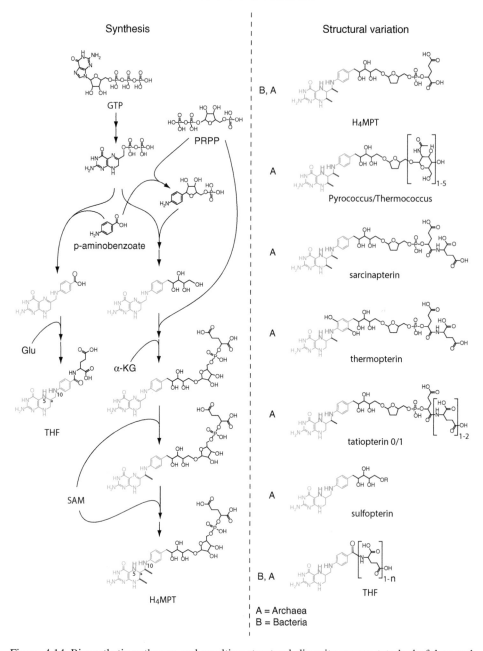

Figure 4.14 Biosynthetic pathways and resulting structural diversity among tetrahydrofolate and the pterin-derived cofactors. Pteridine and benzene groups are shown in blue, nitrogen atoms that carry bound $C_1$ groups in green, electron-withdrawing carbonyl groups in red, and methyl groups that regulate steric hindrance in magenta. Appearance of a cofactor among bacteria or archaea is noted on the right. The simpler folate synthesis uses pABA directly, whereas neopterin synthesis first modifies pABA by substituting a ribitoyl moiety for the carboxylate. Subsequent modifications of pterins include the addition of one or two methyl groups. Their order of addition in biosynthesis is the same as their order of appearance phylogenetically. (After [91], Creative Commons [153].)

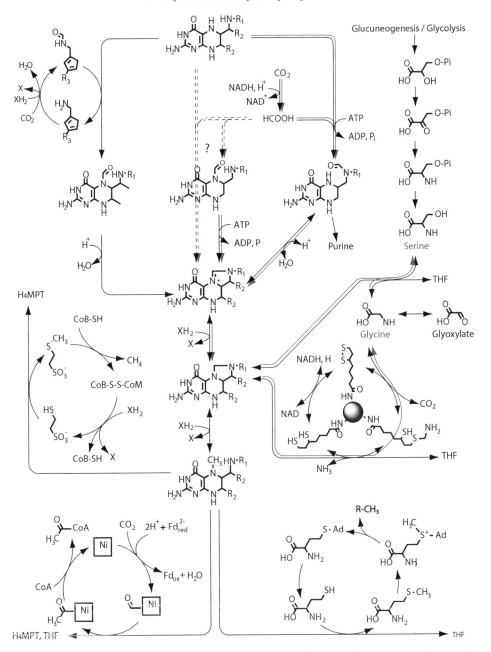

Figure 4.15 The reactions of one-carbon assimilation shown schematically in Figure 4.4, expanded here to show molecular intermediates. The alternative between the characteristic pattern of formylation of $N^{10}$ on folates (upper-right branch) and $N^5$ on pterins (upper-left branch) is shown, along with the proposed scheme of ATP-dependent attachment or cyclo-ligation that may explain the missing $N^{10}$-formyl-THF synthase in bacteria that otherwise appear to fix carbon through this pathway. Reductive glycine and serine synthesis is indicated in red. Note the similarity between bacterial use of methylene groups on lipoic acid (right) and methanogenic (archaeal) counterparts from methyl groups on the CoB-CoM system. Acetogenesis from methyl groups is shown bottom-left, and $S$-adenosyl-methionine chemistry is shown bottom-right. (After [91], Creative Commons [153].)

purines, formyl-tRNA, and formylation of methionine (fMet) during translation, (2) methylene groups to form thymidilate, which are also used in many deep-branching organisms to synthesize glycine and serine, and (3) methyl groups which may be transferred to *S*-adenosyl-methionine (SAM) as a general methyl donor in anabolism, to the acetyl-CoA synthase to form acetyl-CoA in the Wood–Ljungdahl pathway, or to coenzyme-M where the conversion to methane is the last step in the energy system of methanogenesis. The variations among pterin cofactors, shown in Figure 4.14, leave the charge, $pK_a$ and resulting C–N bond energy at $N^5$ roughly unaffected, while the $N^{10}$ charge, $pK_a$, and C–N bond energy change significantly across the family, as we noted in Section 4.4.4.1.

**Flavins and deazaflavins** The flavins are tricyclic compounds formed by condensation of two pterin groups, while deazaflavins are synthesized through a modified version of this pathway, in which one pterin group is replaced by a benzene ring derived from chorismate. Flavins are general-purpose reductants, whereas deazaflavins are specifically associated with methanogenesis.

**Thiamin** Thiamin combines a CN heterocycle common to the purine-derived cofactors with a thiazole group (thus it incorporates sulfur), and shares functions with both the purine cofactor group and the alkyl-thiol group reviewed in the next subsection. The synthesis of thiamin from aminoimidazole ribonucleotide (AIR) is by far the most complex synthesis in this family. In bacteria and archaea the synthesis sequence begins with an elaborate molecular rearrangement, performed in a single step by the enzyme ThiC [405]. The ThiC enzyme is currently unclassified, and its reaction mechanism incompletely understood, but it shares characteristics with members of the 3.5.4 cyclohydrolase family. As in the first committed steps in the synthesis of folates and flavins from GTP, both a ribose ring and a five-member heterocycle are cleaved and subsequently (as in folate synthesis) recombined into a six-member heterocycle. The complexity of this enzymatic mechanism makes a pre-enzymatic homologue to ThiC difficult to imagine, and suggests that thiamin is both of later origin, and more highly derived, than other cofactors in this family. Unlike reactions carried out on folates, reactions involving thiamin take place not on the pyrimidine ring, but rather on the thiazole ring to which it is attached, and which is created in another elaborate synthetic sequence.

Figure 4.16 shows the substrate rearrangements in the subnetwork leading from GTP to methanopterins, folates, riboflavin, and the archaeal deazaflavin $F_{420}$. In the pterin branch, both rings of neopterin are synthesized directly from GTP, and an aryl-amine originating in pABA provides the second essential nitrogen atom. The flavin branch is characterized by the integration of either ribulose (in riboflavin) or chorismate (in $F_{420}$) to form the internal rings. The cyclohydrolase reactions are the key innovation enabling the biosynthesis of this family of cofactors, and also of purine RNA. The heterocycles that are formed or cleaved by these reactions provide the central structural components of the active parts of the final cofactor molecules. Except for TPP, the distinctions among purine-derived cofactors are secondary modifications on a background structured by PRPP and CN heterocycles.

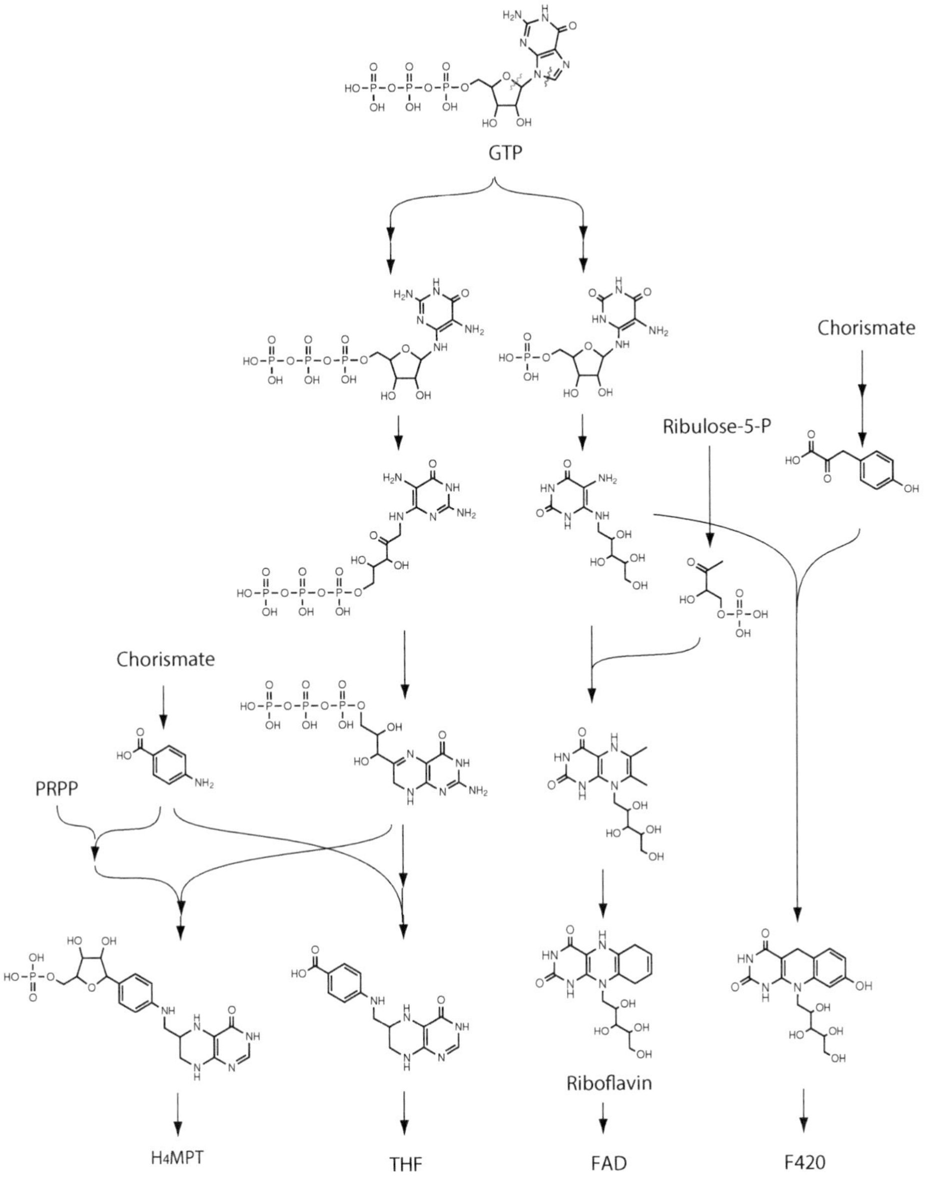

Figure 4.16 GTP is converted into four major cofactors and close derivatives: tetrahydromethanopterin, tetrahydrofolate, riboflavin, and the archaeal deazaflavin called $F_{420}$. Chorismate is a source of benzene rings in three of these. (After [91], Creative Commons [153].)

**Histidine** The last "cofactor" in this group is the amino acid histidine, synthesized from ATP rather than GTP but using similar reactions. Histidine is a general acid-base catalyst with $pK_a$ of 6.5 not found among RNA and well suited to catalysis at the physiological (cytoplasmic) pH of most organisms. In addition to serving in proton abstraction, histidine forms a variety of coordination compounds with metals in redox reactions [829], and the $\pi$-bonds in the imidazole ring can participate in stacking interactions with other planar molecules. We return to functional similarities to cofactors, in other complex amino acids, in Section 4.5.3.

### 4.5.2.2 The alkyl-thiol cofactors

The major chemicals in this class include the sulfonated alkane-thiols coenzyme-B (CoB) and coenzyme-M (CoM), cysteine and homocysteine including the activated forms $S$-adenosyl-homocysteine (which under methylation becomes SAM), lipoic acid, and pantetheine including pantetheine phosphate. The common structure of the alkyl-thiol cofactors is an alkane chain terminated by one or more sulfhydryl (SH) groups. In all cases except lipoic acid, a single SH is bound to the terminal carbon; in lipoic acid two SH groups are bound at sub-adjacent carbons. Differences among the alkyl-thiol cofactors arise from their biosynthetic context, the length of their alkane chains, and perhaps foremost the functional groups that terminate the other ends of the chains. These may be as simple as sulfones (in CoB) or as complex as a peptide chain esterified to an ATP "handle" (in CoA).

Cofactors in this class serve three primary functions, as reductants (cysteine, CoB, pantetheine, and one sulfur on lipoic acid), carriers of methyl groups (CoM, SAM, one sulfur on lipoic acid), and carriers of larger functional groups such as acyl groups (lipoic acid in lipoyl protein, phosphopantetheine in acyl-carrier protein). A highly specialized role in which H is a leaving group is the formation of thioesters at carboxyl groups (pantethenic acid in CoA, lipoic acid in lipoyl protein). This function is essential to substrate-level phosphorylation, and appears repeatedly in the deepest and putatively oldest reactions in core metabolism. A final function closely related to reduction is the formation and cleavage of S–S linkages by cysteine in response to redox state, which is a major controller of both committed and plastic tertiary structure in proteins. The dative bonds to Fe and Ni centers in enzymes, from the terminal sulfur atoms on cysteine, distinguish these bonds from the nitrogen dative bonds (sometimes also to Fe or Ni) in tetrapyrrole cofactors. The following comments note key patterns or roles from this cofactor class.

**Convergent evolution of carrier/co-reductant pairs** An important motif involving all three of S–C, S–H, and S–S bonds, seen in several independently derived cofactor sets, is the use of one sulfur to carry a transferable carbon group and a second sulfur to provide a co-reductant when the carbon is transferred. Coenzyme-B (CoB) and coenzyme-M (CoM) act together as methyl carrier and reductant to form methane in methanogenesis [261]. In this complex transfer, the fully reduced ($Ni^+$) state of the nickel tetrapyrrole $F_{430}$ forms a dative bond to $-CH_3$ displacing the CoM carrier, effectively re-oxidizing $F_{430}$ to $Ni^{3+}$. Reduced $F_{430}$ is regenerated through two sequential single-electron transfers. The first, from CoM-SH, generates a $Ni^{2+}$ state that releases methane, while forming a radical $CoB^{\cdot}-S-S-CoM$ intermediate with CoB. The radical then donates the second electron, restoring $Ni^+$. The strongly oxidizing heterodisulfide CoB–S–S–CoM is subsequently reduced with two NADH, regenerating CoM–SH and CoB–SH.

A similar role as methylene carrier and reductant is performed by the two SH groups in lipoic acid. CoM is specific to methanogenic archaea [48], while lipoic acid and *S*-adenosyl-homocysteine are found in all three domains [90, 175]. The reconstruction of *Aquifex aeolicus* in Section 4.4.5 has led to the proposal that lipoic acid may have originated in its role as the methylene transfer cofactor for reductive glycine synthesis, and later been extended through gene duplication and divergence to serve in a host of other $C_1$ and $C_2$ group transfers. Lipoic acid is formed from octanoyl-CoA, emerging from the biotin-dependent malonate pathway to fatty acid synthesis [495], and along with fatty acid synthesis, may have been present in the universal common ancestor. The universal distribution of the glycine cycle, for which we have argued above, supports this hypothesis.

**Role in the reversal of citric acid cycling** Lipoic acid becomes the electron acceptor in the oxidative decarboxylation of $\alpha$-ketoglutarate and pyruvate in the oxidative Krebs cycle, replacing the role taken by reduced ferredoxin in the rTCA cycle. Thus the prior availability of lipoic acid was an enabling precondition for *reversal of the cycle* in response to the rise of oxygen.

**Carriers of acyl groups** Transport of acyl groups in the acyl-carrier protein (ACP) proceeds through thioesterification with pantetheine phosphate, similar to the thioesterification in fixation pathways. In fatty acid biosynthesis acyl groups are further processed while attached to the pantetheine phosphate prosthetic group.

**The parallelism of reductive glycine synthesis and methanogenesis** The similarity between the glycine cycle and methanogenesis in Figure 4.15 emphasizes the convergent roles of alkyl-thiol cofactors. In the glycine cycle, methylene groups are accepted by the terminal sulfur on lipoic acid, and the subadjacent SH serves as reductant when glycine is produced, leaving a disulfide bond in lipoic acid. The disulfide bond is subsequently reduced with NADH. In methanogenesis, a methyl group from $H_4$MPT is transferred to CoM, with the subsequent transfer to $F_{430}$, and the release from $F_{430}$ as methane in the methyl-CoM reductase, coupled to formation of $CoB-S-S-CoM$. The heterodisulfide is again reduced with NADH, and employs a pair of electron bifurcations (discussed in Section 4.5.5.1 below) to retain the excess free energy in the production of reduced ferredoxin ($Fd^{2-}$) rather than dissipating it as heat. The striking similarity of these two methyl-transfer systems, mediated by independently evolved and structurally quite different cofactors, suggests evolutionary convergence driven specifically by properties of alkyl thiols.

### 4.5.3 The complex amino acids as cofactors

In Chapter 5 we will classify amino acids into three groups – simple, sulfur, and complex – according to shared characteristics in their biosynthesis which are also reflected in their functions and their positions in the genetic code. The two sulfur-containing amino acids are cysteine and methionine. The complex amino acids are phenylalanine and tyrosine, tryptophan, histidine, and lysine. They have in common considerably larger numbers of biosynthetic steps than the simple amino acids, and syntheses that typically draw from multiple disconnected regions of the metabolic chart. Examples include the incorporation of aromatic rings from the chorismate pathway in phenylalanine and tyrosine, and indirectly via indole in tryptophan, the synthesis of histidine from ATP, and the synthesis of lysine from three of four TCA precursors (oxaloacetate, pyruvate, succinate). For some purposes it will be appropriate to group methionine with these complex amino acids, and to group cysteine in the biosynthetically and functionally simple class.

The functions of the complex amino acids are often very similar to those of cofactors in that they are stereotyped and in some cases highly tuned. In cases where these acids draw functional groups from remote regions in the metabolic chart, they may also share functions with conventionally defined cofactors produced from the same precursors. A good example is tryptophan, which draws its aromatic group from chorismate, the common precursor to quinones. The delocalization of electron density around the aromatic ring is one of the features that permit quinones to carry radical intermediate states, an unusual role for non-metal metabolites. A similar electron delocalization provides tryptophan one of its important roles in catalytic centers: the non-local transport of electrons between substrates [460, 744]. Histidine originates from ATP (by reactions considered further below) and is only converted into amino acid form in the last biosynthetic step [776, 872]. Methionine has a relatively limited although crucial role in genetic encoding (acting as a start codon in most organisms), but as part of the cofactor cycle (homocysteine/$S$-adenosyl-homocysteine/$S$-adenosyl-methionine/methionine) it is essential for the transport of methyl groups throughout reduced-$C_1$ chemistry.

In understanding control over metabolism it will often be sensible to regard the six complex amino acids as functionally being cofactors, synthesized in such a way that they can be incorporated into polypeptide chains.

### 4.5.4 Situating cofactors within the elaboration of the small-molecule metabolic substrate network

The close relation of many cofactors to deep core pathways suggests an alternating co-emergence between substrate molecules and the cofactors derived from them. Most deep core reactions require electron or group transfer from one of a few key cofactors, and many of these cofactors are in turn synthesized at surprisingly low levels in intermediary metabolism, comparable to or below the complexity required to reach nucleotides. The following list provides a few representative examples.

1. Nicotinate, the active group in the ubiquitous redox cofactors NADH and NADPH, is synthesized from aspartate and dihydroxyacetone, two simple metabolites in the early arc of rTCA and the first part of the gluconeogenic pathway [143].
2. The active thiol in coenzyme-A is the thiol of a cysteine in a polypeptide, the other monomers in which are $\beta$-alanine and a branched-chain amino acid derived from valine.

3. As diagrammed above in Figure 4.13, even the remarkable family containing the flavins (redox), folates ($C_1$ metabolism), and thiamin (carboxylation and numerous other $C_1$ and $C_2$ transfers) is synthesized from GTP in a network where many of the reactions are performed by a functionally closely related family of cyclohydrolases. These are the same enzymes required in the synthesis of purine RNA. A related set of reactions produces histidine from ATP.

If the biosphere emerged by feedback, as a sequence of organic molecules took over control of pathways from earlier mineral or metal-ligand systems, it should not make sense to regard "metabolism" (in the sense of the network of substrates) as a single system, which could have emerged whole, to be followed by cofactors and catalysts as other unified stages which brought control and eventually molecular evolution. More plausible sequences should be those that interleave the elaboration of metabolic networks with transitions in which new cofactors are introduced. The cofactors feed back soon after their synthesis first becomes possible, to enhance the flux that produces them. In this picture, the emergence of metabolism consisted of a sequence of relatively short self-amplifying feedbacks between organosynthesis and catalysis. Cofactors such as NADH or simple thiol precursors to CoA should have been early, with folates and flavins entering at a later stage, and thiamin or biotin later still. The status of lipoic acid is interesting, as it mediates a variety of functions both directly and in acyl-carrier proteins. Braakman [92] has suggested based on the *Aquifex* reconstruction of Section 4.4.5 that the earliest *genes* for lipoic acid synthesis are likely those associated with reductive glycine synthesis. While the genes cannot be as early (within metabolism-first assumptions) as the first synthesis of glycine, it is interesting that the genetic patterns place both the cofactor and reductive glycine synthesis as the earliest cellular variants.

This proposal for the incremental incursion of cofactors has the difficulty of all metabolism-first proposals: we cannot suggest specific mechanisms by which a network of increasingly complex molecules would be selected among possible chemistries and how the mechanisms of selectivity would have been endowed with persistence. However, to the extent that any such mechanism would have depended on differential growth and chemical competitive exclusion, the network of short flux-enhancing feedback loops involving simple cofactors should have served as a hierarchy of "attractors" in a context of otherwise non-specific catalysts.

### *4.5.4.1 Was an original folate "selection" the platform that enabled later base-pairing functions of RNA?*

In a recursive elaboration of the metabolic network and the cofactor inventory, folates would arise at times comparable to or earlier than monomer purine RNA, if these were all synthesized originally by cyclohydrolase reactions similar to those that they share today. In such a contemporaneous production network, however, folates would have immediate functions as monomers in lifting $C_1$ reduction and incorporation "off the rocks" (that is, off dependence on mineral-surface group transport), whereas the functions of RNA are limited and uncertain. With our colleague Shelley Copley, we have proposed [158] that small dimer RNA molecules are plausible intermolecular and intramolecular catalysts, because of their inventory of acid and base groups, exocyclic amines, and hydroxyl groups (on the PRPP base) amenable to esterification. Any functions of RNA that relied on template-directed replication using base pairing, however, would have required many further selective filters to take place, including selection for homochirality and restriction of polymerization to oligomer synthesis.[84] We entertain the possibility that folates first provided the network feedback that selected their chemistry kinetically, and thereby formed a foundation for RNA as a by-product. The stability of that generative chemistry would have allowed a period of subsequent exploration in which the other problems along the route to RNA base pairing could have been solved.

The cofactors could also have participated in the transition to oligomerization and base pairing. As is hardly surprising (given the way they are synthesized), both pteridine derivatives and alloxazine derivatives (flavin relatives) have faces capable of base pairing with nucleotides. Both can be incorporated into polynucleotides,[85] either filling gaps in double helices, where they may be bound solely by Watson–Crick base pairing and hyrophobicity to nucleotides in the complement strand [340], or through formation of nucleotide substitutes. Flavin nucleotide analogues have been made and incorporated into DNA with high incorporation yield and at least moderate stability [718], and libraries of pteridine and alloxazine derivatives with variable exocyclic groups have been shown to base pair selectively against the standard nucleotides [663].

### 4.5.5  Roles of the elements and evolutionary convergences

Many cofactor functions derive from properties of the chemical elements. In this way part of the structure of metabolism directly reflects order in the periodic table. Three examples that we consider in more detail include the role of transition metals as mediators of radical chemistry with important ramifications for redox energetics, the use of heterocycles with conjugated double bonds incorporating nitrogen, and functions that exploit special properties of bonds to sulfur atoms.

The grouping of cofactor function according to element properties constitutes an additional distinct form of modularity within metabolism. Like the substrate network, cofactor groups often share or reuse synthetic reaction sequences. However, unlike the small-molecule network, cofactors can also be grouped by criteria of catalytic similarity that are independent of common biosynthetic descent or repetition of reaction sequences along parallel synthetic pathways.

Other patterns exist between cofactor classes and distinct subnetworks of core metabolism, which are probably entailed by the relation of elements to categories of reactions used by these pathways, but which have consequences for the staging of

---

[84] We return to these problems, and what is known of mechanisms to address them, in Section 6.5 and Section 6.6.
[85] DNA is generally used as the study system, due to its greater stability.

biogenesis. For example, the $C_1$ reduction superhighway employs generalized folates that are CN heterocycles tuned with oxygen or methyl groups, but without sulfur. Only transfers of $C_1$ groups off folates, and out of the superhighway to serve other one-carbon metabolism, involve sulfur-terminated cofactors, such as lipoic acid, *S*-adenosyl-methionine, or coenzyme-M. In contrast, both carboxylation reactions in the rTCA cycle require sulfur cofactors (thiamin and biotin), though in neither of these is the sulfur atom the carrier of the transferred carboxyl group. This difference may separate the stages of biosynthetic complexity at which cofactors could first have fed back in to support $C_1$ reduction (slightly earlier) versus rTCA cycling (slightly later).

### *4.5.5.1 Transition-metal centers and electron bifurcation*

Section 4.3.6 and Section 4.3.7 noted the highly conserved status of metal-center enzymes involved in carbon fixation, and certain associated cofactors and small cofactor-like peptides such as ferredoxins. A range of metal-center gas-handling enzymes, particularly nitrogenases and hydrogenases, are similarly conserved and associated with difficult reactions [261, 262, 294, 437]. Transition metals – principally iron sometimes substituted by nickel, and more rarely cobalt – have a special role in reactions involving radical intermediates, because the readiness with which these elements can exchange single electrons makes pairs of metal atoms well suited to carry radicals formed by separating electrons in a covalent or dative bonding pair.

An especially important use of radical-pair intermediate states is a process known as **electron bifurcation**, which is coming to be understood as a central mechanism in cellular redox energetics [103, 353, 415, 713, 851]. Bifurcation solves one of the deepest and most universal problems of life: coupling an exergonic reaction to an endergonic reaction to drive an uphill process. This may take the form of generating reductants with midpoint potentials below those of environmental electron donors [284, 415, 484, 714], or it may be used to "titrate" cellular redox, combining low- and high-potential reductants to produce reductants at an intermediate potential to drive a desired reaction with little free energy wasted as heat [851].

One of the most often invoked problems in explaining the emergence of life is explaining how exergonic and endergonic processes could come spontaneously to be coupled. Modern organisms construct molecular machines that perform these transformations, but we must explain how the endergonic reactions were driven before the creation of these machines, and in what sense such coupling mechanisms could have been "necessary" in a prebiotic context. Bifurcation accomplishes this elegantly by employing *the structure of quantum state space and aqueous energetics* as the source of coupling.[86] This does not free modern biochemistry from the problem of exploiting electron-transfer centers with desirable properties, but it means they do not need to *create such state*

---

[86] It thereby removes the need to first evolve complex protein engines that couple mechanical work to pair transfer or group transfer. The problem of discovering an electron-transducing machine is potentially simplified by reducing the complexity of the rearrangements needed, though it does not change the fundamental requirement for a cycle.

*spaces* within CHON chemistry alone, a difficult problem for *p*-orbital elements. It also opens the possibility that contexts available prior to life could have coupled to the same quantum state spaces without the construction of machinery requiring natural selection.

## How bifurcation works

Electron bifurcation is possible because electron bonding pairs – spin-antisymmetric electron pairs in a common spatial orbital – are stable (low-energy) states in aqueous solution while radical pairs are high-energy states and thus difficult to excite. Bonding-pair exchanges such as oxidations, reductions, and group transfers are therefore relatively facile in water, while transitions that require separating a bonding pair into a radical pair are not. If, however, two metal centers can jointly present low-potential and high-potential single-electron transfer sites, one electron of the bonding pair can be driven to the low-potential site, in favor of being left on the organic donor as a radical. When two bonding pairs are split in this way, the two high-potential single-electron holding sites can donate their electrons to a high-potential paired-bonding state, and likewise for the low-potential single-electron holding sites, forming a pair-bonding state at lower midpoint potential than the original donor-bond. Often *ferredoxins* or *flavodoxins* are the single-electron transfer centers to the terminal electron acceptors in bifurcation reactions.

Because electron bifurcation can create low-potential reductants, it opens sectors of organosynthesis that cannot proceed spontaneously with common molecular reductants such as $H_2$. Therefore many authors have proposed a central role for bifurcation in a mineral context as a driver of the earliest organosynthesis [96, 523, 592, 593]. If plausible precursors to metabolism could not be generated without some form of bifurcation to generate low-potential reductants, then a search for mineral contexts in which bifurcation *must* occur because it provides the only relaxation channel for redox energy may provide a pruning rule to identify plausible geochemical contexts and reaction systems.

Although the most active current research concerns bifurcation mediated by transition-metal centers, the phenomenon was first understood for *quinones*, which support hydroquinone (reduced), semiquinone (radical), and quinone (oxidized) states [95]. Although the semiquinone is highly unstable, it persists long enough in lipid environments to mediate the sequential exchange of two single electrons to heme groups in cytochromes. The other major group of organic cofactors that pass through radical intermediate states are *flavins* [103]. In electron-transport chains, flavins that remain permanently bound within flavoproteins pass electrons from cytosolic NADH or $FADH_2$ to Fe-S centers in the flavoproteins and ultimately to quinones. The protein environments tune the midpoint potentials of attached flavins to create a bucket brigade [406] between cytosolic reductants and quinones.

### 4.5.5.2 Nitrogen in heterocycles

The distinctive features of biochemical $C_1$ reduction are the attachment of formate to tuned heterocyclic or aryl-amine nitrogen atoms for reduction, and the transfer of reduced $C_1$

groups to sulfhydryl groups (of SAM, lipoic acid, or CoM). Under the hypothesis that biological carbon fixation was preceded by something like a Fischer–Tropsch reaction, carbon monoxide would have adsorbed onto metal sulfide or metal oxide surfaces, where it would have been reduced (probably with solution-phase $H_2$ or $H_2S$). The use of alkyl-thiol cofactors as transfer carriers for $C_1$ may show continuity with reduction on metal sulfide minerals. However, the mediation of reduction by nitrogens appears to be a distinctively biochemical innovation. An important question to the formulation of a consistent metabolism-first theory of origins is why nitrogen heterocycles were recruited to this role of $C_1$ carriers, and what this tells us about the more specific mineral contexts that preceded them.

### 4.5.5.3 Bonds to sulfur

Sulfur is a "soft" period-3 element [314] that forms relatively unstable (usually termed "high-energy") bonds with the hard period-2 element carbon. It is a less electron-withdrawing element than oxygen, so thioesters create less-oxidized carbon centers than carboxyl groups. We suggested in Section 4.3.3.2 that the ability to fix carbon through carbonyl insertion at thioesters may be understood in part as a retreat from an oxidation boundary that leads to decarboxylation, toward a more neutral state of carbon (neither too oxidized nor too reduced) in which stability and reactivity are both achieved.

For the alkyl-thiol cofactors in which sulfur plays direct chemical roles, three main bonds dictate their chemistry: S–C, S–S, and S–H. Sulfur can also exist in a wide range of oxidation states, and for this reason it often plays an important role in energy metabolism, particularly for chemotrophs, and due to its versatility has been suggested to precede oxygen in photosynthesis [358].

Although not alkyl-thiol compounds as categorized above, two additional cofactors that make important indirect use of sulfur are thiamin and biotin. In neither case is sulfur the element to which transferred $C_1$ groups are bound. For reactions involving TPP, the $C_1$-unit is bound to the carbon between sulfur and the positively charged nitrogen, while in biotin, $C_1$-units are bound to the carboxamide nitrogen in the (non-aromatic) heterocycle opposite the sulfur-containing ring. The importance of sulfur to the focal carbon or nitrogen atom is suggested by the complexity of the chemistry and enzymes involved in its incorporation into these two cofactors [65, 405].

Sulfur centers also carry particular roles in electron bifurcation. While metal centers must stabilize radical intermediates in electron bifurcations, the bonding pairs which are split during single-electron transfers often come from small metabolites including heterodisulfides of cofactors. These exchange single electrons with Fe-S clusters (typically via flavins), and they are essential sources and repositories of free energy in pathways using bifurcation. For example, the heterodisulfide bond of CoB–S–S–CoM has a high midpoint potential ($E_0' = -140$ mV), relative to the $H^+/H_2$ couple ($E_0' = -414$ mV), and its reduction is the source of free energy for the *endergonic* production of reduced ferredoxin ($Fd^{2-}$, $E_0'$ *in situ* unknown but between $-520$ mV and $-414$ mV [415]), which in turn powers the initial uptake of $CO_2$ on $H_4$MPT in methanogens.

## 4.6 Long-loop versus short-loop autocatalysis

The foregoing summary of the kinds of feedback provided by cofactors enables us to compare the contribution they make with the contribution made by the topology of the small-molecule substrate networks, to the self-renewal and self-amplification that all biological systems must perform.

The metabolic substrate, the system of cofactors, and the macromolecular catalysts all depend on one another in all extant cells. Therefore, subsystems smaller than integrated cells can only be considered "autocatalytic" in proposed contexts where some supporting functions of modern cells were provided by pre-cellular surrogates. Often these surrogates will be incompletely known or inadequately defended, so the best we can do is to try to make a set of consistent assumptions and to apply them uniformly to all feedback loops. We will work within the premise stated in Section 4.2.4: that the earliest stages in the elaboration of metabolism were governed less by innovation of entirely new classes of chemistry, than by an incremental replacement of geochemically mediated mechanisms by organic counterparts, as increasing pathway-concentration and flux made these available. Under this interpretation, the shorter loop lengths and simpler reagents in the small-molecule substrate should place them earlier in the history of emergence, and qualify them as initially more robust determinants of metabolic architecture, with feedback from cofactors coming later and depending on the pre-existence of significantly ordered organosynthetic pathways.

The context in which loop length, and the complexity of the intermediates through which the loop passes, are informative is precisely the context of *non-specific or unreliable catalysts* that we have reconstructed at the base of the phylogenetic tree, and which should have been even more fundamental to the LUCA and earlier stages. In the earliest times, when the networks of intermediary metabolism were first being elaborated, these catalysts presumably were limited to what was available in geochemistry. Unreliable catalysis (more generally, any loss of function) reduces pathway output with each additional step. As the number of steps required to close a self-amplifying loop increases, the alternatives to the useful outcome (created by alternative branchings at each step) grow and the stringency of chemical selection required to complete the loop becomes greater. By the measure of the number of steps required to close a self-amplifying loop, the feedback of the loop-fixation pathways in the substrate network of Figure 4.6 might be called **short-loop autocatalysis**, while the feedback from completion of the biosynthetic network for cofactors could be called **long-loop autocatalysis**. To maintain output in a much longer pathway requires more selectivity of the reactions and so we associate it with later stages of emergence.

Self-amplifying pathways concentrate flux through few metabolites and therefore function as *compositional and evolutionary modules* in a sense that we will develop further in Chapter 8. Their self-renewal is a source of robustness, and their concentrating effect is a kind of dynamical attraction for trajectories of matter. Together these properties provide a kind of buffering against perturbations that makes self-amplifying pathways easier building blocks from which to assemble robust hierarchical systems.

The lower the level at which a pathway can achieve self-amplification, the fewer distinctive environmental supports we need to propose for it to first become a mechanism of selection, and the earlier this selection is expected to have acted. This premise helps restrict plausible sequences in the emergence of core functions such as carbon fixation, in particular for the important cooperative action of autocatalytic loops such as rTCA and direct $C_1$ reduction in W-L. While both W-L and rTCA would require abiotic counterparts to modern co-factors and enzymes to exist as pathways, the topology of feedback is achieved in lower-level structures in rTCA,. Therefore we expect that rTCA would have begun to feed back to influence its own growth and chemical competitive exclusion at an earlier stage than W-L, which could do so only after organic cofactors could begin to displace whatever were the pre-organic counterparts.

The next level (by the criterion of path length) of potential self-concentration would come from the successful generation of cofactors in the course of elaboration of the metabolic network. This feedback too is expected to have a sequence. For instance, the pterin cofactors on which the W-L pathway principally depends are the products of simpler and shorter biosynthetic pathways than the sulfur-containing cofactor thiamin (also generated from AIR in a related sequence of elaborations) on which the carbonyl insertions of rTCA depend. Whether biotin with its unusual integral sulfur needs to be considered for $\beta$-carboxylations, or where it should be placed, we do not propose. By comparisons such as these, *we may presume a sequence for stages of self-amplification, in which metabolism incrementally took over control of its own fluxes from environmental supports.* However, assigning times to these stages, or even relating them to the other elaborations required for molecular replication or cellular integration remains a difficult problem to constrain.

## 4.7  Summary: continuities and gaps

The logic of chemotrophic metabolism suggests that determination from constraints of low-level chemistry can still be seen beneath four billion years of evolutionary optimization. It also suggests affinities to geochemistry in many of the deepest and most invariant features. Still, we are left with an important gap between what we infer to have been essential from the top down (in microbiology) and what we can show to be essential from the bottom up (in laboratory models of geochemistry). The gap should serve to focus the next generation of research questions.

Biochemistry seems to suggest, from many directions, a unique central place and organizing role for what we have called the "universal covering network" of metabolism, in Figure 4.5, and within that, a preferred status for the rTCA reactions and the $C_1$ reduction sequence now performed on folates. Not only is the universality of this network attested phylogenetically; what evidence we can compile about the direction of evolutionary change suggests that the redundancy of its functional-group chemistry was even more essential to the possibility for its existence in the era of the earliest cells with their non-specific and perhaps unreliable catalysts. A subset of pathways within the network is our best current candidate for metabolic processes in the LUCA, and the ease with which parallel pathways in the covering network have been reached by few evolutionary innovations recommends them as the kinds of variants we would expect to find in a fluctuating "ensemble" of pathways in pre-cellular eras.

However, current experiments in organosynthesis in geochemically plausible environments do not yet generate the conclusion that *chemically* the universal network was also favored or expected [614]. Partial overlaps have been achieved, which we review in more detail in Chapter 6. However, the main addition reactions of the rTCA cycle, currently dependent on their low-potential ferredoxin reductants, thiamin cofactors and Fe-S center enzymes, do not have understood geochemical analogues. Mechanisms for suppression of side-reactions that would make such a cycle autocatalytic also have not yet been demonstrated. Thus key conceptual features that make the core network seem *biochemically* essential do not yet appear within geochemical approaches.

This is not a crisis – much about the energetic and catalytic functions of metal centers either in solution or in minerals has not yet been systematically studied. It does, however, suggest that either the gulf between geochemistry and biochemistry is larger than we have supposed, or that biochemistry carries clues about energetics and environments in geochemistry that have not yet been recognized and fully utilized.

## 4.8 Graphical appendix: definition of notations for chemical reaction networks

The suite of biochemical pathways in an organism, and even more so the known pathway variants in the biosphere, create networks in which key metabolites may be inputs or outputs to many reactions. In order to represent these compactly in the text, and to emphasize homologies that result from redundant functional-group chemistry, we adopt a graphical notation slightly different from the traditional multi-arrow notation used in much of organic chemistry. We also desire a notation which emphasizes that the microphysics of reactions is inherently reversible: the direction in which a reaction proceeds is governed by the free energies of its inputs and outputs. These may be handled separately from the reaction's conservation laws and may vary across conditions even for the same reaction.

Chemical reactions have the property of **stoichiometry**, meaning that fixed proportions of reactants and products are interconverted in each turnover of a reaction. Stoichiometric relations cannot be represented by simple graphs, and require *directed hypergraphs* [64]. It is possible to display the hypergraphs representing chemical reactions as *doubly bipartite* simple graphs, meaning that both nodes and edges exist in two types, and that well-formed graphs permit only certain kinds of connections of nodes to edges.

In this appendix we define the graph representation used in the text. A variety of graph manipulation rules, which translate directly into operations on the rate equations of chemical reactions and can be used to aggregate reactions, are derived in Appendix B of [91]. These are the basis for the reduced-form kinetic model of the rTCA cycle presented in Section 4.4.4.3.

### *4.8.1 Definition of graphic elements*

- Filled dots dots represent concentrations of chemical species. Each such dot is given a label indicating the species, such as

$$\overset{\text{ACE}}{\bullet} \leftrightarrow [\text{ACE}]$$

used to refer to acetate in the text.
- Dashed lines represent transition states of reactions. Each is given a label indicating the reaction, as in

$$-\,-\,\overset{b}{-\,-\,-}$$

- Hollow circles indicate inputs or outputs between molecular species and transition states, as in

Each circle is associated with the complex of reactants or products to the associated reaction, indicated as labeled line stubs.
- Hollow circles are tied to molecular concentrations with solid lines

$$\underline{\qquad \text{ACE} \qquad}$$

one line per mole of reactant or product participating in the reaction. (That is, if *m* moles of a species A enter a reaction b, then *m* lines connect the dot corresponding to [A] to the hollow circle leading into reaction b. This choice uses graph elements to carry information about stoichiometry, as an alternative to labeling input or output lines to indicate numbers of moles.)
- Full reactions are defined when two hollow circles are connected by the appropriate transition state, as in

describing the reductive carboxylation of acetate to form pyruvate.

- The bipartite graph for a fully specified reaction takes the form

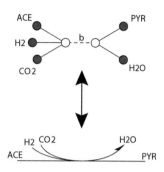

$$(4.4)$$

where labeled stubs are connected to filled circles by mole-lines. The bipartite graph corresponds to the standard chemical notation for the same reaction as shown.

# 5

# Higher-level structures and the recapitulation of metabolic order

Metabolism exists on Earth today only in a context formed by multiple higher-level structures, which are themselves constructed by living processes. In this chapter we consider four kinds of higher-level living order: ribosomal translation, some broad classes of oligomer catalysts, the tripartite bioenergetic system of redox couples, protons, and phosphate esters, and cellular compartmentalization including the association of cells with genomes. Like metabolism considered in Chapter 4, these higher levels show modular architecture and suggest a history of independent subsystems that were brought together to form extant cells. Many of the module boundaries follow divisions already seen in metabolism. Biosynthetic pathway patterns and a layered structure in the genetic code may reflect layers in the accretion of components of the translation system. The unification of bioenergetics by cells mirrors hierarchy in biochemistry, with a core of redox and thioester activation, and large-scale incorporation of phosphates only later, perhaps enabling the rise of an oligomer world. The cell is not merely one kind of compartment but at least three. The three core functions of cellularization – unification of bioenergetics, catalytic rate enhancement, and homeostatic regulation of the cytosol – may have come at different stages of separation from mineral-hosted environments. The resulting picture of life is of a confederacy of subsystems, which retain some distinct identity even in their current union. During biogenesis these gained autonomy from the environments that drove them into existence by becoming more dependent on each other.

## 5.1 Coupled subsystems and shared patterns

Universal metabolism at the ecosystem level is the chemical source of life, and (as we will argue in Chapter 8) in many respects its informatic foundation as well. However, the distinctive chemistry of living systems only occurs on Earth today in a context of elaborate higher-level structure. Biosynthesis depends in essential ways on catalysis by oligomers synthesized through ribosomal translation, on integrated and regulated bioenergetic systems, and on containment in cells. The maintenance and optimization of all these systems takes place through selection on genes and genomes, which must then be transmitted together during descent to maintain their functionality. These higher-level systems present whole webs of interdependency and entwined chicken-egg paradoxes.

To understand the origin of life we cannot only pursue the relations of metabolism to geochemistry. We must also decipher the relations among these higher-level structures into patterns of serial dependency. We saw in Chapter 4 that the architecture of metabolism is not a simple nested hierarchy that may be diagrammed as a tree of dependencies, and the same will be true at higher levels as well. We should not ask for one sequence of contingencies that led from geochemistry to cells. Rather, we are looking for elaboration in parallel systems that could have been maintained concurrently by geochemical environments, and in some circumstances brought into contact where they became interdependent.

The complex intricacy of the molecular biology of the cell might make reconstructing its formative stages seem impossible – the evolved interdependencies have accreted in too many intervening layers. Yet the cell possesses a modular architecture, as indeed it must to function and to be maintained by evolution. We believe that this architecture preserves enough primordial functional distinctions to suggest the sequences and contexts through which they emerged.

As Chapter 4 did for metabolism, this chapter will concern patterns of partial autonomy among cellular systems, and disaggregated contexts in which they could have originally emerged. A remarkable property of the functional modules in the world of oligomers and of the cell is that their boundaries often retrace boundaries we have already seen in the architecture of metabolism. For some of these we argue that the metabolic modularization is functionally and therefore also temporally prior. Either the abilities or the needs of metabolism defined many boundaries to which the emergence of higher-level structure was subject as it transpired.

The resulting picture of cellular life is of a confederacy. Cells are unified and heavily interdependent systems-of-systems. The subsystems they bring together are of many kinds and show order at many scales of aggregation and timescale. Their interdependencies are bounded, though: they are not so extensive as to have eliminated the internal identity of modules or their conditional independence from one another given a small number of restricted molecules or structures that serve as interfaces. In Chapter 8, where we bring together this repeated signature from all the previous chapters, we will argue that the maintenance of such partial autonomy is essential to the robustness and evolvability of such hierarchical systems, and also that ultimately their structure appeals to constraints at the lowest levels for its long-term stability.

### 5.1.1 *Shared boundaries: correlation is not causation*

Of course it is to be expected that hierarchically organized life would be modularized in a way that aligns across levels in the hierarchy. Metabolism, bioenergetics, compartmentalization, heredity, ecosystem assembly, and selection all affect each other. Modularization at any level should tend to imprint its architecture on other levels in a variety of ways. To understand where system structure originated, and which levels act as fundamental constraints on the others, requires further evidence about circumstances or mechanisms.

### *5.1.2 Different kinds of modularity have changed in different directions under evolution*

Section 4.2.2 introduced the argument that evolutionary optimization over time can replace early forms of modularity anchored in external conditions, with later forms defined by integration and modularization of biological subsystems. Section 4.4.5 illustrated a case of this kind, which could be directly supported by evolutionary reconstruction, for the enzymes of rTCA cycling in *Aquifex aeolicus*. Similar changes in the character of modularity are found in higher-level structures.

The richest case of this kind is the emergence of the translation system culminating in the ribosome, the amino-acyl tRNA synthetases and tRNAs, and the universality of the standard genetic code. The code contains a wide variety of metabolic regularities, which in greater or lesser degrees seem to require the explanation that *the associations among components that now participate in translation originated in an era when these components were much more functionally integrated within metabolism.* In the ancient system that gave rise to the components of the translation apparatus, the organization of metabolism was expressed but translation was not yet a separate function. In contrast, the modern system places remarkable barriers between the architecture of metabolism and the world of selection on proteins. It has evolved to eliminate functional cross-talk between coding and metabolism, making the translation system a much more tightly integrated module in its own right.

Similar interpretations can be made for the origin and integration of bio-energetics and for the multiple functions of cellular compartmentalization. Bioenergetic systems probably once followed the modularity of geochemistry, driving biosynthetic reactions involving redox, thioester, and phosphate activation from independent free energy sources. Today the integration of bioenergetics as a cellular system, which provides multiple-carrier support for metabolism as an integrated network, is made possible by cellular interconversion of redox and phosphate energy carriers. Processes of catalysis and chemical homeostasis may once have been tied to different geochemical settings, allowing subnetworks in organosynthesis to operate independently of one another. Today they are dependent on one another but are buffered from dependencies on environmental details through compartmental integration within cells.

The interpretation of higher-level modularity in living systems thus depends on their larger evolutionary context. Often the creation of one kind of modularity has entailed the elimination of others. Sometimes, as in the case of translation, evidence has been left behind. Translation systems could function just as well with many (though by no means all!) permuted codes besides the one we observe. Refining the function, robustness, and autonomy of the translation system has required eliminating active roles for its components within metabolism, but it has not required eliminating all *patterns* from active roles that they have played in the past.

### 5.1.2.1 Metabolism as the context for early modules at higher levels

Modularity in higher-level systems sometimes reflects patterns first seen within metabolism. In the genetic code, these take the form of relations between coding assignments and amino acid biosynthetic pathways that are more precise and more restrictive than the relations between the coding assignments and the amino acids themselves, which are the termini of the pathways. Some other regularities simply result from differences among classes of reactions, which are inevitably reflected in their catalysts. For instance, the enzyme forms for small-molecule organic reactions and for addition of activated monomers are qualitatively different: one class finely tunes the transition state and the other merely orients the substrate so that its activated bond can react spontaneously. The evolutionary histories of conservation or convergence in these two classes of enzymes differ accordingly. Still other regularities reflect the core/periphery distinction in the metabolic network: enzymes for core reactions tend to be more conserved than those for peripheral reactions.

The question we set out to address in this chapter is whether metabolism was a pre-existing context, which supported the emergence of higher-level structures piecemeal and prior to their integration with one another, or whether the imprinting went the other way, from control systems onto metabolic architecture.

### 5.1.2.2 Three reasons metabolic constraints may be recapitulated at higher levels

Metabolic constraints could appear in higher-level organization for at least three different reasons. First, absolute limits on the small molecules that can be synthesized may constrain all higher-level systems that can be assembled from them. We have argued that the universal core of small metabolites and a few of the very low-level biosynthetic pathways impose limits of this kind. Second, some pathways may not be the only possibilities, but they may be the most energetically efficient, or the most catalytically facile, or in some other way easier to afford or to maintain, and these differences may be reflected in long-term evolutionary convergence. A third reason metabolic signatures may appear in subsystems where they have no apparent consequence for function is that they were unique or were selected while the system was forming, and they are sufficiently close to neutral that there has been no selective force to replace them.

### 5.1.2.3 Signatures of the direction in which constraints have acted

Section 4.2.5 reviewed mechanisms by which constraints from a low-level system such as metabolism can propagate "from the bottom up" to impose limits on higher-level systems such as pathway organization, even within contexts where "top down" information flow from genes to pathways is the proximal mechanism for the generation of phenotype. One way to assess the direction of causation is to look for patterns in one subsystem that are not necessary to its function or that it lacks a mechanism to imprint on other subsystems. These are the patterns most likely to have been received rather than to have been donated. The patterns from biosynthetic pathways in the genetic code, which supersede the properties of the terminal amino acids, are of this kind. They are not properties of the function of the translation system, because all of its properties are conditionally independent of biosynthetic pathways given the amino acids actually coded. Therefore *the function of* translation, as it is currently accomplished, cannot have imprinted this structure on the biosynthetic pathways.

Most of the shared patterns that we review in this chapter are of a more ambiguous character. They could have been caused either by constraints from low-level chemistry or by selection from higher-level functions, because the criteria in the two cases are broadly compatible. The crucial thing that modularity at higher levels provides is a way to *parallelize* the emergence of life into subproblems, and to arrange subsystem dependencies into sequences that are not complicated and hence not unrealistically improbable to form.

## 5.2 Metabolic order recapitulated in higher-level aggregate structures

This chapter describes four subsystems within cellular life that are instances of higher-level organization relative to metabolism: the genetic code, some broad functional and evolutionary distinctions among oligomer catalysts, the coupling of redox/phosphate energy systems, and cellular compartmentalization.

**The genetic code** The genetic code is a nearly universal biological pattern resulting from highly restricted relations among tRNAs, the aminoacyl-tRNA synthetases (aaRS), and to a lesser extent the ribosome.[1] Ribosomal translation is itself the outcome of a long transition period in the production and use of RNAs and perhaps peptides, about which some of the history can now be reconstructed, but many questions of system context and evolutionary cause remain open. The convergence on a universal code must have been an even later process, roughly concurrent with the later stages in the optimization of ribosomal translation for accuracy. Yet the code is imprinted with multiple signatures of metabolic order reflecting the deepest stages in amino acid biosynthesis, a stage of metabolic elaboration that could only have occurred long before the components of the translation apparatus were aggregated into their current system.

**Catalysis by "tweezers" and "hands"** Oligomer catalysts can be classified roughly into what we will term "tweezers" and "hands." The tweezers catalyze organic reactions, while hands catalyze oligomerization reactions from phosphoryl-activated or adenylyl-activated monomers. The former alter microscopic transition states and show patterns of conservation in highly specific residues in the active site; the latter mostly provide binding pockets to orient activated monomers, and under either conservation or evolutionary convergence, the gross geometry is maintained but specific residues are not. Thus catalysts mirror the bimodal distribution between organosynthesis in small molecules and a wholesale conversion to phosphate-mediated dehydration for large molecules.

---

[1] The lesser extent is the role the ribosome plays in proofreading aminoacylated tRNAs for the correct charges.

**Bioenergetics** Bioenergetics consists of a tripartite system of chemical free energy carriers: redox couples, protons, and phosphate esters. Redox and phosphates are coupled through two channels. A more limited and possibly older direct chemical channel is mediated by thioesters. A much more elaborate channel connects electron transport to phosphorylation through proton transport under a trans-membrane potential, which today depends on complex, biologically constructed protein assemblies and topologically closed cell membranes. The partitioning of metabolic pathways between redox, thioester, and phosphate support may be among the most important features enabling the emergence of metabolism. Redox couples and perhaps thioesters are plausible geochemically, and perhaps all small-molecule organosynthesis can be driven by these. Whether coupling of redox energy to proton-motive force also has geochemical roots is currently an open question, to which we return in Section 6.2.4. The current biological use of anhydrous phosphates in numerous activation and dehydrating-polymerization reactions remains problematic. The ability to recycle phosphates may have marked the boundary between the small-molecule and oligomer worlds, and may be closely associated with the earliest functions of cellularization. It may also have been the innovation that first made ribosomal translation useful.

**The relation of genomes to cells** Finally, we consider a modularization that is not yet fully integrated. This is between molecular replication and cellular containment. Metabolism, catalysis, and bioenergetics are so interdependent and so reliant on cellular aggregation that it is difficult to envision earlier stages when they were autonomous or incomplete. Molecular replication of genomes is not nearly so closely tied to the cell. Horizontal gene exchange – common in the present and inferred to have been extensive in the past – along with the virosphere, characterize a world where genomes are in a strategic relation with cells, competing with each other, and caught between advantages of service and exploitation of cellular processes.

### 5.3 Order in the genetic code: fossils of the emergence of translation?

The translation system is a complex multi-component system that is a higher-order structure with respect to metabolism. In the modern world of DNA, RNA, and proteins, translation can be abstracted as the process by means of which information flows from RNA to proteins. If we take this abstraction to define the system boundaries, the translation apparatus includes the genes for transfer RNA (tRNA), the aminoacyl-tRNA synthetases, and the ribosome including ribosomal RNA and ribosomal peptides. An important property of the relations among components in that system is the mapping from triplet RNA codons to amino acids known as the genetic code.

The standard genetic code [854] and modifications in variant codes known from the vertebrate mitochondrion and many bacteria are shown in Figure 5.1. Each codon consists of three nucleotides. Major rows indicate the nucleotide in first position; columns indicate

| First Codon \ Second Codon | G | C | A | U | Third Codon |
|---|---|---|---|---|---|
| G | Glycine | Alanine | Aspartate | Valine | U |
| G | Glycine | Alanine | Aspartate | Valine | C |
| G | Glycine | Alanine | Glutamate | Valine | A |
| G | Glycine | Alanine | Glutamate | Valine | G |
| C | Arginine | Proline | Histidine | *Leucine* | U |
| C | Arginine | Proline | Histidine | *Leucine* | C |
| C | Arginine | Proline | Glutamine | *Leucine* | A |
| C | Arginine | Proline | Glutamine | *Leucine* | G |
| A | *Serine* | Threonine | Asparagine | Isoleucine | U |
| A | *Serine* | Threonine | Asparagine | Isoleucine | C |
| A | *Arginine* (Serine/Stop) | Threonine | Lysine | Isoleucine (Methionine) | A |
| A | *Arginine* (Serine/Stop) | Threonine | Lysine | Methionine | G |
| U | Cysteine | Serine | Tyrosine | Phenylalanine | U |
| U | Cysteine | Serine | Tyrosine | Phenylalanine | C |
| U | (Tryptophan) | Serine | | Leucine | A |
| U | Tryptophan | Serine | | Leucine | G |

Start
Stop

Figure 5.1 The standard genetic code. Large, medium, and small fonts distinguish simple, sulfur, and complex amino acids, which we argue entered the code in that order. Italics indicate positions suggested to be "codon captures." The column with second-position A contains the most hydrophilic acids by many measures; the column with second-position U contains the least hydrophilic. White indicates modifications found in the invertebrate mitochondrial code.

the nucleotide in second position, and minor rows indicate the nucleotide in third position. The amino acid assigned to each codon is inscribed in the corresponding box. Colors, typeface, and font are used to express a few of the many regularities that we will describe in detail in the remainder of the section.

The emergence of the translation system brought into existence a role for proteins as the effectors of molecular control, and permitted a contraction of the (still considerable) roles of RNA to memory and regulation.[2] Translation makes Crick's abstraction in the Central Dogma [167], of a one-way flow of information from gene systems to proteins and from there to metabolism and structure, a good practical approximation in many settings. From the Central Dogma, a paradigm has emerged of genes and enzymes as not only controllers but to some extent creators of metabolic order.

From the perspective of this "downward" information flow from macromolecules to small molecules, it has been expected that metabolism should reflect patterns in

[2] Those roles were contracted further, from another direction, with the emergence of DNA and the transcription system, removing the need for RNA to serve as the primary long-term memory system in cells.

the macromolecular world. The Central Dogma does not obviously suggest that the macromolecular world should carry patterns that are best explained in terms of limitations on the form of metabolism. Yet such patterns not only exist; they are pervasive and some appear quite law-like.

Several confounds must be addressed before we can claim that a pattern exists in the code and that it is of metabolic origin. The first is to define the model of chance that constitutes the *null hypothesis*. Random shuffles of code assignments are a common null model, but we will argue the real significance of patterns in the code extends beyond the amino acid assignments to codons, to the identities of the biological amino acids as well. We return to this argument in Section 5.3.4.

The second confound concerns the *direction of causation* for patterns. It is to be expected that some correlation will exist between small-molecule biosynthetic systems and macromolecular systems that interact with them. One method of imprinting, from homologies in the small-molecule substrate to patterns in the evolution of their catalysts, arises from the context-dependence of pathway completion, explained in Section 4.2.5. The regularities in the genetic code are not of this kind, because the translation system itself does not relate to metabolism in the same way. Translation appears to have evolved to minimize the degree of direct connection between the highly structured world of organosynthesis, with its many mechanistic constraints, and the combinatorial world of oligomers. We will focus on the code because, perhaps to the greatest degree among macromolecular systems, certain kinds of small-metabolite patterns appear to have no reason to appear in it, yet they do.

A large literature now exists on the patterns in the code (which reaches back to the decade before the mechanism of coding was even understood [887]). Some patterns are very well defended as being unlikely by definitions of chance that are conservative and robust. Likewise, some explanations of evolutionary causation, particularly those emphasizing robustness of the translation system, seem likely to be valid under very broad assumptions. We will cite a richer and more extensive system of regularities than the most conservative of this literature, so we must be careful to separate the argument that the pattern is "real" from hypotheses for how it might have been created or interpretations of what it means.

Therefore we will approach the problem of order in the genetic code in two stages. Initially we *describe* major patterns, emphasizing the *kinds* of relations they capture, and the kind and amount of arbitrariness each removes from the amino acid inventory and assignment positions. We emphasize the fact that the most comprehensive and specific statement of many patterns is intrinsically metabolic, meaning that it transcends the abstraction of translation as a self-contained functional "black box." At points we will mention broad classes of interactions that we think might give rise to such patterns, but in the descriptive phase we will avoid digressing to propose mechanisms or interpretations.

When we initially characterize the arbitrariness that a pattern removes from a null model of a translation system assembled "by chance," these measures of arbitrariness will need to be considered in relation to the complexity of the descriptions of the patterns themselves.[3] Patterns that can be simply expressed and that preclude many alternative codes are likely to be informative under a broad range of assumptions, and not only in connection with a particular explanatory hypothesis. When the patterns express constraints within multiple subsystems – not only the assignment or function of the code but also the biosynthetic network for the amino acids that fill it – the scope of variation on which they impose restrictions, and hence the information in them, becomes even greater.

We will show that *large parts of the genetic code can be read as almost an instruction set for the biosynthesis of the amino acids that are assigned in particular blocks of codons.* The codons are "instructions" in the sense that they may be associated in a regular way with specific and more or less exclusive steps in biosynthesis. Why those associations should exist, and whether they reflect immediate causes or more indirect cases of historical "freezing in" to the structures of a translation-system-in-formation, is a separate and more difficult problem to be addressed in different terms.

Having described several strong patterns in the code, we then return to reasons that have been given for it to be so ordered. They fall broadly into two classes.

1. Non-randomness may affect the function of the translation system (particularly its capacity to buffer errors in its own execution or in the memory role of the genome), and may have been created by Darwinian selection as the code was forming, and maintained by purifying selection in the period since it became established. The kinds of non-randomness that might confer error buffering are often implicitly defined and apply to the code as an aggregate entity. Most importantly, they arise *from the function of translation*, and depend on the structure and function of the translation system and the amino acid inventory as those properties exist while selection is acting – today this means that they apply to the standard code and the standard 20 biological amino acids.

2. Some patterns in the code are more completely and specifically stated with reference to *early steps in amino acid biosynthetic pathways,* which are masked to some extent by subsequent elaborations to form the modern coded amino acids. These patterns seem necessarily to refer to transitions that occurred in the past but are not active today. These are most interesting when they are not implicitly defined patterns that apply to the code as an aggregate (as the patterns responsible for error buffering often are), but biochemically explicit *constructive* rules that can be used sequentially to assemble aspects of the encoding system from subsets of its degrees of freedom, such as the identities of different nucleotides at different positions in codons.

Often regularities of the two kinds are not exclusive. The early pathway homologies associated with specific bases in a code may produce amino acids with similar properties in codons that are adjacent by single-base substitutions, which minimizes the impact of translation errors. Despite this partial confound of explanations, however, enough instances exist

---

[3] For readers familiar with Rissanen's *minimum description length* [674] as a measure of the complexity in symbol sequences that may be partly random and partly algorithmic, or with the Akaike [13] or Bayesian [716] information criteria for when an additional estimated model parameter is warranted by the regularity it captures in an ensemble, these are the kinds of approaches to assigning a complexity to descriptions that we have in mind.

of patterns which easily fit within one category but not the other, to suggest that sources of order of both kinds are represented in the modern code.

It is easier to formalize conservative null models of information for the genetic code than it is for metabolism, and this is one of the reasons the study of regularities in the code has remained an active area of research. A few general points from standard information theory [164] will be all we use in the formalization of this chapter.

1. Information is inherently a distributional concept [288].
2. The range of processes that must be referred to in order to state a pattern indicate the scope of variations that should be considered to define an information measure for that pattern. (This statement pertains to the "kind" of information in a pattern.)

The scope of variation indicated by some regularities in the code may extend well beyond the processes that would ordinarily be included within the abstract function of "translation" regarded as a black box. Indeed, only patterns that did violate this abstraction could enable us to situate translation within a larger context both in biochemistry and in history.

Where we use notions of "information" or "meaning" informally, we do so deliberately rather than offering simple combinatorial factors and referring to them as "the information" in each pattern.[4] The assignment of information measures to patterns in the code, even for simple permutation and error models, requires *many* assumptions about the range of possibilities and the appropriate distributions to reflect different levels of constraint. Translated proteins confer fitness through many properties, in many contexts, and on many timescales. The notion of "error" that may be buffered by a translation system is therefore complex and multi-factorial, and identifying and weighing these factors is a complex problem. To construct corresponding models for the even larger distributions that are appropriate, if biochemical variation is included in the scope of a pattern's "meaning," leads us into such a large combinatorial space of chemical and network possibilities [25, 677], that the modeling problem is similar to the problem of determining whether self-organizing metabolism is possible or unique. In treating these problems qualitatively, we acknowledge and even emphasize that formalizing them may be a difficult problem of empirical characterization.

The genetic code is a pattern made up from the relations of a set of **aminoacyl-tRNA synthetases** (aaRSs) to a set of **tRNAs**. For most amino acids in most organisms, one aaRS charges all tRNA codons for that acid.[5] tRNAs, like ribosomal RNAs and ribosomal proteins but to a lesser extent, show signatures of variation that largely follow those of the 16S rRNA or consensus phylogenies of proteins. In stark contrast, genes for the aaRSs are among the most freely horizontally transferred of any essential cellular component [266, 891, 898]. On the one hand, the transferability of aaRSs is not surprising, since their function is highly modular and the genetic code is now nearly variationless among

---

[4] A few combinatorial examples are used as bounds, to illustrate how they can be defined and the assumptions inherent in those definitions.

[5] For exceptions, and the patterns in them, see Woese *et al.* [898].

organisms. On the other hand, this fact makes it seem unavoidable that the modern suite of aaRS homologues is a result of many waves of replacement. It has also led Carl Woese and colleagues to argue [898] that the current generations of aaRS proteins cannot be the ones responsible for the assignments in the genetic code.[6] Woese argued more generally [887] that the aaRSs – although they appear to mediate the arbitrary pairing of amino acids with anywhere from two to six tRNAs – were not the source of the major regularity in the coding assignments, but that this source was the tRNAs themselves. That argument was the basis for his early *stereochemical hypothesis* [895] of a selective chemical affinity between tRNA molecules and amino acids. The associations we will point out, between blocks of tRNAs and early biosynthetic steps for amino acids, give even more strongly the impression that coding assignments at or near the observed ones existed as reference patterns in some other system and were ratified by aaRSs as the transfer of these slowly homogenized the code among organisms.

Therefore we study the order in the code as an opportune window on the actual problem of interest, which is the fundamental but obscure emergence of translation.

### 5.3.1 Context for the code: the watershed of the emergence of translation

The emergence of translation was quantitatively and qualitatively a watershed in the evolution of biomolecular memory systems. It was not the same as the emergence of polymerization (much later), and probably was not the first source of peptides. Thus, although it was by all evidence a product of an RNA World, it did not grow out of a simple RNA World or an RNA-*only* World. At the same time, given the seemingly exclusive role of proteins in the coupling of proton and phosphate energy systems in extant life, and the large demand of nucleotide polymerization for structural phosphate, and of polymerization in general for dehydrating potential, it is not apparent what sources or volumes of polymerization would have been available at the origin of translation. The advent of coded proteins seems likely to have had a large impact on phosphate recycling, and depending on how early this impact was felt, it could have fed back on the polymerization processes driving the emergence of translation.

The inevitability of all these constraints, and our uncertainty about their form, must be present as we consider the emergence of the particular relations that introduced translation as a new biological function.

#### 5.3.1.1 The major transition that culminated in ribosomal translation

The lynchpin that solidified the polynucleotide and polypeptide worlds, and bound them to each other and to the control of metabolism, was the establishment of the ribosome as a translation apparatus. The ribosome's own proteins and the enzymes for the synthesis of

---

[6] Evidence supporting this view has greatly expanded within the past decade in both diversity and specificity. It derives both from increasingly sophisticated comparative sequence analysis to reconstruct the evolutionary history of aaRSs in relation to the use of amino acids in peptides [265, 266], and from multifactor evidence about the pre-LUCA history of the components that aggregated to form the ribosome [85].

ribosomal RNA, as well as the replicases and transcriptases, are all produced within the translation system. It is therefore the essential constructor of the components in all systems required for memory in the form of molecular replication, and for control and regulation.

The translation system is formidably complex, involving multiple polynucleotide and polypeptide components and interdependent with both genome replication and transcription. Enough can be reconstructed about the history of the ribosome (see Section 6.8) to show that it was a descendent of a collection of subsystems, which probably formerly served diversified functions in condensation catalysis and template-directed RNA replication. These must already have been refined and stabilized by some kind of evolutionary dynamic, and the emergence of a ribosome from them was a joint process of assembly and innovation.

### 5.3.1.2 The genetic code as a window on translation

It is in the context of understanding the emergence of ribosomal translation that we consider the meaning of metabolic patterns in the genetic code, because genetic encoding emerged in the same transition, and the code like the ribosome is likely a construct of many layers. Early investigations into the order in the code were conducted when the mechanism of coding was not known, and the code was seen as a source of clues to that function. Today studies of this kind focus on stochastic effects and *deviations* from the primary mode of function.[7] As we describe below, however, the patterns in the code that appear to be most universal, most specific, and most informative are in many cases links to metabolic order that are at best partly expressed in the modern amino acid properties and assignments. These patterns that appear to require going beyond a characterization of extant function are most interesting to us as evidence about the past.[8]

### 5.3.1.3 Estimates of information in codon assignments alone

The information within the assignments of the code alone is not actually large. Regarded only as a combinatorial entity, the genetic code consists of the assignment of 20 amino acids to 64 three-base codons. The exact number of possible alternative assignments depends on the constraints we assume, but as a starting null model we may consider the number of simple permutations of the existing assignment. Referring to the number

---

[7] Some early efforts to deduce the code assignments as part of inferring the mechanism of translation are recounted by Carl Woese [887]. The theoretical efforts were overtaken by advances in measurement that made the code assignments a matter of (laborious, but unproblematic) observation. Once the assignments were known, they seemed to shed little light on how translation "must" work, and as molecular biology gradually came to describe how translation *does* work, the mechanism seemed compatible with quite freely permutable assignments. Not only had predictions not anticipated the function of the ribosome; from the observed function such predictions seemed to have no hope of contributing to understanding. Woese's summary of this era, though not complimentary, is a constructive call to frame such questions of pattern in the evolutionary and system-level contexts including especially the role of stochasticity and error, which were themes through all of Woese's work.

[8] Patterns in the genetic code are probably not the only source of evidence about the emergence of translation. The ribosomal RNAs and peptides, the aminoacyl-tRNA synthases, and enzymes for RNA or DNA replication or transcription may also carry signatures that span the same transition. Gustavo Caetano-Anolles and collaborators have presented phylogenomic reconstructions of structural motifs in ribosomal RNA and peptides [112, 337], as well as in other classes of protein enzymes [111], which they argue show not only existence but preserved evidence of relations in macromolecule families that span the period of emergence of the ribosome.

of times each acid currently appears in the standard code (including stop codons) from Figure 5.1, the number of permutations is given by the multinomial

$$n_{\text{perm}} = \frac{64!}{4!\,6!\,6!\,2!\,1!\,4!\,4!\,4!\,2!\,2!\,2!\,2!\,2!\,2!\,4!\,6!\,3!\,1!\,2!\,3!} \approx 2.3 \times 10^{69}$$

$$\approx 20^{53.3} \approx 4^{115.2} \approx 2^{230.4}. \tag{5.1}$$

The figures in the second line give the same number as the count of distinct polypeptide sequences of an appropriate length with 20 amino acids, of polynucleotide sequences with four bases, or of binary strings. The information needed to fully specify the extant code within the possible permutations is comparable to the Shannon entropy in the sequence diversity of a single short peptide of around 53 residues, under the most generous assumption that the shuffles of the code are weighted uniformly.[9]

We think that not only the positions and degeneracies but also the identities of the 20 biological amino acids are the referents of the information in the code. The information that is "about" the translation system does not factor independently from the information that is "about" the biosynthetic network, and precisely that lack of decomposability reflects the way the translation system itself was originally entangled in the structure of biosynthesis. The possibilities for amino acid/codon associations then become much more numerous because the inventory of possible alternative amino acids not used by the biosphere may be larger. Still, the upper limit of information in the code is likely to remain much smaller than the upper limit in the comparative analysis of all informative RNA oligomers and proteins in the translation and RNA/DNA replication and transcription systems.

### 5.3.2 The modern translation system could be a firewall

The problems of inventing structure and function in the world of organic chemistry are heavily circumscribed by constraints of geometry, atomic and molecular bond properties, thermodynamics, and reaction kinetics. We have argued in Chapter 4 that these constraints are so limiting that the necessity and form of the earliest stages in biogenesis may be predictable from first principles of chemistry, energetics, geochemical context, and an appropriate dynamical statistical mechanics.

Once a monomer inventory has been established, however, and the arena of invention has shifted to phosphate chemistry and dehydrating polymerization, the constraints that had limited organosynthesis become irrelevant except insofar as they are frozen into the structures and properties of the established monomers. A striking feature of extant life is how sharply it appears to have defined this transition. The abandonment of organosynthetic modification for molecules larger than simple cofactors is almost total. The adoption

---

[9] To define a more realistic information measure we would not like to assume that permutations are weighted uniformly, but to favor some permutations over others to reflect various other regularities or constraints. The information measure that results for such a weighted distribution is not the Shannon entropy which is simply the logarithm of the multinomial (normalized by the size, 64), but the *relative entropy* or Kullback–Leibler divergence of the multinomial from the prior distribution for sampling of permutations. See [164] and [764, 765] for examples of the use of relative entropy and its interpretation.

of oligomers as the biological macromolecules introduces a taxonomy of secondary structure and other folds as the new higher-level building blocks [189], but forecloses many other options for complexity that would exist if organosynthesis continued to be used to macroscales.

Even the nucleobases and standard Watson–Crick base pairing seem to have been selected to provide nearly uniform geometry and energetics.[10] As we noted, the apparatus of translation requires the parallel synthesis of a large number of components. Most known cells generate at least 45 distinct aminoacylated tRNAs, and the *E. coli* genome encodes 87 tRNA genes.[11] This price and the genomic and biosynthetic costs of other such parallelisms seem to be paid to permit the translation system to function as independently as possible from the constraints of low-level organic chemistry that would inevitably arise if the translation system reused some components for multiple functions. Note the striking contrast of this genomic largesse with the tight economy of enzyme reuse in metabolism described in Chapter 4 (Section 4.4.5), seemingly wherever local functional-group homology permitted it.[12]

The monomer-to-oligomer transition created sufficiently new organizational abstractions that the translation apparatus could nearly function as a "firewall," insulating the problem of sequence selection for macromolecular function from the rich but arbitrary structure of constraints in the underlying metabolic network. Not only could it function in this way, selection appears to support a significant genomic and biosynthetic cost to approximate this function.

Patterns in the code reflect imperfections in this firewall, either present or past. In the present, they come not from the ideal function toward which selection pushes the translation system, but mostly from the small rate of errors that it has not managed to eliminate. The structure of errors betrays the fact that the translation apparatus is a collection of chemical machines, and the kinds of errors they commit reflect the molecular implementation of their abstract functions. In the parlance of modern computer science, the details of implementation leak through the firewall that selection has worked to erect between the "hardware" layer and the "abstraction" of translation. Patterns in the code that reflect the past more plausibly than they reflect errors in present function may go further: they may reflect stages in which the ancestors to the translation system were not even a good approximation to a firewall, but instead were embedded components within a complex

---

[10] The exception that proves the rule comes most obviously from the use of inosine in some tRNAs, and of non-standard base-pairing rules such as the wobble-base non-standard pairs, discussed further below. Inosine base pairing is energetically and geometrically close enough to Watson–Crick base pairing to be compatible with much RNA secondary structure and energetics, but different enough that the use of inosine is limited to a few roles where the differences it creates from standard Watson–Crick pairing are its main contribution.

[11] Source, the GtRNAdb database [125]. This includes one tRNA for selenocysteine but excludes a possible pseudogene.

[12] This contrast casts early efforts to "deduce" the genetic code in a kinder light than they often get from molecular biologists [887]. These efforts failed – as the parallel architecture of the coding mechanism ensured that any such deduction would fail – because they were based on the premise that the coding system should express a kind of economy, a reuse of either mechanisms or relations. In other areas of biology, this premise is valid and causally important. The fact that a costly and complex translation apparatus involving many parallel components has evolved, and that it creates a specific and neutral configuration space from chemistry that is overall highly structured, is a significant surprise. It suggests that factors such as regulatory flexibility, certain forms of robustness made possible by redundancy, or the design simplicity of uncoupled interchangeable parts, have outweighed whatever economies might have arisen from reuse of components or symmetry of assignments.

chemical system in which the abstractions of metabolism and oligomerization were not yet well insulated from one another.

### 5.3.3 What kinds of information does a pattern contain, and how much?

The notion of "information" can be applied to a patterned system such as the genetic code in both informal and formal ways. We will use the term in both ways. It is desirable to formalize where possible, but it is also desirable to retain meanings that, for one or another (often technical) reason, we cannot yet formalize. We will, however, intend that all our uses of the term "information" be in line with the technical usage, and refer at least implicitly to a formalization. The technical sense of information is important not only because it makes possible quantitative comparisons of the amounts of information in different patterns, but because it imposes a structured way to think about the differences in *kind* of the information that different meanings for a pattern may carry.

#### 5.3.3.1 The necessary role of distributions of possibilities in the definition of information measures

In technical usage, we always intend "information" in the sense introduced by Claude Shannon [732].[13] We start with a question such as "In how many ways might a functional translation system have been assembled?" This formalizes our state of prior understanding by admitting some distribution of alternatives that we consider possible.[14] Among these, the observed translation system and code are particular instances. Then we have the following definition.

> **Definition** The *information* inherent in whatever constraints are needed to single out the extant code from among the prior possibilities is the reduction from the entropy of the prior distribution to zero, which is the entropy of a posterior distribution in which the observed form is the only one supported.

It is not essential to this definition of information whether we use the Shannon–Gibbs–Boltzmann functional or some other functional of the distribution to define the entropy; what is essential is that *information is an inherently distributional concept*. To formalize or quantify the "amount" of information in a rule, we must make some explicit commitment to the distribution on which that rule places restrictions.

We also will purposely avoid the premise that some forms of information in the code are "semantic information" that falls outside the scope of Shannon's approach to formalization. Shannon's original distinction between the combinatorial information captured in his entropy measure, and other associations with a pattern that might be "semantic"

---

[13] Chapter 7 provides a more systematic treatment of the combinatorial origins of Shannon entropy than is required for this section.

[14] In the terminology that has become conventional in discussions of entropy, such a distribution of possibilities is termed an *ensemble* [289].

and excluded by the combinatorics, was meant to emphasize that formalizing information required an explicit commitment to ensembles, counting, and distributions. The distinction is not an inherent property of meaning or the nature of pattern. It is, rather, merely a practical distinction between those aspects of pattern that we have framed precisely enough to consider in distributional terms, and those for which we do not yet have as precise a framing.

Any pattern may exist with respect to several different distributions. For example, a pattern in the code may be special because most permutations would not preserve it, and thus it singles out those that do. An elementary information measure may be assigned to this classification of permutations by comparing the Shannon entropies of uniform distributions over permutations that do preserve the pattern relative to all possible permutations.

The same pattern may *also* make manifest reference to biosynthetic pathways and chemical properties, which are not measured by shuffles in the assignments of amino acids to codons. With respect to the measure of permutations and its associated Shannon entropy, these other associations are "semantic." This only means, however, that they are outside the narrow distribution obtained by eliminating all forms of variation besides permutations. In a wider distribution that includes variations in the redundancy, form, or metabolic situation of the amino acids, the same associations may naturally inherit a measure of Shannon's distribution-based information.[15]

The most important reason to insist on distribution-based thinking about information is that it forces us to recognize when joint variations in multiple systems are the domain within which an informative criterion or rule acts. It is tempting in reference to extant life to abstract metabolism, translation, and selection dynamics as functionally separate black boxes, and to attempt to assign a limited information measure, such as an entropy of assignment permutations, as "the" information content particular to the genetic code. However, we will see that many regularities in the code take much of their meaning from the restrictions they imply about amino acid biosynthesis. Quantifying information measures can lead us to an understanding of the history of translation, but we must permit ourselves to recognize when the information they carry is about the coupling of metabolism to emerging translation, and not about variations in either system by itself.

### 5.3.3.2 *Functional criteria and construction rules: two (non-exclusive) classes of informative constraints*

The constraints on the code that will interest us are of two main kinds, which are not mutually exclusive. (That is, the same pattern may be of both kinds.) One group may be called *functional criteria*: a single measure is introduced which ranks different alternatives for the code. As in a one-sided statistical significance test, the measure of uniqueness for the extant code is the number of alternative codes that rank equally high or higher by the functional criterion, out of the total number of alternatives. The logarithm of this ratio is an entropy difference if the uniform distribution is used on both the pre-selected and

---

[15] An immediate candidate is the *mutual information* between the distribution over assignments and the distribution over biosynthetic alternatives, which is obtained from the joint distribution over both biosynthetic pathways and assignments. This approach to defining semantic information is explained in the work of Christoph Adami [11, 12].

post-selected sets of codes. Functional criteria applied to genetic codes have mostly been measures of robustness of physical properties of coded proteins under empirically derived rates of coding error.

A very different way to restrict codes is to *construct* their assignments by successive application of a set of *rules*. The rules may apply to the synthesis of amino acids jointly with their assignment to codons. These rules need not carry a functional interpretation as conferring some advantageous property on the code, and they need not create rankings for most codes. They may simply exclude codes that cannot be constructed according to the rules. Most of the metabolic patterns that we list in the genetic code will be expressible as rules of assembly, applied jointly to amino acid biosynthesis and to codon assignments.

### 5.3.3.3 Information and compression of description length

The amount of information in one or another pattern in the code determines the extent to which the pattern shortens, or *compresses*, the shortest description that will specify the code entirely.[16] For example, if the amino acids and their redundancies are given, and only the possibilities of assignment to codons is considered, the number of possible distinct permutations is given in Eq. (5.1). A specification of any particular code such as the standard genetic code requires about 230 bits of information to simply identify one permutation from a list. A rule which restricts the permitted distributions, and which is not itself complex to state, can reduce the information required to identify the actual code within the resulting shorter list. It is easiest to illustrate this with a particular example.

The assignments of amino acids to the standard genetic code are shown in Figure 5.1. We notice immediately that five amino acids[17] are assigned in blocks of four with the third codon position making no distinction. We could state – as an uninterpreted rule – that the proper ensemble in which to look for the actual genetic code is not the ensemble counted in Eq. (5.1), but an ensemble in which these five acids are always assigned together in blocks of four. Other rules may also apply, but for the sake of illustration we consider only the information in this restriction, and permit the other acids to be assigned in all the remaining possible ways.

The number of possible assignments with priority given to these first five blocks is[18]

$$n_{\text{perm}} = \frac{16!}{5!\,11!} \cdot \frac{44!}{6!\,6!\,2!\,1!\,2!\,2!\,2!\,2!\,2!\,2!\,2!\,6!\,3!\,1!\,2!\,3!} \approx 1.7 \times 10^{45}$$

$$\approx 2^{150.2}. \tag{5.2}$$

---

[16] The proposal to measure the complexity of a pattern in terms of the description length of a compressed algorithm to produce the pattern was first advanced by A. N. Kolmogorov in 1963 [449], and has since been developed by many authors; see [124, 485, 674]. As Cover and Thomas [164] explain, the description length in a compressed symbol string is also the entropy of an ensemble from which the string is a random sample, so algorithmic complexity is a measure of the entropy of algorithmic selection. Different measures of complexity vary in whether they call for complete or partial descriptions. Gell-Mann and Lloyd define a measure called *effective complexity* which is to measure the description length only of the regularities in the algorithm but not the parts that are random [289].

[17] They are Gly, Ala, Val, Pro, and Thr.

[18] One first assigns the five blocks out of the 16 possible pairs of the first two codons. The remaining 15 acids are assigned, with their previous degeneracies, at random among the remaining 44 positions.

Only about 150 bits are required to identify the code within a list of those permitted by the rule for block assignments, so the description length of the *random search part* of the specification of the code has been reduced by 80 bits relative to the permutations (5.1) in an ensemble unrestricted by block assignments.[19]

A glance at the code in Figure 5.1 makes clear that many more groups such as this could be singled out. The task in searching for informative and meaningful rules is to identify those that are suggested to be meaningful according to many criteria, and that do not include a large number of arbitrary conditions and special cases. Where a rule provides no more restriction than simply choosing an entry from a list, especially if the rule is complex or apparently arbitrary to formulate, it is preferable to treat the corresponding assignments as random, admitting our ignorance about whether they are significant, and if so, what they signify. Clearly, whether a rule provides a sufficient reduction in entropy to justify its complexity will depend on the domain in which it acts: rules that make only a few restrictions about assignment positions are not very informative with regard to permutations, but they may become very informative in a larger context, if they place strong restrictions on biosynthetic chemistry.

Rules may act probabilistically, capturing most of the instances in some group. They may still be good and informative rules if the exceptions can be treated in sufficiently simple ways, including treatment as random exceptions if no simple explanation can be proposed. Many of the rules we state below will have this character.

### 5.3.3.4 Compression is informative with or without mechanism

The fact that the modern translation apparatus could work in most instances with any permutation of the actual code implies that *the existence of rules that compress the description of the code is informative in itself*, whether or not those rules constitute a mechanistic explanation for the order in the code. In some cases convincing explanations have been given for order in the code, for instance as a consequence of selection for error buffering. In other cases, we can propose mechanisms but cannot attach high confidence to them. As we emphasized at the beginning of the section, the existence of a compression rule is simply an observational fact, which stands alone as a constraint on all subsequent mechanistic interpretations or theories of causation.

### 5.3.4 Part of the order in the code is order in the amino acid inventory

The most efficient path to present the order in the genetic code of Figure 5.1 begins with a description of the order in the amino acids that are encoded. That is, it begins not within the translation system at all, but within metabolism.

---

[19]  To obtain a full description of the code relative to our original ensemble, it is necessary to assign a length to the statement of the block-rule itself. Formalizing this step is a problem usually studied in the theory of algorithms, which is outside the scope of our discussion, and not obviously essential to a first exploration of the biochemical order in the genetic code. We will therefore restrict the few calculations we carry out to the description of the random search left after rules have been assigned. We chose the block-assignment rule as an example because it is context free and constitutes one of the simpler rules that organize the code.

### 5.3.4.1 Three classes of amino acids: simple, sulfur, and complex

The amino acids are not all alike in their biosynthetic complexity, the locations of their pathways in the metabolic network, their frequencies of use or their functions in proteins. At the same time, they are also not so different as to be non-comparable. They can be broadly grouped into three classes, and the same groupings are respected by many of the criteria we have just mentioned. The three classes and a few of their shared biosynthetic characteristics are as follows.

**Simple** *Gly, Ala, Asp, Glu, Ser, Asn, Gln, Thr, Pro, Arg, Val, Leu, Ile.* The amino acid is synthesized from a backbone in the core CHO network, typically by a sequence of steps that build upon and modify the core backbone, but do not draw large assembled units from remote subnetworks within metabolism. Much of their structure is therefore dictated by the available core backbones and a few frequently used modifying reactions.

**Sulfur** *Cys, Met.* The only sulfur-containing acids. Cysteine in particular has a variety of roles in proteins that depend specifically on properties of S and SH groups. Methionine has a prior role as the methyl-transfer cofactor *S*-adenosylmethionine.

**Complex** *Phe, Tyr, Trp, Lys, His.* Except for histidine, each biosynthetic pathway combines synthesis of a primary backbone from the core CHO network with parallel synthesis of a large component from either the core or a remote subnetwork. The aromatics phenylalanine, tyrosine, and tryptophan incorporate (respectively) benzene, phenol, and indole derived from chorismate. Lysine synthesis is initiated by condensation of pyruvate with aspartate semialdehyde, by a ring-forming enzyme related to the synthase for the cofactor pyridoxal phosphate.[20] Histidine, as noted in Section 4.5.2.1, is in a class more closely related to folates and RNA than to other amino acids.

The structural, biosynthetic, and functional differences of amino acids are *strongly* reflected in their assignments in the genetic code (both the standard code and all variants).

### 5.3.5 Four major forms of metabolic order in the code

Much of the possible permutation-symmetry in the code can be removed with four strict or nearly strict rules of codon assignment. We focus on those which are relatively simple to formulate and which exclude many variants. All four appeal to amino acid groups defined either by molecular properties or by biosynthetic relations, and seem likely to have historical information, as well as consequences for performance of the contemporary translation system. One, which we consider first, is related to the mechanism of translation itself.

---

[20] Source: MetaCyc [412].

## *5.3.5.1  Third-base redundancy*

The most obvious feature of the code, mentioned briefly as a compression rule in Section 5.3.3.3 and described more completely here, is that the third codon position carries no distinction at all in eight of sixteen blocks indexed by the first two codon positions. In the eight blocks where the third base is distinctive, the distinction is made only between purine (designated R) and pyrimidine (designated Y) in six of these. (In the mitochondrial code, only the R/Y distinction is made in all eight blocks.)

An immediately striking pattern is the *dominance of the remaining two-base blocks by biosynthetically simple amino acids*. In every block where the third base makes no distinction, the codon assignment at all positions in the block is to one of the simple amino acids listed in Section 5.3.4.1. In six of the eight blocks where a third-base distinction exists (excluding GAN), it is between a simple acid and either a complex acid or a stop codon (or both). (Here we group Cys with the simple acids and Met with the complex acids, for brevity, and for the moment we do not regard the sulfur acids as a separate class.) The simple amino acid in a block is never assigned to fewer than two codons; the complex acid or stop codon is never assigned to more than two. Two blocks require further refinement beyond an R/Y distinction in the standard code. At AUN, Ile is assigned to three positions and Met is assigned to AUG. At UGN, Cys is assigned to UGR, Trp to UGG, and stop to UGA. Two complex acids are never assigned to codons within the same block. Except for the block with aspartic acid (GAY) and glutamic acid (GAR), and the block combining serine (AGY) and arginine (AGR), two simple acids never share a block. In the case of AGN, its two simple acids also occupy all four codons in different blocks (respectively UCN and CGN), and the AGN assignments are known to be highly variable in numerous species [700].

These observations may be summarized by saying that thirteen of the sixteen possible pairs of first two codon bases may be assigned to a unique biosynthetically simple amino acid, excluding only GA, which may be assigned two simple acids, and AG and UA, which contain only potentially re-assigned amino acids (known as "captures") or else complex/stop assignments in the standard code. Figure 5.2 shows this two-base association. Complex amino acids may be regarded as a secondary group, occurring in only six of sixteen blocks. They always require the third codon position to designate them, and in the standard code two of these have been reduced to representation by single codons. The complex acids are almost comparable to stop codon assignments in these respects, with the difference that in both the standard and the mitochondrial code, stop codons appear in two distinct blocks.[21]

Francis Crick [165] was the first to identify the third-base redundancy as a consequence of the chemical implementation of the translation system itself, in what is known as the **wobble-base explanation**. In the ribosome, the first two bases in the codon and their anticodon partners are sterically constrained, so that only the standard Watson–Crick

---

[21] Which blocks they appear in differ, however, among different non-standard codes [443].

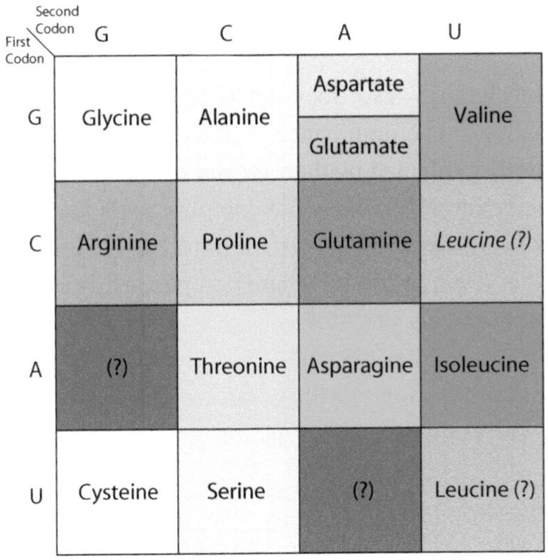

| Second Codon | G | C | A | U |
|---|---|---|---|---|
| **First Codon** | | | | |
| G | Glycine | Alanine | Aspartate / Glutamate | Valine |
| C | Arginine | Proline | Glutamine | *Leucine (?)* |
| A | (?) | Threonine | Asparagine | Isoleucine |
| U | Cysteine | Serine | (?) | Leucine (?) |

Figure 5.2 The assignment of simple amino acids to two-base projections of the code from Figure 5.1. AG (a pair of putative "captures") and UA (no simple acid) are unassigned, and GA is assigned to two acids since we have no natural criterion to exclude either. Leu is provisionally allowed in either UC or UU blocks, though in later sections we will argue that UU is the more natural assignment.

patterns of hydrogen bonding between bases are low-energy configurations. The third-position codon is "loose" and can wobble on the ribose phosphate backbone, enabling other hydrogen-bonding patterns that are made possible by the distribution of N and O atoms on nucleobases but would be sterically forbidden in the other positions. Using only the standard nucleobases, only the G-U pair is introduced as a new possibility, but if inosine is used in the tRNA anticodon, three new possibilities I-A, I-U, I-C are introduced.

The admission of alternative base pairings by permitting wobble enables as few as 45 anticodons to match to all 64 possible codons (a positive adaptation to reduce redundant biosynthetic cost?), but it precludes using the third position to make finer distinctions than purine/pyrimidine reliably, so it requires coding assignments that do not depend on this distinction.

Wobble-base pairing creates stochasticity in one degree of freedom in the translation process. If that stochasticity is not to create large error rates, it places a constraint on coding assignments. Beyond those two factors, however, the wobble-base description is not directly an explanation of *cause* of either this feature of the ribosome or the degeneracy it forces on coding assignments. Several features may indicate that both wobble-base pairing and the degeneracy of the code had an origin earlier than modern coding. G-U pairs are widely used elsewhere in RNA self-interactions, particularly in RNA catalysis that makes use of bound metals [827]. By eliminating an amino group in the major groove of double helices, which all other Watson–Crick base pairs introduce, G-U pairs permit phosphate groups in the backbone to create uncanceled negatively charged pockets that bind divalent cations. Inosine (in which the exocyclic amine of guanine is removed altogether) may be put to similar uses.

The wobble-base mechanism also does not predict which amino acids are relegated to third-codon designations. The dominance of fully degenerate codon blocks by simple amino acids may be partly explained by their greater frequency in proteins: simple acids in degenerate blocks receive more codons while complex acids identified only in the wobble position receive fewer.[22] However, the pattern of third-base redundancy we have described is more explicit than that: even in blocks where complex acids are coded, they share these blocks with simple amino acids and not with other complex acids, reducing the third-position redundancy that could have been allotted to simpler amino acids. Below we will propose an alternative interpretation in which the directly observed property of biosynthetic simplicity associated with coding blocks is the informative property.

### 5.3.5.2  *The Wong correspondence: the first base and biosynthesis*

A very strong pattern in the first base of the codon assignments to amino acids, which captures three of the bases (C, A, U) and 15 amino acids (in all three groups: simple, sulfur, and complex), was first published by Tze-Fei Wong in 1975 [901].[23] These first three bases index the principal carbon backbone (the part containing the $\alpha$-amino and carboxyl group) from which the assigned acid is synthesized. Wong presented the regularity as a relation of codons to the first *amino* acid in the biosynthetic pathway for each assigned acid, and also argued for a mechanistic interpretation in terms of coevolution of the aminoacyl-tRNA synthetases with enzymes in the biosynthetic pathways.

Morowitz [572] reformulated the same pathway regularity as a relation to the universal precursors in core metabolism by recognizing that the first-codon positions in the Wong correspondence can be taken to index the $\alpha$-*ketoacid* precursor in the biosynthetic pathway, as one of three TCA intermediates. Figure 5.3, from Copley *et al.* [158], shows the molecular forms of the simple amino acids in the two-base assignment of Figure 5.2, along with the TCA intermediates that are their $\alpha$-ketoacid precursors. At several positions, a biosynthetic precursor to the terminal amino acid is shown to illustrate homologies under other rules, to which we return in the following subsections.

First-position G in codons does not correspond to a single carbon backbone across synthesized amino acids, but Copley *et al.* observed that the G-first series can be given a regular biosynthetic rule as all acids in this group are made by a reductive amination of a CHO $\alpha$-ketoacid as their terminal (or in most cases, their only) biosynthetic step. Again, three of these precursors are the universal TCA intermediates, and a fourth is

---

[22] Here again, the claim that amino acid frequency implies use of more codons quickly becomes an appeal to complex system-level properties. Whether multiple codon usage is the best strategy to support high frequency of an amino acid in protein synthesis depends on tRNA production rates, the availability of gene duplications, and consequences for the impact of mutations, and so rapidly becomes a question about complex system-level trade-offs. Most organisms use different codons for the same amino acids with different frequencies, a phenomenon called *codon bias*, suggesting that multiple factors drive the tuning of the "standard" translation system into slightly different operating ranges to adapt to local conditions.

[23] See also [797] for a related review.

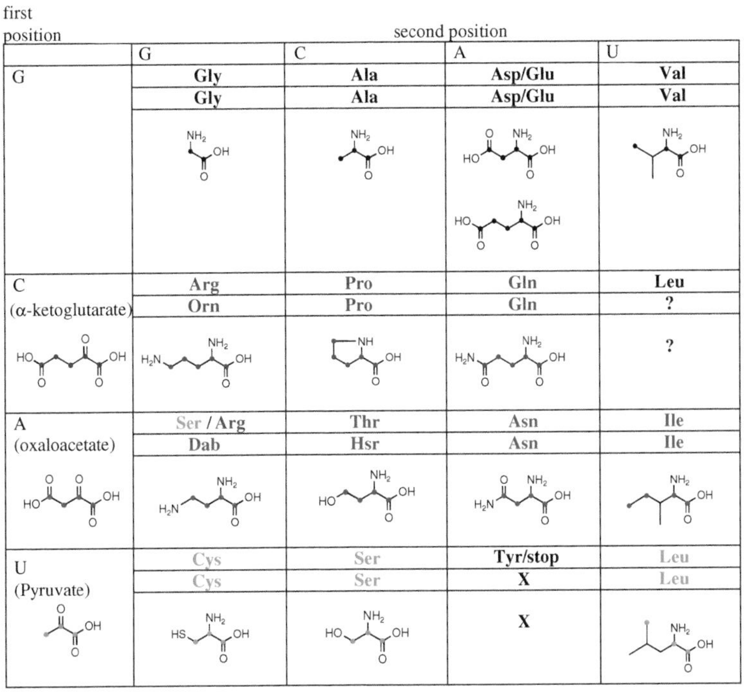

first position

second position

| | G | C | A | U |
|---|---|---|---|---|
| **G** | Gly<br>Gly | Ala<br>Ala | Asp/Glu<br>Asp/Glu | Val<br>Val |
| **C**<br>(α-ketoglutarate) | Arg<br>Orn | Pro<br>Pro | Gln<br>Gln | Leu<br>? |
| **A**<br>(oxaloacetate) | Ser / Arg<br>Dab | Thr<br>Hsr | Asn<br>Asn | Ile<br>Ile |
| **U**<br>(Pyruvate) | Cys<br>Cys | Ser<br>Ser | Tyr/stop<br>X | Leu<br>Leu |

Figure 5.3 The principal association of dinucleotides with simple amino acids, with molecular forms and TCA precursors shown, from [158]. Colored dots indicate positions of atoms from the input backbones. Only one principal assignment is shown for each acid. Positions of codon captures, or positions occupied by only complex acids, are left unfilled and marked with (?). Where biosynthetic precursors display pathway homologies, the precursor rather than the terminal acid is shown. Above each graphic the terminal acid, and the stage showing the homology, are labeled. (After Copley *et al.* [158].)

glyoxylate (accessible from the TCA cycle through retro-aldol cleavages from malate or isocitrate[24]).

These two extensions of the Wong correspondence are compatible with all amino acid assignments except histidine (which is an outlier from all strict rules we will exhibit), though the G-first rule admits many additional acids beyond those actually assigned to G-first codons. The four rules, along with the amino acid assignments each would accept from the standard biological amino acid inventory, and the exceptions found in the code, are shown in Table 5.1.

The first-base rules can potentially admit biological amino acids outside the sets actually assigned within their corresponding rows of the standard code. Because each amino acid has a unique primary backbone, the Wong correspondence provides a partition which is unique for **CN** and **AN**, and admits only Val as an excess member from the actual **UN**

---

[24] Glyoxylate is also accessible from several other aldol cleavages of nearby reactions in the universal covering network of Figure 4.5, which branch from arcs that are homologous to the TCA arcs.

Table 5.1 *Rules from first-codon positions*

| Rule | Context | Description | Admits | Violates |
|------|---------|-------------|--------|----------|
| 1a | **GN** | Reductively aminate a CHO $\alpha$-keto acid and stop; source ketoacids are drawn from the TCA cycle, the aldol network that includes glyoxylate, and the branched-chain amino acid biosynthetic pathways | Gly, Ala, Asp, Glu, Val, Leu, Ile | |
| 1b | **CN** | Synthesize from $\alpha$-ketoglutarate backbone | Arg, Pro, Gln | Leu, *His* |
| 1c | **AN** | Synthesize from oxaloacetate backbone | Thr, Asn, Ile, *Lys*, Met | Arg, Ser |
| 1d | **UN** | Synthesize from pyruvate backbone | Cys, Ser$^o$, Leu, Val, *Phe*, *Tyr*, *Trp* | |

All standard biological amino acids admitted by a rule are indicated in the fourth column, even if they are not assigned within that row in the standard code. Italics indicate biosynthetically complex amino acids, by the criterion that they contain a second component synthesized from a distinct subnetwork in the metabolic chart from the primary backbone synthesis. Ser$^o$ indicates serine synthesized via the oxidative pathway from phosphoglycerate, on the gluconeogenic pathway from pyruvate. (The enolpyruvate group in Phe and Tyr comes from PEP in chorismate synthesis. Although chorismate is also used in the synthesis of indole in Trp, the primary backbone reflecting the Wong regularity comes from a serine added in the last step.) Histidine does not appear within this system in any natural way, since it has no primary backbone as an amino acid.

row. Our extension to **GN** is a more permissive rule, and admits Leu and Ile (not assigned within the G-first row) on an equal footing with Val (assigned).

The compression of the first-base assignment rules is similar to that illustrated for third-base redundancy in Section 5.3.3.3, except that it applies to entire rows of sixteen entries, with multiple amino acids admissible in each row, rather than the simpler third-base rule that refers to four-codon blocks and single amino acid assignments. Like the third-base rule, it greatly reduces the number of admitted permutations by reducing the large factorials in these multinomial formulae by forbidding mixing across rows.

Because each first-base rule applies to an entire row of codons, it makes no further specification of their order within the row. Amino acid assignments may be further restricted, either by the application of rules for other base positions independent of the first base, or by contingent applications of other rules given the first base. The independent application of other rules may be mechanistically uninterpreted; a contingent application generally requires some mechanistic justification or else we regard it as arbitrary and not informative beyond simply declaring the assignments. One such mechanistic interpretation based on intermolecular catalysis by dinucleotides, advanced by Copley *et al.* [158], is reviewed in Section 5.3.7.2 below.

### *Exceptions to the first-base rules*

Very few assignments in the standard code violate the Wong correspondence, and all of these except His are duplicate assignments, in which the amino acid is also found assigned in a different row that falls within the Wong correspondence. Duplications are eligible to be interpreted as instances of "codon capture": a secondary modification of a pre-existing more regular code to form the less regular modern code. Captured positions could be modifications that increase the number of codons assigned to amino acids used with high frequency in proteins (Arg and Ser), or they could reflect a compensation if some codons in an original position are re-assigned to complex amino acids (Leu possibly replaced by Phe at UUY).[25]

In some instances, such as the assignment that splits the four-codon **A**GN block between Ser at **A**GY and Arg at **A**GR, we favor the interpretation of capture because these "secondary" assignments both fail to respect the Wong correspondence, and neither provides a unique association of a simple amino acid with the **A**GN block. Both amino acids in their other positions (Arg at **C**GN and Ser at **U**CN) are unique and are also within the Wong correspondence.[26]

The third exception, Leu at CUN, could be a capture from adjacent **U**UR, where it satisfies the Wong correspondence, but in this case the "captured" position is a majority position, and Ser shares the putatively ancestral codon block with Phe. Since Phe is complex it could be interpreted as a later addition to the correspondence, making an ancestral assignment of Leu to **U**UN plausible. However, using contingent rules this way, to construct a relatively complicated narrative for the history of codon assignments, is *ad hoc* and we do not assign it much weight unless it is supported by mechanistic or other justification.

For the branched-chain amino acid Isoleucine, which has variant synthetic pathways, the Wong prediction holds only for a subset of these. The particular subset for which it holds is curious. Threonine is synthesized from oxaloacetate via aspartate and homoserine, satisfying the Wong prediction. Threonine *deamination* (and dehydration) to 2-oxobutanoate initiates one pathway to isoleucine, as shown in Figure 4.8 of Chapter 4, consistent with the Wong prediction for Ile codons AUY/AUA. While the deamination pathway is more common, the reconstruction of the most plausible pathway in *Aquifex aeolicus* reported in Section 4.3.6.9 is instead the reversed arc of part of the 3-hydroxypropionate carbon fixation pathway, through citramalate. While this pathway is longer, it uses enzymes homologous to those in the TCA cycle and is carried out entirely within a CHO network, with reductive transamination only in the last step. The *Aquifex* pathway is more plausibly the ancestral one in bacteria, and along this pathway Ile does not fit within the Wong pattern.

### 5.3.5.3 *Hydrophobicity measures, second-codon positions, and pathway homologies*

A second set of rules relates the second base in codon assignments to close similarities of side-chains, which derive from homologous sequences of biosynthetic reactions.

---

[25] The complication of the code by capture, which tunes properties of translation but reduces the simplicity of block assignments relative to a presumed ancestral version, is reminiscent of certain enzymatic optimizations in metabolism which we considered in Section 4.4.5.

[26] A capture of Arg from adjacent CGN might have buffered some proteins against point mutations or tRNA/mRNA mismatches, but we do not place much emphasis on *ad hoc* explanations of this kind for particular cases, unless they can be subsumed in rules that apply to many assignments. A capture of Ser is not even easily argued to be a buffer against point-mutation or matching errors, since the assignment of Ser within the UN row is at the non-adjacent positions UCN.

Because side-chain forms significantly affect physical properties of the amino acids, these second-base rules have historically been regarded as **physical property rules**, particularly affecting the hydrophobicity or hydrophilicity of coded amino acids. Physical property rules were the basis of the original stereochemical hypothesis [895], with which Woese attempted to account for amino acid/codon correspondences in terms of chemical affinities of amino acids for tRNA stems as a prior asymmetry present during the formative stages of translation. Woese used the concentration dependence of partitioning in pyridine-water mixtures known as the **polar requirement** as a proxy for an affinity measure between amino acids and RNA. We return in Section 5.3.7.4 to another use of the polar requirement [108, 833] as the basis for a robustness criterion, which could have shaped the code in the modern era of translation, independent of the status of the stereochemical hypothesis or other pre-coding chemical affinities.

The polar requirement gives one measure among many for *hydrophobicity* [327]. In the standard code of Figure 5.1, the amino acids in the **NU** and **NA** columns are assigned by most such measures as the most hydrophobic and most hydrophilic groups, respectively.[27] The relative ranking of amino acids in the other two columns depends on the hydrophobicity measure used. A correlation to which we do assign large significance is that in almost all cases, including hydrophobicity but going beyond it to more detailed chemical patterns, similarities of physical properties result from homologies in biosynthetic sequences that modify side-chains from the $\alpha$-ketoacid precursors, as can be seen very clearly in Figure 5.3.

As for other property regularities – whether size, hydrophobicity, or more detailed properties affecting particular catalytic functions – "similarity" is only a coarse-grained and approximate relation among different amino acids, whereas pathway homology is strict. Just as the Wong correspondences provided logically precise rules to partition amino acids, pathway homologies provide similar rules to partition acids assigned to codons with distinct second bases. Therefore again in the second-codon position, we find that codons are much more precisely correlated with biosynthetic pathways than they are with the physical properties or even the specific forms of the final amino acids.

A set of rules to partition amino acids according to second-codon positions is given in Table 5.2. Two of the rules in this table are defined in terms of modification of terminal carboxyl groups, and so they cannot be applied to carbon backbones that lack these groups. Since the first-base rule for the **UN** row requires synthesis from pyruvate, which has no terminal carboxyl, the second-base rules for **NG**, **NC**, and **NA** are undefined in this row and we treat them as making no assignments.[28] In some cases, such as **UA**, the standard code in fact has no simple acid assigned; the **UAR** positions are used as stop codons and the **UAY** positions are occupied by the complex acid Tyr. The second-base rule for **NU**

---

[27] Haig and Hurst [327] note that it is the second base that determines the polar requirement of amino acids to a strong first approximation.

[28] We will later motivate many of these rules as reflecting the *presence* of biosynthetic pathway sequences. From this perspective it is natural that rules which cannot be applied simply leave codon positions open to be filled in other ways.

Table 5.2 *Rules associating side-chain pathways with second-codon positions*

| Rule | Context | Description | Admits | Violates |
|------|---------|-------------|--------|----------|
| 2a | **NG** | Aminate and reduce the terminal carboxyl or dead end | Arg, Dab | Ser |
| 2b | **NC** | Reduce the terminal carboxyl or dead end | Pro$^c$, Thr | |
| 2c | **NA** | Aminate the terminal carboxyl or dead end | Gln, Asn | *His, Lys* |
| 2d | **NU** | Condense a pyruvate moiety with the primary backbone and isomerize with acetohydroxy acid isomeroreductase, as shown in blue in Figure 4.8 | Val, Ile, Leu, AMC | *Phe, Met* |

Pro$^c$ refers to the step to glutamate semialdehyde, which spontaneously cyclizes to proline. Rule 2b produces homoserine, which is isomerized to threonine, possibly as a later modification. The application of rule 2a to Arg is through its biosynthetic precursor ornithine. Condensation of pyruvate with $\alpha$-ketoglutarate homologous to acetolactate formation would produce 2-acetyl-2-oxoglutarate. The corresponding keto-isomerization would produce 3-methyl-2-oxo-5-carboxypentanoate, and the amino acid would be 2-amino-3-methyl-5-carboxypentanoate (AMC), which is not a biological amino acid.

is compatible with the **GN** rule (1a) because it produces the $\alpha$-ketoacid 2-ketoisovalarate, which is reductively aminated as its terminal step.

When the physical property rules apply, they reflect *homologies among reactions within the first three steps of amino acid biosynthesis.* For most of the amino acids captured within these rules, further steps significantly modify the properties of the terminal amino acid assigned to the codon block, either by biosynthetic elaboration or by isomerization. Examples include Pro (cyclization of glutamate semialdehyde), Thr (isomerization of homoserine), and Arg (by the equivalent of dehydrating addition of urea to ornithine). For ease of recognition in the table, we have listed the terminal amino acids as "admitted" by these rules, though the rules in fact only *restrict* the biosynthetic pathways as far as the precursors. (In Figure 5.3, the terminal amino acid is listed first, followed by the name and molecular form of the precursor that exhibits the pathway homology.) For these rules particularly, we regard the correlation with early biosynthesis, which is not only an exact property but also a much stronger regularity than exists for the terminal acids, as indication that selection for any function that depends only on the currently assigned amino acids is very unlikely to be the primary causal explanation.

Like the Wong correspondence with first-codon positions, the second-base physical property rules also admit amino acids that are not part of the standard biological set and are not precursors to coded amino acids in biologically known pathways, such as diaminobutyrate (Dab, at **AGN**) and AMC (aminomethyl-carboxypentanoate, at **CUN**). In both cases where these exceptions occur, the current codon blocks are occupied by amino acids that we have labeled as potential "captures" by the Wong correspondence in Section 5.3.5.2.

Interestingly, Arg at the putative "capture" position A**G**R is compatible (via ornithine) with rule 2a, though it violates the Wong correspondence 1c. The Ser "capture" assignment at A**G**Y, does violate rule 2a, but we return to a third rule by which it sensibly fills this position in Section 5.3.5.4. It is impossible from current evidence to know whether we would be justified to speculate that diaminobutyrate and aminomethyl-carboxypentanoate were once synthesized, and perhaps assigned in an aboriginal version of the code, or in whatever chemical association preceded it.

*Exceptions to the second-base physical property rules*

All of the biosynthetically complex amino acids that occupy positions for which the second-base physical property rules make assignments, violate those assignments.[29] The acids Tyr and Trp are not listed as violations because they occupy the U**N** row where the rule 2 system cannot propose assignments. This was perhaps to be expected, since when these rules apply, they govern the earliest biosynthetic steps, while the complex acids are distinguished by their late or parallel pathway segments.

Lys at AA**R** violates rule 2c as a specification of a joint reaction of amination coupled to ATP hydrolysis. Yet even here the first step in lysine biosynthesis from aspartate is activation from ATP to form aspartyl phosphate, perhaps mirroring the synthesis of asparagine with either ammonia or glutamine as amine donor, which involves hydrolysis of one ATP to AMP and pyrophosphate. Apart from the presence of two additional aliphatic carbon atoms in its carbon backbone, which are added through a seemingly Byzantine synthetic pathway, lysine is equivalent in structure to diaminobutyrate which would be naturally assigned at A**G**N by both the rule 1 and the rule 2 systems. Given its complexity and the way it partly but not fully respects multiple rules that are otherwise quite regular, we regard the Lys assignment as a plausible outcome of either coevolution of synthetic enzymes with aminoacyl-tRNA synthetases, or selection for property similarity from the coding function, by robustness arguments of the kind we review in Section 5.3.7.4 below.

### 5.3.5.4 GTP and correlation with the reductive glycine/serine synthetic pathway on folates

A third rule, given in Table 5.3, connects patterns in the code to the quasi-freestanding reductive pathway to glycine and serine biosynthesis, which was discussed at length in Section 4.4.3.2 as an outcome of reconstruction of early carbon fixation pathways. The association is between second-position **G** in codons and the biosynthetic origin of THF in GTP biosynthesis. This is an *alternative* regularity to rule 2a, conditioned on the same context but specifying a different and also alternative biosynthetic pattern. Whereas rule 2a accounts for only one actual assignment (Arg at C**G**N and its adjacent "capture" at A**G**R), this third rule incorporates *all* other assignments in column N**G** of the standard code.[30]

---

[29] Here we refer specifically to our rule for side-chain modification. Although that rule correlates with hydrophobicity, it is not the only mechanism to produce polarizabilities within a given range. Notably, many of the complex amino acids are quite close matches to their neighbors in the polar requirement, as noted by Carl Woese.

[30] This regularity was first observed by Rogier Braakman.

Table 5.3 *Correlation of NG codons with amino acids synthesized reductively via glycine or serine on the THF pathway*

| Rule | Context | Description | Admits | Violates |
|------|---------|-------------|--------|----------|
| 3 | NG | Synthesize from a backbone in the reductive glycine/serine pathway from folates | Gly, Ser$^r$, Cys, *Trp* | Arg |

Ser$^r$ refers to this pathway to serine, to contrast with the placement of Ser$^o$ in the UN pathway.

Rule 3 is similar to the mechanistic interpretation by Copley *et al.* [158] of the Wong first-base rules (Table 5.1) and the physical property second-base rules (Table 5.2). The proposed relation is based on catalysis of the biosynthetic pathway by a molecule (THF) related to the codon (**G**) at which that amino acid is currently assigned. The glycine/serine association is less comprehensive than the previous two rules, but accounts for many assignments in the first column that appeared as outliers with the previous two rules, most glaringly serine at **AG** which violates both rule 1c and rule 2a.

Because this rule concerns one base only, it again admits a group of amino acids without further restricting their assignments. In the case of assignments such as glycine at **GG**, this ambiguity may be reduced by looking for compatibility of an assignment with more than one rule, but because reductive synthesis in modern organisms is an alternative to transamination of glyoxylate (an offshoot of the TCA cycle), the status of such redundancies is difficult to interpret.

A weak "echo" of rule 3 may be the assignment of His at **CA**, in the sense that histidine is associated with ATP through reactions of cyclohydrolases in the same family as those that associate folate biosynthesis with GTP. The functional interpretation is not the same in the two cases: the **NG** rule 3 suggests a catalytic function related to **G**, like the property rules 2, whereas the **NA** assignment of His reflects its biosynthesis *from* an ATP precursor. However, this is the only one of the rules that may apply to His jointly with other amino acids; apart from this, His appears to be an outlier among the amino acids both biosynthetically and in assignment.

Rule 3 captures Trp at **NGR** in a way that rules invoking TCA precursors to biosynthesis do not. By the Wong correspondence, the branched-chain acids Leu and Val, and the aromatics Tyr and Phe, which are synthesized directly from pyruvate, are grouped together with Ser and Trp (which is synthesized from serine), because Ser$^o$ is a pathway from pyruvate. Rule 3 distinguishes Ser$^r$ as an independent construct unconnected to pyruvate, and synthesized directly as an $\alpha$-amino acid without a precursor $\alpha$-ketoacid. It therefore excludes both the branched-chain and the aromatic amino acids but retains tryptophan.

If this rule is to be interpreted as a reflection of an early association between biosynthetic pathways and small-RNA systems, it must have reflected a very early association. Surely the rTCA cycle or closely related pathways must have existed and provided alternative routes to glycine and serine, by any era in which an amino acid/RNA association was forming which anticipated translation and the code. It seems plausible (on the grounds that both aldol cleavages and reductive transaminations are facile chemistry for which enzymes exist elsewhere in core carbon fixation and elementary amino acid biosynthesis), that the transamination pathway from glyoxylate also existed in this era, though we find no direct support for this from the phylogenetic reconstructions reported in Chapter 4 (Section 4.4). The partial complementarity and partial overlap of Rule 3 with the TCA denominated rules 1 and 2 in the code seems to recapitulate – almost as a kind of analogy – the complementary but compatible relation of the folate-$C_1$ and rTCA carbon fixation pathways in core metabolism.

### 5.3.6 Rule combinations

Each of the preceding rules, taken alone, only associates a set of amino acids with a set of codons. Because each rule treats a particular base as a "letter" that contributes a distinct meaning to the codon "word" in which it appears, combinations of rules can act together to produce much more specific restrictions. Set theoretically, these are simply intersections, but because the rules themselves refer to biosynthetic steps which also have semantic content in the context of pathways, we will distinguish three kinds of rule combinations, which suggest different relations of the code to metabolic order.

**Intersections** combine assignment rules that are biosynthetically independent, or which respect the decomposition of biosynthetic pathways because the molecular structure of intermediates dictates the order in which the rules apply. They narrow the possible assignments to codons (often to unique amino acids), but they do so by building up a sequential description of the synthetic pathways. The rule intersections are the patterns in the code that we have characterized as being like an "instruction set" for amino acid biosynthesis.

**Redundancies** are defined as positions where a coded acid satisfies more than one rule, but in which the rules refer to alternative pathways. Redundancies could plausibly reflect either multiple reasons for an early chemical association to have existed between amino acids and RNA dimers, or flexibility that allowed an early code to accommodate multiple biosynthetic chemistries. They are not naturally interpreted as "instruction sets," however, because the rules do not act sequentially to assemble pathways.

**Property similarities** are patterns that can be applied to terminal amino acids which are specifically *not* consequences of pathway homology. They are the most unambiguous candidates for patterns that were selected into the code to enhance error buffering, which derive from the function of translation itself and the properties of the currently assigned amino acids.

Although this approach to combining rules respects their associated pathway-semantics, we emphasize that still it is *not a mechanistic explanation* for codon assignments. Hypotheses about mechanisms involve many more assumptions both about what is chemically possible and about what was historically relevant, and they are much more likely to be wrong. Independently of whether any particular mechanistic explanation appears to be plausible (or whether any good candidate mechanisms are even known), the rules and rule combinations in this section provide a *compressed description* of the amino acid assignments in the code – compressed both in *where* they are and in *what* they are – and thereby characterize the structure of its non-randomness in relation to biosynthesis.

### 5.3.6.1 Intersections

The most informative rule combinations, which by themselves come close to giving unique amino acid assignments for a large fraction of codon positions, are intersections of the wobble-base rule for preferring simple acids, and the extended Wong correspondence (1) and physical property rules (2). Recall that the wobble-base rule provides a rationale for associating a biosynthetically simple amino acid with every codon according to its first two bases; in blocks where a complex acid is assigned, it may secondarily displace the simple acid at either the **NNR** or **NNY** codons, but not at both. Complex acid assignments are always secondary to distinctions in the first two positions, and so respect codon blocks defined by the first two bases. Usually they also respect the Wong correspondence.

Here is a list of rule intersections that greatly restrict or uniquely identify the simple amino acids in most blocks ($\wedge$ stands for logical-and: take the set intersection of admitted acids from the appropriate rows in Table 5.1 and Table 5.2).

**1a $\wedge$ 2d GN $\wedge$ NU**: Val, Leu. Correct for Val at codons **GUN**; unable to exclude Leu (which is not coded in this block).

**1b $\wedge$ 2a CN $\wedge$ NG**: Arg (via ornithine). Unique and correct for all four codons **CGN**.

**1b $\wedge$ 2b CN $\wedge$ NC**: Pro$^c$. Unique and correct for all four codons **CCN**.

**1b $\wedge$ 2c CN $\wedge$ NA**: Gln. Unique and correct at codons **CA**Y; complex His at **AAR** is consistent with the wobble-base restriction but not with either of the other block rules.

**1c $\wedge$ 2b AN $\wedge$ NC**: Thr (via homoserine). Unique and correct for all four codons **ACN**.

**1c $\wedge$ 2c CN $\wedge$ NA**: Asn. Unique and correct at codons **AA**Y; complex Lys at **AAR** is consistent with the wobble-base restriction and the Wong correspondence.

**1c $\wedge$ 2d AN $\wedge$ NU**: Ile. Unique and correct at three codons {**AU**Y, **AUA**}. Met at **AUG** is more restrictive than the wobble-base restriction and satisfies the Wong correspondence.[31]

**1d $\wedge$ 2d UN $\wedge$ NU**: Val, Leu. Correct for Leu at codons **UUR**; unable to exclude Val (which is not coded in this block). Phe at **UUY** satisfies the wobble-base rule and the Wong correspondence.

Several observations or comments are raised by the rule intersections.

---

[31] In the mitochondrial code and in some bacterial codes, Met is coded at both **AUR** codons, and so follows the simple wobble-base rule.

**Partitioning ambiguous assignments**   The very simple early step rules 1a and 1d, and
the pyruvate-condensation rule 2d, apply equally to Val and Leu, as almost any simple
rules seemingly must, since the biosynthetic pathways of these two branched-chain amino
acids (shown in Figure 4.8) are the same until quite late stages. While this leaves us unable
to choose between Leu and Val in any rule intersection, we may still expect that they
will be assigned homogeneously to two *different* blocks, as indeed they are. Block-wise
uniform codon assignments would follow either from early associations of pathways with
dinucleotides, or from selections for robustness in which the wobble-base was the most
frequent source of tRNA/mRNA mismatch errors. The rule intersections therefore leave
these two *groups of* assignments ambiguous only up to one binary choice.

Other similar partitionings could be invoked to separate Ser and Cys, which are unassigned within the rule 2 system, and if we can suggest a breaking of this ambiguity for
Ser, as in Section 5.3.6.3 below, partitioning may then explain the position of Cys by
exclusion.

**The rule 1a ambiguities**   Because rules 2a–2c cannot apply in the context of rule 1a, these
rules leave the relative assignments of Ala, Asp, and Glu unspecified. If Val is given at **GU**,
the others may be restricted to the remaining three **GN** blocks by partitioning. We will note
in Section 5.3.6.2 below that Gly at **GG** may also be preferred on the grounds of a rule
redundancy. Nonetheless, rule 1a (reductive transamination, then stop) is very different in
character from the Wong carbon-backbone rules. It seems to associate biological amino
acids with one-step departures from the CHO world, but as a result it is incommensurate
with any other rules that stipulate multiple steps beyond amination. Asp (at **GA**Y) and Glu
(at **GA**R) are both simple acids, distinguished by a wobble-base, so that neither is preferentially associated with **GA**. The complexity of keto-isovalarate synthesis, its comparative
remoteness from the TCA precursor network, and the distinctiveness of the TPP-dependent
acetohydroxy acid isomeroreductase responsible for the key condensation in Figure 4.8, are
all overlooked in regarding valine at **GUN** as "just another" one-step transamination of an
$\alpha$-ketoacid. The unusual degree of ambiguity left in **GNN** codon assignments by rule 1a
suggests that they contain further information about early stages in formation of the code
which we have not recognized.

**Stretching the rules for complex amino acids**   It goes beyond the legitimate use of
the rules as we have stated them – as rules applying to primary backbone synthesis –
to note that 1d $\wedge$ 3 is a non-conflicting rule intersection that singles out Trp at **UGG**.[32]
The chorismate component is synthesized from PEP, while Trp as noted differs from the
aromatic amino acids Phe and Tyr in being synthesized from serine rather than from pyruvate (thus respecting the rule 3 association of **NGN** codons with Ser$^r$). Trp at **UGG** thus
has some characteristics of a rule redundancy. Its association with Ser$^r$ is unique under
rule 3, but it also falls under the Wong correspondence if the backbone is interpreted as

---

[32] As for Met, this extends to the full wobble-base set **UGR** in the mitochondrial code.

Ser$^o$ synthesized from pyruvate. Alternatively, the association of the major row **UNN** with pyruvate could also apply through the secondary synthesis of chorismate (originating in erythrose-4-phosphate in the pentose phosphate network derived from PEP) rather than to the primary backbone.

**The recurrent association of UTP with pyruvate** Rules 1d and 2d seem to indicate an association of UTP with pyruvate that spans many contexts. Pyruvate is the precursor to the primary backbone in rule 1d, and the source of the branched chain in rule 2d. As noted, it is also the precursor to chorismate in the aromatics and Trp. This regularity extends beyond the code, as UTP rather than ATP is the primary activating nucleoside-triphosphate in polysaccharide synthesis.[33] The entire sugar phosphate network, as noted in Figure 4.5, is a branch from the central network of carbon fixation via pyruvate. We do not know how to express this association as a biosynthetic rule, but it appears to be a strong enough co-occurrence within metabolism to be regarded as more than a result of observer bias.

**Wobble-base ambiguities for complex amino acids** The wobble-base distinction and the Wong rule are generally compatible with the assignment of the complex amino acids, but do not otherwise predict them specifically, either with respect to columns or with respect to which of R/Y will receive the complex acid. In general we do not expect metabolic predictions for coding of complex amino acids to be as strong as those for simple acids, on the grounds that the simple acids were more plausibly part of an organosynthetic gemisch[34] imprinted on a nascent code, whereas the complex acids could have been post-coding additions that enhanced the functionality of the polypeptides formed by translation.

### 5.3.6.2 Redundancies

Some amino acids satisfy more than one pattern, via pathways that could be alternatives rather than complementary steps within a single synthesis. Here are two such redundancies.

**1a ∧ 3 GN ∧ NG**: Gly. Unique and correct for all four codons **GGN**. Gly$^r$ is the reductive synthesis, while Gly$^t$ is a standard transamination from glyoxylate off the TCA cycle.

**1d ∧ 3 UN ∧ NG**: Ser, Cys, Trp, via either Ser$^r$ or Ser$^o$. Correct for Cys at **UG**Y and Trp at **UG**R; unable to exclude Ser (which is not coded in this block). This redundancy is all the more enigmatic, as 1d ∧ 3 could be an intersection instead of a redundancy if chorismate synthesis is considered informative, and because Ser$^o$ is adjacent at **UC**N and Ser$^r$ is adjacent at **AG**Y.

---

[33] That is: glucose is activated to UDP-glucose for assembly into polysaccharides, whereas most other monomers activated with nucleosides are adenlyated.

[34] German for "mixture." We adopt this term from Christian de Duve, who used it to refer to early unsupervised geochemical organosynthesis.

### 5.3.6.3 Property similarities not reflecting biosynthetic homology

Most intersections, because they define strict repetitions of biosynthetic steps as is seen in Figure 5.3, also produce amino acids with similar properties. For this reason imputations of cause based on pathway-RNA connections, and alternative explanations based on coding robustness against mutations or tRNA/mRNA mismatches, are inherently confounded for many code regularities. In this section we mention property similarities that can be clearly identified because they only appear as a result of late synthetic steps within pathways that do not initially produce similar forms. If these unambiguous property regularities applied to the *terminal* amino acids at all positions, they would be the best candidates for explanation in terms of error buffering by translation. The interpretation of property similarities is complicated, however, by the fact that some describe relations between pairs of terminal amino acids, while others better describe relations of a terminal amino acid at one position to a biosynthetic precursor at another.

**1d ∧ 2b**  **UN** ∧ **NC**: Ser$^o$. Unique and correct for all four codons **UCN**. Recall that **UN** is a row for which rules 2a–2c do not define outputs, because pyruvate lacks a terminal carboxyl that can be modified by reduction or amination (or both). Therefore the combination 1d ∧ 2b as an *intersection* leaves block **UC** unassigned. However, Ser shares the form of an $\alpha$-aminoacid which is aliphatic except for a terminal hydroxyl, with the two precursors homoserine (at **ACN**) and glutamate semialdehyde (at **CCN**), which are valid instantiations of rule 2b. Whereas this hydroxyl is obtained directly by reduction of a terminal carboxyl group in homoserine and glutamate semialdehyde, it is obtained indirectly via a phosphoenolate in the Ser$^o$ oxidative pathway.

This "property similarity" is not an elementary candidate for explanation in terms of coding robustness, because the amino acids to which it refers (Thr and Pro) have both undergone secondary modifications from the precursors that express the pattern.

A second very marginal property similarity may be the aminated-aliphatic amino acid lysine at AAR, with our putative assignment of diaminobutyrate at AGN. This similarity, however, is far from being expressed in any of the rules we have given, and we mention it only in passing.

### 5.3.7 Accounting for order

Compared to the striking empirical observation that a small number of simply stated rules can remove nearly all the permutation-degeneracy from what could have been a random code, and possibly restrict key early steps in amino acid biosynthesis at the same time, our ability to interpret this regularity will be limited. The explanations that are easiest to formalize and in which it is easiest to have confidence involve physical property similarities among amino acids at adjacent codons, and the degree to which these would be lost if the codon assignments were shuffled. Such explanations, however, only apply once the function of translation is well established, and they invoke only coarse measures of similarity. For the much more detailed and explicit rules that invoke metabolic order, and which we will argue must have acted before the translation system had attained its modern form, we can classify interactions and suggest a few scenarios, but not much more.

### 5.3.7.1 The earliest translation and a context in statistical proteins

In modern organisms we take it as a starting assumption that proteins are specified by their amino acid and nucleotide sequences, and define fitness for those sequences in terms of relative protein functions. For the evolution that shaped the genetic code, however, this is not likely to be the correct starting premise.

Carl Woese argued over a span of four decades [887, 892] that both the emergence of the genetic code, and the shift between an era of horizontal gene transfer and the era of modern phylogeny must be understood together in the evolutionary context of the refinement of ribosomal translation of proteins. Through this transition, gene transfer and unreliable translation[35] were synergistic in the early stages, while inhibition of gene transfer and reliable translation were synergistic in the late stages. Moreover, gene transfer was as essential to the kinds of evolutionary innovation that fixed the code in the early stages, as it was deleterious to the further refinement of translation in the late stages.

The emergence of the ribosome as a translation apparatus could not have been driven by the functionality of proteins (since that did not yet exist), but must have been an exaptation[36] of some other process that was under selection for a different function. Any early translation-like use of RNAs must have been extremely error prone. Because translation to proteins and protein-mediated replication of genes are mutually interdependent processes, they inherently possess *self-amplifying feedback:* large error rates in either process militate against the assembly of systems capable of reducing the error rates. The only robust notion of phenotype in such a world would have been one that was not associated with particular sequences, but with typical properties in distributions of sequences (Woese termed these "statistical proteins"), both of polypeptides and of polynucleotides. This was the statistical world in which genes did not yet form groups identified with particular species, and in which additive fitness at the gene level was a good approximation.[37]

Codon shuffling provided a way to improve the reliability of translation of statistical proteins without requiring the prior availability of reliably translated proteins in the replication, transcription, and translation systems. This would have broken the chicken-egg dependency of a reliable translation apparatus on having reliable components, and would

---

[35] Reliability of genome replication and transcription are also relevant. Though governed by different molecular systems and perhaps undergoing innovation in different periods, they would have been part of the broad interdependent transition from more stochastic to more deterministic dynamics of polynucleotides and polypeptides.

[36] The term was introduced by Stephen Jay Gould and Elizabeth Vrba [313] to refer to traits that are under ongoing selective pressure, but not always with respect to the same function. Relations of this kind play a central role in evolutionary innovation.

[37] This is equivalent to saying that the phenotypic effects of genes, while they would generally depend on interactions within an individual, would be averaged over many contexts in the population. The distribution of fitness effects that would determine the propagation of genes would thus have factored into a product of gene-level distributions. Factorization defines the relevant notion of absence of "long-range" order, and produces additive fitness in the Price equation.

have begun the bootstrapping process of refining the translation system. Selection in a statistical world where fitness can be evaluated at the level of the gene is short ranged and "thermodynamic" in character. We review in Box 8.3 (in Chapter 8) a formal model showing how this domain of selection leads to robust convergence on a unique, optimal genetic code [833].

Only the kind of canalization[38] that could protect phenotype in a world of statistical proteins could have allowed for codon change. However, the more reliable translation bootstrapped into existence by an error-buffering code created a Wittgenstein's ladder [886]: having climbed up on the robustness conferred by statistical proteins, species were required to throw it away. More reliable translation enabled more precise translation and more precise DNA/RNA replication. Fitness would accrue to organisms that used this greater precision to make finer gene distinctions and carry more information in genomes. Relying on precision in turn shifts the level at which functionality is selected *away from the protein distribution and toward particular protein sequences*. Shuffling of the code becomes impossible, and – crucially – even mixing components of the ribosome from lineages that have drifted apart becomes detrimental.

The evolution of more accurate replication and translation systems involves the satisfaction of large numbers of simultaneous constraints on many interacting components.[39] Under conditions of even modest drift among separated populations, components of refined translation systems become mutually incompatible. Horizontal exchange of incompatible ribosomal RNA and peptides is actively deleterious because it disables the cell's own metabolically costly products. Inhibiting horizontal transfer shifts evolutionary pressure onto the refinement of intermolecular relations that were previously established at the communal level in a statistical world. Thus the refinement of translation both enabled, and required, a shift in the modularity of the genetic system away from horizontally transferred individual genes and statistical proteins, and toward integrally coevolved genes and (nearly) deterministic proteins.

This transition in the level and character of selection, to which we return in Section 8.3.4.4, is expressed in the emergence of phylogenetically resolvable lineages, first in components of the ribosome and the proteins involved in transcription, and later in other cellular systems as these come to be integrated in more detail with core processes. Woese coined the term *progenote* to refer to the earlier, genetically communal stage of cellular life. The transition between the eras of mostly horizontal and mostly vertical exchange he termed the *Darwinian Threshold*.

---

[38] *Canalization* is a term introduced by Conrad Waddington [846], to refer to evolution of genotypes, not to change phenotype, but to make the generation of phenotype robust against uncertainty or error either from mutation or from environmental factors. The early examples concerned developmental regulatory systems that could have evolved in response to environmental stimuli, but in which regulation was later transferred to internal controls to make them robust or anticipatory with respect to environmental variation. A key insight is that redundancy evolved to resist environmental perturbation can also confer robustness against genetic variation, analogous to the dual role of statistical translation and variation in the genetic code considered here.

[39] Such networks of constraints within an *intramolecular* sequence are characteristic of the compact folding domain of any globular protein. In Section 6.8.1.3 we review evidence that such constraints affect *intermolecular* sequences in the interaction of peptides and RNA in the ribosomal core.

The routes to reliable translation from unreliable components may have been very limited, and especially in the early phases before codon-shuffling had conferred what robustness it could, other scaffolds from the organization of the chemical substrate may have played a crucial role. Woese originally proposed the stereochemical hypothesis as one such scaffold [895]. We will suggest that even stronger and more specific associations (although perhaps of different origin) are suggested by the foregoing rules for codon assignments.

### 5.3.7.2 *Possible mechanistic interpretations of pathway order*

A translation system that formed through progressive refinement of statistical associations could display regularities originating from many sources. It might reflect structure from the original chemical scaffolding, either because that structure became "frozen in" to an accreting hierarchy of dependencies as the translation system took form, or because it led to a system that continued to be well matched to the chemical inputs on which it acts. Regularities of this kind might not reflect the obvious action of selection in a Darwinian era, but they could readily reflect bulk-process kinetic selection of the kind we have invoked to account for the earliest stages of metabolism. A translation system also might display regularities reflecting the modern function it has evolved to serve. Abstractly, this function is to couple a highly constrained organosynthetic world to a neutral world of sequence combinatorics. Regularities of the latter kind might well be best explained with models of Darwinian competition, if not among genomically integrated individuals, at least among variant genes for components of the translation system itself.

It is natural to suppose that regularities of a biochemical nature would be imprinted on the system in its earliest phases, to be later augmented or perhaps replaced by regularities associated with the modern function of coding. The important questions are to what extent the two processes lead to the same structure, where they differ, and in cases where they differ, whether remnants of the earliest forms of regularity could survive later evolution.

We have argued in Chapter 4 in favor of a relatively early role for cofactors as participants in the elaboration of metabolic complexity. It is not clear when, in the sequence of coevolution of cofactors and metabolic pathways, components that would later be used in translation began to form. The qualitative distinction between the assignment patterns of simple, sulfur, and complex amino acids could be interpreted as evidence that only some amino acids were synthesized in eras when dinucleotides, and then a reading frame, and finally a full translation system were forming. Similar limitations might have applied to the components of the translation system itself. Chiral selectivity of monomers, regioselectivity during ligation, and perhaps even the phosphate potential to form large polymers at all, could have forced components to be introduced in a particular order. At a higher level of aggregation, limitations in the availability of peptides to contribute to ribosomal function and reliability may have led to a bootstrapping sequence of classes of proteins, with the establishment of a crude class required to raise the performance of an early ribosome to the point where a more refined class could be translated.

## *The coevolutionary hypothesis for rule-set 1*

Wong in [901] originally interpreted the first-base correspondence that gives our rules 1b–1d in terms of **coevolution** of the enzymes catalyzing amino acid biosynthetic pathways with the aminoacyl-tRNA synthetases (aaRSs). Some version of this argument may be valid over a limited scope, but even current evidence shows that the relations of aaRS divergence and amino acid incorporation are diverse,[40] and it is unlikely we will be able to piece together even the history of that coevolution, much less its causality, from the extant enzymes alone. The comparative analysis by Woese *et al.* in [898], based on length, alignment, and clade-specific innovations, suggests a complicated history of descent and replacement for known aaRS. They relate maximum-likelihood trees[41] for each aaRS homology class to all others and to the phylogenetic tree for 16S rRNA – Woese terms the latter the *canonical tree* – which provides a proxy for the consensus tree of descent of many large and ancient protein families. While, for most aaRS families, some congruence with the canonical tree is seen, about half of the families also show large and idiosyncratic departures. The patterns of clade-specific innovations are not merely excess sequence overlaps; often they include specific inserted domains with well-conserved distinctive sequences, or other large-scale rearrangements.

The most striking feature of the phylogenies of different aaRS classes is that their divergences from the canonical tree fall on clearly resolved branches that by any account must be much later than the divergences of major bacterial, archaeal, and even eukaryotic lineages, and much later than the establishment of the universal code. The most striking case is that of the glutamine and asparagine RNA synthetases. The asparagine aaRS clusters within the archaeal/eukaryotic group, and the glutamine aaRS solidly within the eukaryotic branch. Both are derived from aspartate and glutamate aaRSs (respectively), which are ancient.[42]

Woese *et al.* conclude that evolution of the aaRSs that we now observe cannot have substantially determined the amino acid inventory or the standard and universal assignments in the genetic code, and that they must reflect waves of horizontal transfer that were made

---

[40] Recent excellent work by Gregory Fournier and colleagues provides windows on pre-LUCA divergence of aaRSs and amino acid incorporation for a few of the more complex amino acids. For the branched-chain amino acids, the authors argue [266] that their incorporation must have preceded the divergence of the current specific aaRSs, while for tyrosine and tryptophan, a specific order of incorporation must have followed the divergence of the aaRSs and neofunctionalization of TrpRS [265].

[41] The method used by Woese *et al.* employed a combination of parsimony and likelihood rather than an explicit forward model. Block-substitution models for alignments were used to produce the 1000 most parsimonious trees, to which amino acid substitution models were then used to assign likelihoods for observed sequences. The maximum-likelihood tree from these 1000 samples was then reported for each case. A similar analysis by Wolf *et al.* [900] which extends to subunit comparisons employs neighbor-joining and Fitch–Margloiash tree-building methods.

[42] Many detailed observations about mechanisms and groupings of innovations that lie behind the conclusions of Woese *et al.* [898] are biologically interesting and may contain clues about the biochemical context in which the innovations occurred, but they are too numerous for the scope of our treatment. Examples include the following. Differences of functioning between Class-I and Class-II aaRS: ligation by Class-I to the 2′ hydroxyl and by Class-II to the 3′ hydroxyl of tRNA. The existence of several groups of mechanistically distinct synthases. The "gemini" group (Cys, Ser, Lys, Gly, Asn, Gln) for which two unrelated charging systems exist; for Gln and Asn, one method is direct attachment of the fully synthesized amino acid, and the other is alteration of glutamic or aspartic acids already attached to the Gln or Asn tRNAs, respectively. Gly, Ser, and Lys have two normal aaRS charging systems, but the relation of the two charging systems for the same acid is not closer than the relations of either to the aaRS for other acids.

We find the case of Gly and Ser interesting because, as reviewed in Section 4.4.3, these two acids have independent and apparently ancient biosynthetic pathways. For SerRS, the minor variant is apparently of archaeal origin while the dominant variant is of bacterial origin.

possible by the universality of a code that was already established at the time the sweeps of replacement occurred. The two parts of the argument are mutually interdependent. Since the code is more universal than any branching in the canonical tree or any of the aaRS trees, there is no indication that innovation in the aaRS sector that we now know added to or modified the code. If it had done so, the patterns of aaRS innovation should at least somewhere be accompanied by patterns in code innovation, but apart from a few special amino acids we do not see the latter. If we conclude that a standard code was established, it then also becomes unsurprising that aaRSs should have undergone frequent and widespread waves of horizontal gene transfer relatively late in the deep tree of life. Against the background of a universal code, their function becomes among the most modular and isolated from other details of cell molecular biology. If the amino acid inventory and tRNAs are given, any aaRS that provides specific charging can be functional in any cell. If charging rates are comparable, their exchange may even be roughly neutral. Cells have no strong reason to become refractory to exchange of aaRSs.

This conclusion attributes the retention of the standard code to the refinement of the translation system that shifts phenotype from being specified at the level of distributions of proteins onto being specified by individual sequences, and makes most code changes lethal. It is the same refinement of translation that leads to inhibition of horizontal transfer of components of the ribosome, as discussed in Section 5.3.7.1. An important question is whether the coevolution of tRNAs and possibly early aaRSs or their predecessors took place (1) jointly with the elaboration of metabolism, (2) as causes of that elaboration, or (3) as a system in which the fitness function was defined by its ability to make use of regularities already present in metabolism.

### Biosynthetic roles for precursors to coding RNA

A very different possibility is that the chemistry that makes the modern translation system possible could have participated in the formation of the amino acid biosynthetic pathways. The ribozymes that led to the study of RNA catalysis [318, 458] are relatively large oligomers synthesized by sophisticated organisms. While many model systems for RNA catalysis are still large enough that their emergence seems implausible in any geochemical setting that was not well on the way to becoming biochemistry, a trend in research has been toward the identification of catalytic function in smaller and smaller RNA sequences.[43]

Active functional groups in ribozymes include general acids or bases [402, 586], or exocyclic amines on adenine, guanine, or cytosine that (in other contexts) are known to form adducts with small organic substrates [548]. Copley *et al.* [158] have proposed that the same groups are eligible to be intermolecular catalysts even on RNA monomers or dimers, essentially a cofactor-like function for these small molecules. Not only free intermolecular catalysis, but also catalysis in bound complexes, should be considered. The platform of

---

[43] The discovery of small RNA sequences capable of folding to form binding pockets, known as *aptamers*, was an early step in this direction [132]. Recent research by Rebecca Turk and collaborators produced transacylation ribozymes with active sites only 3 nt long [813].

phosphoribosyl-pyrophosphate (PRPP), on which all nucleotides and a number of cofactors are assembled, provides three hydroxyl groups (counting the activated $1'$ OH bound to $N^9$ of purines or $N^3$ of pyrimidines), to which ester or ether bonds can be formed. Interaction of the attached groups then creates a possibility for intramolecular catalysis within the covalent assembly.

The possibility for monomer or small-polymer RNA catalysis has precedent in the increasing number of examples of small-molecule organocatalysis.[44] For instance, the tRNA itself appears to act as a catalyst in the formation of peptide bonds in the P-site of the ribosome [865]. If the $2'$ hydroxyl of the terminal adenine is replaced by either H or F, the rate of peptide bond formation is reduced by a factor of $10^6$. In two other examples of intramolecular catalysis, hydrolysis of the anilide group from succinanylic acid occurs $10^5$ times faster than hydrolysis of acetanilide due to the additional carboxyl group of succinate, and hydrolysis of tetramethyl-succinanylic acid is accelerated by a further factor of 1200 relative to succinanylic acid due to steric hindrance by the added methyl groups [355].

The proposal for catalysis (intramolecular or intermolecular) by monomer or dimer RNA is also consistent with our suggestion in Chapter 4 that the functionality of folates may have provided a source of selection favoring reaction systems that possessed cycloligation chemistry, with RNA as a by-product. If small-RNA catalysts could have supported organosynthetic reactions, this might explain their early selection, before chiral selection, oligomer ligation, and other problems had been solved that are required for RNA secondary structure and base pairing.

With our colleague Shelley Copley, we have suggested a mechanism [158] by which dinucleotides could have participated directly in the selective synthesis of different amino acids. The key observation this mechanism seeks to explain is that rules 2–3 above can be assembled to make a "decision tree" for amino acid biosynthesis, as shown in Figure 5.4. The bases at different codon positions determine which branch on the tree is taken at each successive step. The tree represents the conditional application of the set of rules, and defines the sense in which codons are converted into "instruction sequences" for amino acid biosynthesis. It is a dynamically more explicit interpretation of the rule intersections described in Section 5.3.6.1 above.

The approach of our model was to explain the branches in the tree as a reflection of alternative chemical modifications made possible by the detailed interaction of carbon backbone length, functional groups, and steric orientation of nucleobases in dinucleotides, if $\alpha$-ketoacids were covalently bound to the $2'$ OH of the $5'$ ribose in a dinucleotide. The general framework of covalent assembly and process is shown schematically in Figure 5.5.

Proposed examples of the way different nucleobases offer distinctive opportunities for distinctive biosynthetic steps are shown in the next two figures. Figure 5.6 suggests

---

[44] Well-known examples of small-molecule organocatalysis of synthetic reactions for molecules of biological relevance include early demonstrations by Sandra Pizzarello and Arthur Weber [638] that alanine and isovaline not only catalyze, but transfer chirality to, tetroses formed by aldol condensation from formaldehyde and glycolaldehyde. Weber and Pizzarello [860], Córdova *et al.* [160], Zou *et al.* [911], and others also demonstrated catalysis of aldol condensations by proline and by a variety of dipeptides and tripeptides. In particular, Barbas with numerous colleagues (see [51], 15th citation, pertaining to Class-I aldolases) demonstrated the similarity of the transition states and reaction sequence in proline-catalyzed aldol condensation to those in aldolase enzymes, which had been worked out fully 30 years earlier by Rutter [695]. More than 130 distinct reaction types now have known small-molecule organocatalysts [508].

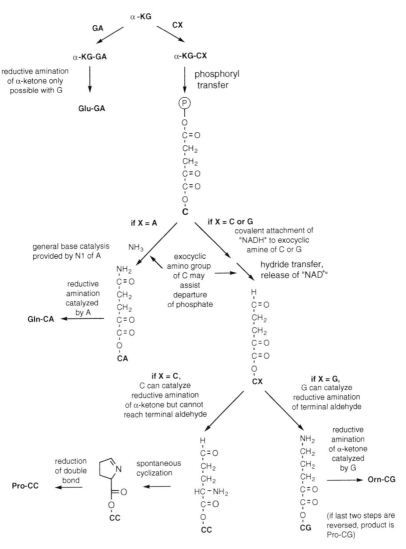

Figure 5.4 A decision tree for amino acid biosynthesis, in which different dinucleotides are associated with different branches. Branching in the tree is contingent, so that some early branches lead directly to the terminal amino acid, while others lead to further options for biosynthetic elaboration. The goal is, first, to understand how rules and rule intersections of the kind described in Section 5.3.5 might have been produced, and second, to propose more refined and more restrictive versions of these rules that might remove the few remaining ambiguities in the rule intersections presented above. (After Copley *et al.* [158].)

that only GTP possess an exocyclic amine in the correct position to facilitate reductive amination at $\alpha$-ketones. Figure 5.7 is an attempt to associate uracil (though in this case, not UMP) with the decarboxylating condensation of two molecules of pyruvate to form acetolactate, or similar reactions with homologous functional-group rearrangements.

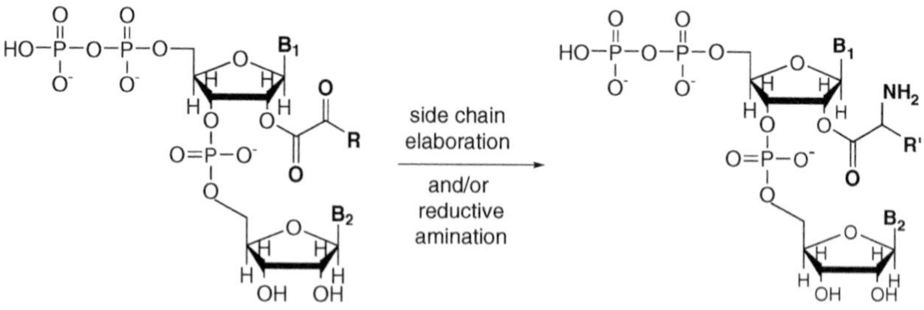

Figure 5.5 The proposed general model of catalysis of amino acid biosynthetic steps by dinucleotides. (After Copley *et al.* [158].)

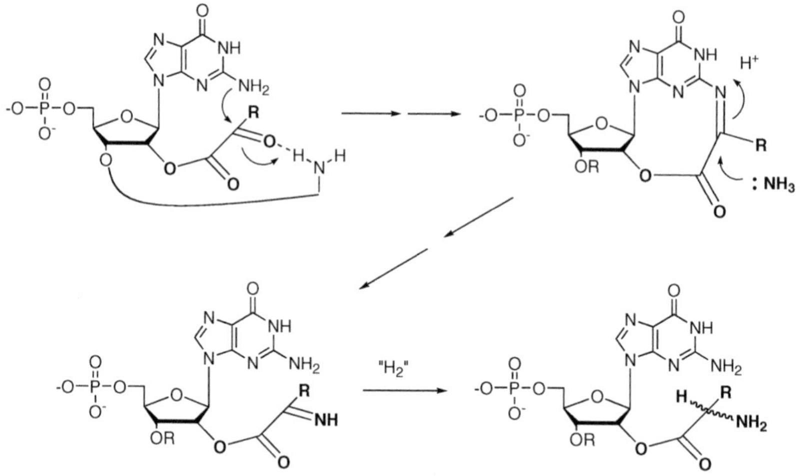

Figure 5.6 Proposed catalysis of reductive amination from ammonia, using the exocyclic amine of GTP. (After Copley *et al.* [158].)

The foregoing reaction schemes are predictions at the levels of diagrams and models, and have not been validated in laboratory experiments. Therefore we do not know if the proposed schemes can be made to work in the laboratory. Even if they can, we are far from having strong circumstantial support that they were important historically. We present them as specific, operational examples of the class of small-molecule RNA/ketoacid interactions that have the correct molecular complexity and the correct relation to pathways to parsimoniously explain the rule 1b–1c associations.

### *Physical property constraints during the evolution of translation*

A third possibility is that order could have been selected as a physical property rule during earlier stages of metabolic elaboration and a precursor to the extant translation system. In the earliest small-molecule era, the components ancestral to the translation system must have been deeply embedded in network relations with other small

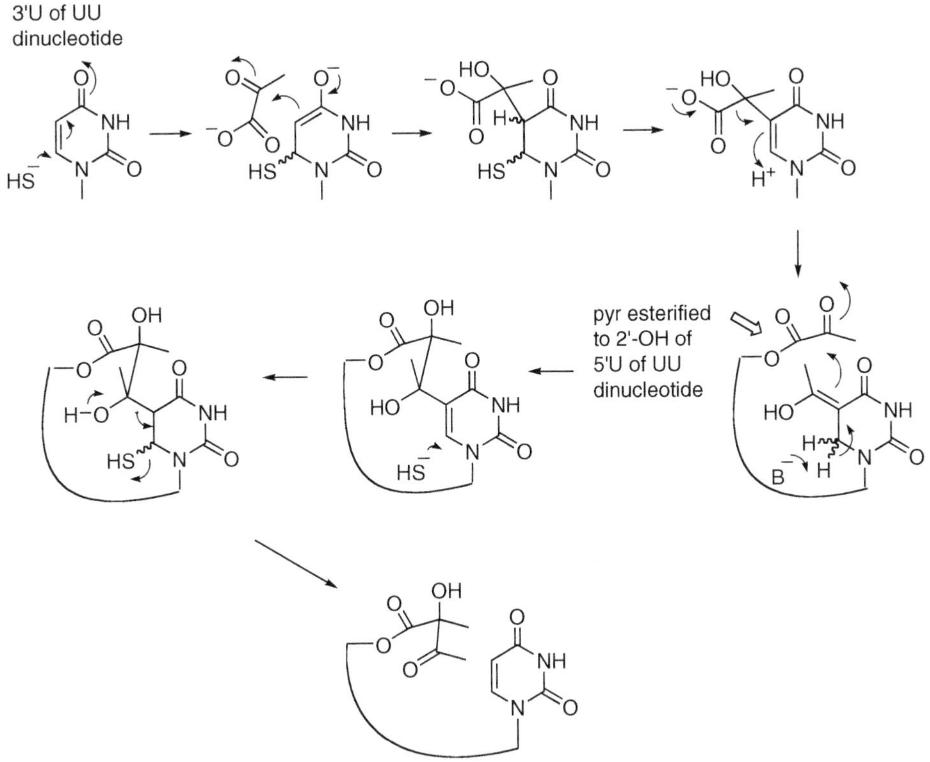

Figure 5.7 Proposed catalysis of decarboxylating condensation of two molecules of pyruvate using uracil. Note that this model does not use uridine-monophosphate. (After Copley *et al.* [158].)

molecules. In the transition phase toward the function of translation as a well-integrated system, when error rates were still very high, qualitatively different kinds of order could have been imprinted on the code by selection for its function as a buffer of errors. These later regularities in the code need not have reflected order in its earlier chemical scaffolding, and they need not have taken the form of any rules of construction defined either from codon positions or from biosynthetic steps. The new order would have appealed only to the codon adjacency relations that determine the distribution of coding errors, and to the physical properties of the amino acids coded at the time.

Order of this kind, which buffers errors, is the most extensively quantitatively studied kind of regularity in the genetic code [108, 276, 277, 442, 833]. It has been easy to formalize, and it affects the reliability of the translation system into the present day. The mechanism by which it is refined is ordinary Darwinian evolution, and the distribution of coding errors and properties of the amino acids can be empirically calibrated. We close the section by reviewing the general properties of error buffering, and then consider a special case in which the polar requirement (introduced in Section 5.3.5.3) is reconsidered, not as a source of early chemical scaffolding, but as a later criterion for minimizing the impacts of coding error.

### 5.3.7.3 *Transcription and translation error, mutation, and the role of the code as a buffer*

Errors in translation that result from mismatch of tRNA/mRNA pairing are rare, but when they occur they create amino acid substitutions in synthesized proteins which reflect the *neighborhood structure* brought into existence by the use of a three-base code in the translation system. A related class of errors comes from single-base mutations in genes. The latter are not characteristics of the translation system and they produce systematic rather than idiosyncratic changes in proteins, but the errors they create reflect the same neighborhood structure in the code.

It is important to appreciate that the reduction of coding error rate is an outcome of intensive selection. We can see this because it is not the same in all organisms: numerous viruses and bacteria across a range of genome sizes have widely varying error rates in transcription and (in the case of bacteria, which encode their own ribosomes) in subsequent translation [131, 170]. For many of these the error rate is close to the Eigen error threshold [213] (see Section 4.2) for the genome length. Error rates seem sure to have been systematically higher in the distant past than they are in modern cells, and very high in the earliest phases of cellularization. At every stage of life we must ask [887] what structures and functions could be produced from unreliable synthetic steps, and what system architectures optimize this reliability. This conception of functions such as translation requires an understanding of the capacities *at the system level* for the order in one subsystem to buffer the consequences of errors in the same *or other* subsystems.

While an idealized translation system – without rate limits, transcription or translation errors, or mutations – could produce the same output with many different coding assignments and redundancies, a carefully selected coding assignment in the actual translation system can reduce the impact of both tRNA/mRNA mismatch errors (within the same system) or single-base mutations (in the genomic memory system), by placing amino acids with similar properties at the adjacent codon positions likely to be substituted by either class of error.

### 5.3.7.4 *The polar requirement and robustness of information flow from genomes to proteins*

The context in which the polar requirement was introduced – as a chemical reason for specificity in the early association of amino acids with tRNA before a functioning system of aminoacyl-tRNA synthetases existed – is somewhat like our proposal for direct connections between RNA and amino acid biosynthetic pathways in Figure 5.4. A quite different interpretation of the polar requirement is as a criterion of similarity of amino acids not to RNA, but *to each other*. Amino acid similarity affects the consequences of errors in translation or of mutations in genes, and the polar requirement furnishes one scalar measure of similarity under which the code can be graded as an error-buffering system. In this later, evolutionary role the polar requirement has continued to enjoy acceptance as an unusually discriminating criterion for distinctiveness of the standard genetic code.

Hydrophobicity is a major determinant of the effect different amino acids have on the kinetics of protein folding, the stability of folds, and the microenvironment created by the folded structure. Therefore it is among the most consequential physical properties for protein function. Because it affects all proteins, it should focus selection especially strongly on the translation system, more so than properties that affect only particular protein families or genes. Therefore robustness of hydrophobicity against point mutations or tRNA/mRNA single-base mismatches also offers a strong and justifiable criterion for robustness of the code *as a component within* functioning translation.

The polar requirement is not only a measure of hydrophobicity – the planar $\pi$-bonding resonance and heterocycle nitrogen which mimic properties of nucleobases may have differential affinity for several properties of amino acids – but whereas the relevance of other properties is most easily understood in terms of amino acid/RNA interactions, the correlation of polar requirement with hydrophobicity is readily interpreted in terms of a first-order consequence for folding and function in every protein. In this association, preservation of the polar requirement under single-base mis-codings is a reasonable criterion under which to test the code for optimality as an error buffer.

The standard genetic code has been recognized for decades to be an unusually good buffer against changes in polar requirement by single-base errors [276, 327].[45] Recent molecular dynamics simulations to compute more refined values for polar requirements than were available from Woese's initial paper chromatography experiments indicate both that the code is more unique than "one in a million" [108], and that this property is an unusually discriminating measure for uniqueness of the modern code.

Why the code should be such a good buffer against changes in polar requirement, and why this particular, rather complicated and even obscure property should be so unusually good to assign the standard code a high rank, is a mystery. It seems implausible that pyridine was a high-concentration solvent during the formative era of translation (so much as to dominate the impacts of all other organics that must also have been present), and pyridine is not an exact or ideal model for all properties of all RNA. It also seems unlikely that using paper chromatography in the early 1960s, Woese through genius or luck somehow identified a physical property of interactions that was singularly important to the forming genetic code. The mechanistic significance of the stereochemical hypothesis or something like it is also still an open question, and the degree to which the polar requirement "best" captures the complex mix of constraints of stereochemical affinity is also not known. Therefore we are not inclined to attach large significance to the modest excess rankings that the polar requirement gives to the standard code, versus alternative measures of hydrophobicity that would be equally relevant to the folding and function of proteins. This does not

---

[45] Steven Freeland and Laurence Hurst [276] showed that under permutations weighted according to the probabilities of different translation errors, fewer than one permutation per million produces an alternative to the standard genetic code that is a more effective buffer of the average value of amino acid polar requirement.

diminish the importance of buffering the hydrophobicity among amino acids as a function of the standard code, and a function that seems surely to have been under heavy selective pressure.

### 5.3.7.5 Why physical property criteria must be conditionally independent of pathway given the acid

The coordinated use of amino acid similarity and the neighborhood structure of three-base codons, to minimize the impact of tRNA/mRNA mismatches, may be understood as a **robustness** mechanism for the function of translation.[46] Although not the only such mechanism, mismatch buffering implies a form of conditional independence that should allow us to distinguish patterns due to evolution of the coding function in its modern form, from historical forces (of the same or different type) that acted in some context before the modern one. The argument for distinguishability may be framed as follows.

> *Any* form of selection that has optimized or that maintains the form and function of the modern code can only depend on the processes of translation themselves and on the inventory of coded amino acids. These system components and their dynamics carry all consequences from lower-level processes that survive to affect the translation system. More formally, any process of selection for properties of translation must be *conditionally independent* of lower-level properties of biochemistry, given the translation system and the terminal amino acids.

The rules 1–3 listed in Section 5.3.5 are correlated with properties of the terminal amino acids, but they are more regularly represented in many positions by early biosynthetic stages. At many positions, subsequent biosynthesis diverges from regularity to produce the terminal acids. We think such regularities can only have arisen from eras before the translation system had taken its modern form.

This is, of course, an *interpretation*, and as such subject to confounds that must be assessed. It cannot be ruled out that, in any codon position, the terminal amino acid, while less regularly related to other terminal amino acids than their mutual early biosynthetic stages were, might still be more closely related than any other amino acid would be in the same position. However, to suppose that by this argument the code can be explained entirely by selection for robustness, rather than by historical biosynthetic commitments, requires complicated assumptions about why different amino acids should not be considered, and depends on the measure of similarity adopted.

It seems to us to require fewer and weaker assumptions to suppose that the early pathway homologies appear in the code because *they were the focal properties* at the time the

---

[46] Generally we expect that robustness makes a positive contribution to fitness, as in other arguments for the emergence of canalization. However, a system-level buffer that makes a certain "wild-type" phenotype robust against small errors in the genome also shields these errors from selection and can permit them to accumulate. Canalization could therefore be advantageous or deleterious depending on the circumstances [455].

pathway and coding assignment were made. The fact that the four rules of Section 5.3.5 restrict the code *so much more* than the criterion of preservation of the polar requirement does, suggests to us that *the biochemical embedding of RNA systems in amino acid biosynthetic chemistry left a larger lasting mark on the code than evolutionary optimization in the last stages of emergence of the ribosome.* The existence of a confound expresses the essential **compatibility** of multiple sources of order. Assignments made originally for pathway homology can precondition a code to reduce the consequences of errors, and thus be preserved under Darwinian optimization.

A variant interpretation can be given, which is historical but does not include an active relation between biosynthesis *per se* and the emergence of translation. It is that the same mechanism of error buffering that operated to standardize the contemporary code also operated in earlier ages when fewer and simpler amino acids were assigned codons, but in which *a well-defined function of translation already existed.* The code would then have buffered errors by preserving the strict homologies that were expressed in shorter amino acid biosynthetic pathways. As amino acid biosynthesis was elaborated, and as the terminal acids attached by the aminoacyl-tRNA synthetases changed, the code could have incorporated these earlier pathway regularities like ancient insects trapped in amber, either through lock-in at the system level, or simply because no alternative assignment was preferable to one that simply followed the evolving pathways. The main difference between this interpretation and one with a more direct chemical involvement of RNA in amino acid biosynthesis is the degree of advancement of the translation system relative to the elaboration of biochemistry.

### 5.3.8  A proposal for three phases in the emergence of translation

We believe that the joint structure in the genetic code and the biosynthetic relations of amino acids indicates three phases in the emergence of translation. Most details in the sequence are obscure, and notions of cause for the transitions are extremely tentative. Nonetheless, a grouping of this kind accounts for regularities across systems, and situates the emergence of translation in relation to the elaboration of metabolism.

#### 5.3.8.1  Simple amino acids associated with dinucleotides in an era when the three-base frame did not exist and the ribosome was not even a ribosome, much less a translation system

In the first stage neither the ribosome, nor a three-base reading frame, and certainly nothing like a modern translation system existed. Significant metabolic order existed, producing both rTCA precursors and also branched-chain $\alpha$-ketoacids in a CHO+S network. This metabolic order made heavy use of promiscuous catalysts, and was possible only because of the high pathway homology among these organosynthetic pathways at the local-group level. RNA was generated from within this organosynthetic network, and folates were active as $C_1$-carrier cofactors in reductive pathways that include reductive glycine and serine synthesis. The rTCA cycle or some close variant, also fed by direct

$C_1$ reduction and synthesis of acetyl-thioesters, was the major source of carbon backbones, and a sugar phosphate pathway resembling gluconeogenesis and some version of the pentose phosphate network was established sufficient to produce phosphoglycerate and ribose phosphates. Reductive amination of ketones and carboxyls was possible, as was some activating step which is currently performed by phosphorylation.[47]

Dinucleotides had regular associations with $\alpha$-ketoacids and with the low-order amino acid biosynthetic steps including reductive aminations of $\alpha$-ketones and terminal carboxyl groups, and some reductions. UTP was in regular association of some kind with pyruvate. Whether RNA directly enabled biosynthesis as a small-molecule catalyst, and whether such catalysis if it occurred was through covalent associations or between unbonded molecules, we do not strongly presume. Whether subsequent stages of amino acid elaboration took place beyond those associated with dinucleotides we also do not presume. Polyphosphates (organic or inorganic) may have been present but perhaps were scarce, and this may have limited the length and complexity possible for oligomers.[48]

It is not essential that dinucleotides have been the only stable groups associated with amino acids or that they have been the only or even the principle organocatalysts. Tiny RNAs, perhaps metal-binding aptamers, and other small oligomers also must have existed in a system that had established cofactor functions and was on the threshold of generating longer RNA oligomers. However, the dinucleotide/amino acid association and the UTP/pyruvate association were the two regularities strong enough to be maintained.

### 5.3.8.2 Early emergence of an ancestor to the ribosomal small subunit as an RNA triplicase, and the emergence of a three-base reading frame

It is against this background of small-RNA/amino acid associations that ancestors to components of the ribosome most plausibly emerged. A trigger might have been increased availability of polyphosphate, whether from an inorganic source or through the earliest harnessing of redox energy via thioesters, as originally proposed by Christian de Duve [183]. The RNA ancestral to the ribosome was not a single system, but probably at least two distinct systems ancestral to the large and small subunits [85]. The ancestor to the large subunit was probably an iron-RNA condensation ribozyme, for reasons we review in Section 6.8 of Chapter 6.

As an ancestor of the small subunit, some form of RNA replicase seems the most likely proposal, and in particular, the function of adding RNA trimers via *template-directed ligation* may have been favored. The arguments for such a "triplicase" function of the small subunit (which are much more speculative than those for the catalytic function of the large subunit), are reviewed in Section 6.8.3. Template-directed ligation is a more primitive method of RNA ligation than single-nucleotide primer extension, and although it requires

---

[47] We wish to leave open the possibility that this activation was performed by thioesters.

[48] If phosphates were very scarce or absent, they may not even have been present as structural elements in cofactors in this era, although as noted this is not a severe restriction on cofactor form or function as usually they are remote from the active part of the molecule.

a source of previously synthesized fragments (which may be wasteful if not all these can be used), it depends on less sophisticated catalysts to perform the ligation.

An important question is what role peptides could (or must) have played at this stage that would have led to their association with ribosomal RNAs [85]. Equally important, if they played a role, is how specific they needed to be, and how the necessary constraints on sequences were preserved through time. There is good evidence for a very deep association of non-globular peptides with the oldest core domains of the ribosome, but much less clarity on how they could have been produced, and how their functional information was preserved in the RNA-peptide system.

### 5.3.8.3 *The sulfur amino acids coincided specifically with the advent of translation*

We will propose that the sulfur amino acids cysteine and methionine came into association with RNA concurrently with the first emergence of the ribosome as a peptide-translation apparatus.

*S*-adenosyl-methionine was likely already present in its role as a cofactor, but its association with the start codon seems to indicate an explicit regulatory function of the code. This need not have been in a system that produced peptides, but it was a system in which controlling movement along the three-base reading frame became important. The emergence of a regulatory role for methionine out of the *S*-adenosyl-homocysteine/SAM/methionine reaction cycle, possibly involving interactions of the adenylyl group with the ancestral small subunit RNA, seems reasonable to propose.

Cysteine is one of the major innovations that can be used to form and regulate tertiary structure in proteins, and also to coordinate catalytically active metal centers. Most known RNA enzymes are metalloribozymes, and all evidence about the core in the ribosome indicates that metal-mediated catalysis was also prevalent among the earliest ribozymes. Depending on the relation of organic reactions to mineral-hosting environments at this stage, RNA-metal catalysis could have been co-extant with small-molecule metal-ligand complexes and surface interactions. The transfer of metal catalysis to proteins, however, seems likely to have been associated quite specifically with the ability to incorporate cysteine in a controlled way into polypeptide sequences.

Thus, Cys contributed two major classes of functions to proteins that would have made ribosomal translation enormously more advantageous than it could have been with only simple amino acids, while Met was introduced as part of a system for regulating this translation.

### 5.3.8.4 *The complex amino acids were added for catalytic refinement in a system with an established translation function*

Many approaches to assigning a temporal sequence to the adoption of different amino acids have been taken,[49] and (not surprisingly) in almost all of these, the acids we have termed "complex" in Section 5.3.4.1 tend to be the latest additions. Their biosynthetic complexity,

---

[49] See Trifonov [812] for a review of primary literature and a summary of consensus and divergent conclusions.

their comparatively low-frequency use in most proteins, and their assignment to codons that in all cases require a third-base distinction lead us to suggest that, not only were they later additions, but they were incorporated in a system that already had an established three-base reading frame and a translation role for the ribosome.

We have noted that some complex amino acids have a resemblance in form and function to cofactors. For histidine this affinity is especially emphasized by its biosynthetic pathway, which converts an ATP-derivative in the last steps to an amino acid that can be incorporated into polypeptide chains. A similar but less striking observation may be made about tryptophan, which in some active sites provides non-local electron transport between substrates [744], using the resonant $\pi$-bonds of indole somewhat as they are used in quinones, another major class of chorismate-derived cofactors. Other similarities could be drawn between the aromatic acids Phe and Tyr, the indole acid Trp, and the imidazole acid His to nucleobases in RNA. Frequently these acids with planar groups are also used in stacking interactions with RNA. The presence of $S$-adenosyl-methionine (SAM) in the cofactor cycle $S$-adenosyl-homocysteine/SAM/methionine which is a central nexus for methyl-group exchange was part of our motivation to group methionine with the complex amino acids and cysteine with the simple acids, when making the two-base association in Figure 5.2. A parsimonious interpretation of all these facts would be that the complex amino acids were recruited by the translation system to bring into existence a new large class of organo-catalytic active sites, augmenting the catalytic sites of a previous generation of catalysts, most of which had relied on coordinated metal centers.

The improvement in reliability of translation may have occurred in stages as a kind of bootstrapping process. The extended structure of core ribosomal peptides and the predominance of amino acid/RNA interactions in the ribosomal core suggest that the earliest function of translated proteins, synthesized from simple amino acids, was not catalysis but rather stabilization of the ribosome itself. Although the three-base reading frame may have been established in an RNA triplicase ancestral to the small subunit, third-base distinctions need not have been reliable enough to support the kind of refined protein synthesis that places metabolically costly complex amino acids at particular places in enzyme active sites. Only once first-generation peptides of simple amino acids had made the ribosome more reliable, and also made the wobble-base codon position more useful, did it become possible to synthesize a new class of sophisticated organocatalysts by incorporating complex amino acids. Aromatic amino acids play a significant role in interactions with RNA in modern ribosomal proteins, and they could have been the first generation to be added, contributing to a self-reinforcing cycle of improved reliability from which their own use and that of Trp and His benefited.

## 5.4 The essential role of bioenergetics in both emergence and control

Order in the biosphere, as in any non-equilibrium system, arises out of the relation between the space of possible processes and the asymmetric boundary conditions that drive flows through them. In Chapter 4 we reviewed recurrent molecular structures and specific networks in organosynthesis that support the main flows of matter and energy in biochemistry. That chapter also introduced the basic mechanisms of dynamical control: the creation of catalysts either in the form of catalytic molecules or in the form of cofactors that mediate group transfer and thereby complete self-amplifying networks.

In this section we consider the energetic boundary conditions that drive pathways, the other dimension of dynamical order. They concern principally three species: electrons, protons, and phosphates. For the biosphere in its planetary-geochemical and geoenergetic context, the main sources of potential are a diverse range of redox couples and (mediated through much more complex mechanisms) light energy, as summarized in Chapter 3.

Within the environment of the cell, the sources of free energy are much less diverse and reflect the universality of the metabolic reactions in which they participate. They consist of a small collection of universal or near-universal redox cofactors and metal centers, phosphoryl donors and acceptors of which nucleoside triphosphates are the most important and inorganic polyphosphates often serve as repositories [100, 454, 461, 462], and proton transfer channels across membranes. These three chemical subsystems jointly maintain chemical organization. Bioenergetic systems furnish an example where three separate classes of processes, all with the abstract interpretation of energy delivery systems, arise from the coherence of low-level chemical details.

Because redox and phosphate energy sources couple to different groups of reactions and are non-interchangeable, they must be jointly maintained, a process performed by complicated systems coordinated at the level of the whole cell. The selective coupling of energy sources via catalysis partitions flows among networks and determines the directions of pathway fluxes, and so constitutes part of the control system for biochemical order. Unraveling the chicken-egg paradox of the mutual interdependence of parallel low-level chemical energy systems and high-level cellular organization is one of the important problems for placing metabolism and cellularization in the correct stages on the path of biogenesis.

### 5.4.1 Energy conservation, energy carriers, and entropy

Biochemistry is a complex collection of near-equilibrium and far from equilibrium processes in chemical thermodynamics. As such, it inherits both very general laws common to all physical systems, and also very particular structural limitations imposed by the constraints of chemistry. The direction of reactions is governed by general relations of energy and entropy common to all thermodynamic processes, but because different chemical inputs cannot be used interchangeably due to the constraints of reaction mechanisms, biochemical thermodynamics factors into distinct sectors in which energy-entropy relations largely act independently. We will introduce the key abstractions in this chapter, using a conventional terminology of "energy" relations that is common in biochemistry, but we caution readers that the terminology is not thermodynamically proper, and ultimately should only be understood as a shorthand for the more careful use of concepts developed in Chapter 7.

### 5.4.1.1 *Roles of energy and entropy in thermodynamics*

The role of energy in all thermodynamic systems is to determine the set of configurations that the system can take *at all*. The conservation of energy, from mechanics,[50] is one of the most fundamental inputs to thermodynamics because it requires that adding energy to one subsystem to make more states available entails taking energy away from other subsystems, and thereby making fewer states available to them. It is in this role that chemical bonds can be rightly said to "carry" or "deliver" energy: the internal energy needed for an electron or bonding pair to exist in a particular bonding state is not available to other electron pairs or the thermal surroundings, to open available states for them.

Energy conservation does not, however, determine the direction of thermodynamic processes; that is dictated by entropy. Among the available states in a whole system, the distribution of energy among its subsystems will be, on average, the one in which the largest number of microscopic configurations overall is made available. This volume of microconfigurations is measured by the entropy. Whether the internal energy in one bonding configuration or another is "high" or "low" is not determined by the amount of energy itself, but by the quantity of entropy it "buys" in the whole system. Therefore the common language of "energy systems," or "high-energy bonds" in biochemistry must always be understood as a shorthand for systems or bonds that, in order to be occupied, sequester energy from the whole system that would buy more entropy elsewhere than it buys in the "high-energy" subsystem.

### 5.4.1.2 *Non-interchangeability of chemical energy carriers*

Conservation of energy is a fundamental and important relation in physics precisely because it is so general: in systems where energy may be transferred in unknown ways among any of a large collection of internal degrees of freedom, the conservation of total energy may be the only certain constraint on the long-term behavior of the aggregate whole. In systems where energy mixes readily, sometimes this total-energy constraint is sufficiently important by itself to predict important system properties. Equilibrium thermodynamics comprises the systems for which that is the case.

Biochemistry, however, operates on finite timescales in which energy is directly exchanged through rapid chemical reactions. Even if, "in the long run,"[51] energy could be exchanged among arbitrarily different sectors through rare or circuitous connections, the fluid energy exchange is governed by reactions that are directly accessible through chemical bonding.

The non-interchangeability of electron transfers, group transfers, or proton transfers, causes each biochemical reaction to be open to coupling only with a limited class of

---

[50] Conservation of energy is a result of a mathematical relation known as *Noether's theorem* [309]. The precise statement is that for systems that do not have explicit time dependence in the laws or parameters, a quantity (the whole-system energy) is conserved and its value dictates the states that it is possible for the system to take. It is most correct to think, not of "conservation of energy" as a law of nature, but of the *theorem* that relates conservation of energy to a symmetry (time invariance) as the law of nature.

[51] Here we recall Keynes' comment from *A Tract on Monetary Reform* [431] (Chapter 3), noted in Section 3.5.5.6.

energy carriers. Not only the total energy matters, but also the degrees of freedom that carry it. In biochemistry, where energy is partitioned and separately regulated, we shift from the more general paradigms of equilibrium thermodynamics to the more particular paradigms of engineering. The structure of biochemistry reflects a complex three-way trade-off: (1) maintaining sufficient diversity of chemical reactions to support complexity; (2) driving these reactions with a few "energy sources" of high chemical potential, thus simplifying and modularizing the system; and (3) minimizing the mismatch between the chemical potentials of the drivers and those of the reactions, to capture as much of the energy transfer as possible in the form of chemical work, relative to the amount dissipated as heat.

### 5.4.2 Three energy buses: reductants, phosphates, and protons

The great diversity of reactions in biochemistry is sustained by three primary energy-carrying subsystems. These are the transferable electrons in oxidation/reduction couples, phosphoryl groups that supply energy to other leaving groups or drive dehydrating polymerization, and membrane-exchangeable protons that may be used to couple the other two systems or may do mechanical or chemical work directly. Although biochemistry may be partitioned into subsystems that require several different "activated" transferable groups (typically carried on dedicated cofactors), the activation energies for the diverse subsystems are provided by either reductants or phosphoryl groups, distinguishing the latter as energy "buses" that deliver free energy to all living processes.

The status of the biological energy buses follows from the very simplest laws of physics and the aqueous medium of life. Electrons are the mobile charges in chemical reactions and in conjugated double-bond systems in organic molecules, and protons are the most mobile species in water.

Of the three energy-carrying species, only redox couples are known to exist in geochemical environments in states of disequilibrium capable of sustaining organisms. (Anhydrous phosphates are common as equilibrium species in environments which are already dehydrated, but there they are not a source of free energy. Proton gradients exist in natural hydrothermal systems, but to be used by cells they require transport through an enclosing envelope. We return to the possibility that proton-motive force was once important in Section 6.2.4.) No known environments currently provide disequilibria in all three systems. Therefore, the integration of energy systems in cells is essential to the function of the cell as a platform capable of supporting metabolism in any known terrestrial environment.

The independent physical character of the three energy systems, their distinctive roles in biochemical reactions, and the partial separation of the subsystems that provide them in cells, suggest some modularization in the emergence of biochemistry, but the interwoven dependence of almost all biochemical systems on two or even all three of these poses one of the largest challenges to unraveling the hierarchy of biochemistry and the process by which it became integrated in cells.

### 5.4.2.1 *Redox energy accompanies the alteration of most chemical bonds*

A chemical reaction is the shifting of electron distribution and spin correlation among molecular orbitals, generally also mediating the physical relocation of nuclei in space. A reaction counts as an **oxidation** or **reduction** if it involves a change in the asymmetry of the electronegativity of the bond partners before and after the reaction. One speaks of redox reactions as *electron transfers*, but only some redox reactions result in a change in the electric charge of a free-standing reactant or product. Those cases tend to involve either metal centers, dissociated protons, or soluble ionic partners. Most electrons are only "transferred" in the sense of partial charges, as they reside more or less asymmetrically on nuclear centers before and after the reaction.

Almost all organic reactions involve some degree of electron transfer in the course of bond rearrangement, so oxidation/reduction and the free energy changes that accompany it are generally among the properties determining whether a reaction can proceed. The set of oxidation/reduction potentials (ORPs), which are the chemical potentials for electron transfer between each reactant and a reference standard, together govern the free energy of the overall reaction.

### 5.4.2.2 *Selection of redox partners is a key biological mechanism to determine reaction directions*

For reaction free energies $\lesssim k_\mathrm{B} T$, reactions are effectively reversible. A few important arcs within core metabolism approximate reversibility, but most do not. At least a subset of reactions in any pathway possess free energies $\gg k_\mathrm{B} T$, which govern the direction of the overall pathway. The sign of the free energy is often determined by the participation of selected electron donors or acceptors as reactants, among sets spanning a range of ORPs. Substitution of redox couples[52] is one of the most important biological ways to reverse pathway directions.[53] By controlling which electron carriers are synthesized, where they are located, and which enzymes are synthesized to use them, cells energetically open or close pathways for flows of matter and also other energy sources.

### 5.4.2.3 *The phosphate ester and dehydration in aqueous medium*

Group transfer is a more complicated aggregate concept, which subsumes redox energy and a variety of quantum selection rules into higher-order functions. In a transferred group, clusters of atoms retain their internal bonding configurations (though often with deformations and energy perturbations). The aggregate electronic properties of the group contribute to the exchanges in which it can participate. Among the transferred groups, phosphates

---

[52] *Redox couple* refers to an electron carrier that exists in one of the oxidized or reduced states as a reactant and in the other as a product, in an electron-transfer reaction [571]. The mid-point potential of a redox couple, the chemical potential of a standard with which equal concentrations of the oxidized and reduced forms would be in equilibrium, determines the contribution of that redox couple to the free energy of an electron-transfer reaction.

[53] An example noted in Section 4.4.6.2 is the direction of citric acid cycling. When low-potential reduced ferredoxins are coupled to the carbonyl insertion reactions at acetyl-CoA and succinyl-CoA, and when fumarate is reduced by menaquinones, the reductive cycle fixes carbon, generates reduced CoA-SH, and oxidizes the quinones. When the pyruvate decarboxylase and $\alpha$-ketoglutarate decarboxylase are coupled to $NAD^+$ and succinate is coupled to reduced ubiquinone, the oxidative cycle produces NADH, acetyl-CoA and succinyl-CoA, and oxidized quinone.

have a special importance in biochemistry. Phosphate occurs universally in biochemistry as $PO_4^{n-}$ or $-PO_3^{n-}-$ (where $n$ depends on context). Frank Westheimer wrote elegantly [870] about the unique capacity of the phosphodiester bond to perform dehydrations in water, making it a source of activation energy for both redox transformations and polymerizations.

### 5.4.2.4 Non-overlapping chemistries

The redox cofactors in cells – NAD, FAD, folates, quinones, and metal-center tetrapyrroles, along with small peptides such as ferredoxins and occasionally metabolites such as succinate – supply energy in almost all low-level organosynthetic reactions. Phosphoryl donors, principally nucleoside diphosphates and triphosphates, supply energy to a small but crucial set of activations in organosynthesis, and control almost the entire sector of oligomer synthesis. Only a few reactions fall in the intersection of phosphoryl activation and oxidation/reduction, and we will have more to say about those in a moment. In most reactions, phosphoryl donors and redox cofactors are non-interchangeable because they are quantum mechanically incommensurable. Even among the redox cofactors, the ORP and the favored number of transferred electrons strongly restrict the range of reactions in which they can participate.

### 5.4.2.5 Proton-motive force: the cash currency of cells

The third principal energy store in the cell is proton-motive force between proton reservoirs separated by cell membranes. Proton-motive force is a sum of chemical potentials from voltage drop and from differential proton activity, the logarithm of which is pH. Proton-motive force can be coupled both to electron transport and to the turning of molecular motors. Its most universal use in cells is to accept energy from redox reactions in electron-transport chains, and to drive ATP synthesis through the $F_O$–$F_1$ ATPase.

Protons are unique in cellular energetics, in constituting a kind of "continuum cash" energy currency. This is why they are key components in membrane-mediated energy transduction systems that couple reductants and phosphates, which have few interconversion channels in small-molecule chemistry. Like electrons or phosphate groups, protons must be exchanged in integer numbers. Unlike electrons or phosphates, however, protons that cross membranes have a potential that is not tuned by quantum mechanics on either the donating or receiving side, but instead by the dimensions of the cell membrane and volume, and by the capacitance of the membrane, which is a nearly universal $1\,\mu F/cm^2$ [572]. Since the cell surface and volume are macroscopic quantities, the proton-motive force can vary much more nearly over a continuum, than the energies of either transferred electrons or phosphoryl groups.[54] When proton transfer is mediated by the ATPase, the relation between the stoichiometry of proton exchange and ATP synthesis, while not continuously variable, may at least be varied in fractions smaller than 1, by varying the number of ports in the rotor component of the motor.

---

[54] This is true only for membranes and enclosed volumes that are sufficiently large. For very small membrane areas, the integer counting of charges may make possible only very coarse discrete steps in voltage. For very small enclosed volumes, pH may change in large increments for single proton transfers.

### *5.4.3 The cellular energy triangle*

Metabolism depends on the coordinated exchange of energy by means of both redox couples and phosphate esters; the lack of either renders the entire system non-functional. Cells rely on the coupling of redox and phosphate energy buses, and on active regulation of energy transfer between them, to ensure that the chemical potentials and availabilities of different energy sources are properly titrated. The same requirement becomes even more complicated when light energy, instead of environmental redox couples, is used as a cell's energy source, because protons as well as charge separations (possibly at multiple potentials [77]) become primary energy carriers.

The principal activity of cellular energy conversion is to recycle anhydrous phosphate using redox energy, enabling the cellular fluxes such as carbon fixation and primary biosynthesis that require both, as well as providing a store of phosphates for polymerization, signaling, and other higher-level processes. Figure 5.8 shows the three energy buses and their three principal modes of coupling in cells. Redox and phosphate energy carriers are coupled through two parallel systems, one chemical, microscopic, and probably primordial, the other mechanical, mesoscopic, and drawing on the full range of molecular innovations used in the cell. The microscopic channel is the direct link through the process called *substrate-level phosphorylation*, indicated with the high-energy thioester bond in the figure. The mesoscopic channel consists of two links: the coupling of electron transfer to proton pumping through the electron-transport chain mediated (in most cases)[55] by quinones, and the coupling between ATP synthesis or hydrolysis and proton transport by the ATPase. Both electron transport and ATP synthesis are mediated by membrane-bound systems in which macromolecules are essential transducers of energy.

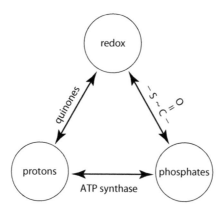

Figure 5.8 The three major energy buses that drive biochemical reactions, and the three links by which free energy is transduced between each pair of subsystems, ensuring that they are maintained in balance in the cell.

---

[55] Some methanogenic archaea and acetogenic bacteria are known which lack cytochromes [773].

### 5.4.3.1 Substrate-level phosphorylation and thioesters

Substrate-level phosphorylation (SLP) connects redox and phosphate energy directly through the modification of individual chemical bonds. It requires bonds that can participate in both electron transfers with redox couples, and group transfers with phosphoryl donors or acceptors, which most organic functional groups cannot. SLP was discovered as a direct source of ATP in the Embden–Meyerhof pathway of glycolysis in heterotrophs [257], but it is not a sole source of ATP in any known organism. In most organisms it is not even the primary source, which is provided by a combination of electron transport and proton pumping at membranes.

The most common reaction responsible for SLP is substitution of a high-energy thioester bond (indicated by S ∼ C in the figure) for a phosphate bond. Phosphate and thiol are both good leaving groups from a carbonyl, and acyl phosphates and thioesters may be substituted nearly reversibly under physiological conditions. Reduction of a thioester to an aldehyde is also approximately reversible using either NADH or NADPH. Thus thioesters form a bidirectional bridge for the transduction of chemical potential between the redox chemistry of nicotinamide cofactors, and the dehydration chemistry of phosphoanhydrides, shown as the direct link in Figure 5.8.

In the rTCA cycle introduced in Chapter 4, ATP hydrolysis to produce an acyl phosphate supplies the energy to form both citryl-CoA and succinyl-CoA. Reduction of the thioester to produce CoA-SH as a leaving group then provides the free energy to enable carbonyl insertion to form (respectively) pyruvate or $\alpha$-ketoglutarate, which would be so endergonic as to be effectively impossible at a carboxylic acid.

### 5.4.3.2 Oxidative phosphorylation

The combination of electron transport with proton-driven ATP synthesis is known as **oxidative phosphorylation**, or "oxphos." Oxphos requires complex multi-protein complexes – the cytochromes and ATP-synthases spanning cellular membranes – as well as a variety of special cofactors. It relies on a mediating proton-motive force maintained by the capacitance of a relatively proton-impermeable membrane and a membrane topology that separates water into "inside" and "outside" reservoirs capable of holding different proton potentials.

Cells use the quantization of energy in electron-transfer and proton-transfer reactions to give a directionality to proton circulation through the electron-transport chain and the ATP synthase. In ordinary oxidative phosphorylation, proton pumping into periplasmic or other intermembrane spaces is irreversible. Heat is dissipated for each electron that is transferred from a reduced metal center to a terminal acceptor such as oxygen, carrying along a proton in the process. At physiological proton activities, protons cannot drive the re-formation of reduced metal centers. In order to return to the cytoplasm they must follow a different route, which in cells is the mechanical route provided by the $F_O$–$F_1$ ATPase.[56]

---

[56] As noted, by varying the number of subunits in the FTPase rotor, the ratio of protons transported per ATP can be used to match the trans-membrane proton potential to the free energy of phosphoryl transfer.

The $F_O$–$F_1$ ATPase, perhaps only second behind the ribosome, forces us to recognize the magnitude of the gap between the era of cellular evolution that led to the LUCA, and the oldest innovations amenable to comparative study through phylogenetic reconstruction. The ATPase which may be driven by $H^+$, $K^+$, or $Na^+$ ions [578, 579, 580]) is a complex, multi-component molecular motor, related to flagellar motors.[57] It is homologous in all known species, and so appears to have been present in the envelopes of the gene-swapping cellular community that constituted the LUCA, from which all domains of life descended. The seeming dependence of the ATPase on quite sophisticated ribosomal translation, and the association of translated oligomers with phosphates rather than thioesters, suggests a complex and perhaps protracted period of coevolution between the emergence of the ribosome as a part of the translation system, and oxidative phosphorylation.

### 5.4.4 Geochemical context for emergence of redox and phosphate energy systems

To us the most likely fundamental source of energy for the earliest geochemical organosynthesis that eventually became metabolism is the redox disequilibrium between reduced metals refreshed at the near-surface through mantle convection, and seawater. The primary locations would have been peridotite-hosted sub-crustal convection zones at and near oceanic spreading centers or volcanic upwellings on continental margins.

The same environments also generally host pH disequilibria in the mixing zones, between hydrothermal alteration fluids and seawater, which opens the possibility for coupling of proton-motive force to other electron or group transfers in the manner of a fuel cell, to which we return in Section 6.2.4. In comparison to redox couples, which reflect a cosmopolitan disequilibrium between the bulk Earth and atmosphere, proton potential differences result from smaller-scale, more local disequilibria of element abundances in mineral phases, and accordingly are more complex and varied.

Polyphosphates are unstable in water against hydrolysis to orthophosphate. Although they can be produced in high concentration in dehydrating environments such as surface volcanos, on both the present Earth and by estimation on the early Earth, the known sources of redox and phosphate energy are not obviously collocated. For this reason surrogates for phosphate esters as early dehydration agents, or perhaps an alternative coupling between redox and phosphorylation potential through reduced-phosphorus salts or acids, would increase the plausibility of early self-maintaining metabolism. Mechanisms to interconvert redox and phosphate energy become essential at some stage before the formation of the cell as system capable of unifying metabolism, oligomer synthesis, and bioenergetics.

Christian de Duve has proposed [183] that substrate-level phosphorylation via thioester intermediates, operating in the opposite sense from the rTCA cycle to produce acyl phosphates from primary thioesters, was the original source of anhydrous phosphate in prebiotic chemistry. This reaction is only one part of his larger **thioester world** picture of an

---

[57] The ATP synthase itself may also be operated in reverse serving as a pump [292].

organosynthetic world not dependent on phosphate-mediated dehydration reactions for either organosynthesis or polymerization. The formation of a thioester is assumed to be driven by redox energy from some very low-potential reductant such as reduced iron, enabling attachment of a thiol group from $H_2S$. Once the thioester has been formed, it can be reduced by attack of an orthophosphate to form an acyl phosphate. Phosphoryl group transfer from acetyl phosphate to polyphosphates is exergonic, so acetyl phosphate could serve as phosphoryl donor for the accumulation of anhydrous phosphate. Acetyl thioesters, as precursors to acetyl phosphate, are of particular interest because they are plausibly produced in carbonyl-insertion reactions from carbon monoxide on metal sulfide surfaces [145, 146, 377].

The use of thioesters as the leaving groups for small non-ribosomal peptides such as gramicidin [298, 775] suggests that thioesters could have fueled an early era with some concentration of oligomers, though in extant cells these peptides appear to be an evolutionary specialization rather than relics from the last common ancestor. Substrate-level phosphorylation remains a plausible source for the earliest *in situ* production of phospho-anhydrides far from equilibrium with the hydrolytic environment, and we return to the status of the thioester world in the context of a larger problem of geochemical one-carbon reduction in Section 6.3.1.9. However, the use by all modern cells of high-energy phosphate esters for ribosomal oligomer synthesis, and the very high fluxes of ATP generation these require, make it difficult to envision a progression of life as far as the ribosome without coevolution of some mechanism to regenerate phosphate esters. It is therefore important to search for geochemical mechanisms that could either have produced phosphate directly from some other chemical disequilibrium (proton-motive force across mineral membranes has been proposed as such a source), or perhaps could have transported anhydrous phosphate from other starting materials where it was an equilibrium species (though this seems more difficult to envision as a steady source). Perhaps this key innovation enabled the transition of the ribosome from decoupled RNA-replicase and general condensation ribozymes to being a peptidyl synthase, and the fixation of the complex catalytic amino acid residues in the genetic code.

## 5.5 The three problems solved by cellularization

E. B. Wilson espoused [883], more pointedly than most writers, the perspective that the cell is the fundamental unit of life. We will argue in Chapter 8 against thinking of life as being defined by any one "fundamental unit," but putting aside for now the problems of multilevel analysis that lead to that argument, we agree that the overwhelming role of the cell as a unit of organization must be recognized and understood. Granting Wilson's point, however, it is important to recognize that the cell as it is understood in modern organisms is not only one kind of compartment or living "granule," but at least three. Three distinct *categories of function* provided by the cell are the following.

1. Integration of redox and phosphate energy systems, through the oxidative phosphorylation link of the "energy triangle" shown in Figure 5.8.

2. Collocation of reaction partners, or of catalysts and their substrates, leading to enhancement of rates for reactions that are second-order or higher-order in biotically created components. Related to this is topological separation of the external milieu from the interior milieu, in which the chemical properties of the interior are much more standard than those of the exterior, and are tightly and dynamically regulated.[58]

3. A zone of control by the genome over metabolic and bioenergetic processes that feed its replication (note that this may be the host genome but may contain other components as well, which are not restricted to descend vertically with an intact cytoplasm).

### 5.5.1  Distinct functions performed by distinct subsystems

These three functions are qualitatively distinct. Although they depend on one another in the context of the modern cell, it is possible to think of alternative environments that would have substituted for these interdependencies, in which each could have been supported independently. The cellular subsystems of macromolecules and compartments that provide the three functions are also sufficiently non-overlapping to suggest that in earlier ages and different contexts, they may have been autonomous from each other, though each would have been more dependent on supports from its local environment. This is a hypothesis for the earlier autonomy, and progressive integration, of higher-level functions of cellularization, similar to the hypothesis we raised for metabolic subsystems in Chapter 4. The argument for autonomy, and evidence that suggests its form, is important because it simplifies the problems of emergence and stabilization.

#### 5.5.1.1  Alternative environments for each function

First we note that the reagent-concentrating function of cells is not inherently tied to the coupling of energy buses, or to the exclusive control by a particular genome.[59] Concentration is thus the function most readily proposed to have been provided early on by environments other than cells.

Prior to having cellular envelopes to enclose and concentrate reactants and catalysts, separated phases, porous media, or adsorbing surfaces could perform similar functions for at least some subnetworks and their catalysts. Proposals have been put forth that either coacervates such as micells [721, 722, 723], mineral foams [524], or fracture systems with large surface areas [843] might have been plausible precursors to cells.

Containment of reaction partners will generally correlate with some degree of stabilization of reaction conditions, but may stop short of the tight regulation that becomes possible with full containment. Biofilms, poly-electrolyte gels, or percolating mineral foams differentially limit diffusion, and may concentrate reaction partners, but they may not regulate pH, ionic strength, ORP, or a host of other chemical potentials in the reaction medium.

---

[58] When the uniformity is maintained in time, the condition is called *homeostasis*. We emphasize both homeostasis and the uniformity of cellular interiors across species.

[59] It is only somewhat tied to the maintenance of a homeostatic chemical milieu.

The latter regulation relies on diffusion barriers across a range of sizes and chemical compositions, topological enclosure to make these barriers effective, and generally a suite of active and regulated transport processes to return the interior milieu to a target condition, which today are provided by the topologically closed cell envelope and its active transport systems.

### 5.5.1.2 Partially non-overlapping subsystems fulfill each function in modern cells

The functional independence of energy coupling, rate enhancement, and chemical homeostasis is underscored by the fact that the functions are not necessarily performed in the same physical compartments in cells. For energy coupling, the clearest example comes from the Gram-negative bacteria and their descendants, the mitochondria and plastids. In these cells, protons are pumped into the periplasmic space, into christae, or into internal volumes such as the thylakoid lumen [155]. The redox couples that drive this pumping, and the ATP produced by the proton return and used in metabolism, are all localized in the cytoplasm (or the homologous mitochondrial or plastid volumes) away from the low-pH proton reservoir.

Functional independence and partial spatial partitioning are important because they provide specific suggestions about where to disaggregate the complexity of cellular processes without qualitatively changing their nature. These divisions can be used to test scenarios for emergence in which different combinations of functions could have been provided by geochemical environments in different periods, and incrementally been brought under control of self-generated organic systems as the latter emerged.

### 5.5.2 An exercise in transversality

Cellularization by the era of the LUCA had incorporated so many refined forms of order, that long times or large populations seem to be required for each of them to have arisen through ordinary stochastic events. An integrated cell, in which all these subsystems require each other's presence to function and so to be maintained by selection, is an impossible model for the selective context in which they first arose. It would require that many rare events of emergence overlapped in time and in space. Recognizing modularity in subsystems allows us to reduce the number of interdependent components to enable each function in an appropriate pre-cellular environment. If few, or even single, rare stochastic events are required to produce a function that can be selected, these can be separated in time and among populations, and preserved until they come into contact. Once in contact, they may form mutual interdependencies that out-compete the autonomous forms, and only the links by which they are coupled need evolve as innovations.

### 5.5.2.1 The concept of transversality in topology

A useful concept in thinking about the separation of selective forces in time or in space comes from differential topology. It is called *transversality* [319]. A topological event, such as the meeting of two lines in a plane, has measure zero relative to both the lines

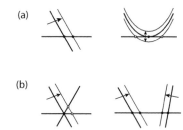

Figure 5.9 Transversal and non-transversal intersections. (a) On the left, a transversal intersection of two lines (black dot) is preserved if either of the lines is displaced. On the right, the tangent intersection of a circle with a line is fine-tuned; a small displacement of either the circle or the line either removes the intersection or produces two intersections (open dots). (b) On the left, a three-way intersection (black dot) is not transversal, as any displacement of any of the lines splits it into three two-way intersections. On the right, a topology that only keeps a set of two-way intersections is transversal.

and the plane.[60] Some intersections of lines, such as crossing at a non-zero angle, have the property that the position of either line may be slightly changed without destroying the property that they intersect. (The place where they intersect will change, but not the property of intersection itself.) Other intersections, such as the point of tangency of two curves, are infinitely sensitive to position. Almost all displacements of either of the curves will lead to either no intersection or two points of intersection. Intersections that are robust in the manner of the two transverse lines, by virtue of the topology of their kind of intersection, are referred to as *transversal*, while intersections that are fragile like the tangent intersection are *non-transversal*. Figure 5.9 illustrates transversal and non-transversal intersections. Note that a three-way intersection of lines is non-transversal, even if each of the two-way intersections is transversal on its own.

The emergence of a rare structure through stochastic fluctuations is an event that may have small measure in time or in population, akin to the intersection of lines. If an environment exists where it confers functionality on its own, and so can be maintained by selection for some extended interval, allowing time for the chance emergence of other structures with other functions, then the sequence of emergence and maintenance by selection is transversal, like the sequence of intersecting lines. If multiple coincidences are required to produce a selectable function, then the event of joint emergence becomes like the three-way (or multi-way) crossing, a non-transversal point in the space of possible histories of the world.

Transversality in parameter space may be as important as the freedom to experiment over extended time. In order for the emergence of biochemistry to have been a likely event in the history of a terrestrial planet, it is *not* necessary that the prebiotic chemical stages have been independent of chemical details in the environment. It is only necessary that, if they depend on details, the variability of the environment made it likely that the necessary

---

[60] The least fine-tuned meeting of two lines is coincidence at a single point. The point is zero-dimensional, whereas each line is one-dimensional and the plane is two-dimensional. In any set with a Euclidean dimension, the measure of a subset with lower dimension is zero.

conditions arose *somewhere*. Consider, for instance, the dependence of organosynthetic reactions on pH. Certain reactions of interest may produce high yield only in limited pH ranges. They would not be likely under all terrestrial conditions. If, however, rock/water chemistry creates local environments with extreme pH values from 3 to 10, and if fluids from those environments are connected and mix through the crustal thermal-flow system of cracks and pores, then it may be very likely that an optimal pH for the reaction yield occurs somewhere if that pH is between 3 and 10.

It is important to the basic structure of our argument that robustness is a system-level property, which need not inhere in any one component alone if it can be argued to inhere in the system.[61] Our main message is that life on Earth may be inevitable, but that at the same time the nature of the life we know is closely tied to the particular chemical structure and energetic dynamics of this planet. The two premises are not inherently in contradiction, but they do require that *life and Earth covary as a joint system, while occupying a range of parameters that is likely to be produced by the dynamics of planetary formation and subsequent evolution.*

A popular example of using transversality to make a prediction that is far from obvious is known as the *Meteorology Theorem*, shown in Figure 5.10. It is a consequence of the fact that vector fields cannot be made smooth everywhere on a sphere without passing through zero somewhere. The non-intuitive result is that at any time, two exactly antipodal points

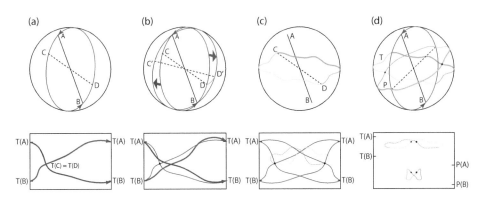

Figure 5.10 The Meteorology Theorem: at any time some pair of antipodal points on Earth must have both the same temperature and the same pressure. This is an application of the Borsuk–Ulam embedding theorem from topology. (a) On any contour between a pair of antipodes (A and B) with different temperatures, there must lie a point (C) with the same temperature as its antipode (D). (b) If the contour is displaced, a curve of antipodes with equal temperatures (cyan, dotted) is traced out. (c) If all contours are sampled, the curve of antipodes makes at least one loop around the sphere. (d) The same procedure if followed for pressure yields another loop; two curves of antipodes on a sphere must intersect somewhere, and at the intersection both pressure and temperature are equal.

---

[61] Albert Eschenmoser, one of the great organic chemists of our time, has argued for a similar point and made specific applications to problems in the chemistry of core metabolism, in [231]. We return to discuss this particular chemistry in Section 6.4.1.2.

on the surface of the Earth have both the same temperature and the same pressure. Thus, a coincidence that would require fine-tuning at any *particular* point is assured to occur always at *some* point.

### 5.5.2.2 The order of stages in the path to cellularization

Our appeal to transversality at this stage is more metaphorical than practically applied. It provides a concrete concept and a visual example to think about the pervasive problem of separating complex evolutionary sequences into independent, robust steps. Below we suggest stages in emergence that seem to serialize the dependence among innovations as far as possible.

**Full cellularization late**  The containment of driven organosynthetic systems is a complicated problem. It requires at a minimum the transport of inputs and wastes, synthesis and retention of catalysts, containment of catalysts and substrates, and support against osmotic pressure. To the extent that specific macromolecules are needed, it requires a memory system. To work out solutions to these many problems in the confines of a self-synthesized cell seems to us very difficult. It is easier to imagine using mineral pores as sources of containment, catalysis, and support for as long as possible, and to accumulate much of the apparatus of molecular biology before enclosing it in membrane-bounded compartments. Hence cellular enclosure seems most plausible to us as one of the later stages of biogenesis.

**Vesicle production relatively early**  This is not to argue that vesicles themselves are late structures. Many lines of evidence [344, 501, 502] (see the summary by David Deamer in [184]) indicate that spontaneous vesicle formation should have been common in geochemical organosynthesis. The inputs to coacervates – the fatty acids or isoprenes – are the waste stream from reduction of carbon and thus likely to have been continuously produced, irrespective of whether they could be exploited for later service. Phase-transfer catalysis [532, 619, 849] of organosynthetic reactions themselves could have augmented this production simply as a mechanism of redox relaxation.[62] Finally, of course, the specific form of phase separation is an essentially deterministic problem in physical chemistry. It is not the formation of vesicles that poses a problem, but the exploitation of them in sustainable systems.

**Oxphos as an early exploitation of vesicles**  If the problems of transition to an oligomer world were worked out gradually in mineral environments, then the assembly of pumps and transporters from these components may not have been a more complex molecular problem than other necessary steps such as refinement of the ribosome, charging of tRNAs, or evolution of DNA replication systems.

It seems possible to us that, as a problem of whole-system integration, oxphos could have been one of the *simpler* uses of membranes to evolve. This is because it can occur

---

[62] We discuss phase-transfer catalysis in the context of dynamical phase transitions in Section 7.6.3.

"inside-out," using vesicles as proton capacitors with biochemistry and molecular insertion taking place outside the vesicle.[63] An inside-out vesicle suffers no osmotic pressure and requires no wall. If its biochemistry is outside (hosted by a mineral enclosure), it requires no pores or pumps. The pH shift in the acid volume, which is likely to be large because it occurs within the small interior of the vesicle, occurs away from the protocytoplasm where proteins must fold and function, much as it does in the periplasmic space of Gram-negative bacteria. Thus the evolutionary problem of protein function is isolated from extreme and potentially fluctuating physical conditions produced by the proton-transport system. If vesicles were once used as proton capacitors, the function of oxphos could have provided a selective context for the evolution of the machinery needed to synthesize and situate complex trans-membrane proteins, which would then have been available to enable more tightly closed and regulated cells.

### 5.6 The partial integration of molecular replication with cellular metabolism

The architecture of the cell, bringing together as it does the functions of energy coupling, reaction rate enhancement, and homeostasis, can be rationalized as a necessary solution to enable a nearly universal metabolism to exist as an autonomous process across a range of very different environments. The distinctive geometric and topological features of the cell as a multi-compartment, and a set of its molecular mechanisms, are particularly accounted for as necessary parts of an autonomous platform for metabolism.

The cell is also, through a less obligate union, a meeting place for metabolism with the conceptually distinct process of molecular replication. Genomes depend on metabolism for materials, and metabolism depends on the genome to preserve and realize a memory of catalysts and structures that when assembled are viable and fit. However, multiple lines of evidence suggest that metabolism and molecular replication do not always need to be bound together in the same place throughout their cycles of competition and replication. Genes or genomes seem to have a native ecology that is independent of, and partly defined by different rules from, the ecology of metabolising cells. Jean-Michel Claverie has proposed [141] that viral lineages have existed in parallel with cells possessing obligatory genomes, throughout the history of life and putatively during the era of separation of the three domains.

Even in the modern world, genetic material enjoys a more complex relation with metabolism, involving considerably greater autonomy, than the metabolic functions of energy coupling, catalysis, and homeostasis have relative to each other. Plasmids, although not freely exchanged, are still quite often exchangeable among prokaryotes. Where bacteria and archaea have erected barriers against lateral transfer of DNA or RNA, viral infection mechanisms can sometimes thwart these defenses. Thus viral lineages define a different

---

[63] This "inside-out" vesicle geometry was the experimental form produced in the famous experiments of Efraim Racker and Walther Stoeckenius [656], first demonstrating the coupling of electron and proton transport, and their connection to the synthesis of ATP.

kind of lifecycle than cellular lineages, with the dispersal of DNA being more or less tightly coupled to metabolism and its cellular containers in different stages. The defenses of eukaryotes against gene exchange are even more elaborate, and yet they also exhibit the most spectacular instances of breaks. The largest mass importations of genes have been the instances of endosymbiosis and subsequent gene transfers to host nuclei [472], but endocytosis, and viral exchanges remain important as well [425].

The modern ecology of cells and the genes that on various occasions govern their activities is a complex interplay of competition by primary productivity, efficiency, or outright conflict, of negotiations of fidelity in which packages of metabolism are a common-pool resource for genes, and of the fine distinction between resourcefulness and banditry. The stability of the ecological dynamics, from what we have currently measured, appears to be dominated by a "normal" marriage of vertically transmitted genomes to the metabolisms they both support and draw from, defining a third fundamental function of the cell as a locus for co-inherited and co-selected genes. This function is less categorical, however, than the two metabolic functions with which we began the section.

## 5.7  Cellular life is a confederacy

*e pluribus unum*

> *de facto* motto of the United States of America from 1782
> until 1956, when it was replaced by "In God we trust"

The observations of modularity and partial autonomy in this chapter and the last do more than suggest rough outlines for a plausible path to emergence of the biosphere. They should fundamentally alter the way we think about the nature of the living state and the processes that contribute to it, as well as its deep history. We find it impossible imagine a single linear chain of dependencies leading from geochemistry to the modern cell. Not only the whole cell, but each of the subsystems we have considered within it, is an assembly of smaller subsystems. Whether these are the network and catalytic motifs in metabolism, the biosynthetic layers and stages in the code and translation system, the three carriers of bioenergetics, or the multiple functions of cellular compartmentalization, the modules within each subsystem suggest slightly different dependencies on a geochemical scaffold, or on intermediate groupings of their own lower-level components. The emergence of the biosphere is not a path through time but rather a tree. It is a tree viewed from the roots, which are anchored in diverse chemistry and energetics, and which coalesce at the apex in the integrated cell.

In Chapter 6 we will sketch some of the sequences in emergence that we believe are suggested by the architecture of modern life. We return in Chapter 8 to a more mechanistic consideration of modules. The emergence of the biosphere as a tree may be the only kind of emergence that was possible. The diversity of living processes that capture opportunities to channel energy and build structure is all localized in the cell today, but it was not localized in the abiotic environment of the early Earth. The problem of emergence was to respond to local energy stresses and exploit local catalysts and compartments, and then gradually transfer them from dependence on the abiotic environments that had stabilized them, to depend on a microenvironment that they created for each other. Chapter 8 will also argue that the separate stabilization of subsystems was not only the only possible route to emergence; the preservation of some degree of separation is likely essential to the preservation of order in the aggregate. This is why four billion years of evolution have not more thoroughly eliminated the module identities or mixed their inter-workings. It is also the reason functional distinctions that may reach as far back as geochemical origins can have been conserved into the present.

Cellular life, even today, is a confederacy. It is both a unified system and a collection of subsystems that retain meaningfully distinct identities and partial autonomy. They interact through simple interfaces that insulate the internal functional details of each subsystem from details in others. Redox and phosphate energy buses support organosynthesis without detailed dependence on the extracellular energy source or its carriers. They are coupled to each other through a uniform "cash currency" of protons. The translation system insulates selection on protein sequences from the architecture of metabolism, but uses patterns in that architecture to compensate for errors, making use of the molecular implementation of translation itself to enhance buffering. The goal in understanding the origin and the nature of life is to identify the sources of opportunity for order, to explain why their coalescence is possible, and to understand where and why coalescence eventually stabilizes at intermediate states of unification.

# 6

# The emergence of a biosphere from geochemistry

The major divergence in approaches to biogenesis turns on the source and mechanisms of the first *selectivity*: whether it came from small-molecule chemistry in a geochemical environment, or from a model of hierarchical control by macromolecules along the lines of Crick's Central Dogma. We argue that the emergence of life followed a path of least resistance along the "long arc of planetary disequilibrium," involving the atmosphere, oceans, and dynamic mantle, described in Chapter 3. The major temporal stages followed the architectural layers of biochemistry described in Chapter 4. The first carbon fixation was mineral hosted. Feedbacks, initially via cofactors and later via oligomer catalysis, lifted core metabolism "off the rocks." The emerging identity of the biosphere reflected the growth of autonomy as much as of chemical invention. Passage to an oligomer phase corresponding to the "RNA World" was a complex and heterogeneous transition, which transpired, and froze into place, in an already ordered organosynthetic context. We propose that cellularization occurred relatively late, and relied on functions of oligomers established in a mineral-hosted environment. The emergence of ribosomal translation originated in two parallel worlds of iron-RNA condensation-catalysis and template-directed ligation, which came together to form the first translation apparatus from mRNA to peptides. The refinement of translation fidelity, together with more precise RNA or DNA replication, ushered in the era of vertical descent along lines first appreciated by Carl Woese. Even in the era of evolution of effectively modern cells, many of the major transitions have been determined by biogeochemical reorganizations.

## 6.1 From universals to a path of biogenesis

In this chapter we suggest a sequence of stages in the emergence of life, based on the biological universals and the connection of life to the geochemical world, reviewed in the previous five chapters. We will propose that the hierarchy in extant life records a *temporal* sequence of stages, and that core biosynthetic pathways of living systems today reflect the pathways by which the same components first entered the incipient biosphere and dictated its form.

The emergence of a biosphere was an extended, and we believe a multistage and hetero-geneous, sequence of transformations. In the early stages of this sequence, the processes that would become core biochemistry took shape but were not yet separate from a geo-chemical background. This *ordered* synthetic network provided a foundation from which higher-level systems could grow in stages, giving rise to the cofactors, then molecular repli-cation, and ultimately cellularization. The possible forms for each new level that emerged were limited by the materials supplied by the lower levels. However, as each new level of components accumulated, they fed back to free the lower levels from dependence on their geological supports. Through successive stages of emergence and feedback, biochem-istry gradually became distinct from geochemistry, and the biosphere became a distinct planetary subsystem.

Thus, we argue, the origin of life began in the emergence of metabolism, and it was a progression through stages of increasing *control* over metabolism and *autonomy* from geochemistry that led to replicating oligomer systems, cellular integration, and eventually Darwinian evolution as a distinct mechanism to produce order.

## Outline of the chapter

The remainder of this section acknowledges the unavoidable role of assumptions in making claims about the origin of life, and introduces the considerations of robustness in hierarchi-cal systems that lie behind our assumptions. We also consider the context and role of RNA, and explain how we believe it should relate to the sources of order in early metabolism.

Section 6.2 argues that biogenesis was scaffolded by the long arc of planetary dise-quilibrium introduced in Chapter 3, involving the Sun, atmosphere, oceans, and dynamic mantle – the same arc of disequilibrium that supports chemolithotrophic life today. The way this arc of disequilibrium was assembled from internal dynamics in the atmosphere, hydrosphere, and lithosphere suggests especially important roles for some compounds and locations in shaping the chemistry of life.

The following five sections then partition the problem of biogenesis into the emergence of a small-molecule substrate and the first organic feedbacks, the first roles for replication in macromolecular systems, and the later stages of integration into cells.

In the last section we consider the major transitions that occurred within the era directly accessible to us through phylogenetic reconstruction. We note that these remain biogeochemical transitions.

### 6.1.1 On empiricism and theory: evaluating highly incomplete scenarios

Wovon man nicht sprechen kann, darüber muss man schweigen.

– Ludwig Wittgenstein, *Tractatus Logico-Philosophicus* [886]

This chapter does not attempt to frame a theory of the living state. Like the first five chap-ters, it is concerned with empirical claims about historical states of the world – albeit proposed states rather than settled facts. However, it is impossible to make empirical claims about biogenesis that do not depend in essential ways on assumptions, either about the interpretation of existing evidence or the relevant directions for experiment. The role of theory is to provide ways to reason systematically about assumptions, when part of their justification can only be arrived at circumstantially.[1]

The need of circumstantial reasoning is pervasive for the origin of life, because the transformation from a lifeless Earth to the first cells proceeded through a large number of transitions, all qualitatively different. Most involved the interaction of multiple different media, each of which we observe after extensive evolutionary refinement, entangled in chicken-egg interdependencies. Accounting for each of these transitions will require imagination and insight, derived from thorough empirical understanding of several kinds of processes. Most will depend on contexts that can only be inferred through incremental stages of reconstruction. For a few of the major transitions to life, compelling proposals exist. For some, there are fragmentary suggestions. Most are still completely obscure.

One approach to defending the validity of an idea for a major transition is to try to embed it in a connected narrative or **scenario** that reaches from prebiotic chemistry to some proxy for complex life. The motivation – to show consistency among the input and output conditions across all links in a chain – is fair and may someday be realizable, but at present we find most such efforts at best marginally helpful. Most steps in such narratives are too weak to add anything to the overall plausibility of the one or two good ideas that the author has to present. We will do better to acknowledge the need to judge the plausibility of ideas for individual steps in the transition to life from a context where connected scenarios are not achievable, and will not be for some time.

For the early, chemical stages of biogenesis especially, we think the greatest contribution from theory will come from understanding but then looking beyond particular materials or mechanisms, to aspects of architecture that are common to all cases of spontaneously emergent, hierarchical, complex order, and which are essential to both their robustness and their comprehensibility. The theoretical criteria that lie behind our proposed stages of biogenesis are summarized in the following three points. We return in Chapters 7 and 8 to more systematic and formal explanations of these ideas.

1. **Strong constraints come from transition points that were gateways** The modular architecture of life, and particularly of metabolism as shown in Chapter 4, suggests that it accreted by passing through key transitions that were *gateways*. The concept of a gateway is a generalization of the familiar example of a reaction transition state: it is an essential configuration or process through which trajectories of change are concentrated. Typically a gateway is associated with a local region of minimum resistance in some barrier that maintains the free energy differences driving the change. We argue

---

[1] In the absence of theory, one does not reason "from evidence alone." Rather, one reasons *non-systematically* about assumptions, or implicitly from a theory that goes unexamined. It is generally agreed [682] that the arguments by Sir Francis Bacon [44] and the early empiricists, that sufficiently many facts would become self-interpreting, are invalid, though well-chosen facts can be very *constraining* on potential theories. Formal models of the role of prior assumptions and the constraints provided by evidence are discussed in Box 8.1 in Chapter 8.

that many gateways we see between modules in extant life mark threshold stages in emergence that were more or less pre-determined by limitations from chemistry or requirements for system stability. These are the points at which we are most likely to correctly infer constraints on the path of biogenesis despite having an incomplete knowledge of how to assemble them into a connected scenario.[2]

2. **Some gateways must have been crossed in a particular order** In some cases it can be argued that gateways in biogenesis must have been crossed in a particular order, because later stages depend on inputs that could only plausibly have been supplied by earlier stages. Such an ordering would not be suggested if dependencies could easily have been back-filled by later evolutionary replacement, but it is implied if the gateways themselves were necessary as sources of robustness for the emergent order. Our proposals for serial dependence among gateways need not assume a single overall hierarchy. Some distinct subsystems of biological order could have emerged in parallel in different environments, and subsequently been brought into interaction and interdependency.

3. **The property of intermediate stability will be key in arguing that gateway transitions existed** Gateways can be necessary, and not readily substituted or replaced, because they define islands of probability for otherwise improbable elements of order. An important property of robust, hierarchical ordered states that emerge through gateways is that the stages that form between the crossing of one threshold and the next tend to be plateaus of incremental stability. (Examples are the stable distributions that form for populations of molecules between rate-limiting steps in a reaction pathway.) Hierarchical complexity is most likely to form along paths where it can accrete through a sequence of incrementally stable stages [754, 755]. The reasons to assume this can be stated in many ways, which we explore at different points in the book, but all of them are variants on the theme that in very complex assemblies, the improbabilities of individual events compound so quickly that in the absence of intermediate stages where cooperative effects provide system-level stability, no plausible theory of emergence exists at all. Conversely, such islands of stability are rare in the space of all possible assembles, so when they exist they tend to dominate the probability of emergence, which occurs by "island hopping." In assuming this we go beyond most geochemical theories of biogenesis, which focus on the existence of possible routes to organosynthesis; we make the stronger and more difficult proposition that contributing to incrementally stable subsystems should be one of the defining criteria for relevant pathways.

These considerations summarize many of the key elements that define a *systems perspective* toward the interpretation of evidence and the choice of assumptions about the origin of

---

[2] An extensive body of experience and a well-developed mathematical theory exist explaining why threshold stages occur in the formation of robust system order. This is the theory of phase transitions, which we review in Chapter 7. An important feature of thresholds is that they separate stages of emergence, so that the dynamics beyond a threshold is mostly governed by the properties of the threshold and not by the details of what happened before it, or "below" it. In that sense thresholds in emergence act like floors and ceilings that divide dynamical regimes into chambers, a key cause of the robustness of emergent order and also of our ability to infer valid theories of systems for which we possess only partial descriptions.

life. That perspective will be central to the way we think about one of the most crucial but also the most complicated questions about biogenesis: what role has RNA played throughout its course?

### 6.1.2  The functions versus the systems chemistry of RNA

Any effort to understand the origin and early evolution of life must draw on the unique properties of RNA and its central place in the structure and function of living systems today. Deciding what the known properties of RNA imply about the stages of biogenesis raises complex issues about the way causation is inferred from functionality, and how wide a system scope must be considered to understand the causal role of any particular component.

Divergent points of view have formed about how to extrapolate backward in time from the known functions of RNA in extant life to a theory of its roles in the origin or early evolution of life. The differences carry important implications for most of the fundamental problems of the origin and nature of life. These include whether RNA was the creator or the inheritor of an ordered organosynthetic context, whether extant biochemistry is continuous or discontinuous with geochemistry, and how early and in what ways a Darwinian evolutionary dynamic was responsible for the selection of biochemistry.

We consider first the way the discovery of RNA catalysis has led to the proposal for an "RNA World," and how various interpretations of this proposal have changed research directions and views about the origin of life. Much in the resulting interpretation of the relation between RNA and the small-molecule substrate of metabolism turns on whether one emphasizes the *ex post* functionality of RNA or the *ex ante* context in systems chemistry for its formation. We return in the following sections to argue that the fundamental difference lies in where one assigns sources of *causation* for the selection and stabilization of the chemical form of life: either in the chemistry of metabolism itself or in the mechanisms of hierarchical control.

#### 6.1.2.1 The introduction and ascendancy of the "RNA World" approach to origins

A striking change has occurred in perspectives on the origin of life since the mid-1980s. The period from the 1953 experiments of Stanley Miller [557] through the 1970s was marked by diverse explorations of the landscape of organic chemistry [280]. Many investigators who were specialists in peptides, sugars, or lipids, sought to frame the chemistry of the molecule classes in which they specialized as the crucial entry point to life. While some claims by individual investigators seem Quixotic from a modern perspective, they jointly accumulated an understanding of the reactions and physical chemistry of a wide class of biomolecules (and of candidates for their geochemical precursors). Still in 1967, the 1920's-era "primordial soup" proposals of Alexander Oparin and J. B. S. Haldane [329, 611] were a cornerstone of thinking about the seed chemistry for life on Earth, recapitulating Darwin's "warm little pond"[3] in the much more extensive twentieth century knowledge of organic chemistry.

However, even in aggregate, the chemical investigations into the origin of life were unsatisfying and conceptual progress seemed to be stalled. They provided essentially no resolution of life's many chicken-egg problems, between the chemistry of metabolism and control by enzymes and genes. They made no contact with hierarchical organization, with the beginnings of evolutionary dynamics, or with the capacity for open-ended variation, which are seen as foundation concepts within biology [828].

The interpretation of the Origin of Life problem, and a very large part of its research efforts, were changed suddenly and permanently by the discovery of **RNA catalysis** by Thomas Cech in 1982 [458] and Sidney Altman in 1983 [318]. It is important to appreciate that RNA catalysis was unexpected at the time of its discovery [613, 675]. Although proposals that the first genomes had used RNA rather than DNA existed in papers by many authors as early as the 1960s,[4] these remained within the accepted division of hereditary from functional molecules in the cell. Once RNA catalysis was recognized, however, the solution to a fundamental problem of hierarchy and complexity in biology immediately presented itself. If RNA could be catalytic, then both the hereditary function of nucleic acids and the controlling functions of catalysts would be present in the same molecules. Walter Gilbert made the first explicit proposal in 1986 that life had passed through an **RNA World** stage [302], in which both heredity and catalysis had been provided by RNA, and not by proteins and DNA.

The RNA World cut what had seemed a Gordian knot: the interdependence of proteins and nucleic acids in the processes of genome replication, transcription and translation, and catalysis. The enormously complex translation system of the ribosome, tRNAs, and the aminoacyl tRNA charging enzymes (themselves proteins) could be relegated to a later age of evolution, now supported by RNA and not needing to occur within small-molecule chemistry. The origin of translation was of course not in any way solved, but it could for the first time be confidently displaced to an era much later than the emergence of life.

The RNA World also offers several other metaphors, which seem to suggest that other problematic aspects of biological hierarchy can be presaged in much simpler chemical systems, or bypassed altogether. A catalytic RNA sequence is both **genotype** and **phenotype**, obviating the need to understand the generation of a genotype-phenotype map, or reducing it to a problem of folding that is much simpler than the construction of an

---

[3] The original passage, in a letter to J. D. Hooker, reads:

> It is often said that all the conditions for the first production of a living organism are now present, which could ever have been present. – But if (& oh what a big if) we could conceive in some warm little pond with all sorts of ammonia & phosphoric salts, – light, heat, electricity &c present, that a protein compound was chemically formed, ready to undergo still more complex changes, at the present day such matter wd be instantly devoured, or absorbed, which would not have been the case before living creatures were formed.
>
> *Letter to J. D. Hooker, 1 February 1871* [178]

It is as close as Darwin came to suggesting a scenario for the emergence of life, as opposed only to its subsequent evolution.

[4] See the reviews by Cech [122] and Neveu *et al.* [589] for history and further references.

entire cell.[5] In combination with the "primordial soup" image of a resource-rich early Earth ripe for exploitation, one might envision self-replicating RNA strands as a kind of naked genome, a *reductio ad infimum* of the viral lifecycle or the horizontally transferred plasmid. Further analogizing strand copying to genome duplication, and analogizing genome duplication to the reproduction of cells (or of organisms more generally), the RNA strand is envisioned as a competitive, Darwinian individual instantiated in the chemistry of a single molecule [122, 488].

We call these metaphors because they map an image of (demonstrated or imagined) RNA dynamics to abstractions distilled from much more complex cellular processes, but the complex requirements for stability in actual modern cells are not mapped to any corresponding dependencies in the RNA world on plausible geochemical settings. We will argue repeatedly in this volume that *any account of the primary architecture of biological systems must include the mechanisms that confer robustness on that architecture.*

Since the discoveries of Cech and Altman, and the articulation of the RNA World vision (at varying levels of metaphor), the bulk of work on the Origin of Life, and by far its most widely recognized framing, has shifted away from heterogeneous chemistry and onto the properties of RNA [297], particularly the study of RNA catalysis and the quest for an RNA replicase that can copy arbitrary sequences (including its own) through template-directed ligation.

We will not review the RNA World position in this volume beyond the current section and some commentary in Chapter 8 on the questions it raises about the flow of information and error correction in hierarchical control systems. Several good modern references exist [185, 280, 750] that summarize both the existing beliefs about the relations of RNA to the origin of life, and the current state of experimental work. Our own position is that some version of an RNA World is very likely to have characterized *early life*, but that the evolutionary questions relevant to this stage are not the same as the questions needed to frame the Origin of Life.

The paradoxes of hierarchy that are almost surely resolved by the RNA World are enormously important, because they open the transition to cellular life into an extended sequence, and they provide a medium for catalysis and control in which the later, more complex layers can more plausibly have evolved. However, the early stages in this sequence remain problems of heterogeneous chemistry. The deficiencies in our understanding, which the Origins community had faced until the 1970s because it could see no way to ignore them, still apply and still reflect hard questions about the earliest origin of structure and order. We think the RNA World has reduced the extent of these problems by moving some transitions to later eras, but it has not changed the original problems of order in early organosynthesis. The heterogeneity of life's chemistry is not an accident but rather is an essence; a projection onto the chemistry of RNA will not replace the need to understand the wider system.

---

[5] The opportunity to project the genotype-phenotype map onto a minimal form that remains non-trivial has been valuable in the study of the evolution of development [258], for other purposes. Here we only argue against over-interpreting a minimal model as a representative model of historical stages.

### *6.1.2.2 Weak versus strong RNA World*

As the "RNA World" has become a dominant framework for discussion it has also become a shorthand. It can be invoked as a context in the report of particular experiments on RNA chemistry as if it referred to a known and standard set of assumptions. However, the assumptions that constitute an RNA World view vary widely and the variations make a difference between a view that claims to be about the origin of life or one that merely characterizes early life. We find it useful to distinguish more conservative from more presumptive interpretations of the RNA World, to be explicit about which assumptions we consider very well motivated and which we consider unjustified from current evidence.

**Weak RNA World** The minimal set of assumptions defining the RNA World, we will term the *conservative RNA World*, or *weak RNA World*. They are: (1) that life passed through a stage in which proteins did not yet exist to carry the function of catalysis, and DNA did not carry the function of heredity; and (2) that in this stage both of these functions were carried by RNA.[6] The weak RNA World may describe much of cellular life before the advent of the ribosome as a translation apparatus from RNA to proteins.

The relatively conservative interpretation of an RNA World is as well motivated as any proposal for early life, and only becomes stronger and more appealing as our understanding of RNA expands. Today it would not be an exaggeration to describe the cell as an RNA-regulated machine in which a very narrow set of functions has been transferred to DNA,[7] and a less narrow but still restricted set of functions has been taken on by peptides.[8] RNA has retained functions in small-molecule catalysis, interference-based regulation, and to a limited extent, large-molecule catalysis [203]. Had the RNA World concept arisen today rather than against a history of cell biology in which RNA had been regarded largely as a "helper molecule," we might refer to the modern era still as an "RNA World,"[9] and to the pre-DNA and pre-translation era simply as less differentiated stages in the complex systems biology of RNA.

**Strong RNA World** A much stronger interpretation and set of beliefs attached to the existing evidence for RNA catalysis both *in vitro* and in the cell is what we will call *strong RNA*

---

[6] Examples of assumptions that are *not* implied in a conservative interpretation include:

1. that peptides were absent or that they were not functional; the only assumption is that they were not produced by translation and thus were not genetically encoded catalysts;
2. that copying of RNA was carried out in an RNA-only system, or that the preservation of RNA reflected competition in the rates of copying;
3. above all, that RNA replication was the first process to distinguish geochemistry from life, or that Darwinian competition among RNA replicators was a necessary prerequisite to the formation of metabolism.

[7] DNA is the "queen bee" of the cell, specialized for preserving and transmitting hereditary information but atrophied in most other functions possessed by RNA.

[8] These include not only catalysis, but also structure, traction and trafficking, and all-important, the regulation of gene expression itself.

[9] This richer understanding of the roles of RNA is captured by Cech in his reference to a "second RNA World" of the present [122].

*World*, or *RNA-first* [589]. This premise adds to the premises of the weak RNA World the claims that (3) RNA was a *self-replicating* molecule,[10] and its competitive replication was a Darwinian process responsible for sequence selection; (4) RNA self-replication defines the point of departure from geochemistry to life; (5) replication and selection of RNA was a necessary mechanism for the emergence of metabolic order mediated by RNA catalysis; and (6) RNA competitive replication and catalysis brought biochemistry into existence as a replacement for whatever "prebiotic" chemistry had provided the inputs to the RNA World. The replacement, in this view, could have been wholesale, or in its later strata it could have left imprints of earlier forms incompletely replaced, in the manner of a "palimpsest" [60].

An important corollary of the strong RNA World premise is that no reason exists for biochemistry to be continuous with the geochemistry or astrochemistry that produced the organic compounds that seeded life. Not only should RNA catalysts have overwritten pre-biotic geochemistry; subsequent waves of evolutionary replacement may have qualitatively altered the architecture of the small-molecule chemistry of the RNA World.

We think that omitting **biochemical continuity** as a source of constraint within a strong RNA World view is unjustified. Its assumptions about the capabilities of RNA heredity and catalysis at the system level extend far beyond any current evidence or even models. It also seems to us to underestimate the severity of the problem of selecting the substrate of biochemistry. Finally, in projecting the problems of biological order almost exclusively onto the properties of RNA, it makes neither analytical nor predictive contact with the remarkable structure and history of biochemistry and cell organization that was reviewed in Chapters 4 and 5.

### 6.1.2.3 Good for doing or good at being?

The RNA World clearly has the status of a *paradigm* [459] for origin of life thinking. It is a framework for structuring investigation and interpretation, which combines evidence on a few points with large, unavoidable assumptions about how these bits of evidence fit together as a system.[11] As in any system where belief is unavoidable and has important

---

[10] Here we mean a class of molecules capable of replicating the members without the aid of other kinds of oligomers. Proponents of RNA-first accept both *strand*-self-replication [488] and inter-strand "hypercycle" replication [823].

[11] Some important assumptions include the following.

The assumption that the study of RNA replication by RNA in protein-free systems is an informative model for the emergence of life on Earth. No such protein-free molecular replication systems exist in extant biochemistry, and it is unlikely that protein-free RNA media could have existed on the early Earth. The relevance of the class of mechanisms must therefore be assumed, and it must also be assumed that the simplification of a protein-free medium does not alter the conclusions so far that it invalidates the study system.

The existence of a geochemically plausible mechanism of RNA synthesis and concentration is currently a matter of assumption. The celebrated synthesis of Powner *et al.* [647] provides a finely tuned mechanism to produce pyrimidines, but one that involves careful sequencing of broadly reactive species, and that is qualitatively unlike biosynthesis of pyrimidine RNA and thus bypasses other molecular species that are closely related to RNA in extant biochemistry. If this synthesis is used as a model, it must be assumed that the synthetic pathway was replaced wholesale within the RNA World that claims the Powner *et al.* synthesis as its basis.

Many further complex, and hard to quantify, assumptions are made about the capacities of RNA catalysis. The goal of designing a free-standing ligase capable of copying general RNA seems nearer [488, 676], but the hypothesis that RNA self-replicases, under Darwinian selection, somehow not only stabilized themselves but catalyzed an entire organosynthetic network into existence to substitute for their original sources, is an enormously complex requirement that presently falls entirely in the realm of speculation.

consequences, it is worth trying to state carefully what lies behind the acceptance of a particular set of beliefs.

The motivation behind the RNA World paradigm clearly derives from RNA's two functions – heredity and catalysis – and from the unusual combination of both in a single molecule [70, 589]. In short, it comes entirely from what RNA is *good for doing*.[12] To the extent that biosynthetic chemistry enters the RNA World paradigm, it enters as a thorny problem [613]. The difficulty of synthesizing RNA in geochemically plausible settings seems to discourage the assumption of its availability, on which the RNA World depends.

From a systems chemistry perspective, as we currently understand the early Earth, RNA would not be high on the list of plausible chemicals to be synthesized and selected, and to accumulate to high concentrations. However, we will argue that, if the role of RNA in extant life is to be truly explained, it must have been a likely product of the systems that gave rise to life. It must have been *good at being* a significant fraction of the system's organic carbon. If RNA was not first a stable component of the particular driven systems in which biochemistry took form, its potential later functions would be too contingent and too improbable to have reached back and pulled it into existence. The seeming chemical implausibility of RNA is a clue to errors in our understanding of the early Earth, and that problem must be overcome *in its own geochemical terms* before an appeal to RNA's functionality becomes plausible.

We assert, however, that RNA should be germane to the context of a coherent *systemic* account of biochemistry. The pathway for RNA synthesis is not an isolated problem unconnected to all other biosynthetic pathways in extant life, and there is no good reason to think it was an isolated problem in geochemistry. If, however, the mechanisms of RNA organosynthesis are considered in relation to a larger biochemical context, accounting for RNA entails solving many of the problems of synthesis and selectivity to produce an ordered protometabolism.

If one goes further, to acknowledge that RNA is an important molecule but not *the only important molecule* in biochemistry, the priorities are reversed. The well-motivated goal is understanding the organization of organic geochemistry into a protometabolism, out of which RNA was a by-product with unusual properties and singular downstream consequences.

### 6.1.3 An emergent identity for metabolism or the emergence of a control paradigm?

Part of the difficulty of judging and integrating different points of view on the origin of life is expressing them in terms of criteria that can sensibly be compared.

The convenient shorthand that opposes "RNA-first" to "metabolism-first" emphasizes the relative order of appearance of two *classes of molecules* in the actual historical

---

This is not a criticism of assumptions *per se*; they are unavoidable in any domain of fragmentary evidence. It is important, however, to gauge their severity relative to the strength of the circumstantial evidence that motivates their acceptance.

[12] Here, "good for" is a shorthand for the standard functionalist perspective in evolutionary biology. Materials and structures are selected "for" the functions they perform, and selection is the arbiter of what is "good."

sequence on Earth. While concrete and material in its reference, this framing poorly expresses the reason to assume that any such well-resolved historical ordering existed, or to regard either order as necessary. Referring instead to "genes-first" versus "metabolism-first" (as is sometimes done) makes clear that the *functional role* abstracted in the concept of the "gene"[13] motivates the assignment of priority to RNA,[14] and calls out the need for a clearer statement of what function of metabolism this is to be compared against.

We will characterize the essential distinction as being between **metabolism-first** and **control-first**, where we mean, more precisely: must one of the following two systems have existed first, in order to create a stable context in which the other could form?

1. The precursors to metabolism, initially a subsystem within geochemistry, but heavily pruned through the collective effect of small-molecule reaction kinetics in an environment dictated by geochemically available catalysts?
2. A "genic" layer of discrete, variable "replicators" capable of Darwinian competition and selection, which must have been interposed between the criteria that dictated what ordered states could be maintained, and the extraction of a biochemical system satisfying those criteria?

Organosynthesis, selectivity, and the emergence of Darwinian evolution all occur within either assumption, but their sequence, their relation to the origin and nature of the living state, and their consequences for the degree of continuity between geochemistry and biochemistry, are very different, as summarized in Figure 6.1.

Starting from the core assumption – that either collective small-molecule kinetics in geochemistry or a hierarchical gene-like control system must be invoked to account for the stable foundation of biological order – the many other, more particular assumptions about structure and function that distinguish metabolism-first from control-first points of view can be derived as corollaries. In the following three subsections we attempt to frame each in the form of a question that is sensibly posed within both the metabolism-first and control-first perspectives, and toward which the two perspectives assume alternative answers. They affect the classes of molecular size that are carriers of essential constraints, the nature and available mechanisms of selection, and whether biogenesis was an extended transformation or a discrete transition.

---

[13] The challenge in using the "gene" concept as a reference for function is that it is derived from the complex behavior of modern cellular systems and involves several abstractions that themselves are sometimes problematic to justify [193]. In practice, mathematical abstractions such as the Mendelian replicator or more recently Manfred Eigen's *quasispecies* are used as guiding abstractions, and then matched as far as possible against particular processes that can be realized in RNA or other molecular systems.

[14] Reference to genes in the abstract opens the discussion to other template molecules besides RNA (see Ref. [466] for a discussion of peptide-nucleic acid, or PNA, in context), though in practice, these alternatives are usually put forth to answer difficulties in accounting for RNA organosynthesis, and it is difficult to decide whether the problems of identifying an entirely different non-biological molecule class, and then explaining how it was replaced by RNA, are less or greater than the original problem of synthesizing RNA.

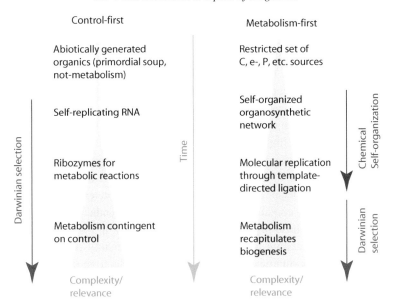

Figure 6.1 Distinctions between control-first and metabolism-first views of the origin of life. Both recognize stages of organosynthesis, the origination of memory and control mechanisms, and the advent of Darwinian competition and selection. The control-first perspective associates the living state more exclusively with the onset of Darwinian dynamics, and downward-acting control roughly as conceived in the Central Dogma. It expects less complexity and little selection in pre-Darwinian chemistry, and regards that phase of organosynthesis as "prebiotic." That chemistry is not strongly expected to show continuity with biochemistry, because the Darwinian horizon is seen as fundamentally innovative. Metabolism-first perspectives see contributions to the nature of the living state beginning in organic geochemistry and continuing as higher-level structures form. They expect greater complexity and selectivity in the geochemical setting, and also expect continuity from geochemical to emerging biochemical pathways. They regard Darwinian transitions as later regime shifts within the unfolding nature of life, which are only partly innovative and largely change modes of control.

### 6.1.3.1 Whether the first constraints on form were carried structurally in the small-molecule or the oligomer sector

The first implication of adopting either a metabolism-first or a control-first premise concerns the size of molecules that are regarded as carriers of essential constraints on the form of life. Metabolism-first assigns a causal role to the constraints on organosynthesis from the system of small-molecule precursors, while control-first assigns that causal role to kinds of selection that in present life depend specifically on informational oligomers.

- **Metabolism-first prioritizes synthesis, asserting that a network of available small metabolites was a prior source of constraints** The early steps in any particular organosynthetic pathway starting from $C_1$ inputs creates some molecules and not others. Clearly this places constraints on possible ways the pathway could be continued to produce higher-level structures, such as macromolecules. In biochemistry the synthetic pathways are organized into nested systems of related monomers, and then larger units made by ligation of the monomers. There is no evidence to suggest that any other class of macromolecules, or any other sequence of macromolecular synthesis than dehydrating assembly from monomers, ever provided large molecules that were used by life.

The metabolism-first premise asserts that the pathway architecture of known metabolism – both the organization into stages and the particular systems of monomers and ligations – is not arbitrary and was not amenable to wholesale replacement historically. Therefore the limits imposed by the small-molecule inventory within the *particular* pathways we know reflect limits on the chemistry of life as a whole. They are regarded as true law-like constraints rather than historical accidents because it is assumed that small-molecule reaction systems are largely determined by the chemical and energetic boundary conditions where they arise. The synthetic routes to a particular metabolite should be those in which the metabolite's role in a small-molecule system explains why the metabolite is special and how the difficulty of deviating from that system constrains opportunities for further organosynthesis from its constituents.

- **Control-first privileges a particular kind of selection in which the dynamics of populations of large molecules is the source of essential constraints** The canonical control-first view is exemplified by what we have termed the "strong RNA World" or "RNA-first" points of view. These assign priority to the role of genes as replicators, of catalysts as controllers, and of Darwinian selection as a mechanism for imparting information in systems (at both metabolic and higher levels).

Strong RNA World hypotheses presume a particular *method* of control, which in the simplest form is captured in Francis Crick's *Central Dogma* of molecular biology [167]. An oligomer functioning as a "gene" must provide a fundamental unit of replication, which stands between the environmental forces of selection, and the ongoing construction (via catalysis) of a physical and chemical "phenotype" that is adaptively shaped by those forces. The control-first premise regards structural limitations on the synthetic pathways for monomers before the era of sequence evolution as "prebiotic" and not as essential constraints on the nature of life. The essential constraint is believed to come from macromolecules as a necessary system to implement a Darwinian/Central Dogma flow of information and control.

### 6.1.3.2 *Whether selection must be Darwinian in kind and mediated by macromolecules in mechanism*

The second place where the implications of metabolism-first and control-first diverge is in the assumptions they make about selection. It is not the case that metabolism-first merely emphasizes the provision of the material substrate while control-first emphasizes the function of selection: both require highly ordered chemical states and thus quite stringent selection. Metabolism-first and control-first differ in what they assume to have been the possibilities for *sources of selection*, and the extent to which these originate in the rules of organosynthesis versus the sequences and population dynamics of macromolecules.

- **Metabolism-first asserts that significant selection is essential in a pre-Darwinian bulk phase** The metabolism-first premise is defined by its assumption that order – entailing *selectivity* – in the small-molecule organosynthetic network is possible using only geochemical mechanisms. We will go further to assert that selectivity from bulk kinetics is *necessary* as a foundation for any further accretion of complexity. We are skeptical that an unsupervised, complex organic mixture with no strong resemblance to biochemistry could be pruned to extract extant biochemistry, solely by means of Darwinian selection on RNA catalysts. We argue that if the basic outlines of metabolism had not been to a large degree prespecified at the level of bulk kinetics,[15] the diversity of states and the update rate required for polymers to correct the errors of unproductive or unlimited reactions would exceed the rate of which Darwinian selection was capable.

- **Control-first asserts that replication and Darwinian selection are necessary in order to achieve selection** The assumption that seems to drive proponents of control-first to accept the many difficulties of a hypothesis such as strong RNA World is that Darwinian selection, acting through the Central Dogma, is the only plausible mechanism capable of imparting information into molecular systems of the kinds we see in the biosphere. We think that this near-exclusive emphasis on replication and Darwinian selection derives from its historical importance [312, 534, 652] in the world of integrated organisms, and from the still relatively poor integration of concepts of causation between biology and the other sciences. The part played by replication in the persistence of the living state is diminished but also made more complex if replication acts in concert with other kinetic mechanisms that stabilize order at other levels than the (genomic or cellular) individual.

### 6.1.3.3 Whether the origin of a living state was demarcated by some particular transition, or transcended many transitions

A third difference, of interpretation, that follows the choice between the metabolism-first and the control-first premises concerns whether the emergence of a distinct living state was an extended planetary transformation or a singular event. We will argue that metabolism-first, if it is to assemble a collection of mechanisms into a theory of causation, should identify the nature of life with alterations of energy flow and matter cycling on a planetary scale, which can accumulate through a series of stages. Control-first, in identifying the nature of life with a capacity for Darwinian evolution, must reify this in a particular material substrate, making the emergence of life a discrete event (at least on geological timescales).

- **Metabolism-first fits within a conceptualization of the living state that can accrete gradually through many changes of organization** Metabolism-first positions are notable for their heterogeneity. Many very different small-molecule systems have been proposed as possible points of departure toward a living state [280]. The lack of a

---

[15] This may mean as little as occupying reaction domains where unusually high reactivity along the pathways is accompanied by unusually low densities of diluting side-reactions. We return in Chapter 8 to attempt to quantify the nature of selection and the problem of performing it solely within a hierarchical genetic control system.

dominating paradigm reflects a view of life in which it is not essential to mark some particular stage of molecular order as a discrete departure from geochemistry to life. We will develop an argument, in Chapters 7 and 8, that the defining character of life is not to be found in particular subsystems, but in the emergence of new channels for energy flow and material cycling, carried by the biosphere as a whole. This criterion is statistical and kinetic in character: it can be applied to the biosphere across time despite many changes of internal organization of the way energy flow and material cycling are supported. The biosphere persists continuously, though its specific components and processes can change through time. The question of distinctness of living systems is, for us, of a different kind than the question when a particular molecule class or selective process contributed to the aggregate dynamics of that system.

- **Control-first defines the advent of "life" in terms of the arrival of some first population of evolving replicators** The control-first view usually places a boundary for the emergence of life at some first achievement of Darwinian evolution in a particular place and medium. We have noted in Chapter 1 that the advent of Darwinian dynamics presumes an associated form of individuality; in scenarios such as the strong RNA World, the individuals are RNA oligomers. We return in Chapter 8 to discuss the complexity of the concept of individuality, and explain why we think attempting to locate an "essence" of life in Darwinian processes is a conceptual error.

### 6.1.3.4 Analogies to open-loop, feedforward, and feedback control systems

An analogy can be drawn from the distinction between sources of constraint assumed in the metabolism-first and control-first views, to different architectures distinguished within mathematical control theory [604] (Chapter 2). The standard control model is that a system subject to disturbances (the *load*) has behavior that also responds to a signal from a *controller*. The signals sent from the controller act as sources of constraint on the load to limit its operating range. The way the signals are chosen or modulated determines how effectively the controller can compensate for the effect of the disturbances (as well as the load's internal dynamics), as part of its larger function of placing constraints on the load's operating range.

In **open-loop** control systems, signals are sent to the load without regard to the disturbance, and the load's behavior is governed by the independent influence of both the control signal and the disturbance. In chemistry, reaction networks with externally given and fixed rate constants are analogous to systems with open-loop controllers. Fluctuation effects of all sorts are sources of disturbance to the reaction network, which result in diffusion of its trajectories within the distribution established by the kinetic constants.

**Feedforward control** requires sensing the disturbances that act on the load, and modulating the control signal using a model of the load's response. If the model of the disturbance's effect on the load is accurate, the controller may be capable of damping the effects of many disturbances. However, if the response to disturbances is difficult to model, or even if the state space for (load × disturbance) interactions is merely larger than the controller's internal state space, the ability to compensate for disturbances will be incomplete, because the load will eventually respond in some way that the controller fails to anticipate. Feedforward control systems are fast and robust, but they are not generally flexible because they are limited by the complexity and precision of the controller's model. Allosteric response of enzymes to the concentrations in a reaction mixture is an example of feedforward control, as is the action of the genome on phenotype in the idealization of the Central Dogma.

Our argument for the causal role of the network structure of organic chemistry and geochemical conditions in selecting the small-molecule metabolic substrate is analogous to some combination of open-loop and feedforward control. It is reliable and robust, but not suited to open-ended variation. In that respect it matches the phenomenology of core metabolism, which does not show evidence of more than limited variation. The fact that this lack of variation has persisted even in the presence of much more complex genetic and enzymatic control systems suggests to us that a large part of the robustness of core metabolism is still owed to its following kinetic and network "paths of least resistance," though we are far from being able to predict why such preferred paths would be the ones we observe.

**Feedback control** requires sensing the response of the load to the disturbance, and then modulating the control signal according to the deviation of the actual state of the load from the target state. Feedback controllers can maintain much better control with much simpler models than feedforward controllers. As a result they are much more flexible, but they are also generally slower. Feedback control systems also tend to be much more fragile than feedforward systems, because the controller can amplify its own errors as these create deviations from the intended state of the load. Darwinian selection, with control mediated by genomes, is an idealized feedback control system. *A priori* variation in genomes is unconstrained; it generates a fully developed disturbance in the phenotype, and only afterward does selection for longevity or fecundity filter the population for fit to the environment.

We will return in Chapter 8 to consider more complex, hybrid models of control, in which the robustness obtained by following paths of least resistance can provide generic error buffering in some degrees of freedom, permitting open-ended feedback controllers to provide flexibility in other properties. We think this is the correct model for the cooperative interactions among metabolites, catalysts, and genes in cellular life.

## 6.2 Planetary disequilibria and the departure toward biochemistry

The metabolism-first view proposes that order arose from collective rearrangement of material and processes, stabilized at the scale of systems rather than individual entities. The origin of life, from this perspective, was most fundamentally a planetary-scale rearrangement: the formation of an accumulating path of molecular synthesis, of accreting complexity in both chemical and physical organization, and of progressive selectivity. This premise must be translated into a concrete list of places, processes, and components that determined the essential structure along the accretionary path.

It is important to avoid too narrow a conception of either location or mechanism for the origin of life, for the same reason it is an error to focus on particular entities (genomes, cells, organisms) as possessing some inherent organizational character that defines the difference between life and non-life. Chapter 3 showed that the support of chemotrophic life on Earth today depends on the integration of stresses generated across systems ranging from the mantle to the atmosphere, even though the organisms themselves are hosted within narrow confines of hydrothermal systems where those stresses are focused. While mechanisms and environments are specific and local, the networks of interaction that determine which of them played a causal role in selecting biochemistry, seem likely to have a *planetary scope*.

Not every disequilibrium process that occurred on the early Earth would have contributed, either, to the necessity or the form of emerging biochemistry. As in any very complex system, most processes of excitation and relaxation would have aggregated to nothing more than dissipative futile cycles, leaving no imprint or framework of causes that would drive the accumulation of particular organic species or their assembly into a system of increasing complexity and hierarchy.

Therefore, after mechanisms of possible organosynthesis have been found, the problem remains how to judge which were *relevant* to the emergence of metabolism from geochemistry and its eventual maturation into a distinct system. This problem arises not only for pathways to arbitrary simple organics, but also for pathways to known biomolecules. Context matters in biology today, and it likely mattered in the course of biogenesis. Multiple synthetic routes in organic chemistry often exist to the same compounds, but only some of these have been incorporated within biochemistry. As Chapter 4 emphasized, the meaning of biological compounds lies as much in the conditions under which they are produced as in the functions that they subsequently confer.

### 6.2.1 The partitioning role of the abiotic geospheres

The possible roles of different mechanisms of planetary disequilibrium in dictating the form of life should have been determined partly by the roles they played in dictating the non-equilibrium states of the abiotic geospheres. Different mechanisms of disequilibration are particular to the atmosphere, hydrosphere, or lithosphere. Each geosphere has an ordered internal state produced by both fast and slow processes, but only a sparse subset of residual species produced within each internal network are stable enough to be transported across geospheres where they can form connections of chemical potential and reactivity at a planetary scale. Such bridging species are likely to have been key sources of constraint on the opportunities to drive a distinct biochemistry into existence.

We may thus partition the problem of constraining theories of biogenesis by first characterizing the major non-equilibrium chemical states of the three geospheres, then identifying the key species that transmit disequilibrium across the geosphere boundaries, and finally identifying key modes of reactivity that are seeded by those connecting species.

### *6.2.1.1 Atmospheric redox state and photochemistry*

Planetary atmospheres on small planets tend to be driven out of redox equilibrium with their rocky cores, if the atmospheres are not lost altogether. As noted in Chapter 3, *hydrogen escape* is the primary process that keeps the atmosphere far from redox equilibrium with the mantle. The atmospheres of Venus, Earth, and Mars all exhibit this disequilibrium.

The Earth's atmosphere is unusual in the solar system due to the retention of a liquid ocean over the planet's history,[16] resulting in the persistence of water vapor that is confined mostly within a troposphere. $CO_2$ and water were the dominant gases controlling the redox state of Earth's atmosphere through most of the Archean. The photochemistry of $CO_2$ and water are coupled in complex ways, which on Earth both reduce the degree of surface oxidation compared to that on Venus or Mars, and at the same time also reduce hydrogen escape and contribute to preserving the oceans and the atmospheric water they supply. Three major species will concern us.

**Photolysis of atmospheric water vapor** This initiates a variety of cycles involving the atmosphere and oceans, as well as liberating a fraction of $H_2$ from seawater, which can escape to space. In the long term the oceans are not stable, as shown by the loss of an ocean from Venus, and the loss of surface waters with attendant mineral oxidation on Mars.[17] On Earth, however, the restriction of most atmospheric water vapor to the troposphere, leaving a relatively dry stratosphere, shields most water from photolysis and slows the rate of hydrogen loss [420]. In the presence of $CO_2$ and $SO_2$ photolysis, scavenging of oxygen radical returns some $H_2$ to water, further limiting its escape by the Jeans mechanism.[18]

**Photolysis of $CO_2$** A consequence of the escape of hydrogen is that starting from mixtures of oxidized and reduced carbon species, released as volatiles by volcanic outgassing, $CO_2$ builds up as the major carbon reservoir in the atmosphere. This is an oxidized carbon pool relative to the mean oxidation state of carbon in the mantle and of magmatic volatiles. The main process of removal for $CO_2$ from the atmosphere is sedimentation or hydrothermal precipitation, which is mediated by the existence of oceans. The removal rate also depends on the activity of divalent cations such as $Ca^{2+}$, and thus is sensitive to the exposed continental surface area and to crustal composition and weathering rates of continents.

Ultraviolet light below 150–210 nm [703] can excite the double bonds of $CO_2$, photodissociating it to form CO and oxygen radical. The same UV spectrum cannot, however,

---

[16] The retention of an ocean is believed to have depended on Earth's being outside the critical radius that permitted relatively rapid cooling from an early "magma ocean" phase [219, 220, 221, 222, 331] that quenched volatiles within the mantle rather than outgassing them in an early hot phase where hydrodynamic escape was extensive. The preservation of surface water has far-reaching consequences for mantle and crustal dynamics through its effects on the strength of subduction zones [453], and through the effects of subducted water on magma formation and resulting arc volcanism [347].

[17] It has been projected that the Earth will lose its ocean to Jeans escape of hydrogen, exacerbated by increasing brightness of the Sun, over the course of the next four billion years [121].

[18] The rise of oxygen and the formation of an ozone layer further increase the shielding of water vapor from photolysis, so the biosphere has fed back to increase the stability of water on Earth.

excite the stronger CO triple bond. If sufficient reduced species exist to scavenge the oxygen radical before it can re-convert CO to $CO_2$, CO can remain in persistent disequilibrium and can even accumulate. Some current models of the early Archean with plausible $CO_2$ levels cannot escape a *runaway of CO* without invoking sinks outside the atmospheric system [432, 814]. Both $CO_2$ and its photolysis products are soluble in seawater and may carry the relatively high redox potential of the atmosphere into contact with molecular species at much lower potentials characteristic of the mantle.

**Photolysis of $SO_2$**  Like $CO_2$, $SO_2$ is produced by oxidation from a range of outgassed magmatic volatiles, and is then subject to photolysis by solar ultraviolet flux. Because it participates in many of the same oxidation cycles as $CO_2$, it provides isotopic signatures that can be informative about the atmospheric redox state, which we consider in detail in Section 6.2.2.2. A complicated network of photochemical and redox reactions determines the ultimate isotopic partitioning of sulfur species in the atmosphere. In addition to $SO_2$, SO, $SO_3$, and OCS also undergo photolysis, and $SO_2$ undergoes photo-de-excitation that is isotope sensitive [139, 876]. Shielding effects from solar flux [814] – both from other atmospheric gases and self-shielding by $SO_2$ – is believed to have had an important effect on isotope enrichment signatures observed from the rock record, and modeling it correctly requires incorporating the vertical structure of the atmosphere with respect to sulfur and carbon species as well as water [139]. Efforts to obtain calibrated, quantitatively reliable models of this atmospheric chemistry are a forefront of current work.

### 6.2.1.2  Transport and reaction of oxides in the hydrosphere

The hydrosphere on long timescales can be divided into two subsystems: the bulk ocean that reflects (within a few log units) the oxygen fugacity of the atmosphere, and continuously generated hydrothermal alteration fluids that reflect the oxygen fugacity of the crust.[19] An important class of reversible reactions that couple the ocean to the atmosphere is hydration of oxides to the conjugate bases of acids. Examples include bicarbonate (from $CO_2$), formate (from CO), bisulfite and sulfite (from $SO_2$), and sulfate (from $SO_3$). Under appropriate conditions such as interaction with transition-metal sulfide surfaces, the hydration equilibrium can be reversed, delivering oxides such as CO to metal centers where they can react. Through the equilibria of dissolved gases and their hydrates, the ocean may be a long-range coupler of atmospheric disequilibria to the water/mineral interface.

Sub-crustal water, whether circulated hydrothermally or entrained during subduction and then trapped, becomes altered to conditions more nearly in equilibrium with the crust or mantle. The resulting disequilibrium with the ocean is focused when hydrothermal fluids are remixed with seawater in vent systems, or when subducted water that induces melting under continental margins is re-released by arc volcanism.

---

[19] Note that shorter-term disequilibria, such as those that occurred during the great oxidation event or the early Cambrian [423, 743], may occur over timescales comparable to mixing times and diffusion times for gases through the ocean volume.

### *6.2.1.3 Reductants, pH, catalysts, and directionality from the lithosphere*

We noted in Section 3.4 and Section 3.5 that plume convection and plate tectonics renew the crust and upper mantle on macroscales, refreshing the supply of reduced metals out of equilibrium with the atmosphere, and recycling weathered oxides to (gradually and minutely!) raise the $f_{O_2}$ of the solid Earth. As important as renewing the stock of redox couples against relaxation is renewing the stock of catalysts against surface oxidation and other weathering processes.

Working to explain the existence of bubble-like Fe-S formations found in orebodies such as the Tynagh iron mines in Ireland, Michael Russell first proposed [683] the existence of alkaline hydrothermal venting systems driven by alteration of mantle peridotites rather than basalts, in locations such as fracture zones off-axis from mid-ocean spreading centers. Russell and Allan Hall [687, 688] argued that alkaline hydrothermal fluids rich in sulfides, when mixed with acidic Hadean oceans rich in dissolved ferrous iron, would precipitate mineral foams of the kind observed at Tynagh. The constant precipitation of new Fe-S surfaces would continually refresh the supply of catalysts with large surface area and important impurities including nickel or molybdenum, on the small scales of the precipitate mound surrounding the vent.[20]

Russell, with several collaborators [96, 524, 525, 526, 540, 689, 691, 693], has gone on to argue that as important as the catalytic chemistry of Fe-S precipitates is the *geometry* they impose on the interactions between vent fluids and seawater. Alkaline vents offer a second potential source of free energy besides redox potential to do chemical work, which is the pH difference between the vent fluids and seawater. As an alternative to simple mixing of fluids, which would produce the chemistry of reaction-diffusion systems, membranes formed as interfaces between vent fluids and seawater would impose a directionality on proton transport as well as potentially electron transport, and could focus pH differences into discrete trans-membrane potentials rather than macroscopic gradients from mixing. The pocket-like enclosures formed by mineral precipitates inflated by the pressure of venting fluids would also provide hemi-compartments. We return in Section 6.2.4 to an extensive set of hypotheses that have been offered about the roles of catalytic Fe-S membranes, proton potentials and translocation, and compartments, as a scaffolding for the emergence of biochemistry and later of cells.

### 6.2.2 *Species that bridge geosphere boundaries to form the great arcs of planetary chemical disequilibrium*

> yea, they have slain the servants with the edge of the sword, and I only am escaped alone to tell thee.
>
> – Job 1:15 (King James translation)

---

[20] These predictions were made before the discovery of the Lost City hydrothermal field, which produces alteration fluids with high $OH^-$ concentration due to $Ca^{2+}$ dissolution, as Russell had argued.

In order for a species to connect chemical disequilibria between two geospheres, it must first survive long enough to exit the environment where it is produced. Stability of a molecular species in a given context reflects the lack of reaction pathways to consume the molecule in that context; thus stability and reactivity are complements or alternatives. The long-lived species that couple chemical potentials across geospheres tend to be the "dead ends" of the networks that produce them.

The view that biochemistry emerged from organized geochemistry asserts that part of the selection of metabolites originated in geochemical kinetics before the emergence of hierarchical feedbacks and Darwinian selection. Selection – by whatever mechanisms – is the process that imparts the "information" in matter that defines the living state.[21] In a metabolism that began as a subsystem of geochemistry, part of that information would have been imprinted "from the bottom up" (or *ex ante*) in geochemical conditions and the rules of organic chemistry, rather than "from the top down" (or *ex post*) via selection of controllers such as oligomers or cells.

The *bridging molecules* that link chemical disequilibria on a planetary scale should be particularly informative because they reflect a particular class of constraints. Their semantic content – what one might say they are *informative about* in biochemistry – is carried in the patterns of reactivity of their functional groups. It attests to the kinetic barriers, or "missing" processes to consume such functional groups in one domain, which were present in other domains.

From the non-equilibrium states and processes within the atmosphere, hydrosphere, and lithosphere described above, we can identify a subset of molecules that on circumstantial grounds seem likely to have played a causal role in driving prebiochemical order into existence.

### 6.2.2.1 The carbon-oxygen system

$CO_2$ was likely to have been the major carrier of redox disequilibria between the neutral atmosphere and the reduced mantle. Section 3.6.6.3 noted the high concentration of mantle carbonates in Archean greenstones exposed in Western Australia, and the interpretation that oceanic concentrations of $CO_2$ could have remained as high as 100–1000 times present levels into the early Archean. $CO_2$ is not, however, a reactive carbon species in most environments [542, 546], even with molecules such as $H_2$ with which a large heat of reaction is possible. Significant reaction rates usually depend on the presence of some mineral catalyst, and when they occur $CO_2$ may first be converted to a more reactive species such as formate or CO, consuming part of the available free energy of redox reactions.

Carbon monoxide, in contrast, may have been the primary *reactive* carrier of oxidation disequilibrium. CO provides one of the best illustrations among atmospheric gases of the fact that reaction conditions are key to determining which species are short-lived

---

[21] The formal basis for statements of this kind is developed in Chapter 7 and Chapter 8. The central observation is that when we say life (and as part of life, biochemistry) carries "information," we mean that some more random chemical mixture serves as a prior expectation or *null model*, and that the actual chemistry of life is sampled from a narrow subset of the null model, restricted by some constraints against variation. The measure of random variation that is excluded by those constraints is then a measure of the information in the living state, and the dimensions in which variation is forbidden describe what this information is "about."

intermediates and which are long-lived carriers of energy stresses, and of the way these roles can change dramatically between geospheres with different reaction conditions.

The CO radical is highly reactive at transition-metal/sulfide mineral surfaces, a class of reactions known as *Fischer–Tropsch-type* (FTT) reactions [31, 651, 734], to which we will return in Section 6.3.1.1. This is a stark contrast to the atmosphere, where as noted in Section 6.2.1.1, solar UV readily photodissociates $CO_2$ to form CO, but does not activate the higher-energy triple bond of the CO radical, which can accumulate in the atmosphere if oxygen radicals are efficiently scavenged. In the atmosphere, then, CO is a "dead end" of reactivity that can accumulate and diffuse into the oceans, possibly being transported into into sub-crustal fluid systems.

The facility of carbon addition from CO through FTT reactions has been understood for decades, and has often been proposed as the first source of organic compounds in the emergence of metabolism [145, 146, 249, 350, 377, 525, 690]. This proposal suffers, however, from the problem of accounting for a significant concentration of CO in a water environment. The positive free energy of CO relative to $CO_2$ in the **water-gas shift** (WGS) reaction [315]

$$CO + H_2O \rightleftharpoons CO_2 + H_2 \tag{6.1}$$

suggests that little CO would be maintained in an ocean near equilibrium.[22] At high temperatures and high $H_2$ activity, the WGS reaction can be driven toward the production of CO,[23] and this has been the basis of proposals for abiotic production of methane by high-temperature FTT reactions. However, at low temperatures, which seem to be required for further elaboration of protometabolic reaction systems, equilibrium favors CO consumption and makes FTT organosynthesis problematic.

A photochemical disequilibrium in the atmosphere may make CO a more plausible source for low-temperature organosynthesis by FTT reactions. Even though the WGS reaction favors $CO_2$ at low temperatures, the sluggish kinetics of the reaction may maintain a concentration of formate in a non-equilibrium steady state for a cool ocean if CO delivery rates are high. Determining whether long-range coupling to the atmosphere would have resulted in significant delivery of formate in hydrothermal systems requires a coupled kinetic model for the joint atmosphere/ocean/subsurface systems.

### 6.2.2.2 The sulfur-oxygen system as a reporter of atmospheric redox state and photochemistry

Sulfur oxides, like $CO_2$ and water, are photoreactive compounds, but their importance to us is different from the importance of carbon species. They are neither primary determinants

---

[22] Note that the photochemical formation of CO, followed by dissolution in the ocean and equilibration under the WGS reaction, forms a thermodynamic *futile cycle* that recycles hydrogen within the atmosphere/ocean system. It is $H_2$ scavenging of oxygen radical that permits CO to accumulate and prevents some hydrogen from escaping the atmosphere, returning it instead to the surface as water. Subsequent WGS reaction of CO that diffuses out of the atmosphere liberates $H_2$ in the ocean, from which it diffuses back to the bottom of the atmosphere, returning the associated oxygen to $CO_2$.

[23] The equilibrium constant decreases by five orders of magnitude between 300 K and 1300 K, and passes through unity at approximately 1100 K.

of physical-chemical conditions in the atmosphere and oceans, nor are they key reactants
in early organosynthesis. Instead, they are useful tracers of atmospheric properties that are
difficult to constrain from other available forms of evidence.

Section 3.6.6.2 introduced the fractionation of stable sulfur isotopes, and noted that the
patterns of fractionation in Archean sediments and hydrothermal precipitates may provide
some of the most detailed known constraints on oxidation state and vertical stratification for
the Archean atmosphere. The existence of mass-independent fractionation (MIF) of sulfur
isotopes in sulfate and sulfide minerals has already been used to infer limits on oxygen levels before 2.45 Ga of $<10^{-5}$ PAL [627], a more stringent bound than those available from
other currently known geochemical proxies. Distinguishing MIF signatures from mass-
dependent (MD) signatures requires an element with three or more stable isotopes. For
sulfur, which possesses four, correlations among multiple isotope fractions, taken together,
can provide multidimensional constraints on atmospheric chemistry [815].[24] Model-based
inference about the Earth's early atmospheric composition is a complex problem and a
forefront of current work, with only a few conclusions established with much confidence.
However, the current modeling efforts suggest that sulfur isotopes may provide sensitive
proxies for key properties of interest such as $CO/CO_2$ ratios, and may attest to screening
of the UV solar spectrum in which both the spectrum and the molecules which may screen
it are of potential prebiotic importance [139, 814].

Atmospheric $SO_2$, supplied by volcanic outgassing, may photodissociate by vibronic
absorption either in a "photolytic" band from 180–220 nm, which leads directly to dissociation, or in a "photoexcitation" band from 250–350 nm, which carries too little energy
to enable direct photodissociation, but which creates excited states that can subsequently
dissociate through collisional interactions [339, 876]. Dissociation by either mechanism
results in an oxidized sulfur pool derived from $SO_3$, and a reduced sulfur pool derived
from SO.

The reason $SO_2$ photolysis does not simply lead to oxidation/reduction futile cycles in
the atmosphere is that the oxidized and reduced sulfur pools *exit the atmosphere* independently and are preserved in independent components of the rock record [139, 237, 627].
$H_2SO_4$ exits as sulfate aerosol, and reduced sulfur is believed to exit principally in $S_8$
aerosols.

In the atmosphere of the Phanerozoic, subsequent to the rise of oxygen, $SO_2$, which
is mostly concentrated below 15 km altitude [139], is shielded from photodissociation
by stratospheric ozone. Any trace SO formed is rapidly re-oxidized to $SO_2$, erasing isotopic fractionation signatures of photolytic interactions [815]. In contrast, the pre-oxygenic
Archean atmosphere (lacking ozone) was more nearly transparent to UV. Photolysis and
photoexcitation of $SO_2$ both occurred, and the resulting oxidized and reduced sulfur pools
remained segregated, allowing the two *exit channels* of acid rain and $S_8$ precipitation to
direct isotopically fractionated sulfate and sulfide into differentiated mineral reservoirs, as
was shown in Figure 3.10.

---

[24] See Box 3.3 for notations and a list of long-term observed patterns.

Three main classes of sulfur MIF signatures appear in both sediments and hydrothermal deposits throughout the Archean from almost 3.9 Ga to 2.45 Ga [815]. This is a remarkably large interval, with respect to the gradual increase of solar luminosity, the decrease of solar angular momentum and UV emission from the upper solar atmosphere, and also biological diversification and impact. Its beginning is suggested from phylogenetic reconstructions to have been a time of minimal microbial diversification (supposing that fully cellular life was even established by this time), whereas the late Archean was a time of extensive evolutionary diversification when microbial metabolisms must have had significant impacts on transport from the atmosphere through the oceans to sediments or hydrothermal systems. The apparent robustness of certain kinds of sulfur MIF signatures under such varied circumstances must therefore be a key outcome of any atmospheric mechanisms and models that are proposed to account for them. The three robust signatures are as follows.

1. **Sign of $\Delta^{33}S$**   Sulfate mineral samples from both sediments and hydrothermal veins seem to require that seawater sulfate throughout most of the Archean was depleted in $^{33}S$ relative to mantle sulfur (so $\Delta^{33}S_{sulfate} < 0$) that provided the major atmospheric source of both $SO_2$ and $H_2S$ through volcanic outgassing. Among pyrite samples (representative of sulfides), those that do not show mass-dependent signatures of subsequent microbial alteration are typically enriched in $^{33}S$ ($\Delta^{33}S_{pyrite} > 0$) compared to mantle sulfur.[25]
2. **Correlation of $\Delta^{33}S$ and $\delta^{34}S$**   The comparison of worldwide samples between pre- and post-rise of oxygen at 2.45 Ga shows that overall MD and MIF-altered sulfur samples partition into two clear bands, with nearly independent patterns of variation in $\Delta^{33}S$ and $\delta^{34}S$.[26]
3. **Anticorrelation of $\Delta^{33}S$ and $\Delta^{36}S$**   A third strong and consistent signature of Archean sulfur minerals is a negative dependence of the form $\Delta^{36}S/\Delta^{33}S \in [-0.9, -1.0]$.[27]

Current efforts to interpret sulfur MIF signatures are based on a variety of computational network simulations of radiation and relaxation chemistry in either well-mixed or vertically stratified atmosphere models. Well-mixed, or "one-box" atmosphere models [814], calibrated to empirical absorption cross sections, can be directly compared to analytic experiments, but they omit aspects of screening and segregation of reactants that are likely to be fundamental to chemistry in actual planetary atmospheres. Vertically stratified models [201, 432, 628] offer more detail but also more parameter uncertainty, and are difficult to directly test.

---

[25] Notable violations of this pattern are often suggested by more than one factor to have resulted from microbial sulfate reduction. An example is pyrite grains included in barite veins in the 3.5 Ga old Dresser formation, which show both $\Delta^{33}S_{pyrite} < 0$ comparable to the surrounding sulfate, and a mass-dependent $\delta^{33}S$ shift; see Ref. [815], Figure 8a.

[26] See Ref. [239] Figure 2. However, for a subset of samples older than 3 Ga, a positive correlation is consistently observed between $\Delta^{33}S$ and $\delta^{34}S$. See Ref. [237], Figure 3a where this is reported as an increase in the slope of $\delta^{33}S$ versus $\delta^{34}S$ to 2.17 rather than the MD reference value of 1.9, or Ref. [815], Figure 7a from samples of the 3.5 Ga old Dresser formation, which show a regression of $\Delta^{33}S$ on $\delta^{34}S$ with the even larger slope 0.56.

[27] See Ref. [239], Figure 4d and Ref. [815], Figure 7b.

The sign of $\Delta^{33}S$ appears to indicate a need for **radiative screening** of wavelengths longer than 202 nm. Under broadband irradiation simulating solar actinic flux, the net fractionation produces $^{33}S$-depleted sulfate from wavelengths below 202 nm, and $^{33}S$-enriched sulfate from the photolytic band above 202 nm. Because plausible solar spectra during most of the Archean have decreasing intensity at short wavelengths, some mechanism of screening appears to be needed to robustly account for seawater sulfate with $\Delta^{33}S < 0$. Carbonyl sulfide (OCS) has been proposed as a candidate for such a screening model, which also absorbs from 8–13 $\mu$m and so contributes to the Archean greenhouse, and which also has natural formation pathways in atmospheres rich in CO [814]. However, the rapid photolysis rate of OCS has led to doubts about its plausibility as a primary screening molecule [139].

Photolysis of $CO_2$ can provide oxygen radical that converts $SO_2$ to sulfate even in an atmosphere with low $O_2$ activity, potentially making the sulfur MIF signature a sensitive proxy for atmospheric $CO/CO_2$ ratios. Atmosphere models that are well mixed (and concomitantly uniformly irradiated) require $CO/CO_2 \gg 1$ to preserve a sulfur MIF signature against oxidation of $SO_2$ to sulfate. While such a conclusion is compatible with the arguments for atmospheric CO accumulation reviewed in Section 6.2.2, its status as an explanation for the robustness of the sulfur MIF signature throughout the Archean is problematic. The same mineral sulfate signature persists into the middle Archean when atmospheric $CO_2$ concentrations had decreased and microbial life was presumably a more important sink for CO from the atmosphere. More fundamentally, in a stratified atmosphere where a 1% $CO_2$ concentration would be optically thick [139], most $CO_2$ photodissociation would have been limited to altitudes above the domain of $SO_2$ photolysis, rendering the constraint against cross-oxidation from the well-mixed model inapplicable. Thus the intriguing possibility remains that sulfur MIF signatures could provide evidence about $CO/CO_2$ ratios, which may be used together with evidence from crustal carbonatization and constraints on the atmospheric greenhouse, to reconstruct the concentrations of CO and $CO_2$ that are of fundamental interest. However, before computational models can be extrapolated to the Hadean eon for which we have no direct mineral evidence, a reliable understanding of the mechanism must be obtained from robust accounts of signatures in the Archean, which is beyond current capabilities.

### 6.2.2.3  The problem of anhydrous phosphates in water: a transporter of exotic dehydration potential or in situ production?

Dilute aqueous solutions, including seawater and probably also hydrothermal fluids under most conditions, are hydrolyzing environments that militate against both primary organosynthesis, and the dehydrating polymerization reactions that produce most complex biomolecules. In biochemistry the phosphoanhydride bond is the source of dehydrating potential responsible for almost all polymerization, and also of activation energy in many carbon fixation and core anabolic reactions [870]. Therefore the problem of obtaining far from equilibrium anhydrous phosphates in a water environment is central to any theory of biogenesis.

Many metabolism-first hypotheses assume *in situ* generation of phosphoanhydride bonds as part of a kinetic cascade involving other primary sources of free energy. We will return to discuss these in Section 6.2.3. Here we note in passing the alternative possibility that anhydrous phosphate is generated remotely in environments where it is in equilibrium, and subsequently trapped and transported into aqueous environments. The hot, dry conditions created by surface volcanism readily create mineral polyphosphates, which may be trapped in volcanic glasses that fall to the seabed. In these glasses, not only the phospho-anhydride bond of polyphosphate, but also the Si–O–P bond, become eligible donors of phosphate to organic compounds as the glasses weather [144, 361, 583, 584].

Although phosphates can be trapped and released far from equilibrium by this mechanism, it is more difficult to argue that they would be delivered into ongoing contact with organic compounds in environments that also provided strong oxidation-reduction disequilibria, such as sub-crustal hydrothermal fluid mixing zones. Hence their importance as drivers of system organization in organic geochemistry is questionable.

A different entry of phosphorus into early organosynthesis, which has no close analogy to its use in biochemistry, could have been via **reduced oxidation-state phosphorus anions** such as the $P^{3+}$ compound phosphite ($HPO_3^{2-}$) or the $P^{1+}$ compound hypophosphite ($H_2PO_2^-$).[28]

The original proposal by Addison Gulick [322] to investigate more reduced phosphorus compounds was motivated principally by the problem of availability of phosphorus: in the presence of $Ca^{2+}$ ions, phosphate precipitates in the very low-solubility apatite minerals, whereas phosphite and hypophosphite remain soluble. Although direct reduction of phosphate is not regarded as a likely source of significant reduced phosphorus compounds under plausible partial pressures of hydrogen on Earth by the end of the Hadean, these oxides are known to be produced through weathering of the *phosphide* ($P^0$) mineral schreibersite $((Fe, Ni)_3P)$ in water, with dissolution timescales $\sim 10$ y. Schreibersite is extremely rare on the present Earth, but it is known to be a minor but ubiquitous component in meteorites and would have been delivered in large quantities during the later stages of infall toward the end of the Hadean. Remarkable recent evidence [625] from Archean marine carbonates shows phosphite concentrations as high as 40–67% in some samples, indicating that reduced phosphorus was an abundant dissolved species in the neutral to reducing oceans before 3.5 Gy.

Hypophosphite readily undergoes addition reactions with organic compounds such as pyruvate, and has been shown to produce lactones that in the presence of aminated compounds form peptide bonds under mild aqueous conditions [101]. Hydrolysis products of $Fe_3P$ also produce both reduced and oxidized anhydrides $P_2O_6^{4-}$ and $P_2O_7^{4-}$ (pyrophosphate) and even triphosphate [624, 715], and so may be plausible donors of poly phosphorus chains which can serve as leaving groups for reactions such as dehydrating

---

[28] These are contrasted with most biological phosphorus, which is in the form of $P^{5+}$ anhydrides of phosphate ($PO_4^{3-}$). Until recently it had been believed that all biological phosphorus occurred only as $P^{5+}$, but a diverse array of recent evidence has shown the presence of $P^{3+}$ and even $P^{1+}$ at levels reaching several percent both within organisms and in ambient environments [626].

polymerization.[29] While a redox cascade from reduced to oxidized phosphorus on the early Earth thus seems plausible as a component in early organosynthesis and a possible source of order-inducing free energy, it remains a very different dimension of disequilibrium from the persistently maintained hydration/dehydration disequilibrium of phosphates used by present biochemistry. Understanding whether reduced phosphorus would have produced an ordered organic system context that continued to biochemistry remains an open problem.

### 6.2.3  Mineral-hosted hydrothermal systems are pivotal in the sense that they are key focusing centers for chemical disequilibria

Thinking of biogenesis as a rearrangement in the organization of matter and reactions, which was scaffolded by a planetary arc of disequilibrium, subtly changes the way we assign causal significance to particular locations. We do not think of vents as locations "where life arose," much as we do not think of particular populations of organic molecules or even cells as sufficiently reflecting the nature of the living state. Rather, vents are important because they are *focusing centers*. They permit chemical potentials, which accumulate gradually on large scales, to be juxtaposed on small scales where part of the free energy is captured in chemical work, along the path to its ultimate fate of dissipation as heat. The initial accumulation of potentials requires either spatial separation or barriers to flow, while the later juxtaposition requires gaps in the barriers. This combination of circumstances is the gating condition for the formation of order, and if such constellations are sparse, they govern the sensitivity of biogenesis to details of planetary context.

The two major chemical potentials for which hydrothermal systems serve as focusing centers are **oxidation/reduction (or redox) potential**, abbreviated ORP or $E_h$ (for electrons) and **pH** (the entropic component of the motive force for protons).[30] Redox potential carries the dominant disequilibrium between the mantle and atmosphere, while pH reflects more context-specific disequilibria of minor-element inclusion in mineral phases and their interactions with water (see Section 3.5.3.3).

The geological asymmetry between $E_h$ and pH is recapitulated in their roles in biochemistry: electron transfer is the fundamental process in the alteration of all chemical bonds, and orbital energies determine all reaction equilibria. Proton transfer plays three more limited though still essential roles. First and most basic, proton abstraction or incorporation

---

[29] Section 6.3.4 notes that phosphorylation plays other roles in biochemistry besides activation, for some of which spontaneous phosphorylation would have been sufficient, so even monophosphorylation activity of phosphite or hypophosphite could have been significant in early organosynthesis.

[30] Both $E_h$ and pH are properly defined state variables only in equilibrium. Because proton transfers equilibrate in water much faster than characteristic timescales for fluid transport, pH typically provides a good characterization of hydrothermal fluids. Because electrons are insoluble in water, electron exchange in aqueous solutions requires collisions of solutes and tends to be slower. Therefore $E_h$ provides only a more approximate characterization, applicable to species that equilibrate within timescales comparable to fluid transport times. We will usually refer to specific chemical composition in hydrothermal fluids, and the redox potential of a pair of reactants is a well-defined property of the molecules, whether or not they are in equilibrium with a bulk value for $E_h$.

governs the accessibility of electron orbitals in many reactions. Second, through the Nernst equation, pH can affect the midpoint potential for electron transfers that involve deprotonated species. Third, proton-motive force at membranes has been harnessed by evolution to provide one of the major bioenergetic transduction systems.

The particular carrier species and reaction centers in the vent environment further transduce and diversify electron (and perhaps proton) potentials into numerous context-dependent **group-transfer potentials.** Two of these, of particular importance for biochemistry, are the thioester bond for which SR (R some organic group) is the leaving group, and the phosphoanhydride bond, for which orthophosphate is the leaving group. Both serve as general sources of activation energy, and in particular they mediate the elimination of water (in aqueous phase!) during ligation reactions.

The central questions about possible roles of vent chemistry in establishing a template for biochemistry concern the couplings of energy flow and chemical complexity in rock/water systems, and the extent to which specific couplings among redox, protons, thioesters, and phosphates in extant metabolism trace continuously back to hydrothermal processes.

The generation of chemical order where subsystems with very different chemical potential profiles are brought into contact is a kinetically governed phenomenon. In hydrothermal environments, the relative rates of different reactions are expected to be pervasively influenced by available metals serving as catalysts, whether as dissolved and chelated ions or at mineral surfaces. George Cody and collaborators have studied variations among the effects of numerous transition-metal sulfides on carbonyl insertion and reduction reactions [145, 147]. More recently, Yehor Novikov and Shelley Copley performed an extensive analysis [597] of the kinetic partitioning of reaction products of pyruvate with reductants and ammonium chloride by a suite of sulfides.[31] Their results show that most metal environments prune the reaction networks to subsets of major products, and that the pruning can be strikingly different for different metals and different mineral stoichiometries and crystallographies. Three of the most important questions about the role of hydrothermal systems as focusing centers, in our view, are how extensively such systems can prune organosynthetic networks, how large a part of the difference between the complexity of equilibrium ensembles and the sparseness of biochemistry can be accounted for with merely geochemical kinetic selection, and whether there are "robust" metal environments in which the product suite that results is not too sensitive to details of mineral composition and suggests a non-arbitrary template for the observed universal core metabolism of extant life.

---

[31] Novikov and Copley [597] studied reactions of pyruvate with $H_2$, $H_2S$, or $NH_4Cl$ at mildly acidic conditions in slurries with several Fe, Cu, and Zn sulfides representative of precipitates at acidic vents. In addition to aldol reactions, they report reductions of ketones to alcohols and reductive aminations, which are the primary departure from the ketoacid CHO world of Figure 4.5 to the amino acids and subsequently nucleobases and related heterocycle cofactors. Lactate, propionate, alanine, and various thiolactate sulfides were common products, but the partitioning of material among these differed starkly with the mineral used, indicating kinetic selection among competing reactions. Minerals studied included pyrite ($FeS_2$, cubic), marcasite ($FeS_2$, orthorhombic), arsenopyrite (FeAsS), chalcopyrite ($CuFeS_2$), sphalerite ((Zn, Fe) S), pyrrhotite ($Fe_{(1-x)}X_x$ with $0 < x < 0.2$), and troilite (FeS).

### 6.2.4  The alkaline hydrothermal vents model

What I relate is the history of the next two centuries.

– Friedrich Nietzsche, *The Will to Power* [914]

In a series of hypothesis papers beginning in the 1980s, Michael Russell with several collaborators has argued that among hydrothermal systems, there is an important asymmetry between the contributions of acidic and alkaline vents[32] to the global structure of geochemical disequilibrium that drove the emergence of life. The authors argue that the role of acidic vents is primarily to establish the background chemistry of the ocean: its content of metals and particularly of iron [463], and (in interaction with atmospheric gases) its pH. They propose a qualitatively different role for alkaline vents, *spatially localizing* complex organosynthesis, proton energetics, and the origins of cells within the precipitate mounds formed around fluid exit channels. The resulting collection of hypotheses have come to be known as the *alkaline hydrothermal vents* (AHV) model. Principal elaborations of these hypotheses have been given by Russell, William Martin, Nick Lane, Allen Hall, Wolfgang Nitschke, and Elbert Branscomb.[33]

The AHV hypothesis papers address many of the problems we have raised about arriving at a selective, ordered organic geochemistry, and they propose geochemical precursors for many of the regularities of metabolism that we have noted in earlier chapters. Many of the papers posit quite detailed scenarios for the emergence of life, some attempting to forge a continuous chain from vent chemistry to the emergence of the first free-living cells and even the ancestors of the major prokaryotic clades. The attempt at narrative continuity across such a large sweep of qualitatively different transitions seems to us tenuous at the present time, but it is also not central to evaluating the merits of some of the important proposals about energy-coupling mechanisms offered in these papers.

An emphasis that sets the AHV model papers apart from much work on prebiotic chemistry is that they seek to account for explicit *patterns within proton and redox bioenergetics*, as an integral part of accounting for pathways in organosynthesis. They extend the continuity hypothesis from geochemistry to life, to apply not only to reaction mechanisms and network structures within anabolism, but also to the use of proton-motive force, and (in several recent papers) to the role of electron bifurcation (see Section 4.5.5.1) as a generator of diverse and low-potential reductants.

---

[32] Today, the distinction between acidic and alkaline vents is for the most part a distinction between basalt-hosted (sometimes called "magma-hosted") on-axis vents, and off-axis peridotite-hosted hydrothermal fields in faulting systems. The major assumption made in the AHV models is that high pH is associated with copious $H_2$ production through serpentinization, while low pH (and resulting dissolution of metals) occurs in hot, on-axis systems. We saw in Section 3.5.3.2 that, even among present-day vents, the relation between $H_2$ production and pH can vary widely, and we noted in Section 3.6.6 that, due to differences in the composition of basalts and also seawater in the Archean (and more so, the Hadean) relative to the present, the character of on-axis systems could have differed still further. Some of these differences and their consequences for the AHV model are explored in [739].

[33] Multiple collaborations contribute to this literature, and several partly independent ideas are now being pursued. The reader will find a summary of most of the development and currently active ideas in [52, 53, 96, 471, 473, 474, 475, 523, 524, 525, 526, 527, 593, 683, 684, 685, 686, 687, 688, 689, 690, 691, 692, 693, 694].

How far this can be done in a natural way, and at what levels continuity should be sought – in the roles of chemical elements, or crystalline unit cells, or even macrostructures such as cell-membrane geometry – are matters of disagreement among the AHV scenarios presented by different authors. Even so, numerous explicit mechanisms are proposed within a broadly coherent emphasis on problems of energy coupling, which may be fruitful to pursue experimentally.

Three properties of alkaline hydrothermal vents, in the context of presumed Hadean ocean chemistry, figure centrally in all attempts to link them continuously to biochemistry and bioenergetics.

1. Peridotite-hosted vents generally – whether alkaline, as at Lost City, or even acidic, as at Rainbow or Logatchev – generate the highest concentrations of soluble reductants, particularly $H_2$ and $H_2S$, among hydrothermal systems [427, 651] (see Section 3.5.5). Low-temperature vents for which serpentinization is the main alteration process and crustal cooling is the main heat source are long lived (tens to hundreds of thousands of years [500], compared to individual vents in magma-hosted systems, which may be active for only tens to hundreds of years [427]). Because the AHV scenarios localize chemistry within the precipitate mound at the surface venting site, they attach significance to the longer stable intervals of chemical disequilibria at low-temperature vents.

2. The pH difference between alkaline vents and a presumed acidic ocean is the largest among hydrothermal systems. (pH values typically assumed are in the range 9–11 for primary alteration fluids and 5–6 for the Hadean ocean [690].) As noted in Section 6.2.1.3, if the FeS films that precipitate under such conditions focus the pH difference so that it produces not merely a gradient under mixing but a discrete trans-membrane potential, then a proton-motive force of $\sim 150 - 300$ mV [473] is available alongside the free energies of redox reactions.

3. The sign difference between the pH of vent fluids and of seawater creates an analogy between the gradient of proton-motive force from the ocean to the hydrothermal mound, and the trans-membrane difference (exterior-to-interior) in archaea and Gram-positive bacteria [96, 475]. If mineral membranes did once perform free energy transductions analogous to those performed in cells, the geometry of chambers and bubbles in the mound could suggest a step-wise progression in which the structure and the energetics were transferred from open mineral compartments to enclosed lipid vesicles.

   Suggestions for the way this continuation may have played out mechanistically have led to more difficulty and disagreement, because of the variety and sophistication of the microbial bioenergetic systems available to serve as models. At the least, the AHV scenarios all propose that organosynthesis depended continuously on trans-membrane potentials, and that organic systems capable of escaping vents must have been selected first for mechanisms to control geochemical trans-membrane potentials, and second for autonomous systems to produce them as a precondition for the utilization of topologically closed vesicles.

In the next section, we sketch a sequence of stages through which we think an ordered geochemistry must have passed, and selective mechanisms that made it progressively more sparse, on the path of becoming metabolism. We emphasize the universal hierarchies reviewed in Chapters 4 and 5, and aspects of geochemical context from Chapter 3. At many points experimental results are available demonstrating key reaction mechanisms in plausible mineral contexts, although many important gaps remain, as well as difficulties reconciling experimental models with chemical observations of systems on Earth today.

For the problems of early partial reduction of $C_1$ units and the first formation of C–C bonds and thioesters, we draw on the AHV hypothesis papers with their extensive use of microbial models. Such models, in the same stroke, emphasize the importance of an integral treatment of bioenergetics and biochemistry, and also the difficulty and ambiguity of finding continuity from highly evolved cellular systems back to mineral analogues. Whether or not the hypotheses in the current AHV scenarios will survive experimental testing in their current detailed forms, they introduce useful motifs such as the architecture of fuel cells, which capture common elements in the way proton-motive force and electron transfer can be coupled across either mineral or cellular membranes.

### 6.3 Stages in the emergence of the small-molecule network

In its earliest stages the emergence of the skeleton of metabolism could not have been considered distinct from geochemistry. It involved organosynthesis and the first steps toward complexity. It could not be regarded as biotic because it was not autonomous. Each reaction sequence depended in detail on the environment and energetics of some hosting system, which we assume was mineral-bounded hydrothermal fluid. Although this stage was merely a subsystem within geochemistry, it nonetheless established some of the defining regularities of metabolism because the chemistry was *selective*.

The interpretations of metabolic architecture given in Chapter 4, including the ongoing constraints it imposed on evolutionary diversification well into the era that can be reconstructed phylogenetically, suggests a decomposition into *layers of accreted complexity*.

The lowest layers are defined by reactions in the dilute-solution limit, probably involving C–H–O structures with sulfur leaving groups, and depending on environmentally supplied transition metal catalysts. Initially we consider only reactions among environmentally supplied molecules, excluding even feedbacks of the reaction products through network-autocatalytic loops. This limit reflects the kinetics of the catalyzing environment in its simplest form,[34] which acted as a conduit directing the flow of matter and the bound free energy in functional groups. It is the counterpart – in the complex hypergraph of chemical reactions – to passive Laplacian diffusion of heat or matter from a concentrated source to a diffuse surrounding.

---

[34] Recall the analogy in Section 6.1.3.4 to the kinds of information imposed by an open-loop control system.

A second stage reflects the lowest level of feedbacks: first-order reactions of products from the dilute-solution limit with environmental molecules. This is the level at which the instability toward self-amplification of network-autocatalytic loops can be seen. If the first network autocatalysis involved C–H–O reactions, the incorporation of environmentally supplied nitrogen would have become important upon the accumulation of CHO backbones.

Further increases in concentrations of products from the first two stages introduce the possibility for new feedbacks in which organic molecules complete networks as group-transfer or redox cofactors, or affect reaction rates as molecular catalysts. The structural incorporation of nitrogen was a precondition to all of these functions and the structural incorporation of sulfur seems likely to have been essential to some of them.

It is tempting to think of the layers of metabolism as reflecting a temporal sequence, but here we say only that they reflect a dependency sequence: second-order reactions require material from first-order reactions. On a young Earth very far from equilibrium, a vigorous organometallic chemistry could have brought several layers of complexity into existence essentially contemporaneously.

### 6.3.1 Carbon reduction and the first C–C bonds

The arguments in Chapter 4 suggest that the first phase in the transition from geochemistry to biochemistry was the emergence of a template carbon fixation network. The essential problem of early organosynthesis is forming partially oxidized organic compounds from $C_1$ inputs, which include high-energy leaving groups and preserve bond configurations that enable continuing reactions. Solutions to all three of these problems are seen – in probably the simplest chemical substrates where they can be instantiated – in the reactions that lead to acetyl thioesters, the most basic and possibly the oldest reactions in biochemistry. Therefore we begin with the mineral-hosted processes that seem most likely to have instantiated this general class of processes on the early Earth.

#### 6.3.1.1 $C_1$ reduction and CO addition at minerals: emphasis on acetate and thioesters

The analogy between the **Fischer–Tropsch** reaction from CO and methyl groups [31], and the **Wood–Ljungdahl** pathway of carbon fixation in acetogens and methanogens, has been noticed by many authors [377, 689, 737], and is believed among many proponents of metabolism-first origin to have been the first route to carbon fixation [63, 282, 523]. CO is the preferred carbon substrate for the carbonyl-addition step in this pathway,[35] but as we noted in Section 6.2.2.1, it is unstable under the water-gas shift (WGS) reaction at low temperatures. Therefore, a hypothesis of FTT synthesis under mild conditions introduces the problem of either transporting CO to a mineral environment or producing it locally in a

---

[35] FTT reactions with $CO_2$ as the carbon source have been demonstrated, but with low yields [267, 544]. $CO_2$ participation in FTT reactions appears to occur primarily through metastable interconversion to formate or CO [543], and whereas starting CO solutions readily produce methane and higher hydrocarbons, starting $CO_2$ produces little or none under mild conditions [546].

kinetic trap that sidesteps the WGS reaction. A third possibility is that CO can be produced at depth and high temperature, where the equilibrium constant of the WGS reaction is closer to unity and the activity of $H_2$ is high, and then transported out of equilibrium to the surface. Evidence has been presented by Lang *et al.* [476] that formate produced abiotically by some mechanism involving metastable transport from deep fluids[36] is present at low concentration in the primary alteration fluids of the Lost City hydrothermal field. Thus the preservation of this unstable species even at high temperature is possible at some level; whether the circumstances in which it occurs are general, and whether they could lead to sufficient concentrations of formate to drive a robust organosynthesis, are not evident from field observations at present, and will likely require models of Hadean hydrothermal systems to deduce.

Other problems in realizing a geochemical $C_1$ reduction pathway involve reduction from $CO_2$, formate, or CO to methyl groups. Full reduction from $CO_2$ to $CH_4$, occurring at high temperatures and pressures in the deep subsurface, is believed to be responsible for a significant fraction of the methane in the upper mantle [31, 68, 127, 267, 369, 428, 429, 543, 737, 738, 826], and even for the synthesis of small hydrocarbons found in fluids from the Lost City hydrothermal field [651].

It is known from experiments that $CO_2$, formate, or CO can be reduced using magnetite [68, 588], a variety of metal sulfide surfaces [146], or the $FeNi_3$ alloy awaruite [760, 761] as catalysts, using $H_2$, $H_2S$, or the metal centers themselves [544] as reductants. In an important early paper for the theory of a geochemical origin of metabolism, Wolfgang Heinen and Anne Marie Lauwers [350] demonstrated reduction of $CO_2$ with $H_2S$ on a substrate of FeS, to form $FeS_2$, $H_2$, $CH_3SH$ and a variety of other minor products.[37] Thus, under some conditions, not merely partially reduced carbon, but partially reduced *activated* carbon in the form of $CH_3SH$, can be produced from geochemical precursors at temperatures below $100\,°C$.[38]

Starting from inputs of $CH_3SH$ and CO, in environments containing precipitates of FeS and NiS, Claudia Huber and Günter Wächtershäuser [377] demonstrated the insertion of the carbonyl to form methyl-thioacetate ($CH_3CO\sim SCH_3$), with part of the thioester hydrolyzing to form acetate ($CH_3COO^-$). Formation of acetate at either high or low pH was observed when NiS alone was present, but co-precipitated mixtures of FeS and NiS led to good yields of acetate at near-neutral pH. Huber and Wächtershäuser proposed that a nickel-bound thioacetate was formed as an intermediate in the synthesis of methyl-thioacetate, a mechanism analogous to that of the CODH/ACS enzyme, but this has not

---

[36] See [666] for further discussion.
[37] Heinen and Lauwers also showed that $CH_3SH$ can be produced directly from $H_2S$ and $H_2$ with input of carbonyl-sulfide (OCS), but that this process is less efficient at generating thiols, with a larger fraction of carbon going into $CS_2$.
[38] Thiols have long been of interest in origin of life studies, because of their central role in extant biochemistry, to which we return in the discussion of thioesters in Section 6.3.1.9 below. They were known even in the 1960s to form from $H_2S$ in a variant on the Miller–Urey experiment [137] (see further results reviewed in [622]), so their possible relevance to prebiotic addition reactions has long been of interest. Fritz Lipmann and colleagues had shown by 1970 [298, 299] that amino acid thioesters could spontaneously polymerize – the $NH_2$ group provides the reductant to return the thiol as the peptide bond forms – and that the non-ribosomal small bacterial peptide gramicidin S is produced in this way.

been verified in a systematic analysis.[39] Cody *et al.* [145, 147], in a much more complex study of the hydrothermal degradation pathways of citric acid,[40] observed the hydrocarboxylation of olefins from CO to form carboxylic acids in the presence of a wide variety of metal sulfide catalysts including FeS and NiS, along with the production of several different thioacids and thioethers.

An empirical caution to all these results is that methyl compounds of demonstrably abiogenic origin, including methane thiol, have not been found in hydrothermal effluents,[41] and the thiols that have been observed appear to have been produced by microbial interaction with subsurface methane [666]. Despite the consensus that Fischer–Tropsch-type synthesis occurs at high temperatures and pressures in the subsurface, it is a separate problem to understand whether intermediates along a $C_1$ reduction sequence are obtainable under mild near-surface conditions. The important difference is that the reduction to partially oxidized $C_1$ intermediates becomes endergonic at low temperatures. This is the problem of the equilibrium of the water-gas shift reaction introduced in Section 6.2.2.1, to which we return in Section 6.3.1.5 below.

### 6.3.1.2 Microbial models of one-carbon reduction and the role of energy coupling among multiple reactions

The literature on alkaline hydrothermal vents, introduced in Section 6.2.4, has made elaborate use of the details of microbial metabolisms to address problems of obtaining partially reduced one-carbon units under mild conditions, and the possible role of coupling to proton-motive force, or perhaps depending on geochemical mechanisms of electron bifurcation.

The early papers made extensive use of acetogens and methanogens as quite literal models, based on their presence in deep-branching clades and the similarity of the Fischer–Tropsch reaction to the W-L pathway. The diversity and also the evolutionary refinement of the energy-coupling mechanisms across this group of organisms, however, has made proposing geochemical analogues difficult, and more recent papers in this literature have explored the use of a wider suite of redox couples in vent environments, and a more complex array of carbon sources from the surface and subsurface. We therefore digress briefly from experimental results on organosynthesis to consider some of the proposals that have been made on the basis of more detailed analogies to microbial metabolisms.

### 6.3.1.3 Acetyl-CoA pathway organisms as models

In early papers developing the AHV model, William Martin and Michael Russell [524, 689] argued for quite extensive correspondences between the geochemistry at alkaline vents leading to life, and metabolisms based on the acetyl-CoA pathway. Martin has continued

---

[39] They also demonstrated that, as an alternative to hydrolysis of this thioester to acetate, a peptide bond to aniline (aminobenzine) could be formed under appropriate conditions.
[40] The motivation for this experiment was to identify kinetically accessible transitions that might be reversed, and relevant addition reactions, to understand organosynthesis to compounds found in intermediary metabolism.
[41] See [773], page 13.

to argue [523, 527] that the acetyl-CoA pathway acted continuously from geochemistry to the first cells, and that it was instantiated through independently evolved $C_1$ reduction branches as acetogenesis in the common ancestor to bacteria and as methanogenesis in the ancestor to archaea.

The degree to which the W-L pathway is analogous between acetogens and methanogens,[42] which bears on the ability to use these organisms as models to make specific predictions about early energetics of carbon fixation, depends on which details one is willing to absorb into abstractions. Making this judgment is difficult because, in different ways, both groups are highly optimized.

Acetogens and methanogens both face the limitation that the reduction of $CO_2$ with $H_2$ yields relatively small net reaction free energies at the environmental concentrations of $H_2$ where these organisms live. This leaves little margin for the dissipation of heat from mismatched redox couples at intermediate steps in their core metabolic pathways.[43] The interaction of biochemistry and bioenergetics in acetogens and methanogens is striking both for its elegance and efficiency at working within this limitation, and for the indirectness with which free energy is partitioned, captured, and shuttled around among different components and processes. The complexities of their pathways reflect the combinatorial problem of either matching, adding, or subtracting free energies from different reaction segments to enable all pathway steps to run forward while keeping dissipation of heat within the tight bounds of viability. Figure 6.2 shows examples from the acetogen *Aceto-bacterium woodii* [640] and the methanogen *Methanothermobacter marburgensis* [415]. When these two groups are compared, the most striking feature is how qualitatively different the energy systems are that achieve this efficiency.

Both methanogens and acetogens form acetyl-CoA through the reaction

$$4\,H_2 + 2\,CO_2 + CoASH \rightarrow CH_3CO{\sim}SCoA + 3\,H_2O \qquad (6.2)$$

with $\Delta G^{\circ\prime} = -59.2\,kJ/mol$ [282]. Although exergonic, as emphasized by Shock and others [21, 22, 545, 746, 749], this reaction free energy is too small to support a cellular bioenergetic system. More importantly, because some intermediate oxidation states are formed endergonically, the net free energy to synthesize acetyl-CoA does not reflect the full suite of constraints on carbon fixation.

The distinctive characteristic of the acetyl-CoA pathway organisms is that they use the same $C_1$ reduction pathway to bypass carbon fixation, excreting reduced carbon as a waste product. In methanogens, the waste is methane, produced in the reaction

---

[42] The CODH/ACS enzyme is homologous among all W-L organisms. The synthesis of the common precursors to folates and methanopterins, 6-hydroxymethyl-7,8-dihydropterin and its diphosphate, are also shared, with some pathways identical in bacteria and archaea at the substrate level, and homologous enzymes for the initial cyclohydrolase reaction from GTP in some pathway variants [773]. (Cleavage of part of the ribose phosphate moiety is the step in which most pathway variability is found.) The later synthesis of folates and pterins involves structurally different pathways in the modification and attachment of the pABA moiety and subsequent reduction and methylation of the pterin (see Section 4.5.2.1), and accordingly involves non-homologous enzymes.

[43] To give a flavor of the pressure these organisms are under to operate efficiently, Lane and Martin [473] note that methanogens may generate about 40 times as much mass of methane in the "energetic bypass" use of one-carbon reduction as they produce in cell biomass.

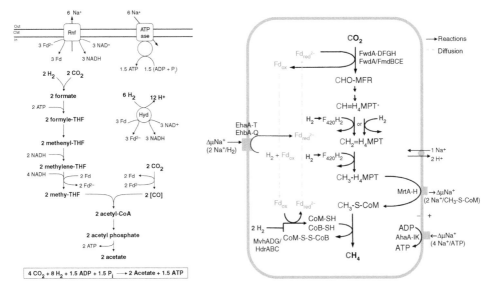

Figure 6.2 Bioenergetic systems enabling carbon fixation by the Wood–Ljungdahl pathway. Left: in the acetogen *Acetobacterium woodii* (after Poehlein *et al.* [640]). Right: in the methanogen *Methanothermobacter marburgensis* (after Kaster *et al.* [415]). In this acetogen, both ATP and ferredoxin (Fd) are recycled internally in the reduction of $CO_2$ to acetate. Ferredoxin is independently recycled between the electron-bifurcating enzyme (Hyd) and the proton-pumping enzyme (Rnf). The net flux between the two systems is a pool of NADH, reduced by $H_2$ and consumed in $C_1$ reduction. ATP from the membrane potential must, however, be consumed to produce acetyl-CoA, where it is not recycled by release of acetate. In the methanogen, Fd is again recycled within the branch of $C_1$ reduction to methane, and the net free energy of reduction is captured in the transfer of methyl groups from methyl-$H_4$MPT to methyl-CoM to produce a membrane ion potential. Part of this ion potential must be fed back to recycle low-potential reduced ferredoxin ($Fd_{red}^{2-}$) in the carbon-fixing branch where methane is not released.

$$4\,H_2 + CO_2 \rightharpoonup CH_4 + 2\,H_2O \qquad (6.3)$$

with $\Delta G^{\circ\prime} = -131$ kJ/mol [801]. Acetogens operate a similar bypass, excreting the less-reduced carbon in acetate, using the reaction

$$4\,H_2 + 2\,HCO_3^- + H^+ \rightharpoonup CH_3COO^- + 2\,H_2O \qquad (6.4)$$

at $\Delta G^{\circ\prime} = -104.6$ kJ/mol [281].[44]

In both acetogens and methanogens, a **membrane ion potential** (in Figure 6.2, for $Na^+$ ions) is the interface that transfers free energy from the carbon-excreting (bioenergetic) branch to the carbon-fixing (acetyl-CoA) branch of the $C_1$ reduction pathway. The free energy injection in carbon fixation also occurs at the same point in the two groups –

---

[44] Note that scheme (6.4) is a two-carbon reaction, so the free energy per carbon is less than half that for scheme (6.3).

to power the reduction of $CO_2$ to formate or a bound formyl group. This implies an energetic dependency hierarchy: $C_1$ reduction in the bioenergetic pathway is energetically self-maintaining; harnessing of free energy from this pathway powers the membrane ion potential; and finally, the ion potential provides free energy to the endergonic carbon fixation branch.

The division between carbon fixation and bioenergetics – with the membrane providing energy transduction – is the qualitative aspect of microbial models that the AHV scenarios attempt to map backward onto vent chemistry. Mineral membranes are to have provided the initial separation, and the membrane potential was to have been generated geochemically, before biochemical sources evolved to supplant the vent-fluid/seawater disequilibrium.

The claim of continuity becomes difficult, however, when attempts are made to interpret it in mechanistically explicit terms, because the mechanisms used even within organisms as biochemically similar as acetogens and methanogens differ not by mere enzyme phylogeny, but at the grossest levels of architecture. In methanogens, electron bifurcation[45] at a soluble enzyme (MvhADG/HdrABC) enables the bioenergetic pathway to be self-maintaining from $H_2$. The high-potential heterodisulfide oxidant, CoB–S–S–CoM, formed upon the release of the methyl group from methyl-CoM, is jointly reduced by $H_2$ along with a ferredoxin that may have a midpoint potential $\sim 100\,mV$ below the midpoint potential of the $2H^+/H_2$ redox couple [415]. The net free energy of $C_1$ reduction is coupled to the membrane energy system at the methyl transfer step from methyl-$H_4$MPT ($CH_3-H_4$MPT) to form methyl-CoM ($CH_3-CoM$). Finally, the membrane potential powers carbon fixation via reduced ferredoxins ($Fd_{red}^{2-}$), at the same entry point as in the bioenergetic branch. ATP, also produced with free energy from the membrane potential, is used outside the $C_1$ system.

In contrast, in acetogens, the bioenergetic branch is self-maintaining from NADH – a lower-energy reductant than $H_2$. Ferredoxins again provide the coupling between late-stage $C_1$ (in this case methylene) reduction and $CO_2$ reduction to the bound CO group at the acetyl-CoA-synthase enzyme. In a separate cycle, ATP synthesized using the free energy of substrate-level phosphorylation from acetyl-CoA is fed back to enable the reduction of $CO_2$ to formyl-THF from $H_2$ (a more endergonic reaction than its methanogenic counterpart on $H_4$MPT). In this system, net $C_1$ reduction does not energize the membrane potential directly. Rather, by drawing electrons from NADH and returning the high-potential oxidant $NAD^+$, the net $C_1$ reduction pathway enables the soluble electron-bifurcation enzyme (Hyd) to reduce Fd with $H_2$ in the presence of $NAD^+$; the subsequent reduction of $NAD^+$ with $Fd_{red}^{2-}$ energizes the membrane ion potential. ATP produced by the membrane potential drives carbon fixation, again at the same point of entry as the methyl branch of the bioenergetic pathway, though using a different free energy carrier than methanogens use at this point.

**In summary** The acetyl-CoA pathway organisms, then, form a clear family in regard to the one-carbon reduction path and the role of nitrogens from pterin and amino-benzoate

---

[45] See Section 4.5.5.1.

in carrying the intermediates. Yet in the biosynthetic enzymes for their specific pterin cofactors, which tune the energetics of the reduction sequence, they seem independent and parallel. Finally, in optimization of the modularity that makes them self-maintaining energetically, they are nearly incomparable.

### *6.3.1.4  Ancient and central roles for other redox couples?*

Citing these difficulties, a recent alternative model by Wolfgang Nitschke and Russell [593] eschews a close association with either acetogens or methanogens. It proposes, instead, an early carbon-fixing system that was not unidirectional in either reducing or oxidizing carbon, but involved both reduction from $CO_2$ with $H_2$ and oxidation from $CH_4$ with atmospherically generated oxidants such as $NO_2^-$. The resulting models are very complex, drawing from pathways in denitrifying methanotrophs [233] and even methylotrophs using the serine cycle (see Section 4.4.6.3). We do not know how to evaluate proposals for models that are at the same time complex and specific in this way, but we will review two points from the arguments behind them that bear on concerns we believe are likely to be general.

### *6.3.1.5  The problem of in situ reduction of $CO_2$ to CO or $H_2CO$ under mild conditions*

The chemical problem that motivates Nitschke and Russell to abandon acetogens or methanogens as model organisms is that their ability to execute $C_1$ reduction *as a pathway* seems to depend in essential ways on complex and refined networks of free energy exchange of the kind seen in Figure 6.2. Direct reductions from $H_2$ are impeded by the positive free energy of formation of the partially oxidized intermediates CO and formaldehyde ($H_2CO$).[46] The first endergonic step, as noted above, is the WGS reaction (6.1). Figure 6.3 shows these free energies of formation both for free $C_1$ groups in solution and for the reduction sequence on THF and $H_4MPT$. In solution, the reduction of formate to formaldehyde with $H_2$ is even more endergonic than the reduction of $CO_2$ to formate, though on pterin cofactors the formyl group can be reduced exergonically to the less-oxidized methylene group with $H_2$ or NADH.

Acetogens and methanogens circumvent the equilibrium of the water-gas shift reaction by coupling the initial reduction of $CO_2$ to exergonic reactions, though as Figure 6.2 shows they do so in different ways. Lacking a mechanism by which either to harness proton-motive force or to use electron bifurcation to carry out a similar reduction geochemically, Nitschke and Russell propose instead to arrive at methanol and formaldehyde by *oxidizing* methane, using the strong oxidant $NO_2^-$.[47] The assertion that, environmentally, both $CO_2$ and methane would be available, but not the intermediates that connect them, is effectively the bistability argument already suggested by the free energy plot of Figure 4.2. Nitschke and Russell interpret the absence of methanol and methane thiol in Lost City

---

[46] It is because the midpoint potential of the $2H^+/H_2$ redox couple is too high to reduce $CO_2$ to CO, that both methanogens and acetogens must either use a reduced low-potential ferredoxin or couple the reduction to hydrolysis of an ATP.

[47] Mechanisms to produce nitrite in the Hadean atmosphere are known (see Section 6.3.3 below); whether they would produce sufficient nitrite to be geochemically important is not clear.

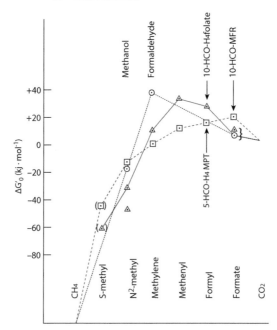

Figure 6.3 Energetics of direct one-carbon reduction, in solution and on THF and $H_4$MPT cofactors. (From Edward and Maden [509].)

fluids as evidence that, where reduction of $CO_2$ can occur, at depth and high temperature, it tends to run to completion, but not to preserve partially oxidized or activated intermediates. Depending on conditions, only methane may be produced [369], or carbon addition may run away to higher alkanes [546], in either case yielding highly reduced and relatively non-reactive compounds.

### 6.3.1.6 Fuel cell models of proton transfer, and bifurcation

Whatever the status of the intricate model of Ref. [593] will turn out to be, it is instructive to review the way Nitschke and Russell, Laura Barge, and collaborators propose to couple proton-motive force to electron transfers, by segregating oxidation and reduction reactions in space with a conductive Fe/Ni membrane.[48] This separation is the design principle behind the **fuel cell** [53].

The two environmental molecules $H_2$ and $NO_2^-$ are assumed to interact with metal centers that can alternate between 2+ and 3+ states. Because both reactions involve ions, the free energies are sensitive to pH. In the example below we use iron as the active metal center.

---

[48] The reader is referred to Figure 5 of [593].

The reduction of metal centers by $H_2$ is to be carried out in the alkaline hydrothermal alteration fluid,

$$H_2 + 2\,Fe^{3+} \rightleftharpoons 2\,H^+ + 2\,Fe^{2+}, \tag{6.5}$$

where the chemical potential of the liberated $H^+$ is low. The complementary oxidation of metal centers by $NO_2^-$ is to be carried out in more acidic seawater,

$$2\,NO_2^- + 4\,H^+ + 2\,Fe^{2+} \rightleftharpoons 2\,NO^\cdot + 2\,H_2O + 2\,Fe^{3+}, \tag{6.6}$$

where the chemical potential of the donated protons is high. The net reaction includes the transfer of an electron pair from $H_2$ to $2\,NO_2^-$ but also the effective translocation of $2\,H^+$ from a fluid at pH 5.5 to a fluid at pH 10.

A final proposal in Ref. [593] *et seq.* is that molybdenum, acting as an obligatory two-electron exchange ion, in a matrix of FeNi or FeFe centers, could accept paired electron states from donors such as $H_2$, and then couple the single-electron transfer to high-potential oxidants such as $NO_2^-$, to the production of low-potential reduced metal centers, implementing a form of electron bifurcation in the mineral matrix. A useful research topic in mineralogy could be a systematic survey of plausible Hadean minerals and precipitates to search for paired metal centers, in which reduction of the two metals occurs at very different midpoint potentials. A second search could then be carried out for impurity metal centers (such as molybdenum) that would have pair-transfer potentials straddled by the midpoints of the metal centers in the primary matrix. Transfers within unit cells would provide the most robust mechanisms of bifurcation, but interface reactions in mineral assemblies could also be considered.

### 6.3.1.7 Selectivity within and beyond the Wood–Ljungdahl pathway

It is worth noting that *all* of these arguments from the AHV literature bear only on the problem of partial reduction of $C_1$ units and the synthesis of the first C–C bonds and thioesters. They offer no basis for specific predictions about how these compounds would be extended to more complex molecules by downstream anabolic pathways. The early AHV papers [524, 689], arguing from the premise that organisms using the acetyl-CoA pathway occupied the base of the tree of life, extended the literal interpretation of these microbial models to propose that the split arcs from acetyl-CoA to the precursors of anabolism would have been somehow an inevitable elaboration of a vent analogue to the W-L pathway. We see no basis for that extended claim of continuity. We believe that the selection of the universal precursors requires an independent explanation, which is best provided by autocatalysis.

### 6.3.1.8 The carbon addition reactions associated with the rTCA cycle

Feedbacks through short autocatalytic loops are the most plausible sources of robust self-amplification that concentrate energy flows and select metabolic precursors. Chapter 4 showed that feedbacks of just this kind lie at the very core of extant metabolism, and that they center on reductive citric acid (rTCA) cycling. We suggest here that they were preceded by the same or similar feedbacks in geochemical organosynthesis.

Before any network-autocatalytic carbon fixation such as rTCA cycling could have occurred, the key reactions must first have been supported. We suppose that, as for the first $C_1$-addition reactions, the first reactions analogous to the rTCA fixation steps depended on catalysts of mineral origin, which were particular to the local functional-group chemistry of these reactions, but not otherwise specific at the level of whole molecules.

The distinctive fixation reaction that sets the rTCA cycle apart from other pathways is the **carbonyl insertion** reaction at a thioester (either acetyl-CoA or succinyl-CoA in the modern cycle), using a thiol as a leaving group. While the modern reaction employing $CO_2$ is both chemically complex and endergonic,[49] requiring a low-potential reductant, an analogue using already reduced and reactive CO as the carbon source is much more plausible geochemically, and not very different from the CO insertion reactions demonstrated by Huber and Wächtershäuser that were reviewed in Section 6.3.1.1.[50]

Following the work of Heinen and Lauwers, and of Huber and Wächtershäuser, George Cody and colleagues performed similar experiments [146] using only FeS rather than FeS-NiS co-precipitates, but working at much higher temperatures (250 °C) and pressures (50–250 MPa). For experimental and diagnostic ease, they used nonyl thiol rather than methane thiol as the organic sulfide input, but CO (from formate) was again the oxidized carbon source. Two of the striking results of their experiment were a demonstration that many Fe centers were heavily carbonylated, and the unexpected formation of *pyruvate*. It is possible, though it was not shown in that work, that an addition reaction similar to carbonyl insertion was responsible for the synthesis of pyruvate, though it is also possible that some more complex intermediate was synthesized and subsequently hydrolyzed or cleaved by an aldol reaction to form pyruvate as one product. Cody and colleagues later performed a more extensive assay of the role of multiple transition-metal sulfide catalysts in related reactions [145].

The reactions in this section and in Section 6.3.1.1, although they differ in their exact products and in the metabolic pathways to which they are most nearly analogous, involve many common mechanisms in C–O–S chemistry and suggest some homology between the beginnings of the rTCA pathway and the Wood–Ljungdahl pathway. This is striking as the two pathways are often regarded as alternative models for the earliest chemotrophic metabolism [63, 282, 473, 523].

The proposal of a deep connection between direct $C_1$ reduction/addition and rTCA cycling, reaching back to a geochemical era before the function of carbon fixation had been transferred to organic cofactors, was made by Günter Wächtershäuser [845], who suggested that a mineral analogue to the Wood–Ljungdahl pathway provided a

---

[49] The reaction from $H_2$ is endergonic by $\sim$18.8 kJ/mol = 0.195 eV, and therefore requires a reduced ferredoxin in the rTCA cycle. The reverse reaction in oxidative TCA reduces $NAD^+$ (via lipoic acid).

[50] Whether the CO must have been supplied exogenously to the reaction fluid, or could have been created from $CO_2$ through interaction with highly reduced metal centers taking the place of low-potential ferredoxins in modern biochemistry, we leave as an open question.

feedstock of acetyl thioester for an autocatalytic loop.[51] This is essentially the relation we will propose, in this section, to have begun in a mineral era and then *persisted* into the eras of cofactors and eventually fully formed cells. It draws on the argument by Braakman and Smith [90, 91] from Section 4.4.4, that the two pathways were not *alternatives*, but rather were *linked*, at the time of the Last Universal Common Ancestor.

The second reaction that must have been supported in order for rTCA cycling to have been possible is the $\beta$-**carboxylation** of ketoacids (pyruvate or $\alpha$-ketoglutarate in the modern cycle) using bicarbonate as the $C_1$ source. This reaction is kinetically facile even without enzymes; the problem for its plausibility is energetic. The energy in modern organisms with efficient enzymes can be provided in two ways by a single hydrolysis of ATP to ADP and a transferred phosphate. The important activated intermediate in either case is carboxyphosphate, together with the enol form of pyruvate. The first method begins by phosphorylation of pyruvate to form PEP, which acts as both phosphate carrier and stable enol-pyruvate form. The phosphate group can be donated by PEP to bicarbonate to form carboxyphosphate, which then transfers the carboxyl group to enol-pyruvate to form oxaloacetate.[52] Alternatively, the activated carboxyl group of carboxyphosphate can be formed directly and transferred to a biotin cofactor to form carboxybiotin, from which it can then be donated directly to pyruvate through the action of the enzyme *pyruvate carboxylase*. Four possibilities for geochemical analogues to this reaction are: (1) that some activated bicarbonate was present in hydrothermal fluids or mixtures; (2) that strong reducing potential led to reduction of small amounts of $\alpha$-hydroxyacids (such as malate or isocitrate) from the very small concentrations of $\beta$-ketoacids that would be formed spontaneously; (3) that some geochemical substrate facilitated the coupling of the free energy of reduction to the process of $\beta$-carboxylation;[53] or (4) that some energy source distinct from redox potential also coupled to the $\beta$-carboxylation of ketoacids.

At present we do not know of experimental results indicating which if any of these is plausible, so our argument for their presence depends on the circumstantial argument for an early role for rTCA cycling. It is important to note, however, that $\beta$-carboxylation reactions or other additions of activated bicarbonate are the least specific of carbon fixation reactions to rTCA cycling. All cyclic carbon fixation pathways employ such reactions, and they are also parts of the anaplerotic reaction segments for the acetyl-CoA pathways. Therefore explaining how bicarbonate addition entered biochemistry will be a crucial problem for any metabolism-first theory that purports to account for the selection of the biosynthetic precursors in life as we know it.

---

[51] The proposal of Wächtershäuser differed from the extant rTCA cycle by proposing sulfur-substituted intermediates, and invoking the oxidation of FeS to pyrite $FeS_2$ as an essential source of free energy.

[52] This pathway is used for gluconeogenesis in *Methanobacterium thermoautotrophicum*; see the MetaCyc [412] pathway gluconeogenesis II.

[53] Recall from Section 4.4.5.1 the fact that the isocitrate dehydrogenase (ICDH) in most modern bacteria binds oxalosuccinate in this manner, combining the endergonic $\beta$-carboxylation with an exergonic reduction to produce a nearly reversible net reaction.

### *6.3.1.9 Substrate-level phosphorylation versus membranes as early sources of phosphoanhydrides*

Christian de Duve had proposed in the early 1990s [183], that **thioesters** were the first energetic leaving groups in protometabolism, enabling not only continued carbon addition, but also dehydrating polymerization and the production of phosphoanhydrides through substrate-level phosphorylation (SLP). The proposal is appealing in many ways, but at the time, for energetic reasons,[54] de Duve could not offer a plausible source of thioesters. If a geochemical analogue to the Wood–Ljungdahl pathway did exist (in alkaline vents or elsewhere), then SLP from species such as methyl-thioacetate could have produced acetyl-phosphate, and transphosphorylation could then have produced phosphoanhydrides, as the same steps do in the acetogenic pathway in Figure 6.2, making de Duve's *thioester world* much more promising. Knowing when the first phosphoanhydrides became available, from what source, and in what concentration, would also improve the precision of the questions that can be posed about $\beta$-ketoacid formation, and an increasing set of roles for phosphate groups that seem necessary by the time sugars entered protometabolism.

An important question is whether SLP was the only source of phosphoanhydrides over the long period until the ribosome had reached sufficient complexity to produce trans-membrane proteins capable of ion-driven phosphorylation, or whether a geochemical analogue to that process operated in parallel to enable the first rise of an oligomer world. Among the concerns of the AHV model of Section 6.2.4 are the early sources of phosphoanhydrides, and depending on whether analogies to acetogenic or methanogenic phenotypes are pursued, either an early or late role for vent pH differences in the production of phosphoanhydrides can be proposed. Continuity with acetogenic phenotypes would suggest an early role for polyphosphate-mediated coupling between proton potential and carbon fixation, whereas continuity with methanogenic phenotypes would require coupling only through the production of low-potential reductants, and would allow membrane generation of ATP to enter biochemistry quite late. Arguments in favor of an early role for proton-driven phosphorylation are made in [53, 96, 593, 690], and arguments against are made in [473, 523].

The question whether the coupling between proton-motive force and phosphorylation might have a basis in condensed matter has been asked ever since Peter Mitchell first proposed [559] that proton translocation across membranes served as the intermediary between redox and phosphate energy systems, and before a reconciliation had been reached between proton translocation and Paul Boyer's 1977 proposal [88] that conformational change in the ATP synthase was key to the release of ATP. Morowitz in 1978 [569] – concerned with the problem of avoiding dissipative losses from a diffusive transport of protons across membranes, and drawing on both the glass-electrode architecture of pH/ORP meters and Lars Onsager's recent explanation of the source of excess proton mobility in ices –

---

[54] de Duve proposed (as one mechanism) the formation of thioesters by thiol attack on carboxylic acids. However, the de Duve hypothesis is endergonic at low temperatures and neutral pH, and overcoming this problem was one motivation for his proposal of a hot, acidic vent environment for the initiating reactions in a thioester world. Wächtershäuser [845] therefore suggested instead a substitution reaction on already-formed, mineral-bound thioacids.

proposed that proton translocation occurred not by classical conduction but by *semiconduction*, whereby protons would be delivered, with the full free energy of the pH difference across the membrane, to the surface where a chemical reaction with orthophosphate might lead to its polymerization.[55] Current experimental work with FeS precipitates [52] demonstrates facile transfer of $P_i$ from acetyl phosphate to $P_i$ to form $PP_i$, but so far no evidence for a coupling of proton-motive force to *de novo* formation of phosphoanhydride bonds.

### 6.3.1.10 Aqueous-phase reactions that are the common currency of carbon and redox rearrangement in biochemistry

The aqueous-phase reactions that make up much of the redundant structure within and across carbon fixation pathways seem less problematic than the initial carbon incorporation steps considered in the previous two subsections. These are the reactions characterized as facile in Section 4.3.7, and catalyzed today by widely diversified enzyme families. They include aldol reactions (both condensation and cleavage) hydrations/dehydrations, and reductions of double bonds. Soluble transition metals bound in metal-ligand complexes with carboxylic acids are either known or suspected catalysts for aldol reactions and reductions (V. Srinivasan and S. Ohara, unpublished). Geochemically plausible reductants include $H_2$, $H_2S$, and $Fe^{2+}$.

Some bonds in metabolites appear easier to dehydrate than others, and this difference deserves to be understood. The bonds that are dehydrated when alcohol groups are displaced by means of aconitic intermediates are also bonds at which aldol cleavages are performed in biochemistry, as shown in Figure 4.9. A surprising observation (G. Cody, personal communication) is that aldol condensations of pyruvate seem to permit facile dehydration of the created C–C bond following the condensation.

### 6.3.1.11 The first stages are characterized by limited kinetic selection

The first layers of organosynthesis as we have described them here should correspond roughly to processes that take place in the parent bodies of carbonaceous meteorites [156, 704]. Side-reactions should produce complex, heterogeneous mixtures of small molecules, and probably keep flow in autocatalytic loops below the threshold for self-maintenance.

*We must emphasize that the plausible early networks – even if they were assembled from reactions that later became the foundation of biochemistry – probably did not possess topologies that looked at all like the biological carbon fixation pathways. Moreover, the transition from these early networks to the first proper cycles may not have looked at all like the assembly of linear pathway segments.* More likely, it came through the addition of non-specific catalysts that would have altered local fluxes at several network positions characterized by similar functional groups, resulting in adjustments of flows across the network.

---

[55] In this, Morowitz was also following ideas of Peter Mitchell, who had proposed that protons played a role at the catalytic site for ATP formation. However, Boyer had established by 1977 that a direct involvement of protons in the catalytic site was not consistent with data on exchange reactions [16], so the Morowitz model would already have required an interpretation as a condensed matter *precursor* to the cellular enzymatic system, not a part of its mechanism.

Only through the incremental accumulation of this patchwork of catalysts would the coalescence of cycles and increasing levels of loop autocatalysis have occurred. This does not mean that the existence of autocatalytic topologies was unimportant: even below the threshold for self-maintenance, flows around loops with amplification could have concentrated carbon in cycle intermediates, as argued in Section 4.4.4.4.

We propose that on Earth, differently than in planetesimals, organosynthesis did not end at the first stage. Whether this can be explained simply because the Earth was larger, more stratified, tectonically active, and overall further from equilibrium, is an important technical problem about whether for planetary organosynthesis, "more is different" [30].

### 6.3.2  rTCA: the potential for self-amplification realized and the first strong selection of the metabolic precursors

We argue that the second stage in the emergence of a robust metabolism must have been the integration of elementary carbon addition reactions and other aqueous-phase reactions into cycles capable of self-maintenance and true self-amplification through network autocatalysis. The threshold for self-maintenance marks a transition between small finite concentration enhancement, and the first true chemical competitive exclusion and strong selection of core metabolites. The rTCA cycle is the most compelling candidate in extant biochemistry for a self-maintaining cycle.

#### 6.3.2.1  The circumstantial argument that early network autocatalysis is needed: kinetic selection of the core metabolites

The observations in Section 6.3.1 suggest that the carbon addition reactions in the rTCA cycle fall in a range of marginal plausibility. Carbonyl insertion at thioesters is a demonstrated chemistry but has not been verified for rTCA intermediates. $\beta$-carboxylation of ketoacids is kinetically facile but energetically disfavored from $HCO_3^-$ in solution. Many intermediates in the cycle have been demonstrated in one-pot reactions by Cody and collaborators, but nothing approaching the entire cycle has been demonstrated. Even if all reactions could be demonstrated in isolation, the same reactions that are required within the cycle seem likely to produce side-reactions that would prevent the cycle from reaching its autocatalytic threshold [614]. If a truly self-maintaining and self-amplifying cycle existed within early-Earth geochemistry, it seemingly must have been part of a larger network that has not been recognized, or to have relied on more selective catalysts than those studied thus far.[56]

The argument that an autocatalytic rTCA was a geochemical input, rather than a subsequent evolutionary optimization from the era of enzymatic catalysts, is therefore at this

---

[56] Leslie Orgel argued, in his last published paper [614], against the naive picture of an autocatalytic rTCA cycle. He framed the counter-argument in terms of a useful list of the laboratory demonstrations that would be required to justify a claim of a simple self-amplifying *in vitro* cycle. The objections concerned both feasibility of reactions – on the kinetic or energetic grounds we have reviewed above – and the control of side-reactions from mechanisms that are feasible. We believe that the feasibility problem is addressed by exploration of energy sources and catalytic environments beyond those currently recognized. The selectivity problem is more likely solved by expanding our discussion from the naive picture of a simple cycle to a more realistic picture of a cross-linked network. It is possible that, in place of the progressive assembly that occurs within the simple rTCA cycle, realistic networks proceed by "over-building" complex molecules, and then fragmenting those and recycling the fragments through various pathways.

stage largely circumstantial. It is an argument that the number of bits of information in oligomer sequences – even in modern organisms, but more so in earlier stages of life – and the slow rate at which these can be selected, is insufficient to allow such catalysts to be an asymptotically self-maintaining control system over small-molecule reaction networks unless those networks are somehow already limited in the diversity and rate of errors they can generate. Self-amplifying cycles are then proposed as a necessary mechanism to focus and simplify reaction networks, to a point where they are amenable to control by selection on genes and enzymes.[57] Cycles that are short and redundant at the functional-group level, which employ relatively facile chemistry, and which are not too productive of side-reactions,[58] are the most likely sources of kinetic selection, as was argued in Chapter 4.

The argument for rTCA in particular is driven by the need to account for the synthesis of all of biomass from the universal precursors: acetyl-CoA, pyruvate or PEP, oxaloacetate, $\alpha$-ketoglutarate, and perhaps glyxoylate. Among these, only acetyl-CoA seems to be a likely molecule from reactions of the Fischer–Tropsch type. To explain why the other universal precursors would be not only possible, but *necessary* invariants of metabolism, we propose that

1. populating the molecular species that are cycle intermediates was the most facile or most robust way to create a redox relaxation channel,
2. the robustness of the cycle resulted from self-amplification, which caused the cycle to dominate carbon entry and electron flux through a process of chemical competitive exclusion, and
3. the same self-amplification made the channel self-focusing so that its intermediates represented a large excess fraction of the organic carbon in the network relative to an equilibrium Gibbs ensemble.

The cycle intermediates would then have been the species available in excess supply, and they would have been the substrates most frequently encountered when cofactors and later oligomers arose in catalytic roles. Because the evolutionary selection of catalysts is a process of inductive inference, the most frequently encountered substrates in a structured milieu provide the most robust statistics to support induction. The most certain evolutionary reward would come from enhancing the flux around the cycle itself, which supplies the precursors from which the catalysts are made.

### 6.3.2.2 A mineral-hosted network and Gibbsian distribution of reactions and species

Mineral or soluble transition metal catalysts for rTCA reactions would likely have been selective only at the scale of the transition state itself or the functional group, not at the scale of the whole molecule. With such limited catalytic selectivity, any reaction network including the rTCA loop would, like a kinetic counterpart to the Gibbs distribution in

---

equilibrium thermodynamics, have contained many more molecules and reactions than the rTCA intermediates. We imagine that it might have resembled the universal covering network of Figure 4.5, since that network is produced by the redundant action of only a few "generators" defined at the level of functional-group transformation, as shown in Section 4.3.7. In addition to parallel pathways to the rTCA arcs, cross-cutting networks such as the glyoxylate bypass should have been kinetically accessible as well.

An important property of such Gibbsian ensembles is that although they contain many molecular species, the distributions satisfy a criterion of maximum entropy constrained by only a few parameters, which are the central tendencies of the distribution and which contain all the "information" in it.[59] For a driven network, the central tendencies that contain the network's information are the set of net fluxes driven by independent chemical-potential differences in the boundary conditions. The measure of "order" in a network with re-entrant cycles, which could be said to exceed the order pre-existing in the environmental chemical potentials that drive it, comes from the fluxes in any pathways that are self-focusing and self-amplifying. Side-reactions from these would be parasitic but would contain no information, like fluctuations around a mean in a thermodynamic distribution.

### 6.3.2.3 Crossing the threshold to self-amplification induces a non-equilibrium phase transition

In such a network with re-entrant pathways amid a cloud of side-reactions, a qualitative transition would result when the rTCA (or related) pathway crossed through its threshold for autocatalysis. This is an example of a non-equilibrium phase transition, as explained more fully in Section 7.6. Crossing a threshold to self-amplification would have led to an initially exponential growth of reaction flux in the self-amplifying pathway that gave order to the network, and to accumulation of carbon in the pathway intermediates and possibly in molecules close to them in the network.

Only the rise of second-order reactions, which could be ignored in the dilute-solution limit,[60] would have altered the dynamics from an initially exponential growth, into a saturated steady state at a new plateau. They would therefore have provided the starting conditions for organic feedbacks through longer loops such as cofactor or RNA biosynthetic pathways.

### 6.3.2.4 Incrementally advancing plateaus rather than combinatorial explosion

The dynamic of incrementally accreting steady states that we propose here is different from the image of combinatorially increasing network complexity proposed by Kauffman *et al.* [367, 368, 422] based on abstract models of molecular catalysis in reaction networks.

---

[59] The sense in which the information in a thermodynamic distribution is all encoded in its central tendencies is explained in Box 7.2 and Box 7.3 in Chapter 7. The current through the self-amplifying part of a network such as the rTCA cycle is what we term an *order parameter* for a dynamical phase transition. The order parameters in phase transitions determine the structure of the distributions, in such a way that fluctuations about them can be considered independent and random.

[60] Aldol addition of pyruvate to itself is a strong candidate [597]. This particular reaction is interesting because the subsequent hydrolysis products revisit elements of the rTCA cycle by means of networks that cut across the cycle (S. Ohara, personal communication).

In constructions of this kind,[61] expansion of the network topology opens new possibilities for material input with each node added to the network. As a result, larger networks always offer more potential for both growth and increased complexity than smaller networks.

In the actual small-molecule biochemistry that is the basis for our proposed network, the reactions that input carbon and the ways of catalyzing them are strictly limited by the orbitals of the $C_1$ inputs and the functional groups to which they can attach. Expanding network complexity introduces opportunities for new *second-order* reactions (as in Kauffman's construction), but in our stoichiometric networks, these are all parasitic side-reactions that drain cycle intermediates and impair carbon intake into the original pathway. Therefore each new generation of side-reactions tends to become self-limiting, until either increases in efficiency of the core network, or pruning of the side-reactions, restore growth in the whole network's flux, permitting the elaboration of a further level of complexity. The kinds of higher-order feedbacks that contribute to network growth therefore tend to be those that enhance the primary input reactions specifically, as cofactors do. These are considered in Section 6.4.

### 6.3.3 Reductive amination of $\alpha$-ketones and the path to amino acids

Neither the production nor the incorporation of reduced nitrogen into biochemistry poses a serious problem of geochemical plausibility. Making nitrogen available for organosynthesis divides into a "hard problem" and a collection of "easier problems." The hard problem is breaking of the $N\equiv N$ triple bond of dinitrogen, a process that occurs abiotically in gases through both shock heating (as occurs in the passage of meteorites through the atmosphere) and electric discharge.[62] It can also occur in hydrothermal fluids through interaction with both iron oxide and iron sulfide minerals [94, 761] at moderately high temperatures and pressures.[63] The easier problems are interconversion of $N_1$ compounds among oxidation states, ranging from ammonium $NH_4^+$ to nitrate $NO_3^-$, which occur through a variety of abiotic and biotic processes.

On the early Earth, most $N_1$ compounds would either have accumulated as ammonia or been returned to $N_2$ [761]. Ammonia in the atmosphere is unstable due to photolysis [251, 696, 697], so the route to organic nitrogen is initiated through oxidized compounds. Shock heating or electrolysis produce NO in the reaction

$$CO_2 + \frac{1}{2}N_2 \rightharpoonup NO + CO. \tag{6.7}$$

---

[61] Because this construction does not include the constraints of stoichiometry that are so fundamental to the structure of chemical reaction networks, it is a more applicable model to gene regulatory networks than to small-molecule geochemistry.

[62] Electric discharges through mixtures of $N_2$ and methane produce cyanoacetylene ($HC\equiv C-C\equiv N$). This is the origin of suggestions that cyanoacetylene is a plausible prebiotic input to pyrimidine synthesis [248].

[63] The $N\equiv N$ triple bond is also reduced biotically by the highly conserved, molybdenum-containing nitrogenase complex in autotrophic bacteria, where it is referred to as *nitrogen fixation*. Extant biochemistry attests to the advantage that accrues from avoiding nitrogen fixation if possible. Even though the nitrogenase complex is widely distributed among autotrophs, and many ecosystem relations exist to exchange the $N_1$ groups they produce, it is estimated that still roughly 1/3 of the nitrogen from natural sources used by plants and bacteria today does not come from biological nitrogen fixation, but originates in atmospheric nitrogen oxides generated by lightning [235, 665].

This NO is readily dissolved in seawater, where it is converted to nitrite or nitrate [784]. In the presence of ferrous iron, both nitrite and nitrate are reduced to $NH_4^+$. Alternatively, in hydrothermal fluids, the $N_2$ triple bond can be reduced directly to ammonia in the presence of various iron oxides and sulfides, with or without auxiliary reductants such as $H_2$ and $H_2S$.[64]

On the current Earth, with its oxidizing ocean and diverse ammonium-oxidizing bacteria, most hydrothermal fluids contain low concentrations of ammonia (micromole concentrations). A few sites contain much higher concentrations, including the Main Endeavor Field on the Juan de Fuca ridge (0.5–0.6 mmol/kg [809]) or the Guaymas basin (>10 mmol/kg [837]). These are believed to derive from decay of organic nitrogen [427], though the possibility of a contribution from mineral processes is not excluded.

In environments with ammonia, reductants, and ferrous minerals, the non-enzymatic reductive amination of ketones is a facile reaction. Claudia Huber and Günter Wächtershäuser [378] experimentally demonstrated reductive amination of $\alpha$-ketones in mixtures containing FeS and Fe(OH)$_2$. It seems likely therefore that any hydrothermal environments capable of generating carbon fixation pathways would concurrently have produced the simple amino acids of Section 5.3.4.1 derived from rTCA intermediates, or at least their early biosynthetic precursors.

### 6.3.4 A network of sugar phosphates and aldol reactions

It is difficult to envision an early role for sugars in an emerging metabolism without some degree of phosphorylation. Sugars that form from formaldehyde addition, in the network known as the *formose reaction*,[65] create extremely complex and non-selective mixtures, which have no analogues in biochemistry.[66] In contrast, *all* sugars in biochemistry are phosphorylated where they participate as intermediary metabolites [776]. If the original

---

[64] Brandes et al. [94] have demonstrated mineral-catalyzed ammonia formation from $N_2$, $NO_2^-$, and $NO_3^-$ under conditions typical of the suboceanic crust and hydrothermal systems. $Fe^0$ in the presence of water acts in some combination as reductant and catalyst, reducing $N_2$, $NO_2^-$, and $NO_3^-$ and at the same time forming iron oxides. Magnetite (Fe$_3$O$_4$) also acts as a catalyst for reduction of $N_2$ in the presence of formate. Both pyrrhotite (FeS) and pyrite (FeS$_2$) readily reduce $NO_2^-$, and $NO_3^-$ at pH values higher than about 7, with nitrite reduction proceeding more rapidly.

[65] Formaldehyde addition to form sugars was discovered by Butlerow in 1861 [109]. While its importance is difficult to assess because of the complex and non-selective character of the products, and the uncertain status of formaldehyde as an important molecule on the early Earth, it still stands as the unique demonstrated example of network autocatalysis within small-molecule organic chemistry. It is interesting and important that autocatalysis in the formose reaction (for instance starting from a glycolaldehyde seed) is in kinetic competition with the disproportionation reaction known as the *Cannizaro reaction*, producing methanol and formic acid [613] page 101, reiterating the very large free energy of formation of formaldehyde discussed in Section 6.3.1.5.

[66] Steven Benner has emphasized [671] that borates can prune this network to pathways which, intriguingly, include ribose. It is difficult, however, to situate the borate pruning mechanism within a system description that meets other requirements for a driven origin of life. Boron is a low-abundance element, its salts are highly soluble in seawater, and the mechanism by which it prunes the formose network has no apparent analogues in biochemistry. In contrast, phosphorylation of hydroxyl groups is ubiquitous. One can invoke extraordinary contexts such as lunar borate flats followed by meteorite spallation that transports sugars from the Moon to the Earth, to try to skirt these difficulties, but such solutions bring in other problems of explaining how small and intermittent supplies of organics could have nucleated a process as robust as life if it were not already a statistically favored outcome. Invoking a borate mechanism seems to raise more difficulties than the mechanism alone solves in favoring ribose, and we regard the association of borate with ribose as a coincidence.

On the other hand, the *principle* exemplified in the borate/formose system seems to us a fundamental one. It is that when large and unruly networks are generated by a small number of functional-group rearrangements, a few ligands to those configurations can drastically change the whole-network topology. Attempting to apply this lesson to the more heterogeneous

geochemical routes to RNA and related cofactors proceeded through a modular network analogous to the network of extant biochemistry, in which these are synthesized on a backbone of PRPP [783], then an early network of sugar phosphates must have preceded the synthesis of nucleotides and their related cofactors.[67] Interestingly Mueller *et al.* [576] showed that the network of aldol reactions seeded with glycolaldehyde and glyceraldehyde are pruned if monophosphates instead of simple carbohydrates are used. Phosphorylation prevents tautomerizations that are responsible for creating many reactive forms of each molecule and generating much of the branched, complex topology of the formose network. It seems most likely that at least substrate-level phosphorylation therefore provided anhydrous phosphate prior to the emergence and use of much of the functionality of a cofactor world, including the use of monomer as well as oligomer RNA.

## 6.4 The early organic feedbacks

The cofactors[68] seem to be the first products of a geochemical organosynthetic network that could have fed back to enhance the throughput of that network. The most important consequence of this feedback would have been to transfer deep core reactions from mineral surfaces onto organic carrier molecules, potentially expanding the range of settings and the combinations in which they could occur. We envision this transfer of core metabolism from mineral to organic carriers as having happened in stages. The argument that these stages were interleaved with the elaboration of metabolism, and that they were earlier than feedback from oligomers, is largely circumstantial, as many of our proposals concerning the small-molecule substrate network itself have been.

Recall from Chapter 4 several properties of the cofactors as a class that distinguish them from oligomers, and which suggest that they could have been employed earlier.

- The biosynthesis of many metabolically central cofactor classes is comparable to or simpler than that of nucleotides, and several cofactors share synthetic pathways with nucleotides.[69]
- Cofactors provide function directly as monomers, unlike polypeptides or polynucleotides, whose functions require chiral selection and dehydrating polymerization in addition to biosynthesis. This property together with the last makes the feedback channel through cofactors a shorter and more robust one than the channel through oligomers.

---

chemistry of a plausible early Earth should provide a helpful pruning rule in the search for relevant geological environments and reaction systems.

[67] Such a network would also then have preceded the electron-transfer cofactors derived from chorismate.

[68] As in Chapter 4, here we use the term *cofactor* to refer to a characteristic class of molecules, in the upper size range made by direct organosynthesis and below a transition to phosphate-mediated oligomerization, and responsible for a group of functionally precise roles in group transfer or electron transfer and occasionally catalysis (often through mechanisms related to group or electron transfer). This usage differs in emphasis from the conservative usage in biochemistry though it refers to the same set of molecules. In standard biochemical usage, the *term* "cofactor" is defined partly in terms of a relation of cofactors to protein enzymes either as prosthetic groups or as coenzymes. Our concern with cofactors lies mainly with the possibility that they had early roles *antedating* oligomer catalysts. We also generalize from the narrowest conventions by not intending exactly and exclusively the molecules known from extant biochemistry. In keeping with all arguments made thus far about metabolism-first, we suppose that in systems with loosely defined order, organosynthetic processes would have generated families of structurally similar molecules, some of which would have also carried related functions.

[69] Cofactor biosynthesis is not simpler than synthesis of the simplest amino acids, but this is not a direct comparison because amino acids do not perform the functions that cofactors do in small-molecule metabolism.

- The organic structures of cofactors tend to dictate some aspects of their functions within the molecule (though quantitative properties may be mediated by context), in contrast to amino acids or nucleic acids, the functions of which (at least in the context of oligomers) tend to be generated only jointly with the surrounding sequence. Thus cofactor feedback requires preservation of the biosynthetic pathway but not additional mechanisms to preserve and re-instantiate sequence information.

Unlike oligomer catalysts, which have a nearly open-ended range of functions, cofactors play a restricted role in completing biosynthetic pathways. They are stereotypical carriers of (often activated) functional groups, each serving a restricted set of group-transfer reactions. In cases where cofactors form parts of the active sites of enzymes, they may determine the reaction transition state.[70] Therefore cofactors have the potential to do what is most essential in a network where higher-order reactions are generally harmful to pathway flux: they can provide targeted support to core reactions associated with carbon and energy intake, while limiting parasitic side-reactions to those involving the same functional-group transfers. The cofactors associated with carbon addition and redox reactions in the universal network in Section 4.3.5 will therefore be of most interest to us, though this set already contains most of the widely used cofactor classes.

The proposition that cofactors fed back early to enhance flux through geochemical carbon fixation pathways is open to the same objection as all metabolism-first propositions: how are the biosynthetic pathways of cofactors or their early analogues to be explained in unregulated geochemical settings? It must be appreciated, however, that this problem affects the production and use of *all* complex biomolecules. The remaining question is whether the orders of difficulty can be ranked. Some specific problems are the following.

- Catalyzing and controlling the biosynthetic pathways for cofactors involves the same difficulties that were seen for small metabolites. Because cofactor synthesis is more complex, these difficulties are generally greater. Two possible resolutions to this problem are that (1) some cofactors are simpler to synthesize than they appear and form spontaneously, as shown in the Oró synthesis of adenine [615] or the non-enzymatic synthesis by Heinz *et al.* [351] of both pterins and flavins from amino acids,[71] or (2) that extant cofactors are evolutionary refinements of related, simpler molecules that carried out similar functions (albeit more crudely) and were parts of the small molecule inventory.
- While the restricted chemistry of cofactors limits the diversity of side-reactions that they enable, it also facilitates the reactions within that class. Therefore the problems that side-reactions pose for directing significant pathway flux into any one channel remain.
- It is difficult to marshal direct historical evidence for the sequencing of cofactors in biochemistry, by methods such as phylogenetic reconstruction, so the best arguments we

---

[70] The possibility that cofactors could have acted as small-molecule organocatalysts on the early Earth is suggested by similar functions for amino acids [638] and an increasing range of other known small molecules [51, 508].

[71] See Sousa and Martin [773] for a more thorough discussion of this point.

can make are still highly circumstantial. Due to their stereotypic functions, cofactors are among the most readily exchangeable components of biochemistry. Genes for interchangeable components tend to be the last to become refractory to horizontal transfer, and undergo recurrent waves of replacement. Thus one reconstructs the history of innovations in biosynthetic enzymes rather than of the cofactor's core function.[72]

Acknowledging these difficulties, we put forth in the following sections what seems the most plausible order for the first entry of cofactors as a causal influence on the structure of metabolism. It may be read as a relative order among the cofactors, whether they arose early or late. However, because both the biosynthetic pathways and the functional roles of cofactors seem to interleave naturally with the dependency sequence among the small metabolites, we propose that feedback from cofactors was interleaved with the elaboration of metabolism.

The order below is based on both the simplicity and size of the small-molecule networks from which each cofactor class could first have been produced, and the centrality of the metabolic reactions it supports. It is encouraging that these two independent criteria suggest similar orderings, with a few important exceptions such as thiamin, which is anomalously complex in several respects (see Section 4.5.2.1). Both criteria divide the arrival sequence for cofactors into two stages. In the first stage, simple redox cofactors and folates, which lack sulfur and are biosynthetically more primitive, would have altered $C_1$ reduction and CO addition. In the second stage, one or more complex cofactors, involving sulfur, would have altered the reactions in the rTCA cycle.

### 6.4.1 C–N heterocycles

The first C–N heterocycles would most plausibly have served as electron-pair carriers in redox reactions, and later as $C_1$ group carriers. The particular combination of conjugated double bonds in the heterocycle, with the alternation between the ion and the non-bonding pair of N, places a two-electron transfer at accessible energies for aqueous-phase redox reactions.[73] The first heterocycles would, however, have lacked the particular energetic tuning that becomes possible when sulfur can be incorporated in combination with nitrogen.

#### 6.4.1.1 Nicotinamide cofactors

A cofactor similar to nicotinamide mononucleotide (NMN) is a natural candidate for the first important heterocycle. Nicotinamide itself can be synthesized readily in a non-enzymatic reaction from dihydroxyacetone phosphate (DHAP) and aspartate [143], which are the same inputs used in *de novo* biosynthesis. The overall biosynthesis of NADH

---

[72] This difficulty has been pointed out, as it affects efforts to reconstruct the incorporation of amino acids from study of sequence relations among aminoacyl-tRNA synthases, by Gary Olsen and Carl Woese [605].

[73] This is almost the unique motif used in non-metal-center redox cofactors, the major exceptions being the class of quinones. It reflects a quite particular interaction of atomic configurations and molecular orbitals in the non-metal sector of the periodic table. The more flexible suite of electron-transfer reactions requires the larger number and much more complex energetics of the partially filled $d$-orbitals of the transition metals.

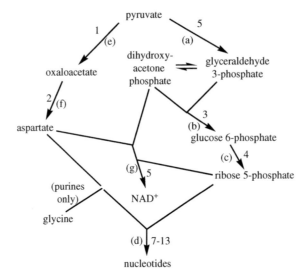

Figure 6.4 The small network connecting pyruvate to NADH and nucleotides. Numbers by the arrows indicate the number of steps to produce each new product. Letters label inputs to those synthetic steps in extant biochemistry: (a) 2 ATP, NADH; (b) none; (c) 2 $NAD^+$; (d) variable amounts of ATP, $CO_2$, $NH_3$, formate, $O_2$, and $NAD^+$; (e) $CO_2$; (f) $NH_3$, NADH; (g) 3 ATP, $NH_3$, $O_2$. Where $O_2$ is used as a terminal electron acceptor, we imagine other oxidants might have been substituted on the early Earth. (After Copley *et al.* [159].)

is not complex, and involves some of the earliest inputs in a network that could reach the amino acids and nucleotides, as shown in Figure 6.4.[74]

Experimental evidence of a non-enzymatic route to a cofactor similar to NMN, however, is still lacking. In NADH, the capacity of the nicotinamide group to exchange electron pairs depends on its ligation to the ribose phosphate base-plate, which has not been achieved non-enzymatically. $N_1$-methyl-nicotinamide (without a ribose phosphate base) has served as a model for $NAD^+$, which can undergo one-electron reduction to form radical states, and these in turn can serve as electron donors to metal centers even in extant biological proteins [240, 241]. Hence some electron-transfer processes could be supported by molecules related to nicotinamide. The context and mechanism by which such molecules could have become coupled to ribose and capable of pair transfer remain important open problems.

### 6.4.1.2 Nucleobase and nucleoside organosynthesis

Unlike nicotinamide synthesis, synthesis of purines and pyrimidines is currently a problem for which the synthetic chemists' best solutions do not closely resemble synthesis in

---

[74] One of the five steps in NADH biosynthesis involves the attachment of an AMP moiety, which is a "handle" for binding the cofactor in an enzymatic world, but not central to its function as an oxidant/reductant. The only other step beyond synthesis of nicotinamide is transfer to a PRPP platform, though this is an important and possibly problematic step, as discussed below. The $O_2$ used in NADH synthesis is used in the first step, to convert aspartate to iminosuccinate. A relevant question is then whether iminosuccinate could have been produced directly with a different nitrogen donor, or whether other oxidants could have been used to produce it from aspartate.

biochemistry. The essential difference is that biochemistry synthesizes from activated $C_1$ and $N_1$ groups, attached serially to a starting backbone (glycine for purines and aspartate for pyrimidines), whereas all current laboratory syntheses are based on condensation of reactive $C_1N_1$ pairs from nitrile ($-C{\equiv}N$) groups.

This presents a problem for understanding continuity: biological nucleobase synthesis grows hierarchically out of amino acid synthesis and the background of activated $C_1$ and $N_1$ exchange chemistry used throughout core metabolism, whereas synthetic approaches rely on diversely reactive functional groups that biochemistry avoids. Accepting that puzzle, however, it is notable that a number of different synthetic pathways from small, reactive CN compounds yield detectable levels of purines and pyrimidines.[75] These heterocycles are thus much more likely products than would be expected from counting the number of combinatorial possibilities for C, N compounds of comparable size.

As we noted in Section 3.6.6.5, photochemistry in planetary atmosphere may yield small but significant concentrations of HCN, so it is possible that HCN, like formate, is a prebiotically relevant input. The question therefore is whether products of HCN condensation chemistry, if they were produced on the early Earth, were part of the causal system that drove the emergence of biochemical order. We offer a brief summary of work to date on nucleobase synthesis. The reader is directed to reviews of Orgel [613] and of Powner *et al.* [648] for more extensive discussion and references.

**Biological mimics** A direct mimic of biological nucleobase synthesis might proceed from condensation of ammonia and carboxyphosphate onto amino acid "seed" molecules. Carboxyphosphate is a common activated intermediate in addition reactions from bicarbonate, as noted in Section 6.3.1.8, and it is used as a carbon source in nucleobase synthesis in *E. coli* [412]. Serial condensation without enzymes, however, would not be selective. Simple graph-grammar models of the allowed reactions[76] yield enormous numbers of products and do not suggest a mechanism to select nucleobases (Gil Benkö, unpublished). To our knowledge experiments have not been carried out to analyze what would be the products of mixtures of this kind.

**HCN polymerization and purines** An extensive literature on the synthesis of heterocycles from HCN polymerization dates back to the work of Juan Oró and collaborators in the 1950s [615, 616, 617, 618]. Most current work on abiotic synthesis of nucleobases and nucleosides derives from this line of investigation.

The formation of purine heterocycles starts from inputs of cyanide and ammonia in any of several forms. The original Oró synthesis [616] used ammonium cyanide; alternative syntheses have used HCN and liquid ammonia [847] or (in solid state) ammonium

---

[75] Percent levels are not uncommon in the simplest reaction systems. Yields in the tens of percents can be achieved (see numerous references in [613, 648]), but the associated reactions are much more carefully staged and controlled.

[76] These models used only a simple representation of the functional groups participating in reactions, and did not attempt to model kinetics or energetics beyond the hydrolysis of the phosphate leaving group. Many of the products were large dendritic molecules, which in actual systems would have been pre-empted by cyclization earlier in the synthesis.

formate [356, 912]. The important polymer that establishes the course of the synthesis is the HCN tetramer diaminomaleodinitrile (DAMN). This tetramer will spontaneously polymerize, and from the resulting precipitate adenine can be recovered by hydrolysis, though the pathway is not well understood for this route. Alternatively, single HCN tetramers can be cyclized with formamidine (a product of ammonia and HCN), and from the resulting imidazole (either directly or by hydrolysis) adenine, guanine, and other derivatives of adenine or hypoxanthine are produced.[77]

For each of these products, particular addition sequences have been suggested in the experimental literature. A computational study by Andersen *et al.* [24] of the HCN polymerization and hydrolysis network using graph grammars has shown that multiple alternative paths should exist to the same molecules. A full kinetic analysis is required to interpret the topological results proposed by Andersen *et al.*, but their work suggests that purine synthesis is not only possible but in some sense generic over a broad class of related pathways.

Orgel notes [613] that production of purines in dilute solutions is difficult because hydrolysis of HCN to ammonium formate (a first-order reaction) competes with tetramer formation (a fourth-order reaction) [613]. Therefore much research on purine synthesis focuses on concentrating mechanisms or alternative routes, some involving photochemistry [246].

**Connections between HCN and carbohydrate chemistry** Although not directly pertinent to nucleobase synthesis, it is interesting to note relations between aldehyde hydrolysis products of HCN polymers and the CHO chemistry of the rTCA cycle, which were proposed by Eschenmoser [231].

We observed in Section 3.6.6.5 that a similar photochemical asymmetry between double and triple bonds leads to the accumulation of CO and HCN in planetary atmosphere models, though to different degrees.[78] Following this similarity through condensation reactions produces an analogy between HCN dimer and glyoxylic acid, a condensation product of CO and formic acid which is also a potential hydrolysis product of HCN dimer. Extending the analogy further, HCN tetramer is analogous to dihydroxymaleic acid and its stereoisomers, which are formal dimerization products of glyoxylic acid and potential hydrolysis products of HCN tetramer. Eschenmoser proposes that all four $\alpha$-ketoacids in the rTCA cycle, as well as simple sugars, are likely products of HCN-polymer hydrolysis. Even more important, he suggests specific short autocatalytic cycles employing glyoxylate and oxaloglycolate, which concentrate reaction flux onto known core metabolites or closely related molecules. Other potential, simple small-molecule organocatalysts may be found in this network.

---

[77] The order of polymerization and hydrolysis need not be strict. Formamide, the direct hydrolysis product of HCN, is by itself a sufficient source for diverse C–N heterocycles under many conditions. A wide variety of purines and pyrimidines have been reported in syntheses from formamide activated by heat, photoexcitation in the presence of $TiO_2$, or dielectric breakdown, or have been catalyzed with clays, phosphate or borate minerals [250, 698, 699].

[78] The similarity of these two triple-bonded compounds is also responsible for their related toxicity, as both bond irreversibly to similar metal centers in biological enzymes.

The work of Eschenmoser suggests a particular path through plausible chemistry linking HCN to glyoxylate and downstream rTCA cycle intermediates. It furnishes an ideal example of the way expert knowledge can be used to suggest relevant reactions, and a skeleton through a potentially very complex network, which can then serve as a seed for computational models that can reveal network properties not readily pursued by the chemist working manually. Andersen *et al.* [28] used their generative algorithm based on graph rewrite rules to explore the neighborhood of Eschenmoser's skeletal pathway, which is produced by recursive application of the same 18 reactions that generate the skeleton pathway.[79] They find three important properties of the surrounding network. First, multiple plausible alternative autocatalytic pathways to the one proposed by Eschenmoser exist from glyoxylate to core metabolites. This does not in any way contradict Eschenmoser's proposal; rather, it potentially increases the confluence in the actual network relative to the predictions from a single pathway. Second, they show that a network of partly overlapping pathways has the topology of a lattice, because the order of many independent reactions can potentially be shuffled. Flux remains within the same class of similar molecules, but the pathway multiplicity increases the aggregate multistep reaction rate. Third, they propose undiscovered autocatalytic pathways producing glyoxylate from HCN. Some of these bypass rate-limiting reaction steps in the original proposal of Ref. [231]. Eschenmoser had suggested that such bypass pathways would be important if they existed, but his analysis did not extend far enough to identify them.

We expect that in the coming decade, this kind of collaboration between expert knowledge, computational network generation, and heuristic-aided search, will shift our conception of small-molecule network chemistry away from one based on single pathways, and toward a more realistic picture of bundles and lattices of dominant pathways, in which network as well as reaction properties govern throughput and flux focusing.

**Cyanoacetylene and pyrimidines** Laboratory synthesis of pyrimidines has centered on the condensation chemistry of cyanoacetylene ($HC{\equiv}C-C{\equiv}N$) or its hydrolysis product cyanoacetaldehyde ($HOC-C-C{\equiv}N$), with cyanate ions ($NCO^-$) or urea ($H_2N-CO-NH_2$). Cyanoacetylene and two molecules of isocyanic acid (HNCO) can condense to form cyanovinylurea, which cyclizes to form cytosine [248]. Alternatively, cyanovinylurea is formed directly by condensation of cyanoacetaldehyde and urea.

**From nucleobases to nucleosides** Purine condensation onto D-ribose can be driven by dry heating in the presence of MgCl or mixed natural sea salts at modest yields (8%) [283].[80]

---

[79] Even for this simple generator set, the systematic elaboration of the network of HCN condensation and polymerization products produces a space that is too combinatorially large to be searched, before it arrives at interesting pathways or molecules. However, by beginning with a proposed skeleton pathway, and elaborating neighborhoods of that pathway, computational methods can expose network properties that are of conceptual as well as practical importance.

[80] It is difficult to assess the importance of this observation for the origin of life; it adds the coordination problems of hydration/dehydration cycles to the other system requirements of reactive intermediates and conditions that can synthesize both heterocycles and sugars. However, if this complex context is not considered too much to accept, the reaction is otherwise consistent with a hypothesis that nucleosides could have formed geochemically from modular reaction systems somewhat like those in biochemistry.

However, even under such seemingly favorable conditions, pyrimidine condensation onto ribose has not proved successful [647].

This difficulty has prompted John Sutherland and colleagues to adopt a variety of step-wise assemblies. Ingar *et al.* [387] replaced the urea + cyanoacetaldehyde synthesis of cytosine with condensation of a cyanamide ($H_2N-C\equiv N$) directly onto a sugar phosphate platform, followed by condensation with cyanoacetylene ($HC\equiv C-C\equiv N$) to form cytidine arabinose-3′-phosphate. In a later synthesis, Powner *et al.* [647] abandoned the modular synthesis entirely, synthesizing cytidine ribonucleotide from aminooxazoles which combine part of the carbohydrate and pyrimidine rings. The roles of cyanamide and cyanoacetylene are similar in the two syntheses, but in the synthesis of Powner *et al.* the ribose moiety is built step-wise starting from glycolaldehyde.

A common theme in these approaches is that individual synthetic steps are demanded to be selective and produce good yields. Synthetic chemists refer to these criteria together as making a pathway *robust*. Whether robustness of individual pathway steps is an achievable or even an appropriate criterion under plausible geochemical conditions, however, is questionable. Albert Eschenmoser [231] suggests that robustness at the network level may make essential use of individual reactions or subnetworks that are not "robust" at the pathway level, but which provide entry routes to self-amplifying loops, or control points for small-molecule catalysts, that can eventually come to control the flow in the larger network. The parallels he proposed between HCN oligomer chemistry and rTCA intermediates were pursued in an effort to look for such entry points.

Eschenmoser's proposal for the dynamic reorganization of chemical networks brings to mind a mathematical analogy with the role of rare events in the physical chemistry of individual reactions: the configuration that forms the transition state is not common – in fact it is almost never realized – but the conversion between reactants and products is dominated by the rare moments when that state is traversed. Whether such gateway processes exist in systems chemistry relevant to the origin of life, and how to find them if they do, are among the most important but also difficult problems in extending what has been learned from reactions to reaction systems. We will look at the context for RNA synthesis in biochemistry for clues, although the attempt to interpret them as geochemical fossils raises more questions than it answers.

### 6.4.2 Cyclohydrolase reactions: purine nucleotides and folates

We believe an important observation about cofactor biosynthesis in extant biochemistry concerns the extensive overlap of biosynthetic mechanisms and even pathway segments between the folate, flavin, and thiamin cofactors and purine RNA, reviewed in Section 4.5.2.1.

A closely related class of reactions leading from aminoimidazole ribonucleotide (AIR) to the purines, and also to the folate and flavin cofactors, are those performed by the **cyclohydrolase enzymes** with EC numbers beginning 3.5.4. Their role in purine synthesis is to close the formyl and formamide[81] groups of formamidoimidazole-4-carboxamide ribotide (FAICAR) to form the pyrimidine ring in inosine monophosphate (IMP). Related roles starting from GTP or ATP are more complex, involving breaking of either the imidazole or the pyrimidine rings (respectively) and also the ribose ring. In the case of GTP conversion to folates, these rings are also re-formed by the same enzyme (EC 3.5.4.15). (Recall Figure 4.13 for an illustration of these conversions.) The cyclohydrolases interconvert C–N heterocycles – particularly heterocycles linked to ribose moieties – to the two most metabolically central cofactors used in $C_1$ group transfer (folates and thiamin), as well as to the flavin (redox) cofactors, RNA, and histidine.

### *6.4.2.1 Which came first, the pathway or the products?*

Any theory of the origin of life will face the difficulty of explaining how the extant biosynthetic pathways to RNA came about. Conversely, under any solution to that problem, the parsimonious explanation for the overlap of RNA and related cofactor biosynthetic pathways is that the two coevolved from a common set of catalysts, which initially were promiscuous and later became diversified and specialized. The first decision that must be made to give an interpretation to the extant pathway is whether

1. one or more of RNA or its associated cofactors already played a role in the earliest life, and the extant pathway to produce it was later evolved and displaced whatever had been the original source of the molecule, or
2. the extant pathway preserves in essence the original reaction system that produced RNA, and its catalysts recapitulate catalysts that were present in cruder form on the early Earth.

If the first interpretation is chosen, it is difficult to believe that both RNA and other cofactors were produced independently and played roles in the earliest life, and that a highly parsimonious pathway was subsequently evolved to produce them all. Rather, one of these functional molecules must have been the reference for evolution of the pathway, and the others were then derived by exaptation. In that case, if RNA is supposed to have been the originally essential molecule, it is difficult to explain how the RNA-derived cofactors took on exclusive roles in all of the very deepest core pathways, unless the late evolution of catalysts was responsible for the whole elaboration of the universal core, leaving us with the most problematic version of RNA-first hypotheses. If, instead, one of the cofactor classes was the originally essential molecule, the difficulties are even more apparent, given the very special properties and singular role of RNA.

Interpretation 2 avoids this dilemma, though it opens the problem of what system of catalysts should have preceded the enzymes in the current pathway. It suggests, however, that if this problem can be solved, folates and flavins (less apparently thiamin) would have been part of any chemical system capable of producing purine RNA, and their functionality in

---

[81] This formamide group comes from carbamoyl phosphate. The formyl group is donated from formyl-THF. Its role in the synthesis of folate cofactors is therefore a cycle-within-a-cycle in organic feedbacks onto small-molecule metabolism.

metabolism might also have been available. This opens a second question of interpretation, subordinate to option 2 above. Either

2a. the functions of RNA were the first target of selection for the refinement of cata-lysts in the biosynthetic pathway, and the diversification and coevolution of the related cofactors followed, or

2b. one or more of the cofactors provided the first function that was a target of selection, and *this stabilized the biosynthetic network that made RNA available.*

The functionality of cofactors, carried at the level of the molecule rather than the oligomeric sequence, makes the cofactors attractive targets of selection for the earliest stabilization of a protometabolism. Folates, feeding back to modify $C_1$ organosynthesis, would have transferred the direct $C_1$ reduction sequence from transition-metal sulfide mineral surfaces to the N atoms of heterocycles.[82] They are the simplest purine-derived cofactors, and notably lack sulfur, a characteristic of all the cofactors in the direct $C_1$ reduction sequence. If it is correct that folates or flavins were present and were somehow selected (kinetically, at the reaction-system level) in the earliest elaboration of metabolism, this suggests that the first roles for RNA may have been analogous to cofactor roles in catalysis, before oligomer RNA took on a different class of functions.[83]

### 6.4.3  Thiamin-like chemistry: lifting rTCA off the rocks

Thiamin is a seemingly unavoidable but also problematic cofactor to consider in the context of early organic feedbacks. The early stages in its biosynthesis are closely related to that of folates and purines, so our suggestion that those reactions should have been present applies to thiamin as well. However, the later stages of thiamin biosynthesis are among the most complex known among cofactors, and are still incompletely understood [405].

Yet thiamin seems impossible to escape in a theory of the transfer of the carbonyl-insertion reactions of TCA cycling from a mineral substrate onto organic controls. It is used in both reductive and oxidative TCA cycling, even as other reagents such as redox partners (ferredoxins versus NADH) are substituted. The finely tuned electronic properties of a carbon positioned between a positively charged N and negatively charged S provide a solution to the transfer of a carboxyl group to a thioester for which biochemistry has never evolved an alternative. Finally, both thiamin and biotin – the two $C_1$ transfer cofactors for rTCA reactions – incorporate sulfur, in contrast to the cofactors of direct $C_1$ reduction which do not.

For these *very* circumstantial reasons we suppose that some important transition in early organosynthesis, perhaps later than the formation of folate-like cofactors but at a

---

[82]  Recall from Section 4.3.4 and Section 6.3.1.1 that the $C_1$ reduction sequence is the likely original source of both acetyl-thioesters and also glycine and serine by direct transfer of methylene groups. It may also have mediated formyl group transfer in the synthesis of FAICAR, thus catalyzing its own synthesis.

[83]  For independent reasons of accounting for the amino acid biosynthetic pathway signature in the genetic code, we have proposed [158] a small-molecule organocatalyst role for RNA dimers, which would be analogous to a cofactor role.

comparable stage in the elaboration of the metabolic substrate, led to the incorporation of sulfur and a shift from CO insertion in thioesters on surfaces, to a $CO_2$-derived mechanism mediated by organic cofactors. The chemistry of thiamin would seem to be the best source of clues for what this transition may have entailed. This transition would have "lifted rTCA off the rocks," and brought the second major carbon fixation pathway under organic control.

### 6.4.4 Biotin: uses in rTCA and in fatty acid synthesis

The other sulfur-including cofactor associated with rTCA cycling is biotin, which serves as a carrier of activated carboxyl groups between carboxyphosphate and an enol-pyruvate acceptor. Biotin synthesis is also complex, but unlike thiamin, biotin may not be essential to the first $\beta$-carboxylation of ketoacids in an rTCA cycle. As we noted in Section 6.3.1.8, alternative routes for $\beta$-carboxylation exist, and stabilization of the enol form as phosphoenol-pyruvate (or homologously, $\alpha$-ketoglutarate) provides both a phosphoryl donor to form carboxyphosphate, and the acceptor for the carboxyl group. Hence the PEP-carboxylase analogue reaction may have been original, and the use of biotin later.[84]

The entry of a biotin-like carboxyl-transfer function still appears to be essential in fatty acid synthesis via malonyl-CoA. It is universally required in the acetyl-CoA carboxylase enzymes, where its function is analogous to that in pyruvate carboxylase. By the arguments we have advanced for other cofactors, this would seem to suggest that despite its complexity, the emergence of a cofactor functionally similar to biotin may have been necessary to enable the transition of fatty acid synthesis away from a Fischer–Tropsch-like mechanism on a mineral surface, to a process under the control of organic cofactors.

### 6.4.5 Alkyl thiols

Perhaps in parallel with the incorporation of sulfur in heterocycles, alkyl-thiol cofactors and amino acids appear as plausible early sources of feedback that would have changed protometabolic throughput and extended the synthetic network. Biosynthetically the alkyl thiols are simpler than heterocycles incorporating sulfur, but their roles seem more generic and less linked to very central core pathways, so it is difficult to argue whether they should have entered earlier or later than analogues to thiamin or biotin.

Cysteine and homocysteine, formed from serine and homoserine, seem likely to be early among the alkyl thiols, and they provide ideal examples of the crossover between functioning as cofactors and functioning as coding amino acids. Cysteine would have been the gateway to three key functions. As the active functional group on pantetheine (after an important decarboxylation) it forms thioesters (in coenzyme-A). In the later context of coded polypeptides, the same group serves as acyl carrier in the acyl carrier proteins. Finally, and crucial to the evolution of stable and controllable folds in

---

[84] This sequence, however, then fails to shed light on the use of sulfur in rTCA enzymes, which if it is not an accident and requires an explanation, must be explained at some other stage.

the protein world, it forms redox-sensitive disulfide bridges. The last function, although very different in context and importance, is mechanistically similar to the functions carried out by pairs of SH groups that jointly serve as $C_1$ carrier and co-reductant (as in the coenzyme-B/coenzyme-M system or in lipoic acid).[85] Homocysteine, in the presence of nucleotides, would have offered a starting point for the methyl-group exchange function of the *S*-adenosyl-methionine cycle (shown in Figure 4.15).

## 6.5 Selection of monomers for chirality

Organosynthesis through the stages of small metabolites and even cofactors,[86] if it could have proceeded at all, could presumably have proceeded in reaction systems where all intermediates existed in mixtures of stereoisomers. Either before, or as part of, the rise of oligomers, however, a mechanism for chiral selection of small molecules must have imposed further order on geochemical organosynthesis.

### 6.5.1 Degree of enantiomeric selection at different scales

It is interesting that only a few chiral centers are found in the network of core carbon metabolism shown in Figure 4.9. Within the rTCA cycle, only malate and isocitrate are chiral. The absence of stereoisomers of most compounds results from a combination of small molecular size, absence of heteroatoms in the CHO sector, and the presence of double bonds in many compounds.

In contrast, the monomers in all the major oligomer classes are chiral. All $\alpha$-amino acids are chiral at the $\alpha$ carbon atom, as are all sugars that retain the chiral center of glyceraldehyde which is an entry point into the pentose phosphate pathway. Ribose in the furanose (five-membered ring) form has three chiral centers, with C-4 determining the designation as D- or L-ribose. Deoxyribose retains only two of the three chiral centers of ribose, and due to their position in the molecule, it has only two enantiomers. D-Fructose (a ketohexose), is one of eight stereoisomers possible from its three chiral centers. Glucose (an aldohexose) has four chiral centers, and is one of the sixteen possible stereoisomers.

We assume that at least some enantiomeric selection must have been performed at the level of bulk processes by geochemical mechanisms, for the same reason we assume that pruning of the metabolic network itself must have occurred: forming chirally selected oligomers from racemic populations of monomers requires rejecting an entropy that grows as a multiple of the oligomer length and the number of chiral centers per monomer. Yet chirality of monomers is essential both in the formation and in the functionality of most oligomers. For example, the secondary structure of many folds of polypeptides requires

---

[85] The biosynthesis of CoB and CoM, while linked to ketoacid and amino acid biosynthesis, is remarkably complex and like many features of methanogens, seems more easily understood as an evolutionarily derived character. The core function is nonetheless simple, and informative about the range of related functions that may have contributed to a selective context for early sulfur amino acids. It may be relevant to note that the CoB/CoM pair act jointly to reduce fumarate to succinate in the enzyme fumarate reductase in some methanogens [349], an alternative to menaquinone-based reduction in bacteria.

[86] Note that this follows as a corollary of the monomer-level action of cofactors, the second property listed under Section 6.4.

homochiral amino acids, and the formation of double helices of polynucleotides requires homochiral ribose groups in the backbone. The incorporation of monomers of the wrong chirality into oligomers can lead to loss of function of the whole fold, making high-fidelity sorting essential. Arriving at oligomer secondary structure therefore requires not just an enantiomeric bias, but *reliable* selection, which seems implausible in a world of crude organic controls, and in any case costly in free energy because an entropy proportional to the synthesized molecular mass must be rejected. Therefore, whatever enantiomeric selection can be provided through cooperative effects at the system level greatly increases the plausibility of a transition to the oligomer world.

### 6.5.2 Mechanisms and contexts for stereoselectivity

Listed below are two general classes of mechanisms that may provide stereoselectivity, under the kinds of steady-state reaction conditions we have proposed for the other stages in the emergence of metabolism. Many more mechanisms can be proposed, under cyclic conditions or transport through heterogeneous environments, but we will not attempt to cover them here.

**Stereoselectivity at mineral surfaces** Most solution-phase synthesis from small-molecule precursors yields racemic mixtures. However, numerous organic molecules are known to bind non-racemically to the crystal faces of common rock-forming minerals including quartz, alkali feldspar, and clinopyroxene [342, 346]. The possible role of mineral faces as a source of selection of biological homochirality has been suggested since the 1930s.[87]

The reason the interactions of many organic molecules with mineral surfaces are stereoselective is that most crystal faces have chiral centers, with the same chirality across a mineral face. These centers can serve as sites for adsorption and perhaps synthesis of organic compounds. Although most minerals are intrinsically achiral in the bulk, cleavage faces at angles to the principal axes of symmetry will generally leave surface "kinks" where the cleavage plane intersects the space-group lattice of the mineral [341]. Growing faces also frequently contain arrays of such chiral centers, as these are sites where the crystal lattice grows by deposition of elements.

Crystal faces are not expected to be a solution to global enantiomeric selection of organic molecules, as the distribution of left- and right-handed chiral faces among crystal assemblies is about even. However, a crystal face that is homochirally selective provides a region that is *macroscopic* on a molecular scale, and can thus create pools of chiral small molecules amenable to condensation to form higher-order compounds.

**Propagation of chirality through small-molecule organocatalysis** Cooperative synthesis in networks of small molecules may have the property of propagating chirality from

---

[87] See [346] for a review and references.

some small molecules which act as organocatalysts, to the products of the reactions they catalyze. Sandra Pizzarello and Arthur Weber demonstrated such propagation of chirality for synthesis of threoses using amino acids as organocatalysts [638]. This mechanism, like mineral surface enantioselectivity, is unlikely to produce global chiral selectivity. Unlike crystal surface selection, however, it is not inherently limited to the volume of an externally provided selective environment. In any networks for which chirality-propagating reactions form loops, the mechanism is potentially susceptible to positive feedback, like synthesis in autocatalytic networks. Thus the extent of selection will depend on the degree of selectivity over links, and the topology of the network.

### 6.5.3 The redundancy of biochemical processes simplifies the problem of chiral selection

For enantiomeric selection as for pruning of the organosynthetic network itself, it is important to recognize that the constraints may have been imposed at the level of *processes*, rather than as an an *ex post* filter on pre-existing molecular populations. Also important, the part of organosynthesis that has survived to become biochemistry, as we have noted, makes heavy use of redundant synthetic reactions. Because of the way biosynthetic pathways ramify from a few starting backbones to many products, and because of the reuse of homologous reactions, selection of key chiral centers in a large family of molecules can readily be produced by selection of a few key catalysts.

Morowitz [572] has emphasized that in extant biochemistry, the D-homochirality of sugars all traces to the chiral center of their common glyceraldehyde precursor, while the L-homochirality of the 19 chiral amino acids derives from their use of a common aminotransferase mechanism. The former is a constraint that propagates through conserved chiral centers in the small-molecule substrate, while the latter propagates through homology of an enzyme family.

Inferring implications of amino acid chirality today for prebiotic chemistry is challenging because amino acid biosynthesis involves modern enzymes but also low-level mechanisms and simple cofactors. Reference [572] emphasized the unifying role of the $\alpha$-amine of glutamate as the amine source for 16 of the 20 amino acids, including (nonchiral) glycine. The amino group transfers that form these amino acids are catalyzed in an almost universally homologous family of transaminases in what is termed a "ping-pong" reaction.[88] In such reactions, the donor molecule (glutamate) transfers its amino group to an enzyme- bound pyridoxal phosphate cofactor (forming pyridoxamine), after which the donor exits as an $\alpha$-ketoacid. The ketoacid receiving the amine then enters the site previously occupied by the donor, and the amino transfer proceeds in reverse from pyridoxamine to the ketoacid. The use of the same binding site by the donor and acceptor preserves chirality from glutamate to the 15 out of 16 transaminated ketoacids that are chiral. The enzymatic source of chirality is more general than the glutamate

---

[88] Other widely studied reactions of this type include the transthiolases associated with the metabolism of acetyl-CoA.

donor, however, applying also to the *ketoglutarases*, which form arginine, proline, and glutamate itself. They also use enzyme-bound pyridoxamine as the amino donor to $\alpha$-ketoglutarate, but the pyridoxamine is formed directly from ammonia and pyridoxal phosphate.[89]

One possible interpretation is that the extant, enzymatic transamination from glutamate reflects an early cofactor-like role for glutamate as a general-purpose amino donor, and that reuse of an active site analogous to a ping-pong reaction was also the original mechanism to propagate chirality. A simpler interpretation may be that universal one-way transfer from an amino-group carrier (now provided by pyridoxamine) was the original mechanism, and imprinted the same chirality on all transaminated $\alpha$-ketoacids.[90] If either interpretation is correct, it suggests that the homochirality of the 19 chiral amino acids does not reflect the solution to 19 independent evolutionary problems, but the adoption of one overall chirality, propagated to 19 instances through a shared mechanism.[91]

## 6.6 The oligomer world and molecular replication

The emergence of an oligomer world was a veritable forest of transitions, heterogeneous in kind and all interacting, to which we cannot begin to do justice here even in summary. We must refer the reader to diverse and rich reviews in the specialist literature [122, 223, 269, 613, 648, 791].

Key transitions in this wholesale reorganization of organic chemistry on the early Earth (in something like a sequential order) include: the incursion of extensive dehydrating polymerization alongside the synthetic chemistry of mixed small-organic monomers; the agglomerative dynamics of small assemblies of amino acids, simple peptides, and RNA; interactions that assort monomers according to chirality and that are regioselective of ligation reactions; the emergence of a "division of labor" between polymerization that favored oligomers (rather than heteropolymers as in a cofactor world), and non-covalent associations of oligomeric polypeptides and polynucleotides; the emergence of a memory system for sequence information using RNA base pairing but perhaps also assortative folding in complexes of RNA and small peptides (see Section 6.8.1.3); the beginnings of selection of small RNA oligomers (and perhaps through non-covalent associations, small peptides?) for functionality; the (probably parallel) emergence of general condensation catalysts and RNA replication enzymes; the (probably two-stage) emergence of tRNAs and aminoacyl

---

[89] Glutamine is also aminated directly from ammonia, but this (along with the homologous amine of asparagine) is a terminal amine rather than an $\alpha$-amine.

[90] The reader will note that this interpretation – that the extant ping-pong reaction, with its input and output legs, and its universal pyridoxamine core, reflects an earlier one-way reaction in which the core was the starting point and all transaminations were outward-going – is similar in spirit to the interpretation we assigned to modern metabolic "bowties" through the universal rTCA core, in Section 4.3.1.

[91] Reference [572] further observes that the use of pyridoxal phosphate as a cofactor provides a kind of "check-point" for the joint L-chirality of amino acids and D-chirality of sugars. The non-chiral pyridoxine-5-phosphate is formed by condensation of 1-deoxy-D-xylulose-5-phosphate with 4-phosphohydroxy-L-threonine. Non-homochiral sources of xylulose-5-phosphate and threonine would reduce the strength of this feedback by failing to produce pyridoxine-5-phosphate from some substrate combinations. The one-off character of this particular molecular relation, however, makes it difficult to assign a significance to the observation.

tRNA charging; the aggregation of replicase, catalyst, and tRNA to create the ribosome; and, growing out of the functioning of the ribosome, the emergence of a distinct abstraction of translation and the apparatus that serves it, including mRNA as a messenger, diversified aminoacyl-tRNA synthetases defining the first genetic code, and the beginnings of a formally distinct RNA genome. Finally, a rapid refinement of ribosomal function, occurring within an evolutionary dynamic of agglomerated fragments that is likely to be very complex and conceptually different from most of our current evolutionary models, created the need for the first partitions against horizontal gene transfer, the emergence of prokaryotic domain distinctions, and the beginning of the phylogenetic era. Each of these transitions, in turn, is made up of a collection of subtransitions, about some of which, remarkably, insights are available.

In this section we will touch on a few of the major transitions, emphasizing connections to low-level chemistry that may have constrained the possible forms of higher-level architecture. The most basic physical questions about the rise of an oligomer world are what kinds of disequilibrium forces were responsible for the emergence of metastable populations of oligomers in water, and what kinds of interactions, shaping the kinetic landscape, determined which oligomer populations emerged and persisted. It is important to understand the role of constraints or selection criteria that originate in low-level chemistry or the synthetic networks of monomers and cofactors, because the order in which these acted on oligomers would have determined whether the emerging oligomer world augmented and locked into place the synthetic patterns from the earlier ordered geochemistry, or else had the power to overwrite the earlier order.

### 6.6.1  Increased need for dehydrating ligation reactions

The rise of an oligomer world from a world of small molecules must have depended on a greatly increased capacity for dehydrating polymerization. The problem of polymerization is not principally to arrive at a particular group of molecules, but *to arrive at a particular functionally important state of disequilibrium*. Both monomer synthesis and polymer functions in life are carried out in aqueous solution, where polymers are only metastable against hydrolysis. Several ideas exist for the way such metastable populations may have been produced on the early Earth. They may be grouped according to the assumptions they make about the following three alternatives.

- **Whether non-equilibrium was achieved with activating groups in a steady state, or by altering the activity of water in the whole system** The most basic distinction is physical: whether polymerization occurred prebiotically as it occurs in present life, by coupling the ligation step so that its dehydration supplies the water for a hydrolytic cleavage of an activating group with a larger free energy of hydrolysis than the polymer bond, or instead by lowering the water activity in the whole system so that the polymer state was energetically favored in some conditions. In the former case, polymerization can occur in steady state with high water activity. In the latter case, hydration/dehydration cycling is required either in space or in time.

- **If activating groups were used – whether thioesters or phosphates are geochemically more plausible** Thioesters and phosphate groups are the two most plausible activating groups to drive dehydration. Each is defined directly from the properties of an element (S or P) rather than a higher-level organic assembly; both are used in extant biochemistry; the thioester and phosphoanhydride bond can in many cases be exchanged nearly reversibly, and plausible generating mechanisms for thioesters or their mineral analogues exist in environments of transition-metal sulfide minerals or $H_2S$, as discussed in Section 6.3.1.

- **If phosphates were the activating groups – whether the pre-biotic source was analogous to substrate-level phosphorylation or analogous to membrane-mediated, proton-driven phosphorylation** Christian de Duve argued that thioesters acting directly to drive polymerization were the most plausible source of dehydrating potential in an early world, and that through substrate-level phosphorylation, they were also the most plausible source of phosphoanhydride bonds, kinetically maintained in a persistent disequilibrium with water. The alternative, an analogue to membrane-mediated, proton-driven phosphorylation, at first seems problematic as all such systems in extant life rely on complex systems of not only macromolecules, but complex coded proteins. However, Michael Russell and collaborators have proposed that this chicken-egg dependency can be broken if mineral membranes and hydrothermally generated proton-motive force coupled proton transport to phosphorylation in the earliest stages of organosynthesis. (For details, see Section 6.3.1.9.)

Among these hypotheses, we favor those using activating groups to maintain non-equilibrium steady states over those suggesting hydration/dehydration cycles. Although the latter are simpler to study in the laboratory, and control the process by creating a variable equilibrium target, they are harder to understand as part of the systems chemistry needed to account for the hierarchy of chemical transformation between geochemistry and life. They introduce yet another dissimilar environment from the aqueous or lipid-phase environments that are essential to all extant living processes. More fundamentally, we think the reason invoking dehydration seems conceptually simple is that it attempts to account for a material (polymers) in terms of an equilibrium condition (dehydration), rather than attempting to understand the specifically *non-equilibrium context* of polymers as part of the larger kinetic organization of the rest of biochemistry.

### 6.6.2 Coupling of surfaces and polymerization

Dehydrating polymerization is inherently a kinetically controlled effect. Formation of anhydride bonds is an energetic relaxation process, which competes (for inputs) with hydrolysis if activated monomers are used, and (for persistence) with hydrolysis of the polymer once it has been formed. Therefore the extent and kind of polymerization will generally depend on the environment. Environments that have the effect of organizing monomers, so that polymerization is favored over hydrolysis, act as catalysts by reducing configuration entropy of the inputs. Two such classes of environments that have been extensively studied are clay minerals and multilamellar lipid bilayer stacks. Both create two-dimensional reaction environments with charged surfaces and large surface-area/volume ratios. Clays have largely been studied in the context of polymerization catalyzed by activating groups, while lipid stacks have been studied in contexts of hydration/dehydration cycles.

### 6.6.2.1 Clays and polymerization of activated monomers

An early survey of many clay minerals by Mella Paecht-Horowitz, J. Berger, and Aharon Katchalsky [620] studied the competition between polymerization and hydrolysis of amino acids activated with adenylyl leaving groups. They found that non-swelling clays such as apatites and kaolinates were not active as polymerization catalysts, but swelling clays such as illite and (even more so) montmorillonite[92] were very active, suggesting that organization was performed between layers.

For this system, in a solution phase without clay, hydrolysis of the monomers dominated relaxation through the polymer state, so that only small quantities of dimers and trimers were formed. For the same system in a slurry of clay particles, hydrolysis of monomers was inhibited, oligomers of length 10–50 were formed, and these were stabilized against hydrolysis by association with the clay surface.

James Ferris with several co-authors [245, 247, 375] has extensively studied the polymerization of activated RNA monomers[93] in slurries of montmorillonite. Both poly-merization and hydrolysis of RNA oligomers occur at the clay interlayer, but in the studies reviewed in [247], polymerization was an order of magnitude faster than hydrolysis. As in the study of Paecht-Horowitz *et al.*, without the clay, hydrolysis becomes the dominant process from the same reagents. Polymerization in clay environments is sometimes found to be regioselective, but the degree of selectivity and whether the $2'–5'$ or $3'–5'$ linkage is favored depends on both the nucleobase and the activating group used. In these experiments both spontaneous activation and template-directed synthesis were observed.[94]

### 6.6.2.2 Lipid stacks and hydration/dehydration cycles

Polymerization using hydration/dehydration (HD) cycles creates a different kind of dis-equilibrium from polymerization using activating groups. Instead of creating a steady non-equilibrium polymer population through leaving-group kinetics under constant con-ditions, the polymerized state becomes a target equilibrium during the dehydration phase,

---

[92] Montmorillonite is a very high-swelling aluminosilicate clay, in which silica and alumina lattices make up opposing faces of each platelet. Substitution defects in the two surfaces lead to charge centers on the platelet, and it is here that many polymerization reactions are believed to occur.

[93] Here the activated monomers were nucleoside phosphorimidazoles 1-methyladenine, not nucleoside triphosphates. See [613], page 110.

[94] The proposal that clay minerals may have aided the concentration and polymerization of small organics dates back to a lecture by J. D. Bernal in 1947 (see [247] page 100), appearing in print in [66]. Graham Cairns-Smith has argued [113, 114, 115] for a very elaborate role for clays in the origin of life, in which the clay surfaces themselves were the original carriers of hereditary information, which was subsequently transferred to a subset of RNA oligomers that formed in the clay environment.

toward which the system moves even with non-activated monomers. The kinetic competition then occurs between the transient period during dehydration and the transient period during rehydration, and depends on the HD duty cycle.

The catalytic role of systems such as lipid stacks remains that of separating timescales. In the HD context, they "organize" the system by preserving mobility of the monomers over the surface so that configurations favorable to ligation can be found, in contrast to simple dehydration in bulk, where the formation of disordered solids reduces mobility in unreactive configurations, leading to a very long timescale for metastability.[95] For a given rate of polymer hydrolysis in the aqueous phase, a duty cycle can be found such that hydrolysis dominates polymerization in bulk dehydration, but polymerization dominates dehydration in the lipid stack.

Veronica DeGuzman, Wenonah Vercoutere, Hossein Shenasa, and David Deamer [187] studied the long-term oscillatory states of RNA polymerization during HD cycles in multilamellar stacks of phospholipid bilayers.[96] Inputs were RNA nucleotides at millimolar concentrations. They used a duty cycle in which 8% of pre-existing duplex oligomer RNA was lost in the system per drying cycle. They found that several cycles of hydration and dehydration produced quasi-steady populations of nucleotides from ∼10–50 nt length. They do not report regioselectivity of the condensations, but they report evidence for some duplex structure in the polymers formed.

### 6.6.3 Distinguishing the source of selection from the emergence of memory in a ribozyme-catalyzed era

It is plausible and even likely that base pairing played a role in the earliest RNA polymerization. Disordered condensations involving many different bonding sites can occur from phosphorylated RNA in solution, but base-pairing interactions that lead to template-directed ligation bias the bonds formed in ways that favor formation of secondary structure. (We return to a specific prediction of the consequences of template-directed ligation in Section 6.8.3 below.) It is even possible that RNA catalysis was relevant to the early condensation reactions involving RNA or other activated monomers.

However, there is an important distinction in the roles assumed for base pairing and catalysis, between the metabolism-first and the RNA-first views of the origin of life. The RNA-first view supposes that the hereditary function of template-directed synthesis was also the basis for the most important form of Darwinian fitness in the RNA system. In contrast, the metabolism-first view separates the roles of template-directed synthesis as a memory mechanism, from catalysis of *other, non-RNA-related* reactions as the determinant of which sequences were preserved in reaction systems. This difference has important consequences for whether the oligomer world would have replaced, or would have preserved, the foregoing organic reaction networks.

---

[95] This distinction is mathematically similar to the distinction between an *anneal* and a *quench* in the cooling of liquids to form crystalline solids.
[96] The phospholipid used was 1-palmitoyl-2-hydroxy lysophosphatidylcholine.

In a view that places RNA before the selection of biochemistry, RNA-catalyzed replication (either of the same sequence as the catalyst or of other sequences) is assumed to be the mechanism by which sequence information persists. The speed and fidelity of replication, which depend on the catalyzing sequence, are *also* assumed to be the filtering criteria that determine which sequences come to dominate populations. Thus, *it is assumed that the mechanism that brings memory into existence also introduces the criterion for selection.*[97] This view focuses entirely on the properties of a single molecule: selection arises out of RNA's interaction with itself or other RNA, and the mechanism of selection is competitive copying.

If one supposes instead that a context of kinetically selected geochemistry precedes RNA as a foundation for chemical order, it is more natural to *split the emergence of memory and the source of selection.* The primary mechanism of memory is still assumed to be through RNA base pairing,[98] and the role of RNA catalysts is de-emphasized though not ruled out.

The key assumption is that the polymerization and base pairing which keep sequence information in the population depend only on background chemical processes already in place, and *not* on the functions of polymers produced through the memory mechanism.[99] The earliest replicating processes need not have been reliable: a process qualifies as a memory mechanism as long as the distribution of oligomers in later generations has an excess probability to contain sequences (or their complements) or phenotypes present in earlier generations, compared to a random sample of other sequences. The memory mechanisms are assumed to preserve whatever sequences or phenotypes are in existing populations, without regard to what those sequences do.

### 6.6.3.1 Service to metabolism first determined preservation

The selection that acts as feedback, determining the RNA sequences that are retained, is assumed to act at the system level, through enhanced production of metabolites, cofactors, and oligomers.[100] The RNA sequences that contribute to greater organosynthesis are preserved by virtue of being passed down in systems that provide more inputs, rather than through more effective individual-level competition for scarce inputs.[101]

---

[97] If RNA sequences are supposed to *self*-replicate, then this survival and proliferation rate satisfies the definition of a true Fisherian fitness [273, 274]. If RNA sequences are supposed to aid each other's replication in a system-level hypercycle [823], then a more complex statistical formulation than Fisherian additive fitness is required.

[98] Peptides could also be selected through mechanisms such as cooperative folding with RNA, discussed in Section 5.3.7. This is an instance of selecting directly on properties of a distribution rather than on particular sequences, a feature that we believe characterized sequence selection before the emergence of precise ribosomal translation more generally, as discussed in Section 8.3.4.4

[99] Martin Nowak and Hisashi Ohtsuki [598] study a very minimal formal model with the property that selection is separated from the mechanisms that produce memory. The idea was later applied to template-directed ligation of RNA by Michael Manapat *et al.* [515].

[100] Sara Walker, Martha Grover, and Nicholas Hud [850] developed an explicit toy model of this form, in which differential diffusion in space imposes selective feedback from primary production onto random *de novo* sequence generation and template-mediated but neutral sequence replication.

[101] See, for example, the review of Szostak [791]. Martin and Russell [452, 524] proposed a mechanism by which differential chemical growth takes on a "Darwinian" character of competition between different pockets in a mineral foam. The reactants in more productive pockets displace reactants in less productive pockets through a process similar to invasion percolation.

This proposal is very general, and does not presume that RNA catalysis is limited to large oligomers. Catalysis of substrates covalently linked to RNA dimers was proposed in [158]. Section 6.4.2 proposed that RNA may first have been stabilized at the network level collectively with folates or flavins synthesized by similar chemistry, in the era before enantiomeric selection and base pairing enabled it to act as a macromolecular catalyst. Working from the other direction, ("downward" from large molecules), Rebecca Turk, Nataliya Chumachenko, and Michael Yarus [813] have synthesized a ribozyme for transacylation that is only 5 nt long, with a 3 nt active site. Shelley Copley has organized known and proposed catalytic functions into a progression in which catalysis might have been extended and in some cases transferred from smaller to larger molecules through an entire sequence reaching from basic organosynthesis to major motifs associated with an RNA World [159], shown in Figure 6.5.

Finding selective forces that do not derive from RNA copying, and copying mechanisms that do not rely on selection to preserve them, increases the likelihood of a transition to a functional oligomer world. The two functions of selection and memory arise and are stabilized through different processes, and may occur at very different times rather than necessarily being contemporaneous. Decompositions of complex mechanisms, which break evolution down into more simple, independent steps, satisfy our criterion of *transversality* introduced in Section 5.5.2.

### 6.6.3.2  *The resultant freezing in of metabolic patterns – consequences for the continuity hypothesis*

If a synthetic organic network was pre-existing, selective, and robust in its form, RNA catalysts would have been selected for compatibility with that (continually unfolding) network. Many RNA catalysts[102] could have been selected in parallel to one another, for compatibility with a geochemical protometabolism, before the concentration and diversity of RNA oligomers led to significant RNA-RNA interactions. Such a dynamic would be expected to re-enforce and "freeze in" the patterns of the small-molecule network by the stage when oligomers directly assisted each other's replication, rather than overwriting the network with patterns derived from competitive self-replication as in the RNA-first view.

Thus the argument to separate memory mechanisms from selection criteria, which is suggested by transversality to be a more plausible route to the oligomer world, is an important part of the premise for continuity from geochemistry to the earliest metabolism.

### 6.7 Transitions to cellular encapsulation in lipids

Some time during the emergence and refinement of an oligomer world, the transition to encapsulation in lipid vesicles also occurred. Encapsulation is a complex, multifunctional concept, as pointed out in Section 5.5, and must have been the culmination of several stages

---

[102] Again, remembering Woese [887, 890], we should more properly say many *distributions* of related RNA catalysts preserved under noisy copying processes.

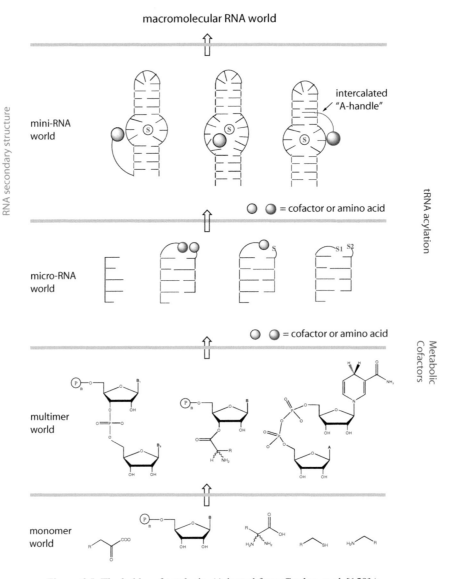

Figure 6.5 The ladder of catalysis. (Adapted from Copley *et al.* [159].)

of increasingly diversified *use* of lipid vesicles, which individually placed lower demands on the interdependence of many molecular systems than are required by fully developed cells. The ambiguity in the way these stages were traversed is reflected in the variety of proposals for this sequence in the literature.

Most authors agree that complex molecular systems including membrane-spanning protein complexes and even the $F_O$–$F_1$ ATPase were in place by the stage of the LUCA, but there consensus ends. Building on the earlier mineral-foam model of Russell and

Hall [687], Koonin, Martin, Russell, and Lane in several publications [452, 473, 524] have argued for an entirely vent-hosted LUCA which employed the ATPase in non-enclosing membranes. They propose that the escape from dependence on vent mineral membranes and proton-motive force occurred in two separate instances of coupled emergence of encapsulating membranes and DNA replication systems. They also associate the transition from pre-cellular molecular populations to cells with Carl Woese's Darwinian Threshold (see Section 6.8.4) and the emergence of the bacteria and archaea. Purificación López-García and collaborators [496] argue, to the contrary, that the enzymes for glycerol-3-phosphate and glycerol-1-phosphate (the backbones to which bacterial and archaeal lipids, respectively, are attached) both reconstruct to the LUCA, and that a population of fully membrane-enclosed cells underwent considerable evolution before the emergence of stable bacterial and archaeal domains. Carl Woese [887, 892] associated the Darwinian Threshold specifically with the transition from stochastic to much more reliable ribosomal translation, in a population of molecular entities sufficiently integrated to be describable in terms of competition and selection, but sufficiently open to horizontal gene transfer that lineages were not stable. Enough proteins specifically associated with membranes and membrane-mediated bioenergetics appear to have been fixed as of this transition (or the set of transitions for different domains [892]) that we are not inclined to regard the pre-threshold populations as non-membrane-bound molecular assemblies, but rather entities that already possessed at least part of the architecture of cells.[103]

The central question is how the adoption of vesicles as compartments was interleaved with the elaboration of macromolecular complexity and the emergence of systems to control and evolve macromolecules. Lipids serving functions such as phase transfer catalysis or polymerization, or non-enclosing membranes supporting a proton-motive force, could have provided selective context for the evolution of molecular complexity and could even have created some integrated subsystems before full cellular autonomy.

In reducing environments, aliphatic carbons are the low-energy end-products of carbon reduction. They are found in meteoritic carbon deposits[104] and are expected to have been common in those environments on the early Earth where organosynthesis occurred. Numerous experiments, including those by David Deamer, Pier-Luigi Luisi, Robert Hazen and others [344, 501, 502] suggest that stacked lamellar lipids and disordered arrays of vesicles would have been present in most systems where organosynthesis produced aliphatic compounds. Therefore the *availability* of lipid structures for use in the early elaboration of biochemistry seems unproblematic.

Adopting a vesicle as a compartment, however, requires providing several complex functions that in extant biochemistry are provided by coordinated, complex *systems* of macromolecules. These include osmotic containment, improvements in proton or ion

---

[103] We particularly wish to avoid regarding the cell merely as a compartment to sequester genomes. We argued in Section 5.5 that the quite independent functions of cells as platforms for metabolism and integrated bioenergetics are essential whether or not genes are readily exchanged. Indeed, relatively rare gene exchange on a generational scale can still be quite sufficient to remove the possibility for stable lineages to remain distinct in a shared environment.

[104] See [75, 184] for accounts of the history and the sequence of publications through which the study of lipids from meteors developed.

impermeability of membranes, transport of nutrients and wastes, and transduction of proton or $Na^+$ chemical potential to reductants or phosphoanhydrides.

The two-way interdependence between compartments and the macromolecular systems they employ is *asymmetric*. The functions provided by macromolecular systems are needed to make encapsulation in vesicles advantageous rather than detrimental (in the case of transport), or even to make encapsulation possible (in the case of support against osmotic rupture). In contrast, compartmentalization does not directly create the functions of catalysis, molecular replication, or translation that supply macromolecular systems; its role is rather to couple them and to change the forms of interdependency that govern their evolution.

Thus, compartmentalization contributes mechanistically toward increasing the *autonomy* of the molecular functions of biochemistry from the geochemical background, but it contributes to the innovation of those functions only indirectly through the context it creates and the effect of that context on evolutionary dynamics.

Therefore, following Russell, Martin, Koonin and others we are inclined to look for innovation of significant molecular complexity within the geochemistry of the rock-hosted hydrothermal fluid, before the transition to the use of vesicles as cellular compartments. It seems to us most plausible that oligomerization and formation of secondary structures involving RNA, and possibly also peptides, would have been reached before cellularization.

Scenarios that propose the opposite order – in which early compartmentalization is needed to evolve molecular complexity – again violate the criterion of *transversality* from Section 5.5.2. If vesicles are assumed to create the selective context for the evolution of molecular functions, but the molecular functions are required for the viability of vesicle-hosted chemistry, then multiple order-forming transitions must be accomplished concurrently. We propose instead a sequence in which macromolecular complexity was selected first for other functions, and then enabled the adoption of vesicles as compartments.

### 6.7.1 Contexts that separate the aggregate transformation into independent steps

An important distinction exists between the complex, *aggregate* problem of adopting vesicles as containers for metabolism, and the simpler problems of evolving solitary functions such as electron, proton, or ion transport, or osmotic containment. The coevolution of molecular and vesicle functions is most plausible in contexts that separate each innovation into an independent evolutionary problem. The following examples are representative of the kinds of separation that have been proposed.

**Breaking the chicken-egg dependency of carbon fixation and bioenergetics** Section 6.3.1.2 noted the energy-dependence hierarchy created by interposing membrane ion potential between the carbon excretion and carbon fixing branches in acetyl-CoA organisms: self-maintaining carbon excretion enables a membrane potential, which then enables carbon fixation. The AHV model argues that the only natural evolutionary order for these three stages is the *reverse* order, where they become stages through which the acetyl-CoA pathway attained autonomy. One suggestion for a way to separate these stages, from Section 6.3.1.5, is to couple proton-motive force to $CO_2$ reduction from $H_2$ in alkaline vents.

**Osmotic containment and polyelectrolyte-gel "biofilms" before cells** Without cell walls to resist osmotic pressure, membranes cannot concentrate catalysts and substrates unless they are in environments where the exterior osmotic strength is already very high. This may not be an insurmountable problem for the earliest life, and in particular it encourages us to consider the role of early polyelectrolyte gels as pre-cellular "biofilms."[105] A compatible but slightly different hypothesis derived from the AHV model [524, 688] proposes that a mineral-hosting environment in the form of an iron sulfide colloid could have extended directly to the earliest cell membranes, first being coated and then supplanted by lipid/protein assemblies.

**Sodium and protons for early membrane potential** Certain problems of membrane evolution would be simplified if $Na^+$ ions rather than protons were the first charged species for which trans-membrane proteins coupled electrochemical potentials to formation of either phosphoanhydrides or low-potential ferredoxins. Unlike protons, which are the most mobile charge carriers in water through hopping mechanisms, $Na^+$ is a normal ionic solute which forms solvation shells. These nano-scale molecular assemblies diffuse much less readily through the ephemeral defects and gaps that form in crude amphiphilic lipid bilayers [641], and which permit both water and proton diffusion.

Armen Mulkidjanian in particular has argued [578, 579, 580] in favor of a prior role for $Na^+$ pumps and ATPases, on phylogenetic as well as functional grounds. Lane and Martin [473] counter that the phylogenetic evidence points, at most, to non-specific use of either protons or $Na^+$ in the earliest ATPases, and on the premise that membrane chemiosmosis grew continuously out of alkaline vent chemiosmosis, they argue that the early selective pressures on the evolution of membrane bioenergetics were created from functions involving proton translocation.

## 6.8 The advent of the ribosome

The translation system from RNA to proteins introduced fundamental new boundaries in the organization of life, both functional and phylogenetic. Functionally, translation stands between the detail and particularity of the biosynthetic pathways, and the comparatively neutral landscape of sequence space.[106] Phylogenetically, it forms a horizon between the

---

[105] We are cautioned, however, that even refined modern biofilms have not spared bacteria or archaea the need for advanced cell walls to support osmotic stress, so some more modest biochemistry must have been supported in any early, wall-free systems.

[106] By this, we mean that the replication system and the translation system permit nearly any protein-coding sequence to exist as a gene and to be expressed, translated, and submitted to selection for evaluation. The functions of the resulting proteins, and their contributions to fitness, are of course not neutral.

modern protein world and whatever RNA/peptide world went before, across which much of the information about how the earlier world functioned has been lost.

Remarkably, however, sufficient variation exists in ribosomal RNA and peptides that through a combination of comparative and functional analysis, it has been possible to reconstruct the evolutionary history of the ribosome back to much simpler origins that clearly antedate the function of translation. This reconstruction provides probably the most specific window that we have into the components and functions of the RNA World [85].

Three conclusions that have emerged with reasonable confidence are the following.

1.  The original component of the ribosome was the ancestor of the peptidyl transfer center (PTC) in what is now the large subunit (LSU), and its function was catalysis of some form of (probably non-specific) condensation reactions.
2.  Pre-ribosomal RNA carried out many of its catalytic functions using ferrous iron as a counter-ion where modern RNA uses magnesium. The catalytic activity of iron-RNA complexes was in many cases higher, it was functionally more diverse, and surprisingly many components of the translation system including components of the LSU and even tRNAs were catalytic.
3.  Subsequent evolutionary elaboration and variation of the ribosome LSU has occurred by the introduction of loops that are rooted to the core through double helices, and by helix extensions, but these have largely preserved the fold and function of the core.

We begin with a review of the remarkable evidence for this history of the LSU and the catalytic functions of the ribosome. Less is understood about the history of the small subunit (SSU) which governs the interaction with tRNA and mRNA, and is believed to have added the "translation" function to the older function of the LSU which was always essentially catalytic. Even for the SSU, however, a partial historical reconstruction is suggested. It is also possible that essential features of the encoding system such as the three-base reading frame originate in very low-level properties of chemical kinetics, which pre-determined the possibilities for encoding systems before the actual code was populated with the existing amino acid assignments.

### 6.8.1  The core and evolution of catalysis

The proposal that the ribosomal LSU originated in the PTC can be followed in a sequence of increasingly specific and methodologically refined analyses in the work of George Fox and others over the course of the last decade [80, 269, 270, 271, 372, 385]. Early assignments of secondary structure for the 23S RNA of the LSU [595, 596] provided a decomposition into domains to which a chronology could be assigned, but did not identify the most ancestral core region as itself a structurally defined domain. More recent analysis [634, 635] combining phylogenetic comparison and recent high-resolution three-dimensional structures from X-ray crystallography identify the core explicitly as a domain, and round out a consensus picture of the origin and early evolution of the LSU.

### 6.8.1.1 Phylogenetically identified secondary structure and six domains

The first comprehensive secondary structure[107] and assignment of domains to the 23S RNA of the LSU was derived by Harry Noller [595] on the basis of phylogenetic comparison of conserved sequence ranges, and even more conservative motifs in the minimum-energy base pairings for those sequences. This structure showed six domains, which surrounded the geometric center. The oldest, indexed Domain V, contained the A-site and P-site constituting the peptidyl transferase center of the LSU. The PTC itself consists of a pair of pseudo-symmetric strands believed to have formed by duplication of a roughly 110-unit oligomer [80]. The geometric center itself was originally represented in secondary structure as if these duplicated strands were extended segments without base-pairing assignments, from which loops forming the other six domains branched.

A remarkable observation by Konstantin Bokov and Sergey Steinberg [80] has enabled the modeling of the ribosome as an accretion of "expansion segments" that accreted onto the ribosome in layers, around the ancient central region. The observation involves an RNA interaction known as the *A-minor interaction* [591], in which a stack of unpaired adenosines on one segment pack into the minor groove of a double helix elsewhere in the sequence. A-minor interactions are not required to stabilize the helix into which they pack, but they do govern the tertiary structure of the segment that donates the stack of adenosines. This gives each A-minor interaction an inherent directionality, in which the A-donating segment depends on the receiving helix but not the converse. In the ribosomal LSU, these directional links are almost perfectly cycle-free, meaning that the expansion segments can be removed in an order that leaves the remaining assembly stable with respect to A-minor interactions. Bokov and Steinberg showed that the LSU can be modeled by this partial order as the result of 12 layers of accretion around the catalytic core.

### 6.8.1.2 Identification of a new domain corresponding to the core

More recent work by Anton Petrov, Loren Williams, and collaborators [634, 635], combining expanded phylogenetic analysis with three-dimensional structure from high-resolution X-ray crystallography, shows that the central region from which the six previously recognized domains grow is itself a compact and double-helical domain structure. This core region, which Petrov *et al.* term "Domain 0" is both the topological center for the tree of branching double helices seen in the secondary structure, and the geometric center in three-dimensional structures. They present evidence that it is an autonomous folding unit, which is compact and which also possesses the greatest sphericity (as defined by three-dimensional surface/volume ratio) of any of the domains.

Domain 0 incorporates the strands of the PTC that are the ancestral core of the LSU in Bokov and Steinberg's model, and contains a cleft that holds the A-site and the P-site in proximity to one another. The fold surrounding the P-site, which clearly antedates

---

[107] *Secondary structure* refers to the topological representation of an RNA fold defined by base pairing. It omits three-dimensional spatial embedding and non-base-pair interactions such as those of adenosine with minor grooves in double helices, known as *A-minor interactions*.

encoded proteins, is reconstructed as the oldest part of the ribosome, and is among the oldest polymeric sequences retained in extant biology.

### 6.8.1.3 Peptides in the ribosomal core

The character and role of peptides in the pre-ribosomal RNA World can also be dimly seen in retained properties of ribosomal proteins. It has long been recognized [269] that, in contrast to most protein enzymes which fold into globular structures, ribosomal proteins especially in the core are extended strands. The functions of their sequences are defined by amino acid–RNA interactions rather than amino acid–amino acid interactions as for globular proteins.

It is interesting to speculate that the melting/crystallization dynamics of RNA-peptide assemblies may have provided a mechanism for the selection of peptides before translation arose. In a world where the compaction of large RNAs into tertiary structures was aided by peptides, a dynamic of intermittent melting and re-crystallization could have enabled folded RNA to sequester peptides with compatible sequences against dilution, performing a kind of population level selection for compatible peptide and RNA sequences.[108]

### 6.8.2 Catalytic RNA and iron

As remarkable as the ability to trace a catalytic function of the ribosome back to a pre-translational era, is what has been learned about the character of RNA catalysis and its implications for this era. Both the compaction of RNA into tertiary structures, and the catalytic function of ribozymes, depend on the presence of divalent cations. In both naturally occurring RNAs and those adopted as laboratory study systems, the ion used has almost always been $Mg^{2+}$. However, on the Archean Earth before the great oxidation event, another prevalent divalent cation would have been $Fe^{2+}$,[109] and it would have been difficult to exclude from associations with RNA, whatever its impact on compaction and catalysis. In recent studies of RNA in anoxic solutions with dissolved ferrous iron, Loren Williams and co-workers have shown both that $Fe^{2+}$ readily replaces $Mg^{2+}$ as the charge-neutralizing cofactor enabling the compaction of large RNAs,[110] and that it is in many cases a better and more diverse catalytic center.

$Mg^{2+}$ has conventionally been regarded as an essential "cofactor" for RNA catalysts. Hsiao *et al.* [371] have shown that $Fe^{2+}$ can also function in this role, and that it both

---

[108] Note that this is a form of selection on phenotype that may admit a whole distribution of sequences with equal likelihood. We return to discuss the selection of distributions of sequences below in Section 6.8.4. Jessica Bowman, Nicholas Hud, and Loren Williams [85] make this argument explicitly, offering it as an interpretation of Carl Woese's earlier proposal [892] that evolution before the advent of translation was carried on "supramolecular aggregates" where today it is carried by genes, chromosomes, or organisms as units of selection.

[109] Recall from Section 3.5.3.3 that Ca–Mg exchange reactions in hydrothermal fluids deplete $Mg^{2+}$, an affect that is believed to have reduced $Mg^{2+}$ abundance in early Archean seawater relative to the present day [40]. The effect at acid venting sites is even more dramatic: when investigators attempt to empirically determine alteration fluid chemistry from samples that are invariably contaminated by some seawater mixing, they use the *zero-Mg* extrapolation from a distribution of samples to estimate the native alteration fluid composition [837]. In the early Archean and Hadean eons, when sulfur was less prevalent to precipitate $Fe^{2+}$ from solutions, the concentration of soluble reduced iron in vent effluents should have been higher than it is today [463].

[110] Systems studied in [40] were the P4–P6 domain RNA in the large self-splicing RNAs known as *Group I introns*, the L1 ribozyme ligase [676], and the hammerhead ribozyme.

improves the activity of some ribozymes (e.g. phosphoryl transferases) and expands the repertoire of RNA-catalyzed reactions to include *single-electron transfers*. Single-electron transfer is impossible for Mg but central to the role of Fe in numerous metal-centered protein catalysts. It is one of the key functions that we have argued made transition metals, and particularly iron, essential to the initial elaboration of organic geochemistry into a proto-metabolism. Hsiao *et al.* hypothesize that $Fe^{2+}$ was the essential metal center in RNA folding and catalysis not only in early life but throughout most of the Archean, and that the great oxidation event, which removed soluble iron from the oceans, led to a wholesale replacement of $Fe^{2+}$ by $Mg^{2+}$ in most RNA.

In addition to the diversity of catalytic mechanisms, a further surprise is the diversity of components in the translation system that are catalytic with $Fe^{2+}$ cofactors. Both the 23S ribosomal RNA and transfer RNAs[111] are efficient electron-transfer ribozymes in the presence of $Fe^{2+}$ under anoxic conditions and in the absence of $Mg^{2+}$.

### 6.8.3 The origin of translation and the three-base reading frame

The origin and evolutionary history of the SSU of the ribosome, consisting of the 16S rRNA and ~21 ribosomal peptides, are less well understood. Most investigators believe the 16S rRNA originated as a catalyst for RNA replication [269], and that the association with the LSU and processing of aminoacylated RNAs originally evolved to augment this replicase function in some way. Only later did polypeptide synthesis become the function for which the ribosome as a system was selected.

One of the "magic numbers" of biology, which appears with the emergence of the ribosome, is the three-base reading frame and the resulting triplet code. It is the kind of property that calls out for an explanation, because from a chemical perspective it at first seems arbitrary, yet it has profound consequences for the informational properties of translation. Hints have been recognized [549] that the three-base reading frame actually could have originated from low-level chemical properties of RNA, and that its explanation is linked to the explanations for the use of the $3'$–$5'$ phosphate ester bond and the existence of secondary structure. As usual, the explanation begins indirectly – starting with a much more general and seemingly independent problem – and then surprisingly reveals connections between low-level chemistry and higher-level informatics.

#### 6.8.3.1 Secondary structure and regioselectivity of ribose phosphate ester bonds

The more general puzzle is how $3'$–$5'$ phosphate ester linkage could first have been enriched, and why the formation of secondary structure was ever possible at all. Ribosyl-nucleoside triphosphates have two nucleophiles – the $2'$ OH and the $3'$ OH – which can attack the $5'$ triphosphate and form phosphodiester linkages. The $2'$ OH is 6–9 times more nucleophilic toward activated phosphate esters [679], and without other constraints, ligation of activated RNA monomers yields a random sequence of $2'$–$5'$ and $3'$–$5'$ linkages with

---

[111] The specific examples exhibited were 23S rRNA from *Tetrahymena thermophilus* and yeast tRNA[phe].

a high proportion of the former. However, the $2'-5'$ linkage interferes with the formation of double helices and thus stable secondary structure.[112] In this particular respect, RNA at first appears an unlikely monomer to have become the main informational molecule of life.

However, the fact that $3'-5'$-linked RNA sequences can form helices provides a direct source of feedback that can select for this linkage, through differential rates of both ligation and hydrolysis. David Usher and Angelika McHale [820] showed that, in long helices,[113] a single inserted $2'-5'$ linkage was hydrolyzed at 900 times the rate of a $3'-5'$ linkage at the same location. Jack Szostak and collaborators [678, 679, 791] have shown how regioselectivity of the ligation reaction, along with subsequent differential hydrolysis rate, imposes a filter on plausible routes to early RNA replication, which jointly selects for $3'-5'$ linkage and secondary structure.

Szostak [791] has argued that processive, single-nucleotide extension of RNA primers, a favored laboratory study system and a qualitative analogue for the function of the modern RNA polymerases, is not a plausible mechanism for the earliest RNA replication. A more plausible mechanism is **template-directed ligation** (TDL),[114] in which short activated oligomers that attach by base pairing to adjacent regions on a template strand are subsequently ligated to one another. TDL can proceed in the absence of a ribozyme or with low-selectivity ribozymes. Although activated monomers can also be ligated, these form large fractions of $2'-5'$ linkages and are readily hydrolyzed.[115] If, however, both bound oligomers are long, they form stable double helices with the template strand, orienting the nucleophilic OH groups so that $3'-5'$ linkages are strongly favored.[116] These linkages are then also slower to hydrolize. Rohatgi *et al.* [679] suggest that the first ribozyme polymerases were oligonucleotide condensation catalysts. To the extent that trimers represent a kind of threshold to the formation of stable double helices, three becomes a magic threshold also for the onset of regioselective ligation.[117]

### 6.8.3.2  *The early ribosome as a "triplicase"*

An independent explanation for the three-base reading frame, also originating in RNA replication but not directly connected to the regioselectivity of triplets for $3'-5'$ linkages, was originally due to Anthony Poole, Daniel Jeffares, and David Penny [644]. They argue

---

[112] References [225, 735] report that aptamers can be evolved which will tolerate interspersal of $2'-5'$ linkages and will still fold and function, so the exclusion of $2'-5'$ linkages need not be total.

[113] The length used was 12 bp, but this appears to be a suitable proxy for arbitrarily long double helices.

[114] See also reviews by Ellington [223] and James and Ellington [390] on TDL and its transition to catalyzed, single-nucleotide primer extension.

[115] An extensive literature on ligation of activated monomers, in solution, on templates, and with ribozymes, is reviewed in Orgel [613]. The dependence of the rate and linkages formed on the activating group, the divalent cation used as a counterion, and the sequence itself is complicated and in some cases unpredictable. In the experiments of Rohatgi *et al.* [678, 679], both imidazoles and triphosphates were studied as leaving groups.

[116] It has been suggested that these longer duplexes could serve as models for template-directed ligation of nucleoside triphosphates [549].

[117] Note that this threshold is bidirectional. If the oligomer RNAs that are inputs to TDL are to be formed by random ligation, then both the action of hydrolysis and the random interspersal of $2'-5'$ linkages cause the concentration of $3'-5'$-linked oligomers to decay exponentially with their length. This exponentially decaying tail in the supply of oligomers must be multiplied by the probability of their incorporation into longer, surviving strands through TDL, which has a lower threshold at length-three.

that high-fidelity RNA polymerization would have been favored if the first polymerase incorporated triplets rather than monomers. They propose that the 16S rRNA began as a "triplicase" for what amounts to TDL.[118] This argument is not exclusive of the arguments of Rohagti *et al.* for the importance of regioselectivity, and it is possible that a selective filter for triplets that was already present in the non-enzymatic world was recapitulated for different mechanistic reasons in the evolution of the first replicases.[119]

### 6.8.4 Reliable translation and the birth of phylogeny

Carl Woese coined the term "the Darwinian Threshold" [887, 892] to refer to the transition – within populations of small molecular assembles that could already have had some of the complexity of cells – from an era governed by communal genomes and extensively shared genes, to an era in which cells became progressively more refractory toward gene transfer, and lineages of vertical descent began to emerge. While we recognize a wider set of transitions as introducing novel levels of "Darwinian" competition and selection, the particular transition called out by Woese remains special because it was also the beginning of a phylogenetic record of history – the other fundamental element of Darwinian dynamics besides a theory of evolutionary causation.

The causal argument that Woese put forth for the existence of a Darwinian Threshold grows crucially out of the system-level integration of the ribosome. We reviewed briefly in Section 5.3.7 the problem of bootstrapping a translation system into existence, when its components – ribosomal RNAs and proteins – are unreliably replicated and translated in the early stages, and the need that this creates for the kind of buffering we see in the redundancy of the genetic code. We will return in Section 8.3.4.4 to argue that the evolution of precise translation was a phase transition[120] that shifted the unit of selection carrying evolutionary information. In the age of unreliable translation this unit was the distribution of genotypes replicated with high error, and the distribution of statistically translated proteins; in the age of reliable translation the informational unit became the individual sequence. The characters selected for in these two regimes are qualitatively different.

Here we remark that the complex, hotch-potch interaction of RNA and peptides seen in the core of the ribosome is exactly the kind of *non-modular* architecture that would be expected to lead to a transition of the kind proposed by Woese. While the LSU of the ribosome is reasonably well resolved into domains at the level of RNA secondary structure, the interaction with peptides near the core occurs along extended strands, and creates very non-local interactions of amino acid and RNA residues. These are the interactions that would have required evolutionary fine tuning for the ribosome to become a precise

---

[118]  Eric Ekland and David Bartel [217] demonstrated that an RNA polymerase which incorporates mononucleotides could also function as a triplicase, but this was an instance of non-specificity for length rather than specificity for length-three.

[119]  The possibility that two functions at successive hierarchical levels may have led to selection for the same magic number is reminiscent of our arguments in Section 5.3.7 for the selection of assignments in the genetic code: an early chemical criterion may create a pattern that is also well (pre)-adapted to a later criterion of robustness or fidelity.

[120]  The same concept is invoked by Woese [892], though we will describe the system that undergoes a shift in organization in terms that are not exactly those he used.

translator. They are exactly the kinds of refinement, however, that accumulate along highly contingent histories, and lead to components that are very difficult to exchange among different evolutionary lineages.

Arriving at an analytic "theory" of the logic that shaped the earliest interaction of RNA and core ribosomal peptides may be one of the most difficult problems in understanding the early history of the ribosome. Quantifying the approximate range, multiplicity, and historical depth of the correlations involved, and their implications for the transitions that excluded horizontal transfer, may however be tractable in statistical terms.

## 6.9 The major biogeochemical transitions in the evolutionary era

The emergence of a ribosome capable of high-fidelity translation produced many changes in the character of evolutionary transitions, but the most major evolutionary transitions retain a simplicity and lawfulness at their foundation that is continuous with the pre-Darwinian era. The important change is that, upon crossing Woese's Darwinian Threshold, the organisms in the biosphere could for the first time have possessed genomes long enough that the diversity of maintainable sequences outnumbered the population sizes of organisms that could instantiate them. Whether the stable, evolvable forms in the earlier, statistical era could have been diverse enough to permit an unlimited accumulation of adaptations is unclear, but by the time the Darwinian Threshold was crossed, life had certainly entered a regime of open-ended adaptation.

At the base of many major evolutionary transitions, however, is a biogeochemical re-ordering that does not admit many forms and is not open ended. With respect to this foundation, the complexity of the Darwinian era leads to new methods of exploration and selection, but the change accomplished has limited freedom. The following three examples are among the most major evolutionary transitions, and each is manifestly biogeochemical in character.

**Innovations in carbon fixation in a reducing world** The innovations of the six known carbon fixation pathways reviewed in Chapter 4 all took place in the deep phylogenetic tree, in the early to middle Archean. This was a period of rapid change for genotypic and phenotypic diversity, and for the construction and exploitation of geochemical and ecological niches. In important respects, life was expanding into a "new world," in comparison to the complex jockeying for ecological position in today's "full world." Yet despite the opportunity that one might expect in a period of early and large qualitative change, the innovations in carbon fixation were few and were bounded in time. Each explores a basic mechanism in carbon addition chemistry, and apart from the first exploitation of autocatalytic feedbacks, even network innovation consisted of variations on a conserved architecture. We believe that this small number of innovations covered the major domains of chemotrophy that planetary geochemical variability provided, and that after each had been populated in some form, evolution turned to a secondary level of pathway refinement.

**Phototrophy** All of our discussion of the first drivers of biogenesis has emphasized the role of redox potential,[121] a free energy source that couples directly to chemical bond formation. In the great planetary arc of redox disequilibrium, light energy participates indirectly as a driver of atmospheric escape and atmospheric ionization chemistry, but despite its enormous free energy potential, light is not used directly by chemotrophic life. Today the major disequilibrium for which biological primary production creates a dissipation channel is between visible and thermal-infrared light, and it is captured by phototrophs. The emergence of phototrophy opened a direct transduction channel for light energy and entropy in parallel to the redox-relaxation channel that (we argue) must have existed first to provide a foundation for the evolution of the complex mechanisms of phototrophy.

**The rise of oxygen** As long as a permanent ocean and plate tectonics persisted on Earth, it does not seem that oxidized surface conditions could have been produced by geochemical mechanisms, akin to those at work either on Venus or on Mars. The establishment of a strong redox disequilibrium between an atmosphere and ocean concentrated with a strong oxidant such as $O_2$, and continually refreshed reduced crust, was the great chemical opportunity on a tectonic Earth, but it could not occur until a process much faster than tectonic recycling could sequester carbon and release oxygen. The emergence of the biosphere created this new timescale and the resulting qualitatively distinct, kinetically maintained redox disequilibrium. As we observed in Section 3.6.7, the rise of oxygen changed the mineral inventory of the Earth's surface, and as we observed in Section 4.4.6, it changed the phenotypic character of evolution in metabolic pathways and ecological exchanges.

### 6.10 Tentative conclusions: the limits of narrative and the way forward

> Finally: It was stated at the outset, that this system would not be here, and at once, perfected. You cannot but plainly see that I have kept my word. But I now leave my cetological System standing thus unfinished, even as the great Cathedral of Cologne was left, with the crane still standing upon the top of the uncompleted tower. For small erections may be finished by their first architects; grand ones, true ones, ever leave the copestone to posterity. God keep me from ever completing anything. This whole book is but a draught - nay, but the draught of a draught. Oh, Time, Strength, Cash, and Patience!
>
> – Herman Melville, *Moby-Dick*, Chapter 32 [553]

We have seen in this chapter that an extensive though still fragmentary body of evidence exists about the sequence and causes of stages in the emergence of the biosphere. Some

---

[121] The role of pH has also been considered, but pH disequilibria are localized both geographically and mechanistically, in contrast to redox disequilibrium which exists at a planetary scale.

proposals (such as those concerning the history of the ribosome) employ few signatures and address quite specific problems, and seem unlikely to be seriously in error. Other arguments are much more circumstantial, and attempt to search a large and poorly characterized chemical space for relevant mechanisms that can be assembled into complex hypothetical scenarios (such as the AHV model). A third group provides a careful characterization of a limited number of demonstrable pathways to products that we know are important to the actual history of biogenesis (selective synthesis of ribose or *de novo* synthesis of nucleobases or nucleoside phosphates), but they provide limited systematic consideration of the context for the demonstrated pathways. A fourth body of method, still in its early stages, seeks to combine expert knowledge (for instance, of chemical mechanisms) with computational search over large combinatorial spaces, to lift our current reasoning from the level of pathways to a more inclusive network view.

Where current efforts seem most seriously lacking is in ways to place these fragments into a system-level context. We have reviewed numerous mechanisms that are known to exist and that have been proposed as sources capable of producing order, but for which it is very difficult to judge whether they were *relevant* to the actual path of biogenesis. Where complex and particular scenarios have been offered as hypotheses, although we believe the actual unfolding of biogenesis was also complex and particular, we doubt that from current knowledge the particularity of a whole scenario can be guessed with much hope of being correct; nor do we believe that completeness should be needed to judge key insights about problems of organization that are likely to be general. Yet the lack of a systems context makes it difficult to frame such general problems in ways that do not seem to depend on particular scenario assumptions.

Continuations at all levels present problems. Perhaps geochemical order could once have been created by proton-motive forces across mineral membranes, or by a cascade from reduced to oxidized phosphorous. Biochemical order today is maintained either by different chemical potentials or by the same chemical potentials contained and gated through machinery that is in most respects qualitatively different. How can we determine when one set of mechanisms generates systemic order that continues into *and serves as a cause for* systematic order maintained by other mechanisms? More generally, in thinking about causation, how do we know when we have identified a sufficiently broad system context to capture all the essential constraints on the emergence or persistence of order – but no more (that is, we also need to know how to weed out detail that is present but non-essential). We have been willing to propose that such causation is at work between geochemistry and metabolism, and between metabolism and higher-level architecture, but our proposals like others leave much to be desired of mechanistic specificity, integration, and certitude.

These problems are typical of the frontier where we characterize systems as being *complex*. On the basis of experience with other systems that were once beyond the horizon of complexity and are now considered understood, we can anticipate that the needed innovations for understanding biogenesis will be of two broad kinds. The first kind may not involve problems of *complication* – that is, our current difficulties may not arise from the variety and heterogeneity of basic generating mechanisms – but may come from the need to account for long-range feedbacks at the system level, which are not apparent from the low-level mechanisms. The recent (twentieth century) understanding of collective and cooperative effects is an example of an innovation of this kind. The second kind are the problems of dealing with complication itself: our capacity to derive relevant scientific abstractions is much weaker for systems where many heterogeneous low-level mechanisms interact to create order, than for systems with more symmetry. We are alerted to problems of complication when, recognizing that existing abstractions are inappropriate or inapplicable, we take recourse in narrative which seems to retain the richness of empirically known heterogeneity,[122] but then find that narrative is a poor guide to abstraction, and does not offer methods of back-tracking to identify incompatible assumptions or missing information when different scenarios conflict.

We think that these problems define the scope and the role of theory in Origins research. Theory should help us identify the scale at which system-level feedbacks remain important to understanding causation, when this scale falls far outside the scale of the basic generating mechanisms, or is qualitatively different from them in kind. Theory should suggest ways to abstract, from the diverse interactions in complicated systems, the decomposition into levels, modules, or other kinds of subsystems that not only make the whole comprehensible, but more importantly are essential to its own dynamical capacity to persist in a patterned state.

The working out of a theory of the origin of life will not be a single effort or separate project; it will proceed in slow steps as an integral part of the reconstruction of history, the search for mechanisms and the suggestion of scenarios. It will not be done by a few researchers or on the basis of a few conceptual insights. We think it will consume the inventive imagination and hard work of the collective Origins community for a considerable time. In the last two chapters of this book, we will however suggest some of the criteria of robustness and hierarchical decomposition that we believe any theory of biogenesis must include. We will also show that a body of concepts and methods exists to formalize at least some parts of this problem. The advance of theory will bear a loose analogy to the advance of life: it need not be a process of pure invention; the core concepts for some parts of the study of robustness and hierarchy have already been learned in our studies of simpler systems. What theory must do to advance is to develop a branch of these methods that is *autonomous* from the supports of our old equilibrium thinking, and correctly matched to the structure of the non-equilibrium processes that are our subject.

---

[122] Notably, we do not take recourse to narrative for problems of the first kind, even when we do not understand them, because the kinds of narratives that could be generated would not be *interesting*. For example, coherence over all possible paths is essential to the stability of an atomic orbital, or a laser, or an atomic clock, yet one would never tell a story about each path, since in relation to the underlying dynamics they are all degenerate. A narrative about the differences between one path and another would be a recitation of random numbers.

# 7

# The phase transition paradigm for emergence

This chapter asks not how, but *why* a non-equilibrium system like the biosphere should have emerged, and how such a state can be stable. A mathematical theory exists which addresses such questions, familiar from condensed matter and particle physics as the theory of phase transitions. We show why some theory of this kind is needed to make sense of the complex facts and contradictory interpretations that have been given for the origin of life, and then introduce the fundamental concepts of phases and phase transitions in forms appropriate to the phenomena seen in earlier chapters. We introduce phase transitions as a class of mathematical phenomena, which frees us from introducing the subject by analogy, and provides a conceptually more fundamental expression of the main ideas. We show how the concepts of phase transition generalize from familiar equilibrium systems to more complex non-equilibrium cases, and summarize which lessons from equilibrium systems can be expected to generalize and where new ideas will be needed. The thermodynamic theory of stability is an application, in matter, of the method of inference known as Laplace's *principle of insufficient reason*, which states that when inferring from any observation, we should be as uncommittal as possible given what we have seen. From Laplace's principle, we show how the theory of optimal error correction is of the same kind as the thermal theory of stability of matter. These different versions of the stability perspective are assembled in this and the next chapter to propose that the emergence of the biosphere must be understood as a cascade of non-equilibrium phase transitions away from a lifeless Earth, and this is the origin of their necessity.

## 7.1 Theory in the origin of life

Many problems in reconstructing the origin of life center on searches for mechanism. We must search large parameter spaces for domains of relevance, with respect to geochemical energy sources, mineral structure and contents, organometallic and organic reactions, pathway completions, and physical structures of many kinds. The last six chapters have mostly been reviews of facts about the life we know, chosen to inform our problems of search.

A different part of understanding, alongside reconstructing the path of biogenesis, originates not in search but in conceptualization. What kind of event (or more properly, sequence of events) was the emergence of the biosphere? Is it to be understood merely as some assembly of reactions that arrive at complex molecules? Is any reaction sequence plausible, or are some preferred? Among reactions inferred to occur in meteorites or comets, on interplanetary dust grains, in the atmosphere or oceans, or in laboratory models of any of these, are all of them together, or is any combination, equally relevant as a model of incipient life? Related to this question is the problem of the likelihood or uniqueness of the path the Earth actually took in forming a biosphere. If we cannot, in some principled way, say that one chemical environment or mechanism is more plausible than another, we also cannot judge whether the set of alternatives taken together offer some path that was more likely to be taken than the eventuality that the Earth had remained lifeless.

One of our main theses in this monograph is that the origin of life cannot be understood as a compounding of rare or arbitrary events, but must be understood as a cascade of *system rearrangements* that were in certain essential ways robust, and at least locally, necessary. Two lines of argument lead to this conclusion, one empirical and the other from the theoretical perspective that we introduce in this chapter. From the empirical side, we have exhibited a number of specific subsystems in which patterns appear to proceed upward from low-level organic chemistry or geo-energetics and not downward from controlling macromolecules or structures. These suggest that constraint and stability flowed from what was generic and necessary rather than from what was particular and perhaps accidental. From the theoretical side, independent of the particular cases in which we happen to have functional or historical reconstructions of living subsystems, we know that living matter obeys certain mathematical laws governing fluctuation and stability, which include the familiar laws of thermal physics but extend also to higher-level structures. The most important problem these address is that many small-scale degrees of freedom have been entrained in states of order that are robust on the timescales we observe directly, and these ordered states have persisted for billions of years. Whatever chemicals and contexts conducted the biosphere on its path of emergence, a theory of biogenesis can only make sense within the larger framework of physical and mathematical law if almost all intermediate stages were robust, and at least most transitions were likely.

The notion of robustness or likelihood that we invoke is not a rejection of stochasticity, but rather a particular expression of it. Local micro-events contributing to biogenesis were surely stochastic and in that sense random, as they remain in all living and non-living matter today. However, the system states in which these events took place were those that survived the filter of constant perturbation. The dynamics of system rearrangement on the macroscale comprised those sets of microscopically random events that, in aggregate, were most likely to occur. It turns out that robust states and system rearrangements have certain properties that distinguish them as a category from arbitrary micro-events, and by these properties we distinguish the scientifically sensible emergence of life from accident.

In this chapter we introduce the main concepts that distinguish stochasticity in micro-events that makes them "accidental," from the more orderly and predictable behavior that arises when many such events are aggregated. The key concept of a robust state is the notion of a thermal *phase*. The key concept of system rearrangement is that of *phase transition*. Both concepts were worked out, and are best known, from the domain of equilibrium physical thermodynamics, and some of the applications we will need to understand the emergence of life use these equilibrium concepts directly. However, the mathematical ideas that underlie phases and phase transitions are far more general than equilibrium thermodynamics. They apply to non-equilibrium and even non-mechanical processes, and to problems of inference, and they underlie important theorems about optimal information transmission and error correction. These latter applications will be central to what is new about the biosphere, besides its particular chemistry, that must be captured both formally and conceptually to understand biogenesis as a planetary transformation.

The current chapter introduces phases and phase transitions as one of the theoretical pillars in established science, to parallel the established facts from geochemistry or biology in the first five chapters. Like the earlier empirical chapters, we do not give an extensive interpretation of the role of phase transitions in the understanding of the living state until some brief sections at the end of the chapter, leaving most of the overall integration for Chapter 8. The material in this chapter is different from that in the empirical chapters in that it does not emphasize a system of observations, but rather a system of mathematical relations that have been used to unify observations into a coherent theoretical system. The point to appreciate about the theory of phases and phase transitions is that it is our current most basic theory of *robust states and processes*.

### 7.1.1 *What does a phase transition framing add to the search for relevant environments and relevant chemistry?*

Whereas microscopic fluctuations are ubiquitous and continual, the more deterministic regime shifts that qualify as phase transitions are comparatively rare and restricted. They typically require collective or cooperative effects among many small-scale degrees of freedom, which are sufficiently strong to drive the system into a small "corner" of its configuration space, thereby circumscribing further fluctuation and creating new forms of order.

Asserting that the origin of life must have comprised a cascade of non-equilibrium phase transitions does not by itself solve the problems of search for relevant environments or relevant chemistry, but it may help to narrow them. The environments, boundary conditions, and chemical mechanisms along the path of biogenesis must both have possessed self-amplifying cooperative effects, and have occurred within a configuration space where self-amplification could drive matter and energy flows into restricted corners or channels. This is the mechanism by which self-organization is *selective* (synonymous with creating order by pruning possibilities), without necessarily being Darwinian. It is the appropriate form of selection to invoke for bulk processes that produce universal background patterns, which then serve as references for the more contingent forms of order created in later evolutionary eras.

The reason we need a *theory* of phase transition and not just a descriptive category, and the reason it matters that the theory is connected to *information and error correction*, is that a theory is meant to quantify the relation between the volume or rate of disorder constantly generated by fluctuations, and the volume or rate at which disorder is rejected by self-amplifying cooperative effects. The conditions under which the two effects are just balanced are the "tipping points" between differently ordered regimes. As balance points between two aggregate-level effects, they tend – like the crossing point between two lines – to be discrete and separate events, explaining why phase transitions are rare and restricted compared to small-scale fluctuations which remain pervasive in all phases and through the tipping points.

In living systems where order carries function, fluctuations that disrupt order become errors. The cooperative effects that reject disorder perform the function of buffering against errors. The concept of buffering, and the way buffering reduces the requirements for error correction and information input in control systems, will be fundamental to our argument in Chapter 8 that both emergence and evolution in the biosphere follow paths of opportunity laid down by phase transitions and the ordered states they create.

## 7.2 Arriving at the need for a phase transition paradigm

Before giving a more technical introduction to the concepts and mathematics behind the theory of phases and phase transitions, we review some of the problems and disagreements that show why a theory of the origin of life without these concepts is inadequate. In the first five chapters of the book we emphasized environments, structures, and functions in extant life. All these can be reported as facts. However, facts are not self-interpreting,[1] and the facts we have reviewed were selected because they bear on an interpretation in terms of metabolic origin that we believe is most compelling. Other researchers recognize these facts, but assign less importance to them as evidence about origins because they are comfortable supposing large and even qualitative discontinuities [60, 589] between the early material and processes that enabled a biosphere to emerge, and the processes that sustain it today. To move beyond exhibiting facts and proposing scenarios for origin, toward a proper theory, we must have further criteria for judging the validity of a model system or the adequacy of an explanation.

---

[1] The balance of theory and empiricism has a long and tortuous history in science, which we cannot review here. In a legitimate effort to oppose theory-as-prejudice, Sir Francis Bacon argued [44] for the possibility of induction from facts, effectively arguing that with sufficient data, facts *are* self-interpreting. It is now generally agreed that the problem of induction has no general solution [682], and that to do science we must characterize our own participation as observers in order to filter biases that the process of observation may impose on the phenomena observed. In fields such as the origin of life, which are severely under-constrained by data, the role of theory looms large. If thoughtlessly chosen, it may be a damaging source of prejudice and oversight. If well motivated, it may be the primary source of well-chosen experiments in a large and otherwise intractable search space.

In this section we argue that the first question to be asked must not be about particular mechanisms to reach life as a kind of pre-defined goal, but rather why any state besides lifelessness could be a goal. In posing this question for chemical systems, we recognize from a modern perspective (really made possible only with the techniques of the past 20 years) that the main difficulty is not accessing reactions, but combining yield with selectivity. The need for chemical concentration and selection, in a chemical era that seems best regarded as pre-Darwinian, leads us to the essential properties provided by thermal phases. We characterize this as the "stability perspective" on the origin of life. Once the need for a theory beyond brute facts and intuition has been articulated, we proceed in the later sections to introduce the concepts behind the stability perspective more formally and to develop applications.

### 7.2.1 Why there is something instead of nothing

> ...the first question that should rightly be asked will be "Why is there anything, rather than nothing at all?"
>
> – G. W. F. Leibniz [864]

#### 7.2.1.1 Metabolism-first or RNA-first: different directions from simplicity to complexity

If we are to understand and adequately address the difference in points of view between proponents of an origin in metabolism and proponents of an origin in genetics, we must fairly present the substance of the arguments for both. As we noted in Section 6.1.3, the metabolic-origin view interprets the facts in the first five chapters as suggesting a sequence in which *the elaboration of metabolism and the accretion of control* progressed from simplicity to complexity. Arguments for early heterotrophy (fed by a primordial soup) or RNA-first also claim a progression from early simplicity to later complexity, but they emphasize other criteria. A proposed world of competing, Darwinian RNA molecules presents a very simple early model of *selection* (see Section 6.1.2.1), whereas in the metabolism-first view selection – meaning the pruning of organosynthesis – is a multi-factorial and distributed function carried out in possibly diverse geochemical environments. In the RNA-first view it is catalytic function and the criteria for fitness that grow from a simple early criterion of replication rate (see Section 6.6.3) to the later support of organosynthesis and network assembly. To acknowledge both problems of finding early simplicity, and identifying a likely path to later complexity, and to judge what constraints each places on any theory of origins, we must start from some framework for plausibility that does not originate in an account of only a subset of life's properties.

#### 7.2.1.2 The first question that should rightly be asked

Questions about the origin of life can be approached from two perspectives, which, borrowing language from probability theory, we might call "conditional" and "unconditional." In the conditional framing, one takes as a starting point the fact that the Earth has a biosphere, and then asks what is the most plausible route by which such a complex system could have formed.[2] In other words, we ask about chemistry, control, or selection, on the condition that the answers account for some aspects of life. The conditional frame provides a way to make use of circumstantial evidence from modern life, but also anchors the question in a living state which is long diverged from the lifeless state out of which the biosphere arose.

The alternative to the conditional frame is a version of Leibniz's "first question that should rightly be asked": why do the patterns realized in the living state exist at all, as opposed to no patterns beyond those realized in non-living matter.[3] The unconditional frame, rather than asking for the most likely path to a pre-specified outcome (a biosphere's having emerged), asks first whether any outcome besides lifelessness is probable. A host of rocky planets and asteroids, many with water, metals, and radiogenic heating, some such as the Murchison parent body even manifesting low-level but quite complex organosynthesis [704], did not develop complex biospheres like that on Earth. As far as current evidence goes, they may not have formed cellular life or even distinctively prebiotic chemical networks. In taking lifelessness as the null model, the unconditional framing asks what is "wrong" with a non-living state, for some bodies such as the Earth, which might render that state unstable. It then asks what the directions of departure from lifelessness should have been.

The overlap between the first directions of departure from lifelessness, and the late-stage convergences to the modern living state, may be only partial. In particular, we argue that notions of individuality, replication, and Darwinian competition have no natural place in early organosynthesis, and are better understood as emergent motifs in an intermediate era between geochemistry and the modern world. The concept of a thermal phase is a potential source of constraint, selectivity, and information that can be dynamical without being Darwinian. This argument is developed in Chapter 8. By asking, for each pattern, in what context its emergence becomes more likely than its absence, we recognize that multiple transitions from simplicity to complexity must be explained. What we avoid doing is treating abstractions such as replication or competition as self-contained or free-standing. The problem is instead to arrange the contexts and constraints from multiple systems so that each of the elementary transitions is locally simple, and can be understood as a necessary departure from the state that preceded it.

---

[2] A particular case of the conditional frame, in which one begins with the existence of people who can pose the question, is familiar as the "anthropic principle" [120].

[3] Our version of the "first question" differs from the original in important ways. For Leibniz, who had limited scientific tools through which to understand causality and who carried an explicitly theological goal, this was an almost syntactically motivated question about the problem of infinite regress in causal explanations. Stephen Maitzen [512] argues that this is an instance where the purely syntactic motivation leads to a semantically meaningless pseudo-question. Some echo of the original problem of regress may remain a legitimate part of science as it concerns problems of initial conditions in cosmology. For us the problem of why patterns can exist, first as mathematical descriptions and second as attainable ordered states of matter, is a very concrete though still surprisingly subtle one. It is interesting that, in our discussion of why something rather than nothing exists, we will be led to Laplace's "Principle of insufficient reason" [478], also called by Keynes [430] (Chapter IV) the "Principle of indifference," and apparently given the name "insufficient reason" in part to refute Leibniz's "Principle of *sufficient* reason," which was meant to account for the existence of entities on theological grounds.

### 7.2.2 Selecting pattern from chaos in organosynthesis

Before Friedrich Wöhler demonstrated the synthesis of urea by artificial means in 1828 [899], the organic compounds were believed to be produced only by living systems (hence their name). Synthetic organic chemistry became a thriving field in the century following Wöhler's synthesis, but something of the earlier perception of its separation from the inorganic world persisted. If not necessarily the embedding in living organisms, then at least the precise control of conditions by chemists was regarded as a precondition of organosynthesis. Alexander Oparin (in 1924) [610, 611] and J. B. S. Haldane (in 1929) [328] proposed that organic compounds were not only possible to produce by abiotic mechanisms, but abundant on the early Earth, in what Oparin termed a "primordial soup."[4] The watershed that removed the perceived need for finely controlled boundary conditions was Stanley Miller's 1953 demonstration with Harold Urey [557] that abundant and complex organics could be formed simply with gas-phase free radical activation, and that the molecules synthesized included many biological monomers.

A host of subsequent experiments and natural observations – from Miller's ongoing experiments to work on the formose network [857], the work by Heinen and Lauwers [350], Huber and Wächtershäuser [377], and Cody and collaborators [145, 146] on mineral-mediated synthesis, early and ongoing experiments with pyruvate [597], spectroscopic analysis of chemistry in the interstellar medium, or increasingly sensitive assays that reveal the complexity of organic products in carbonaceous chondrites [156, 704] – have made clear that gaining access to organic reactions is not the fundamental problem for understanding the origin of life. In most cases, excepting those with the most restricted classes of reactions, we may not understand how the synthesis proceeds, but empirically it is clear that relaxation of redox disequilibria and a variety of other kinds of excited states proceeds at least in part via complex organic intermediates.

The problem for the origin of life is rather to understand *what makes a redox relaxation orderly*, so that matter and energy flux are concentrated into a small subset of compounds, which are produced in significant concentrations. The compounds capable of seeding downstream networks of increasing molecular size, and therefore relevant to life, are complex. Structurally and energetically, they contain "internally stored" disequilibria from adjacent oxidized and reduced carbon centers, which would not typically arise within long-lived chemicals. If we think of the biological monomers as selected (by whatever means), they seem to require very fine specification among the huge number of apparently similar compounds that can be retrieved from databases with similarity searches to biotic compounds [574]. The tens or hundreds of thousands of compounds estimated to be in the Murchison meteorite are certainly more diverse than the compounds that make up the foundation of metabolism, but from a process perspective, they may be hard to exclude. They likely result from the application of a small collection of basic reactions, unrestricted by the whole-molecule specificity that enzymes confer on metabolites. Understanding the role of this kind of abiotic organosynthesis in prebiotic chemistry is thus a problem that combines yield with selectivity.

---

[4] Haldane's term for the same proposal was "primitive soup."

If one thing more than others is surprising across the range of experiments, it is that despite the stark differences in their sources of activation energy and reaction conditions, many products found at significant concentration are *common to multiple reaction systems*, and a subset of these are also known metabolites.

The problem of concentrating selection may itself decompose to smaller problems of order. Mechanisms that favor concentration should probably not be understood as acting only on particular molecules or at particular reactions, but rather as acting on ordered non-equilibrium states to which those molecules or reactions open control points. Therefore even in cases such as meteoritic organosynthesis which are not as selective as we imagine the path to the biosphere must have been, the reaction networks and products may be reproducible and robust, as a result of being ordered around certain main fluxes.

### 7.2.3 The phase transition paradigm

What kinds of processes could explain a planet's making a rapid transition away from a lifeless, chemically relatively inactive state, into a state containing a biosphere? These require changes in the planet's chemical composition and its characteristics as a free energy input/output system. After forming, the altered planet must then remain stably in that more ordered state, as far as we know, indefinitely.

Any explanation must recognize the Gibbs equilibrium state or weakly driven near-equilibrium states as null hypotheses, since many planets and asteroids remain in these states. It must cast biotic order as a deviation from the equilibrium state, and then account for the spontaneous emergence and indefinite maintenance of that additional order in the face of constant perturbations in all planetary components at all scales of space and time. This last property turns out to be the difficult and subtle one, and it will be our emphasis in this chapter and in Chapter 8.

The surprising fact about the answers to questions of this kind is that, so far within our understanding of statistical mechanics, they are all of the same kind. The processes that spontaneously generate and then preserve new forms of order in the face of pervasive disrupting perturbations are the **phase transitions**.[5] Phase transition is a mathematical concept, with origins in the laws of large numbers. Systems that undergo spontaneous transitions to order are both distinctive mathematically, and broadly similar to one another in the character of their ordered states and fluctuations.

Systems of very diverse constitution can become ordered through phase transitions. The reason they are not just a collection of special cases, each resulting *ad hoc* from its particular dynamics and unrelated to the others, is that most known kinds of macroscopic order must emerge and persist against the disordering effects of continual microscopic disturbances. Although it took nearly a century to fully understand, this property – that order must be asymptotically stabilized in the presence of disorder – causes the robust physical transitions to order to share the following five characteristics.

---

[5] Note that Darwinian selection is a *mechanism* that contributes to making populations more ordered, and so is a concept of a different kind. Phase transitions are defined by what they accomplish – the asymptotic suppression of fluctuations away from a spontaneously generated ordered state – and not by the particular mechanisms they use to curtail or correct these fluctuations.

1. Transitions whereby systems form macroscopic order are vanishingly rare against the background of situations in which no new order forms. The cases of order depend on the action of mutually reinforcing collective effects among many small-scale degrees of freedom.
2. The collective variable that carries the macroscopic order is robust due to this highly parallel mutual reinforcement, and the residual fluctuations in the microstate, conditional on the background order, are independent.
3. Becoming ordered entails making a large volume of the microscopic state space inaccessible to (or through) fluctuations. The (logarithmic) measure of this forbidden volume in state space is an entropy. The form of the entropy reflects the basis in which fluctuations are independent.
4. The sharp reduction in the volume of available state space results from a kind of competition among multiple distinct but equivalently stable ways of becoming ordered. The possible ordered states are mutually exclusive, so the formation of order through (a large class of) phase transitions is accompanied by the unpredictability of some properties of that order.
5. All these properties of macroscopic order are consequences of averaging under the laws of large numbers. Because many details of the fine-scale dynamics cancel under averaging, it is possible for microscopically distinct systems to have similar kinds of aggregate order. A corollary is that quantitative features at the macroscale can sometimes be obtained from incomplete or imprecise knowledge of microscopic interactions, though when this is possible is a technical and often difficult question.

Our hypothesis becomes that *the lifeless Earth was not only an energetically stressed Gibbs state, but a **metastable** or **unstable** state, and that a process of relaxation through a sequence of progressively more stable dynamical states was what we recognize as the emergence of the biosphere.*

The phase transition paradigm provides a structured framework within which to assign meaning to the observations of modularity, universality, hierarchy, redundancy, and context presented in the first five chapters. It provides a route to formal abstractions of biological order in terms of asymptotically reliable error correction and the requirements for stability in hierarchical control systems. Within the joint perspective of stability, error correction, and control we may justify the sequence of biogenesis proposed in Chapter 6 as one in which the intermediate stages are most plausible as stable states, following the evidence from extant modularity, and in which the transitions require the fewest rare or unlikely events.

### 7.2.4 Developing the stability perspective

In searching for a theory of biogenesis, we must remember that before it is anything else – before it uses TCA precursors or RNA, before it is cellular, before it is Darwinian – the biosphere is a dynamical ordered state of chemistry that has persisted on Earth for almost four billion years. The property of *asymptotically stable dynamical order, as a concept,* must be a central criterion for any explanation. The task of this chapter is to bring this premise from statistical physics together with the evidence from geochemistry in Chapter 3, from metabolism, organic chemistry, and biochemistry in Chapter 4, and from the genetic code, bioenergetics, and cellularization in Chapter 5.

Each scientific discipline has distilled certain abstractions as its most basic elements that must be respected in any sound explanation, and therefore each discipline leads us to recognize information in different features of the descriptive account of life. The challenge for an adequate theory of origins is to concurrently respect the insights carried by them all. The unique insight from the stability perspective, beyond the instantiations of it that are already subsumed in the equilibrium theory of minerals or mass-action chemical kinetics, is that the problem of containing fluctuations exists at *all scales* of material and processes. The implications of thermodynamics are not exhausted when we have defined the Gibbs free energies of molecules or transition states; they continue to apply at an ascending series of scales through population genetics, ecological dynamics, and long-term evolution. The mathematics of phase transitions provides both a formulation of the stability problem as a problem of joint error correction at multiple systemic levels, and a description of the kind of cross-level interactions that enable optimally efficient and effective error correction. It therefore offers a unifying description of stability capable of bridging the heterogeneous, multilevel, coupled mechanisms that have jointly contributed to the emergence and persistence of life. After introducing the basic concepts and some worked examples in this chapter, in Chapter 8 we provide a synthesis from the stability perspective and discuss the ways in which it leads us to reconceptualize the nature of the living state.

As we address the requirements for stable order as a problem in its own terms, which supersedes specific mechanisms, we do not mean to suggest that the mechanisms are unimportant. Indeed, particular mechanisms will be essential to inducing the order formed in each case. The message distilled by appropriate abstraction is that all understood cases of asymptotically stable order result from certain mathematical consequences of aggregation, which apply across the spectrum of mechanisms. Understanding that these exist, and knowing what they are, can provide a context for the study of mechanisms. They can help select, among the wilderness of approaches of which most are dead ends, those that will ultimately integrate into an adequate theory of biogenesis.

### 7.2.4.1 The four core ideas to be introduced

In the following four sections we introduce four core concepts related to the nature of steady thermal states, fluctuations, stability, and error correction.

Section 7.3 introduces the basic consequence of the combinatorics of large numbers known as *large-deviations scaling*, and explains how the thermodynamic concepts of state variables, the entropy, and the independence of fluctuations follow from this scaling. The large-deviations perspective emphasizes that the entropy is defined from properties of fluctuations. From this starting point, its other properties follow, including entropy maximization as the condition for stability, the informational properties of entropy, and the roles of other state variables as boundary conditions or sources of constraint. Our approach may seem somewhat formal at first, because we are careful to avoid analogy or any association with equilibrium statistical mechanics except as a source of familiar examples. However, by staying focused on core concepts, we not only recover all the familiar roles of entropy in classical thermodynamics: With no confusion, we may then directly pass to entropy principles for non-equilibrium processes, in the last section of the chapter.

Section 7.4 reviews the mathematics and physics of phase transitions in equilibrium systems, where they are most exhaustively understood. We start with a worked analysis of the simplest (binary) phase transition, and demonstrate how all basic quantities and core ideas follow using only elementary calculations. We devote some time to explaining how the phase transition paradigm has provided a systematic way to understand the relations among reductionist description, emergence, and predictability. Emergence and contingency in the phase transition paradigm are neither anti-reductionist nor even particularly surprising, and can be understood quite matter of factly in terms of ordinary ideas of symmetry and degeneracy. We also summarize the relations between stability and long-range order which are important in physical phases and are likely to be fundamental in understanding the much more complicated character of historical contingency in biology. It is not as widely appreciated as it should be that a hierarchy of phase transitions currently provides our best theory of all states of matter from elementary particles to correlated-electron states in ultra-cold condensed matter, so we briefly review the phenomenology of this theory which also suggests crudely the kind of hierarchical description we expect should arise for living order.

Section 7.5 offers a parallel introduction to thermodynamic phases that starts, rather than in theories of matter and mechanics, in Claude Shannon's theory of information and optimally robust communication in the presence of noise. Not only does Shannon's information theory lay bare the principles of large-deviations scaling and thermodynamics without the distractions of mechanics, also the canonical examples of asymptotically reliable communication offer models of multilevel error correction that provide a starting point to understand selective functions in biology.

Finally, Section 7.6 introduces the theory of thermal stability and phase transition appropriate to *non-equilibrium processes* rather than merely equilibrium states. Because we work directly with the mathematical principles that define phase transitions, rather than appealing by analogy to phenomena in equilibrium, we have no difficulty showing that the principles of entropy maximization and phase transition apply away from equilibrium as well, though one must replace the equilibrium entropy with a properly constructed non-equilibrium entropy function that is more complicated and generally harder to compute. We begin the section with a discussion that we hope will dispense in a constructive way with a certain discourse on the role of entropy in living systems that has been rehearsed without making any really new contribution since Boltzmann's time. We hope it will become so obvious as to be disappointing that, when one distinguishes the *concept and function* of entropy (which are general), from particular *forms* such as equilibrium entropy functions (which are constructed within equilibrium contexts), equilibrium thermodynamics can stand in no conflict with the emergence of a biosphere, but also provides limited help beyond the mechanisms of basic chemistry in accounting for that emergence. The section closes with a simple worked example showing how appropriate non-equilibrium functions are constructed, and suggested analogies to a more integrative description.

### 7.2.4.2 *Using the core ideas in a synthesis*

Chapter 8 attempts to apply the ideas to the interpretation of metabolism and biological hierarchy. The synthesis is, again, not merely a matter of claiming analogy but a problem of identifying the important principles of organization. The new ideas in Chapter 8 that go beyond thermodynamic states and phase transitions concern modularity, hierarchical control, and the emergence of individuality. We hope that giving a more thorough understanding of the phase transition paradigm will provide principles for assigning importance to facts, and a basis for interpretation in which our assertions about life fit consistently within a larger scientific system for understanding order in nature. Origin of life studies is a field with an extraordinary diversity of proposals, but few good ways to systematically judge the framing of questions and the prioritization of evidence. While we may never be as systematic or efficient at choosing good questions as we can be at validating laboratory results, the role of theory should be to make it easier to ask questions whose answers will remain useful over time.

### 7.2.5 *A chapter of primers*

The academic disciplines have become fragmented, constrained by limits of human attention and effort, but the emergence of the biosphere was not fragmented. Scientifically grounded treatments of origin must be chemical but we believe they must also be framed at the level of networks and distributions and statistics, because processes defined at these levels are the only kind that seem possible on an early Earth. However, the disciplines of organic chemistry, geochemistry and biochemistry, statistical mechanics, and error correction each have such a weight of concepts and applications that lie behind their habits of thought that they are far from having grown together into a unified and accessible understanding of nature.

We expect that for most readers of this book, organic and biochemical knowledge will be the most familiar and large-deviations theory, statistical mechanics, and communication theory the most distant. Even for physics readers versed in thermodynamics, which for a century has been introduced from a starting point in mechanics and conservation of energy (rather than from counting and the properties of entropy), our routine use of "entropy" to refer to quantities that are *not the equilibrium entropy* may seem strange and even jarring. Therefore, in this chapter more than the others, we expand several sections into small primers on basic concepts, methods, and examples that we want to be available to all readers to draw from. We introduce the entropy from a large-deviations perspective explicitly to show a formulation in which *constraint and boundary conditions* are important, but apart from being one of the more common constraints, energy plays no further special role. We develop two key examples at length with algebraic steps filled in to a level that can be followed by early graduate students or advanced undergraduates. One is a phase transition and the other is an optimal error-correcting code. Both examples are core course material within respective disciplines, but they are too rarely taught together. Within each section, we try to emphasize the key concepts beyond the arithmetic. These are the ideas that make each discipline a coherent landscape to its practitioners, but too often in coursework they are lost in the rigors of examples. Some of the smaller and more purely technical constructions, included to give precise meaning to ideas mentioned in the main text, are included in a sequence of boxes.

We must reiterate a point raised in the preface. To try to do justice to a process that was at once geological, chemical, physical, and incipiently biological, we have had to draw from the knowledge of many communities and to impose on many generous colleagues to try to educate and referee us. If the juxtaposition of this chapter with others in the book leads to reading that is uncomfortable at some point from any angle, we hope the points of discomfort can be starting points to bring together readers with complementary strengths. The origin of life may be a phenomenon that cannot even be *thought about* correctly from within the knowledge and perspective available to any individual, but it may be within reach of the right kinds of collaborative community.

### 7.3 Large deviations and the nature of thermodynamic limits

Standard introductions to thermodynamics such as Enrico Fermi's classic 1956 text [244] typically start with the deterministic maximization of classical entropy and its consequences as the starting point to frame the theory. From this starting point one departs through studies of fluctuations toward more and more microscopic levels of description. Even textbooks on statistical mechanics [373, 441] stress the coherence of thermodynamics as a free-standing theory irrespective of its justification in microscopic models. Indeed, Fermi's goal in summarizing classical thermodynamics without statistical foundations was to emphasize that the classical theory could be entirely constructed from empirical foundations in *thermometry* and *calorimetry*.[6]

Here we will take a very different line of approach into the subject, in which fluctuations and statistical inference are the starting points which *define* the entropy, and in which entropy maximization and the resulting structures of classical thermodynamics follow from application of the fluctuation theorems to interacting systems. We take this approach partly because the taming, control, and use of fluctuations are the concepts from thermodynamics of direct interest to us in understanding the biosphere's emergence, function, and ongoing evolution. But we also take this approach because it is more fundamental to the essence of thermodynamic limits. The energetic approach of most texts, exemplified by Fermi, is well suited to the study of equilibrium systems coupled to mechanical controllers, in

---

[6] He might, as well, have added barometry, magnetometry, and a variety of empirical studies of the forces that govern the stability and control of chemical reactions. Listing the eclectic inventory of empirical disciplines that identify different thermodynamic forces would have diminished the importance that energy had for historical reasons, but would have increased the centrality of entropy and opened a more symmetric treatment of the many constraints of which entropy is a function.

which work, heat, and other energy exchanges are the main quantities of interest. However, by seeming to root thermodynamics in energy conservation rather than in statistical inference, it obscures the generality of the theory and the straightforward (if sometimes technically difficult) extensions of the core concepts away from equilibrium mechanics to non-equilibrium processes, information theory and error correction, and problems of inference that arise in hierarchical control systems.

### 7.3.1 The combinatorics of large numbers and simplified fluctuation-probability distributions

The "laws of large numbers" [164] refer to a variety of consequences of the exponential character of combinatorial counting that may come to dominate the statistics of assemblies consisting of many components or events. Because the combinatorics of counting may be universal even when the things counted are not, laws of large numbers can lead to universal behavior in aggregate systems that is simple and predictable, even when the small-scale details of those systems differ or are partly unknown.

The consequence of laws of large numbers that leads to the theory of thermodynamic limits is known as *large-deviations scaling* [811] of the probabilities of macroscopic fluctuations in aggregate systems. The requirements for large-deviations scaling to even be expressible as a property of some system are first that the extent of aggregation can be characterized by some **scale**, and that the **structure** of fluctuations in aggregate properties can be characterized in some common way across a range of scales. For example, a box of gas may have a scale characterized by the volume it occupies or the number of particles, or its internal energy. Fluctuations of density by 1% compared to the average, over some fixed fraction within the total spatial volume, are patterns formulated in terms of relative scales, which are well defined and comparable, whether the box and its contents are large or small.

In systems that show large-deviations scaling, probability distributions for fluctuations simplify to the following form [811]:

$$P_{\text{fluct}} \sim e^{-Ns} \tag{7.1}$$

in which $N$ characterizes scale and $s$ characterizes the dependence on structure.[7] The tilde $\sim$ indicates that this relation is true to leading exponential order; that is, there is a limit $s$ for which:

$$\frac{1}{N} \log (P_{\text{fluct}}) + s \to 0 \tag{7.2}$$

as $N \to \infty$. The sub-leading contributions must scale at least as functions of $N$ that grow more slowly than $N$. In many practical cases they actually scale as $\log N$.

---

[7] For the sake of introducing the concept, we are *defining* $N$ to be whatever scale parameter appears in a relation such as Eq. (7.1). Hugo Touchette [811] emphasizes that in many cases where a system may have a natural scale factor such as component number, mass, etc., the large-deviation relation (7.1) may require a scale factor that is different from the linear power of this system size.

The important feature in Eq. (7.1) is that **scale separates from structure**, because of the log-probability factors. In Eq. (7.1) the factor $N$ reflects only system size, and scales the probabilities for all fluctuations, while $s$ depends only on the structure of fluctuations but not on $N$. $s$ is called the **rate function** of the large-deviations limit. Again, for example, in a box of gas, fluctuations in density and fluctuations in internal energy will have probabilities related through the dependence of the same rate function $s$ on their form and relative magnitude, at any scale, while both kinds will be suppressed in absolute probability by the same overall particle number $N$.

The rate function in a large-deviations limit is also called the **entropy**. In ordinary thermodynamics it would be referred to as the *specific entropy* or entropy per particle, and the total system entropy $S = Ns$ would denote the product of scale $N$ and specific entropy $s$.[8] In everything that follows, *we take the rate function in a large-deviations limit to be the definition of the entropy* relevant to the system. (The usage of the term "entropy" in large-deviations principles is very wide: it encompasses the usual equilibrium-specific entropy and its non-equilibrium generalization, but also a variety of generating functions that capture system context, which in equilibrium thermodynamics would be interpreted as free energies.)

Exponential suppression of fluctuations with increasing scale is the source of stability in thermodynamic systems. The separation of scale from structure results in an enormous simplification of probability distributions for aggregate fluctuations, and enables the use of increasing aggregation scale as a source of greater stability.

The most important point to appreciate about large-deviations scaling is its generality. In any system where aggregation leads scale to separate from structure, we have the possibility to identify thermodynamic limits and exponential regression toward a central tendency. By taking the large-deviations rate function to define the thermodynamic limit, we obtain a principled definition of entropy that is not linked to any restricted context or mechanism, or even to the distinction between dynamics and problems of pure statistical inference.

### 7.3.1.1 Canonical example: the central limit theorem

Perhaps the best known application of large-deviations scaling is the explanation it provides for the ubiquity of Gaussian probability distributions in statistics. This example illustrates both the suppression of fluctuations in large systems, and the flow of statistics for a variety of microscopically different processes toward a common aggregate distribution, which is the basis for the technical theory of *universality*.

---

[8] We must apologize to readers for a terminology that suffers many collisions because the same word has been used to stand for related but different quantities in fields with separately evolved traditions. The famous Shannon entropy of a probability distribution $p$, written $-\sum_i p_i \log p_i$ for discrete distributions over an index $i$, or $-\int dx\, p_x \log p_x$ for densities over a continuous variable $x$, corresponds to an entropy per sample or a specific entropy $s$. The entropy of Boltzmann, which is the log of the count of typical microstates for an $N$-particle system compatible with a given macrostate, multiplies the sample-probability entropy by the number of particles, and so corresponds to $S = Ns$. As long as one makes clear, in any context, whether single samples or aggregates are being considered, the usage of the term "entropy" can be disambiguated.

Suppose that some random variable $X$ has a probability density to take values $x$, given by

$$P(x \leq X \leq x + dx) = p(x)\,dx, \tag{7.3}$$

with mean $E(X) = \mu$ and *finite* variance $E(X - \mu)^2 = \sigma^2$.

Index successive independent samples of $X$ as $X_1, X_2, \ldots$, and define a new random variable $X^{(N)} \equiv X_1 + \cdots + X_N$. $X^{(N)}$ is one choice of what is called a *summary statistic* for the collection $\{X_1, \ldots, X_N\}$, which reflects simple aggregation.[9]

We now introduce a rescaled variable $Y^{(N)} \equiv X^{(N)}/N$ to remove the leading effect of aggregation on scaling, which comes through the mean. One of the most important theorems in mathematics for the understanding of scaling and universality – the *Central Limit Theorem* proved by Pierre-Simon Laplace in 1810 [479] – states that $Y^{(N)}$ has a probability distribution that converges, in the limit of large $N$, to the distribution

$$P\left(x \leq Y^{(N)} \leq x + dx\right) \rightarrow \frac{1}{\sqrt{2\pi\sigma^2/N}} e^{-N(x-\mu)^2/\sigma^2}\,dx, \tag{7.4}$$

with mean $E\left(Y^{(N)}\right) = \mu$ and variance $E\left(Y^{(N)} - \mu\right)^2 = \sigma^2/N$.

Equation (7.4) shows an example of large-deviations scaling. The scale factor $N$ is simply the number of samples in the aggregate. The rate function $s = (x - \mu)^2/\sigma^2$ depends on $x$ which characterizes the structure of fluctuations about an invariant mean $\mu$ in aggregates of any scale. The normalization constant proportional to $\sqrt{N}$ is a sub-exponential prefactor that is absorbed in the leading-exponential approximation $\sim$ in Eq. (7.1). Much of the technical theory of universality in statistical mechanics may be understood as a generalization of the way the central limit theorem collapses an infinite diversity of higher-order moments in the microscopic density $p(x)$ into the two-dimensional family of Gaussian limit densities.

### 7.3.2 Interacting systems and classical thermodynamics

From a starting point in the macroscopic fluctuation distribution (7.1), we may immediately reconstruct the entropy-maximization principle and its corollaries in classical thermodynamics by considering the joint fluctuations of multiple systems, each independently described by large-deviations scaling, when these are brought into contact to become subsystems within a larger aggregate system.

#### 7.3.2.1 Constraints on the entropy and state variables

The large-deviations scaling limit for the distributions of macro-fluctuations results from averaging over those internal states of a system that are left free under the specification of the macrostate. They therefore correspond to the variables that would not be

---

[9] See Box 7.2 for more on summary statistics.

controlled by a boundary condition put in place to *force* the macrostate into a shifted configuration. Thermodynamic limits are typically useful when the internal, uncontrolled degrees of freedom are many and the controlling boundary conditions are comparatively few.

The properties of a macro-configuration that determine the distribution over microstates, and which therefore appear as arguments of the entropy function are called the **state variables** of the system. Which state variables define a system's macrostate will generally depend on the kinds of boundary conditions imposed on the system. In particular, when multiple subsystems are brought into contact to form a larger, aggregate system, the relevant state variables for the aggregate will depend on the way the interaction causes each system to impose boundary conditions on the others.

### 7.3.2.2 *Entropy maximization at the most probable state*

The important assumption about aggregation of multiple systems which each obey large-deviations scaling is that their interaction takes place only through low-dimensional boundary conditions, leaving the internal degrees of freedom unconstrained except through these variables. If this assumption of independence holds, the probability for a fluctuation in a joint system composed of two interacting subsystems is given by the product $P_{\text{fluct, joint}} = P_{\text{fluct,1}} P_{\text{fluct,2}}$, in which the subscripts 1 and 2 label the subsystems. From Eq. (7.1) – which we take to *define* the relevant entropy function – the most probable configuration is the one that maximizes $S = S_1 + S_2$.[10]

If joint fluctuations are *constrained* because some quantity $X = X_1 + X_2$ takes a fixed value in the aggregated system, and only its apportionment between the subsystems can fluctuate, then the jointly most probable configuration is the maximizer $\delta S = 0$ along the surface of constraint $\delta X_1 = -\delta X_2$. As $\delta X_1 \to 0$ this vanishing variation becomes a condition on the first derivatives of entropy in the two subsystems:

$$\frac{\partial S_1}{\partial X_1} = \frac{\partial S_2}{\partial X_2}$$
$$= \frac{\partial s_1}{\partial (X_1/N_1)} = \frac{\partial s_2}{\partial (X_2/N_2)}. \tag{7.5}$$

In the original formulation of statistical mechanics from Hamiltonian mechanics, by Boltzmann and Gibbs, the quantities that macroscopically constrain interacting systems either are geometrical or are the *conserved* quantities of the mechanical motion, because conservation is not affected by aggregation under laws of large numbers. Therefore geometric quantities such as volume ($V$), constants of motion such as internal energy ($U$), or

---

[10] Whether we call this configuration a "fluctuation" is a matter of terminology. The most-likely configuration could be considered the "zero fluctuation" or degenerate case of non-fluctuation. Since, however, the most likely configuration internal to a system will generally change if it is brought into contact with other systems, all such configurations are deviations from pre-contact configurations, and are in that sense equally eligible to be considered fluctuations.

conserved particle numbers of various kinds, generally arise as the state variables which are arguments of the entropy.[11]

We will draw on examples of this kind for illustration, but our concern is with the behavior of arbitrary Markovian stochastic processes with large-deviations scaling limits. Therefore we will think of entropy maximization as occurring along any surface of constraint which is a consequence of the transition probabilities of the stochastic process.

### 7.3.2.3 Scaling of conserved quantities and entropy with system size

In many systems, and in all the examples that will concern us, the constraining quantities all scale in linear proportion to each other. Hence any of them may be used as a reference for system scale, and alterations in the ratios $X_i / N_i$, used in the second line of Eq. (7.5), define what we mean by the "structure" of fluctuations independently of overall system scale.

For example, in a pair of coupled systems with fixed numbers of particles, $X$ might be the internal energy $U$, or the volume $V$, and the specific entropy $s \equiv S/N$ would be a function of $X/N$ interpreted as the energy density $U/N$ or volume per particle $V/N$. In a different case, in which particles are allowed to diffuse between compartments of fixed volume, we might take volume $V$ as an invariant measure of each subsystem's size, and let $X$ be particle numbers which are exchanged, so that specific entropy per unit volume $s = S/V$ would be a function of the density of particles $X/V$.

Because many combinations of scaling are possible, it is more common to work directly with the entropy $S$ and the conserved quantities $X$. While doing so more directly expresses the character of constraints on system states, it tends to mask the separation of system scale from structure that defines the large-deviations limit.

### 7.3.2.4 Extensive and intensive state variables

The extensive state variables are those that scale with system size in the same way as the entropy in the large-deviations function. The derivatives of entropy with respect to these quantities, in which the scale factor cancels, are termed intensive.

Intensive and extensive state variables provide *dual* characterizations of a system as long as the entropy function is convex. Introduce the notation

$$\lambda \equiv \frac{\partial S}{\partial X}, \tag{7.6}$$

to refer to the intensive state variable obtained as the gradient of $S$ with respect to $X$. The *Legendre transform* of entropy $S$ with respect to an extensive state variable $X$ may be given a name such as $F$, and is constructed as

---

[11] It is interesting that quantities such as electric charge, which in early twentieth century physics were taken to be conserved for unknown reasons, much as atomic elements are conserved in chemical reactions, took on the explanation in quantum field theory of constants of motion resulting from symmetries. The latter description is essentially geometric [293, 863], in an enlarged configuration space. We therefore regard the roles of all constraints on the entropy as being conceptually equivalent, and resist giving energy and volume different roles in thermodynamics only because the "reason" they are conserved may appear different in some historical period.

$$F \equiv X \frac{\partial S}{\partial X} - S \equiv X\lambda - S, \qquad (7.7)$$

which is to be evaluated as a function of $\lambda$. $S$ continues to take $X$ as its argument, but now $X$ is to be evaluated at a point $X(\lambda)$ where $\partial S/\partial X = \lambda$. It is for this evaluation to be unique that we require $S$ to be convex, so that each unique derivative requires a unique argument.

The gradient of $F$ with respect to $\lambda$ recovers $X$, as

$$\frac{\partial F}{\partial \lambda} = X + \frac{\partial X}{\partial \lambda}\left(\lambda - \frac{\partial S}{\partial X}\right)\Bigg|_{X(\lambda)} = X(\lambda). \qquad (7.8)$$

Hence, given an entropy function from a large-deviations scaling limit, we may describe a system either by the quantity $X$ it contains, or the gradient $\lambda$ of $S$ at which that quantity $X$ is maintained.

The illustration here treats $S$ as having only a single argument $X$, but the construction extends to any collection of extensive variables which act independently as boundary constraints on the system state. In the more complex case, the partial derivatives in Eq. (7.6) refer to gradients in which all extensive variables independent of $X$ are held fixed. A variety of mixed Legendre transforms of $S$ provide an array of potentials through which different combinations of extensive and intensive state variable values may be obtained by differentiation. Box 7.1 below provides an example.

The argument for maximum probability that leads to Eq. (7.5) may then be expressed as the classical thermodynamic equilibrium condition that *coupled systems take on their most likely macroscopic configuration when the intensive state variables dual to any exchangeable quantities are equal.*

### 7.3.2.5 The historical energy perspective versus the more natural entropy perspective

The explanation of classical thermodynamics from statistical mechanics originally developed as an outgrowth of Hamiltonian mechanics in a many-particle setting. It therefore emphasized conservation of energy, and placed the problem of statistical inference in a secondary role. Entropy appears mysteriously in this treatment as an explanation for the contribution of *heat flow* to a system's internal energy, alongside more "intuitive" contributions such as work, and also mysterious *chemical potentials* associated with particle exchange including chemical reactions.

Almost all presentations of thermodynamics and statistical mechanics have followed this historical tradition. Continuing to follow the energy perspective, however, has had several unfortunate consequences.

1. It obscures the central role of the entropy as the *defining* property of thermodynamic limits, and hence the ready generalization of the principles of thermodynamics beyond mechanics.

2. It masks the essentially symmetric role that energy, volume, particle numbers, or other conserved quantities have as *constraints* on the set of configurations over which entropy is maximized. Entropy can naturally depend on quite diverse and heterogeneous clusters of constraint variables in the large-deviations analysis of general stochastic processes, which will be our concern as we move away from equilibrium statistical mechanics and into the more general problems of inference about living systems.

3. The concept of "heat" is often used ambiguously to refer to any contribution to the internal energy change which is not accounted for in work flow. A proper usage would distinguish those changes which act continuously as constraints on entropy from those which do not, as this determines which transformations even *possess* well-defined intensive state variables of temperature, pressure, chemical potential, etc. Heat flow, like mechanical or chemical work, can then be restricted to characterize those changes of constraint for which intensive variables are continuously defined. All these constrained changes are distinguished from transfers of energy, volume, or particles over trajectories that are momentarily out of equilibrium, during which the exchanged quantities do not act as constraints on the entropy.

4. It has needlessly mystified generations of students about the nature and role of entropy, and the structure of thermodynamics reflected in its suite of thermodynamic potentials dual to the entropy, and the meanings of their gradients.

In Box 7.1 we provide a very brief review of the most standard classical thermodynamics as presented, for example, in Fermi's classic introduction. From the entropy perspective, the standard constructions become elementary.

---

**Box 7.1    Familiar thermodynamics from the entropy perspective**

All the standard relations of classical thermodynamics, usually introduced through a "conservation of energy" relation, may be alternatively and more transparently constructed by starting with entropy. The only difference between systems described by one state variable and systems described by many comes from the number of arguments of the entropy function, expressing the set of constraints on which a macrostate depends. Here we present a small representative example. The construction extends immediately to more equilibrium variables such as magnetization with its dual magnetic field, but also to non-equilibrium contexts in which the constraints are functions of time rather than fixed quantities.

The classical thermodynamics of systems capable of exchanging heat, volume, or particles with their environment[a] results when large-deviations scaling leads to an entropy $S(U, V, N)$ which is a function of internal energy $U$, volume $V$, and the number $N$ of some kind of particle, where we consider only one kind for simplicity. The gradients of $S$ with respect to these arguments are

[a]    We therefore include directly systems that undergo chemical reactions.

**Box 7.1    Familiar thermodynamics from the entropy perspective (cont.)**

$$\frac{\partial S}{\partial U}\bigg|_{V,N} \equiv \frac{1}{T}$$

$$\frac{\partial S}{\partial V}\bigg|_{U,N} \equiv \frac{p}{T} \tag{7.9}$$

$$\frac{\partial S}{\partial N}\bigg|_{U,V} \equiv -\frac{\mu}{T}.$$

The familiar *energetic* intensive state variables are temperature $T$, pressure $p$, and chemical potential $\mu$,[b] but note that the natural dual variables to $U$, $V$, and $N$ are various combinations of these.

The variation in the available log-volume of state space $\delta S$ created by a joint change in energy $\delta U$, volume $\delta V$, and particle number $\delta N$ is then given by

$$\delta S \equiv \frac{1}{T}\delta U + \frac{p}{T}\delta V - \frac{\mu}{T}\delta N. \tag{7.10}$$

Expression (7.10) is usually introduced as an empirical "conservation of energy" relation

$$\delta U = T\delta S - p\delta V + \mu\delta N, \tag{7.11}$$

in which a mysterious "heat flow" $T\delta S$ must somehow be added to the system's energy to enable an increment $\delta V$ of volume and $\delta N$ of particle content. Note how much more intuitive as well as how much more direct it is to recognize Eq. (7.10) as a *definition* of the intensive state variables, and $\delta S$ as the change in entropy *made available to the microstates* by increments $\delta U$ of energy, $\delta V$ of volume, or $\delta N$ of particles.

A few standard Legendre transforms of the entropy may be given straightaway. The Helmholtz free energy $F$ comes from the Legendre transform with respect to internal energy, though the combination that arises as the dual to entropy is $F/T$:

$$U\frac{\partial S}{\partial U}\bigg|_{V,N} - S = \frac{U}{T} - S \equiv \frac{F}{T}. \tag{7.12}$$

$F$ satisfies

$$\frac{\partial (F/T)}{\partial (1/T)}\bigg|_{V,N} = U, \tag{7.13}$$

---

[b] Note that if $\mu < 0$, the system can increase its entropy by gaining particles. Particles from an environment with a higher chemical potential may then spontaneously fall into this system to produce a joint state of higher entropy, analogous to particles spontaneously falling into a "potential well." This is the reason for the negative sign convention for chemical potential.

**Box 7.1    Familiar thermodynamics from the entropy perspective (cont.)**

and the derivatives with respect to $V$ and $N$ are the same as those of $-S$. Equation (7.12) is written in energy units as the more familiar $F = U - TS$. Equation (7.13) is readily converted with a little symbol manipulation to the conventional expression in energy units $\partial F / \partial T|_{V,N} = S$.

The Gibbs free energy is similarly constructed as a Legendre transform on two variables:

$$U \left. \frac{\partial S}{\partial U} \right|_{V,N} + V \left. \frac{\partial S}{\partial V} \right|_{U,N} - S = \frac{U}{T} + \frac{pV}{T} - S \equiv \frac{G}{T}, \qquad (7.14)$$

satisfying

$$\left. \frac{\partial (G/T)}{\partial (1/T)} \right|_{p/T,N} = U$$
$$\left. \frac{\partial (G/T)}{\partial (p/T)} \right|_{1/T,N} = V, \qquad (7.15)$$

and the Landau "grand potential" is

$$U \left. \frac{\partial S}{\partial U} \right|_{V,N} + N \left. \frac{\partial S}{\partial N} \right|_{U,V} - S = \frac{U}{T} - \frac{\mu N}{T} - S \equiv \frac{\Omega}{T}, \qquad (7.16)$$

satisfying

$$\left. \frac{\partial (\Omega/T)}{\partial (1/T)} \right|_{\mu/T,N} = U$$
$$\left. \frac{\partial (\Omega/T)}{\partial (\mu/T)} \right|_{1/T,N} = -N. \qquad (7.17)$$

Again the derivatives with respect to the extensive arguments are the same as those of $-S$.

From the standpoint of mere manipulation of partial derivatives, as long as the entropy is convex, it is a matter of convenience whether one chooses entropy, internal energy, or some other quantity to be maximized along a surface of constraint determined by the other quantities. For many mechanical problems in which thermal systems are coupled to heat baths or movable walls, the energy approach is the most direct in which to calculate quantities of interest. The entropy approach is more natural for many contexts of statistical inference, in which it is manifest that $F/T$, $G/T$, and $\Omega/T$, are cumulant-generating functionals for the moments of $U$, $V$, or $N$.

### 7.3.3 Statistical roles of state variables

Because the macroscopic state variables *determine* the constraints on microstates, they are also the most useful quantities in *inference* about the values of the microstates. This property will be very important when we consider the role of phase transitions in living systems. The exponential suppression of fluctuations in thermodynamic stable states provides a form of *buffering* for unwanted microscopic noise in a hierarchical complex system. Not all fluctuations are removed; those mediated by the state variables that define the stable states remain available to serve multilevel system dynamics. For this dynamics to make use of the microscopic variables despite their uncertainty, it must be possible to make reliable inferences about those properties through which they influence their surroundings. It is this inference that is supported by the state variables.

The concept of an optimally informative quantity in a problem of statistical inference is captured in the notion of a **sufficient statistic**, summarized in Box 7.2. The state variables are sufficient statistics for maximum-entropy distributions in which they are the arguments of the entropy.[12] We return to the reason state variables arise as the natural control parameters in the evolution of hierarchical complexity in Section 8.2.5.

---

**Box 7.2   Sufficient statistics**

Suppose we have a system that we name $\mathcal{S}_x$. Suppose that the state of $\mathcal{S}_x$ is a joint state of $n$ random variables $(x_1, \ldots, x_n)$. We will think of $\{x_i\}$ as the microscopic state variables.

Next suppose that each random variable $x_i$ takes values from a set $X_i$. Denote by $|X_i|$ the cardinality of (number of elements in) $X_i$. For convenience, take $n$ and all the $|X_i|$ to be finite natural numbers. Then the number of possible configurations that can be defined for the system $\mathcal{S}_x$ is

$$\prod_{i=1}^{n} |X_i|.$$

A *summary statistic* [655][a] for the state of the system $\mathcal{S}_x$ can be any function $\sigma_x(\{x_i\})$ of the values of the microscopic state variables. We will be interested, however, in functions that are many-to-one, so many values of $\{x_i\}$ are equivalent under $\sigma_x$. Therefore we let $\sigma_x$ take values $\sigma_x(\{x_i\}) \in \Sigma_x$, for some set $\Sigma_x$ whose cardinality

$$|\Sigma_x| \ll \prod_{i=1}^{n} |X_i|. \tag{7.18}$$

[a] Summary statistics, including subclasses such as sufficient statistics, are part of the larger category of *descriptive statistics*. For an introduction see [667].

---

[12] Benoit Mandelbrot [516] has developed formally the role of state variables as sufficient statistics, and their relation to invariants of the kind we have invoked here as constraints between coupled subsystems. Mandelbrot goes on to show that the Gibbs exponential distribution is, in fact, the *unique* distribution that has a single (scalar) sufficient statistic corresponding to each invariant on the macroscopic state.

---

**Box 7.2    Sufficient statistics (cont.)**

Now suppose there is a new random variable $y$ that we wish to sample, predict, or estimate, and a stochastic process that generates a probability distribution over values of $y$ from the value of the state $\{x_i\}$. In an abbreviated notation we denote this distribution by $P(y \mid \{x_i\})$. Typical situations are that $\{x_i\}$ is the state of a system at some time, and $y$ is some function of the state at another time, generated by complex dynamics from the earlier state. Or $\{x_i\}$ might be the microscopic state of one subsystem and $y$ might be a function of the state of some other subsystem coupled to $\mathcal{S}_x$ through a boundary.

The summary statistic $\sigma_x$ is called a *sufficient statistic* for $y$ [781] if $P(y \mid \{x_i\})$ has the property that

$$P(y \mid \{x_i\}) = P(y \mid \sigma_x(\{x_i\})). \tag{7.19}$$

In words, we can know no more about the values of $y$ given knowledge of the whole state $\{x_i\}$ than we know given the value that $\sigma_x$ takes. This also says that the number of possible distinct forms for $P(y \mid \{x_i\})$ is limited to $|\Sigma_x|$, rather than the much larger number $\prod_{i=1}^{n} |X_i|$.

---

### *7.3.3.1 Entropy as a measure of potential information*

The principle of maximum entropy may also be understood as a principle of *maximum ignorance*. The entropy-maximizing distribution over microstates is the least constrained distribution consistent with the values of the state variables. For this reason, any reduction in entropy required, as the boundary conditions on a system change, is eligible as a measure of **information** gained about the system's microstates as a result of the change in its macrostate. The ultimate reduction in entropy comes from a completely detailed *sample* of the system, which identifies a unique microstate. The sample is equivalent to a distribution with only one member, which has zero entropy. Such a distribution might be created by imposing a much more detailed boundary condition than the one that led to the original macrostate. Therefore the entropy of a macrostate has the interpretation of the amount of information *missing* in the corresponding distribution about sampled microstates.

E. T. Jaynes has emphasized [391, 392, 394] the use of maximum-entropy distributions as a solution to the problem of *least-biased inference*. Jaynes interprets the state variables in a statistical mechanical system as measures of what one knows about the microstates of the system. The least-biased inference that can be drawn about variables which are not deterministic functions of the known quantities is made by first computing the maximum-entropy distribution given the known quantities, and then computing expected values of the desired quantities in this distribution. An equivalent understanding of the information capacity in a transmission line as a measure of the variety of signals that could be sent – and hence would be unknown before the transmission but known precisely if the transmission could be carried out reliably – was the basis for Claude Shannon's information theory [732], introduced in Section 7.5 below.[13]

Later we will work out an example showing how both the computation and the interpretation of Jaynes can be justified and used within a framework of Bayesian inference, in specific contexts. However, we do not mean either to endorse or to follow the main program of Jaynes that purports to give a general approach to inference. Two mathematical areas in which we differ are these.

1. Jaynes argues for a general role for the Shannon–Gibbs entropy on axiomatic grounds [394], which is a starting point to define the problem of inference and information. We regard entropy as a derived property of large-deviations scaling when that scaling pertains, and do not commit to any invariable functional form.[14] Where we identify a problem of inference, its structure derives from the underlying large-deviations scaling.

2. Jaynes seeks a general program by which "knowledge" corresponds to constraint functions for the entropy, which then define the macroscopic state variables. Our view is less ambitious and openly dependent on constructive arguments. We define thermodynamic problems by the constraints that a definite set of boundary conditions impose on a system which then still remains partly unconstrained. The state variables reflect what is *controlled* about the system, and the interpretation of constraint as knowledge must be constructed from the specific mode of control.

By making clear how the fluctuation behavior of a system *defines* particular problems of inference, and not attempting to impose entropy maximization as an axiomatic approach or to privilege unique functional forms for entropy, we will avoid claims of an objectively based unique criterion for inference.[15]

A consequence of the macroscopic state variables' being the sole constraints on microscopic states is that the samples of states in a maximum-entropy ensemble do not otherwise constrain each other. The formal statement of this property is that fluctuations in a maximum-entropy ensemble are *conditionally independent* of one another, given the values of the macroscopic state variables as the conditioning context. Conditional independence of fluctuations is reviewed in Box 7.3.

---

[13] The association of entropy changes with information changes *of the opposite sign* has been given a shorthand by Schrödinger [711] in saying that life feeds on "negative entropy," abbreviated further by Brillouin [98] to "negentropy." We revisit in Section 7.6.1.1 the sense in which this shorthand captures an important property of the relation of order in the biosphere and in the environment, but also the limits of using the equilibrium concept of entropy to try to formalize such a relation.

[14] Since all of our examples involve elementary multinomial distributions, the Shannon–Gibbs entropy arises as the logarithm of the multinomial.

[15] No specific program of inference seems to be justified without significant assumptions about context which must be at least in part *as hoc* (meaning, "to the case"). For all presently advanced "consistency criteria," reasonable exceptions can be shown [816], and even the use of sample values as constraints on probability requires other assumptions [817], which become explicit if sampling and probability are used to refer to a system's internal dynamics in place of an external observer's beliefs. Cases are readily obtained in which entropy maximization is inconsistent with Bayesian updating [724, 725], though the latter also requires many assumptions about the relation of hypotheses to evidence, as we suggest briefly in Box 8.1.

**Box 7.3   State variables and independent fluctuations**

The principle of maximizing Gibbs entropy, and the role of state variables as constraints on that maximum, imply that fluctuations in a thermodynamic system about its most likely state are independent in a specific way. We begin with simple entropy maximization, and then consider fluctuations.

The Gibbs prescription for thermodynamics proposes that the probability distribution we use to describe a thermal ensemble should maximize the entropy $S = -\sum_{\alpha=1}^{\aleph} p_\alpha \log p_\alpha$. Here $\alpha \in 1, \ldots, \aleph$ are the possible values that a system state might take. Gibbs entropy maximization is not free from constraint. We *define* the constraints on the system by the expectation of one or more functions of the state $\alpha$. Usually these will be boundary conditions imposed by the environment. Here, to simplify the example, we suppose there is a single function $H_\alpha$ (for equilibrium systems the most common constraint is from a Hamiltonian energy function), and its expectation is $E(H) = \sum_{\alpha=1}^{\aleph} p_\alpha H_\alpha$.

An insight by Lagrange [309] can be applied to the maximization problem as long as the entropy is convex. Lagrange observed that such a constrained entropy is maximized if any non-vanishing gradient of the entropy points in a direction that can only be accommodated by moving the constraint. Therefore we obtain the constrained maximum by solving for the absolute maximum of a *Lagrangian* function

$$\mathcal{L} = S - \beta \left( \sum_{\alpha=1}^{\aleph} p_\alpha H_\alpha - U \right) - \eta \left( \sum_{\alpha=1}^{\aleph} p_\alpha - 1 \right), \qquad (7.20)$$

where $U$ is the value to which $E(H)$ is constrained, and the second term requires that probabilities sum to unity. The "Lagrangian multipliers" $\beta$ and $\eta$ give, at the solution, the proportionality between a given change in the entropy and the movement of a constraint needed to permit that change.

Use $\bar{\eta}$, $\bar{\beta}$, and $\bar{p}_\alpha$ to label the solutions that maximize $\mathcal{L}$. Then these solutions,

$$\bar{p}_\alpha = e^{-[1+\bar{\eta}+\bar{\beta}H_\alpha]} = \frac{1}{\bar{Z}} e^{-\bar{\beta}H_\alpha}, \qquad (7.21)$$

are precisely the probabilities used in the partition function of Gibbsian statistical mechanics [441], which we label $\bar{Z}$.

In elementary statistical mechanics (discussed further in the text), the partition function $\log \bar{Z} = -\bar{\beta}\bar{F} = \bar{S} - \bar{\beta}U$, where $\bar{F}$ is called the *Helmholtz free energy*. We use overbars here on all quantities besides $U$ as a reminder that these solutions are determined by $U$. In particular, the entropy $\bar{S}$ is the maximum possible for a distribution in which $E(H) = U$.

**Box 7.3    State variables and independent fluctuations (cont.)**

Subtracting all boundary terms from the Lagrangian (7.20) is equivalent to subtracting away this maximal entropy. If we evaluate $\beta$ at $\bar{\beta}$ and $\eta$ at $\bar{\eta}$, the remaining maximization problem is over the function

$$\mathcal{L} - \bar{\beta} U - \log \bar{Z} = \mathcal{L} - \bar{S} = -\sum_{\alpha=1}^{\aleph} p_\alpha \log\left(\frac{p_\alpha}{\bar{p}_\alpha}\right) - \left(\sum_{\alpha=1}^{\aleph} p_\alpha - 1\right). \qquad (7.22)$$

$\sum_{\alpha=1}^{\aleph} p_\alpha \log\left(p_\alpha / \bar{p}_\alpha\right)$ is called the *relative entropy* of the distribution $p_\alpha$ from $\bar{p}_\alpha$ [164], and is discussed further below.

**Large-deviations scaling of fluctuations**   Maximizing a constrained Gibbs entropy identifies a single distribution $p_\alpha = \bar{p}_\alpha$. The motivation to perform this maximization, of course, is an implicit understanding first due to Boltzmann that this is the most-likely distribution to be found in samples of a system's state that may fluctuate. Large-deviations theory, from which (as noted in the text) classical thermodynamics can be derived, makes more precise the sense in which the classical distribution is "most likely" among the possibilities.

Large-deviations theory considers systems that are not infinite but only large, and made up of $N$ components that may be said to fluctuate "independently." The scaling variable $N$ determines how likely fluctuations are to survive averaging under laws of large numbers.

To give a meaning to the Lagrangian maximization problem over $\{p_\alpha\}$, we take $n_\alpha = N p_\alpha$ to be the fraction of the $N$ components that we find in each state $\alpha$ in a sample of the system's state. Because the system's state can fluctuate, samples with $n_\alpha \neq N \bar{p}_\alpha$ are possible. The system shows large-deviations scaling if the probability for any joint configuration has the scaling at leading exponential order (denoted by $\sim$)

$$P(\{n_\alpha\}) \sim e^{N(\mathcal{L} - \bar{S})} \approx \left(\prod_{\alpha=1}^{\aleph} \bar{p}_\alpha^{\, n_\alpha}\right) \left(\begin{array}{c} N \\ n_1 \ldots n_\aleph \end{array}\right). \qquad (7.23)$$

Because we fix $N$, all possible values of $\{n_\alpha\}$ satisfy $\sum_{\alpha=1}^{\aleph} p_\alpha = 1$, so only the relative entropy term in Eq. (7.22) survives. In the second expression we have used Stirling's formula to express some terms in the exponential as logarithms of the multinomial coefficient, given by

$$\left(\begin{array}{c} N \\ n_1 \ldots n_\aleph \end{array}\right) \equiv \frac{N!}{n_1! \ldots n_\aleph!}. \qquad (7.24)$$

---

**Box 7.3    State variables and independent fluctuations (cont.)**

Equation (7.23) says that the fluctuations in the macroscopic system are those that would be produced by independent sampling of a random variable $N$ times, where the values taken by the variable depend only on $\{\bar{p}_\alpha\}$. That is, if we write the sample of a system's state as a random variable $a_i$, for $i \in 1, \ldots, N$, then any joint probabilities for outcomes of such a sampling process factor as

$$P\left(a_i = \alpha_1, a_j = \alpha_2, \ldots\right) = \bar{p}_{\alpha_1} \bar{p}_{\alpha_2} \ldots = P(a_i = \alpha_1)\, P\left(a_j = \alpha_2\right) \ldots. \quad (7.25)$$

The probabilities $P(a_i)$ are really conditional because all $\bar{p}_\alpha$ depend on (but *only* on!) on the constraint $E(H) = U$. Therefore the factorization (7.25) defines the *conditional independence* of samples $\{a_i\}$ given the state variable $U$.

   Whether one thinks of this combinatorial independence as the justification for Gibbs entropy, or thinks of Gibbs ensembles as treating systems "as if" they are composed of independent samples, is not essential in this discussion. What matters is the form of conditional independence, given the state variables, which is inherent in large-deviations scaling of fluctuations.

---

### *7.3.4 Phase transitions and order parameters*

The entropy function can become non-convex, and when it does so, thermodynamic systems may become ordered in a way that is either less fully determined by the boundary conditions than a typical intensive/extensive dual pair of state variables is, or else opens the possibility for both intensive and extensive boundary conditions to be imposed as constraints independently. The first case describes the *second-order phase transitions*, which are associated with the spontaneous formation of macroscopic ordered phases that fail to manifest some symmetries present in the underlying dynamics. We provide a full worked example of a second-order phase transition, and explain its relation to symmetries, in Section 7.4.1. Systems possessing second-order phase transitions are an extremely important class of thermodynamic cases, because it is through these that asymmetry, multiscale dynamics, and unpredictability or historical contingency enter the world of stable entities.

   The other case describes the *first-order phase transitions*, in which no new asymmetries are created, but in which two state variables that normally would be duals become required as boundary conditions to specify the state of the system fully. We provide a brief characterization of entropy functions which give rise to first-order transitions at the end of Section 7.4.1, and consider the directly applicable empirical case of lipid-water phase separation and ordering in Section 7.4.6.

## 7.4 Phase transitions in equilibrium systems

Thermodynamics first provides foundations for order in the biosphere through the physical substrate of equilibrium or near-equilibrium condensed matter. Living systems use a rich variety of stable thermal states ranging from molecules themselves, to lipids or crystals, to folding macromolecular structures, the intrinsic stability of which comes from physical chemistry prior to active energy-consuming control. As equilibrium matter is also the domain where thermodynamic principles are most widely used and best understood, we develop several examples of equilibrium steady states and phase transitions to make both the mathematics and the characteristic behaviors familiar.

We begin with a mathematical example of what is probably the simplest phase transition, a collection of binary variables with no spatial structure. After illustrating the major concepts of reduction, emergence, and predictability in this system, we review the way the same ideas have been applied to systems with richer state spaces to describe the entire hierarchy of known equilibrium states of matter, from the elementary particles to minerals and superconductors. The mineral history of the planet is itself another hierarchy in microcosm, arising within a relatively narrow band of energy from the interaction of the rotational states of molecular bonds with the problem of forming space-filling lattices. The hierarchy of matter gives a rough model, at equilibrium, for the kind of hierarchy of phase transitions that we propose dictated the steps in the emergence of a biosphere *away from equilibrium*. Before closing the section we describe the phases associated with oil–water separation in amphiphilic molecules. This particular constellation of equilibrium phase transitions is the source of vesicular containment and of a variety of control strategies used by cells.

### 7.4.1 Worked example: the Curie–Weiss phase transition

We wish to show with a concrete example, which can be fully solved by an elementary calculation, how cooperative effects among many degrees of freedom can compress the fluctuations in a system's microstates down into what might be called a "corner" of its configuration space. This is the modern understanding of what happens when a system freezes.[16] Reducing temperature, increasing pressure, or in some other way reducing the range of microstates available to the system entails a reduction in its entropy. This reduction in turn defines a useful measure of the *information*, or the excess order that is intended when people refer to self-organization, in the frozen state compared to the melted state.[17]

It is important that the control parameter be something blunt, such as a reduction in temperature, so that the emergent order reflects hidden constraints from these corners of

---

[16] In this usage "freezing" is a general term, applicable not only to the conventional usage of cooling-induced crystallization, but to magnetization, superconductivity, and the host of other changes in state that share the mathematics of second-order phase transition. A related set of concepts with somewhat different mathematical details handles other phase-transition phenomena such as condensations. For simplicity, here we limit the discussion to second-order transitions, which have a particularly important relation to the presence and the representation of *symmetries*.

[17] Unlike a direct sample of a microstate, a phase transition does not reduce a system's distribution to a point, but only to a narrower distribution. Freezing provides an intermediate amount of information corresponding to reduction in freedom of the system's internal state dynamics.

the state space or the structure of the microscopic interactions. Order that merely reflects a complicated boundary condition imposed by a controller does not require a special explanation. The formation of internal order from blunt boundary conditions is the property that gives phase transitions their appearance of creating "emergent" complexity within a system, beyond whatever appears to be imposed by its environment.

Although the frozen phase of a system may be more ordered than its melted phase, the excess order never violates maximization of the entropy. Both in the disordered phase and in the ordered phase, an equilibrium thermal system has a distribution over microstates that is as undetermined as it can be, given the restrictions of access to these states that its internal energy or the values of its other state variables impose.

A further important property of phase transitions is they do not rely on the control parameter's being taken to an extreme value (such as temperature taken to zero) to remove all accessible fluctuations. More subtly, an internal competition among multiple possible, but mutually exclusive, responses to tightening environmental constraints causes the system to rapidly compress into one or another subdomain of its configuration space while still permitting residual fluctuations. This competition is expressed in the shape of the macroscopic entropy function, which determines the system's response to its environment. The multiplicity of solutions leaves a macroscopic variable (the order parameter) indexing the ordered state, which is stable against internal fluctuations and provides a low-dimensional interface of the system's order with the environment.

The simplest possible example that demonstrates all these phenomena with no extra baggage is the two-state phase transition introduced by Pierre Curie and first analyzed by Pierre-Ernest Weiss [866].[18] A large set of symmetric and uniformly coupled microscopic variables can take only binary values. At high temperature, the macroscopic state is "melted" and is unique. At low temperatures a "frozen" state spontaneously forms which, like the microscopic variables, can take only two symmetric values, although these values depend on the temperature.

The entropy in the model is the logarithm of a binomial distribution, which is the simplest example that shows all important properties of large-deviations scaling in thermal systems. The relation to binomial (or multinomial) counting remains the most basic justification for use of the Shannon–Boltzmann–Gibbs entropy, and illustrates the sense in which fluctuations about the average in maximum-entropy states are independent of one another.

### 7.4.1.1 Don't think about magnets

> Don't think about elephants.
>
> – Attribution unknown

The Curie–Weiss phase transition was developed as a microscopic theory of ferromagnetism, and it will be convenient for us to use physical terms such as "magnetization" or

---

[18] This analysis introduced the *mean field approximation*, which has become a standard first level of analysis for problems in statistical mechanics and field theory.

"spin" to refer to mathematical quantities in the model. For some readers, it may be helpful to have intuition from a physical system. For others it may be a distraction, because we do not intend that *physical magnetization* was an important factor in the emergence of the biosphere. We develop the Curie–Weiss model here only as a self-contained introduction to all the important mathematical features of phase transition. We may, nonetheless, use the physical properties of magnets to make some of the important roles of symmetry in this phase transition more accessible.

Ferromagnets are materials such as metallic iron or magnetite (($FeO$) ($Fe_2O_3$)), which do not produce a macroscopic magnetic field at high temperatures but upon cooling through a system-dependent critical temperature, spontaneously create a magnetic field and magnetization. The critical temperature for ferromagnets is named the *Curie temperature* or "Curie point" in honor of Pierre Curie.[19]

The important feature of second-order phase transitions illustrated by magnetization is that *the state adopted by a magnet has less symmetry* than the boundary conditions imposed on it by its environment.

The simplest environment is one that imposes no preference on the direction in which a magnet is oriented. Yet magnetization is a directional quantity that cannot be non-zero except by pointing in some direction. Therefore, becoming magnetized requires choosing a direction. Which way the magnet orients is an essentially arbitrary choice. Thus, the magnetized state both hides the underlying symmetry of orientations permitted by space, and makes the direction selected upon freezing unpredictable in a specific way.

A magnet has a strength as well as a direction, and unlike the direction, the strength is fully predictable from the properties of the configuration space and the microscopic physics. Thus the magnet illustrates the complementary character of reduction and emergence, or of necessary and chance properties.

Because of the directional character of magnetization, it was natural for Curie and Weiss to propose that the existence or non-existence of overt (macroscopic) magnetization might reflect a problem of *organization* rather than a *de novo* creation of a new property. It was possible to imagine that a macroscopic chunk of magnetizable material consists of a collection of atomic centers that are individually magnetic but whose directions of magnetization can have different orientations. Since the atoms are small and their number is many, if their directions are not aligned, no magnetization is visible on the macroscale. By far the largest number of microscopic configurations have roughly equal numbers of spins pointing in opposite directions (as we will show). Ferromagnets have the unusual property that their microscopic magnets (known as "spins" because the magnetization results from a property of valence electrons related to the rotation group) weakly affect each other through their

---

[19] Although we do not propose that magnetism was an important factor in the emergence of the biosphere, it is rare that life passes over opportunities to exploit an ordered phase if one is provided by physics at accessible energies. Magnetotactic bacteria [480] are known to use aligned grains of biotically produced ferromagnetic minerals to sense geomagnetic fields. The known minerals, produced by secretion in these bacteria, are magnetite or greigite (($FeS$) ($Fe_2S_3$), introduced in Chapter 3). Unlike migratory birds [582], magnetotactic bacteria do not use the horizontal component of magnetic field lines for navigation. Instead they use the vertical components to travel over macroscopic depths in oxygen transition zones in water or sediments, presumably to seek desirable environmental oxygen fugacities.

magnetic fields, in a way that favors their aligning in the same direction. Whether a macroscopic material becomes magnetized or not depends on whether the disordering effect of thermal vibrations, which favors filling the state space, or the mutual aligning effect of the spins that reduces their available configurations, is stronger.

### 7.4.1.2 The binary-binomial phase transition

In more realistic models of physical magnets, spins may be taken to point in any direction in three-dimensional space, but in the Curie–Weiss model, they are regarded as discrete entities with only two possible states, called "up" and "down." The state space for a set of $N$ such spins is just an $N$-dimensional cube. In this model all spins affect each other equally and weakly, so that no complications of spatial dimension enter. The magnet effectively exists as a point in space.[20]

To make a formal model, we introduce a label $i \in \{1, \ldots, N\}$ for spins, and a dynamical variable $s_i$ for each spin $i$, which can take the two values $s_i \in \{-1, 1\}$.

The important property of the system that makes it magnetic is the assignment of energies required to populate each joint state of all the spins.[21] The microscopic energy function for the Curie model is[22]

$$H = -\frac{\epsilon}{2} - \frac{\epsilon}{N} \sum_{\langle i,j \rangle} s_i s_j - h \sum_{i=1}^{N} s_i. \tag{7.26}$$

The notation $\langle i, j \rangle$ indicates the pair of spins $i$ and $j$ with $i \neq j$, and the sum is over all pairs. Each pair that point in the same direction reduce the energy by an amount $\epsilon/N$, where $\epsilon$ is some energy scale that characterizes the interaction strength of the spins. In a system/environment complex, energy not used in the system's state may move to the environment, increasing the volume of its accessible states and thus its entropy. Each pair that point in opposite directions raise the energy by $\epsilon/N$, so that configuration can only be occupied if less energy is available to environmental states. We scale the interaction strength by $1/N$ so that we may consider a thermodynamic limit of large $N$, while keeping the interaction strength on any individual spin $s_i$ finite in the limit. $h$ is an optional externally applied magnetic field, which reduces the energy in proportion to the net magnetization $\mu \equiv \sum_{i=1}^{N} s_i$.

Since the state space for the joint spins counts each configuration equivalently, the entropy of the environment, which depends on the energy left over from the system's state, determines the likelihood of each system microstate. The assignment (7.26) of interaction

---

[20] A more thorough pedagogical review of the Curie–Weiss model is provided in [445].
[21] The energy function is conventionally called the *Hamiltonian*, with reference to the Hamilton–Jacobi function of mechanics [309], from which statistical mechanics originated. For simple equilibrium models of condensed matter, a full dynamical specification is not usually given; rather the sampling of the system's states is supposed to be governed by thermal forcing by complex environmental factors, possibly together with chaotic internal dynamics, in a manner that is well approximated by replacing the dynamics with a random sample. Therefore the "Hamiltonian" in this usage is simply an energy function over states.
[22] A constant offset to the energy is added from the start to make the forms of later equations simpler.

energies to spin microstates makes states with aligned $\{s_i\}$ more likely, and among these it favors states with $\mu h > 0$.[23]

Because both internal and external interactions are symmetric, the energy (7.26) is a function only of the magnetization $\mu$, and may be written

$$H = -\frac{\epsilon}{2N}\mu^2 - h\mu. \tag{7.27}$$

The magnetization when $n$ spins point up (and so $N - n$ point down) is given by

$$\mu = 2n - N. \tag{7.28}$$

The number of otherwise equivalent microscopic configurations that all have this same value of magnetization is the binomial coefficient

$$\binom{N}{n} \equiv \frac{N!}{n!\,(N-n)!}. \tag{7.29}$$

The spin interactions and counting of microscopic spin states are shown in Figure 7.1. Configurations with all spins pointed either up or down are the top and bottom corners of the hypercube. Neighborhoods of these corners have few vertices compared with the equator,

(a)                    (b)

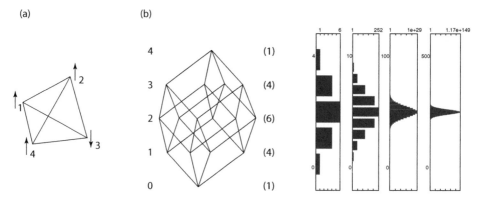

Figure 7.1 (a) The all-to-all coupling of ($N = 4$) spins that can take values $s_i \in \{-1, 1\}$, indicated by up or down arrows. Each link between a pair of spins $\langle i, j \rangle$ represents a contribution to the energy of $-\epsilon s_i s_j / N$. (b) The possible spin configurations for $N = 4$ make a four-dimensional hypercube. Vertices in the same row have the same number $n = |\{i \mid s_i = 1\}|$. $n$ is shown on the left beside each row, and the number of vertices at that $n$, given by the binomial coefficient (7.29), is shown on the right in parentheses. Binomial coefficients rapidly become concentrated near the "equator" of the hypercube as $N$ grows large. The third panel shows bar graphs counting the number of vertices at each level $n$ for $N \in \{4, 10, 100, 500\}$.

---

[23] Here we choose units in which the magnetic moment of each spin is set to unity, and $h$ and $\epsilon$ are measured in energy units that incorporate these moments.

where equal numbers of spins point up or down. As $N$ grows large, the concentration of almost all vertices near the equator of the hypercube becomes more extreme. This is why, if cooperative effects can restrict possible microstates to be distributed near either the top or the bottom corner, they can drastically reduce the number of such microstates available (they can reduce the entropy), and increase the information about the microstate in the thermal distribution.

### 7.4.1.3 The partition function determines thermodynamic properties

The traditional approach to thermodynamic modeling, which we will use here, introduces the notion of an *ensemble* [289] of microscopic configurations, each having some probability, and derives macroscopic properties as those that are expected in the ensemble. All microscopic configurations are sampled, and the probabilities assigned to them are those that maximize the entropy of the resulting distribution.[24]

The sum over all configurations of the exponential function that maximizes entropy – *unnormalized* – is called the *partition function* for the ensemble. The equilibrium partition function for a spin system with all-to-all coupling is

$$Z \equiv e^{-\beta F} = \prod_{i=1}^{N} \sum_{s_i \in \{-1,1\}} e^{-\beta H}. \tag{7.30}$$

Here, $\beta = 1/k_B T$, where $T$ is absolute temperature, converted to energy units by Boltzmann's constant $k_B$.

$Z$ may be thought of as a count of the number of likely system states, each weighted by the exponential term in the probability distribution. Recall that this term $e^{-\beta H}$ in turn reflects (up to an overall normalization) a count of states in the environment's heat reservoir after the spin system has accounted for energy $\beta H$. Hence, apart from an overall normalization, we may regard the partition function as a count of the number of states available in the system/environment pair, constrained by the total energy of which most is held in the environment. If there is no constraint so that the distribution is uniform, the partition function is normalized so that it simply counts the total number of microscopic system states.

The temperature is the boundary condition imposed by the environment that will determine how strongly the system's internal energy skews the distribution on states. $F$ in Eq. (7.30) is the *Helmholtz free energy*, introduced above in Box 7.1. If we limit our consideration to boundary conditions with $h \equiv 0$, $F$ is a function only of temperature.

---

[24] The justification for the ensemble and its probability distribution may come from several premises. A strong assumption is that the microscopic dynamics is sufficiently random, and that the coupling of the system to its environment is sufficiently slow, that the complex microscopic dynamics may be replaced by a random sample from the ensemble. This is known as the *ergodic hypothesis*, and may or may not be true depending on the system studied. An alternative argument is that the least-biased inference that can be drawn from the system's boundary conditions is one that samples the maximum-entropy distribution consistent with the boundaries. This justification was advocated by E. T. Jaynes over many decades [394]. While we have said we do not follow Jaynes in regarding entropy maximization as a general solution to the problem of inference, we recognize that it is a *weaker* assumption than ergodicity because, in emphasizing that state variables reflect our *control* over the system by coarse boundary conditions, for many problems of inference we should average over system preparations with distinct initial conditions that may supplement the incomplete ergodicity of its own dynamics. We show in Box 8.2 how this interpretation embeds naturally within a framework of Bayesian updating.

It is related to the internal energy $U$ and entropy $S$ as $-\beta F = S - \beta U$, with $U$ evaluated where $dS/dU = \beta$.[25] Thus in the partition function $e^{-\beta F} = e^{(S-\beta U)}$, one factor $e^S$ is the count of typically occupied states, and the other factor $e^{-\beta U}$ mimics, for the system as a whole, the exponential weights $e^{-\beta H}$ given to the microstates.[26]

We may immediately make contact with the entropy of microscopic configurations because Stirling's approximation for factorials in Eq. (7.29) approximates the natural logarithm of the binomial coefficient as

$$\log \binom{N}{n} \approx -N \left[ \frac{n}{N} \log \frac{n}{N} + \left(1 - \frac{n}{N}\right) \log \left(1 - \frac{n}{N}\right) \right]. \tag{7.31}$$

The right-hand side is simply the entropy of the normalized distribution on two elements $(n/N, 1 - n/N)$, which attains a maximum of $\log 2$ at $n/N = 1/2$.

Using the above expressions for magnetization, energy, and entropy, and introducing a rescaled magnetization variable $x \equiv \mu/N$ (the fraction of the possible magnetization), we may approximate the partition function (7.30) in forms that are easy to compute:

$$
\begin{aligned}
Z &= \sum_{n=0}^{N} e^{\beta\left[(\epsilon/2N)(2n-N)^2 + h(2n-N)\right]} \binom{N}{n} \\
&\approx 2^N \sum_{n=0}^{N} e^{(N/2)\left[\beta\epsilon x^2 + 2\beta hx - (1+x)\log(1+x) - (1-x)\log(1-x)\right]} \\
&\approx \frac{N2^N}{2} \int_{-1}^{1} dx \, e^{(N/2)\left[2\beta hx + (\beta\epsilon-1)x^2 - x^4/6\right]} \\
&\equiv \frac{N2^N}{2} \int_{-1}^{1} dx \, e^{-N\Phi(x)}. \tag{7.32}
\end{aligned}
$$

In the third line of Eq. (7.32) we have replaced the sum by an integral. The argument of the exponential integrand is called an **effective potential**, given in the fourth line by

$$
\begin{aligned}
\Phi(x) &\equiv \frac{1}{2} \left[ -2\beta hx - \beta\epsilon x^2 + (1+x)\log(1+x) + (1-x)\log(1-x) \right] \\
&\approx \frac{1}{2} \left[ -2\beta hx + (1 - \beta\epsilon) x^2 + x^4/6 \right]. \tag{7.33}
\end{aligned}
$$

---

[25] In Box 7.1 we introduced the convention of expressing boundary conditions in terms of derivatives of the entropy, rather than following the tradition of computing derivatives of the internal energy $U$. Readers interested in a more systematic development of this point of view can see [767] (Appendix B).

[26] This manner of weighting that treats aggregated and disaggregated states equivalently is the justification to use the exponential terms $e^{-\beta H}$, rather than fully normalized probabilities, to define the partition function. In the modern understanding of the structure of matter as a hierarchy of phase transitions, there is no unique microscopic level and no unique macroscopic level. The output $e^{-\beta F}$ from one form of aggregation may be used as the elementary weight in a larger ensemble with many copies of the smaller one. In the case that the smaller ensemble has no entropy response to energy changes, the microscopic weight simply becomes $e^{-\beta H}$, providing an alternative route to the use of the exponential probability distribution.

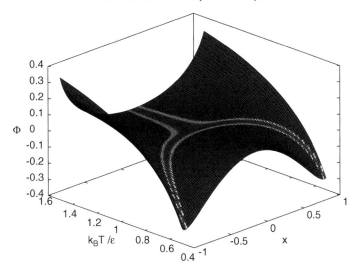

Figure 7.2 The effective potential $\Phi$ of Eq. (7.33). Potential surface shows the change of curvature from attractive toward $x = 0$ to repulsive as $k_B T/\epsilon$ passes below 1, and the emergence of new minimum values $\Phi < 0$, where the cooperative effect in the energy dominates the combinatorial effect in the entropy. Colormap shows the set of $x$ values at each $T$ for which $\Phi$ is within $1/N$ of its minimum value at that $T$, as the non-red colors. Only these $x$ make a significant contribution to the partition function (7.32).

$\Phi$ contains the terms in the original energy function (7.27) (multiplied by $\beta$), together with the microscopic entropy.[27] Values of $x$ contribute to the partition function if they are within roughly $\sim 1/N$ of the minimum of the effective potential at a given temperature. The profile of the effective potential, which determines the trade-off between entropy-dominance and energy-dominance, and the ranges of $x$ that contribute to the partition function at $N = 500$, are shown in Figure 7.2.

At large $N$, fluctuations in this fully coupled system are small, and the magnetization is dominated by the maximizer of the exponential, where the first derivative of the effective potential vanishes, given by

$$0 = \frac{1}{2} \log \left( \frac{1+x}{1-x} \right) - \beta \left( \epsilon x + h \right)$$

$$\approx \frac{1}{3} x^3 - (\beta \epsilon - 1) x - h \tag{7.34}$$

and the second derivative is positive, given by

$$0 < \frac{1}{1 - x^2} - \beta \epsilon$$

$$\approx x^2 - (\beta \epsilon - 1). \tag{7.35}$$

---

[27] The polynomial expansion of effective potentials such as $\Phi(x)$, shown to fourth order in the second line of Eq. (7.33), was proposed by Lev Landau as a phenomenological model for general second-order phase transitions, including ferromagnetism, in 1937 [468]. We will see a similar expansion used to quantitatively analyze phase transitions in lipid systems in Section 7.4.6 below.

The first line in each expression is the exact evaluation from the effective potential (within the approximation of Stirling's formula), and the second line is the algebraically simpler polynomial expansion to lowest non-linear order. The first line of Eq. (7.34) is the transcendental equation for the entropy gradient in terms of the self-generated magnetization $\beta \epsilon x + h$, treated in [441].

The cubic polynomial approximation in Eq. (7.34) can be solved for $x$ in closed form at general $h$, but if we are not concerned with the response to external fields near the critical point, we may study the behavior of energy and entropy, and the spontaneous breaking of symmetry, by setting $h = 0$. The resulting dependence on temperature, of this *stationary-point approximation* $\bar{x}$ for the magnetization, is given by

$$|\bar{x}| \approx \max \left\{ 0, \sqrt{3 (\beta \epsilon - 1)} \right\}. \tag{7.36}$$

The magnetization remains zero for all temperatures $k_B T \geq \epsilon$, which is the *Curie point* for the ferromagnet.[28] For $k_B T < \epsilon$, $\bar{x}$ – the *order parameter* of this phase transition – departs from zero as either a positive or negative square root and hence a *non-analytic* function of $\beta$. Non-analytic dependence of the order parameter on the control parameter $\beta$ is the distinguishing feature of a second-order phase transition [885], and square-root dependence is characteristic of this class of approximations for discrete-spin ferromagnets.[29]

In the same stationary-point approximation, the internal energy is given by

$$\begin{aligned} U \equiv E(H) &\approx -\frac{N\epsilon}{2}\bar{x}^2 \\ &= -\frac{3N\epsilon}{2} \max (0, \beta \epsilon - 1), \end{aligned} \tag{7.37}$$

where $E(*)$ denotes expectation in the exponential distribution of Eq. (7.30). The entropy is simply the log of the binomial coefficient, evaluated at $\bar{x}$, given by

$$\begin{aligned} S &\approx N \left[ \log 2 + \frac{3}{4} - \frac{1}{12}\left(\bar{x}^2 + 3\right)^2 \right] \\ &= N \left[ \log 2 - \frac{3}{4} \max \left(0, (\beta \epsilon)^2 - 1\right) \right]. \end{aligned} \tag{7.38}$$

In this simple approximation, all macroscopic state variables are given by the stationary points of the integrand in Eq. (7.32). To determine the stationary points we set the gradient of $\beta H$ equal to the gradient of the microscopic entropy (7.31) with respect to $x$. As a

---

[28] The Curie point is simply a freezing temperature. As the applicability of phase transition principles continues to extend to a growing list of systems, it becomes more natural to use "freezing" as a generic term, and not to insist on a special name for each transition point unless one exists for historical reasons.

[29] This approximation is only valid for sufficiently small magnetizations that Stirling's formula for the binomial coefficient is a good approximation. Here we limit to $|\bar{x}| \ll 1$ and $\beta \sim 1/\epsilon$.

result, the estimates (7.37) for internal energy and (7.38) for entropy automatically satisfy the required relation $dS - \beta dU = 0$ when they are varied as functions of the argument $\bar{x}$. In the symmetric phase neither internal energy nor entropy changes with temperature in this approximation.[30] If we use the dependence of $\bar{x}$ on $\beta$ in the magnetized phase, the internal energy and entropy do change, at the rates

$$\beta dU = dS = -\frac{3N\epsilon^2}{2}\beta d\beta = \frac{3N\epsilon^2}{2k_B^2 T^3}dT. \tag{7.39}$$

Equation (7.39) is often written $dU = k_B T\, dS$ and referred to as "conservation of energy" between internal energy and heat. The simpler and more literal reading is that $\beta$ is simply the gradient of the log-volume of state space $dS$ gained as $dU$ passes from the environment to the system. The environment and system jointly maximize entropy when their values for these gradients are equal, so that no total entropy can be gained by redistributing internal energy between them.

### *7.4.1.4 The shape of the entropy function*

The distinction between the disordered and the ordered phase of the ferromagnet results from the contrast between persistence of the average magnetization $E(x) \sim 0$, near the maximum of the microscopic entropy, at all temperatures above the Curie point, and its sharp departure from zero below the Curie point. The departure of typical values of the magnetization $x$ from zero compresses the likely microstates more and more tightly into either the upper or lower corner of the hypercube in Figure 7.1, with decreasing temperature. Referring to the right-most bar graph in that figure, we illustrate the behavior of the entropy here for $N = 500$. The sharp drop of the entropy reflects the exponentially (in $N$) smaller number of vertices in the tail of the distribution compared with near its center.

The "shape" property of the entropy function $S(U)$ that describes this compression is shown in Figure 7.3. The qualitative nature of the distinction is most easily expressed using the simple stationary-point approximation for the mean magnetization $E(x) \approx \bar{x}$ from Eq. (7.36). The exact evaluation of the partition function at $N = 500$ smoothes this distinction near the transition point, and makes the characterization of this shape more intuitive, but does not otherwise change the qualitative behavior of the entropy.[31]

In the stationary-point approximation, both the internal energy and the entropy saturate at their maximum values $U \sim 0$, $S \sim N \log 2$,[32] everywhere above the Curie point. The derivative $dS/dU = 1/k_B T$ "pivots" through the whole range $k_B T \in (\epsilon, \infty)$ at this fixed

---

[30] Only higher-order fluctuation-related corrections to the energy depend on temperature, since the mean magnetization is always zero.

[31] The simple form of the partition function (7.32) makes it easy to sum directly with a computer, and in this sum fluctuations around the most likely configuration also make small contributions to the thermodynamic properties.

[32] Here $\sim$ denotes the linear-order dependence on $N$.

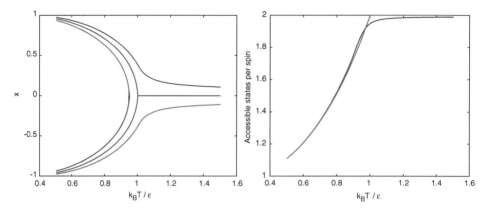

Figure 7.3 Left panel: magnetizations $x$ occur with non-vanishing probability in the region where the effective potential remains within $1/N$ of its minimum. The region between the blue and green boundaries corresponds to the non-red region in Figure 7.2. The stationary-point approximation $E(x) \approx \bar{x}$ is shown in red. Right panel: accessible states per spin defined as $\exp\{S/N\}$ in the partition function (7.30), evaluated by direct computation of the sum, versus $k_B T/\epsilon$. Blue is the exact evaluation; green is the stationary-point approximation.

value for $U$ and $S$. Only for $k_B T < \epsilon$ does a change in $T$ lead to a loss of $U$ and hence of $S$ as given by Eq. (7.39). A loss of $S$ linear in $N$ describes a loss of state-space volume exponential in $N$, for $T$ below the Curie point.

In the exactly computed partition function, average energy and entropy show a smoother transition, but still approach the $N$-independent and $N$-linear behaviors above and below the Curie point, respectively. In the ordered phase, the range for likely values of $x$ is a single interval, and stochastic trajectories for $x$ can drift anywhere within it in short times. Below the Curie point the allowed region for $x$ splits into two intervals separated by a region of very low probability. Stochastic trajectories in one or the other interval remain within that interval for a time that grows exponentially with $N$ relative to migration times within the interval.

Near the Curie point the fluctuations in $x$ briefly widen as two asymmetric intervals below separate from the one symmetric interval above. Even though this gives a growth in fluctuations $\sim N$, it must be remembered that this is only in *one dimension*. The total volume of accessible state space continues to decrease monotonically through the transition.

### 7.4.1.5 Domain flips and large-deviations scaling of fluctuations

The partition function (7.32) contains within it a distribution over all macroscopic configurations and not only the most likely one. Therefore we may use it to estimate the probabilities for fluctuations in the macrostate as well as the probabilities of particular microscopic configurations. The probability of a macrostate $x$, rather than being the probability of a particular microstate, is the weight of a distribution over all microstates for which $x$ is a typical value.

The evaluation of macrostate fluctuation probabilities follows the construction of the large-deviations function introduced qualitatively in Section 7.3. Readers interested in the technical methods used to extract leading exponential approximations to fluctuation probabilities are referred to [224, 767, 811]. Recall that large-deviations scaling arises if the leading exponential dependence of macroscopic fluctuation probabilities factors into a function of $x$ and a function of scale $N$. The dependence on $x$, which characterizes the "structure" of fluctuations (in this case, changes in magnetization) will then identify the rate function which is also the per particle entropy.

In the Curie–Weiss model the leading exponential dependence of the un-normalized distribution in the partition function (7.32) is given by $e^{-N\Phi(x)}$, in terms of the effective potential, and the leading exponential dependence of the partition function $Z$ is given by the minimizing value $e^{-N\Phi(\bar{x})}$, in terms of its minimum. Therefore, a normalized probability for the system to fluctuate to a macroscopic magnetization $x$ is given by

$$P(x) \sim e^{-N[\Phi(x)-\Phi(\bar{x})]}, \tag{7.40}$$

in which $\sim$ indicates that only the leading exponential dependence is kept.

Now we recall where the effective potential came from. In the stationary-point approximation, $-N\Phi = S - \beta U$. That is, the effective potential equals the entropy of the system minus $\beta$ times its internal energy. But $-U$, up to a constant, is the energy that the system has lost to an environment at temperature $T$, and $-\beta U$ is therefore the entropy it has made available in the environment. In other words, the effective potential (with a minus sign) is none other than a combined entropy of the system plus environment:

$$-N\Phi = S - \beta U$$
$$= S_{\text{tot}} \equiv S + S_{\text{Environment}}. \tag{7.41}$$

Equation (7.40) therefore shows the two most important properties of large-deviations scaling for fluctuations. First, the probability of a fluctuation to a macrostate $x$ is exponential in the difference between the entropy $S_{\text{tot}}(x)$ and the maximizing value $S_{\text{tot}}(\bar{x})$:

$$P(x) \sim e^{-[S_{\text{tot}}(\bar{x})-S_{\text{tot}}(x)]}. \tag{7.42}$$

Second, both of these entropy terms scale linearly in $N$, and their dependence on $x$ is through the effective potential $-\Phi = S_{\text{tot}}/N$, which does not depend on $N$.

A particularly important fluctuation is the one that causes the magnetization to escape from one stable state and go to the other. The least likely point along such a trajectory is the passage through $x = 0$. For the constant offsets we have chosen, $\Phi(0) = 0$, and the probability of a flip is given by

$$P_{\text{flip}} \sim P(0) \sim e^{-N[-\Phi(\bar{x})]}. \tag{7.43}$$

The probability to lose a stable macrostate decreases exponentially with the number of spins in the system, at a rate determined by the minimum value of the effective potential, attained at $\bar{x}$.

The two equations (7.40), (7.43) illustrate the most important consequences of the laws of large numbers for the stability and memory of macrostates in ordered thermal systems. We will see identical relations again in Section 7.5, where we review asymptotically optimal error correction as another application of the phase transition paradigm.

### *7.4.1.6 Second-order and first-order phase transitions*

The phase of a system entails many properties, including its internal organization and also its response to boundary conditions. The notion of phase transition is similarly general, describing changes in a system's response not only along single contours as some control parameter is changed, but over entire surfaces in which multiple internal and external control parameters may vary. Within these surfaces, the change in the degree of order may be continuous or discontinuous, and the order itself may spontaneously break a symmetry or it may simply alter the response of the system to symmetries that are already broken by boundary conditions. Characteristic classes of these behaviors are associated with the **order** of a phase transition, a convention introduced by Paul Ehrenfest [210] to refer to the lowest-order derivative of a response function that becomes discontinuous along the curve of some control parameter at a transition point. A single system that has multiple phases may undergo either second-order or first-order transitions along different contours in the space of control parameters.

The Curie–Weiss model is a model of a second-order phase transition, which forms order by breaking symmetry. The second-order phase transitions are those in which the order parameter passes continuously – though with discontinuous derivative – from the value zero reflecting disorder, to non-zero values reflecting order. Not all phase transitions are of this type. The **first-order phase transitions** exhibit *discontinuous* order parameters at the transition point. They do not generally spontaneously create new asymmetries, but they do generally produce strong non-linearity in the magnitudes of system asymmetries induced by asymmetric boundary conditions.

The approximation of the partition function $Z$ in Eq. (7.30) using the leading exponential dependence on $\Phi$ allows us to understand the relation between the effective potential and the **Helmholtz free energy.** Summing over the degeneracy of states at each total $n$ (hence, each value of $x$), but leaving $x$ as an argument of the effective potential $\Phi(x)$, allows $x$ to function as a macrostate variable, the fluctuations of which are governed by the sum of subsystem entropies. The summand $S(x)$ in the second line of the formula (7.32) giving $\Phi(x)$ measures the spin system's combinatorial entropy at any $x$, while the summand $-\beta H(x)$ measures the environment's entropy at that $x$. The total entropy is maximized at the condition $\beta = \partial S/\partial U = (dS/dx)/(dH/dx)$. If we wish to consider the aggregate system incorporating the spin system and heat bath, as a function of temperature $1/\beta$, then $\Phi$ is evaluated along the contour $x(\beta)$ of maximum entropy at each $\beta$, and the resulting function $N(\Phi - \log 2)$ becomes the (descaled) Helmholtz free energy $\beta F$.

The same interpretation of $\Phi$ can be carried through for general values of the external magnetic field $h$. When $x$ is evaluated where entropy is maximized as a function of both $T = 1/\beta$ and $h$, $\Phi$ has the interpretation of a **Gibbs free energy**.[33] Although not as rich as a full model of first-order phase transition, it will allow us to illustrate the kinds of discontinuities that characterize first-order transitions.

The potential $\Phi(x)$, varying both $T$ and $h$, is graphed in Figure 7.4. The minima of $\Phi$ give the "classical" values of the extensive variable $x$ in response to the system's intensive boundary conditions. At $h = 0$ they illustrate the pitchfork bifurcation of the second-order transition from Figure 7.2. For $T < 1$ and non-zero $h$ (right panel), the minimizer which we denote $\bar{x}(h)$ gives the response of the classical magnetization to the environmentally applied magnetic field. Above a critical field strength $h_c$ of either sign, $\bar{x}(h)$ is a single-valued function, but for $-h_c < h < h_c$, two local minima exist. If $h$ is made to oscillate with magnitude larger than $h_c$, the system shows hysteresis. Under a suitable oscillation protocol, it can be trapped in a local but not global minimum which is a metastable equilibrium, and for $N \to \infty$ it can persist there indefinitely.

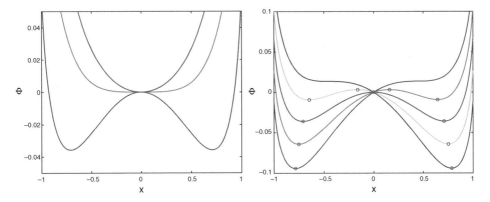

Figure 7.4 The effective potential $\Phi(x)$ interpreted as a large-deviations rate function or joint system/bath entropy as in Eq. (7.41). Left-hand panel: $h = 0$ and $T = \{0.8, 1.0, 1.2\}$, showing the transition through the critical point and symmetry breaking. Right-hand panel: $T = 0.8$ and $h = \{-1, -0.5, 0, 0.5, 1\} \times h_c$, where $h_c \approx 0.06225$ is the critical field strength above which $\Phi$ has a single minimum. Positions $d\Phi/dx = 0$ are marked with circles.

---

[33] Since the minimal Curie model has no spatial extent, there is no volume variable or conjugate environmental pressure. However, $h$ and the magnetization $x$ stand in the same relation as Legendre dual variables, so we will refer to the Legendre transform with respect to $h$ as a Gibbs free energy in this context.

At large but finite $N$, the system can undergo rare spontaneous domain flips when more than one equilibrium coexists. The escape process from a metastable equilibrium produced by a time-varying $h$ protocol is a simple model for nucleation, and creates a finite life-time for metastable states. The lifetime becomes very short for $h$ near $h_c$, smoothing the discontinuity of the hysteretic response to oscillating $h$. Rare spontaneous population of metastable states at small $h$ leads to long-time averages of $x$ that do not show hysteresis but only a strongly non-linear dependence on $h$. Over sufficiently long times, the ratio of the time spent in the stable versus metastable equilibrium diverges exponentially in $N$ for $h \neq 0$, so that as $N \to \infty$ the system asymptotically spends all of its time in the true equilibrium. However, at $h = 0$ any mixture in the interval $\left[ \bar{x}(0^-), \bar{x}(0^+) \right]$ provides an equally good approximation to the minimization condition of the free energy.

In a freely fluctuating system, the ratio of residence times in coexisting equilibria is controlled by the escape events which are exponentially rare in large $N$, and thus exponentially weaker than the curvature of $\Phi$ that governs the fluctuations about local minima. Thus any time-averaged magnetization in the interval $\left[ \bar{x}(0^-), \bar{x}(0^+) \right]$ could be maintained by an external perturbation exponentially weaker than the system's internal magnetization but stronger than the spontaneous domain flips. This decoupling of two variables $(x, h)$ which in "normal" phases are Legendre duals of each other, is the characteristic of the *coexistence regime* in a first-order phase transition. An example is the way volume – the Legendre dual to pressure in either a liquid or a gas – can be independently varied at fixed pressure to determine the proportion of liquid and gas at the coexistence pressure. Each of these features will appear in models more directly applicable to life, developed in Section 7.4.6.3 and Section 7.6.4.

### 7.4.2 Reduction, emergence, and prediction

The Curie–Weiss example illustrates the most important relations among reduction, emergence, and prediction. While phase transition does not define the only notion of "emergence" found in informal usage, it is by far the most thoroughly understood sense of the term, and it is the one that we believe has the most to offer to understanding the order of the biosphere. Most important, the example shows that reduction and emergence are not opposing notions: they are two complementary parts of the same description. The reductionist description includes all components and relations needed to construct the system. The emergent properties not predicted from the component view are those that reflect fundamental *degeneracies* in what can be constructed, which result from underlying *symmetries*.

In the ordered states created by phase transitions, chance and necessity are not alternatives. They are brought into existence together and in a sense cause each other. The necessary strength of order creates the framework of variables chosen by chance. In turn, the existence of the degeneracy that makes chance possible creates the "shape of the entropy function" that drives entropy out of the system as it falls into one or another of the ordered states.

### 7.4.2.1 The Curie–Weiss model is a reductionist description

The Curie ferromagnet is a "reductionist" description in the most thorough sense of the term used in physical sciences. Its physical laws are fully determined by the components and their interactions, and no new agency of causation comes into existence with the formation of the magnetized state. The *strength* of the order parameter is completely predictable from the microscopic properties of the system, even though its direction is unpredictable. This strength variable creates the background that also determines fluctuation properties. To bring these relations into everyday life, one may think of frost on a windowpane: the hardness of the ice is a function of temperature determined and computable from the microphysics. Only the orientation of the crystals (if they are on a uniform background) is undetermined.

### 7.4.2.2 The unpredictable variables have both limits and contexts

The emergence of a property (the orientation of the magnetization) that cannot be predicted from the reductionist description defines a very limited form of unpredictability, and one that is linked to other properties of the system. The orientation is only unpredictable because it reflects a hidden but still present underlying degeneracy. In an infinite magnet, this direction is eligible to become a "frozen accident" property of the system. In a finite system, it is a variable that may take on dynamics that become exponentially slower (in the system size) than the scale of microscopic fluctuations, but which are never removed. In the discussion of fluctuations in Section 7.4.1.5, for instance, we may think of samples being randomly drawn from the ensemble to represent random perturbation by thermal vibrations. The shifts of the distribution capable of producing flips in the direction of the magnetization, Eq. (7.43), will be exponentially rare in $N$ and hence domain-flip dynamics will be exponentially slower than the thermal sampling dynamics. Such frozen accident variables in discrete systems are thus a source of dynamics on timescales much slower than the microdynamics.

We have not illustrated an equilibrium example with continuous symmetry breaking here, but a property of such models is that the continuous degeneracy of possible solutions creates new accessible macroscopic excitations. We note an example of shear sound waves in crystals below in Section 7.4.4.4. A second example – noisy clocks – arises for driven dynamical systems that form limit cycles or steady currents.[34]

### 7.4.2.3 Susceptibility to weak forces and historical contingency

The phase transition paradigm gives a formal model of the combination of susceptibility to weak forces, with subsequent long-term persistence of the outcomes, that Stephen J. Gould popularized [311] under the name "historical contingency." In practice, only mathematical models have ideal symmetry and thus perfect degeneracy. Almost all physical systems are subject to interactions across vast ranges of scale from strong to weak. The "unpredictable" directions of the order parameter are in real cases usually set by small asymmetries in the

---

[34] For a worked example in the context of population biology, see [768], Chapter 6.

boundary conditions. For frost on a windowpane these may be slight impurities or dust. Such small effects are not sufficient to interfere with the reductionistically determined properties of order such as the strength of the ice matrix, and for the matrix as for the melted liquid state they would be un-noticeable. Only the degenerate properties can amplify such weak effects into large-scale macroscopic patterns.

In cases where the boundary conditions are dynamic instead of static, the most accessible movements associated with degeneracies are tickled by forces too weak to disrupt order. For instance, the energy in earthquakes (away from the fault) is far too weak to break crystal bonds, but it can propagate through the Earth in the shear waves supported by crust and mantle crystals.

### 7.4.3  Unpredictability and long-range order

The macrostate in a thermal system can only deviate from the symmetry of the underlying laws (or of the boundary conditions) as a result of mutually reinforcing *collective effects* of many small-scale degrees of freedom. Therefore the unpredictable part of an ordered state displays – indeed, it is maintained in the most essential way by – *long-range order* [307] of the micro-variables. From the perspective of statistical inference about samples from the ensemble, the order parameter is the leading estimator for *all* of the small-scale orientations, which describes how they are confined within subdomains of the possible state space.

#### 7.4.3.1  Ordered systems become "less thermodynamic"

Each phase transition that pushes a system further into some corner of its state space and reduces its entropy makes that system a little "less thermodynamic." Even though the fluctuations of the system have less entropy, slightly more about its average configuration is unpredictable or sensitive enough to weak boundary conditions that it is often most convenient to estimate it empirically. Perhaps the most novel aspect of the mathematics of life will be that its forms of long-range order become so numerous and govern so much of the kinetics that we can no longer deal with them as isolated special cases one by one. We must begin to develop a theory of movement within spaces structured by long-range order.

#### 7.4.3.2  Conditional independence given the order parameter

The order parameter is the leading estimator for shared order of many micro-variables because they can be shown to fluctuate about it *independently*. As shown in Box 7.4 for the Curie–Weiss example, the probability distributions over all spins become conditionally independent of each other, given the value of the magnetization, in the large-system limit.

---

**Box 7.4  Conditional independence in the Curie–Weiss model**

Two variables $s_1$ and $s_2$ are said to be conditionally independent of each other, given the value of a third variable $\mu$, under some probability distribution $P$, if

$$P(s_1, s_2 \mid \mu) = P(s_1 \mid \mu)\, P(s_2 \mid \mu). \tag{7.44}$$

**Box 7.4  Conditional independence in the Curie–Weiss model (cont.)**

The Curie–Weiss model shows how the magnetization, which is its only extensive state variable, makes the residual fluctuation probabilities for all spins conditionally independent in the large-system limit.

As in the text, let $s_i \in \{-1, 1\}$ label the collection of microscopic spin orientations. Let the conditioning variable $\mu$ be the magnetization. Although $\mu$ is the convenient macroscopic observable, we will write probability distributions as functions of $n = (\mu + N)/2 = |\{i \mid s_i = 1\}|$, the number of "up"-spins in the microscopic model, for convenience.

For any $n \in 0, \ldots, N$, the number of possible configurations is $\binom{N}{n}$. Of these, the number of configurations with some given spin pointed up, say $s_1 = 1$, is $\binom{N-1}{n-1}$. Therefore the probability

$$P(s_1 = 1 \mid n) = \left( \begin{array}{c} N - 1 \\ n - 1 \end{array} \right) / \left( \begin{array}{c} N \\ n \end{array} \right) = \frac{n}{N}. \tag{7.45}$$

A similar expression for the joint configuration $s_1 = 1$, $s_2 = 1$, gives

$$P(s_1 = 1, s_2 = 1 \mid n) = \left( \begin{array}{c} N - 2 \\ n - 2 \end{array} \right) / \left( \begin{array}{c} N \\ n \end{array} \right) = \frac{n(n-1)}{N(N-1)}$$

$$= P(s_1 = 1 \mid n) P(s_2 = 1 \mid n) \left\{ 1 + \mathcal{O}\left(\frac{1}{N}\right) \right\}. \tag{7.46}$$

The distributions for the two spins become independent of each other, given the magnetization, at large $N$, which we may write

$$P(s_1 = 1, s_2 = 1 \mid \mu) \rightarrow P(s_1 = 1 \mid \mu) P(s_2 = 1 \mid \mu). \tag{7.47}$$

Conditional independence holds if we are given the exact magnetization. In an ensemble defined by the temperature $T$, a small range of fluctuating $\mu$ will be admitted, and the factorization (7.46) will be modified by a weighted sum over $n$. Away from the critical point, these fluctuations remain $\mathcal{O}(1/N)$ and the scaling in Eq. (7.46) still applies. For any finite $N$, however, there is some sufficiently close neighborhood of the critical point that in this neighborhood the $\mathcal{O}(1/N)$ is multiplied by a large prefactor. In this range we cannot speak of the $s_1$ and $s_2$ being conditionally independent given the *average* magnetization $E(\mu)$ in the ensemble, which is the classical state variable. Such joint fluctuation right at the critical point reflects the role of collective effects in causing the phase transition.

This notion of a "best" conditioning variable for inference about a system is the correct statistical concept to put in place of what has sometimes been termed "downward causation"[35] in ordered states. Because the mean magnetization in the Curie–Weiss model determines the distribution for each individual spin, some descriptions seek to reify the magnetization as a new causal entity brought into existence by the formation of the ordered phase, which acts "downward" on the individual spins. The Curie–Weiss model shows that in all phases, no other cause ever exists for the magnetization of any spin than the collective influence of all other spins, microscopic and equal. The order parameter (which for the Curie–Weiss model happens to be simply the sum) is the function of all other spins that reflects all of their joint influence and nothing more.

Finally, note that in both the unmagnetized and magnetized states, *fluctuations about the mean value remain a property of the distribution*. The order parameter determines the mean to which the fluctuations regress, but it does not eliminate them. This concept will be of absolute importance to the description of chemical flows on the early Earth as a dynamically ordered state.

We will invoke the rTCA cycle as a candidate for the order parameter of these flows, but *we will not intend* that those flows resemble the pathways in modern organisms. In particular, we do not assert that only rTCA intermediates and reactions would have been present in the early chemical ensembles on Earth. Modern biochemical pathways have tightly controlled outputs because they have been refined by billions of years of enzymatic evolution. Networks on the early Earth, facilitated by crude and non-specific catalysts, could never have had this selectivity. We expect that they would have been clouds of connected side-reactions, much as the distribution of spin configurations makes a cloud around the mean magnetization. Nonetheless, it will be meaningful to speak of rTCA as a candidate for the unique source of order in that cloud, if the other reactions are conditionally independent of one another given the reactions in the main pathway.[36]

### 7.4.3.3 Limitations in the use of symmetry breaking

Symmetry breaking is important in nature as a source of dynamics across wide ranges of scale, and we expect that it will also be important to enable the hierarchical architecture of living systems, in which higher levels exercise memory and control functions over lower levels. However, we do not wish to overstate the role of symmetry as the sole concept behind phase transition. First-order phase transitions may be nowhere symmetric, yet they describe constitutive rearrangements in the states of systems. Whereas symmetry is a default characteristic in gases, liquids, or crystals, due to their large numbers of

---

[35] The general use of this term in studies of emergent properties appears to follow Roger Sperry's more particular application [774] to consciousness.

[36] To express the same idea in more explicitly information-theoretic terms, a modern chemical pathway, which contains high concentrations of desired reactants and sharp suppression of undesired side-reaction outputs, requires a very peaked distribution with many cumulants, in order to distinguish the enhanced from the suppressed reactions. We may expect that the price of maintaining these cumulants is a large cost in deaths through natural selection, as modeled by Matina Donaldson-Matasci, Carl Bergstrom, and Michael Lachmann [202]. In comparison, distributions on the early Earth would have been less peaked with heavier tails about the central molecules and reactions. Specified by fewer cumulants, these distributions would contain much less information than the pathways that we argue are their descendants.

copies of identical atoms, molecules, or bonds, heterogeneity becomes more common in the chemical systems that make up life.

A second way in which symmetry could be less important in biology than it has been in equilibrium thermodynamics is that the separation between phase transitions may be less sharp. Even in the ordinary physical thermodynamics of biomatter, hierarchies of numerous phase transitions nested closely in temperature or pressure are common, including sol/gel/liquid-crystal transitions in lipid films, or the notoriously complex behavior of emulsions and other "non-Newtonian" fluids [709].

Finally, one of our main uses of phase transition will be to account for the buffering between levels in a hierarchy that makes it possible for a simple controller to direct a controlled system that may have many more internal states than the controller can model, sense, or correct. However, unlike simple physical phase transitions, the phase transitions relevant to life generally leave still complex interfaces for control within hierarchies. Indeed, these are responsible for the common chicken-egg problems of reconstructing the origin of these systems. When a controller can "reach in" and alter the internal dynamics of a controlled system, this interaction may itself reduce the purity of the phase transition that creates the controlled system, or the role of symmetry in that transition.

### 7.4.4 The hierarchy of matter

#### 7.4.4.1 Mere rubbish: the origin of life and the origin of matter

> It is mere rubbish thinking, at present, of origin of life; one might as well think of origin of matter.
>
> *Letter to J. D. Hooker, 29 March 1863* [178][37]

We imagine it would have surprised Darwin how much *simpler* a problem the origin of matter has turned out to be than the origin of life. This might have been particularly surprising in Darwin's day, had it been possible to exhibit the rather indirect and abstract mathematics of quantum field theory that has eventually given rise to a conceptually consistent (and radical!) and yet quantitatively satisfactory theory of matter. How different the language of quantum fields seems, and how indirect from the human scale and immediate perception, than the phenomena and structures associated with life, in which we are immersed.

Yet the origin of matter has proved a solvable problem, while the origin of life is still open, in part for reasons that are straightforward to explain and in retrospect not surprising at all. The hierarchy of phase transitions that has produced the inventory of low-temperature states of matter from the original universal vacuum has been a hierarchy of equilibrium transitions. The states of matter are, by and large,[38] defined by local properties

---

[37] Page 18 of [178], scanned as Darwin Online.
[38] The confinement transition [446] creating hadrons and mesons has a spatial component, as do the vortex states in superconductors [78], and some phenomena such as twinning in crystals.

of the vacuum. The transitions are also, in most cases, separated by large increments of energy or temperature scale. As a result the contributions to the structure of matter have been easily *reducible*: that is, we can study in accelerator laboratories the same phenomena that shaped the universe as a whole. Therefore, while the phenomena of matter themselves sometimes occur at small sizes and high energies that require planning and specialized machinery to reach, they have certain properties that from a mathematical standpoint make them tractable even with manual calculations. It is not accidental that the same phenomena which make the phase transitions in elementary matter tractable to analyze also make the transitions themselves highly regular with few contingencies.

Many, and perhaps most, of these mathematical simplifications are likely to be absent from the driven chemical order of life. That extra freedom makes the emergence of the biosphere a phenomenon in a larger, more difficult space to search, than the hierarchy of matter, even though the constituents of living matter probably require no finer resolution than single atoms to describe. Because they are non-equilibrium phenomena, living states of order can have spatially non-local and historical dependencies [72, 73]. Multiple transitions may occur in close neighborhoods of scale. Life also arises out of condensed matter phases that offer very large starting complexity, even prior to their incorporation into living order. For all these reasons, the phenomena of life are more difficult to handle with reductionist methods; it is more difficult to study the components in isolation without changing the way they work to such an extent that one fails to learn about their contributions *in situ*. The same interdependencies that make components of life difficult to study separately can also lead to dense networks of conflicting pressures in the dynamics of living systems themselves, creating more opportunities for variation, and a new role for historical contingency not only in arbitrary details of structure, but in fundamental dimensions of function and variation.

In this section we will propose that the success in understanding the hierarchy of matter, even though it is on the whole the simpler of the two problems, may still teach us useful lessons for understanding the transitions that led to the emergence of life. The details will be different. The particular symmetries will be different. Perhaps even the organizing role of symmetry will be different, though we are not quick to jump to that conclusion. However, the basic notion that progressive long-range order accretes, with later building blocks being made from cooperative effects among previously formed building blocks, is such a general principle of the laws of large numbers, that we strongly expect that some version of it is the right thing to look for in biogenesis.

### 7.4.4.2 The frameworks for our current theory of matter

Before undertaking a more detailed introduction to the kinds of non-equilibrium phase transitions that would plausibly apply to the emergence of life, it may be helpful to review the way in which our entire understanding of states of matter has been recast in terms of a hierarchy of *equilibrium* phase transitions. The phase transitions that formed the elementary particles and states of condensed matter illustrate many roles of symmetry that are richer than any that the Curie–Weiss model provides. The hierarchy of matter also illustrates the progression from simplicity and high symmetry in a primordial high-temperature vacuum to the remarkable complexity of chemical, mineral, and ordered-electron states at low temperatures. This progression serves as a loose analogy or model to the progression that we propose led from the modestly organized bulk organosynthesis of geochemistry, through primordial metabolism, and ultimately to the multilayered complexity of cellular life and complex ecological communities.

Two closely related mathematical disciplines – statistical mechanics and quantum field theory – were together the source of our current theory of the states and the nature of matter. They share many concepts and methods, and mathematical structures of one can often be continuously transformed into those of the other. The modern understanding of critical phenomena and ordered phases has grown from concepts traded back and forth between the quantum field theory of the vacuum and the statistical mechanics of condensed matter.[39] We argue next that the physical configuration space in which the states of matter form, and the physical and chemical space in which the biosphere exists, have many parallels.

### 7.4.4.3 Configuration spaces with product structure

The most basic point needed to appreciate the modern theory of matter – and also to see parallels in chemical organization – is the way the relation of matter to the vacuum, and the nature of different kinds of matter, are understood in terms of **internal symmetries**. They are the generalization of the simple up/down reflection symmetry of spins in the Curie–Weiss model. We also add the ingredient of physical space, which may or may not be important in particular cases. All spins in the Curie–Weiss model were equivalent and equally coupled, so if one wished to assign them a spatial description, it would be necessary to say they exist "at a single point" in space. For some forms of physical order, and potentially for many kinds of biological order, the existence of an extended spacetime with dynamic degrees of freedom at each point may be important to the kinds of order that can form.

Both the vacuum and any matter in it have the topology of *physical space*, which we may describe with three spatial coordinates and a time coordinate. At every four-dimensional point in that spacetime, however, the vacuum also has a collection of other "internal" coordinates, referred to as *matter fields*, which determine its local state. The full configuration space thus has the structure of a *product* of the spatial and internal degrees of freedom. The important property of these internal variables is that, in unordered phases, they are

---

[39] Important early contributions to the understanding of universal behavior were the work of Lev Landau on phase transitions in equilibrium statistical mechanics [468] and of Steven Weinberg on the concept of *effective field theories* [861] for the quantum field theory of the vacuum. Following a generation of perplexity about the role of the laws of large numbers in the field known as *renormalization*, to which Richard Feynman was a key contributor for the theory of electrodynamics [252], the modern understanding of renormalization began with a paper by Murray Gell-Mann and Francis Low in 1954 [290]. Gell-Mann's student, Kenneth Wilson, taking these ideas further, turned renormalization into both a conceptually coherent discipline and a widely understood and used tool, using thermal states and phase transitions in condensed matter as applications [885]. Joseph Polchinski in turn applied Wilson's description systematically to vacuum quantum field theory [643], emphasizing particularly that this step explained why effective field theories of finite complexity could provide both correct and ultimately unique descriptions of natural systems with infinitely many small-scale forms of interaction.

equivalent physically but still represent distinct states. Their equivalence is expressed as a *symmetry*: all physical properties of the vacuum and matter would be unchanged if the internal coordinates at all points in space were jointly rotated under the action of a certain continuous group.[40] The entities that we recognize as *particles* are spatiotemporally non-uniform ripples in the internal matter-field variables, in which the internal state is not rotated by the same group element at all points.[41]

The important point for our discussion of phase transitions and life is that, in the hierarchy of transitions that form matter, all freezing leads to order in the *internal* fields, producing a state that is *uniform* in spacetime. Not until far down in this sequence – the phase transitions in the strong interaction that produce nucleons and related particles – are phase transitions found in which the internal order is coupled with non-uniform order in space.

We emphasize this point now because later, when we discuss the mathematics of non-equilibrium phase transitions, it will be convenient to illustrate it with models such as dielectric breakdown (electric sparking) or crack propagation, where the order forms in space. Similar examples have been popular as analogues for biological order, including the use of the Belousov–Zhabotinsky reaction or the theoretical "Brusselator" [451] which form spatial/chemical patterns when seeded with uniform inputs. These examples capture part of the relevant mathematics, but in all of them, spatial or even spatiotemporal non-uniformity is essential to the existence of order. The ordering of internal variables in the hierarchy of matter, which in most cases is uniform in space, is a better analogue to the bulk chemical order that we propose for the first stages of biological emergence.

### 7.4.4.4 The freezing hierarchy of matter

The concept of freezing has become the foundation to understand the hierarchical organization of all states of matter, in the following way. At very high temperatures (equivalent to high energy scales) the group of internal symmetries at each point in spacetime is highly dimensional. As temperatures drop – in the early universe this was a result of expansion of the whole universe – the energy density available to excite fluctuations in these many-dimensional matter fields drops, and *a subset* of the dimensions freeze into an ordered state. The remaining, unfrozen variables of the matter fields remain symmetric, but under a new, lower-dimensional symmetry group. The process repeats as temperatures drop further, and more internal fluctuations of the matter fields freeze out to states of order. After several such transitions, at temperatures comparable to those in stars, on planets, and in space today, freezing transitions occur that involve spatial order rather than only order in the internal fields; these are the transitions that create nucleons, nuclei, atoms, molecules, and condensed matter.

---

[40] The continuous groups that form the basis for the theory of elementary particles are the *Lie groups*, named for Sophus Lie.

[41] Here we are oversimplifying the treatment of this topic by failing to explain what it means for the fields to be jointly rotated by the same amount. A proper rendering would take us into the domain of continuous local symmetries known as *gauge symmetries*, which is beyond the scope of this treatment. It is possible that the geometry of driven chemical reactions has parallels to this more complex structure, but we must leave that as an open question.

In each freezing transition, if the frozen state is degenerate, new low-energy degrees of freedom are produced, which reflect the hidden degeneracy. These often serve as the elementary particles that communicate forces, which in turn lead to later phase transitions lower in the energy hierarchy.

We cannot digress to give even a descriptive review of the many phase transitions that are now well understood, and which give a predictive, quantitative account of *all* of the known properties of the elementary particles and their interactions.[42] However, it is helpful to briefly list several major stages in the hierarchy of matter, so that the reader can see how the phase transitions we propose for the origin of life are meant to fit within this much larger context.

1. Above about $10^{25}$ eV or $10^{29}$ K, we have no direct experimental constraints on the symmetry of matter fields, so this is where the current frontier of the unknown lies. Somewhere below this scale, one or more freezing transitions took place that created a symmetry group which is a product of three *special unitary groups*, written as $SU(3) \otimes SU(2) \otimes U(1)$,[43] which are responsible for the properties of known matter. The $SU(3)$ component describes the *strong force* of quantum chromodynamics (QCD),[44] and the $SU(2) \otimes U(1)$ factor described a force termed the *electro-weak* force.

2. At roughly $246 \times 10^9$ eV or $3 \times 10^{15}$ K, a new freezing transition occurred within the electro-weak force, which separated the three heavy particles of the weak force (recall Figure 3.1) from the electromagnetic force.[45]

3. At about $10^9$ eV or $10^{13}$ K, a pair of interconnected phase transitions – this time involving freezing of the internal fields of $SU(3)$ in configurations that also have spatial order – create the nucleons (proton and neutron) and a suite of related heavier particles that are readily produced in accelerators, stars, supernovae, and high-energy collisions in space. An important, relatively low-mass particle created by the degeneracy of the frozen state is the $\pi$-*meson*, which is chiefly responsible for communicating the strong force as we see it in nuclear matter at low energies.

4. A classical *condensation* transition in ordinary thermal matter at $10^8$ eV or $10^{12}$ K (the energy scale equivalent to the mass of the $\pi$-meson) leads to the formation of nuclei from protons and neutrons and the creation of the elements. In a plasma of nucleons and electrons that is still too hot to form atoms, the elements cannot yet express the properties of chemistry, but the die for these properties is now cast in the charges of the nuclei.

5. In a classical condensation transition at 10 eV or $10^5$ K, electrons and nuclei from plasma condense to form neutrally charged atoms. The properties of chemistry are first brought into existence with this transition, and it makes atoms distinguishable by their

---

[42] A readable introduction is provided in [152].

[43] Each factor is something like a spherical rotation group in respectively 8, 3, and 1 dimension. See [293] for definitions.

[44] At very high temperatures, however, it was not yet very strong.

[45] The recent identification of the first measured scattering events mediated by *Higgs bosons* [150] – named after Peter Higgs who was one of the inventors of the mechanism that froze the electro-weak force – further confirms the consistency of this model with new observations. In essence, the experiment delivered energy by colliding particles, in a sufficiently small region of space to *melt* a bit of the frozen vacuum and restore the matter and force fields in their unfrozen form.

spectra when they interact with light. The passage back and forth between plasma and neutral atoms is part of the normal structure of the surfaces of stars, and even occurs briefly during lightning strikes.

6. Further condensation of neutral atoms forms molecules at temperatures at or below the energy of most covalent bonds, of order 0.5 eV or 5000 K. The enormous combinatorial complexity of chemical bonding states is now brought into existence.

7. Physical condensation that breaks the symmetry of spatial motions creates condensed matter from atoms or molecules. Starting from atoms or molecules in gas phase, either liquids or solids may form. Liquids result from only freezing out symmetries of *location*, while solids result from freezing out symmetries of both location and *orientation*. The very complex problem of putting chemical bonds into regular arrangements both in space and in orientation leads to the explosion of diversity encountered in the mineral world, as we discussed in Section 3.3.2.4. The new low-energy excitations that express the hidden degeneracy of the frozen states are *sound waves*. Liquids have only one kind of sound (pressure waves), while solids have two (pressure and shear), reflecting the greater number of fluctuations that freeze out to form them.

8. Many kinds of condensed matter have electronic states in which either charges are mobile, or their spin orientations can fluctuate. A variety of further phase transitions can occur in such systems that remain within the scaffold of their solid condensed matter state, but which generate new electrical and magnetic phases. These include conducting, semiconducting, and magnetic states, and they may be accompanied by a variety of new kinds of low-energy excitations. The Curie–Weiss transition to a magnetized state, from Section 7.4.1, is of this kind. These phase transitions occur at the temperatures of metallurgy or common laboratory refrigeration, and they are a familiar part of engineering and everyday life.

9. The last phase transition we note is one that commonly occurs at tens or hundreds of degrees Kelvin. Pairs of electrons, interacting through any of a variety of kinds of sound waves created by earlier freezing transitions, can condense into special coherent states called *Cooper pairs* in the Bardeen–Cooper–Schrieffer mechanism of superconductivity.[46] In this freezing transition, the photon (the last of the surviving massless elementary particles) undergoes a transition much like the one experienced by the weak force at a temperature that was 14 orders of magnitude higher. Electromagnetism below this scale becomes a short-ranged force within the superconductor, much as the weak interaction became a short-ranged force in the vacuum of spacetime. By this scale, *all* of the symmetry in the original vacuum of the universe has been hidden by freezing into one or another ordered form, and the expressed structure of matter is entirely carried by composite entities.

Many more exotic phase transitions are understood, which contribute to the menagerie of states of matter, but we will stop our list here. An important relation among the freezing

---

[46] For an introduction, see [806].

transitions in the hierarchy of matter is that the forces that enable each new transition are mediated by ordered states brought into existence by earlier transitions. This is the same kind of relation as one sees among *building blocks* in the assembly of complex ordered systems. Small building blocks are assembled in a few particular combinations to make blocks of intermediate size, and these then become the components that are combined to form the next level of structure. Each frozen state of order is, in a sense, a composite built from the elements that were free at higher energies, and it becomes a component in larger composites upon freezing to lower energies.

### 7.4.4.5 Ceilings and floors from phase transitions

The hierarchy of matter illustrates an exceedingly important property of phase transitions, without which the universe might have been a web too tangled ever to be understood. As each phase transition occurs, it brings into existence the states of matter and the elementary excitations that will describe matter everywhere below, until the next transition freezes some of them into a new background. The state of the vacuum, or of condensed matter, does not continuously change character as it cools, because the symmetries available to hide are finite in number. Therefore the phase transitions are discrete events that alter finite subsets of symmetries at particular temperatures.

Every phenomenon in equilibrium matter takes place at some energy scale, and therefore falls between some pair of phase transitions. The transition above brought into existence the excitations available to the phenomenon; the transition below will freeze out a subset of them to create an even more complex ordered state. The transition above is (in energy scale) a **ceiling**, and the transition below is a **floor**. Remarkably, to estimate the leading-order behavior of any phenomenon, we *do not need to know the physics above the ceiling or below the floor*. That is, the excitations that were accessible before freezing happened no longer matter once the frozen excitations exist and are known. Likewise, the long-range structures that will form at the next freezing do not affect the current degrees of freedom while they exist as free excitations.

### 7.4.4.6 Historical perspective: reconceptualizing the nature of matter

The gradual elucidation of the phase transition paradigm, over the course of almost a century, brought about a radical change in what we understand matter to be. It altered the way we ask questions about a fundamental theory of matter, not by arriving at the "true ground layer" which had been the original goal, but rather by showing that this goal was not necessary and perhaps was not even well conceived.

It is fair to say that before the modern phase hierarchy, science had no uniform theory of what matter was. Probably from the time of Democritus and Leucippus to the early part of the twentieth century (though in widely differing forms), some entities have been understood to be composites of other, more "fundamental" components. The fundamental components, however, were always a menagerie, sometimes sharing some physical properties such as electric charge, but in substance fundamentally non-comparable, and simply *named.* The phase transition paradigm replaced the menagerie with a uniform representation of all matter and forces as ripples in matter fields that exist at each point in space and transform under various *representations of particular symmetry groups.*[47]

In each era where some entities were understood to be composite, at some level their components were viewed as "atomic" in the original Greek sense of "indivisible." Each era was brought to an end as its atoms were found to be composites of a new set of components. This situation is unsatisfying in two respects. First, for some entities science is a description of relations while for others it is a mere list of names and properties. Second, each generation's atoms were likely to be divided in the next generation. Physics wanted to know what, if anything, was the "bottom layer" of matter.

The phase transition theory of ceilings and floors has shown that falsifiable theories can exist at intermediate levels, whether or not we know what the bottom layer is, and *whether or not a bottom layer even exists.* A consistent, empirical, falsifiable theory can exist of relations-within-relations, and can expand out from our current middle scale of size and temperature in both directions. It does not matter for the consistency of science whether the expansion ever ends.

The idea that one can do practical theory at middle scales without knowledge of the extremes sounds pedestrian enough, but it is emphatically *not* something which can be taken for granted. For a brief interval around the 1960s and 1970s, the apparent violation of this premise led to a period of crisis for quantum field theory. Models with only finitely many parameters had been successfully used by Enrico Fermi to model weak interactions as early as the 1930s [243]. As later generations tried to include corrections from quantum fluctuations to Fermi's calculations and others of similar type, it appeared that an infinite hierarchy of interactions would always be generated, in which matter at all scales of space and energy would contribute arbitrarily severe fluctuation effects to any observation at any scale. The attempt to organize calculations to tame these effects was the program called **renormalization**.

Properly understood, however, renormalization was a much worse theoretical problem than the taming of uncontrolled fluctuation effects to preserve the simplicity of theories. The correct understanding of such fluctuation terms was that *they should have been part of every theory from the beginning* [643, 861]. Every field theory had, in principle, an infinite hierarchy of types of interactions with an unknown parameter for each. In the finite theories of Fermi and later generations, infinitely many of these parameters had been simply ignored, and so set to zero for no reason. How could it have been that a theory with infinitely many assumptions, each almost surely wrong, could correctly predict the outcomes of experiments?

The explanation of fluctuation effects in terms of a hierarchy of ceilings and floors – through which one did not need to see – by Gell-Mann [290], Wilson [885], Weinberg [861], Polchinski [643] and others, gave a principled understanding of why finite

---

[47] It remains a rather astonishing property of nature that the symmetries required so far have all been low-dimensional finite (usually) simple groups, or (still finitely generated) continuous groups such as displacement in space.

theories could correctly predict nature. Steven Weinberg pointed out that all theories should be understood as **effective theories**, each with an infinite set of unknown parameters. For the so-called *renormalizable* theories, errors even in infinitely many of these parameters could be absorbed by finite shifts in the remaining parameters, as long as the correct finite set of parameters were kept. The parameters of the finite theories could therefore be set empirically at the scales of observation, and could then make accurate predictions for neighboring scales. The theory of gravity, which does not have such protective ceilings, is still not understood as an extension of theories of matter.

### 7.4.4.7 The formation of the planet

The mineralogical history of the Earth, briefly reviewed in Chapter 3, mirrors in microcosm the hierarchy of matter. The separation of core and mantle, the partitioning of iron-magnesium from calcium-aluminum silicates, the formation and later accumulation of continental rafts, and the subduction and regeneration of ocean-bottom basalts, are all structured around phases of matter with very different properties, which emerge in response to boundary conditions that, on cosmic scales, all fall within a narrow range of energy. The incredible complexity of continental crust, formed by repeated diverse forms of volcanism, surface weathering, and collision, creates a puzzle that geologists looking down from the surface continually work to unravel. Unlike the hierarchy of matter, in which phase transitions are cleanly separated by energy scale and nested according to their symmetries, the phase transitions in the mineral history of volcanic, tectonic planets leave a huge diversity of stable forms chock-a-block in most surface rocks.

### 7.4.5 Parallels in matter and life: the product space of chemical reactions

To explain the parallel between the hierarchy of matter and the dynamical hierarchy of living order, we contrast the product-space of elementary particles, with its internal matter fields at each point in spacetime, with the state space in which biochemistry occurs, drawn schematically in Figure 7.5. Biochemistry exists in space and time, and like matter is indexed by their four coordinates. It also has internal degrees of freedom, because at every point in physical space, a copy exists of the network of all possible chemical species and all possible reactions. This network takes the place of the field variables in the theory of matter. Most of the network is unoccupied by actual molecules in realized chemical systems, so it is to be understood as a network of *potential* configurations.

In the hierarchy of matter, phase transitions occur as the vacuum is cooled. As internal energy is removed from a system, the range of fluctuations in the matter fields that can be accessed is reduced. Lowering temperature and removing internal energy is one way in which the distribution of fluctuations is *constrained* by its boundaries, as was shown in the Curie–Weiss example of Section 7.4.1. A freezing transition permits the space of fluctuations everywhere to be greatly reduced all at once, as the system is crowded into a corner of its state space.

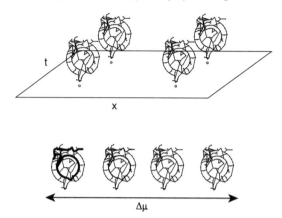

Figure 7.5 Upper panel: the product structure of chemical reaction space. $(x, t)$ stand for coordinates in space and time. The small graphs are the potential reactions of all chemicals, which can occur at each point. Interior nodes on the graphs are connected at neighboring points by diffusion, since chemicals are transported without change through space. Lower panel: phase transitions in the chemical graph are created not by lowering temperature, but by increasing the difference in chemical potential between a source compound and a sink compound. Here we illustrate the onset of autocatalysis through rTCA with a drain through the lipid pathway as the dark line, which becomes more intense as the driving chemical potential is increased.

For driven chemical reactions in Figure 7.5, instead of cooling, the environment increases the inequality between the chemical potentials for some species (such as electron pairs) between two or more nodes in the reaction graph. An environment that can independently impose different chemical potentials at different nodes in the reaction graph can always place the system under *at least as much constraint* as an environment that imposes the same chemical potential at all nodes. The resulting driven system must then always have a smaller accessible configuration space than the corresponding equilibrium system.

Our proposed explanation for emerging biological order, illustrated in the figure, is that the response of geochemistry to this tightening constraint was a phase transition in the chemical network that concentrated matter and energy flow in neighborhoods of a few self-amplifying pathways (heavy black lines). In modern life, lipid/water and protein-aggregate phase separation, and other kinds of spatial organization, play important roles in maintaining the many forms of order in cells. On the early Earth, spontaneous spatial arrangement may not have been as important on the microscale as it is in cells. In addition, some spatial order may have been pre-existing in rock structure.

Although we wish to stress the many parallels between the phase transitions that have produced such a thorough theory of matter, and the raw material from which we may try to build a phase transition understanding of life, the two hierarchies will differ in certain fundamental ways. The most important differences in chemical order will come from the fact that the "internal" degrees of freedom live on the discrete topology of a chemical reaction network,[48] rather than in a simple continuous group manifold, and the fact that the formed order is dynamical rather than static. Both of these differences are aspects of the different roles played by symmetry.

### 7.4.6 Oil and water

The near-equilibrium phase transitions in the lipid–water system are the processes most widely described in terms of error buffering that simplifies the problem of forming living matter. Lipid–water "self-organization" is recognized to limit excursions of molecules away from a central tendency toward order, without active regulation. Life exploits this low-level buffering at many levels of living structure, from chemical synthesis, to compartmentalization, catalysis, and bioenergetics, to mediating the folding, placement, and segregation of oligomers. Within lipid phases, other condensed-state transitions such as liquid/gel/liquid-crystal transitions alter physical properties that are important to the containment functions of cell membranes and the environment they provide for embedded proteins. Oil–water transitions may also have affected the path of biogenesis because structures created by near-equilibrium phase thermodynamics may mediate feedbacks that create dynamical phase transitions, enhancing lipid supply rates on which the equilibrium phases depend.

The array of phase transitions in the lipid–water system, even in chemically purified and homogeneous systems, is remarkably complex. Added to that complexity, in the ascending hierarchy of aggregation and developmental complexity from bacteria and archaea through protists to multicellular organisms, increasing scale brings increasing molecular complexity, as lipid phases progress from incorporating 100 or fewer distinct molecules to as many as thousands. Increasing molecular complexity can smooth transitions and blend properties from multiple phases, creating membrane states that are finely tuned, functionally diversified, and robust under environmental perturbations.

Chemically homogeneous lipid–water coacervates, ranging from soaps (salts of fatty acids) to complex gylcerol-linked biological molecules such as lecithins, have received extensive study as prototypes for the first cells. Their formation, permeability, chemical environments, and division have been characterized and modeled. Robert Hazen and David Deamer [184, 344] have extracted chemically heterogeneous lipid components from the Murchison meteorite and shown that these will spontaneously form vesicles through phase separation. Recent decades have seen many experiments on containment of RNA and other monomers within vesicles to study the relation of polymerization with vesicularization, and the use of coacervates as complex catalytic environments for chemical syntheses of industrial interest is also an active area [404].

Here we review only a few properties of the equilibrium phase system of oils in water, as instructive examples. We return in Section 7.6.3 below to consider ways in which the physical and chemical properties of separated amphiphile systems feed back on chemical synthesis to enable inherently non-equilibrium transitions in dynamical trajectories.

---

[48] The mathematical representation of reaction networks that reflects the stoichiometry of reactions is a *hypergraph* [64].

### *7.4.6.1 Energy scales and phase characteristics*

The main features of lipid–water mixtures are controlled by chemical bonds at two scales.

**Hydrogen bonds and domain separation** Phase separation is possible because lipids (whether simple or complex) are amphiphiles, consisting of a polar head group covalently bonded to one or more long hydrocarbon tails. In pure water, hydrogen bonds with characteristic energies 4–13 kJ/mol (0.04–0.14 eV) form extensive transient networks. Polar head groups can participate in hydrogen bonding and so are soluble, but non-polar hydrocarbon chains cannot participate and therefore disrupt the bond network. The minimum free energy solution for non-polar solutes is phase separation which minimizes the disrupted volume in the water bond network. Amphiphile solutions differ from simple oil–water mixtures because the polar head groups form an interface with water, and the hydrocarbon chains are sequestered to form a quasi-bulk non-polar domain. Although the hydrocarbon chains are anchored at the polar head group, and although the resulting non-polar domain only has a depth less than roughly twice the chain length, many bulk hydrocarbon properties are recovered within this domain, which are well approximated by properties of bulk polyethylene [585].

**van der Waals bonds and interior chain orientation** Space-filling by hydrocarbon chains, sequestered from water by the polar head groups, is mediated by van der Waals forces with characteristic scales of 2–4 kJ/mol (0.02–0.04 eV) per bond between any two $CH_2$ groups in contact, from the same or neighboring chains. Additional interaction energies on adjacent carbons serve to raise rotated orientations known as *gauche* configurations above the straight-chain *trans* orientation by a similar factor $\epsilon \approx 0.5$ kcal/mol [585].[49]

A variety of secondary and case-specific interactions may determine other structures. These include ionic interactions among charged head groups or head group/counter-ion interactions, geometric factors from head group/lipid body relative areas, and excluded-volume effects from chain-length differences (leaving chain tails at the mid-plane of the bilayer that cannot take an all-trans conformation) or unsaturated bonds in the hydrocarbon chains.

Lipid concentration and geometry further govern the phase diagram for forms of macroscopic order. Distinctive alternations of both low-concentration and high-concentration phases exist.

**Low-concentration phases** In aqueous solution at low lipid density the three dominant domains are solubilized lipids, micelles, and lamellar sheets/vesicles [505].[50] The micelle/lamellar distinction is most strongly controlled by amphiphile geometry,

---

[49] For polyethylene, which is mediated entirely by van der Waals forces acting collectively, the bulk melting temperature $T_M \approx 138\,°C = 411$ K corresponding to 3.3 kJ/mol (0.035 eV). Hence the thermal activation energy must be approximately 1.6 times the trans-gauche activation energy and in the upper range of van der Waals bond energies to melt a saturated hydrocarbon lattice.

[50] Many details of the trade-off between surface curvature and edge effects, in static or dynamic settings, determine whether vesicles or laminae are the steady-state configurations. The determination rapidly shifts to fine interactions of equilibrium and non-equilibrium ensembles, and unfortunately is outside the scope of this brief summary and examples.

with roughly cylindrical amphiphiles such as phosphatidyl cholines (PCs) favoring lamellar (double-layered sheet) geometries, and inverse conical geometries such as lysophosphatidyl cholines favoring micelles [757]. Conical lipids such as phosphatidyl ethanolamine with a small head group favor inverse micelles [585], but this geometry is not generally available in the dilute lipid regime.

**High-concentration phase geometries** At higher concentrations of lipid in water, connected lipid phases as well as connected water phases become available. Surface interaction through the water phase can become important, and a variety of new macroscopic phases related to surface geometry and interactions become possible. Inverse micellar as well as micellar phases can exist with close-packing in space. Hexagonal arrays of indefinitely extended lipid rods or water-filled voids can exist, as well as layered lamellar stacks [505]. Finite-length rods terminating in "vertices" with smoothly interpolated surfaces may create inter-networked connected domains of both water and lipid that extend indefinitely. In such geometries, finer distinctions of the average convexity favored by the head group/lipid body radius relations can be accommodated. Correspondingly, the free energy differences per atom that govern the favored arrangement of these coarse-grained "objects" is weaker than the primarily molecular interactions that govern domain separation and internal chain alignment.

### 7.4.6.2 Internal lipid phases and phase transitions

Within a lipid–water coacervate separated by the primary hydrogen bond driven interface formation, three secondary phases arise from van der Waals interactions and excluded-volume effects among the lipid hydrocarbon chains. These are generally referred to as liquid, gel, and liquid-crystalline phases [585]. The gel/liquid-crystal transition has received the most detailed empirical characterization through NMR and X-ray diffraction studies, and is the subject of the most extensive theoretical modeling.

The separation of scales between the layer-forming hydrogen bond energy in water, and the weaker van der Waals forces between adjacent hydrocarbon chains, is revealed by the fact that simple orientation models ignoring most conformational constraints produce reasonable predictions for the dependence of melting temperature on chain length. Over the range of biologically interesting vesicle-forming phosphatidyl cholines from di-lauroyl-($C_{12}$) to di-stearoyl-($C_{18}$) PCs, melting temperature increases roughly linearly with chain length [575], eventually saturating for $C_{\infty}$ chains near the $138\,°C$ melting temperature of polyethylene in bulk phase [585]. For phosphatidyl cholines, with their two chains, melting temperature is maximized where the embedded chain lengths are roughly equal [130].

### 7.4.6.3 Nature of the gel-to-liquid crystal phase transition

The primary order of internal hydrocarbon chains in a bilayer is distinguished as liquid-crystalline ($c$), gel-like lamellar ($\beta$), or liquid ($\alpha$) [505]. In both the $c$ and $\beta$ phases the chains are fully extended, but they may be oriented either along the normal to the layer face (designations $c$ and $\beta$), or at an oblique angle which therefore breaks a circular rotational symmetry (designated $c'$ and $\beta'$). Oblique orientation may permit more complete alignment of multiple chains in a PC bilayer, but it increases the area per polar head group and the degree of hydration, and so may be associated with an increased surface free energy.[51]

In the crystalline phases, chains are microscopically aligned and linked by multiple van der Waals bonds along each maximally extended chain. A limited amount of rotation of the carbon backbone away from the all-trans form by gauche bends, shown in Figure 7.6, gives rise to the lamellar gel phase. Because gauche bends are forbidden by excluded-volume constraints in a crystalline all-trans background, their formation requires loosening of the lipid matrix relative to the crystalline state and breaking of long-range alignment of hydrocarbon chains. If $\theta$ denotes the angle between the main lipid backbone and the normal to the plane of a single $CH_2$ group, the second Legendre polynomial

$$q \equiv \frac{1}{2}\left(3\left\langle \cos^2\theta \right\rangle - 1\right),\tag{7.48}$$

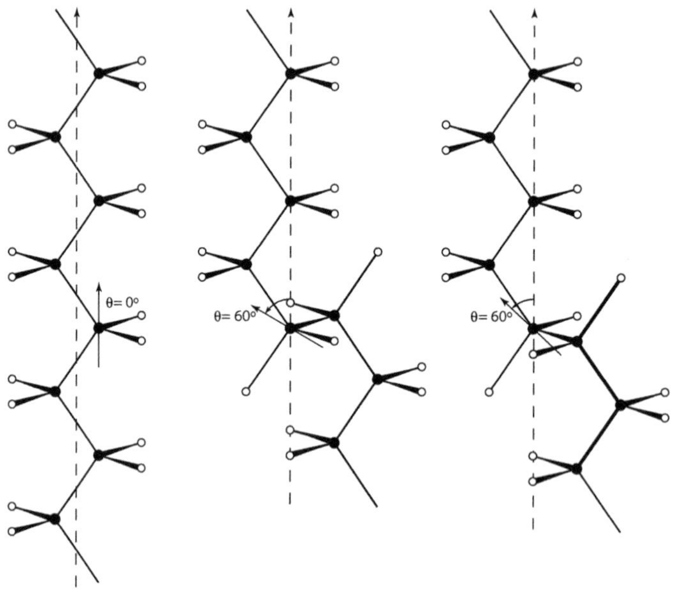

Figure 7.6 A segment of a saturated hydrocarbon chain showing conformations. An orientation angle $\theta$ is defined between the normal to the $CH_2$ plane and the orientation axis of the maximally extended lipid. Left panel is the maximally extended all-*trans* conformation. Center and right panels are the two *gauche* bends obtained by rotating one C–C bond by 120° about the axis of adjacent C–C bonds. Both these rotations, because they are about an off-vertical axis, result in a 60° orientation for the two $CH_2$ groups at the ends of the rotated bond. Filled dots are carbon atoms, open circles are hydrogen. Chain termination is not shown.

[51] An interesting rippled gel phase, designated $P_{\beta'}$ jointly breaks the rotational symmetry in both the lipid orientation and the surface geometry [505]. A ripple with wave vector along the direction of oblique tilt in the $\beta'$ phase permits alternating wave faces to contain normally oriented lipids, and maximally oblique lipids, thus partitioning surface and interior free energies.

averaged over all $CH_2$ groups, is typically used as an **order parameter** $q$ for the liquid-crystal to gel transition. The definition of $\theta$ and its values in various chain conformations are shown in Figure 7.6.

In a strictly ordered phase $q = 1$, while in a fully disordered (liquid) phase $q = 0$. Because more orientational disorder within backbones decreases their projection along the primary axis, more melted states produce thinner bilayers with more area per head group, approximately conserving the volume per amphiphile measured as bilayer thickness times head group area.

Polynomial expansions in $q$ with empirical coefficients have been used to make symmetry-based **Landau models** of the free energy of lipids near the liquid-crystal/gel phase boundary. An example is the form

$$G = G_0 \left[ \frac{1}{4} \left( \frac{q_c}{q} - 1 \right)^4 + \frac{\alpha}{2} (T - T_c) \left( \frac{q_c}{q} - 1 \right)^2 + \alpha (T_M - T) \left( \frac{q_c}{q} - 1 \right) \right] \quad (7.49)$$

given as Eq. (6) in Morrow *et al.* [575]. Here $q_c$ is the (empirically estimated) order parameter at the critical transition, and $T_c$ and $T_M$ are (empirically estimated) temperatures respectively at an inferred critical point and at the melting point.[52] The expansion (7.49) is identical in character to the polynomial approximation for the effective potential illustrated for the Curie–Weiss transition in Section 7.4.1.3, except that temperature governs both the linear and quadratic terms, rather than having each coupled to an independent boundary condition. The parameters at which models of the form (7.49) provide good fits to many characters in both phases indicate that the transition is weakly first order: the system has a latent heat of freezing into the liquid-crystalline phase and a coexistence region with mixed values of $q$ as a function of bilayer thickness or surface area.

A critical temperature $T_c > T_M$ is needed in the model (7.49) to produce the observed discontinuity in the orientational order at the melting temperature $T_M$. Although at $T = T_M$ the system has a phase with $\partial G / \partial q = 0$ at $q = q_c$, this phase is a local maximizer of free energy. It is therefore unstable to decay into a mixture of gel and liquid-crystal phases. For $T > T_M$ the gel phase is the global minimizer of $G$, while for $T < T_M$ the liquid-crystalline phase is the global minimizer. This behavior is similar to the dependence of $\Phi$ on $h$ for the Curie model, shown in Figure 7.2. However, the Landau model (7.49) has only one control parameter ($T$), so the system never passes through the critical point.

### 7.4.6.4 Uses in biology

The first question that arises from simplified phase transition models is whether these reflect physical properties in real cell membranes. Bacteria and archaea, which use relatively simple lipid inventories, are indeed found to retain some of the discrete phase transition character of laboratory lipid mixtures. Melchior *et al.* [551] have shown that

---

[52] In Eq. (7.49) we use the result in [575] that the coefficients estimated for the linear and quadratic terms are approximately equal. In the original model they were independently fitted to data.

both the plasma membranes and viable cells of *Mycoplasma laidlawii* show a cusp in the specific heat at a temperature in agreement with the lipid melting temperature, and moreover that cooling and melting transitions leave organisms viable. The melting temperature appears to be maintained $\sim5\,°C$ below the optimal growth temperature for organisms grown in different settings, and also increases or decreases with the length of the fatty acid chains in the medium on which the bacteria are grown.[53] The suggestion of this work is that bacteria may have mechanisms to tune the melting temperature to provide the required degree of membrane fluidity, though they cannot entirely remove the discrete character of the transition.

For post-oxygenic organisms, the character of membrane fluidity is more complex, and for eukaryotes in particular, the gel/liquid-crystal transition is converted into nearly a continuum. Eukaryotes synthesize sterols, which appear to partly compartmentalize lipid hydrocarbon chains and interrupt long-range order in chain conformation. Their membranes are more fluid than pure lipid below its characteristic temperature, and less fluid above.[54] Eukaryotes also synthesize a wider variety of lipid chains, evidently to provide a more flexible fit to membrane-embedded proteins while protecting the impermeability of the membrane envelope to protons. The bilayer thickness in different cellular compartments has been shown to control the inventories of proteins that embed in each membrane class.

Andrew Pohorille and colleagues [641, 642] have shown that the close proximity of polar and non-polar environments provided by amphiphile surface layers is a natural environment in which to have stabilized the first **peptide secondary structures.** $\alpha$-helices, which alone are not stable in either water or oil, may be stabilized at a membrane surface by alternating hydrophobic and hydrophilic residues periodically. Embedding of the hydrophobic residues with the hydrophilic residues projecting into water stabilizes the helix in addition to its internal hydrogen bonds. Such embedded-helix secondary structure provides a natural route to the formation of channels and pores from multiple helices, which may jointly turn to span the membrane with their hydrophobic faces in contact or surrounding a core of hydrogen-bonded water molecules.

### 7.5 The (large-deviations!) theory of asymptotically reliable error correction

In this section we review the theory of asymptotically robust error correction due to Shannon. The most important observations are that: (1) such a theory *exists*, (2) this is a *large-deviations* theory, and (3) identification and removal of all errors with high probability occurs by a process mathematically identical to regression toward a thermal ordered state. Message systems capable of conveying multiple messages have ordered states of the

---

[53] *Mycoplasma*, a highly degenerated obligate intracellular pathogen, does not synthesize fatty acids and so must take them from the host.

[54] It is noteworthy that, of the roughly 30 steps in cholesterol biosynthesis, all reactions beyond squalene require molecular oxygen. In eukaryotes, where fatty acid desaturation is performed in the endoplasmic reticulum, this step also requires $O_2$. Hence the complexity of regulation of membrane properties in eukaryotes is heavily dependent on molecular oxygen along with the greater cell complexity it enables.

kind produced by phase transitions, and imperfect error correction due to finite correlation length is equivalent to the drift of thermal ordered phases in finite systems.

We will focus on Shannon's formulation of this problem – and within that, on the systematic development of a few of the simplest models as examples – because it allows us to cover with minimal technical overhead the whole range of topics from the optimal use of redundancy in finite systems to the attainment of asymptotically zero error rates. A parallel literature on the control of error through *architectural* redundancy exists, which descends from von Neumann's seminal paper on **probabilistic logics** [838]. It exhibits many of the same mathematical phenomena, including the interplay between "executive" and "restoration" stages (see Section 7.5.7 on repeaters in transmission lines), and an asymptotically exponential dependence of the error probability on a measure of the redundancy scale (a large-deviations limit). In some ways this approach addresses the problems of "hardware" implementation more directly, but the minimal models needed to exhibit these features are a larger topic unto themselves. We expect that isomorphisms exist between many problems of reliable computation and reliable communication, under which the "positions" for signals in a communication are mapped to components that carry signals in a logical device, but we do not try to develop the mappings in this treatment.

### 7.5.1 Information theory as a mirror on thermodynamics

Claude Shannon's information theory was a pure formulation [732] of the combinatorial aspects of the laws of statistical inference. Shannon's entropy was Gibbs' entropy, arising from the combinatorics of symbol strings the same way it arose for Gibbs from the combinatorics of particle counts (see Box 7.5). The two papers of Jaynes in 1957 [391, 392] laid out the map between the constraining roles of state variables and the structure of the inference problem implied by them, making clear how the thermodynamic entropy can be understood as an application of information entropy to systems whose internal states are incompletely determined by their boundary conditions, and shifting the discussion of thermodynamics away from an emphasis on mechanics and toward a more general framing in terms of information and control.

---

**Box 7.5    The combinatorial basis of Gibbs–Shannon entropy**

The use of the Gibbs–Shannon entropy as a measure of information may be justified in many ways.[a] Because our interest here is in large-deviations scaling and the behavior of fluctuations, it is sufficient to use the original justification of Gibbs, based on the simple counting of microstates.

Suppose a system has $N$ (distinguishable) objects, which may take any one of $K$ states. $N$ may be a number of particles and $K$ the states available to a single particle, or $N$ may be a number of symbols in a sequence, and $K$ the size of the alphabet.

[a]    For a systematic review, the reader is directed to Jaynes [394].

---

**Box 7.5    The combinatorial basis of Gibbs–Shannon entropy (cont.)**

If, in some ensemble, every assignment of objects to states is equally probable, the frequency of samples that have $n_i$ objects in state $i$, for $i \in 1, \ldots, K$ and $\sum_{i=1}^{K} n_i = N$, is

$$P(n_1, \ldots, n_K) = \frac{1}{K^N} \frac{N!}{n_1! \ldots n_K!}.$$

The reason the laws of large numbers cause *scale* to separate from *structure* is that, from Stirling's formula for the logarithm, the multinomial coefficient converges on the approximation

$$\log\left(\frac{N!}{n_1! \ldots n_K!}\right) \approx -N \sum_{i=1}^{K} \frac{n_i}{N} \log \frac{n_i}{N} \equiv N h\left(\frac{n_1}{N} \ldots \frac{n_K}{N}\right).$$

$h\left(\frac{n_1}{N} \ldots \frac{n_K}{N}\right)$ is the Gibbs–Shannon entropy function[b] of the frequencies $(n_1/N, \ldots, n_K/N)$, which sum to unity.

Therefore, ensembles at different $N$ produce samples that can be grouped by the same *fractional* occupations $(n_1/N, \ldots, n_K/N)$. At leading exponential order, the frequency of such fluctuations scales as

$$P\left(\frac{n_1}{N}, \ldots, \frac{n_K}{N}\right) \sim e^{-N[\log K - h(n_1/N, \ldots, n_K/N)]}.$$

$\log K$ is the maximum of $h\left(\frac{n_1}{N} \ldots \frac{n_K}{N}\right)$, so the log-frequency of macroscopic configuration $(n_1/N, \ldots, n_K/N)$ is, at leading order, $-N$ times the difference of the entropy of the macrostate from its maximum.

---

[b]    Here we follow the notation in Cover and Thomas [164].

---

### 7.5.2   The large-deviations theory of optimal error correction

In 1949 Shannon proved [731] the remarkable result that with suitable encoding, signals could be sent at a *non-zero rate* through a transmission line, and recovered by a decoder at the other end with an asymptotically zero probability of error, even if the transmission line introduced errors to the signal continuously. The surprising feature of this result is that an expansion of the encoded form of the signal by only a finite factor to add redundancy is sufficient to compensate for corrupting noise as long as the noise is sufficiently uncorrelated and not too strong. The catch to the theorems is that reliable transmission is only attained in a suitable infinite-time limit, so that the redundancy may be used over arbitrarily long stretches of time. Thus the relative redundancy per bit of message information is finite, but the absolute redundancy in the whole signal must grow as the signal length grows, in order to use this redundancy optimally.

These theorems and numerous extensions, along with signal encodings that approximate the bounding transmission rate with low error, are the basis for modern error-correction technology in communication and data storage. As important as the maximum attainable message rate in the asymptotic limit is the performance of coding systems that can be decoded in *finite time* at the cost of a small probability of error. Both results will be of interest to us, as models to understand the apparently asymptotically stable character of the living state maintained by the biosphere as a whole, but the apparently finite duration of most *particular* entities or states of order in which life is carried.

The Shannon theorems for the information capacity of a noisy transmission channel, and the performance of optimal error-correcting codes, have connections at two levels to results familiar from equilibrium thermodynamics. These show the equivalence of robust error correction to regression toward thermal ordered phases.

1. The capacity and coding theorems are thermodynamic results in all important mathematical respects, without making reference to mechanics or physical conservation laws. The channel capacity is a property of an infinite-waiting-time limit equivalent to the infinite-system-size limit of classical thermodynamics; the code reliability for finite-duration transmissions is a fluctuation result equivalent to the macroscopic fluctuation theorems for finite equilibrium systems.
2. Simple models of optimal error correction have direct maps to thermal relaxation processes including first-order phase transition. These serve to show in explicit terms both that robust error correction is a process within the phase transition paradigm, and that thermal relaxation can be used to perform error correction.

The capacity and performance theorems for uncorrelated noisy channels are helpful tutorial examples because they map directly onto familiar results from *equilibrium* thermal systems. However, since the channel models themselves are not equilibrium models (and need not even be understood as mechanical models), the most important point to take from the existence of a mapping is that the interchangeable description of error correction and thermal relaxation applies to any processes with similar large-deviations structure.

### 7.5.3 Transmission channel models

The fundamental object in most basic models of error correction is a **transmission channel** [164]. The channel accepts inputs from a transmitter and then alters them in some way that "corrupts" the transmitted signal. A receiver with a decoder must identify and remove the corruptions to identify the transmitted signal. Since a transmitter of messages must be free to send any one of several signals, the decoder must be able to identify corruption without knowing beforehand what was the intended signal.

In the original applications to telephony, the channel was a physical telephone line that carried electrical voltage/current signals across space with a time delay. The only property captured mathematically in channel models, however, is *mixing* of noise onto the

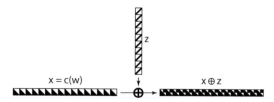

Figure 7.7 Mixing of signal and noise. In a discrete channel, a set of bits (left-hand boxes) are given the values $x$ of a codeword $c(w)$ for a transmitted word $w$. Codewords contain some redundancy among the bits (not all codewords are sent), indicated by half-shading. A noise source $z$ (vertical boxes) corrupts bits of $x$, indicated by the operation $\oplus$. Not all bits of $x$ may be corrupted, so in a maximally compressed representation, the number of boxes of $z$ is fewer than that of $x$. Mixing may be as simple as addition modulo two for binary digits. The output $x \oplus z$ is delivered to a decoder. Because both $x$ and $z$ contribute variation to the received bits, a larger set of sequences may be received than the number of codewords sent. If the code uses the maximum reliable rate, the received bits have maximum entropy for the codeword length, indicated by heavy shading.

transmitted signal,[55] illustrated in Figure 7.7. Therefore channel models serve equally well as representations of persistence through time with progressive corruption of ordered states by fluctuations.

### 7.5.3.1 Correlation length and transmission rate

The fundamental assumption allowing asymptotically reliable error correction is that over some timescale, corrupting noise is *uncorrelated*.[56] The correlation length of the noise defines a fundamental time unit, above which transmitted signals may be chosen freely by the transmitter. Successive freely chosen signals are therefore represented as letters from some (discrete or continuous) "alphabet," and transmitted signals consist of sequences of letters drawn from this alphabet. The most efficient use of an alphabet is one that places no constraints of correlation on the sender in its choice of messages other than those required to overcome channel noise, and possible constraints from the alphabet size. Unconstrained transmission of this kind is *maximum entropy rate* transmission, and we will discuss only this case.[57]

### 7.5.3.2 Block encoding for error correction

Robust error correction is possible because transmitted messages can be assembled from letters into *words*, and the signals that represent them can likewise be grouped into blocks

---

[55] "Mixing" is a standard term from electrical engineering, which can refer to a wide range of computational algorithms that combine signal and noise. Many mixing algorithms are either real addition or the threshold-generalization of this to non-linear systems as flipping of binary digits. We will use only these simple cases in examples.

[56] We work here only with the simplest case in which additive noise is literally uncorrelated in time. More complex cases with correlated noise have received extensive treatment [831]. Their common feature is that noise correlations can be modeled and used to compress a correlated noise stream to an alternative representation in which the compressed signals are uncorrelated.

[57] The theory that relates signal encodings for which the sent messages do not have maximum entropy rate, to maximum entropy rate encodings, is the theory of *data compression*. It too is a thermodynamic theory, and much of what we describe here for error correction has dual results in the theory of compression.

called *codewords*. When block encoding is used for error correction, the number of possible codewords is made larger than the number of distinct message-words that are to be sent. Therefore many possible codewords are never used, and the words that are used can be chosen to carry redundant information about the letters that make them up. The number of codeword letters among which correlation is used to impose redundancy defines a measure of the *correlation length* of the encoding system relative to the reference correlation length given by the channel noise.

The ratio of the number of message-words actually transmitted, to the number of possible codewords that could be transmitted in the same time, defines the *rate* of the coding system. The rate is measured logarithmically, so that for messages with no internal correlation constraints, the number of possible messages grows exponentially with transmission time, as a stationary stochastic process. In examples, we will denote by $D$ the number of letters used in a codeword, so that $D$ is the correlation length of the code. If $n$ is the number of message-words encoded in a code of length $D$, the rate $\mathcal{R}$ of the code is defined as[58]

$$n \equiv e^{D\mathcal{R}}. \tag{7.50}$$

A worked example of a channel with continuous Gaussian-distributed message signals and noise, and optimal codes at all finite correlation lengths $D$, is given in Box 7.6.

### 7.5.4 *Capacity and error probability*

The following is the capacity theorem proved by Shannon. Any channel with uncorrelated noise has a maximal rate of reliable message transmission. This rate is known as the *channel capacity*, and we will denote it by $\mathcal{C}$. The channel capacity is the maximum, over all possible encodings, of the mutual information between the sent messages, and the received messages that they are converted into by channel noise.[59]

Moreover, message transmission at rates $\mathcal{R} < \mathcal{C}$ may be made asymptotically reliable. Perfect reliability,[60] however, is only attained in the limit of infinite correlation length $D \to \infty$ at any rate $\mathcal{R} > 0$.

In the same manner as we have approached the classical limit of thermodynamics by first studying fluctuations in finite systems, we state the capacity theorem as a limit of *non-zero* probabilities of error for codes with finite correlation length. For a code with block length $D$ and rate $\mathcal{R}$, the probability that the decoder will mistakenly assign the received signal to

---

[58] We will use natural logarithms because in general we make no assumptions about whether the alphabet is discrete or continuous, or what is its size. In cases where a binary alphabet is assumed, it can be convenient to replace Eq. (7.50) with its binary equivalent, to write

$$n \equiv 2^{D\mathcal{R}_{(2)}}.$$

[59] With apologies, we must refer the interested reader to Cover and Thomas [164] for a systematic treatment of the full suite of relevant concepts from information theory.

[60] The correct expression is "almost-sure" correct decoding, meaning that the set of errors approaches measure zero asymptotically.

a codeword other than the one transmitted is bounded, at leading exponential order, by the formula

$$P_{\text{error}} \sim e^{-D(\mathcal{C}-\mathcal{R})}. \tag{7.51}$$

Box 7.6 gives geometric arguments for why these specific terms enter.

**Large-deviations scaling and asymptotic limit** Equation (7.51) for coding reliability is a large-deviations result equivalent to Eq. (7.40) for the Curie–Weiss model. It is a general form for combinatorial models, as shown in Box 7.5. The correlation length $D$ is the scale factor for system size (here, the set of codes is the "system"), and $(\mathcal{C} - \mathcal{R})$ is the rate function of the large-deviations scaling. As in equilibrium thermodynamics, the rate function is a difference between two entropies, $\mathcal{C}$ and $\mathcal{R}$. The capacity $\mathcal{C}$ is a maximum (per symbol) entropy for reliable codes, corresponding to the maximum specific entropy for most likely classical states. Smaller per symbol entropies $\mathcal{R}$ characterize "looser" codes that sacrifice efficiency by including fewer codewords.[61] In channel models, if $D$ is made into an explicit correlation *time*, both $\mathcal{C}$ and $\mathcal{R}$ become *temporal* entropy rates for stationary stochastic processes. The ability to map an extended-time process into an ensemble description of independently transmitted code blocks, which is afforded by the finite correlation length of the noise, permits a map from signal transmission to equilibrium thermal ensembles.

### 7.5.4.1 Signal/noise ratio and channel capacity

As Box 7.6 illustrating the Gaussian transmission channel shows, the capacity of a channel with continuous-valued signals and Gaussian noise depends on a competition between the signal power and the noise power. Each of these "power" constraints, in telephony, is a limit on the variance of either the signal or the noise strength, which is the continuous-signal version of an "alphabet size." The signal/noise power ratio, denoted $P/N$, sets the limits within which codewords in a code may be placed far enough apart that excursions due to noise do not cause one to be mistaken for another at the decoder. For the Gaussian channel that sends a scalar signal (e.g. voltage), the expression for capacity is

$$\mathcal{C} = \frac{1}{2}\log\left(1 + \frac{P}{N}\right). \tag{7.52}$$

---

**Box 7.6   Optimal coding for the Gaussian channel**

The Gaussian channel is a minimal model in which to study optimal error correction for signals which are continuous valued, but satisfy mean-square constraints on signal level. In telephony, signal level has the interpretation of a voltage, and the squared

---

[61] We may imagine that, in a system where reliable transmission was a "function" in the biological sense, selected for maximum transmission rate, competition would drive systems toward codes in which the rate approaches the capacity, much as fluctuations drive thermal systems toward the most probable (hence maximum-entropy) attainable states.

**Box 7.6   Optimal coding for the Gaussian channel (cont.)**

level is signal power. Many other interpretations also exist. Signal and noise values in the Gaussian channel are all sampled independently from Gaussian distributions because they are the maximum-entropy distributions constrained by signal or noise variance.

A channel is simply an input/output device that mixes inputs from a *transmitter* and a *noise source*, to obtain an output signal delivered to a *receiver*. All three are modeled as random variables denoted by capital letters. The input signals from the transmitter are denoted $X$, the noise $Z$, and the outputs, which result from addition of inputs and noise, are denoted $Y = X + Z$.

Finite noise correlation length in the Gaussian channel is modeled by dividing time into a discrete sequence of intervals indexed with integers $i$, and sending one signal $X_i$ in each interval. The corresponding noise values $Z_i$, are independently drawn from a Gaussian density with mean zero and variance $N$. Written explicitly, the probability for assignments $z$ to the random variable $Z_i$ satisfy

$$P(z \le Z_i < z + dz) = \frac{1}{\sqrt{2\pi N}} e^{-z^2/2N} dz. \tag{7.53}$$

A shorthand for this sampling relation is

$$Z_i \sim \mathcal{N}(0, N), \tag{7.54}$$

in which $\mathcal{N}(0, N)$ is called the *normal* distribution with mean 0 and variance $N$.

Although this model of the Gaussian channel is discrete in time, the signal levels are continuous, so we define the concept of redundancy in a signal in another manner than counting "bits" as would be done for signals with binary alphabets. We introduce a *code block* consisting of $D$ successive intervals of time, and assign *codewords* to different messages as sequences $X_1, \ldots, X_D$, the values of which may be correlated within a code block. Since noise is autocorrelated within single intervals, $D$ gives the correlation length of a code. For any number $n$ of distinct codewords, writing $n = e^{DR}$ defines $\mathcal{R}$ as the *rate* of the code. We wish to study the reliability of optimal codes that have the same rate $\mathcal{R}$ but different correlation lengths $D$.

If $x_i$ is the value assigned to a symbol $X_i$ in a given codeword, we represent a constraint of variance over the code block by the constraint

$$\sum_{i=1}^{D} x_i^2 \le DP. \tag{7.55}$$

$P$ is the signal power per symbol, which competes with the noise power $N$ to determine the rate at which messages can be transmitted reliably.

**Box 7.6    Optimal coding for the Gaussian channel (cont.)**

From a famous argument first recognized by Shannon [731], we can estimate the upper bound on the rates of coding, which is called the *channel capacity* and denoted $\mathcal{C}$. The range of likely outputs generated from a single input by an uncorrelated noise sequence $Z_1, \ldots, Z_D$ is a $D$-dimensional sphere of radius $DN$; that is

$$\sum_{i=1}^{D} z_i^2 \lesssim DN. \tag{7.56}$$

For uncorrelated signal and noise values, the total range of possible outputs for a power-constrained input block is

$$\sum_{i=1}^{D} y_i^2 \lesssim D\,(N+P). \tag{7.57}$$

The most reliable decoder we can implement for uncorrelated signals and noise is one that divides the $X$-sphere of radius $\sqrt{DP}$ into equally spaced cells, and for each received signal $Y_1, \ldots, Y_D$, maps the received signal to the nearest codeword $w$ by Euclidean distance $\sum_{i=1}^{D} \left( y_i - x_i^{(w)} \right)^2$, where $x_1^{(w)}, \ldots, x_D^{(w)}$ are the levels in the code block for word $w$. The likelihood that the wrong codeword will be selected may be made arbitrarily small if the distances between code blocks for any two words $w$, $w'$, satisfy

$$\sum_{i=1}^{D} \left( x_i^{(w')} - x_i^{(w)} \right)^2 > DN. \tag{7.58}$$

The codewords therefore correspond to a packing of spheres of radius $\sqrt{DN}$ into the $Y$-sphere of possible output values of radius $\sqrt{D\,(N+P)}$. The number of distinct codewords under such a packing scales as

$$\left( \frac{D\,(N+P)}{DN} \right)^{D/2} = \left( 1 + \frac{P}{N} \right)^{D/2} \geq e^{D\mathcal{R}}. \tag{7.59}$$

The channel capacity is the upper bound on the rate, and is therefore given by

$$\frac{1}{2} \log \left( 1 + \frac{P}{N} \right) \equiv \mathcal{C}. \tag{7.60}$$

An illustration of sphere-packing in a sequence of increasing dimensions $D = 1, 2, 3, \ldots$ is shown in Figure 7.8.

Now we may show that, for a sequence of codes at increasing $D$ but *fixed rate*, we may reduce the probability of error asymptotically to zero as long as the rate is

**Box 7.6   Optimal coding for the Gaussian channel (cont.)**

less than the channel capacity. We may also estimate the form and rate of approach to the asymptote at finite $D$.

An asymptotically reliable code is obtained by spacing the codewords a distance $\sqrt{D(N+\epsilon)}$ apart for some positive $\epsilon$, which does not depend on $D$. The increased spacing permits fewer distinct codewords, so the rate of this more reliable code is given by

$$\mathcal{R}_\epsilon \equiv \frac{1}{2}\log\left(\frac{N+P}{N+\epsilon}\right) \approx \mathcal{C} - \frac{\epsilon}{2N},\tag{7.61}$$

where we suppose $\epsilon/N \ll 1$ in taking the $\epsilon$-linear approximation.

Now we may estimate the reliability with which a signal can be sent if its rate is below the channel capacity. The probability of a mis-decode is bounded above by the probability that $z$ falls outside the radius $\sqrt{D(N+\epsilon)}$ from the transmitted codeword (since the mis-decode happens only if some other codeword is closer to $Y$ than the transmitted word):

$$P\left(z_1^2 + \cdots + z_D^2 > D(N+\epsilon)\right) \approx \frac{1}{\Gamma(D/2)}\int_{\frac{D}{2}\left(1+\frac{\epsilon}{N}\right)}^{\infty} du\, u^{(D/2-1)}e^{-u}$$

$$\sim e^{-D\epsilon/2N} = e^{-D(\mathcal{C}-\mathcal{R}_\epsilon)}.\tag{7.62}$$

(In Eq. (7.62) we have reduced the $D$-dimensional Gaussian integral to the $\Gamma$-function form in the descaled radial variable $u \equiv (1/2N)\sum_{i=1}^{D} z_i^2$. The symbol $\sim$ indicates the leading exponential dependence on $\epsilon$ up to constant prefactors of order unity that become independent of $D$ at large $D$.)[a] The probability of a mis-decode converges exponentially to zero with scale factor $D$ and rate function $\mathcal{C} - \mathcal{R}_\epsilon$.

---

[a] Cover and Thomas [164] give a more formal argument that permits almost-all good codes to be generated, letter-by-letter, as independent samples $X_i^{(w)} \sim \mathcal{N}(0, P)$. An upper bound for the probability that a noise-corrupted signal will decode to a different word than the transmitted signal is given by the ratio of the noise-sphere volume to the received signal space, given in terms of their entropy functions by

$$\frac{e^{Dh(Z)}}{e^{Dh(Y)}} = e^{-D[h(Y)-h(Z)]} = e^{-D[h(Y)-h(Z|X)]} = e^{-D[h(Y)-h(X+Z|X)]}$$

$$= e^{-D[h(Y)+h(X)-h(XY)]} = e^{-D\,I(X;Y)}.$$

Here $h(X)$ is the Shannon entropy function of the random variable $X$, and $I(X; Y)$ is the mutual information between two random variables $X$ and $Y$.

From this expression we gain a direct graphical picture of the derivation for the channel capacity. The original received signal results from a fixed but arbitrary $X$; thus $X$ does not appear in the expression for $h(Z)$. The probability of error measures the fraction of events in which $X + Z$ falls within the entire volume of received signal space from randomly generated values $Y$. This value is bounded above by $e^{-D\,I(X;Y)}$.

Since the number of possible codewords that could create such an overlap error is $e^{D\mathcal{R}}$, the upper bound for the error is

$$e^{D\mathcal{R}}e^{-DI(X;Y)} = e^{-D(\mathcal{C}-\mathcal{R})}.$$

The number $e^{D\mathcal{R}}$ of independent other codewords that can therefore be assigned while retaining an asymptotically zero error rate is bounded by $\mathcal{R} \leq \mathcal{C}$.

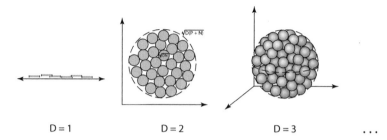

$$\sqrt{D(P+N)}$$

$$\sqrt{DN}$$

D = 1     D = 2     D = 3     . . .

Figure 7.8 Non-overlapping codewords for the Gaussian channel at increasing block size. Dimension $D$ indicates the number of voltage signals in a code "block." Axes are voltage levels at each of the $D$ sample times that define the block. Centers of the gray intervals, disks, or spheres are positions of codewords used by the transmitter. Gray regions indicate the range of signals likely to be received at the decoder after corruption with noise through the transmission channel. The codewords are chosen so that these "noise balls" form a random close-pack of spheres within the region of available power (the interval in $D = 1$, or the circular or spherical region within the thin dashed line at $D = 2$ or $D = 3$).

### 7.5.5 Error correction and molecular recognition in an energetic system

Thomas Schneider has applied the Gaussian model to the problem of optimal molecular recognition [705, 706], an application in which $P$ and $N$ are not *power*, but rather *energy* constraints. He therefore maps the Gaussian model not only mathematically, but *literally*, to a thermal relaxation system.

The problem addressed in the Schneider model is how to optimally use the position and momentum coordinates in $D$ non-covalent associations (e.g. hydrogen bonds) between a protein transcription factor and a bound sequence of DNA, to discriminate a target DNA sequence. Here the number of "letters" in a code block is $2D$ (one position and one momentum coordinate for each bond), $N = k_B T/2$ is the thermal noise per position or momentum degree of freedom, and $P$ is a "priming enthalpy"[62] given to the transcription factor before it is released. In the model, the transcription factor in solution is supposed to attach to the DNA at a random point, and then migrate along the DNA, incrementally converting the priming enthalpy to heat as it finds nearer sequences, until it settles on a single sequence with no priming enthalpy remaining and no ability to migrate further. This process is illustrated in Figure 7.9.

The random attachment point defines an ensemble of sequences that the transcription factor may initially encounter. The optimal-recognition question is then what is the largest entropy of initial sequences that can be completely eliminated by selecting a single target sequence reliably, for a given priming enthalpy per bond. The answer to this question from the Shannon theorems is that the optimal sequence entropy per bond is given by[63]

---

[62] One could imagine, for example, mechanically deforming a protein which would subsequently relax while moving along the target DNA strand in search of the sequence to which it is matched.

[63] The reason this expression is greater by a factor of 2 than Eq. (7.52) is that Eq. (7.52) describes a single degree of freedom.

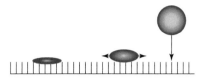

Figure 7.9 The Schneider model for optimal molecular recognition. A "primed" transcription factor with stored enthalpy $P$ (red circle) goes out of solution to attach at a random point on an unpaired DNA sequence (comb). Because the attachment point is random, the possible sequences are all those to which the bonding sites of the transcription factor could attach. The protein migrates along the DNA, and as it finds more favorable sites, it releases some enthalpy (flattens in the figure) to limit migration to fewer sites. When it has released all enthalpy (flat ellipse) it can no longer migrate. Schneider asks, among how large a set of initial attachment points can the protein correctly identify a single target sequence, as a function of the priming enthalpy $P$.

$$\mathcal{C}_{\text{bond}} = \log\left(1 + \frac{2P}{k_{\text{B}}T}\right). \tag{7.63}$$

### 7.5.5.1 Removing "signal" versus removing noise

A difference between Shannon's Gaussian channel and the Schneider recognition model is that, for the Gaussian channel, the decoder's function is to identify and remove the channel noise (of power $N$), and to recover the transmitted signal diversity (of power $P$). In the Schneider model, the noise cannot be removed because it is a property of the finite-temperature bath. Rather, it is the entropy of the initial sequences limited by priming energy $P$ that is to be removed by relaxation. (See Box 7.7.)

---

**Box 7.7   Error-correcting codes and molecular recognition**

The relation between the Shannon capacity model of reliable signal transmission, and the Schneider model of reliable molecular recognition, may be more easily seen by using a formal model of the computational process to which each corresponds. We use a result of Rolf Landauer [469], Charles Bennett [61, 62], and others, that any computation may be modeled as a two-stage process. The first stage is a rearrangement, which may be done within the computer and without coupling to the environment. The second stage consists of *erasure* of some bits of the computer's external state, which requires an input of work to transfer the entropy of the erased bits from the computer to the environment. The Shannon model and the Schneider model share identical computational steps but differ in their erasure steps.

We augment the mixing model from Figure 7.7 with the necessary computational steps to represent a full encode/transmit/decode sequence in Figure 7.10. In this Box and in the figure we suppose a discrete alphabet with binary digits, so that the maximum entropy per symbol is one bit. (This produces a simpler language than the Gaussian model, but the capacity theorem takes the same form.)

---

---

**Box 7.7    Error-correcting codes and molecular recognition (cont.)**

A maximum-entropy ensemble of message-words $w$ is delivered to an encoder, which generates codewords $c(w) \equiv x$ that are input to the channel. Since this process is deterministic, mapping all uncertainty about $w$ to an equivalent uncertainty in $x$, the computation can be done *reversibly* [61] – that is, without coupling to a thermal environment. The entropy of the inputs is denoted $S_w$.

A noise source $z$, with entropy $S_z$, is mixed with the codewords $x$ to produce an output $x \oplus z$. If the rate of the code $c(*)$ is the channel capacity, the output has entropy $S_w + S_z$ equal to one bit per symbol.

The computation performed by the decoder consists of mapping the output $x \oplus z$ to an estimate $w'$ drawn from the same set as $w$ (and therefore having the same entropy), and an estimate $z'$ for the noise. Since this is a deterministic computation on the received bits, it can be done reversibly as well.

Next, however, the Shannon and Schneider models do different things. In the Shannon decoder, the message estimate $w'$ is delivered to another data stream (perhaps the input to another round of encoding and transmission, or to another computer). The noise estimate $z'$ is not needed, but to reset the computer to receive the next codeword, the uncertainty in its noise state must be transferred to some variable, such as a thermal variable, in the environment. To make environmental state space "available" to accept this entropy, if the computer operates at temperature $T$, work equal to $k_B T S_z$ must be provided, which is delivered to the environment as heat $\Delta Q_z$ (per word).

In the Schneider model, the thermal noise $z$ cannot be removed. Instead it is the entropy of the unpredicted DNA sequence ($w$), which is rejected to an environmental bath. The priming enthalpy $P = k_B T S_w$ (for an optimal code) is delivered to the environment as heat $\Delta Q_w$ in reducing the transcription-factor/DNA pair distribution to a single estimate $w'$.

---

The Schneider model is physically equivalent to a microscopic model of the relaxation of a supercooled liquid into a crystal in a first-order phase transition. The ensemble of protein states bound to all possible DNA sequences corresponds to the set of local microscopic configurations in a liquid. The protein state bound to the recognized sequence corresponds to the particular microscopic configuration compatible with the crystalline ordered phase. The enthalpy $P$ is optimally used to carry away the entropy of non-crystalline configurations if the relaxation is adiabatic. The out-flow of enthalpy, as an ensemble of liquid configurations loses its diversity and converges on the single-crystalline form, corresponds to the latent heat of freezing of the transition.

The difference between the Schneider and the Shannon computational models may be understood as a *duality* between representations of a first-order, and a second-order phase transition. In the Schneider model where entropy $S_z$ remains and entropy $S_w$ is rejected,

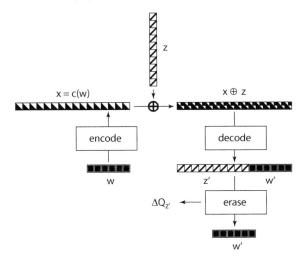

Figure 7.10 Computation diagram for conventional channel error correction. $w$ is a random variable holding a string of maximally compressed input bits (one bit entropy per bit signal, indicated by filled shading). $x = c(w)$ is the data expanded with an error-correcting code, representing the same entropy with more bits (partly shaded). $z$ is a compressed representation of environmental noise, mixed ($\oplus$ in the diagram) by the transmission channel with the encoded signal to yield a string $x \oplus z$. If the code rate equals the channel capacity, the output bits are fully utilized, so $x \oplus z$ contains one bit of entropy per (expanded) signal bit. Decoding may be implemented as a reversible logical operation, whose function is to rearrange the received bits so that estimates $w'$ of the original data bits occupy the leading places, yielding by implication a representation $z'$ of the inferred channel noise in the remaining bits. If encoding is optimal, the representation $z'$ is itself maximally compressed. For conventional channel encoding, the noise bits are not of interest and may be reset by an ideal erasure with rejection of heat $\Delta Q_z = k_B T S_z$ to the environment, where $S_z$ is the entropy in units of the natural logarithm (nats) of the noise ensemble $z'$. In the long-signal limit with optimal encoding, $w'$ differs from the input $w$, hence $z'$ from $z$, only with vanishing probability.

only one final configuration is selected, and rejected heat is the latent heat of a first-order transition. In the Shannon model where the noise entropy $S_z$ is rejected, the system may settle on any one of many codewords. It is likely to settle on the word $w' = w$ (the word that was transmitted), but if it settles on some $w' \neq w$, that outcome is also a stable solution for the decoder. The degeneracy of all possible decoded words $w'$ is equivalent to the degeneracy of ordered states produced by a symmetry-breaking phase transition. We have not said here where the selection of the codewords originates, which is the problem corresponding to the reductionist solution for the ordered phases of a frozen system.

### 7.5.6 The theory of optimal error correction is thermodynamic

These examples, like all of our introductory examples, are chosen because they are *minimal models* in which certain consequences of the laws of large numbers can be demonstrated. Optimal error correction in the Gaussian channel demonstrates the three-way trade-off among message diversity, correlation length (which carries a cost of time and computational complexity both in the design of codes and in the process of decoding), and reliability, under a constraint of noise that is taken to be exogenous. The most important and non-trivial messages we wish to stress are that *a theory of asymptotically reliable error correction exists*, and that *the source of reliability is a suppression of fluctuations equivalent to that found in the thermodynamics of phase transitions.*

The application of these ideas to non-equilibrium systems involving the topology of chemical reaction networks, and possibly spatially structured diffusion, is a more involved problem. Inevitably, the role of the laws of large numbers becomes contextualized in a host of system-specific details and calculation techniques. The technical overhead of performing calculations is likely to be off-putting enough that most of the presentation will be relegated to a specialist literature. We hope, in emphasizing that these minimal models are useful precisely because they contain *only* the laws of large numbers, to avoid the misunderstanding that the relations of error correction to thermal relaxation are mere "analogies," or that there is any obstacle of principle to extending them to the non-equilibrium chemical arena.

### 7.5.7 *One signal or many?*

The most stable thermal systems are those with a single maximum-entropy macrostate and no unpredictability. These must map to the error-correction model if the two are indeed equivalent. Yet unique stable equilibria typically require only a finite correlation length to evolve stochastically. How then are they to be compatible with the fluctuation theorem (7.51)?

The answer is that the unique case in which a decoder can be absolutely reliable with finite correlation length is when the message uses *only one codeword*. A message stream using only one word has rate $\mathcal{R} = 0$ because it has no uncertainty.

In systems that require only a few codewords, regression can be strong and reliable even with short correlation length. However, the non-zero probability of error requires that rare events will cause the decoded estimate to differ from the encoded signal occasionally.

We may represent the persistence of a system in time as a chain of *repeaters* in a transmission line, shown in Figure 7.11. Natural systems can persist because they have fewer "functionally" or "logically" distinct states than the number of possible microstates. Examples of such functionally distinct states are the macrostates in a phase transition, or the codewords $w$ in a code. The physical instantiation of each logically distinct state includes redundancy in the matter and events, which constitute the codeword $c(w)$. This physically instantiated system $c(w) \equiv x$ is "input" to the dynamics of nature at the beginning of a time interval and mixed with noise $z$ over the interval. A surveillance mechanism attempts to identify the error and restore an estimate $w'$ for the input *logical* state. It re-encodes this estimate and transmits it through the next interval of time. If $w' \neq w$ with some probability $P_{\text{error}}$, the state of the system will drift over time. However, if the code is good, the drift rate will be much slower than the rate at which noise $z$ enters the system.

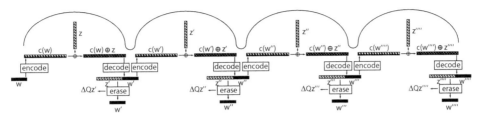

Figure 7.11 Persistence in time modeled as a channel with a "repeater." The *logically distinct* states of a system are physically instantiated with some redundancy as they exist in the world. Over time, the physical instantiation accumulates errors modeled as channel noise. At intervals, a surveillance system uses the redundancy of the physical instantiation (corresponding to the codewords) to estimate the input and the noise, to reject the noise, and to re-encode and re-transmit the input. Since the only estimate of the original codeword $w$ is the decoded estimate $w'$, if $w' \neq w$, the system state will drift without mean-regression among its codewords.

Drift under a repeater is in fact a valid representation of the drift of finite systems among their possible ordered states created by a second-order phase transition. The motion may consist of rare large transitions if the ordered states are discrete, as in the binary (Curie–Weiss) phase transition of Section 7.4.1, or they may consist of slow diffusion along a continuous axis of symmetry. Examples of the latter, in models of evolutionary population processes, are developed in [768] (Chapter 6).

While symbol-sequence error-correction models capture the qualitative properties of drift correctly, they are not appropriate models for all forms of phase transition order, because their symmetry groups are not appropriate to all forms of phase transition. The natural codes for the Gaussian channel are packed spheres. For a two-word code, they may mimic the symmetry of the binary phase transition, but they would fail as models in which the frozen phase was a dynamical limit cycle or some other extended state.

More important, in this section as in most engineering applications of error correction, we treated the machinery of encoding and decoding as exogenous, in order to focus on properties of optimal error-correcting codes. When we consider the emergence of hierarchical structure in biology, we will need to reconsider these questions. In general, we believe that the phase transition paradigm serves the emergence of hierarchical order in the biosphere because simpler transitions, at the lower levels, bring into existence the encoding/decoding machinery and selection of a code, which then defines the regression problem for the next higher transitions.

## 7.6 Phase transitions in non-equilibrium systems

Thermodynamics has seen great progress in two areas in the second half of the twentieth century. Information theory provided a rearview-mirror in which to see that physical thermodynamics had always been about control and inference, not fundamentally about energy. The second advance came from improvements in the mathematical understanding of large-deviations theories as the foundation of thermodynamics. As a result of work by Andrey Kolmogorov [447, 448], Yakov Sinai [758], E. T. Jaynes [393], and a host of modern workers, it is now widely understood that well-defined principles of thermodynamics apply to systems away from equilibrium as well as at equilibrium.[64] The non-equilibrium calculations can be considerably more difficult, and fewer examples are currently well understood, but these are difficulties of technique and approximation rather than of principle. An interesting question, about which only a little is currently known, is what new common motifs of order may appear in non-equilibrium systems that are rare or absent in equilibrium. We expect that the biosphere employs many such motifs.

### 7.6.1 On the equilibrium entropy and living systems

To understand what a non-equilibrium thermodynamics is, and why it is needed, we first summarize some frequently met efforts to understand the biosphere using equilibrium ideas.

#### 7.6.1.1 The Darwin–Clausius non-dilemma

Rudolf Clausius' 1862 memoir [140] "On the application of the theorem of the equivalence of transformations to interior work" introduced the formula $\delta Q/T$ for what we now know to be the change in equilibrium entropy in an adiabatic transformation, as a function of heat flow and temperature. His later compilation of this and other memoirs, *The Mechanical Theory of Heat* [357], published in 1865, just six years after the first edition of Darwin's *On the Origin of Species*, contained the famous Clausius statements of the first and second laws of thermodynamics. (1) The energy of the universe is a constant, and (2) the entropy of the universe tends to a maximum.

Following the first translation of Clausius' *Mechanical Theory of Heat* into English in 1850 by John Tyndall, Tyndall proposed to Herbert Spencer around 1858 that Spencer's view of a complex and stable endpoint of human society as a result of (Spencer's conception of) evolution could not be an expression of a universal natural law because it was in conflict with the Clausius claim that the final state of the universe was heat death. Spencer's reply to Tyndall[65] in a letter in 1858 or early 1859 includes the response

> ...Thus, you see, that my views commit me most fully to the doctrine of ultimate equilibration.
>
> That which was new to me in your position enunciated last June, and again on Saturday, was that equilibration was death. Regarding, as I had done, equilibration as the ultimate and highest state of society, I had assumed it to be not only the ultimate but also the highest state of the universe. And your assertion that when equilibrium was reached life must

---

[64] For an accessible overview, see Hugo Touchette's review [811] for physicists. A more extensive treatment of entropy laws in physics for mathematicians is given by Richard Ellis [224].
[65] Source: David Duncan [204], vol.1, page 136.

cease, staggered me. Indeed, not seeing my way out of the conclusion, I remember being out of spirits for some days afterwards. I still feel unsettled about the matter, and should like some day to discuss it with you.

In these early writings we see the expression of a perceived incompatibility of thermodynamics with biological – and by extension in Spencer's case, social – evolution that has been a topic of ongoing discussion for the succeeding 150 years. Informally, thermodynamics asserts that inanimate matter tends toward a state of disorder, while adaptation through natural selection is held to be responsible for making living systems more complex or functional, or at least more homogeneous as populations come to be dominated by their most adapted forms, and by any of these criteria, more ordered.

We open the discussion of non-equilibrium phase transitions with a review and resolution of this perceived conflict, even though it has been discussed in many earlier treatments, and even though the "resolutions" that have been offered have referred to equilibrium thermodynamic quantities. Our reason is that the substance of the concern is a valid scientific question about the origins of order, and also that the refutations offered so far are at best partial. At worst, they actually stand in the way of a proper formulation. A careful treatment of both the dilemma and its resolution shows why a theory of phase transition is relevant and in particular why this must be an inherently non-equilibrium phase transition.

### 7.6.1.2 No conflict, but neither yet a solution

The perception of a conflict between the orderliness of the biosphere on the one hand, and the association of inevitable or stable states with disorder on the other, is complicated to address. Technically the perception is an error, and there is no conflict. However, the arguments usually put forth to show that it is an error are themselves inadequate because they are conducted in a language that is part of the error. Finally, one does not wish to lose the grain of truth in the perceived conflict, which is that the biosphere is more ordered than many other systems, and that the nature of this excess order must be understood. All of the problems originate in moving from the informal need to explain a pattern to a formal language that is not sufficiently careful. Therefore it is perhaps best to retrace the original problem, and show how in a careful language it is resolved.

The correct insight is that the Gibbs equilibrium state is an appropriate thermodynamic *null model* for the level of order in a stable state. The biosphere is more ordered than the Gibbs equilibrium at the same values of internal energy, volume, and conserved atomic constituents, and this excess order is what must be explained. The proper way to put the question is to ask what *kind* of explanation is required: does one need to go beyond principles of maximum entropy to explain stability, or is the biosphere similar to other cases of phase transition in which excess order is explained within a principle of maximum entropy? (Our answer will be that the latter is correct.)

The difference between converting the informal question into a confusion, and converting it into a statement that contains the seeds of its own solution, turns on a careful use of language. Here is a statement of the maximization principle that living matter **does not** satisfy, in which we are careful to give operational definitions to the terms.

> The state variables of equilibrium thermodynamic systems, such as internal energy, volume, densities of particular types of molecules, etc. – since these simply measure the system's contents – can be computed for the matter in living systems as well as for matter in equilibrium distributions. Since the entropy function for a Gibbsian equilibrium ensemble can be expressed as a function of these other state variables, it can also be computed when the variables are estimated for living matter. If one carries out this calculation it is easy to check that typical biological inventories of molecules, if produced by reversible transformations from Gibbs ensembles, would represent states of much less than the maximum entropy given only their contents of energy, volume, atomic constituents, etc.

Notice that all we know from this statement is that a certain function computed for equilibrium ensembles as an entropy is not maximized in a certain class of non-equilibrium ensembles, if only the equilibrium conserved quantities are taken as constraints on the maximization. Clearly the two possibilities are either (1) that the principle of entropy maximization (which has been the foundation for equilibrium thermodynamics) does not apply to non-equilibrium systems and more particularly to life; or (2) that the function derived as an entropy for an equilibrium ensemble is not the same as the entropy required to describe a living system, in which case the observation of its non-maximization is not enough to tell us whether some other function is maximized, and whether that function is constructed as an entropy appropriate to the ensemble it describes. We claim the correct conclusion is the latter, but it must now be arrived at in two steps.

The first part of the resolution to the perceived dilemma comes from observing that the biosphere is not an isolated system: it is coupled to an environment and is therefore an **open system**. Hence any entropy function describing only the biosphere would not be predicted by thermodynamics to take a maximum value, because it would not furnish a complete description of the relevant constraints and statistical uncertainties. Again, as a pre-formal statement, this observation is valid. The answer that is usually given, however, attempts to form a description of open systems in terms of an equilibrium entropy that could only be well defined if the biosphere were a system formed reversibly, now augmented by additional equilibrium entropy measures for the environment. This formal response only corrects half of the error that led to the perceived conflict. Here again we state the standard answer, being careful to operationally define terms.

> *If both life and its physical environment were described by equilibrium ensembles* then the Clausius law would only require that the equilibrium entropy in the environment increase by a larger amount than that by

which the entropy of living matter decreases. As long as this condition
is met, there is no conflict between a highly ordered biosphere and a
principle of maximum equilibrium entropy.

This is the answer that was given by Boltzmann in 1886 [82], brought into popular
awareness by Erwin Schrödinger in *What is Life* in 1944 [711] with his expression that
life consumes "negative entropy," and further expounded by Leon Brillouin [98], Henry
Quastler [654], and others. The essential insight is sufficient to explain why the order of
the biosphere is compatible with everything understood about thermodynamics, and when
properly used it can also be a source of bounds on biological order.

However, the biosphere is not created by reversible processes, and it is not an equilibrium ensemble. Therefore, the "open system" described in this statement is still not the
same kind of ensemble as the open biosphere. The realization of this difference is the
important thing lost in careless use of the term "*the* entropy," put forth as if the principles
of thermodynamics applied to the form of the equilibrium function without thought for its
its status as merely the *evaluation* of an information measure defined in a restricted context.

The macroscopic entropy that is thermodynamically meaningful must be an information
measure for the distribution from which it is computed. This is the content of the large-
deviations *definition* of the entropy function from Section 7.3. The state variables which
are its arguments must also be appropriate sufficient statistics for that ensemble, as we have
illustrated in the foregoing examples. *The function of these variables (the particular mathematical formula) which is a macroscopic entropy function for an equilibrium ensemble will
not generally be the actual entropy function for a non-equilibrium ensemble.* The equilibrium state variables are not generally sufficient to capture the full set of constraints acting
away from equilibrium,[66] and the actual entropy function will generally have a different
form, which can be reduced to or projected onto the equilibrium entropy in the non-driven
limit.[67]

The equilibrium entropy does, however, provide a bound on the degree of stably
sustained order for the following reason. The kinds of boundary conditions, such as
non-uniform temperature or chemical potentials, needed to drive a system away from
equilibrium always constitute excess constraint and therefore additional information to
that given by boundary conditions that would produce the same matter composition in
an equilibrium ensemble.[68] Therefore the equilibrium-entropy reduction required to construct living matter by a reversible transformation, without taking account of process, is

---

[66] For a simple example in which both equilibrium and non-equilibrium entropy functions are local in time, and in which the
non-equilibrium entropy simply adds *currents* as additional state variables to those in the equilibrium limit, see [762, 763].

[67] As we show below, whereas the equilibrium entropy can be a *function* only of a system's state, the non-equilibrium entropy
will generally be a *functional* – a function of functions – the arguments of which are entire time-dependent histories of states
and the transitions between them.

[68] Note that, to define such an ensemble, we would generally require a host of partitions as barriers between regions with
different chemical potentials. For living chemistry, which is driven largely by oxidation/reduction potential (ORP), sometimes
even different carbon atoms within a molecule have different effective ORP, so that the molecule on its own could relax
through internal redox reactions if reaction mechanisms were available. Hence the definition of boundary conditions required
to assign an equilibrium entropy measure to living matter may be extremely complicated.

a *lower bound* on the entropy that an equilibrium environment would need to accept to produce the actual non-equilibrium system.[69] The Boltzmann–Schrödinger *et al.* statements about consumption of environmental "entropy" then provide valid expressions for what is thermodynamically *forbidden*. However, bounds from the equilibrium entropy may be too loose to reflect the additional constraints and mechanisms that determine which forms of order are possible away from equilibrium.[70] Note that it is not necessary that life be formed via reversible (also called *adiabatic* or *isentropic*) processes for the Boltzmann–Schrödinger bound to apply. It is only necessary that the resulting *states* be suitably approximated by equilibrium ensembles. For many purposes, environmental subsystems such as the solar thermal black-body radiation at 6000 K or the 300 K surface of the Earth are adequately approximated with equilibrium ensembles. Living subsystems bounded by compartments, or simply separated by long reaction times except where they are coupled by molecular gating mechanisms, also may sometimes be well modeled as equilibrium states as long as the origin of the compartments and gating mechanisms is not the phenomenon to be explained.[71]

The property of driven ensembles that is excluded from the local-equilibrium approximation is *the rate structure of transitions* between states. A defining criterion of equilibrium is that all waiting times for transitions, no matter how long, have been surpassed, so that the occupation of states is determined only by the energy, volume, conserved atomic constituents, etc. that each state requires, relative to the total quantities in the system. In contrast, in driven systems, states are occupied only if they can be reached (as, for example, in chemical reactors) by energy or materials between the time these enter the system at the source and the time they leave it at the drain. In some systems, including many popular reaction-diffusion models, the approximate equilibrium entropy at successive times contains enough information to determine dynamics with the aid of a set of transport coefficients. More generally, factors that can affect transition rates and which are not captured in the equilibrium entropy alone can range from the response of reaction rates to catalysts, to the construction – within a system – of far from equilibrium structures such as compartments, surfaces, scaffolds, or turnstyles [96] which may partition reactants into separate near-equilibrium, but decoupled, subsystems. The rate structure away from equilibrium, and the history dependence it induces in the system's response to its boundary conditions, can be considerably more complex than the probability structure at equilibrium.

The final step that is required, after recognizing that the biosphere is an open system, is to provide suitable non-equilibrium entropy functions appropriate to any living

---

[69] Since the non-equilibrium entropy is generally a functional of histories, some coarse-graining is generally required to project it onto any function of states at a single time, which can be compared to an equilibrium entropy. See [768] (Chapter 7) for one example of such a projection.

[70] Reference [763] provides an example in which the state variables that create non-equilibrium order in response to non-uniform boundary conditions are excluded from the set of equilibrium state variables by their behavior under time reversal. Since the equilibrium entropy can only take equilibrium state variables as its argument, it is forced to omit exactly the terms that cause non-equilibrium organization.

[71] These bounds can be used to compute limits for the thermodynamic cost of natural selection, and to map the chemical work relations to the Landauer principle for the thermodynamic cost of computation, showing that the ensemble descriptions of adiabatic chemistry and ideal computation are isomorphic [764, 765, 766].

or environmental subsystems for which the equilibrium entropy is too loose a bound to capture the kinetic as well as energetic constraints on formation and maintenance of order. Fortunately, one of the areas of greatest current progress in statistical physics is the demonstration that such entropy functions for non-equilibrium processes are well defined and constructible. We will provide a simple worked example below, and fuller discussions are given in [71, 73, 190, 278, 465, 522, 767].

### 7.6.1.3 Why it matters to retain entropy principles

Ultimately, terminology should be a convenience and an aid to clear thought. Usually, as long as we can operationalize our expressions in terms of explicit physical processes, we can overcome some terminological inconsistencies that we have inherited from the past. In the case of entropy, one unfortunate inconsistency has been the use of the same word to refer either to the mathematical measure of fluctuation probabilities, or to the particular quantity constructed by Gibbs for equilibrium many-particle systems. The identification of these two uses has led to expressions such as "entropy production" as a quantity at work in irreversible systems, first by Lars Onsager and later by Ilya Prigogine and collaborators.[72]

We do not conflate the two uses. For us, *entropy* is defined by its role in large-deviations expressions for fluctuations, in whatever context pertains. Particular instantiations of the entropy are referred to in association with the systems they describe. Hence the "entropy" that is "produced" in diffusing systems should be called the "equilibrium entropy," to emphasize that this function – drawn from an ensemble slightly different than the actual ensemble of diffusive histories – is not the actual entropy that governs the dynamics of diffusion and may not be sufficient to determine the actual entropy. It is much less jarring, and closer to what physics now understands, to realize that the principle of maximum entropy does not change between equilibrium and non-equilibrium applications; only the function or functional used in calculations changes to properly reflect the change in the inference problem being posed. The theory of stability for driven systems remains one of counting: those macro-trajectories are stable which have the largest number of stochastic ways to be entered or maintained, and the fewest ways to be exited.

The order of the biosphere is excessive relative to a Gibbs ensemble, but only in the same way as the order of other frozen phases is excessive. *The biosphere is driven by "blunt" boundary conditions that are much better approximated by Gibbs ensembles than the biosphere itself is.* We have mentioned solar radiation and the 300 K Earth surface as examples. Other examples include oxidation/reduction disequilibria created when $H_2$ is an electron donor and $CO_2$ or $O_2$ are terminal electron acceptors. Each of these molecules

---

[72] Onsager introduced the term as part of a "principle of minimum entropy production" in two papers [607, 608] on the consequences of microscopic irreversibility in 1931. Even this name for Onsager's principle is misleading. The rate of change of equilibrium entropy in the boundary reservoirs is only minimized relative to a bilinear form in the transport currents, which Onsager called the *dissipation function*, in which the kernel was defined from local, near-equilibrium transport coefficients. Why the entropy production should be minimized along this surface is not contained in the "principle." Similarly, rules for maximization of the rates of change of equilibrium entropies introduced by Prigogine and collaborators [303, 451] are defined with respect to various constraints defined from near-equilibrium transport coefficients.

is a stable dominant component in some Gibbs ensemble, but they can only be described in equilibrium if the ensembles are separated. The chemical problem of understanding life is to explain why disequilibria which originate in the *ensemble* populations become concentrated in disequilibria both between and even *within* organic molecules, along the most probable relaxation pathways.

We close the section with a discursive example that may make some of these ideas more accessible. Attempts to speak precisely about open and closed systems, or equilibrium and non-equilibrium entropy principles, may seem obscure and technical. But in the modern world, where we are surrounded with human-engineered thermodynamic machines, the actual limitations of descriptions that use only equilibrium concepts are seen ubiquitously in everyday events.

### 7.6.1.4 The parable of refrigerators

Out of the understanding of equilibrium thermodynamics created by Sadi Carnot [119], Clausius, Boltzmann, and Gibbs, we have built a world full of refrigerators and engines. Refrigerators consume (usually electrical) work to pump heat out of their contents and into their surroundings. Efficient refrigerators are tolerably well approximated by near-equilibrium ensembles on the inside and the outside, coupled only through the heat pump. The refrigerator locally decreases the entropy of its contents by increasing the entropy of the surroundings. Actual refrigerators receive their energy from engines (the power plants), which produce work by producing net entropy in the environment (again, in a suitable near-equilibrium approximation). The refrigerator/power-plant combination produces a local entropy decrease by cooling food or some other contents, and is driven to do so by the irreversible production of entropy at the plant. This example involves thermal engines using temperature differences, but equivalent processes exist for chemical cycling [765]. The biosphere is a kind of chemical refrigerator, and what it refrigerates by becoming ordered is the matter of which it itself is made [762].

All these facts we take for granted. Yet they essentially tell us nothing about *why* refrigerators exist in the world, how they came into existence, why and how the events in nature transpire which repair or replace them as they wear out, or for how long the presence of such machines is likely to persist as a feature of the world. To answer any of these questions for refrigerators as technological artifacts is of course an unimaginably complex problem: it requires models of social order, and in turn of human agency, design, and intention, and in turn of human evolution of intelligence and technology, and in turn of life. It may be that the least unlikely path to arrive at refrigerators is the one that passes through human design and depends on an enormous number of intermediate levels, ultimately rooted in the existence of the biosphere.

At each level in the technological transition, near-equilibrium thermodynamic models may provide useful constraints on which structures can or cannot exist. Their role is clear not only for the design of the refrigerator and the powerplant themselves, but for intermediate materials designs and many lower levels as well. Yet even if they account for the leading entropic constraints from flows at each level, they omit the crucial coordination processes that structure these flows and determine whether the path to a given artifact is or is not taken.

We invoke the parable of refrigerators because they make the power and the limits of equilibrium thermodynamics familiar, not because their emergence is simple to understand. The emergence of a biosphere, although not yet as familiar, depends on steps much lower in the hierarchy, and thus poses a simpler problem.

### 7.6.2 Ensembles of processes and of histories

The conceptual shift from equilibrium to non-equilibrium thermodynamics was made clear by A. N. Kolmogorov and Y. Sinai in 1959 [448, 758].[73] It does not consist in a shift from computing values to computing time-derivatives of the equilibrium entropy. Rather, it replaces the entire ensemble description, and recognizes that for processes, the relevant entropy of transitions must be an *entropy rate*.

The elementary points to which equilibrium probability distributions are assigned (termed "atoms" in probability theory) are *states* of a system. The states may have momenta as well as positions as their coordinates, but they are defined at a single time. For non-equilibrium systems, in the most general case, the ensembles of states from equilibrium are completely replaced with ensembles in which the atoms are *histories*. (We will sometimes refer to them as "trajectories" when it is clear that we mean the trajectory of the entire system under one unfolding of its dynamics.) Probabilities are assigned to microscopic histories, and entropies are computed and maximized for these probability distributions. Macroscopic "state variables" also become histories. The most likely macroscopic trajectory for a system is the one that maximizes the entropy over micro-histories consistent with the macro-variables. Jaynes termed such entropies-of-histories **calibers** [393], by analogy to the volume in the bore of a rifle. The principle of maximum caliber – still a principle of maximum entropy, though now on a more complex space – replaces maximization of the state-entropy from equilibrium thermodynamics. In constructions of this kind for stochastic processes, the caliber is the large-deviation rate function for fluctuations of macro-histories.

For non-equilibrium steady states, the caliber for a history of finite duration is the time-integral of an entropy rate corresponding to the Kolmogorov–Sinai entropy rate. For more general transients or time-dependent histories such as limit cycles, the caliber takes the form of a non-trivial *action functional*, which we have termed the "stochastic effective action."[74] This effective action both generalizes the entropy of equilibrium

---

[73] Kolmogorov and Sinai worked on the rather complicated problem of defining an appropriate entropy rate function for chaotic dynamical systems evolving on non-smooth attractors. Much simpler counterparts to the Kolmogorov–Sinai entropy exist for discrete-state and discrete-time or continuous-time Markov chains. Although simpler in dispensing with continuous measures and chaotic attractors, these other Markovian systems are more easily extended from the non-equilibrium steady states studied by Kolmogorov and Sinai, to a variety of transient conditions. A variety of non-equilibrium reaction-diffusion models in this class are topics of active current work.

[74] A stochastic effective action for small fluctuations from equilibrium was first computed by Lars Onsager and S. Machlup in 1953 [609], building on Onsager's 1931 work [607, 608] on microscopically reversible systems. A modern tutorial introduction is available in [767]. For correlations of fluctuations in a non-equilibrium ensemble at a single time, the large-deviations function obeys a *Hamilton–Jacobi* equation dual to the stochastic effective action functional [73]. Even for single-time correlations,

thermodynamics to non-equilibrium systems, and also generalizes the Lagrange–Hamilton action functional [309] from mechanics to thermodynamics.

Non-equilibrium ensembles, entropies, and phase transitions are generally much more difficult to compute than their equilibrium counterparts, and we provide only the simplest example in Section 7.6.4 below, which lacks many features of the more general case that we expect to be important for the emergence of the early biosphere.[75] We provide physical examples of the effects of interest, relevant to biological emergence, in the next two sections.

### 7.6.3 Phase transfer catalysis

Section 7.4.6 reviewed properties of equilibrium oil–water phase separation that biochemistry can put toward essentially structural uses. We have also considered one energetically neutral but catalytically important possible role for oil–water interfaces in Section 7.4.6.4, which is to use the dielectric contrast of the interface to guide protein folding. In the latter role, the presence of an equilibrium phase transition can be part of a feedback loop that alters dynamics. Here we note that a dielectric contrast may also provide a free energy differential for dissolution of different molecules, and hence affect the directions as well as the rates of reactions. In this role the equilibrium phase transition may be a mechanistic component in a larger non-equilibrium phase transition, on which the equilibrium materials depend.

The aggregate alteration of reaction kinetics by all mechanisms in phase-separated systems is termed **phase transfer catalysis**. The process is of wide interest for industrial synthesis, where a variety of reactions have been studied that are not directly connected to synthetic pathways that supply the lipids. A representative example is the catalysis of Diels–Alder reactions[76] studied by Otto *et al.* [619], in which not only the water/hydrocarbon contrast, but also the cation Cu (II), serving as a counter-ion to negatively charged carboxyl groups on the lipids, were essential parts of the catalyzed reaction mechanism.[77]

The possibility for catalytic feedback on the lipid synthesis reactions that supply the membranes, resulting in **phase transfer autocatalysis**, has been proposed to be important for the origin of life, and a few experimental models have been studied as proofs of concept.

Walde *et al.* [849] demonstrated a simple feedback system in which membrane-catalyzed hydrolysis of a fatty acid anhydride is provided by vesicles formed from the component fatty acids. In these experiments, a water-insoluble layer of a symmetric anhydride of a long-chain carboxylic acid was overlaid on a water phase which was the reaction phase.[78] In the absence of vesicles, spontaneous hydrolysis of the anhydride was very slow, while in the presence of vesicles it was enhanced to a rate determined by the vesicle area.

---

this Hamilton–Jacobi equation requires reconstruction of the history dependence that is a novel feature of non-equilibrium entropies.

[75] Examples illustrating each of these properties are currently known. For non-locality and history dependence of single-time correlations, see [72]. For phase transitions in both discrete-time and continuous-time molecular or population processes, see [457, 768, 770].

[76] In the Diels–Alder reaction, a diene is condensed with a substituted olefin.

[77] See Ref. [16] of [619] for an elaborate list of similar applications of micellar catalysis.

[78] The long-chain fatty acids studied were caprylic acid or oleic acids.

Fabio Mavelli and Pier-Luigi Luisi [532] have carried out a mass-action network analysis of two variants on this process, and shown good quantitative agreement with the experimental progression of the reaction. In their model, area-mediated hydrolysis was considered the rate-limiting step, with vesicle incorporation, growth, and budding approximated as being maintained in equilibrium at all times at the fatty acid concentration. The system shows the initially exponential growth of reaction rate characteristic of a first-order autocatalytic system, which saturates only when the input supply of anhydrides is exhausted.

It is not evident within current biology that this mechanism of phase transfer catalysis was important to the emergence of lipid metabolism. In cells, fatty acid synthesis by decarboxylating addition of malonyl-CoA is carried out on a single enzyme complex, with the fatty acid esterified to a pantetheine unit of an acyl-carrier protein. These protein complexes are bound in cytosol, so they furnish no direct model of a role for phase transfer catalysis in fatty acid synthesis. Nonetheless, the free energy of decarboxylation of malonyl-CoA should be more negative if the resulting $-(CH_2)_2-$ group can be sequestered in a non-polar environment than if it must occupy an aqueous medium. It might be fruitful to seek precursors to the malonate pathway in which the extending chain embeds in a non-polar environment and thereby enhances the reaction rate.

### 7.6.4 A first-order, non-equilibrium, phase transition in the context of autocatalysis

A self-contained calculation of the phenomenon of network autocatalysis will serve to illustrate the construction and some important properties of non-equilibrium entropy functions. The model we develop below is network autocatalytic, like the lumped-parameter rTCA model of Section 4.4.4.3 (though even simpler in its rate equations). We choose the following model to have a feature the minimal rTCA model lacks: it undergoes a first-order phase transition rather than a second-order transition. This feature allows us to study a model that possesses two alternative metastable states under the same boundary conditions, to formalize the notion of chance nucleation of an ordered dynamical state.[79] The model will illustrate how the richer entropy functional for a non-equilibrium process allows us not only to describe the relative probabilities to occupy two non-equilibrium steady states, but also to compute the most probable *trajectory* that leads from one to the other.

#### 7.6.4.1 A one-variable chemical reaction model

The simplest models of stochastic chemical reaction with autocatalytic feedback are one-dimensional models in which a single species changes concentration through Poisson-distributed reaction events.[80] We will choose parameters for simplicity and in order to produce a structure that illustrates all basic concepts.

---

[79] This model thus duplicates an important property of rTCA *without* a feeder such as the Wood–Ljungdahl pathway, which is the possibility to exist in either a high-throughput state or a collapsed, low-throughput state. It differs from the plain rTCA model because the collapsed state is not permanent.

[80] A richer two-dimensional model with the same stoichiometry is well known as the Selkov model. The richer model illustrates some of the important new features of escape processes in non-equilibrium systems, but requires much more formal machinery

The basic reaction scheme to be used as an illustration is

$$A + 2X \overset{1}{\rightleftharpoons} 3X$$

$$X \overset{2}{\rightarrow} \varnothing \tag{7.64}$$

$$A \overset{3}{\rightleftharpoons} X.$$

Superscripts on conversions label the reaction numbers.

In reaction 1, an environmental source molecule $A$ is converted to an intermediate molecule $X$ in a ter-molecular reaction that requires two molecules of $X$ as input. The phase dependence on the dynamics of $X$ will be the property of interest. Reaction 2 models dilution of $X$ and is therefore shown as unidirectional.[81] Reaction 3 is a spontaneous conversion of $A$ to $X$ which does not require autocatalysis but which we suppose occurs at low rate. The background of reactions 2 and 3 ensures regular solutions, and reflects the non-zero probability of any allowed reaction. Very rare reactions may be treated with limits in which rate constants for either reaction 2 or reaction 3 approach zero.

Since spatial geometry is not of interest, we do not introduce reaction volumes and concentrations, but work directly with numbers of molecules. This approach is also needed to define the stochastic reaction process. Let $n_A$ denote the number of $A$ particles, and $n$ denote the number of $X$ particles.

We begin with the mass-action rate equations because these define the possible averaged dynamics and fixed points of the network. Stochasticity is introduced later. The three mass-action reaction currents for scheme (7.64) are given by

$$j_1 = n_A n^2 k_1 - n^3 \bar{k}_1$$

$$j_2 = n k_2 \tag{7.65}$$

$$j_3 = n_A k_3 - n \bar{k}_3.$$

We suppose that $A$ is buffered to constant $n_A$ so we will not write dynamics or an explicit distribution for it. Although the scheme (7.64) is only a simple toy model, we suppose that it is meant to model some system completely, meaning that no ancillary sources of chemical potential or other energy are present beyond those explicitly noted. Under this interpretation the equilibria for reactions 1 and 3 must be the same because they represent the same net conversion. Therefore

$$K_1 \equiv \frac{k_1}{\bar{k}_1}$$

$$= K_3 \equiv \frac{k_3}{\bar{k}_3}. \tag{7.66}$$

---

to present. Therefore we exploit the simplicity enforced in a one-dimensional model, and refer the reader to Dykman *et al.* [207] for the more complex case.

[81] The reaction could be written as reversible, with the concentration of $\varnothing$ buffered at zero.

The reader may wonder why we have adopted the seemingly greater complexity of the ter-molecular reaction 1, especially since in the minimal rTCA model the autocatalytic uptake reaction is only bimolecular. Here we briefly review the kinetics of linear auto-catalysis, and explain why it does not show a first-order phase transition, and indeed has a unique though non-linear solution except in singular limits. A bimolecular alternative to reaction 1 in scheme (7.64), which we will call 1′, has scheme

$$A + X \overset{1'}{\rightleftharpoons} 2X, \tag{7.67}$$

and current

$$j'_1 = n_A n k'_1 - n^2 \bar{k}'_1. \tag{7.68}$$

The $\left( j'_1, j_2, j_3 \right)$ system produces the mass-action kinetic equation for $n$

$$\begin{aligned} \frac{dn}{dt} &= j'_1 - j_2 + j_3 \\ &= n_A k_3 + \left( n_A k'_1 - k_2 - \bar{k}_3 \right) n - \bar{k}'_1 n^2. \end{aligned} \tag{7.69}$$

If $k_3 \rightarrow 0$, Eq. (7.69) always has a fixed-point solution at $n = 0$. The zero fixed-point is unstable, and the solution has a second fixed point at $n'_0 \equiv \left( n_A k'_1 - k_2 - \bar{k}_3 \right) / \bar{k}'_1$ if this quantity is positive. Otherwise the zero fixed point is unique and stable. This is the kinetic model for the simplified rTCA network shown in Chapter 4. The fixed point depends continuously on $n'_0$ but has discontinuous derivative if $n_A k_3 \rightarrow 0$. If $n_A k_3 > 0$, the unique fixed point has a smooth but non-linear dependence on $n'_0$ in the neighborhood of zero.

### 7.6.4.2 A system with a first-order phase transition

Returning to the $(j_1, j_2, j_3)$ scheme: the mass-action kinetic equation for $n$ is

$$\begin{aligned} \frac{dn}{dt} &= j_1 - j_2 + j_3 \\ &= n_A k_3 - \left( k_2 + \bar{k}_3 \right) n + n_A k_1 n^2 - \bar{k}_1 n^3. \end{aligned} \tag{7.70}$$

Our interest in both the mean and stochastic analysis is the scaling in time and particle number, so we begin by extracting scale factors.

The natural normalization for particle number, to assess the importance of autocatalysis, is the concentration of $n$ that would be in equilibrium with $n_A$ if the autocatalytic reaction were arbitrarily faster than the drain reaction. Introduce the notation (arbitrarily using $K_1$ to label the equilibrium constant)

$$n_0 \equiv n_A K_1. \tag{7.71}$$

The relative particle number is then $x \equiv n/n_0$, and a natural time constant for the reaction rate is $\bar{k}_1 n_0^2$ (the backward reaction rate per particle of the autocatalytic reaction 1 at saturation). Therefore normalize time by introducing a non-dimensional coordinate $\tau$ with

$$d\tau \equiv dt\bar{k}_1 n_0^2. \tag{7.72}$$

The quantities that determine the dynamics are the steady-state values of $x$, which are the roots of the rate equation (7.70). The roots of the general cubic equation are cumbersome to write down, but they are easily approximated in terms of the roots at $n_A k_3 \to 0$. Zero is always a root in this limit, and we denote two other roots by $x_1$ and $x_2$ with $x_1 + x_2 = 1$, taking $x_2 > x_1$. The product of the two roots corresponds to the only term in Eq. (7.70) not removed by scaling:

$$x_1 x_2 = \frac{\left(k_2 + \bar{k}_3\right)}{\bar{k}_1 n_0^2}. \tag{7.73}$$

Equation (7.73) is solved by

$$x_1 = \frac{1}{2} - \frac{1}{2}\sqrt{1 - \frac{4\left(k_2 + \bar{k}_3\right)}{\bar{k}_1 n_0^2}}$$

$$x_2 = \frac{1}{2} + \frac{1}{2}\sqrt{1 - \frac{4\left(k_2 + \bar{k}_3\right)}{\bar{k}_1 n_0^2}}. \tag{7.74}$$

Finally, introduce the notation

$$\epsilon \equiv \frac{k_3}{K_1 \bar{k}_1 n_0^2} = \frac{k_3}{k_1 n_0^2}. \tag{7.75}$$

$\epsilon$ is a "bleed rate" from weak reactions outside the autocatalysis/dilution system, which we will suppose is small compared to $x_1$ and $x_2$. The mass-action rate equation (7.70) in descaled variables becomes

$$\frac{dx}{d\tau} = \epsilon - x\left(x - x_1\right)\left(x - x_2\right). \tag{7.76}$$

### 7.6.4.3 Roots and stability

The structure of solutions is governed by the roots of the cubic when $\epsilon$ is small. In the limit $\epsilon \to 0$ the qualitatively distinct regimes are the following.

- If $\left(k_2 + \bar{k}_3\right)/\bar{k}_1 n_0^2 = 0$ then the rate equation has a second-order root at $x_1 = 0$ and a unique stable root at $x_2 = 1$. From any other starting state besides $x = 0$ the system attracts to $x_2$.
- For $0 < \left(k_2 + \bar{k}_3\right)/\bar{k}_1 n_0^2 < 1/4$ the rate equation has stable roots at $x = 0$ and $x = x_2$, and an unstable root at $x = x_1$.

- If $\left(k_2 + \bar{k}_3\right)/\bar{k}_1 n_0^2 = 1/4$ then the rate equation has a second-order root at $x_1 = x_2 = 1/2$ and a stable root at $x = 0$. The second derivative of the rate equation at $x_1 = x_2 = 1/2$ is negative, so the dynamics is stabilizing everywhere except the roots and marginally stable at $x = 1/2$. Therefore it can only attract to $x = 0$.
- For $\left(k_2 + \bar{k}_3\right)/\bar{k}_1 n_0^2 > 1/4$ the unique real root is $x = 0$, which is stable.

For $\epsilon \ll x_1$ the three roots are $x \approx \left\{ \dfrac{\epsilon}{x_1 x_2}, x_1 - \dfrac{\epsilon}{x_1(x_2 - x_1)}, x_2 + \dfrac{\epsilon}{x_2(x_2 - x_1)} \right\}$.

### 7.6.4.4 Stochastic process and master equation

A stochastic process that will reproduce the mass-action kinetics (7.70) may be defined by introducing a density $\rho$ with probabilities $\rho_n$ that a sample from an ensemble of reactors will be found with n particles of type $X$.[82]

$\rho$ is time-dependent, and evolves under a probability-flow equation known as a *master equation*, written

$$\frac{d\rho}{d\tau} = \mathbb{T}\rho. \tag{7.77}$$

Here $\rho = [\rho_n]$ is regarded as a (potentially infinite-dimensional) column vector on index $n = 0, 1, \ldots$; $\mathbb{T} \equiv [\mathbb{T}_{nn'}]$, called the *transfer matrix*, is a square matrix in which coefficient $\mathbb{T}_{nn'}$ gives the rate at which probability originally in $\rho_{n'}$ is transferred to $\rho_n$. $\mathbb{T}$ is a *stochastic matrix*, meaning that $\sum_{n=0}^{\infty} \mathbb{T}_{nn'} = 0$ for all $n'$. This is needed so that total probability $\sum_{n=0}^{\infty} \rho_n$ will be conserved.

The master equation will give the time progression of any density. Therefore it may be used to study the temporal correlations of histories initially concentrated at a single value of n, including the gradual approach to the steady-state distribution around one or more fixed points. In the case where histories may approach and remain within neighborhoods of multiple fixed points with non-vanishing probability, $\rho$ will be multi-modal, and the relative weights in different modes will be proportional to the frequency with which trajectories are found near each fixed point. By combining analyses of distributions initially concentrated at various points, and of stationary distributions, it is possible to extract persistence times and transition frequencies between fixed points. In this way the master equation can be used to extract the quasi-discrete dynamics of nucleation transitions among different phases, on timescales much slower than the microscopic reaction timescales. If the constitutive parameters are changed on slower timescales still, the phase-change response, and metastable dynamics of the reaction system may also be derived.

A remarkable suite of methods was developed within the second half of the twentieth century, which not only enable all these analyses, but even combine them into a single coherent ensemble analysis so that they do not need to be performed as separate experiments on different instances of $\rho$. The methods are technical, and may be of more interest

---

[82] Here and in Appendix 7.7, where integer values of n will be important, we use Roman script. Math Italic will be reserved for *n* treated as a continuous variable; either the average under $\rho$ or as a field variable of integration used in the generating-function solution of the master equation derived in the appendix.

to specialists in applied mathematics than to general readers, so we have placed our review and computations in Appendix 7.7.

The important general point, and the one we wish to emphasize as a key component of the stability perspective – illustrated by this model – is that the large-deviations theory of non-equilibrium stochastic processes such as the reaction network (7.64) shows the full suite of behaviors of thermodynamic phase transitions, only in a much richer space of functions rather than states. Here we will review a few key quantities from the computation, explaining what fluctuation each describes and the equilibrium-thermodynamic quantity it generalizes.

### 7.6.4.5 Phase transitions in the reaction rate

The property of obvious interest in the $(A, X, \varnothing)$ system with reactions (7.64) is the possibility to switch between low and high concentrations of $X$ with corresponding low and high reaction rates, either in response to changes in the constitutive parameters or through spontaneous transitions when both non-equilibrium steady states are present as fixed points of the rate equation (7.76). This property is the non-equilibrium counterpart to first-order phase transition, including the possibility for metastability and nucleation, and coexistence of two phases at the critical parameters.

The quantities determining the possibility for nucleation or coexistence must be the rates to increase or decrease $x$ in Eq. (7.76) and the scale factor $n_0$, as these are the only quantities appearing in the theory. Qualitatively put, in regions that move from the unstable fixed point to the stable fixed point, motion follows the classical rate kinetics and should have probability $\sim 1$ in the large-deviations limit. Escapes from a stable fixed point, whether these involve accumulation of $x$ through regions where the mean loss rate is larger than the mean gain, or loss of $x$ in regions where mean gain exceeds mean loss, will become exponentially rarer the larger these differences become. The rate equation (7.76) is graphed in Figure 7.12. Escapes require $x$ to travel downward in the upper region where the expected rate is positive, and upward in the lower region where the expected rate is negative.

The question for a large-deviations theory is whether rates of escape can be derived from a potential, and the answer from Appendix 7.7 is that in one dimension they always can be. For a non-equilibrium system, however, where the objects of interest are trajectories, we must explain what class of trajectories arise as arguments to such a potential.

When more than one fixed point exists, most trajectories at large $n_0$ will spend almost all of their time in neighborhoods of range $\delta x \sim \mathcal{O}(1/n_0)$ of one or the other fixed point. We label these fixed points $\bar{x}_\pm$ for the high and low values of $x$ respectively. Excursions away from the fixed points will be rare, and if a macro-fluctuation reaches some value $\bar{x}$, the least unlikely path by which it will have arrived there is one in which the escape is a time-reversed image of the usual mass-action kinetics.[83] That is, it will be an initially

---

[83] This result is exactly true only for one-dimensional systems such as this example. It is often approximately true, however, even in higher-dimensional systems as long as the rate equations produce direct attraction to fixed points, and not limit cycles with caustics or other more complex dynamical behaviors [510, 511].

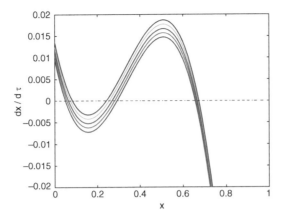

Figure 7.12 The rate equation (7.76) for the $(A, X, \varnothing)$ system with reactions (7.64) with constitutive parameters $x_1 = 0.4$, $x_2 = 0.6$. Five values of $\epsilon \in \epsilon_{\text{crit}} + [-0.002, 0.002]$, where $\epsilon_{\text{crit}} \approx 0.01166$ is the critical value for detailed balance, are shown. Zero-crossings of the rate $dx/d\tau$ are fixed points. Crossings with negative slope are stable, those with positive slope are unstable.

slow exponential divergence, which accelerates toward $\bar{x}$ within a time $\tau_{\text{escape}} \sim \mathcal{O}(1)$ in a history of any duration.

Therefore, the concept of an escape, whether full or only partial, is defined by an *event*. While the escape trajectory itself is a history, it is conditioned on a single observation of $\bar{x}$ with no other stipulation. Thus, although the history has an inherent duration, this duration is not part of the information defining the escape, so a large-deviations potential for escapes can be a function of state variables at single moments, much like potentials in equilibrium systems.

Equation (7.103) in Appendix (7.7) gives the **potential** governing escapes indexed by a limit $\bar{x}$:

$$\Phi(\bar{x}) \equiv \int_0^{\bar{x}} dx \, \log\left(\frac{x_1 x_2 x + x^3}{\epsilon + x^2}\right). \tag{7.78}$$

The lower limit of integration is arbitrary and determines the overall offset of the potential. For convenience we have chosen $x = 0$ in Eq. (7.78). The probability associated with an escape trajectory from an initial stable fixed point $\bar{x}_\pm$ has the large-deviations approximation[84]

$$P_{\text{escape}}(\bar{x} \mid \bar{x}_\pm) \sim e^{-n_0[\Phi(\bar{x}) - \Phi(\bar{x}_\pm)]}. \tag{7.79}$$

---

[84] Although escapes are events, their probabilities must be expressed as *rates* per unit of $\tau$. Appendix 7.7 discusses the sum over escape events and the functional determinant that converts probabilities of the form (7.79) to escape rates. Since the dimensional prefactor that converts $P_{\text{escape}}$ to a rate has less than exponential dependence on $n_0$ we suppress it in large-deviations expressions.

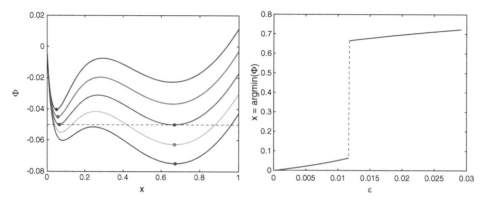

Figure 7.13 First-order phase transition in the $(A, X, \varnothing)$ system with the constitutive parameters shown in Figure 7.12. Left-hand panel: plots of the effective potential $\Phi$ from Eq. (7.103), with global minima marked, for $\epsilon \in \epsilon_{\text{crit}} + [-0.002, 0.002]$. (These are the counterpart, for a specific class of non-equilibrium fluctuations, to the plots of the equilibrium effective potential for the Curie model shown in Figure 7.4.) The horizontal dashed line is tangent to both minima of $\Phi$ at $\epsilon_{\text{crit}} \approx 0.01166$. Right-hand panel: dependence of the minimizing argument $\bar{x}$ on $\epsilon \in (0, 2.5 \times \epsilon_{\text{crit}})$, corresponding to the marked points on the left-hand panel. The discontinuity in $\bar{x}$ at the transition is the distance between the two minima which have equal depth.

The potential (7.78) is graphed for several values of $\epsilon$ at fixed $(x_1, x_2)$ in Figure 7.13. Its minima occur at those $x$ which are stable zeros of the mass-action rate equation (7.76) shown in Figure 7.12. Its maxima occur at the unstable zero.

A "full" escape is a trajectory leading from $\bar{x}_\pm$ to the unstable fixed point, from where ordinary mass-action kinetics can take the trajectory with probability $\sim \mathcal{O}(1)$ to either fixed point. Detailed balance between fixed points therefore occurs where the escape probabilities (7.79) for the two minima are equal, which is the value of $\epsilon$ where the minima of $\Phi$ have equal depth. The condition that $\Phi$ have two equal minima corresponds to the criterion for a critical temperature in a first-order phase transition, that its Gibbs free energy have equal values in both phases.

The sections shown in the left-hand panel of Figure 7.13 may be compared to the Landau free energy in Eq. (7.49) that accounts empirically for the first-order gel/liquid-crystal phase transition in lipids. Apart from the fact that the potential (7.103) is less symmetric than the fourth-order polynomial Landau expansion for the lipid free energy, the two show the same qualitative behavior, with the leakage rate $\epsilon$ in the reaction network serving the role as control parameter that temperature $T$ serves for the lipid phase transition.

In a system with large $n_0$, trajectories will spend almost all their time near the lowest minima of $\Phi$. Therefore, if $\epsilon$ is slowly varied, at the critical value $\epsilon_{\text{crit}} \approx 0.01166$, the reaction system will jump discontinuously from being almost always on one branch of $x$ to being almost always on the other, shown as the dashed transition in the right-hand panel of Figure 7.13. The variation must be slower than the characteristic time $\sim 1/P_{\text{escape}}$, however, in order for the phase dependence to be single valued. Faster changes of $\epsilon$ can leave the system in metastable states, which decay on timescales comparable to the escape rate. The range of $\epsilon$ for which $\Phi$ has two local minima is bounded, just as for the spinodal in a first-order phase transition. Outside these limits, the mass-action kinetics convert the system from any initial condition to its stable phase on intervals of order unity in $\tau$.

### 7.6.4.6 Extended-time trajectories

Most macro-trajectories, whose probability as fluctuations we may wish to estimate, cannot be specified from single events the way escapes can be. For these more elaborate trajectories, we require a large-deviations function that is a time integral of an appropriate **entropy rate** characterizing the instantaneous suppression of increments along the path. Formulae of the latter kind are also derived in the Appendix. To illustrate that a well-known effective action for diffusion systems is in fact a non-equilibrium entropy, we review the effective action of Onsager and Machlup, published in 1953 [609].

We first write any path $x = \bar{x} + x'$ as a sum of a solution $\bar{x}$ to the classical mass-action rate equations (7.76) from whatever initial conditions we are given, and a remainder $x'$ whose likelihood as a fluctuation we would like to estimate to leading exponential order. The probability for path $x$ has the form

$$P[\bar{x}] \sim e^{-S^{\mathrm{OM}}}, \tag{7.80}$$

where now we use square brackets to indicate that the argument of $P$ is a *function $x(\tau)$* rather than simply an instantaneous state. $S^{\mathrm{OM}}$ is the Onsager–Machlup effective action, which for this system has the form

$$S^{\mathrm{OM}} = \frac{n_0}{2} \int d\tau \, \frac{\left[ \left( \partial_\tau - \frac{\partial}{\partial \bar{x}} \frac{d\bar{x}}{d\tau} \right) x' \right]^2}{\epsilon + \bar{x} \, (\bar{x} + x_1) \, (\bar{x} + x_2)}. \tag{7.81}$$

The rate at which a fluctuation history $x'$ may decay toward $\bar{x}$ without suppression is simply the gradient of the classical rate equation (7.76), given by

$$\frac{\partial}{\partial \bar{x}} \left( \frac{d\bar{x}}{d\tau} \right) = \left( x_1 x_2 - \bar{x}^2 \right) - 2 \, (\bar{x} - x_1) \, (\bar{x} - x_2). \tag{7.82}$$

One can check (see Eq. (7.111) in the Appendix) that, if $x'$ is taken to be the difference between any actual solution $\bar{x}$, and the same trajectory displaced by a small time interval $\delta\tau$ (which is automatically also a solution), then the numerator in Eq. (7.81) vanishes on that $x'$. In other words, the most-likely fluctuation $x'$ is a trajectory that would make $\bar{x} + x'$ another solution to the classical rate equations. Any other history is suppressed to the extent that it deviates from this property. The reference scale that determines what counts as a significant deviation is the denominator term $\epsilon + \bar{x} \, (\bar{x} + x_1) \, (\bar{x} + x_2)$ in Eq. (7.81), which is the variance of Poisson fluctuations about the mean reaction rates.

The Onsager–Machlup effective action is only valid to quadratic order in $x'$, and only valid for expansions about classical solutions (which may simply be fixed points). The

Appendix shows how this is generalized to include fluctuations about escape trajectories, or more complex formulae that go beyond quadratic order.

### 7.6.5  *The lightning strike analogy*

We close this section with an analogy, to bring together the various threads of the argument with an intuitive example, and to show that the mathematical relations we are proposing are already familiar from ordinary processes. The proposal is that the emergence of energy flow through metabolic pathways in the early Earth – and over the longer term, life as a whole – was akin to a lightning strike, but that instead of cutting a channel through physical space, metabolism cut a channel through the *graph of chemical reaction possibilities*.

#### 7.6.5.1  *Mathematical similarities*

Lightning is dielectric breakdown of dry air. Various atmospheric processes create a charge separation between a cloud and the ground or between two clouds, and a voltage stress across the intervening atmosphere. This voltage stress is a non-equilibrium boundary condition, because voltage is a kind of chemical potential (for electrons), which could not differ across locations in a fully equilibrated system. A lightning strike shares the following properties of large-deviations origin with the biosphere, though in a much simpler realization.

**The mismatch of equilibrium and transport states**  Dry air is the equilibrium (Gibbs) state for nuclei and electrons at cool atmospheric temperatures, but it is a poor conductor of electricity. Ionized plasmas are far better conductors, but they are not thermal equilibria at atmospheric temperature. This difference sets up the tension between the low-temperature equilibrium of most of the gas, and the non-equilibrium stress in the boundary conditions.

**Disequilibrium boundaries and the potential for order**  The free energy from charge separation offers the potential to maintain a disequilibrium state in some tiny fraction of the atmospheric volume, but not in the whole volume. Therefore, in order for an actual lightning strike to form, a *kinetic* concentrating mechanism must exist to move a large fraction of excess internal energy into a small region of space – a redistribution that would strongly violate maximization of the equilibrium entropy if this were an equilibrium system.

**Kinetic concentrating mechanism**  The ion cascade is such a mechanism. If a small region of gas can become ionized through chance fluctuations, the voltage can accelerate the freed electrons and increase their kinetic energy. These mobile electrons, upon colliding with neutral atoms at the ionization front, create a chain reaction liberating new electrons and forming new ions. The result is a local region of high-temperature, high-conductivity plasma, through which most of the separated charge can flow to neutralize its counter-charge at the other end of the channel.

**Necessary order** The lightning strike, like fractures, hurricanes, or many other kinds of breakdown, illustrates that under sufficient stress, a failure process that deviates from the Gibbs equilibrium becomes not only possible but, if one waits long enough, certain. All three of these examples also show that the earliest failure can occur with far less externally available energy than would be required to change the whole medium to a different Gibbs equilibrium. Only a tiny channel (usually millimeters wide) ionizes in a lightning strike that may be kilometers long. In a fracture, the rupture of crystalline bonds occurs across a single atomic diameter. The deformation at the fracture tip concentrates diffuse stress energy from the entire bulk of the medium onto individual bonds around the tip. In the bulk, the distributed stress is far too weak to break crystalline bonds, but when stress from trillions of displaced atoms is focused at a single link, it becomes strong enough to break a covalent bond. Thus fracture constitutes a very simple instance of concentrating energy from a diffuse background to a density sufficient to do simple chemistry.

**Reductionism and unpredictability** Lightning illustrates the characteristic relation of chance and necessity that we have come to associate with phase transitions. The threshold needed to produce ionization, and the degree of ionization in a given channel, can be well predicted from a knowledge of the density and composition of the gas, the driving voltage, and the store of charge available to flow through the channel. Yet the microscopic position of the channel is highly unpredictable. It forms from locally random processes and is sensitive to tiny fluctuations in density or velocity of the air into which the ionization leader grows. Hence long lightning strikes zigzag, and like snowflakes, each one is different. If you are a rare unlucky person struck by lightning, such differences may be very significant, but the overall contributions of lightning to the electrochemistry and energy distributions of the atmosphere do not depend on them at all, and follow from the reductionist part of their description.

**The use and limitation of equilibrium state variables** We can use equilibrium state variables such as temperature or conductivity to describe the core of the ion channel and the surrounding neutral air, and these variables capture much of the shape of the local phase-space distributions for nuclei, electrons, or atoms in a near-equilibrium approximation. However, to understand why lightning is a common and, ultimately, inevitable consequence if a gas is placed under sufficient dielectric stress, these state variables and the equilibrium entropy computed from them clearly do not provide the correct maximization principle. We must also recognize the electric current flowing through the medium, and the greater likelihood it creates for random collisions to send the whole system into states that preserve and extend the local ion channel, than into states that dissipate it.

**Equilibrium entropy production is necessary but not sufficient** If we take a *very* coarse-grained view, of total charge and energy distributions, and of times long before and long after the lightning strike, we recover the observations of Boltzmann and Schrödinger. The transient lightning strike has increased entropy in the surrounding world by converting voltage to heat. We may even extend this description to the near-equilibrium state variables during the quasi-steady state while charge is flowing (a timescale that may last only a fraction of a second but is long compared to microscopic equilibration times). The "entropy decrease" in forming the spatially uneven plasma channel within a neutral volume of air will be less than the entropy increase from the transported charge that has flowed through the channel to that point.

It is clear, however, that this necessary condition from equilibrium thermodynamics is not enough to predict whether the strike can occur, or what form it will take. The balance that maintains the plasma is between an entropy of the channel/air state, and an *entropy rate* at the charge reservoirs due to the rate of charge transport between them. To balance an entropy against an entropy rate requires the appropriate timescale. This timescale is set by the driving and quenching of the ion cascade, and it is what the non-equilibrium entropy function of histories must supply.

### 7.6.5.2 *Limitations of the analogy*

Analogies must always be used with caution. They can show how new phenomena are partly familiar and thus bring intuition to bear, but ultimately we use analogies as a crutch when we have not yet understood new systems fully in their own terms. Therefore, we note what is valid and what is invalid in the lightning analogy:

**The right laws of large numbers** The lightning strike is a good analogy to the role we propose for early metabolism to the extent that it reflects the appropriate laws of large numbers in a dynamical setting. It makes intuitive contact with the properties of phase transitions that are familiar in equilibrium, but shows the insufficiency of the equilibrium state variables and entropy to capture essential timescales, and the necessity to include kinetics in estimates of most likely histories.

**The wrong space in which order is formed** The lightning model can mislead, however, if we focus on it as an example of order *in physical space*. Chemistry lives in a more complicated space, which we showed in Section 7.4.5 is a kind of Cartesian product. Physical space is one factor in this product, and continues to govern diffusion. However, to model chemistry we must multiply each point in physical space by a copy of the complete hypergraph of possible compounds and reactions. Organization can occur in the reaction graph even if the reactor in physical space is well-mixed. It is not necessary that different chemical potentials exist at different points in space, but that they exist between different molecular species *in the reaction graph*, so that (for example) bonding pairs must find paths between low-potential electron donors and higher-potential acceptors. Thus, we must take the mathematical intuition that the lightning strike provides for physical space, and relocate it into the reaction network.

**Boundary-layer models are ultimately physical models** Reaction-diffusion models such as the Belousov–Zhabotinsky reaction or the Brusselator, in which chemical differentiation occurs in diffusive boundary layers, have been popular examples of non-equilibrium order formation to put forth as analogies to the spontaneous emergence of life. It is important to appreciate, however, how little *chemical* complexity such examples use. Few reactions are possible, and these are tightly limited by the boundary sources that feed the reactor. Structure forms, not as a result of competition among pathways for concentration, but as a balance between reaction times and times for diffusion across spatial concentration gradients. The resulting order is "soft" – no large free energy differences (relevant to the mass-action dynamics) are maintained within a distance of atomic diameters in space – and in that way they resemble other common examples of soft order in physics, such as weather patterns, vortex sheets, sand ripples, etc. Also like other common diffusive patterns, the kinds of hierarchical complexity these dissipative quasi-chemical structures can form is limited. They can only become more complex in space, because no new chemical hierarchy is possible, and the patterns in space are limited by the few low-level diffusive patterns and the large energy dissipation they require to maintain.

The lightning analogy, fractures, or hurricanes, are similarly best regarded as models of organization in physics or physical chemistry. They incorporate a limited set of chemical transformations (ionization, crystal-bond rupture, evaporation/precipitation), but they are all boundary-layer phenomena.[85] Their order forms in physical space, and there is neither complexity nor an opportunity for hierarchy within their chemical possibility graphs.

**Too much symmetry and too little constraint** The low potential for complexity, which makes these physical examples poor analogies for the complexity in the biosphere, ultimately derives from the high symmetry of physical space compared to the reaction graph. The complexity of a lightning strike or fracture path is low because it is generated by a local random process of tip propagation, among positions that are all essentially alike. How different this weakly guided random walk is, from the path of a bonding pair through the diverse reactions of organic chemistry between its low-potential initial electron donor and its high-potential ultimate acceptor!

As important as the excess of symmetry is the dearth of constraint. Failure propagation in physical space can be written for the lightning strike or the fracture, as a random walk on a simple graph. In contrast, the chemical state of metabolism lives on the much more complicated space of a hypergraph. The requirements for concurrency in the hypergraph lead to a much more complex space of constraints than similar transport properties on simple graphs.[86]

---

[85] Even the stress field around a fracture tip is a boundary layer in continuum elasticity.

[86] Here we may directly compare the problem of the search for network autocatalysis in graphs and hypergraphs. Wim Hordijk and Mike Steel [366] have derived a (fast) polynomial-time algorithm to search for network-autocatalytic loops in a graph, when the graph is generated from an actual chemical reaction system by projecting out only the requirement that a species be present, but not keeping its concentration or stoichiometric coefficients. In contrast, Jakob Andersen *et al.* [25] have shown that the same search in a fully stoichiometric hypergraph is NP-complete, and thus admits no fast algorithm. Their example is one of a very large number of NP-completeness results known for hypergraphs.

### 7.6.6  Conclusion: the frontier in the study of collective and cooperative effects

The forms of order we find in the biosphere should define a frontier for the study of collective and cooperative effects in statistical mechanics and information theory. Like any frontier in science, this one should lead us to new concepts and abstractions that are as distinctive from those we now know, as the character of life is from the parts of the non-living world on which science has heretofore gained some traction. At the same time, we are not entirely without maps that may help us to explore this frontier. Not everything we find will be unprecedented, and not everything will need to be built from ignorance. We close with thoughts about what we expect to remain the same across the frontier and what seems likely to be new.

The study of critical phenomena in the late nineteenth and twentieth centuries changed our conception of objects and interactions, and introduced a variety of useful calculating tools and methods. Among these were renormalization and the uses of effective theories. The systems we understood with these tools, however, were all simple in the crucial sense that they were all highly symmetric. The best understood cases – magnets, pure crystals, liquid–vapor transitions, and elementary particles in the vacuum – are all spatially homogeneous. Even the most complex cases currently under study, the glasses and disordered systems [255, 556], while they are spatially heterogeneous, are still *statistically* homogeneous in their disorder, and this is crucial to their tractability with current methods.

The other factor that has contributed to tractability of study cases of phase transitions, in addition to their strict or statistical spatial homogeneity, is that the internal configuration spaces studied (these are termed "fibers" in a construct known as a *fiber bundle*) have largely been simple groups: discrete point groups, the rotation group, the unitary groups, and so on. Through the restrictions that both kinds of simplicity entail, the role of symmetry in organizing collective and cooperative effects across scales has been paramount. Phase transition theory as we understand it is the theory of states defined by symmetries, and mediated by collective fluctuations as these either respect or hide symmetries.

To apply what we have learned about collective and cooperative effects to the emergence and organization of the biosphere, we will need to extend the study of laws of large numbers and robustness into the domain of more highly structured elementary configuration spaces. The simplest phase transitions may be defined within bulk (but driven, non-equilibrium) chemistry, and in that sense may bear a passing resemblance to the phase transitions that define the hierarchy of matter, but most transitions that underpin life have involved structure both in chemistry and in space.[87] The internal chemical configuration space itself, rather than being generated by a small collection of operators with a structurally simple

---

[87] The precedent in particle physics for a phase transition that inherently requires structure jointly in an internal symmetry group and in physical space is the confinement transition in quantum chromodynamics that produces the hadrons and mesons – protons, neutrons, $\pi$-mesons, etc. – from quark and gluon constituents. Even this transition, involving the relatively simple internal symmetry group SU(3), proved very difficult to understand quantitatively, and has only gradually come to be tractable with heavy use of numerical integration combined with the renormalization group as an organizing conceptual framework. A few other combined space/internal ordered states, such as magnetic vortex states in superconductors, have been simpler to understand approximately with elementary methods, though these have the advantage that they are defined through deformations of elementary ordered states that exist as spatially homogeneous bulk phases.

algebra, is generated by as many as perhaps a few dozen elementary functional-group transformations. Which operations can follow which others depends on structural relations within the molecules they create. These kinds of dependencies are not simple to represent in algebraic terms, and require more complex abstractions.[88] Finally, structure in driven systems involves fundamentally non-local interdependencies whereas its antecedents in equilibrium have involved only local quantities, a result that has been elegantly emphasized by Bertini and co-workers [72] even for the simple case of temporally steady-state diffusion in one dimension. Many living processes make use of cyclicity on one or more timescales as an inherent mechanism to create and maintain order, a form of temporal interdependence that introduces yet higher-order structure. We return to this qualitatively distinct aspect of living systems in Section 8.3.3.1.

One of the most central questions about the frontier of collective and cooperative effects that led to the biosphere is whether symmetry will continue to be the paramount organizational concept that it has been for the hierarchy of matter. Already the study of glasses has driven the concept of expressed and broken symmetries into very indirect notions of symmetry that do not reflect equivalence either in space or in the internal configurations, but which refer to the reproducibility of effects in a highly abstract mathematical domain where the dependence on the dimension of replicates *as a scaling variable* [255] is key. Will inhomogeneity require similar inventiveness for biological order? Will the complex structure of molecules and reactions lead to no natural sense in which one pathway can be compared to another that we would recognize as a symmetry transformation? These outcomes seem possible. On the other hand, it must be remembered that in non-equilibrium systems, the sudden expansion of the relevant probability measures from ensembles of states to ensembles of histories introduces an enormous new class of symmetries which are eligible to be expressed or hidden, most importantly the symmetry of time translation itself.[89] While perhaps the familiar internal symmetry groups will not continue to be useful to describe phase transitions in driven chemistry, a large new class of relevant symmetries may take their place. These are conceptual questions of fundamental importance to our understanding of how and why biological order has arisen, but their answers must arise from the technical solutions to a hierarchy of chemically explicit cases.

## 7.7 Technical appendix: non-equilibrium large-deviations formulae

The stochasticity of single reaction events[90] is the source of spontaneous noise in the trajectories described by time-dependent probability distributions via the master equation.

---

[88] The distinctive feature of chemistry, which resembles a property of computation, is that transformations defined from one set of functional groups can *bring into existence* new molecular configurations that are the functional groups or contexts for other reactions. Promising approaches to automating such generative systems include the *double push-out automaton* representation of graph grammars [26, 27] for chemical reaction systems, or the Kappa language [174] for even more complex dependencies in systems biology. (See also www.kappalanguage.org.)

[89] For an exploration of new roles of time-translation symmetry in non-equilibrium entropies, using worked examples from population processes, see [768].

[90] The first systematic development of these methods to transition state theory was made by Henry Eyring. See the classic text by Glasstone, Laidler, and Eyring [304].

However, single reaction events often furnish an inconvenient basis in which to solve for the large-deviations behavior of the reaction network, because it is the rare sequential, cumulative fluctuations by many particles that produce excursions in the macrostate. More direct approaches to extract these sequences are often provided by the Laplace transform of the density $\rho_n$, which is called its *generating function*.

Since the late 1960s and early 1970s, a family of powerful methods has been developed to compute generating functions for both equilibrium and non-equilibrium stochastic processes. The equilibrium moment-generating functions are of course the partition functions, and their logarithms, the cumulant-generating functions, are the free energies. The advancement of these methods for classical stochastic processes has benefited from exchange with a very similar class of algebraic and field-theoretic methods developed for the ensembles describing quantum mechanics and quantum field theory.

In this appendix we provide basic constructions, descriptions of analytic methods, and a few algebraic forms and results, for the non-equilibrium phase-transition example introduced in Section 7.6.4 of the main text. With apologies to the reader, we do not attempt a systematic review of algebraic and field-theoretic methods to analyze generating functions, which would exceed the scope of this discussion. Extensive reviews from a mathematical perspective are available in [224, 278], and reviews from methods more familiar to physicists may be found in [408, 531, 767, 768, 770].

### 7.7.1 Master equation for the one-variable model of autocatalysis

We begin with the reaction scheme and transfer matrix introduced in Section 7.6.4.

The stochastic process is formulated in terms of single-particle events. The discreteness of integer changes, which are Poisson distributed in time, is the fundamental source of stochasticity and the determinant of its magnitude. Therefore we repeat here the mass-action rate equation (7.70) for $n$, using the descaled variables $\epsilon$, $x_1$, $x_2$ of the text as constitutive parameters:

$$\frac{dn}{d\tau} = \epsilon n_0 - x_1 x_2 n + \frac{n^2}{n_0} - \frac{n^3}{n_0^2}. \tag{7.83}$$

The master equation (7.77) in the main text, with indices written explicitly, is

$$\frac{d\rho_n}{d\tau} = \sum_{n'=0}^{\infty} \mathbb{T}_{nn'} \rho_{n'}. \tag{7.84}$$

All reactions in scheme (7.64) change n by $\pm 1$, so the transfer matrix is defined entirely in terms of the rate at which independent events lead to n $\rightarrow$ n + 1, given (in descaled variables) by

$$r_n^{(+)} = n_0 \left( \epsilon + \frac{n(n-1)}{n_0^2} \right), \tag{7.85}$$

and the rate at which independent events lead to $n \to n - 1$, given by

$$r_n^{(-)} = n_0 \left( x_1 x_2 \frac{n}{n_0} + \frac{n(n-1)(n-2)}{n_0^3} \right). \tag{7.86}$$

In these equations, unlike the mass-action rate equations, we make a specific choice that the reactant and catalysts must be *distinct* particles. Therefore higher-order rates such as $n^2$ must be replaced by $n(n-1)$ etc. For $n_0 \gg 1$ this makes no difference in quantitative results, of course, but it produces algebraically simpler forms for later quantities.

For unit changes in n, rather than write the indices explicitly as in Eq. (7.84), we may indicate the downward unit shift operator by $e^{-\partial/\partial n}$, and the upward shift operator by $e^{\partial/\partial n}$, and write the transfer matrix as an operator in the form[91]

$$\mathbb{T} = \left( e^{-\partial/\partial n} - 1 \right) r_n^{(+)} + \left( e^{\partial/\partial n} - 1 \right) r_n^{(-)}$$

$$= n_0 \left\{ \left( e^{-\partial/\partial n} - 1 \right) \left( \epsilon + \frac{n(n-1)}{n_0^2} \right) + \left( e^{\partial/\partial n} - 1 \right) \left( x_1 x_2 \frac{n}{n_0} + \frac{n(n-1)(n-2)}{n_0^3} \right) \right\}. \tag{7.87}$$

### 7.7.2 Ordinary power-series generating function and Liouville equation

The ordinary power-series generating function for $\rho$, also called the *moment-generating function*,[92] is an analytic function $\psi(z)$ of a complex argument $z$, given by

$$\psi(z) \equiv \sum_{n=0}^{\infty} z^n \rho_n. \tag{7.88}$$

When $\rho$ evolves under Eq. (7.84), $\psi$ evolves under a differential equation known as the *Liouville equation*, which has the form

$$\frac{d}{dt} \psi(z) = -\mathcal{L} \left( z, \frac{\partial}{\partial z} \right) \psi(z). \tag{7.89}$$

For classical stochastic processes, the Liouville equation is usually a kind of diffusion equation, and the Liouville operator $\mathcal{L}$ generalizes the Laplacian for uniform diffusion.

---

[91] We can check that $d \sum_n \rho_n / dt = 0$ so probability is conserved. The only shift operator that eliminates a real element of $\rho$ is the term $\left( e^{\partial/\partial n} - 1 \right)$ acting at $n = 0$. Probability is being added from contributions from all $\rho_{n>0}$, but being subtracted from all $\rho_n$ including $n = 0$. However, this is a downshift term, and there is no lower entry into which probability can flow from $\rho_0$. Fortunately, the rate for this term is at least linear in n, so there is no contribution at $n = 0$, and all probability fluxes balance.

[92] We use the term "ordinary power series" (or OPS) here [877] because the analytic functions built as powers of a complex variable are only one form in a menagerie of generating functions that may be constructed to extract correlations of interest from a density $\rho$. The OPS generating function is based on the algebra generated by $(z, \partial/\partial z)$ acting in a space of polynomials. The use of other algebras, which may be crafted to give most direct access to other correlations than those extracted by the Laplace transform, is the domain known as *duality* [822].

$\mathcal{L}$ is determined from the form of $\mathbb{T}$ in Eq. (7.87), and the sign is chosen so that $\mathcal{L}$ will have positive eigenvalues. For this problem

$$\mathcal{L} = n_0 \left(1 - z\right) \left(\epsilon + \frac{z^2}{n_0^2} \frac{\partial^2}{\partial z^2}\right) + (z - 1) \left(x_1 x_2 + \frac{z^2}{n_0^2} \frac{\partial^2}{\partial z^2}\right) \frac{\partial}{\partial z}. \tag{7.90}$$

The Liouville equation (7.89) may formally be integrated (or "reduced to quadrature"), by exponentiating the Liouville operator, as

$$\psi_T = e^{-\int_0^T d\tau \, \mathcal{L}} \psi_0, \tag{7.91}$$

to express the generating function at a time $T$ in terms of the initial function at time 0 (indicted by subscripts to $\psi$). Because this stochastic process has no explicit time dependence, the operator $\mathcal{L}$ commutes at different times, so the operator exponential may be defined as a simple power series in $\mathcal{L}$.[93]

### 7.7.2.1 Quadrature of the generating function as a functional integral

A construction originally due to Masao Doi [199, 200] and Luca Peliti [630, 631], but which has counterparts for stochastic differential equations and many other representations [278, 759, 767, 768], allows us to write the generating function $\psi(z)$ at time $T$ as a functional integral. The idea is to break the quadrature (7.91) into small increments of time $\Delta\tau$, and between each increment of evolution $e^{-\Delta\tau\mathcal{L}}$ to insert a representation of unity in a basis whose elements are the generating functions for an overcomplete set of Poisson distributions. A number field analogous to n is reintroduced (though indirectly[94]), as the mean value indexing each Poisson distribution in the basis.[95] We denote this field by $n$ in math Italic. It is equivalent to the thermodynamic average occupancy that appears in the mass-action rate equation (7.83), except that it is allowed to run over all positive values with a distribution whose first moment will give the classical mean.

A distinguishing feature of this approach to evaluating generating functions is that a *second* field must be introduced, which turns out to be both Legendre dual and also – remarkably – Hamilton–Lagrange dual to $n$. (We return to explain this Hamiltonian duality in a moment.) As a Legendre dual to $n$, the field (which we denote by $\eta$) has the interpretation of a chemical potential. It is related to the chemical potential in equilibrium thermodynamics that defines the chemical work needed to reversibly induce a shift in a concentration of some particle species. In the functional integral, the second field has the effect of extracting the moment from any generating function that defines its correct projection onto the corresponding basis function in the Poisson representation of unity.

---

[93] Under more general conditions of explicit time dependence in the master equation, it would be necessary to use a time-ordered exponential.

[94] One construction begins not with $n$ directly, but with a basis of coherent states [531], which must then be converted to number $n$ by a canonical transformation [767]. A different construction begins with $n$ directly [637, 759].

[95] Since Poisson distributions are indexed by a single parameter, which is both mean and variance, a scalar $n \in [0, \infty)$ is sufficient to index this basis set.

The result of the standard construction is a form for $\psi_T$ at a late time $\tau = T$, in terms of an initial function $\psi_0$ at $\tau = 0$, given by

$$\psi_T(z) = \int_0^T \mathcal{D}\eta \, \mathcal{D}n \, e^{(ze^{\eta T}-1)n_T - S} \psi_0\left(e^{\eta_0}\right). \tag{7.92}$$

The functional measure $\int \mathcal{D}\eta \, \mathcal{D}n$ is a product measure over configurations of $\eta$ and $n$ at a closely spaced sequence of consecutive times. The normalization and other details of its construction may be found in [767].

With these substitutions, the kernel of the functional integral is given by[96]

$$S = \int d\tau \, (\eta \partial_\tau n + \mathcal{L}) - \int d\tau \partial_\tau (\eta n)$$
$$= n_0 \int d\tau \left( \eta \partial_\tau x + \hat{\mathcal{L}} \right) - n_0 \int d\tau \partial_\tau (\eta x), \tag{7.93}$$

in which the Liouville operator (7.90) has become a field-valued Liouville *function* of the form

$$\mathcal{L} = n_0 \left[ (1 - e^\eta) \left( \epsilon + \frac{n^2}{n_0^2} \right) + (1 - e^{-\eta}) \left( x_1 x_2 \frac{n}{n_0} + \frac{n^3}{n_0^3} \right) \right]$$
$$= n_0 \left[ (1 - e^\eta) \left( \epsilon + x^2 \right) + (1 - e^{-\eta}) \left( x_1 x_2 x + x^3 \right) \right] \equiv n_0 \hat{\mathcal{L}}. \tag{7.94}$$

Relative to the Liouville operator on analytic functions (7.90), the complex scalar $z$ is replaced by $e^\eta$ and $\partial/\partial z$ is replaced by $ne^{-\eta}$

Our motivation in writing the transfer matrix $\mathbb{T}$ in terms of shift operators $e^{\pm \partial/\partial n}$ and number indices n was to be able to skip the analytic ordinary power-series generating function altogether. After all the procedure of converting to a functional integral has been carried through, the kernel functional $S$ has the general form (7.93), in which $\mathcal{L}$ is obtained from $\mathbb{T}$ in Eq. (7.87) by changing sign and replacing $\partial/\partial n$ by $-\eta$ and n by $n$. In the second line of Eq. (7.94) the descaled field $x \equiv n/n_0$ in the same way as it was defined in the mass-action rate equation.

The solution for properties of $\psi_T$ now proceeds through a variety of approximations and transformations of the kernel $e^{-S}$ in Eq. (7.92). Methods of steepest descents starting from local minima of $S$ may be carried out by expanding $S$ as a power series in small differences of the integration variables $(\eta, n)$ from a set of stationary-path solutions.[97] In the same way as the argument $z$ allows the generating function $\psi$ to compute moments of $n$ at a single time, introduction of additions to $\mathcal{L}$ known as *source terms* may be used to

---

[96] The total derivative in Eq. (7.93) is one of several kinds of surface terms in the functional integral (7.92) that can be chosen to vanish in all stationary-path approximations and do not affect the subsequent evaluations, so from now on we will drop them from the notation.

[97] Other standard representations, such as the Langevin stochastic differential equation approximation, may also be derived directly from Eq. (7.92). See [767] for a standard construction using the Hubbard–Stratonovich transformation.

convert Eq. (7.92) to a generating functional, from which we may compute moments of the history $n(\tau)$ over intervals of time.

The cumulant-generating functional obtained from Eq. (7.92) is the correct non-equilibrium generalization of the equilibrium Gibbs free energy. Its functional Legendre transform with respect to the sources added to $\mathcal{L}$ is the correct generalization of the equilibrium entropy that determines fluctuation probabilities in the large-deviations limit.

### 7.7.2.2 A Hamiltonian dynamical system from a stochastic process

Although the original stochastic process had no conserved particle number and no kinematic momentum variable, the field integral for the time-dependent generating function has introduced these structures for the pair $(\eta, n)$. The function $S$ in Eq. (7.93) has the form of an *action functional* [309], in which the Liouville operator serves as the *Hamiltonian*. Here $n$ serves as a position coordinate and $\eta$ is its conjugate momentum under $\mathcal{L}$. We may derive the resulting conservation law (and the corresponding conserved phase-space volume element) by solving for the stationary paths of this action. Recalling that the action appears in an exponential in the path integral (7.92), these stationary paths will be important as the starting points for steepest descent approximation of the integral.

*Stationary paths* of the action are paths along which the first variational derivative vanishes. Denoted by $(\bar{\eta}, \bar{x})$, they satisfy the equations of motion

$$\partial_\tau \bar{x} + \overline{\frac{\partial \hat{\mathcal{L}}}{\partial \eta}} = 0$$

$$-\partial_\tau \bar{\eta} + \overline{\frac{\partial \hat{\mathcal{L}}}{\partial x}} = 0,$$

(7.95)

in which overbar denotes evaluation of the partial derivatives of $\hat{\mathcal{L}}$ at $(\bar{\eta}, \bar{x})$. It is immediate to check from Eq. (7.94) that if $\bar{\eta} \equiv 0$, both $\hat{\mathcal{L}}$ and $\partial \hat{\mathcal{L}}/\partial x \equiv 0$ at all $\bar{x}$, so this solution for $\bar{\eta}$ is consistent. At these solutions, the equation of motion for $\bar{x}$ (the first line of Eq. (7.95)) reproduces the mean-field rate equation (7.76) in the main text. Hence the $\bar{\eta} \equiv 0$ manifold contains the solutions to mass-action relaxation from all initial conditions for $\bar{x}$.

All solutions to the stationary-path equations (7.95) satisfy

$$\frac{d}{d\tau} \overline{\hat{\mathcal{L}}} = 0.$$

(7.96)

Since solutions to the classical mass-action equations begin with $\overline{\hat{\mathcal{L}}} = 0$, all smooth solutions to the stationary-path equations preserve $\overline{\hat{\mathcal{L}}} = 0$ through time. Note that for these solutions, since $\bar{\eta} \equiv 0$ also, the leading term in the action $S \equiv 0$. Hence trajectories expanded about any of the mass-action laws are the leading, unsuppressed terms in the large-deviations limit.

### 7.7.3 Escape trajectories and effective potential for non-equilibrium phases

In addition to the solutions to the classical mass-action laws, an important class of *locally least-action solutions* to the stationary-path equations exist, which characterize macroscopic transitions between *different* stationary points of the classical rate equations. These are the non-equilibrium counterparts to nucleation events in equilibrium first-order phase transitions, by which a metastable phase may collapse to the stable phase, or (at the critical point), two coexisting phases may exchange matter in a condition of detailed balance. The non-classical stationary paths still have $\overline{\hat{\mathcal{L}}} = 0$ but have $S = \int d\tau \, \bar{\eta} \partial_\tau \bar{x} > 0$. Therefore they have non-zero exponential suppression in the large-deviations limit, making nucleating transitions rare events.

For one-dimensional systems, including this model with its single dynamical variable $x$, the non-classical solutions turn out to be easy to write down directly. The condition $\overline{\hat{\mathcal{L}}} = 0$ implies that the $\bar{x}$ solutions, which we will refer to as *escape paths* from the attracting stationary points, are the time-reverses of the solutions to the mass-action rate equations. Therefore, along an escape path,

$$\left. \frac{d\bar{x}}{d\tau} \right|_{\text{escape}} = \bar{x}_1 \bar{x}_2 \bar{x} + \bar{x}^3 - \epsilon - \bar{x}^2. \qquad (7.97)$$

The required corresponding solution for $\bar{\eta}$ is given by

$$\bar{\eta}|_{\text{escape}} = \log \left( \frac{\bar{x}_1 \bar{x}_2 \bar{x} + \bar{x}^3}{\epsilon + \bar{x}^2} \right). \qquad (7.98)$$

If we denote by $S^{(0)}$ the solution for $S$ along any of these paths, which is the 0th-order expansion of the field integral (7.92) in fluctuations about its stationary paths, then for any monotonic trajectory with $\overline{\hat{\mathcal{L}}} = 0$ we have

$$S^{(0)} = \int d\tau \, \bar{\eta} \partial_\tau \bar{x}. \qquad (7.99)$$

Equation (7.99) is general, and may be used along any stationary solution. For monotone trajectories, in which $\bar{\eta}$ has a unique value at each $\bar{x}$, the time dependence may be removed and $\bar{\eta}$ may be regarded as a function of a dummy variable $x$ of integration, to obtain

$$S^{(0)} \rightarrow \int dx \, \bar{\eta}(x). \qquad (7.100)$$

For relaxation trajectories $\bar{\eta} \equiv 0$, and for escapes it is given by Eq. (7.98).

Escape trajectories are paths from either of the stable fixed points of the rate equation (7.76) to the unstable fixed point. The evaluation of $S^{(0)}$ along any such path gives the leading term in the log of the rate of escapes, as

$$S_{\text{escape}} = \int_{x_{\text{stable}}}^{x_{\text{unstable}}} dx \, \log\left(\frac{x_1 x_2 x + x^3}{\epsilon + x^2}\right). \tag{7.101}$$

For escapes from the lower stable point, the integral to the unstable fixed point is upward in $x$ and the logarithm is also positive. For escapes from the upper stable point to the unstable point, the integral is downward and the logarithm is negative. Therefore $S_{\text{escape}}$ is positive in all cases.

### 7.7.3.1 Detailed balance of escapes and the counterpart to the Landau free energy for instantaneous non-equilibrium states

The condition for *detailed balance*, that the upward and downward escape rates are (at leading exponential order) the same, is then

$$S_{\text{escape, lower}} - S_{\text{escape, upper}} = \int_{x_{\text{lower stable}}}^{x_{\text{upper stable}}} dx \, \log\left(\frac{x_1 x_2 x + x^3}{\epsilon + x^2}\right) = 0. \tag{7.102}$$

Therefore we may regard

$$\Phi(\bar{x}) \equiv \int^{\bar{x}} dx \, \log\left(\frac{x_1 x_2 x + x^3}{\epsilon + x^2}\right) \tag{7.103}$$

as an *effective potential* or "Landau free energy" governing the relative persistence times in either low or high reaction-rate states.

Note that $\Phi(\bar{x})$ is only defined so far for comparisons among different $\bar{x}$ at the same values of the constitutive parameters $\epsilon$, $x_1$, $x_2$. The appropriate starting point of integration, and hence the relative offset of $\Phi(\bar{x})$ at different constitutive parameters, is not something for which we have given a prescription.[98]

If we denote the high-flux and low-flux fixed points by $\bar{x}_{\pm}$, then $\Phi(\bar{x}) - \Phi(\bar{x}_{\pm})$ is the large-deviations rate function for *instantaneous* excursions from $\bar{x}_{\pm}$ as the most recent past asymptotic fixed point. It is most useful for predicting the relative occupation times at $\bar{x}_+$ and $\bar{x}_-$ in a long time series that may involve several transitions. If we denote these occupation times by $\tau_{\pm}$, the ratio is given to leading exponential order by

$$\frac{\tau_+}{\tau_-} \sim e^{-n_0[\Phi(\bar{x}_+) - \Phi(\bar{x}_-)]}. \tag{7.104}$$

When $\Phi(\bar{x}_+) \neq \Phi(\bar{x}_-)$, one state is exponentially suppressed at large $n_0$, and as $n_0 \to \infty$, only the minimizer of $\Phi$ will be occupied at all times.

The condition of detailed balance when $\Phi(\bar{x}_+) = \Phi(\bar{x}_-)$, in which a long time span may include intervals of both phases even if $n_0$ becomes large, is the non-equilibrium equivalent to the condition of equal free energy minima that defines the **critical temperature** in a first-order phase transition. In the zero-dimensional but extended time reaction

---

[98] In the main text, we set the lower limit of integration at zero, so that $\Phi(0) = 0$ at all values of $\epsilon$, $x_1$, $x_2$.

network, the coexistence of the two phases occurs in time, rather than in space as water and steam or water and ice may coexist.

If the constitutive parameters, $\epsilon$ or $x_1$, $x_2$ are changed slowly, the reaction network may be *metastable* for times less than the escape time $\tau_{\text{escape}} \sim \exp(-S_{\text{escape}})$, where $S_{\text{escape}} = n_0 \left[ \Phi(\bar{x}_{\text{unstable}}) - \Phi(\bar{x}_{\pm}) \right]$.

The exponential separation of timescales between reaction events and escape events, with even slower changes in constitutive parameters, permits us to consistently speak of phase changes between non-equilibrium steady states, even though each steady state is itself determined by minimizing an action functional over trajectories. In general, we do not wish to use the same value $n_0$ to characterize the nucleation time as the value that characterizes the whole-system stability. In familiar first-order transitions, the entire volume need not pass through the unstable configuration between phases. Only a *local* nucleation volume must make the transition. To realistically describe local nucleation, however, requires a richer model than the single-phase model we have developed here.

### *7.7.3.2 A comment on the nucleation potential and energetics*

The nucleation potential $\Phi - \min_{\bar{x}}(\Phi)$ is much like an equilibrium Gibbs free energy in two ways. First, it is a function of state and not an integral of a rate that depends on the time interval considered.[99] Second, it is constructed much as the Gibbs free energy is, which gives the probability of a macrostate fluctuation in an equilibrium ensemble: $n_0 \Phi$ is an integral of a change in particle number $dn$ against a field $\eta$ with the interpretation of a chemical potential shift that would instantaneously maintain that particle number.

However, the $\Phi$ constructed for escapes in this non-equilibrium ensemble *does not reduce* to any simple interpretation in terms of work, dissipation, or "production" of the equilibrium entropy. This is true, even though equilibrium chemical potentials remain approximately valid as state variables for the reactant and product systems. The reason $\Phi$ is not a measure of work is that it depends only on the *aggregate* kinetics (something $\leftrightharpoons X$) – the only information written into the master equation – but it does not disaggregate different sources or sinks. To see the irreducibility explicitly, we rewrite $n_0 \Phi$ in the original $n$ variables and evaluate its components:

$$n_0 \Phi(\bar{x}) = \int^{\bar{n}} dn \, \log \left[ \frac{(k_2 + \bar{k}_3) n + \bar{k}_1 n^3}{n_A (k_3 + k_1 n^2)} \right]$$

$$= \int^{\bar{n}} dn \left\{ \log \left( \frac{n}{n_A K_3} \right) + \log \left( 1 + \frac{k_2}{\bar{k}_3 + \bar{k}_1 n^2} \right) \right\}. \tag{7.105}$$

---

[99] This statement is true because the subtraction of $\min_{\bar{x}}(\Phi)$ effectively allows the integral to start at one of the stable fixed points, which implies an asymptotically slow departure. Any shift to a finite value of $\bar{x}$ is a trajectory with a finite transit time (that is, of order 1 in $\tau$ units), independent of the asymptote. It is the characteristic transit time of escape trajectories that contributes to $\Phi$ and not the asymptote. Because of their finite transit time and their low frequency in the large-deviations limit, escapes of this kind are called *instantons* [149], borrowing a term used for an identical construction in equilibrium field theory methods for evaluating generating functions and functionals.

The first logarithm has the interpretation of $(1/k_B T)$ times the difference of chemical potentials between the $X$ and $A$ system, which integrated $\int dn$ is a measure of chemical work. However, the net rates also receive a contribution from the second reaction $X \to \varnothing$ through $k_2$, which is essential to the existence of three fixed points. The energy dissipation through this reaction is a free variable that does not affect the kinetics, because $k_2$ depends only on the one-particle chemical potential of $X$ and the transition state free energy for the conversion to $\varnothing$.

The important message is that the large-deviations structure of the system is determined by the rates that appear in the master equation, which draw from a combination of equilibrium and transition state chemical potentials. The map between kinetic structure and the equilibrium chemical potentials alone need not be either an injection or a surjection: either system may have parameters left unspecified by the other.

### 7.7.4 Gaussian fluctuations about stationary-path backgrounds

The stationary-path solutions to Eq. (7.95) are the trajectories that are least suppressed in entropy – either globally, in the case of the solutions to the mass-action kinetics for $\bar{x}$ with $\bar{\eta} \equiv 0$, or locally compared to surrounding trajectories, as in the $S^{(0)} > 0$ solutions. All other trajectories are suppressed as macro-history fluctuations by additional terms, which we now compute. We do this in the functional integral, using extended-time sources to make a generating functional, and then taking a functional Legendre transform of this generating functional to define the entropy of histories which is the large-deviations function. Motivations, and details of the general construction, are given in [767].

For Gaussian-order fluctuations, as for the simple binomial distribution of Section 7.4.1, the leading exponential dependence of fluctuations is simply given by the exponential weight function $e^{-S}$ in the functional integral for $x$, after integrals over $\eta$ fields have been performed. We write the fields of integration $(\eta, x) = (\bar{\eta}, \bar{x}) + (\eta', x')$, where $(\bar{\eta}, \bar{x})$ is a solution to the stationary-point equations (7.95), and expand to successive orders in $(\eta', x')$. The zero-order expansion is $S^{(0)}$, the evaluation of $S$ on the stationary-path solution. The linear-order term $S^{(1)} \equiv 0$, because the coefficients of $\eta'$ and $x'$ are the first variational derivatives (7.95).

### 7.7.4.1 Recovering and generalizing the Onsager–Machlup effective action

The term at second order in $(\eta', x')$, which we denote by $S^{(2)}$, after $\eta$ fields have been integrated, may be written

$$S^{(2)} = \frac{n_0}{2} \int d\tau \left\{ -\left( \overline{\frac{\partial^2 \hat{\mathcal{L}}}{\partial \eta^2}} \right)^{-1} \left[ \left( \partial_\tau + \overline{\frac{\partial^2 \hat{\mathcal{L}}}{\partial \eta \partial x}} \right) x' \right]^2 + \overline{\frac{\partial^2 \hat{\mathcal{L}}}{\partial x^2}} (x')^2 \right\}, \qquad (7.106)$$

in terms of the second *partial* derivatives of $\hat{\mathcal{L}}$ evaluated on a stationary background.

Let us denote the full expansion to second order, including $S^{(0)}$ in case it is non-zero, as

$$S^{OM} \equiv S^{(0)} + S^{(2)}, \tag{7.107}$$

with $S^{(0)}$ given by Equation (7.99). Equation (7.107) is a generalization of the form first derived by Lars Onsager and Stefan Machlup in 1953 [609]. The original form, missing $S^{(0)}$ and the term $\overline{\partial^2 \hat{\mathcal{L}}/\partial x^2}$ in $S^{(2)}$, applies to neighborhoods of solutions to the mass-action rate equations, where both terms vanish. The more general form applies also to escape trajectories.

Expressions for the second partial derivatives that appear in Eq. (7.107) are

$$\frac{\partial^2 \hat{\mathcal{L}}}{\partial \eta^2} = -e^{\eta} \left( \epsilon + x^2 \right) - e^{-\eta} \left( x_1 x_2 x + x^3 \right)$$
$$\rightarrow - \left[ \epsilon + x \left( x + x_1 \right) \left( x + x_2 \right) \right], \tag{7.108}$$

$$\frac{\partial^2 \hat{\mathcal{L}}}{\partial \eta \partial x} = -e^{\eta} \left( 2x \right) + e^{-\eta} \left( x_1 x_2 + 3x^2 \right)$$
$$\rightarrow 2 \left( x - x_1 \right) \left( x - x_2 \right) + \left( x^2 - x_1 x_2 \right), \tag{7.109}$$

$$\frac{\partial^2 \hat{\mathcal{L}}}{\partial x^2} = \left( 1 - e^{\eta} \right) 2 + \left( 1 - e^{-\eta} \right) 6x.$$
$$\rightarrow 0 \tag{7.110}$$

(The second line in each of these expressions is the form when $\eta \rightarrow 0$, appropriate to classical solutions.)

Equation (7.108) is negative-definite for positive $\epsilon$ and $x$. This is the source term for fluctuations in the non-equilibrium fluctuation-dissipation theorem [408]. It is the sum of aggregate half-reaction rates for ($X \rightleftharpoons$ anything) in both directions.

Equation (7.109) is minus the gradient of the mean reaction rate, and can be of either sign. It is the regression rate for fluctuations, which is positive near stable fixed points and negative near unstable fixed points.

### 7.7.4.2 Consistency of different expansions

The large-deviations function for some paths may be computed in more than one way, so it is useful to check that different computations yield the same answer. In particular, the escape paths are given an action $S^{(0)}$ from Eq. (7.99) when they are computed as stationary non-classical solutions to the equations of motion (7.95). The same trajectories may be inserted in the Onsager–Machlup expression for $S^{(2)}$ in Eq. (7.106), regarded as fluctuations about a fixed-point steady state. In the latter case, the whole trajectory must be given and not just its endpoint $\bar{x}$. Equation (7.99) is an exact solution, but when expanded to quadratic order in $\bar{x} - \bar{x}_\pm$, it gives the same result as Eq. (7.106). Moreover, if the value of $\bar{x}$ is given as an endpoint constraint, the escape trajectory produces the *minimum* value of $S^{\mathrm{OM}}$, and in this sense is specified by an event.

### 7.7.4.3  Zero-eigenvalue eigenvectors of the Onsager–Machlup integral

It turns out that $S^{(2)}$ has exactly one eigenvector of zero eigenvalue if $(\bar{\eta}, \bar{x})$ contains an escape trajectory from a stable fixed point to the unstable point.[100] All other eigenvectors of fluctuation have positive eigenvalues. The zero mode corresponds to the time derivative of the escape solution. We prove that fact as follows.

If $(\bar{\eta}, \bar{x})$ is a solution, the fact that the underlying dynamics is time translation invariant implies that $(\bar{\eta}, \bar{x}) + \delta\tau\,(\partial_\tau\bar{\eta}, \partial_\tau\bar{x})$ is also a solution in the limit that the time offset $\delta\tau \to 0$.[101] We can therefore expand Eq. (7.95) in powers of $\delta\tau$ for any such solution, and equate terms to leading order in $\delta\tau$ to obtain a set of identities relating the trajectory and partial derivatives of $\hat{\mathcal{L}}$. The equality implied at linear order in $\delta\tau$ is

$$
\left(\partial_\tau + \overline{\frac{\partial^2\hat{\mathcal{L}}}{\partial\eta\partial x}}\right)\partial_\tau\bar{x} + \overline{\frac{\partial^2\hat{\mathcal{L}}}{\partial\eta^2}}\partial_\tau\bar{\eta} = 0
$$
$$
\left(-\partial_\tau + \overline{\frac{\partial^2\hat{\mathcal{L}}}{\partial\eta\partial x}}\right)\partial_\tau\bar{\eta} + \overline{\frac{\partial^2\hat{\mathcal{L}}}{\partial x^2}}\partial_\tau\bar{x} = 0.
$$
(7.111)

If we multiply the first line of Eq. (7.111) by $\partial_\tau\eta$, and the second line by $\partial_\tau x$ and subtract the second line from the first, we obtain

$$
\partial_\tau\left(\partial_\tau\bar{\eta}\,\partial_\tau\bar{x}\right) = -\overline{\frac{\partial^2\hat{\mathcal{L}}}{\partial\eta^2}}(\partial_\tau\bar{\eta})^2 + \overline{\frac{\partial^2\hat{\mathcal{L}}}{\partial x^2}}(\partial_\tau\bar{x})^2.
$$
(7.112)

If $\delta\tau\,(\partial_\tau\bar{\eta}, \partial_\tau\bar{x})$ is used for $(\eta', x')$ in Eq. (7.107), then the (generalized) Onsager–Machlup action evaluates on this fluctuation trajectory to

$$
S^{\mathrm{OM}} = S^{(0)} + (\delta\tau)^2\frac{n_0}{2}\int d\tau\left\{-\left(\overline{\frac{\partial^2\hat{\mathcal{L}}}{\partial\eta^2}}\right)^{-1}\left[\left(\partial_\tau + \overline{\frac{\partial^2\hat{\mathcal{L}}}{\partial\eta\partial x}}\right)\partial_\tau\bar{x}\right]^2 + \overline{\frac{\partial^2\hat{\mathcal{L}}}{\partial x^2}}(\partial_\tau\bar{x})^2\right\}
$$
$$
= S^{(0)} + (\delta\tau)^2\frac{n_0}{2}\int d\tau\left\{-\left(\overline{\frac{\partial^2\hat{\mathcal{L}}}{\partial\eta^2}}\right)(\partial_\tau\bar{\eta})^2 + \overline{\frac{\partial^2\hat{\mathcal{L}}}{\partial x^2}}(\partial_\tau\bar{x})^2\right\}
$$
$$
= S^{(0)} + (\delta\tau)^2\frac{n_0}{2}\int d\tau\,\partial_\tau\left(\partial_\tau\bar{\eta}\,\partial_\tau\bar{x}\right).
$$
(7.113)

[100]  For multiple escapes, the number of independent zero eigenvalues equals the number of escapes.
[101]  More generally, the solution $(\bar{\eta}, \bar{x})$ offset by *any* value of $\delta\tau$ is another solution, but to use it we must keep the entire Taylor series in $\delta\tau$.

If a time-dependent solution $(\bar{\eta}, \bar{x})$ is produced by initial conditions, the total derivative in the second term of Eq. (7.113) may not vanish, but for any internal dynamical solution where $\partial_\tau \bar{\eta} \, \partial_\tau \bar{x}$ either vanishes or is periodic at the boundaries of the $\int d\tau$, the second term vanishes and the fluctuation $(\eta', x')$ is un-suppressed. Two very important classes of solutions possess this property.

1. **Classical solutions to the equations of motion with time-dependent background** Examples include limit cycles or more complex attractors with $\partial_\tau \bar{x} \neq 0$. These solutions have $\bar{\eta} \equiv 0$ and $\partial^2 \hat{\mathcal{L}} / \partial x^2 \equiv 0$ always, so $x' \propto \partial_\tau \bar{x}$ is always a zero-eigenvalue fluctuation of the (original) Onsager–Machlup action. On a limit cycle, or on an attractor over sufficiently long times that an initial condition is closely approached later in the trajectory, the limits of integration may be chosen so that surface terms cancel.
2. **Internal non-classical (local least-action) solutions such as escape trajectories** For these solutions $\bar{\eta} \neq 0$ but $\partial_\tau \bar{\eta} \, \partial_\tau \bar{x} \to 0$ far from the transition. The consequent vanishing of the total derivative in Eq. (7.113), together with the fact that $-\partial^2 \hat{\mathcal{L}} / \partial \eta^2 \geq 0$ everywhere, implies that $\partial^2 \hat{\mathcal{L}} / \partial x^2 \leq 0$ somewhere. However, we are nonetheless assured that $S^{OM}$ has no eigenvectors of negative eigenvalue, because the kernel of $S^{OM}$ is an elliptic differential operator and $\partial_\tau \bar{x}$ is a nodeless eigenvector, so it must be the eigenvector of lowest eigenvalue.

$S^{OM}$ is the functional of trajectories that correctly generalizes the equilibrium entropy as the log-probability of fluctuations. Its scale factor is $n_0$, and $S^{OM}/n_0$ is the large-deviations rate function, which takes the scale-independent trajectory $x(\tau)$ as its argument. If $\partial_\tau \bar{x} \equiv 0$, the large-deviations function has only positive eigenvalues. If $\partial_\tau \bar{x} \neq 0$, it has a single zero-eigenvalue eigenvector $x' \propto \partial_\tau \bar{x}$ which is unsuppressed relative to the background $\bar{x}$.

### *7.7.4.4 The meaning of unsuppressed fluctuations*

All solutions $x' \propto \partial_\tau \bar{x}$ which have zero eigenvalue reflect the fact that time offset is a *symmetry* of the underlying dynamics, which is spontaneously hidden by any non-constant solution $\bar{x}$. The theorem that time-dependent classical solutions may diffuse without mean-regression as an expression of the underlying but hidden time-translation symmetry is the stochastic process version of **Goldstone's theorem**. The theorem is well known from the elementary particle theory of phase transitions as the cause of masslessness (or the approximation of masslessness in models) of particles such as the $\pi$-meson. In condensed matter it is the explanation for the fact that sound waves formed by freezing satisfy a wave equation with long-wavelength solutions. In stochastic processes, it implies that periodic solutions such as limit cycles are *noisy clocks* [768], which attract toward the orbit of the cycle but within that orbit perform a random walk from exact periodicity. Because all solutions within the orbit are admitted equivalently, random fluctuations that advance or retard the solution slightly on the orbit have no reference by which to regress.

The same theorem when applied to escape trajectories shows that a partition function for the non-equilibrium stochastic process must include an integral over all times at which the escapes may occur. This integral, which is finite for any finite $T$, replaces the Gaussian integral over the offset $\delta\tau$ in Eq. (7.113) which appeared to diverge. Sidney Coleman [149] (Chapter 7) shows for equilibrium systems how this integral over a dilute gas of escape trajectories – including a functional determinant from the Gaussian integral, which we do not compute here – exponentially mixes asymmetric initial distributions back to ergodically sampled distributions over fixed points at late times. The counterpart for stochastic processes evolves any initial generating function $\psi_0$ with an exponential decay term in which the leading exponential dependence is the one-time escape (7.101), and the initial rate is given by the functional determinant.[102] By such exponential corrections, a generating function for a distribution initially concentrated around either $\bar{x}_+$ or $\bar{x}_-$ is replaced with a mixed distribution with the relative occupation frequencies of Eq. (7.104) at late times. When the constitutive parameters place the system at its critical point so that $\Phi(\bar{x}_+) = \Phi(\bar{x}_+)$, this is the familiar result that, even when fast dynamics hide an underlying symmetry by concentrating trajectories around a single fixed point, long-term dynamics at the macroscale restores this symmetry.

---

[102] See [767], Eq. (77) for a discussion of this correction.

# 8

# Reconceptualizing the nature of the living state

Here we bring together the empirical regularities from the first five chapters and the stability perspective from Chapter 7 to sketch an integrated theory of the emergence of the biosphere. Further required ideas, taken up in this chapter, include arguments for the importance of modular architecture in hierarchical complex systems, applications of the stability perspective to the problem of hierarchical control, and an explanation of the way modularity supports control by buffering errors. We interpret the subsystem decompositions reviewed in the empirical chapters as modules created by phase transitions, which made possible the emergence of hierarchical control, but dictated the architectures for which stability at the whole-system level was possible. The same basic relation – the affordance of modules that buffer errors – is the mechanism whereby low-level laws have repeatedly constrained the forms and functions of higher-level assemblies in biology. An essential element in the biosphere's emerging distinctness from abiotic Earth systems has been the emergence of individuality, a complex concept instantiated in many ways, and precondition for Darwinian evolution. For us individuality is not primitive, but is an organizational motif that emerges at intermediate stages in systems that already possess significant structure. Therefore Darwinian dynamics also is not a sufficient starting point to define the nature of life. We propose an alternative essence for the nature of life, in which the biosphere as a whole is the defining level of organization, energy transport through covalent bond chemistry is the essential function, and the complexity of the chemical substrate is an essential source of both complexity and stability.

## 8.1 Bringing the phase transition paradigm to life

The patterns presented in the first five chapters suggest that life grew out of geochemistry. Metabolism is the living subsystem closest to the lawfulness of geological processes. The ecosystem is the level of organization that carries the necessary and universal subnetworks within core metabolism, and that has hosted their accretion over geologic and evolutionary timescales.

In Chapter 7 we proposed that biogenesis took the form of a cascade of non-equilibrium phase transitions. Redox stresses, perhaps along with other chemical potentials (protons? phosphates?), forced electron flow through the graph of possible chemical reactions. Auto-catalysis (at both network and single-molecule levels), by a hierarchy of first short and then longer loops, created sufficient positive feedback that the flow was concentrated into selective channels. The two important parts of this proposal are that assemblies of complex organic molecules provided the paths of least resistance for free energy relaxation, and that the effects of positive feedback were mathematically similar to the feedbacks that create a lightning strike, except that the "channel" is selected among paths through the reaction graph rather than among paths through space.

The phase transition paradigm offers concepts of cause or necessity for the emergence of a biosphere on Earth, and it relates these to life's subsequent persistence and its observed robustness under the perturbing influence of fluctuations. A phase transition origin of life has a natural continuity with geochemical processes that are understood in similar terms, and makes it sensible to think of selection, or of metabolism's being a source of constraint and information, without appeal to the more complicated conditions required by Darwinian evolution.

A theory of the origin of life, however, must not only be continuous with geochemistry at its beginning: in its culmination it must arrive at the many characteristics that make living systems distinct from the Earth's abiotic systems. The concept of phase transition is one input to such a theory, but the description so far has been too simple to embed these ideas adequately in biology.

In this chapter we attempt to provide a more complete synthesis of the many roles ordered phases have played, and continue to play, in structuring the origin of life and the organization of the biosphere. Here we make contact with some of the motifs that were central in our empirical chapters, including hierarchical and modular architecture and the propagation of constraints across levels. The attempt to connect origins of bio-logical order to first principles of robustness, and to maintainability and evolvability of hierarchical complex systems, provides part of the motivation behind our choice of pat-terns emphasized in the earlier chapters. Many of the relations among levels in living systems are relations of control, for which a formalism of control theory is available from engineering. Stabilizing hierarchical control systems inherently involves asymptot-ically reliable error correction, bringing in the other perspective on phase changes from Chapter 7.

All these considered together create a view of life that is inverted in three ways from the usual characterization.

1. In several universal properties of metabolism and in some higher-level features of cellularization and molecular replication, constraint and information appear to have originated in low-level law-like processes and not in the more complex higher-level sys-tems that nominally control the lower levels today. (In the understanding of emergence we propose from arguments in Chapters 5 and 6, the need for control originated with

the integration and interdependence of subsystems that were then freed from disparate geochemical settings that had previously sustained them.)

2. In the progression from abiotic geochemistry to modern complex life, the ecosystem is a more fundamental and law-like carrier of metabolic regularities than the individual organism. Ecosystems are not merely assembled communities of individuals, but metabolically integrated systems in their own right, partly independent of the identities of their constituents in some characteristics of input/output, and imposing prior constraints on the ways member species can be assembled to form them.

3. Individuality becomes a complex and constructed form of organization that emerges from a chemical context already structured by kinetic and energetic selection. The Darwinian concepts of replication and competition, which are predicated on the existence of individuals that can serve as units of selection, are likewise delayed in their onset and reduced in importance, to become only one mechanism among several that generate living order and information.

Accepting these three inversions does more than suggest a language and scenario for the origin of life. It forces us to reconceptualize the nature of the living state. Many kinds of order, from different sources, are equally inherent to the function and persistence of the biosphere. Recognizing this makes life more continuous with other material systems, but it emphasizes the insufficiency of a characterization of life based principally on evolution.

Therefore in this chapter we will also propose a characterization of the living state. *Life is not an attribute possessed by individuals, but a role defined through participation in the biosphere. The biosphere exists as an energy-flow channel through organic chemistry, which is only reached through elaborate synthesis of mediating structures. Without a biosphere, these domains of chemical-reaction space would go unaccessed (or accessed only at very low rate) by abiotic processes.* These are the state and the function associated with living matter, corresponding to the three phases of condensed matter and their characteristic chemistries that are associated with the atmosphere, hydrosphere, and lithosphere.

### *Order of the presentation*

The remainder of the introduction summarizes the technical arguments of the chapter. We first motivate the specific need for a theory of robustness from basic observations about the disruptive effects of error on dynamical order. We recall key characteristics of ordered phases from Chapter 7, and then show how these dictate the paths of easiest assembly in hierarchical systems. The key theorems from control theory governing the possibility of control are then introduced, and the contribution of ordered phases to buffering is summarized. We then introduce the idea that it is only the universal features of life, anchored in lawfulness and predictability, that make open-ended evolution compatible with whole-system stability. A formalization of this idea is provided in terms of concepts of asymptotically optimal error correction from Section 7.5.7. The complex nature of individuality – the problem of what it is, why it arises, and what it does – is then argued to underlie some of the most fundamental differences between static and dynamic ordered states, and to be something that must be explained rather than taken for granted or simply assumed to arise in molecular systems. This is the most important argument against the motivation for models of self-replicating RNA considered in isolation from a protometabolic foundation. These observations lead to the reconsideration of the nature of life, and the essential role of energy flow, and particularly of the chemical substrate.

### 8.1.1  Necessary order in the face of pervasive disturbance

> Things fall apart; the center cannot hold;
> Mere anarchy is loosed upon the world.
>
> – W. B. Yeats, *The Second Coming*

In living systems, *every level, from all atomic locations and bonds, to control, replication, and selection, is pervasively and continually subject to perturbing fluctuations*. At a molecular level these include thermal fluctuations; in chemical systems they often include solvent effects and variations in electrical or magnetic environment due to fluctuations of molecular arrangement; in information systems they arise from errors in symbol persistence or symbol transmission; in higher-level processes they may result from aggregate phenomena, including randomness in birth and death, mutations, or fluctuations in the states of environments.

The problem of containing the effects of perturbation so that they do not either accumulate, or worse amplify, to the point of system collapse is *quantitative*. It is not enough to say that one or another mechanism tends to produce order, the way one can say that Darwinian evolution in populations of organisms with *given* ontogeny and *given* heredity can narrow the population composition on a fitness peak. It is necessary to show that a mechanism can produce reliable regression to *all* the requirements of order in realistic environments. The equilibrium phase transition paradigm has taught how difficult a condition this can be to satisfy in systems with many degrees of freedom that produce spontaneous order, but has also provided examples demonstrating that the conditions can be met.

To apply the same ideas to the biosphere we must formulate them for non-equilibrium systems with complex architecture. Even were we only to address the empirically known features of life without trying to explain them, this would require at the least modeling the autonomy and control problems between chemistry, individuals, and ecosystems. However, this multilevel structure may not be merely an empirical regularity. The formation of the motifs of individuality, replication, competition, and selection may be aspects of optimizing error correction in dynamical systems which have no conceptual precedents in equilibrium. In that case, the emergence of architecture as well as optimization in the context of architecture becomes part of the pattern to be explained.

#### 8.1.1.1  Principles from the phase transition paradigm that form a starting point

We summarize the key conclusions from the phase transition paradigm that are general enough that they must apply to the biosphere just as they apply to abiotic systems.

**Symmetry provides the null hypothesis** A system must have some average behavior, even if its components vibrate. The notion of "order," regarded as a surprising condition that requires explaining, needs a null hypothesis for what would count as the absence of order. The null hypothesis for systems that undergo second-order phase transitions is provided by the *symmetry of the microscopic dynamics*. Since fluctuations arise from the microphysics, if it is symmetric they should be symmetric, and the average behavior toward which the system regresses would be expected to show the symmetry of its underlying dynamics. Gibbs equilibrium chemical ensembles satisfy this criterion. Their boundary conditions are contents of energy, volume, and conserved atomic numbers (or indirectly, temperature, pressure, and chemical potentials). Fluctuations into and out of molecular bonding states populate all bonds in proportion to the likelihood that random energy fluctuations and geometric arrangements are sufficient to form them. The concentration of different molecules is determined by the resulting random combinatorics of bond assembly subject only to these global constraints.

**Deviation from symmetry requires cooperative effects** "Surprising" states of order in stochastically perturbed systems are those in which the average behavior *does not* regress to a state with the same symmetry as the microscopic dynamics. These are the states of frozen order. Since microscopic fluctuations affect every degree of freedom, a sufficiently pervasive force to affect every micro-variable is required to lead the system to a frozen ordered state. Since frozen states are (by definition) not those with the symmetry of the microscopic laws, such a force can only have its origin in the system's own dynamics. The name given to the mutually reinforcing interactions that produce regression away from states of highest symmetry is *cooperative effects*.

**Freezing has thresholds: the mere presence of ordering effects is not enough** It is easy to propose microscopic interactions that favor correlated or ordered states. Electron spin alignment through magnetic fields in paramagnets is such a mechanism. So is the van der Waals force that can stick two molecules together in space. Autocatalytic self-amplification in networks is another example, as is Darwinian selection in populations. The important mathematical lesson learned through the study of phase transitions is that the mere existence of such interactions does not ensure the emergence of macroscopic order. Increasing the strength of the ordering effects, or reducing the strength of the fluctuations, does not gradually make a system more ordered. Rather, the system remains completely disordered over a finite parameter range, showing that the presence of order-inducing interactions is not sufficient. At a threshold ratio in the strength of the ordering and disordering effects, even though these interactions have not changed character and have only barely changed in relative strength, the system suddenly becomes macroscopically ordered, entraining infinitely many degrees of freedom if it is infinitely large. The problem of forming and maintaining order is a problem of *global sufficiency of mean-regression*, which must be understood over and above the presence of order-favoring mechanisms.

These properties have their origin in the laws of large numbers, and so they presume a high degree of equivalence among the microscopic interactions. Systems with small-scale heterogeneity may not aggregate in as simple a way, and may show more gradual ordering.[1] The basic implication between stable order and cooperative effects, however, should remain valid even outside the idealized case. The stable order that requires an explanation is order not already written into the microscopic interactions or imposed by the boundary conditions. For this order the necessary explanation can only come from cooperative effects in the dynamics *which are sufficiently pervasive* to produce reliable mean-regression against all the small-scale fluctuations which would otherwise drive the system to the least order required by the microscopic laws.

### 8.1.1.2  Global error accounting and biology

The equivalence of the thermodynamic concepts of joint mean-regression in ordered thermal states to problems of optimal error correction was developed in Section 7.5 on Shannon's capacity theorem and error-correcting codes.

The picture of a distributed error-correcting system, in which individual deviations from a dynamical target state are identified and corrected by reference to the system as a whole, is already used widely in biology. What is missing from the existing biological picture, however, is a mathematical theory of sufficient conditions for self-maintenance.[2]

The existing theory of phase transitions shows that, even in remarkably simple systems with high levels of equivalence among micro-variables, the criteria for order are not simple. In biology the same subtleties should arise, and the technical difficulties will be much greater. First, the ordered living state is dynamic rather than static. Second, both fluctuations and the couplings that can induce order involve multiple timescales and interactions among multiple levels of aggregation. Background reactions among metabolites, effects from correctly formed and mis-formed catalysts, physiology, and selection at multiple levels feed both error correction and error up and down in biological hierarchies. To perform the error accounting in such an enormously complex system from its "engineering" details is at present an unimaginable task. What a paradigm such as freezing allows us to recognize is that such an accounting must exist, that we should expect it to involve feedbacks among many small-scale variables, and that it is precisely the multiplicity of these feedbacks that enable the stability of the ordered state.

### 8.1.2  The role of phase transitions in hierarchical complex systems

Any single phase transition, resulting from the cooperative effects of a large number of similar degrees of freedom, is architecturally flat, like the binary phase transition from

---

[1] Current work on glasses, however, shows that significant disorder can still lead to sharp transitions [255]. Glasses are random in structure as well as in dynamics on microscopic scales. As long as the structural randomness is *statistically homogeneous*, however, they can still show sharp transitions between liquid and frozen states.

[2] Examples exist in very limited domains, such as Manfred Eigen's criterion for *error catastrophe* [213] in genome replication. Such models, however, are extremely simplified and context dependent. A full theory of sufficient conditions for self-maintenance would require as its context only the rules of chemistry and the geochemical and energetic boundary conditions on the Earth.

Section 7.4.1. The space of variation and the interactions of the variables may involve components with many characteristic scales, but with regard to the collective effects that lead to a single kind of spontaneously emergent order, other degrees of freedom are boundary conditions and not participants. When degrees of freedom with many scales may all undergo cooperative formation of order, they do so through multiple elementary phase transitions perhaps influenced by each other's state of order.[3]

If the biosphere were a simple system resembling ordered abiotic systems such as hurricanes, Bénard cells, or lightning strikes, a non-hierarchical notion of phase transition might provide not only a compelling metaphor but also a route to a full explanation for its existence.[4] Some aspects of life almost appear this simple, in that they are small, apparently universal, and apparently indefinitely persistent. The metabolic core network and some aspects of cell form are examples. But these features that are simple, and which we can recognize as simple because of their universality, do not undergo evolutionary change, do not manifest historical contingency, and neither manifest nor by themselves explain most of the hierarchical complexity associated with life. They are part of life but they cannot constitute a complete account of its nature.

In Section 8.2 we consider the consequences of hierarchical organization for the problem of global error correction and whole-system stabilization. We continue to regard collective mean-regression in ordered phases as the most natural framework to explain globally sufficient error correction, but we must develop a richer theory of coupled multilevel systems to explain how we think this mechanism has acted to create and maintain the biosphere.

### *8.1.2.1 Why a phase transition account of metabolism alone is not enough*

Although it is perhaps most intuitive that core metabolism and some physical properties of the cell might be explained as non-equilibrium ordered phases, it is important for several reasons not to conflate the phase transition paradigm merely with energy-driven order or merely with bulk-phase order. The most important reason is that neither autocatalytic metabolism nor any aspects of cell function beyond the simplest lipid/water partitioning exist on Earth today except where they are maintained by the full complexity of biotic structure and dynamics. While core metabolism may not "evolve" in the sense that it may not show variation or ongoing innovation, it nonetheless depends on the full support structure of catalysis and bioenergetics that is maintained by ontogenetic, evolutionary, and ecological dynamics. Whether this reflects an intrinsic role for hierarchy, or a displacement of earlier abiotic supports by more effective biotically self-produced systems,[5] a notion of

---

[3] The lipid phase diagram reviewed in Section 7.4.6 furnishes an example of the rich structure of phases that can result when degrees of freedom characterized by several closely spaced phases are all susceptible to cooperative effects. Other examples include the complex phases of silicate minerals, resulting as these do from combinations of bond orientation and space-group symmetries.

[4] Were that the case, however, the biosphere would then become just another elementary physical system, not demanding new paradigms. One of the challenges in the following sections, if a case can be made that the core layers are explained by simple phase transitions, will be to argue that those early transitions are not "prebiotic" or "abiotic," but that they are the foundation layer within a biosphere which continues to function at other levels by the same rules.

[5] We believe that each of these is important in some domains. We return in Section 8.5 to what it might mean for hierarchy to be necessary to the emergence of ordered phases in dynamical systems.

phase transition that did not both give rise to, and then use, the observed forms of hierarchical order would be missing the central requirement for an explanation of error correction, which is its global closure.

Even if a narrower interpretation of the role of phase transitions – restricting to energetically driven bulk physical processes – were to provide a self-contained account of the geochemical roots of metabolism, and even if we could argue that some of the mechanisms responsible for order in the early world continued to constrain later more complex evolution, it would be desirable not to view this as the essence of the paradigm. Such an interpretation would cordon metabolism off as a geochemical feature of "proto-life," and perhaps regard the persistent mechanisms as a scaffold of "physical constraints" on evolution,[6] leaving the essence of life to be understood in other terms. This situation would be an improvement over "frozen accident" views of metabolism, in that it would give principled reasons for the continuity of early and late organosynthesis, and a frame within which extant biochemical structure and evolution could be interpreted as carrying information about origins. However, it would erect a division between invariant and changeable features, as being respectively "prebiotic" versus reflecting what makes life distinct; we think such a division is false. Invariant and variable features are both inherent in the nature of life, and stabilizing and adaptive roles for selection contribute jointly to the preservation of living order. The appropriate role for a theory of global stability is to encompass and explain the interplay of chance and necessity.

### 8.1.2.2 Error propagation and error correction in hierarchical systems

Hierarchical architecture fundamentally alters the dynamics of error generation, propagation, and correction. For example, within a cell, most thermal fluctuations are harmless because small-molecule geometries regress toward standard states due to the structure and energetics of quantum state spaces. Some fluctuations, however, lead to isomerizations or deleterious (or simply wasteful) side-reactions, whose products must be identified and either repaired or degraded and excreted, and whose reactants may need to be replenished due to loss. Many fluctuations in the molecular inventory of cells may be automatically sensed and regressed by allosterically controlled enzymes [376], but some require active mechanisms of sensing and alteration of gene expression. Even with the high-volume error-correction mechanisms of biochemistry and cell physiology, many small differences are not sensed, and must be left to filtering by natural selection that purges unfit genomes. At each stage in this hierarchy, mechanisms at a coarse, slow scale correct many errors at finer scales. However, error creation as well as error correction can be multiplied in a hierarchy. Mutant genomes can produce harmful or simply non-functional catalysts. Mis-transcribed or mis-translated catalysts may create multiple errors in the flow of their target

---

[6] In evolutionary modeling it is generally desirable to distinguish order that is an *ex post* expression of selection from constraints on the mechanisms that generate variation (though there will generally be a gray area for forms that are marginally viable). Among the constraints, biologists also sometimes separate separate "physical constraints" regarded as abiotic in origin from "evolved constraints," which may have been produced by evolutionary dynamics in the past but serve as hard limits on the variations that can be generated in the future.

substrates before they are identified and removed, or may even kill the organism if they are not removed.

Hierarchy therefore may facilitate the attainment of global error correction by partitioning errors and limiting the domains within which different kinds of error may propagate. At the same time, it may greatly complicate our ability to trace and understand the mechanisms of error creation and error correction. Errors generated at one level may propagate to other levels and either attenuate or multiply before they are finally identified and corrected. The challenge for any paradigm of comprehensive error correction is to identify the conditions for global closure, and their limitations. It is the problem that the simple phase transition paradigm addresses for flat systems, but now complicated by cross-level and multiscale interactions. Therefore we must add to the basic statistical understanding of spontaneously ordered phases a theory of how these interact with architectural motifs in hierarchical systems.

### 8.1.2.3 *Concepts of architecture: modularity and optimal control*

A number of writers on organization in general and biological organization in particular have argued that **modular architecture** is a necessary precondition to hierarchical complexity, and is often enhanced as that complexity evolves. We devote Section 8.2 to the arguments of Herbert Simon [754, 755, 756], which serve both as a framework for the interpretation of facts, and somewhat more formally, as a principled prior expectation for Bayesian model selection in the search for appropriate theories of organization. Simon argues for an empirical generalization that hierarchical complex systems can be constructed and maintained only when large aggregates are made from smaller subsystems that are at least partially self-maintaining. To have this property they must also be partly autonomous, even once they are incorporated into larger systems. As an interpretive frame, these arguments have been our motivation to emphasize features in biochemistry, cellular control, or evolutionary history, that provide quantitative evidence about module boundaries and partial autonomy of subsystems. In a more formal Bayesian context, we may say that if we observe a very complex hierarchical system such as the biosphere, and must look for theories to explain why its existence is possible and how it came about, we have a better chance of finding a correct theory among the set that presume a sequential assembly of previously autonomous *and stable* modular subsystems.

John Gerhart and Marc Kirschner have argued [295, 296], from their efforts to frame a theory of developmental biology and regulation, that modular architectures provide the essential joint property of stabilization and adaptability. By permitting subsystems to be partly autonomous and self-maintaining, such architectures keep the problem of stabilizing the larger system from becoming intractable as the scale of aggregation grows. At the same time, by clarifying the domains of autonomy and interdependency, module boundaries facilitate the variations that are needed for adaptation. They insulate the problems of maintenance within subsystems as far as possible from the effects of adaptive variations in larger-scale assemblies. Gerhart and Kirschner argue that even where modularization is not present as a precondition to the formation of structure, it should be expected to evolve under positive selection for its service to robustness and adaptability. We review some of these arguments in Section 8.3.4.

The arguments both by Simon and by Gerhart and Kirschner assert the need for modularity, but they say less of a general nature[7] about why modularization is possible or under what conditions, and about ways in which limits on available modularization may determine the large-scale structure and dynamics of organizations.

The reason we have introduced phase transitions at some length in Chapter 7, and particularly the correspondence between mean-regression and error correction, is that we believe self-organizing non-equilibrium thermal phases are the correct candidates for many low-level modules that have enabled the emergence and stability of the biosphere. More generally, the phase transition model of jointly determined reliable error correction captures formally many of the mathematical properties attributed to modules in descriptive empirical accounts. In other words, *the availability of transitions into ordered phases underlies the particular hierarchy of modules that make up the biosphere.*

The hierarchy of life is not a passive structure but a dynamically coordinated process, in which components at multiple levels jointly perform complex functions – among them the creation and maintenance of structure. Therefore many of the interlevel links in living systems are links of **control**, in which one level (usually a coarser level) serves as a controller over one or more other levels (usually finer levels). The central concepts from control theory provide a way to understand the kinds of interaction and interdependency that arise at module boundaries. Control functions contribute to the global error-correction system that maintains the biosphere, either by providing part of the guidance that keeps subsystems within their stable operating ranges, or by adapting them to changing contexts at or above the scale of the controller.

An important inherent problem for hierarchical control systems is that the number of misbehaving states on the small scale can grow much more rapidly with aggregation, than the number of distinct states that a controller made from coarse components can distinguish and hence act on optimally. In Section 8.2.6 we review a set of classic accounting identities from optimal control theory which show that the state diversity (known in control theory as *variety*) of a controller can never be less than the variety of the states that it can maintain under control. The difference between the cardinality of potential error states in low-level controlled modules, and the sensing states in the high-level superstructures that control them, must be made up by **buffering** within the low-level modules themselves.

The phase transition paradigm attaches both a conceptual framework and a mathematics of joint error correction to the notion of buffering in systems with many autonomous but coupled degrees of freedom. Both Simon's and Gerhart and Kirschner's arguments that modularity is necessary are based on the need for buffering. A system must not depend on its controller too much in order to return to its own optimal operating range, or it will require more signal variety than the controller can provide. Conversely, it must not

---

[7] This is not to diminish the importance of their abstraction. They say a great deal *in particular* about mechanisms in biology that achieve modularization, which must be the starting point for an empirically grounded theory.

be interfered with too severely by control signals if the controller's own circumstances change, or else the need for adaptation comes into conflict with the need for stability. If either of these needs for insulation is not met, the system as a whole is likely to become unmaintainable.

The first five chapters emphasized facts about geochemical context, energetics, mineral homologies to biological mechanism, and structure in biology and biochemistry, which we believe reflect the most important constraints on the availability of modules capable of some autonomy and of buffering their internal states. In Chapter 6 we arranged these into a sequence of stages – most corresponding to accretions of metabolic or cellular subsystems – which we propose as a sequence through which life emerged. These stages follow a natural module decomposition. Their sequence is one that we think makes the stages most plausible as stable intermediate ordered states, and requires the fewest or simplest innovations to carry out the transitions. In the cascade of (mostly non-equilibrium) phase transitions that made up biogenesis, we suggest these as the points of passage where the combined features of reaction mechanisms, network topology, and in some cases spatial geometry, made modules available from the basic physics. As our understanding of reaction mechanisms and our ability to model and test large networks improve, these transition points should become theoretically predictable.

### 8.1.3 How uniqueness becomes a foundation for diversity

The stability perspective is a general assertion that criteria of global error correction must be met but are not easy to meet. It asserts, more particularly, that they are most readily fulfilled when the degrees of freedom that can take on error states act cooperatively and in large numbers to jointly regress toward a target state of order. Target states of order that can be maintained in this way are rare, and so ordered systems capable of very long-term persistence are rare.

Although this assertion is by no means limited to energy-driven order or bulk-phase order, when it is applied to core metabolism or lipid phases in water, it suggests phase transition models intuitively like those in equilibrium thermodynamics. The ordered phase would be expected to be unique given its boundary conditions, and the independent, small-scale stochastic dynamics would be distinctly non-Darwinian.

Yet order including metabolism at least since the last cellular common ancestor (and probably well before) has been maintained by replicating, competing, and selected organisms, or even by smaller components such as exchangeable genes. Bulk-phase ordered systems do not provide a complete picture of life today because, supposing they did once exist, at some point very early in the Earth's history they gave rise to replicating, competing entities capable of Darwinian evolution, and they now depend on those forms for persistence even when the underlying pathways do not show evidence of adaptive change. The dependence on Darwinian selection is so pervasive and so manifest in extant life that for many researchers, no other paradigm for the emergence and maintenance of order is even conceivable.

We must ask, if ordered metabolism performing the stress-relief function of an energy channel, and various ancillary phases of matter as by-product, could once have emerged as autonomous geochemical forms, *how was it possible for them to support a second emergent Darwinian dynamic, and why are the original forms no longer found in any other context?* From one perspective the answer is facile: modern catalysts and energy systems *have* been extensively refined by molecular evolution, even if the pathways they catalyze have not changed. Modern organisms can consume energy sources to create environments in which their ancestors could not have survived let alone "competed," and the stresses needed to create autocatalytic organosynthesis in geochemistry may not exist anywhere on Earth any more.[8] Moreover, the parallelism provided by partitioning of biomass into organisms provides a platform for the maintenance of diversity, from which selection can cull errors without losing the basic organism template.

Surely this explanation is all correct, but there is also much that it leaves out of account. Selection can refine the content of populations, but it must be given viable individuals on which to act. *Under what circumstances do non-Darwinian systems give rise to individuality?* The independence of individual lives that enables parallelism requires significant autonomy, and hence generates complex new control problems. Under what circumstances do the coevolutionary dynamics of populations of individuals retain the capacity for global stability that is the normal state of order in bulk systems if they can attain ordered states at all. Darwinian selection is a wasteful as well as a slow and coarse-grained filter, discarding whole organisms with a natural scale of generations. When does Darwinian selection have the requisite variety needed to identify and correct the fine-grained and frequent errors that undermine development as well as ecological coevolution? The difficulty of providing adequate answers to these questions suggests that the emergence of a Darwinian world from geochemistry poses genuine problems from the stability perspective: why it is possible, why in aggregate it is advantageous, and what limits the extent and contribution of individual-based evolution.

Our answer to these problems is that the emergence of individuality, and with it Darwinian evolution, introduced *some* important new forms of order into biodynamics, but that *individuality is neither a simple nor a unique property, there is no inherent form that is self-sufficient or preferred, individuals are not the only carriers of biological information, and in particular they do not support globally complete error correction on their own.* This is why the low-level solutions such as core metabolism – the solutions that we propose once existed in bulk phase and are still expressed most consistently in ecosystems – are not prior to life; they are as essential to its nature as the later structures that built on them. The biosphere as an error-correcting system depends on interactions

---

[8] This interpretation merely widens the scope of an argument already recognized in Darwin's 1871 letter to Joseph Hooker [178], with the famous lines

> But if (and Oh! what a big if!) we could conceive in some warm little pond, with all sorts of ammonia and phosphoric salts, light, heat, electricity, etc., present, that a protein compound was chemically formed ready to undergo still more complex changes, at the present day such matter would be instantly devoured or absorbed, which would not have been the case before living creatures were formed.

among all scales, from physical chemistry that makes individual form possible, to the variable "experiments in living" by individuals in populations, and ultimately to metabolic completeness in ecosystems as the source of stabilizing coevolution.

### 8.1.3.1 The advent of individuality

The important new form of order that makes life progressively less like an any ordered equilibrium phase is the emergence of individuality. Individuality, whether in development, transmission, or selection, is a complex concept applicable at many levels. Each new form or kind of individuality is like a new ordered phase, in that it renders some variations independent of others. However, the kind of independence that defines individuals – localized in space and periodic in time – makes it increasingly difficult for collective effects to propagate single solutions through all living matter. The result is that life, partitioned among individuals, maintains a plurality of parallel forms. Throughout the changes that constitute the emergence of each new kind of individuality, the correction of errors remains a global function jointly carried out by bulk chemistry and individual dynamics, so that out of the plurality of forms the aggregate biosphere continues to possess a unified identity.

We emphasized in Section 7.4.2 that the same cooperative effects that can free a system to form new order beyond that imposed by its boundary conditions can also make some aspects of that order unpredictable. When biomass is partitioned among individuals, the basic structure that supports the integration of the unit enables many variants to carry out experiments[9] on ways of being alive. With the emergence of individuality, a new more severe form of unpredictability, and with this, historical contingency and chance, enter the dynamics of the biosphere in an essential way.

In treating individuality as a new form of statistical partitioning, which distills like droplets out of a foundation of metabolic and physical order, feeding back on that background but not replacing its fundamental architecture, we recover a view of individuals more in keeping with the modern understanding that draws jointly from development, population processes, and ecology. In Section 8.3.4 we review some of this perspective as it is understood from the perspectives in embryology, development, the evolution of regulatory strategies, and paleoecology. While all of the evidence in these disciplines comes from an era and a level of complexity of organisms much later and higher than the metabolism that is our main focus, we believe the more detailed and realistic attempt to understand modern evolutionary dynamics suggests joint mechanisms of error correction that are much more continuous with the statistical picture of the early world than a naive distinction of Darwinism from bulk processes would suggest.

### 8.1.3.2 Ecosystems and the universal metabolic chart

Although new forms of individuality change the structure of evolution and the character of error correction in the biosphere, much of living order remains unexplained by organization

---

[9] This characterization was drawn from David Deamer [184].

at any single level. Perhaps the most convincing case for this argument is the persistence of the universal metabolic chart in a Darwinian world.

In modern ecosystems composed of highly evolved and complex organisms, the part of biosynthesis performed within any organism can vary widely from perfect autotrophs to extremely reduced (heterotrophic) obligate parasites. The completeness of the metabolic chart – its sufficiency for the synthesis of all biomolecules in all member species – depends on trophic relations among species that must be preserved as properties of their ecosystems. Ecosystems that are not biosynthetically complete cannot persist, but completeness is a global constraint that only acts on organisms through long-range feedbacks and ecological flux balance of energy, elements, and (in most ecosystems) higher-level organic compounds as well. The subset of metabolic functions under the control of any single organism's genome may experience stabilizing or disruptive selection within that species, but the property of metabolic completeness is only transmitted vertically in the joint community of all species, and is maintained by their coevolutionary dynamics. The locus described by coevolutionary dynamics, including all global constraints (and also internal regulatory interactions among species), is the ecosystem, an entity that is not itself Darwinian.

Ecosystems, then, become the carriers of universal order in the biosphere such as invariant core metabolism, while successive, emergent forms of individuality change the internal mechanisms by which that order is maintained, from what it was in the era of bulk chemistry. We regard the earliest ecosystem formation, not as a problem of community assembly from pre-existing species, but as a problem of distributing the task of maintaining an obligatory metabolic chart to selection acting on multiple autonomous genomes rather than on a single genome or communally traded gene pool.

This approach to individuality provides a coherent way to relate many patterns across a range of both scales and times. The identities of species as units of selection does not obviate the identities of ecosystems as elementary units of order. The traits preserved by ecosystems may be ancient ones such as core metabolism that have unique solutions, while member species preserve or continually adapt new functions that may have more diverse solutions.

### *8.1.4 Beyond origins to the nature of the living state*

By far the most commonly met approach to characterizing the nature of life is one that begins with Darwinian evolution [403, 828]. Within this, the events of selection that lead to adaptation are emphasized over the events of **stabilizing selection**, which may far outnumber the former but more easily go unnoticed. Stabilizing selection is the background life-and-death turnover whose reason for existence is at least as great a puzzle to explain as the small adaptation it enables. Stabilizing selection of features that show no adaptive change is like a great pond on which adaptation forms only small ripples. One notices the ripples and fails to realize the need to account for the medium in which the ripples form.

Conflating life with the capacity for adaptive evolution marginalizes the features that do not change and the mechanisms that maintain them. They become mere "constraints" or "protometabolism." The reason a biosphere formed organisms or lifecycles in the first place, as opposed to remaining a continuum in space and time as most abiotic ordered systems do, becomes an unframeable question, because the nature of life is cast in terms of a dynamic that can only be expressed if individuals are presumed to exist.

Everything we have argued up to this point makes it impossible for us to see the nature of life through such a frame. *Metabolism is not a relic of proto-life but the distributed substrate and the reference for coevolutionary coherence.* The emergence of individuality was a partitioning of the error-correction problem between new kinds of autonomous subsystems and the pre-existing distributed substrate. The substrate is retained at multiple levels, from the appeal to physical chemistry, to the informational capacity of ecosystems to coordinate coevolution. The partial autonomy of individuals introduces new kinds of buffering but also creates the need for new interfaces through which hierarchical control remains sufficient to ensure asymptotically reliable error correction. Selection can filter for high degrees of refinement but it has limited control-theoretic variety because it is coarse in its scale of action and also slow and wasteful. Therefore the allotment of control over processes such as metabolism, first to cofactors, cells, and genomes autonomous from the mineral substrate, and later to multiple ecologically coupled autonomous lineages, was a problem of finding the most effective distribution of regulation and control across levels. The granularization of living processes in both space and time, into organisms and generations, so unlike order in the abiotic world but pervasive in life, is a transformation requiring an explanation, and perhaps more remarkable than anything adaptation does with individuals once they exist. All these questions would demand to become central to our understanding of life, simply to reflect the understanding already within established domains of biology, even if we were not led to them through the effort to understand origins.

In Section 8.4 we offer a conceptualization of the nature of the living state in which the full range of these observations is considered fundamental. Darwinian evolution is not a defining precondition for life, but rather a consequence of the nature of individual-based organization within a living state specified at other levels. The biosphere as a whole rather than any particular kind of individuality or ecological community-structure within it is the locus at which life is identified as a distinctive planetary phenomenon. Life is defined, whether for matter, for events, or for organizational forms, by participation in the biosphere. This definition is not tautological because the biosphere is unified by global closure with respect to the regression to its ordered state. Finally, the ordered state of the biosphere itself is distinguished because it is an order that originates in chemistry, and within chemistry it opens distinctive domains of covalently bonded molecules and energy flow through reactions, which would not be accessed by other geospheres in the absence of life. Far from being an incidental substrate to a "living process" defined by abstractions, we argue that the chemical substrate is the unifying precondition that entails the whole range of other distinctive features of life.

We close the section by revisiting the question of phase transition and robust order in a multiscale setting. Much more can be done with the rich structure of jointly optimal error correction than we have been able to present in the simplest models. In Section 8.5.4 we raise what to us is one of the most important questions to answer generally, because it concerns phenomena that are more general than any level of scale or any particular mechanism. In the abiotic geospheres, durable patterns are maintained because they are realized in durable entities. The dynamic order of life shows the opposite pattern: durable patterns are realized on transient entities, and the most durable pattern of all – core metabolism – consists of the roles of the most fleeting entities – the core metabolites. Yet this dynamical order is arguably the oldest fossil on Earth. The relation of patterns to entities in the biosphere expresses most forcibly of all the need for a theory of phases of processes.

## 8.2  Metabolic layering as a form of modular architecture

Sections 7.4 and 7.6 considered the statistical behavior *within* a single system undergoing a phase transition – the kinds of cooperative effects we have characterized as "architecturally flat." Important concepts illustrated within these minimal models include the system's entropy change, the ambiguity of the order parameter, and the conditional independence of residual fluctuations about the average ordered state. Here we consider the face that such a system presents to other systems within a hierarchical organization. Because ordered phases are internally stable, they can limit error propagation and buffer against errors in a hierarchy. In cases where the ordered state is ambiguous, the alternatives can become dynamical entities or control interfaces in their own right on slower timescales. The conditional independence of internal fluctuations makes the order parameter, statistically speaking, the *best* control variable over the system's internal states.

These are the properties that govern the design of a system from components. Phase transitions provide a natural foundation that creates modularity in design, and provides the services of error buffering, controllability, and predictability that have been argued to make modular architecture a prerequisite to hierarchical complexity. This section explains the theoretical premises from statistical mechanics behind our claim that the modular architecture of life is especially informative about the way it must have emerged.

### 8.2.1  Herbert Simon's arguments that modularity is prerequisite to hierarchical complexity

Herbert Simon was a widely synthetic thinker about the nature of organization in the second half of the twentieth-century. His insights were drawn from a broad empirical knowledge in disciplines ranging from history and business to biology and engineering, and many of them re-framed the way subsequent generations thought about the problems of becoming organized, and how those problems are solved in natural or designed systems.

One of Simon's most important assertions is that modular architecture is a prerequisite to the emergence or maintenance of hierarchical complexity. Complex systems most often or most easily arise if they are assembled from subsystems that have some degree of inherent internal stability, which the larger assembly does not then need to provide. The argument was advanced in a variety of papers on the architecture of complexity in the 1960s and 1970s [754, 755], and given an expanded context in *The Sciences of the Artificial* [756].

The classic **parable of two watchmakers** is the best known version of this argument. It refers to the problem of assembling a complex system.

> There once were two watchmakers, named Hora and Tempus, who man-
> ufactured very fine watches. Both of them were highly regarded, and the
> phones in their workshops rang frequently – new customers were con-
> stantly calling them. However, Hora prospered, while Tempus became
> poorer and poorer and finally lost his shop. What was the reason?
>
> The watches the men made consisted of about 1,000 parts each. Tem-
> pus had so constructed his that if he had one partly assembled and had to
> put it down – to answer the phone say – it immediately fell to pieces and
> had to be reassembled from the elements. The better the customers liked
> his watches, the more they phoned him, the more difficult it became for
> him to find enough uninterrupted time to finish a watch.
>
> The watches that Hora made were no less complex than those of
> Tempus. But he had designed them so that he could put together sub-
> assemblies of about ten elements each. Ten of these subassemblies,
> again, could be put together into a larger subassembly; and a system
> of ten of the latter subassemblies constituted the whole watch. Hence,
> when Hora had to put down a partly assembled watch in order to answer
> the phone, he lost only a small part of his work, and he assembled his
> watches in only a fraction of the man-hours it took Tempus.

In "The architecture of complexity" [754], a variant form of the story is drawn from the **Alexandrian empire.** This form refers more to the problem of maintaining a com-plex organization, and its subtext is that information about the underlying modules can be gained from the fragments that form when larger architectures fail.

> Philip assembled his Macedonian empire and gave it to his son, to be
> later combined with the Persian subassembly and others into Alexan-
> der's greater system. On Alexander's death, his empire did not crumble
> to dust, but fragmented into some of the major subsystems that had
> composed it.
>
> The watchmaker argument implies that if one would be Alexander,
> one should be born into a world where large stable political systems
> already exist. Where this condition was not fulfilled, as on the Scythian
> and Indian frontiers, Alexander found empire building a slippery busi-
> ness. So too, T. E. Lawrence's organizing of the Arabian revolt against
> the Turks was limited by the character of his largest stable building
> blocks, the separate, suspicious desert tribes.

These examples can be read as references to stability, error correction, or control, three technical ideas from physics and engineering that have significant overlap. Modules, whether watch parts or nation-states, may have considerable complexity which it is the function of their own internal architecture to maintain. The stability of the module relies on the stability of its internal structure, and is expressed in the stability of the interface it presents to the outside world. In a hierarchical assembly, it is this interface through which the module shares functions with other modules. In the Alexandrian empire, the interfaces between populations and Alexander were the ruling elites of the nation-states within the empire.

Simon's parables can either be understood as descriptions of the ways extant systems are built and function, or they may be taken as grounds to assign Bayesian priors as we attempt to select models from incomplete evidence. (For a formalization of **model selection** as an application of Bayes' theorem for joint probabilities, see Box 8.1.[10]) If we observe that a hierarchical complex system exists but have only limited information about its past forms, we may give a prior preference to those scenarios for emergence that pass through sequences of stable intermediate stages. From the perspective of maintenance rather than emergence, if we see an extant system and we have an incomplete understanding of the information and control interfaces that it employs, we may give prior preference to models that assume stability is maintained to a large degree within subassemblies of limited complexity, and that control of these assemblies passes through interfaces that are less complex than the full internal states.

It is important to appreciate that a theoretically motivated preference for modularity matters in interpreting data. In the *ex post* description of some system, given by an authoritative historian or a finished theory, the module decomposition may seem self-evident. This self-evidence is often revisionist history – the decomposition was not at all evident before the system was understood.[11] The extant biosphere is a network of chicken-egg paradoxes, as large systems depend on small systems for material, and the small systems in turn depend on the large systems to determine what material is produced and to enable its production. Perhaps the deepest divisions in current origins studies are between those who consider the material dependencies more binding and hence conserved, and control systems as later innovations, and those who consider the emergence of control to be the origin of a living state. Among the many ways one might attempt to cut such cycles, we prefer those in which the data suggest a module decomposition, whether in the dependency for materials or the dependency for control.

---

**Box 8.1 Bayesian model selection**

The modern alternative to a naive empiricist view of science can be at least partly expressed using the framework of *Bayesian model selection*. This framework captures the intuition that while lists of facts can never fully determine a model, increasing the number and diversity of facts may progressively narrow our choices.

---

[10] Bayesian model selection may be understood as the modern formulation of the problem of induction addressed by Sir Francis Bacon in 1620 [43, 44]. Evidence is not self-interpreting, but it can shift weight among competing prior hypotheses. The Bayesian framework requires a formal representation of the prior preferences.

[11] Duncan Watts [855] provides helpful reminders about the fallacy of self-evidence.

---

**Box 8.1    Bayesian model selection (cont.)**

If the prior choice of eligible models is made explicit, this narrowing can be done in a principled way.[a]

Suppose that $M$ is a random variable whose values index some set of possible models to explain a phenomenon. For equation fitting, the interpretation of $M$ is often straightforward; for more complicated problems of interpretation, indexing a model space may become very difficult. Let $D$ be a random variable whose values are the possible outcomes of some datum to be measured experimentally. In general, a model is required to predict at least a definite probability distribution for measured data values, which we denote $p(D \mid M)$. This quantity is called the *likelihood* of the data given a model.

We represent our uncertainty about the correct explanation prior to measuring the datum $D$ with a *prior probability distribution* assigned to models $M$, denoted by $p(M)$. The prior may come from anywhere: poorly justified theoretical preference, assumptions about symmetry [396, 397], or highly refined "laws" from elsewhere in science. The virtue of Bayesian model selection lies in forcing us to recognize that *we have uncertainty about models*, and that the way we represent probabilities over models is a part of the theoretical commitment of interpreting data.

The prior probability for values of $D$ under this distribution is then given by

$$p(D) = \sum_M p(D \mid M)\, p(M)\,.$$

The *Bayesian posterior* distribution for models after we have seen the value of $D$ is denoted $p(M \mid D)$. Supposing we have committed to the prior on $M$ and to a well-defined likelihood function, the posterior is given by *Bayes' theorem*, as

$$p(M \mid D) = \frac{p(D \mid M)\, p(M)}{p(D)} = \frac{p(D \mid M)\, p(M)}{\sum_{M'} p(D \mid M')\, p(M')}\,.$$

We may perform Bayesian updating recursively, using the output $p(M \mid D)$ from one experiment as our prior for the next.

Although as a formal tool for model selection, Bayesian rules have only been found useful for fairly simple problems with well-defined model spaces and tightly predictive models, they provide a helpful framework in which to think about many classic problems in scientific method. In addition to embedding Bacon's empirical stance correctly within the problem of induction, they show where Popper's asymmetry between verification and falsification [646] arises, $p(D \mid M) \equiv 0$ for some $M$ rules out that $M$, and permit some "softening" of the Popperian stance to imprecise models and measurement.

---

[a]  See Andrew Gelman and Cosma Shalizi [291] for a much more comprehensive account of methods and interpretations in model selection.

### 8.2.2  The modularity argument in a dynamical setting

In Simon's argument for modularity, modules are important because they limit the propagation of error, and they make errors easy to identify and to either repair or eliminate. In the watchmaker parable, the effort put into assembling each module is lost if the module must be set down to fall apart. This loss is a pure cost to the watchmakers. Modular design simply limits the number of components to which the error can propagate, and hence the cost of redundant work.

The parable of watches can be taken literally for the simplest mechanical systems, toward which the only actions of interest are those of the watchmaker. This situation is simpler even than the simplest thermal equilibrium, however, because in thermal equilibrium all components are constantly disturbed and jostled. Modules can only be "set aside" if they rarely fall apart when left alone. For driven non-equilibrium systems, the situation is far worse. In addition to the constant thermal disruption of microscopic order, the same random reactions by which order is assembled stands ready to degrade it away. *Unless a driven system is continually self-amplifying, it cannot even persist.* It is as if, in addition to handling the customers, the watchmakers were bedeviled continually by gremlins that disassembled any module not kept in hand.

To apply Simon's arguments to driven dynamical order, it is not enough that modules merely limit the propagation of disruptive error. They must actively provide some degree of error correction within their own boundaries. Only then can a module be stable and thus serve as a foundation for a larger assembly.

### 8.2.3  Error correction from regression to ordered thermal states

In Section 7.5 we showed that a theory of asymptotically reliable error correction does exist, due to Claude Shannon, and that it is a large-deviations theory with the same structure as regression toward ordered thermal states. Reliable error correction is not attained simply by introducing mechanisms that tend to catch and correct errors. It depends on using long-range correlations in an optimal way to ensure that – asymptotically – *all* errors are identified by comparison against a majority of uncorrupted, redundant degrees of freedom somewhere in the system.

#### 8.2.3.1  The paradox of asymptotic stability with finite correlation length

A difficulty in applying the Shannon theory to chemical systems, however, is that asymptotically reliable error correction only occurs for either trivial systems that have only one stable ordered phase, or systems with infinite correlation length and unlimited complexity and compute-time available to the decoder that identifies error states. Real living systems do not have infinite time and complexity to identify and correct errors, and they do have multiple possible states. Therefore some errors do arise, and the patterns that on short terms appear stable tend to drift on longer terms. We must therefore understand how the biosphere can be (apparently) asymptotically stable while its components possess only finite complexity and internal correlation.

This essential question in the theory of error correction is where Simon's arguments about hierarchy enter. Because the emergence of the biosphere was not one phase transition but many, it does not employ only one level of error correction, but rather many levels. Our proposal is that the lowest levels *do* have unique solutions which follow from the structure of chemistry and the Earth's energy systems. These solutions, as results of phase transitions, are not strictly "unique," but rather unique up to unimportant degeneracies like the direction of frost on a windowpane described in Section 7.4.2.1.[12] Biological order at the lowest levels does not build on the noisy variables subject to drift, but only on the reliable aspects of order that are ensured from the reductionist description. These low-level systems such as ecosystem-level primary production, then bring into existence machinery such as cells and molecular replication systems capable of Darwinian evolution. The low-level systems also define the criteria for growth at the ecosystem level which organize individual selection and inter-individual interactions into an "error-correcting code" of co-adaptation toward which the higher-level assemblies regress.

Individual solutions at the higher level do drift. We return in Section 8.4 to propose that this drift is "unimportant" in the sense that the assembly of possible solutions, linked to stability at the lowest chemical level, constitute the law-like and indefinitely persistent order of the biosphere. First, however, we develop the role of chemistry in the lowest modular levels.

### 8.2.4 The universal metabolic chart as an order parameter

We have now built the necessary concepts and language to express the role we believe the universal network of core metabolism – or something closely related to it – played in the emergence of the biosphere. *We interpret the fluxes through the universal core network as the **non-equilibrium order parameters** for oxidation/reduction chemistry of organic compounds on the Hadean Earth as the biosphere formed.*[13] The ensemble theory of phase transitions and ordered states introduced in Chapter 7 implies the following consequences of this interpretation of the early core network.

The discussion of order parameters and fluctuations in Section 7.4.3.2 makes clear that even if Hadean geochemistry came to be ordered around a certain set of cyclic flows within a core network, the intermediates in those cycles would not have been the only chemicals present in significant concentration in the reaction systems. This is especially true at the beginning when catalysts were non-specific, and probably acted mostly at the functional-group level.

It may be correct to propose that, whereas modern enzymatic catalysts both open reaction channels and prune for molecular specificity, in the earliest geochemical era the two

---

[12] Because biological order is dynamic rather than static, a more apt example might be the random walk around a limit cycle performed by a population process in its ordered state, an example that is worked out in [768] (Chapter 6). The chemical reaction counterpart to phase noise in a population process is random noise in the progress along the reaction coordinate in an autocatalytic pathway. It is motion along the direction of symmetry (time translation), but is not part of the definition of the cycle itself.

[13] Other environmental disequilibria, perhaps from proton-motive force or phosphate dehydration potential, may have been important scaffolds for the transition to cellularity, but the status of all of these is currently speculative, and ultimately they have all become internal recycled species driven by net redox fluxes.

tasks were divided. Catalysts would have opened limited mechanisms of reactivity, but kinetic selection through feedbacks at the network level may have been the principal mechanism for whatever pruning was performed. Under this picture, we look for much of the additional information created by selection in the succeeding billions of years not to be reflected in alteration in the main architecture of pathways, but rather in the extensive and demanding task of pruning the network so that only the skeleton remained occupied and deviating fluctuations were suppressed.

If the core network is not to reflect the only chemicals or reactions present, what then is its role? The ability to provide a non-trivial answer to this question defines the essential difference between an order parameter and any microscopic history of events. A non-equilibrium order parameter, like an equilibrium order parameter, is a typical behavior toward which fluctuations regress. In this sense, *the order parameter is the attractor of the fluctuations in a non-equilibrium network,* even though only a part of the whole-network flux may be in the particular reactions that the order parameter denotes. We thus interpret the core network as the set of fluxes that organized the primordial, non-equilibrium, exponentially distributed network.

In addition to being an attractor, the core molecules and fluctuations in this interpretation are sufficient statistics for early chemistry. While many – perhaps millions (see Section 7.2.2) – of other molecules or reactions would have been sampled as random fluctuations in the reactor, those fluctuations would also have been conditionally independent of each other given the fluxes in the order parameter, as we illustrated for the simple order parameter of the magnet in Box 7.4.

One of the aspects of ensemble thinking that takes time to get used to is that its falsifiable claims are all made about incompletely specified observations. The order parameter is not any realized single history of microstates, but a central tendency toward which microstates regress. That central tendency is not even a measurable quantity, but one that must be approximated using sample estimators. Yet despite this non-material status, the order parameter predicts properties that can be tested, and which, if they are not confirmed, can rule out the hypothesis for the system's state of order. To qualify as an order parameter, a pathway flux (in an experiment, or in a statistical model of evolutionary diversification), should be the most predictive summary statistics for the state of the network. Departures from the ensemble-average value of the key fluxes should be mean-regressing, and the expectations for other fluxes should be functions of those fluxes that are taken to constitute the order parameter. Finally, small-scale fluctuations in concentration or reaction flux in the remainder of the network should satisfy a large-deviations principle and should be conditionally independent given the values of the core fluxes. These are the criteria that the phase transition paradigm stipulates for chemical "self-organization."

### 8.2.4.1 Error containment within modules

The metabolic modules from Chapter 4, organized around a collection of core pathway fluxes, are Simon's modules with respect to *partitioning of the error-correction problem among levels of structure.* They do not separate errors in space or in time (that function will arise when we consider individuality below). Rather, the core pathways create mean-regression toward a stable inventory of organic species different from a Gibbs equilibrium distribution. Fluctuations away from the central tendency that supports the pathway flux – including departures toward the equilibrium Gibbs distribution – are "errors" in this dynamical context that are removed by the non-equilibrium reaction kinetics. As higher-level systems come to incorporate the core metabolites that accumulate in the non-equilibrium distribution, new errors can arise through hierarchical interactions. However, secondary fluctuations in biosynthetic pathways will still be shielded from low-level fluctuations, conditional on maintenance of the flux in the core pathway.

The question whether an ordered but complex low-level network will support the emergence of higher-level structure may now be broken into two quantitative parts. The first question is whether the network as a whole possesses sufficient mean-regression to maintain a state of driven order. If this condition is met, the second question is how asymmetric the internal order is – whether it causes a subset of molecules to be produced in significant excess of the rare or trace compounds in the network. In other words, how *selective* is the network for a subset of its molecular species? Species present in excess concentration – which we expect to be the intermediates in the order-parameter skeleton – are most eligible as building blocks for higher-level aggregates. Networks in which the non-equilibrium is not very asymmetric are often referred to informally as suffering a "combinatorial explosion" of chemical complexity under the recursive action of the generating reactions. They may have very small concentrations of millions of species, as the Murchison meteorite has, but perhaps none of these in sufficient concentration to support coherent cells or chiral oligomers with secondary structure. Networks that are more skew are more similar to extant biochemistry, with high concentrations of a few species and a relatively easy problem of selecting and incorporating these into higher-level structures.

### 8.2.5 *The use of order parameters in induction*

When ordered states emerge in thermodynamics, the laws of large numbers average away infinitely many details, and leave only low-dimensional manifolds of possible states of macroscopic order. The reduced dimension of macroscopic order, and the conditional independence of fluctuations about this order, make the distributions that describe thermal ordered systems enormously *simpler* than randomly chosen distributions for the same microscopic states. As a result, ordered thermal systems are similar in many respects. A single theory of phase transition, in which only the symmetries and degrees of freedom differ between one level and the next, governs the whole hierarchy of ordered states of matter, rather than a set of *ad hoc* descriptions, one for each level.

The drastic projection that large-deviations scaling entails, onto simple distributions, also leads to a relatively simple information theory for thermal ordered systems. In an ensemble created by a phase transition, detailed knowledge of the instantaneous state is only useful for predicting states at other times, to the extent that it provides an estimate of the order parameter of the ordered phase. We summarized this formally in Box 7.2

of Chapter 7 by saying that the order parameter is a *sufficient statistic* for states of the system: for samples from many ensembles relevant to long-term dynamics, one gains as much predictive capability from knowledge of the order parameter as from knowledge of the order parameter plus any measurement of the instantaneous fluctuations in the system. In other words, if one's goal is prediction of future states of the system, there is nothing beyond the order parameter that is worth knowing (or estimating) for the sake of prediction.

In the assembly of hierarchical complex systems from modular subassemblies, *the different modules must make predictions about each other's states* in order to jointly perform a coherent function. This is a tautology. Each module's structure defines its input/output characteristics, and therefore the set of distinctions it can sense in its environment and the set of acts it can take that will influence that environment. Simply by *having* a structure, the module is committed to making the prediction that the environmental distinctions it can sense are those that will be produced by the other modules, and the actions that it can take will contribute to a coherent aggregate function. Modules that make wrong predictions in regard either to sensing or to acting do not contribute to functional aggregates.

### 8.2.5.1 Assembly through selection and the problem of optimal induction

All the systems we are considering are selected on the basis of the functions they have executed, in either present or past realized states. Whether the selection occurs by means of thermal relaxation at or away from equilibrium, or later through a Darwinian dynamic, the mechanics of aggregation has only data from the past to draw upon. This is a statement that the processes of assembly and maintenance are *causal* in the physical sense of the term. It can be refined to say that they are **Markovian** – that all information from the past relevant to dynamics is represented in features of the current state – if one adopts a sufficiently detailed description of the current state. Beyond causality or even the Markov property, a stronger claim can be made in most practical cases: since the mechanisms of maintenance and aggregation are dynamical and are continually being degraded by noise, they have limited memory, and in effect can only draw on a *finite* collection of past events and consequences.

The problem of assembly of hierarchical complex systems is therefore a problem of **induction** from finite past data to the selection of modules. Once a particular modular architecture with its supporting interfaces has been constructed, it then entails a commitment to some input/output routine with particular, limited capacities for sensing and reaction. It has now been accepted for most of the last century in philosophy and mathematics [682] that the problem of induction from finite data has no general solution. Yet every assembly of finite systems in an open world is perforce a commitment to some inductive hypothesis – finite systems cannot do otherwise.

Therefore *the problem faced by the assembler of a hierarchical complex system is to base the interactions between modules on those features that best support induction*. The features of one module that another module samples and responds to must not only be informative about the state of the first module, they should be consistent and reliable. The more reliably the parts in the aggregate work together to produce the same input/output behavior, the more effectively a selective filter can choose among alternative forms.

In Box 8.2 we use the Curie ferromagnet to illustrate the sense in which the order parameter is an optimal sample estimator to support prediction. Following Box 8.1, we frame the problem of induction as a problem of model selection that arises when a phase transition has occurred. If a thermal subsystem that forms a module has only one phase, its ordered state variables are stable but have no diversity. Such systems can contribute exogenous constraint but no non-trivial function. This is the model for the simplest and most restrictive role of the Gibbs distribution.

If a phase transition places the system into one among a collection of its possible ordered states, the system may be more predictable (in the sense that its internal entropy is reduced) and may even become capable of macroscopic dynamics, but for another subsystem to use these capabilities the second system must estimate the first system's ordered state which is not fully determined by the microphysics. In such a situation the partition function (7.30) with a bimodal distribution – but with no further discrimination between the modes – provides a natural Bayesian prior, and the magnetization becomes a sufficient statistic to be estimated with a Bayesian posterior from samples. Box 8.2 shows that an actor outside the system can do no better to estimate even complex functions of multiple microscopic spin variables, than to estimate the magnetization and then compute the expectation of the micro-variables in the Bayesian posterior distribution for the magnetization. This Bayesian posterior is a good proxy for the abstraction known as the "symmetry-broken" partition function, in which a modeler removes by hand one or the other mode of the bimodal prior distribution, to reflect the dynamical inaccessibility of one mode to random dynamics that begin in the other mode.[14]

---

**Box 8.2   Conditional independence, the order parameter, and inference**

The existence of an ordered state created by a phase transition can simplify and stabilize the problem of inferring joint distributions for some microstates, given samples of others. The limited values the order parameter may take, together with the conditional independence of microscopic variables given that order, strongly constrains the possible forms of probability distributions.

A nice example, which also illustrates the interplay of reduction and emergence, is again given by the Curie ferromagnet. Suppose we know the microphysics of the Curie magnet so we know the partition function, but we do not know which

---

[14] While this by-hand removal of modes to represent spontaneous symmetry breaking is a commonly used technique in descriptions and even calculations about symmetry breaking, we prefer to avoid it because it has no real grounding in principles and creates needless confusion in presentations. A better solution is to explicitly work with time-dependent correlation functions among multiple observables. The factorization of such correlation functions into fingers, each finger dominated by observations that all occur within the same ordered state, provides both a quantitative characterization of the dynamical decoupling between ordered states in finite systems, and a principled way to describe symmetry breaking and symmetry restoration in large-system limits. See [768] for further discussion and examples.

**Box 8.2    Conditional independence, the order parameter, and inference (cont.)**

magnetization state we will happen to find the spin system in after the temperature has fallen below the Curie point. We must estimate that variable by sampling the actual system. Suppose, for the sake of example, that we need to estimate a distribution for some function that may involve many spins, but that individual samples of spins are costly so the number we can perform is limited. What is the best inference algorithm we can carry out, and how efficient is it?

We may formulate this as a classic problem of Bayesian inference. Our Bayesian prior for states of the magnetization $\mu = 2n - N$ is just

$$P(n) = \frac{1}{Z} e^{-\beta H} \begin{pmatrix} N \\ n \end{pmatrix}. \tag{8.1}$$

$n$ is our "model" in the Bayesian language. Note that, while $P(n)$ may be tightly concentrated about either one or two values for $n$, it is always symmetric under $n \leftrightarrow N - n$. This is the statement that the reductionist description cannot predict the direction of magnetization.

Suppose that the data sample from which we need to infer the magnetization state comprises the values of some set of $M$ spins, where $M \ll N$ (so we make a very incomplete sample of the system). The likelihood function for $m$ of these to be up if $n$ total are up, written $P(m \mid M, n)$, is given by

$$P(m \mid M, n) = \begin{pmatrix} n \\ m \end{pmatrix} \begin{pmatrix} N - n \\ M - m \end{pmatrix} \Big/ \begin{pmatrix} N \\ m \end{pmatrix}$$

$$\approx \frac{1}{\sqrt{2\pi M (n/N)(1 - n/N)}} e^{-M D_{\mathrm{KL}}\left(\frac{m}{M} \| \frac{n}{N}\right)}. \tag{8.2}$$

In the second line we have used Stirling's formula for factorials, omitted terms at $\mathcal{O}(M/N)$, and introduced the notation

$$D_{\mathrm{KL}}\left(\frac{m}{M} \Big\| \frac{n}{N}\right) \equiv \frac{m}{M} \log\left(\frac{m/M}{n/N}\right) + \left(1 - \frac{m}{M}\right) \log\left(\frac{1 - m/M}{1 - n/N}\right) \tag{8.3}$$

for the *Kullback–Leibler divergence*, or *relative entropy*, of the distribution $(m/M, 1 - m/M)$ from $(n/N, 1 - n/N)$ [164].

The relative entropy is a pseudo-distance function for probability distributions. It is strictly non-negative and equals zero only when $m/M = n/N$. It fails to be a proper distance function only because it does not satisfy the triangle inequality required of true distance functions.

**Box 8.2  Conditional independence, the order parameter, and inference (cont.)**

The Bayesian posterior distribution for $n$, given the outcome $m$ of our sample, is the usual expression from Box 8.1:

$$P(n \mid m, M) = \frac{P(m \mid M, n) \, P(n)}{\sum_{n'=0}^{N} P(m \mid M, n') \, P(n')}, \tag{8.4}$$

in which we use the form (8.1) for $P(n)$ and (8.2) for $P(m \mid M, n)$.

Suppose that $N$ is large, and let $|\bar{\mu}|$ be the stationary-point magnitude of the magnetization, which is a function only of $T$ and $N$. Denote by $\bar{n} \equiv (|\bar{\mu}| + N)/2$ the value of $n$ at which $\mu = |\bar{\mu}|$. Then below the Curie point, the Bayesian prior distribution (8.1) is well approximated by two point distributions at $n = \bar{n}$ and $n = N - \bar{n}$.

In this limit we obtain a very simple expression for the Bayesian posterior (8.4). Because the prior has support only for $n = \bar{n}$ and $n = N - \bar{n}$, the posterior has the same property. The two values are given by

$$P(n = \bar{n} \mid m, M) \rightarrow$$
$$\frac{1}{2} \left\{ 1 - \tanh \left( \frac{M}{2} \left[ D_{\mathrm{KL}} \left( \frac{m}{M} \Big\| \frac{\bar{n}}{N} \right) - D_{\mathrm{KL}} \left( \frac{m}{M} \Big\| \frac{N - \bar{n}}{N} \right) \right] \right) \right\}, \tag{8.5}$$

and $P(n = N - \bar{n} \mid m, M) = 1 - P(n = \bar{n} \mid m, M)$.

The posterior probability swings to favor whichever value of magnetization is allowed by the partition function and is closer to the sample fraction $m/M$. The sharpness of the swing is proportional to $M$, reflecting the sample confidence.

Predictions for any other spins besides the ones measured in the original sample of $M$ are conditionally independent of the measured spins (and of each other), given the estimate for the order parameter. The best estimate we can form for the probability that any given spin $s_i = 1$ is given by the usual chain rule, using the Bayesian posterior estimate (8.4) for $P(n \mid m, M)$:

$$P(s_i = 1 \mid m, M) = \sum_{n=0}^{N} P(s_i = 1 \mid n) \, P(n \mid m, M)$$
$$= \frac{\bar{n}}{N} P(n = \bar{n} \mid m, M) + \left( 1 - \frac{\bar{n}}{N} \right) P(n = N - \bar{n} \mid m, M). \tag{8.6}$$

More important, the joint distribution for any collection of such spins, as long as its number remains much smaller than $N$, is simply the generalization of the first line of

---

**Box 8.2    Conditional independence, the order parameter, and inference (cont.)**

Eq. (8.6) to a sum of weighted joint samples. Note that $n$ is a sufficient statistic for all $s_i$, which in this case we cannot directly measure but try to estimate from sample estimators $m/M$ and knowledge of the partition function.

We have cast this example as a problem of Bayesian inference for simplicity, to show how we might use prior knowledge that $P(n)$ concentrates onto two point distributions. The essential point, however, is that the distribution *has* this property. If we replaced the Bayesian reasoner with an evolutionary or reinforcement-learning algorithm, it would be rewarded for those inferences that assumed these properties of $P(n)$. The optimal function of sample estimators would remain the one given by Eq. (8.6).

---

### 8.2.5.2  Interpretations for primordial organosynthesis

The simplification that stable and low-dimensional macrostates afford, in the problem of assembling hierarchical complex systems, is the formal basis of our preference for theories of the origin of life that do not grow from a complex molecular "tar" as their original source of organics. The evolution of catalysts or other hierarchical controllers is a problem of reinforcement learning: the catalyst must first encounter substrates from which it produces an outcome that can be rewarded, and once amplified through selection, the population of favored catalysts should constitute a good out-of-sample prediction for the substrates likely to be encountered in the future and the outcomes of acting on them. A stable state of molecular order, with few dimensions of easy variation, simplifies both the sampling problem to identify the order and the prediction problem for molecules encountered later. The opposite situation holds for a complex tar or combinatorially explosive organic network, with many molecular species present in small numbers.

### 8.2.6  Control systems and requisite variety

So far we have described the Simon argument for modularity in terms of stability and then in terms of prediction, functionality, selection, and induction. A third framework, which is central to almost all descriptions of hierarchy in biology, and which we have invoked in previous chapters, is that of control theory. We have described cofactors and catalysts as the first controllers of the kinetics of organosynthetic networks. The Central Dogma is a control-theoretic concept: that genes via proteins determine the expression of phenotype, and thus transmit the consequences of natural selection. Many of the feedbacks from large systems onto their smaller synthetic precursors, which create the most vexing chicken-egg paradoxes for inferring the origins of living order, are very naturally cast as control relations.

W. Ross Ashby, a contemporary of Herb Simon, was one of the pioneers in the discipline now known as cybernetics [38], and also an important contributor to the engineering paradigm of optimal control theory. In control theory, one subsystem, known as the controller, acts on another subsystem which is being controlled, to maintain the controlled subsystem within a restricted domain of its possible states or dynamics. In a seminal paper, Roger B. Conant together with Ashby showed [154] that as a tautology, any system satisfying the definition of a good controller contains a *model* of the controlled system in some representation. The model is essential, because it is the comparison between the controller's sensory inputs from the real system, and references stored in the model, that determine what actions the controller may take to correct unwanted excursions by the control system and to return it to its desired operating domain.

### 8.2.6.1 The law of requisite variety

Using this notion of an internal model, Ashby introduced the criterion for the sufficiency of control known as the **law of requisite variety** [38, 39]. The law asserts that the number of distinct states that a successful controller can sense, and the number of distinct actions it can take in response to its samples, must at least equal the number of distinct forms of disturbance that can drive the controlled system out of its target domain, minus any self-correction the controlled system can perform on its own. The self-correction is known in control theory as "buffering."

In a hierarchical complex system where the larger system stands in the relation of a controller to one or more of its component modules, the Simon argument for modularity is based on buffering. If a module can contain the propagation of errors in its own internal state, or can actively correct the errors, then the environment need not contain a model of those errors in the controlled module's internal state. A reduction in the requisite variety that results from a module's self-buffering reduces the demands on both the complexity of the environment and the complexity of the couplings through which the environment imposes control on the module. In a system that forms modules through phase transition, the requisite variety of a controller that interacts with the module only through its order parameter may be very low. In such situations error correction remains low cost and predictable, even if the controlled system has many fluctuating internal variables.

It is certainly not an absolute requirement that only the order parameter be used as a controlled variable; that is simply the easiest and most robust approach. A controller may reach into the system's microstates and influence them directly, driving the system far from a Gibbs state characterized by a simple order parameter. This kind of refined control is exercised in modern cells by highly specific enzymes, and it is why the pathways of modern cells are far from the thick networks we suppose existed in the Hadean.

Box 8.2 shows how phase transitions are used to reduce requisite variety in actual magnetic storage devices. The internal state of a tiny magnetizable domain may take $2^N$ distinct configurations, where $N$ can be very large. Yet the order parameter remains binary at all $N$. In actual storage devices, large $N$ is used to provide stability to the net magnetization. The controller places memory into a magnetic domain by coarsely driving the magnetization either positive or negative. If the controller includes active error correction, it need not monitor the exact internal state of a magnetic domain – it senses only the net magnetization and drives it into the appropriate domain if it violates a checksum condition in the error-correcting code.

### 8.2.6.2 When unpredictability is unimportant and left unused in a control system

In general, following Section 7.4.2.3, we think of the complementarity between the low-dimensional space of states of macroscopic order, and the high-dimensional space of buffered fluctuations, in terms of a separation of scales. In a perfect phase transition, the order parameters are exactly degenerate. In the case where these are used as control variables in a hierarchical complex system, the coupling to a controller can be strong enough to break the degeneracy and select particular macrostates, but still much weaker than the mean-regression that ensures the stability of the thermal phase. In Box 8.2 these two subspaces filled complementary roles as the Bayesian prior and the sample estimator in a Bayesian model-selection exercise. Any residual degeneracies that are uncontrolled, along with any imperfections in the control signals, become the most likely directions of historical contingency.

We have proposed, however, that the lowest level of biochemistry anchors the control problem for the living state because it is a *unique* solution. What then happened to the role of degeneracy created during the formation of ordered phases? The answer is that for some phase transitions, the degenerate variables are variables of indifference to the important aspect of order. In the lightning analogy of Section 7.6.5, the particular zigzag path is irrelevant to its role in the atmosphere as an engine of electrolysis. Likewise, for the lipid–water phase separation that forms vesicles, their topological roles in containment, bioenergetics, and phase separation, are invariants under the Brownian fluctuations of the exact spatial geometry of the membrane. Other examples unique to non-equilibrium phase transitions, in which the exact timing in the formation of an ordered state is a variable of symmetry but may be unimportant to the functional role of order, are developed in [768]. All three kinds of "irrelevant symmetry" can render a spontaneously formed state of non-equilibrium chemical order effectively unique.

### 8.2.7 Biology designs using order parameters

Raffiniert ist der Herrgott aber boshaft ist er nicht

– Albert Einstein

These observations about easy and difficult paths of inference, and the nature of evolution as an inference engine, may be brought together to predict the kinds of order we expect to find in the biosphere. Using the metaphor of evolution as producing "design" in the architecture of complex systems, we claim that *the components of biological design should be, primarily, the order parameters made available by phase transitions.*

Paralleling the behavior we expect from evolution as a filter for the material elements that support induction, we can make a similar claim about the appropriate prejudice (in the sense of Bayesian priors) to carry about theories of biogenesis. *The emergence of a biosphere should have been least unlikely along trajectories in which assembly and induction were based on the order parameters of phase transitions*, and our search for theories should favor those with this structure, among hypotheses that are otherwise unconstrained.

To the extent that the introduction of new order parameters through phase transition is a sparse occurrence within the many-dimensional continuum along which boundary conditions may vary, the opportunities for new design are likewise limited. These limits – which, in the equilibrium world are derived from the reductionist part of the derivation of ordered phases – should likewise be the constraints on possible forms for the biosphere that are predictable from principles.

### 8.2.7.1 The biosphere's inference problem and our own

The implications of this argument for the interpretation of structure and variation are the reason we have chosen to model the emergence of the biosphere as accreting layers and modules from the known biosphere, rather than postulating unrelated stages or mechanisms. Recall Simon's observations on the Alexandrian empire in Section 8.2.1. Not only was Alexander's empire assembled from previously autonomous nation-states; as the empire disintegrated, it returned to those lower levels of political aggregation as its fundamental units.

In most respects we cannot study the disintegration of biological order, but we can study small elements of residual autonomy, along with evolutionary variation indicating where innovations were possible in the past. Examples include the substitution of pathway segments in core carbon fixation in Chapter 4, or the partial autonomy of genomes and cellular metabolism in Chapter 5. We suppose that the modules that limit and structure variation in the evolutionary era are plausible as subsystems that stabilized the process of biological assembly to form hierarchies.

We do not assert that no other intermediate states were important in the past which have since been lost in extant living systems. However, it is difficult for us to imagine alternative stepping stones that combine more remarkable features suggesting internal stability than the modules in extant metabolism.

## 8.3 The emergence of individuality

Biology epitomizes the study of diversity within unity. Up to this point we have heavily emphasized the structure of unitary aspects of life such as universal core metabolism, which show few or no variants and extremely long-term persistence, because we believe that structure reflects causation.

One of the most striking aspects of life, however, which sets it apart from ordered states of the non-living geospheres, is the maintenance of incredible biotic diversity – Darwin's "endless forms most beautiful and most wonderful" [177]. Much of this diversity is systemically complex, in the sense that diverse organisms persist because they are adapted to diverse circumstances. But much of the diversity reflects the preservation of about equally good solutions to the *same* problem of remaining alive in any single circumstance. The latter diversity is the standing variation maintained within all living populations. It is the raw material of adaptation as R. A. Fisher and George Price showed formally,[15] but also a source of drift and historical accident.

### 8.3.1 Darwinian evolution is predicated on individuality

*The* paradigm in biology, for the way information enters the biosphere and is maintained through the production and culling of diversity, has been Darwinian evolution ever since Darwin published *On the Origin of Species* in 1859 [177]. The three requirements for Darwinian evolution, using Lewontin's formulation [482] introduced in Section 1.3.2.5, are

1. a unit of selection capable of expressing traits and maintaining them while it exists, and capable of *replicating* to produce another generation of similar units;
2. variations in the trait that the unit of selection is capable of supporting, which are *heritable* when the unit replicates;
3. a mechanism of selection that can discriminate among the trait variants, and alter the composition of populations of these units between generations by affecting either their survival or their fecundity.

We will refer to replicators capable of fulfilling the Darwinian criteria as **individuals**.[16] We distinguish *kinds of individuality* (roughly corresponding to the notion of "units of selection") from *individuals*, which are the entities. Following Leo Buss [106], this usage allows "individual" to coincide with the common-language use of the term in those instances where the individual is an ordinary organism. The important property of individuals is that they are *undividable* with respect to some level of selection that admits a Darwinian description. Individuals can cause population dynamics to differ from bulk-phase dynamics, because they retain identities and do not diffuse into the environment or into one another.

The Darwinian framework requires as its inputs quite complex entities and rather sophisticated functions. These are its starting points for analysis in the same way as atoms and molecules are starting points in chemistry. But to understand the emergence of life, we need a more principled discussion of the role played by individuals and replication, and a realistic assessment of what would be required to explain when or how entities fulfilling those roles can arise. The replicators studied in biology are entities of enormous complexity. When they are maintained in natural systems, they exist in a context of biotic

---

[15] This is the result known as *Fisher's fundamental theorem of natural selection* [256]. It is one term in the more complete accounting identity for population processes known as the *Price equation* [650].

[16] Vigorous debates have been launched about what terminology should be used to refer to the various concepts [537], which we will pass over. Our concern is to appreciate the numerous examples already in biology of different ways the preconditions for Darwinian evolution can be met, and to understand how living order began to deviate from being a bulk geochemical process, and chance entered the biosphere.

complexity that (including their kind and other kinds as well) is even greater. Whether these individuals are organisms or cells (maintained long-term in ecosystems), chromosomes or genes (maintained by the physiology of cells), or artificial replicators (synthesized and purified by chemists working in sophisticated laboratories [488]), their ability to fulfill the Darwinian criteria depends on contexts that are not simple. The scientific work of evolutionary dynamics comes in explaining how the appropriate kinds of individuals come to exist, how their variation and interactions are structured, and how the criteria for their selection are determined.

In this section we characterize the relation of natural selection to the phase transition paradigm, and of the Darwinian world to the pre-Darwinian era of unique solutions to problems of becoming ordered. Our overarching thesis has been that the biosphere draws on eclectic mechanisms for error correction, across the diverse domain of non-equilibrium stochastic processes. Our thesis in this section is that Darwinian selection was brought into existence as a new kind of non-equilibrium mechanism, by the **emergence of individuality** which was a new kind of organization. Our main concern is what sets the Darwinian mechanism apart within the wider class of order-forming stochastic processes, and how the stabilizing mechanisms in the biosphere were in some ways preserved and in some ways changed by the emergence of individuality.

Darwinian dynamics is a special case within the more general class of order-forming, non-equilibrium stochastic processes. To the extent that life ubiquitously draws on other mechanisms to support the formation of order, it cannot be "only" Darwinian. Because it has been common even to *define* life in terms of Darwinian evolution [403], we wish to stress the oversimplification inherent in such a position.[17] There are two approaches to a pan-Darwinian view of origin that we specifically wish to avoid.

**Widening "evolution" to include everything dynamical** First, one can stretch the interpretation of the organism to include any functional constellation of events, and the notion of replication to include any mechanism by which a pattern can persist with amplification, and selection to include any statistical preference for some configurations over others, so that tautologically any order-forming process is Darwinian.[18] For us this dilutes the term and prevents it from making what we consider to be one of the most important category distinctions. Prior theories already exist for equilibrium and non-equilibrium thermal relaxation, which are built around perhaps incomplete, but nonetheless operational and empirically grounded concepts. Imposing Darwinian metaphors at this level does not improve our understanding of these simple forms of order. The very thing that makes Darwinian dynamics important, new, and interesting is the more particular character of selection arising out of properties of individuality, which we will review in more detail below.

---

[17] In Section 8.4 we return to discuss the logical problems with this identification.

[18] Our colleague Michael Lachmann has used this approach to produce minimally structured limiting models for order formation so that, within a single framework, he can study the difference in the dynamic as more complex units of selection are formed. While we understand the motivation for such an approach, we will prefer to use a more restrictive terminology. From a quite different starting point, Lee Smolin has tried to apply the Darwinian framework to the problem of specification of physical constants [771], as an alternative approach to many-universe cosmology from the anthropic principle [120].

**Narrowing "life" to its Darwinian aspects**   Alternatively, one could *define* life to consist only of systems that are maintained by Darwinian evolution, and insist that only entities that replicate and compete can be a basis for models of early life. Strong versions of RNA World fall into this category, as we reviewed in Section 6.1.2.1. We think such Darwinism-by-fiat imposes a somewhat naive model of the replication of organisms artificially on the chemical domain where there is not a compelling motivation to do so. All Darwinian information-capturing processes we know of take place in very complex or constrained systems, where a host of small-scale errors are maintained by processes outside the selective criterion of interest.

In Section 8.5 we pursue the problem of how life achieves global closure with respect to the identification and correction of continuously generated errors at all levels. In science we currently have no theory of global error correction in systems that have both complex states and finite correlation length, and we argue that such a theory must be central to understanding biogenesis. Relevant to the discussion in this section, no current models lead us to believe that a Darwinian dynamic among any known replicators is capable of providing such error closure.

Arguments that life must be Darwinian often point out (correctly) that at present we also have no demonstration of mechanisms capable of maintaining protometabolic order in bulk chemical systems. However, imposing a particular Darwinian model on chemistry because we do not currently have other ideas is a "God of the gaps" argument. It is a reasonable source for hypotheses, but if the Darwinian model depends on as many undefended hopes as the gap it is meant to fill, it should be viewed with similar reservations.

We acknowledge, but do not wish to over-interpret, the role of replicable entities. While the emergence of replicable entities, whether micells, vesicles, or oligomers, was undoubtedly important to the elaboration of living order, these are better regarded as contributions to the emergence of the individuality of organisms – which must yet be explained – than as surrogates for organisms.

When it arose, Darwinian selection introduced new mechanisms to generate future functions, and at the same time changed the mechanisms that maintained existing function. Yet even the most universal metabolic features of life are found only as characteristics of complex organisms that rely on Darwinian selection to maintain their whole intricate physiology. Evolution may create adaptation within the cellular and genomic control apparatus, but with respect to the metabolic background it is merely stabilizing. One of the important questions that a theory of the emergence of a biosphere should at least frame productively is *why unique solutions would come to depend for their maintenance on mechanisms that feed on standing variation.* The deeper question, why the emergence of individuality has been so fundamental in the elaboration of life though it has almost no counterpart in equilibrium ordered systems, may lie at the heart of the distinction between static and dynamic order.

### *8.3.2 How unique solutions give rise to conditions that support diversity*

The common feature in the explanations we have proposed for the *universal and putatively unique* forms of biological order is that the error-correction mechanisms sustaining them arise through system-wide distributed interaction of loosely connected components. This description includes regression of microscopic molecular deviations in a proposed pre-cellular bulk chemistry, but we also mean to include mechanisms such as extensive horizontal gene transfer, or feedbacks through trophic ecosystems which we argue are responsible for maintaining universal metabolism in the post-cellular era. Each of these has only "short-ranged" correlation in the sense that the components are semi-autonomous.

These are the processes in which evolution has a "thermodynamic" character in the colloquial sense: *the system can find its stable phase through the independent adjustment of many small degrees of freedom.* Distributed mechanisms with this character are associated with strong convergence toward a few solutions with only local deviations and little persistent, correlated variation. In one way or another the optima are locally defined (similar to the properties of simple gases or liquids in thermodynamics), whether for genes because these are rapidly transferred among different chromosomal contexts, or for ecological fluxes because the trophic exchanges between individuals are haphazard and fluctuating.

A qualitatively different relation to standing variation is found where a form of individuality has emerged to integrate many components, and where multiple individuals form populations in which each individual is a parallel platform doing whatever individuals of that kind do. In these cases, *a kind of long-range order exists among the components within any individual,[19] which renders the selection of components non-local, and creates a more favorable condition for the maintenance of standing variation.* This kind of aggregation exists among genes linked in a chromosome,[20] cell lineages in a multicellular organism, or populations on patches in island biogeography. Box 8.3 gives two examples that contrast distributed maintenance of a unique solution, with cellular-level maintenance of parallel solutions of the kind that typifies species.

---

**Box 8.3   Global systemic solutions versus parallel solutions**

**The parsimonious tree of early carbon fixation, and the reticulated tree required to explain methylotrophy**  The high parsimony that is possible for early branches in the tree of autotrophy in Figure 4.10 gives no empirical basis to invoke the maintenance of distinct carbon fixation pathways in organisms brought to occupy a common environment. This signature would be expected from an era with facile horizontal gene transfer and limited integration of metabolic genes into species-restricted groups. Organisms would be driven strongly toward unique metabolic optima governed by local geochemical conditions, but they would retain little memory of conditions other than those currently occupied.

The carbon fixation phenotypes that live on the early, high-parsimony branches stand in striking contrast to the methylotrophic phenotype of post-oxygenic $\alpha$-proteobacteria shown in Figure 4.12, which can only be placed on a tree with

---

[19] The long-range order is maintained by whatever forces maintain the existence of that form of individuality.
[20] See Lewontin [483] (Chapter 6).

---

**Box 8.3   Global systemic solutions versus parallel solutions (cont.)**

numerous reticulations. The methylotrophic network calls for many instances of gene transfer for large pathway segments, of a kind that could only have resulted if the donating bacteria and archaea maintained their distinct pathways even as they were brought together into a common environment with the ancestors of the proteobacteria to which these were eventually transferred. This situation would be expected of organisms with more tightly integrated genomes, and less pervasive horizontal transfer at the single-gene level.

**Contrasting the universal genetic code to non-universal 16S rRNAs** The universal genetic code appears to antedate all microbial divisions because it has no variation among most of the major clades in the tree of life. Kalin Vetsigian, Carl Woese, and Nigel Goldenfeld have argued [833], and demonstrated quantitatively in a formal model, that this form of strong convergence is a signature of **innovation-sharing** in a world where standing variation among genotypes was not maintained within populations.

They demonstrate that optimization of the code as a buffer, against changes in the polar requirement from single-base errors in translating statistical proteins, depends on the extensive and facile transfer of genes, following a line of argument put forth by Woese [887, 892] (reviewed in Section 5.3.7.1). Although statistical proteins can (uniquely) accommodate exchange among organisms with different code assignments, they nonetheless benefit when gene transfer is not an additional source of perturbation. Therefore buffering exerts a positive feedback: the more a population has converged on the same code that is a good buffer, the more viable gene transfer becomes, reinforcing the strength of convergence on a universal code. Vetsigian *et al.* argue that *only* communal evolution in a world of gene exchange could have produced a code with the observed levels of both universality and optimality (supposing that the polar requirement is the appropriate criterion of selection).

The universal code, stabilized through a distributed mechanism of gene transfer, stands in contrast to the progressive separation of species first signaled by differences in 16S rRNA. The fine tuning of ribosomal peptides and RNA for precise and reliable translation obviates exchange of components. Ribosome lineages do not separate because their performance differs and is a criterion of selection. Instead they are driven to ramify even if their functionality is essentially identical, because they are made of components that are not readily made interchangeable.

---

A different way to characterize the emergence of individuality is in terms of the change it creates in the *state-space structure of stochastic processes*. As Section 1.3.2.4 noted, all natural phenomena, whether abiotic or biological, are stochastic processes with the *Markov property*: all information from the past that is relevant to the future is incorporated in the system's state in the present. However, the structure of state spaces with this property can be very different from case to case, and this accounts for a large part of the characteristic difference between non-living and living systems.

The whole-system state spaces for systems with few, robust states of order tend to be characterized by extensive mixing among components that have limited complexity or persistence within their internal, component-level states. In contrast, systems that maintain persistent, correlated fluctuations are those in which the components have complex and persistent internal states. These components satisfy the description of individuals. This kind of internal-state complexity is necessary for a component to function as a "platform" for executing an autonomous copy of some process.

In the language of Conant and Ashby [154] such internal states are implicit **models** of an individual's likely future situations within its environment. Often these models are products of extensive selection. The individual contains within itself a representation of its environment (chemical, energetic, ecological, etc.) derived from the statistics of long prior histories. Unlike bulk chemistry, in which order only persists if it is continually reinforced by feedback from the *instantaneously present, local* environment, the biological platforms that allow many parallel solutions to run can support those solutions both with present feedback, and with **memory** of past events incorporated in their internal models.

The emergence of standing, heritable variation, then, is associated with the emergence of platforms capable of maintaining internal models and distilling information from the past into distinctions of their internal states. This property of being partly freed from the need of instantaneous, local reinforcement from the environment is both a precondition to being able to autonomously execute a function, and a form of insulation that keeps idiosyncratic variant models carried by different individuals from being rapidly homogenized.

### 8.3.3 The nature of individual identity

The criteria defining individuality have been realized many times and in many ways in the emergence of life and its ongoing evolution. They are inherently statistical and not associated with a particular substrate or instance. Individuals, of whatever kind, are packages of components that are *collocated in space, that affect each others' states, and that have at least a high probability to be either reproduced or lost together.*

By this criterion, oligomers such as RNA capable of taking on secondary structure and undergoing template-directed replication create a kind of individual association among the RNA bases they contain. (This is the form of individuality that strong RNA World scenarios invoke [122, 488].) Chromosomes create a kind of individual association among genes [483] (Chapter 6). Cells create many kinds of individual associations: redox systems, protons, and nucleoside polyphosphates are coupled by oxidative phosphorylation; the catalysts, substrates, and reactions in the cytoplasm are bound together with each other and with cellular energy systems; replicating chromosomes are coupled to the metabolism they support, and bound at least with some probability to either its death or its replication by cell division.

Whether colonial organisms qualify as individual with respect to their constituent cells depends on whether the collective maintains and transmits traits through its relations that go beyond the traits of the member cells. Multicellular organisms are individuals when they include differentiated tissues and germline/soma populations [106]. The complexity of the concept of the physiological or developmental individual in relation to the chromosomal unit of selection is well illustrated by organisms with complex lifecycles, in which several different unicellular or multicellular phases, potentially with different chromosomal contents, may constitute stages within one replication cycle [354]. In sexual organisms, for some purposes it is the haplotypes transmitted in the gametes that behave as individuals, and the diploid (or higher-)ploid organism is a game played by the haplotypes to produce new haplotypes [768] (Chapter 8).

The important relation in any of these cases is that during life, the components within an individual have stronger or more elaborate material or control dependencies on each other than they have on any particular objects in the environment. Their destruction or reproduction are also correlated both in space and in time. The property of individuality most fundamental to Darwinian evolution, then, is that the correlation of the locations, states, and events within individuals allows many copies to perform quite complex and nearly identical functions and yet be only weakly coupled to each other.

### 8.3.3.1 Granularity and shared fate

We have emphasized the importance of parallelism and conditional independence of individuals as the source of their ability to maintain variation in a population, to a degree that would be impossible if their components were freely exchanged and selected independently in a common-pool environment. Two further properties of individuals that can greatly alter dynamics from the character of bulk physics or chemistry might be termed *granularity* and *shared fate*.

**Granularity** In many modern kinds of individuals, the locations, states, and events of the components are not merely more correlated within individuals than across them, the components within individuals are often tightly integrated and nearly insulated from particular objects or states in the environment. Being present within an individual is often therefore a kind of hard constraint on the states the component can take, which limit its capacity to either sense or respond to the external world.

Such hard constraints are characteristic of granular systems. They can completely alter the qualitative dynamics from forms seen in fluid systems which otherwise are similar in character. Examples of granular systems are sandpiles, ricepiles, and other flowing granular matter [19, 46, 47], or the electrical states of digital computers, as contrasted with those of analogue computers. Granular systems can **jam** [489] – that is, they can be unable to find solutions that fluids would readily find to respond to external pressures – because the internal constraints within grains can forbid steps in the search that less tightly bound components might otherwise have executed. The complex properties of granular systems relative to fluid systems are characteristic of the complexity of search and solution problems in integer arithmetic compared to the simplicity of the same problems for real numbers.

We emphasize jamming here because the inability to respond to external pressures is one mechanism by which individual-based systems may preserve more standing variation than amorphous systems. The example of fixation of the genetic code in a world of statistical proteins from Box 8.3 is such a case. When all aspects from code assignments to protein synthesis are error prone and stochastic, selection for robustness can proceed to produce a high level of correlation in amino acid properties at adjacent codons. In a world with precise protein translation and selection for specificity, the same solutions are not so readily reached, because changes in codon assignments result in systematic alteration of proteins throughout the organism and near-certain fatality. A more refined coding system produces a more integrated translation system and a more "granular" relation among all the individual's proteins and their contributions to phenotype. The same selection pressure under conditions of more refined coding, in the models of Vetsigian *et al.* [833], produced both less redundant coding assignments and more variation between and within populations.

Strong granularity in space leads to granularity in time as well – that is, the whole organism replicates or dies as a unit. This granularity makes the idealization of replication applicable to complex packages of components and traits such as whole organisms. In searching for evolutionary causation, it may sometimes be useful to distinguish the temporal granularity imposed as a *topological* condition by cell division or germline/soma separation, from **senescence** interpreted as an evolved property within the lifecourse.[21] The evolution of senescence, characterized by regularities such as the Gompertz mortality curve [41, 310, 782] that sets rather strict bounds on maximum lifespan, replaces the Poisson statistics of random failure in simple systems with a much more programmatic and predictable behavior of the individual with respect to its own offspring and its ecological context.

**Shared fate** When cells die, very low-level components such as intermediary metabolites may simply disaggregate and be taken up by other cells. Most higher-level aggregates, however, are simply lost. Unlike the nation-states of the Alexandrian empire, most modern components of life have such refined dependencies on their cellular or organismal contexts that not only their state of aggregation but the components themselves are discarded if the individuals disintegrate. As a result, the components within an individual usually have strongly shared fates of death or reproduction.

Shared fate is another mechanism that can lead to increased standing variation. The cause of death of an individual may be one unfit trait, but any other traits not suitably dispersed in the population may be lost if the unfit trait causes the individual to leave no offspring.

---

[21] The most basic way senescence can evolve to create a tightly programmed and predictable schedule of mortality is George Williams' concept of *antagonistic pleiotropy* [879]. Among genes which create a mix of positive and negative fitness effects, those for which the positive effect occurs before reproduction, and the negative effect late in life, can accumulate through selection despite their negative consequences. Because the onset of fecundity is generally tied to body size and so is part of a quite regular growth trajectory for many organisms [868], morbidity and mortality inherit this regular schedule as well.

### 8.3.3.2  *The requisite variety of organism selection*

Here, then, is the relation we believe exists between Darwinian selection in a world of individuals, and the larger universe of error-correcting mechanisms in the biosphere.

Each kind of individuality defines a complex process jointly conducted by some components within the biosphere. The individuals which instantiate each such kind run nearly parallel copies of some collection of the functions of life, including their biochemistry, their physiology, their ecological roles, and so forth. All of these functions are performed with some error, both random error due to stochasticity of events, and error which may differ systematically between individuals due to their inborn differences. The granularity of the individuals both limits this error and incorporates some mechanisms to identify and correct it within the lifecourse, providing a form of buffering.

The accumulated differences in performance that cannot be buffered within individual lifecourses lead to the changes in their situation that are expressed as differences in fitness. These are the differences that are exposed to mechanisms of selection such as spectra of stresses that kill weaker members or cause different individuals to leave more or fewer offspring. The selection mechanisms act only at the events of death or reproduction, in contrast to the buffering mechanisms which act continuously through the lifecourse of the organisms.

The reason we believe that pan-Darwinism oversimplifies the problem of global stabilization is essentially one of counting. Consider, for example, the error rate and requisite variety of selection for a bacterial cell. A cell of *E. coli* that weighs 600 femtograms contains about 4.6 million bases of DNA in its genome but roughly $3 \times 10^{13}$ atoms. The genome is replicated roughly once per 20 minutes in exponential growth phase, while molecular vibration frequencies are in a neighborhood of $10^{12}$ per second, and turnover of small metabolites may be estimated from turnover rates of ATP, which are on the order of 250 per minute [362]. The opportunities for error to be introduced into the molecular inventory of the cell scale with the multiplicity of bonds in the atoms and the vibration rate. Most errors are exponentially rare compared to this elementary scale – which is precisely a form of buffering offered by chemistry – making the reference rate for error difficult to estimate, but we may suppose that the primary exponential suppression is the difference between the vibration frequency and the reaction rate. Relative to the timescales of reactions, the probability of errors is still suppressed by some factor relating to the reliability of catalysts, translation, etc., but not by a very large factor when all the possible sources of error are taken into account. Therefore, as a rough estimate, we may take the possible error rate as the order of the number of atoms multiplied by the reaction frequency, or $(3 \times 10^{13}) \times 250/\text{min} \sim 10^{16}/\text{min}$, divided by some smaller factor which counts the fidelity of reactions and reflects the fact that most molecules do not turn over as frequently as ATP does.

We ask, therefore, whether the selective filter, which can incorporate information at an upper limit somewhere well below megabases per hour,[22] is plausible as a sufficient

---

[22] This limit comes from the requirement that mutation rates must fall below the Eigen error-catastrophe threshold for the cell to remain viable in the long term.

mechanism to identify and correct all errors that can enter the cell through its molecular dynamics, at a rate loosely estimated to be several, but not very many, orders of magnitude below $10^{16}$ per minute.[23] Any excess of the molecular error rate above the reproductive error-correction rate must be buffered by the cell's own organization, above and beyond the part of that organization that is corrected by selection itself. *In other words, cellular organization must take a form which allows the requisite variety of selection (expressed as a rate[24]) to be less than the information in the genome multiplied by the reproduction rate.*

These numbers for a bacterial cell provide figures of merit, but the selection of interest is on the genome of the organism. For multicellular organisms, the error rates must be multiplied by the number of somatic cells in which damage can significantly affect the survival and reproduction of the germ line.

From this perspective, the population-level consequence of Darwinian selection is not qualitatively different from that of local regression toward non-equilibrium states through other mechanisms. Both are opportunistic processes. They create order only in domains of organization-space where the architecture provides sufficient buffering to meet the criteria of requisite variety of the controlling mechanisms. The study of equilibrium phase transitions has taught us that the opportunities to form asymptotically stable order are rare. We expect that the same will be found to hold for non-equilibrium processes – that the order of life clusters closely around these architectures that provide buffering, much as trees in a dry land follow the waterways.

We expect the domain of Darwinian selection to be subject to similar limits. Life is not "only" Darwinian because, even if selection on individuals is used pervasively, it is used within limited locales of opportunity, where manageable forms of individual organization can exist. At low levels such domains reflect structure in chemistry and energetics; at higher levels they may result from the prior action of selection on other units. In all cases, though, the error correction must be closed as a system, so even if the architectures that provide sufficient buffering are biotically generated, they exist within a hierarchy where they may never become fully independent of buffering from the lower organizational levels out of which they first arose. It is in understanding the requirements for such architecture, and recognizing the particular cases in which they are met, that we express much of the lawful or predictable nature of life.

---

[23] The actual error rate is again difficult to estimate because it results from a combination of stochastic and deterministic factors. When catalysts are properly formed, errors are stochastic with a rate determined by the accuracy of the catalyst. But if catalysts suffer a single mutation which is a single event, they may produce systematic errors at a rate comparable to the rates of reactions. The correct accounting for this global error-correction system is a difficult quantity to estimate in the complex context of actual cells.

[24] Requisite variety is often introduced as if it is a static property, counting the number of distinct internal states among which a controller can discriminate. It is expressed as a rate if we multiply the number of distinct states by the rate of events at which the controller can respond to them. To some extent these quantities can be interchanged. For example, the controlled system may be able to deviate only in a limited number of modes, but the amount of the deviation may depend on the time between corrections from the controller. In this case, the controller may require a larger internal state space in which the magnitude of the control response as well as the mode of deviation is distinguished. In general we suppose that any control system can be reduced to a minimal representation in terms of a *generator* [206] which is rate valued, in the same way as a continuous-time Markov chain can be represented by its generator. The generator expresses the least qualitative diversity by which all responses of the controller can be generated by recursive application, and the intensity of control can then be modulated by setting the characteristic timescale on which the events induced by the generator occur.

The concept of individuality is broad and permits many realizations. Jointly, the emergent forms of individuality can have complex consequences for the character of biological dynamics and error correction. If the emergence of a Darwinian world was the emergence by layers of this complexity, that transition should have not been either simple or unitary.

### 8.3.4  Individuality within the structure of evolutionary theory

In this book we have mainly considered the way an ordered metabolism could have emerged, and how it would have supported and constrained the emergence of molecular replication, vesicles, and bioenergetic systems, and their later assembly to form cells. The stage when these integrations were complete marks the beginning of increasing diversity and historical contingency, and at this stage we leave the argument to be carried on by others.

However, even these early considerations lead to a view of emerging Darwinism as a complex constructive process with many stages. We think that the difficulty of reconstructing the emergence of individuality is an *appropriate* difficulty, and we have rejected approaches that try to bypass it and yet use Darwinian evolution as a point of departure to life. We think this view meshes well with the careful and constructive approach to levels of development and selection taken in modern evolutionary biology and ecology, more so than a simplified view of Darwinian transitions would do.

#### 8.3.4.1  Individuality in developmental complexity and differentiation

A landmark book integrating evolution and development from the perspective of embryology is Leo Buss' *The Evolution of Individuality* [106]. A theme that runs through the book is that developmental complexity arises as living units that were formerly autonomous are brought together into new associations where their autonomy changes. Each new association marks a new kind of individuality and a new unit of Darwinian selection. The components, which may once have been fully autonomous Darwinian units in another context, do not necessarily cease to evolve once they combine to form a new kind of individual. More often it is the context for their evolution that changes. The aggregate system may alter or limit the mechanisms that produce variation in the components, it may restrict the forms of competition they can undergo, or it may change the criteria that define their fitness. Sometimes the components continue to undergo direct selection by environmental factors, and in such cases their success as autonomous agents may support or may conflict with the success of the larger assembly of which they are parts.

#### 8.3.4.2  Variation and modularity

In *Cells, Embryos, and Evolution* [295], Gerhart and Kirschner emphasize that while the variation that is the raw material of evolution is not goal directed, it is nonetheless highly *structured*. They provide a detailed and empirically rich review of the many mechanisms by which the assembly of components into an aggregate may change the kinds of structure the components may generate. They are particularly interested in positive selection for increasing modularity as a facilitator of adaptive variation that at the same time enhances stability by insulating the internal components of modules from detailed interference by their environments.

The Gerhart and Kirschner argument for the ongoing evolution of order echos Herb Simon's argument for modularity as the prerequisite to the emergence of hierarchical complex order. The basic mechanisms of error buffering and the reduction in requisite variety of control systems are the same as those already listed in Section 8.2. Again, the themes we have used to understand pre-cellular life are recapitulated in later eras.

### 8.3.4.3 Genes, chromosomes, and kernels

Even in the narrowed context of molecular replication, we may suspect that the emergence of a natural unit of selection in the early world was not simple on the grounds that it is not simple today. A useful illustration that individuality must ultimately be handled statistically, and that even then it may present a complex problem, is provided by debates over the gene versus the chromosome as a unit of selection.

For Gregor Mendel [553] the gene was inherently statistical, reflecting the partitioning of traits in strains of plants for which these statistics were particularly simple and unambiguous. In the decades following the discovery of DNA structure by Franklin, Watson and Crick [275, 853] and the inference of the way genes influence phenotype, there was a progressively more materialist reification of Mendelian statistical notions, culminating in the use of "a particular copy of a non-recombining sequence at some locus in some individual" as a *definition* [439] of the gene. Although this characterization has suffered many setbacks and undergone many qualifications as increasingly many forms of epigenetic and post-transcriptional modification have been discovered [193, 203, 778], the abstractions it presumes are still widely used and influential. They have led, among other consequences, to a view of evolution in which "the gene" is the only fundamental unit of selection [181, 880], both indivisible within itself and autonomous from other genes through the agency of crossover (in sexual organisms) or other forms of DNA rearrangement.

Against this view, Richard Lewontin created an elegant model of **linked heterosis**,[25] in which weak heterosis at many loci in a chromosome could be amplified if the complementary alleles at many positions occurred together to form complementary chromosomes. Even in the presence of crossover, if selection was sufficiently strong, linkage disequilibrium could be maintained in populations. The Lewontin model illustrates that *covariance*, as an independent property of chromosomes from the contribution of individual genes to phenotype, could become a selected character because it could be transmitted if crossover was sufficiently infrequent. The difference between even idealized genes, and other units with weaker covariance, was only a matter of degree.

---

[25] See the integrated treatment in [483] (Chapter 6) and references therein to the development of the model in the primary literature.

The very nature of translation requires that, where genes do correspond to non-recombining DNA sequences, the explanation must be found elsewhere than in the genetic system. Perhaps no other components in the cell have evolved to be as interchangeable as the successive phosphodiester bonds between DNA bases. Crossover within genes is only suppressed to the extent that it is likely to produce a lethal phenotype, which never survives the lowest-level filters in ontogeny to be presented for selection. This criterion is merely a stronger form of the fitness deficit in Lewontin's linked heterosis model when two jointly homozygous chromosomes are paired, but it is not different in essence.

Two factors that tend to cause Mendelian genes to correspond to regions of DNA are the need to preserve the **reading frame**, and use of open reading frames to encode **folding motifs** in proteins. The reading frame is imposed by the ribosome and tRNA, for reasons considered in Chapter 5. Protein folding is a source of multiple long-range constraints, which are difficult to preserve in crosses within an exon. Numerous, parallel constraints that a successful protein folding motif has evolved to satisfy render the gene a "granular" unit of selection, compared to softer correlations such as linkage. *The granularity of the gene is thus an epiphenomenon* of modularity in coding and protein synthesis, each reflecting constraints on the kinds of modularity available to make physiology robust and evolvable.

As a final example, we consider a case in which collections of genes forming whole regulatory circuits become so tightly integrated that they admit no non-lethal variants and so are removed from the spectrum of evolutionary variation. They are like super-genes, protected not only from crossover but also from non-silent mutations.

Eric Davidson and Douglas Erwin have coined the term **kernel** [180, 227] for early developmental regulatory networks that show this property. They are particularly interested in regulatory genes such as the endomesoderm regulatory network for which the sea urchin provides a thoroughly studied model system [179]. The diversification of this and closely related early-developmental networks was a large factor in the diversification of animal forms in the Cambrian explosion [228]. Davidson and Erwin argue that as organisms became more complex, so many secondary developmental pathways eventually came to depend on the early regulatory networks that almost any change in the early networks led to multiple downstream differences and almost sure fatality. The argument recapitulates, at the level of developmental regulation, Carl Woese's arguments for the freezing in of the genetic code, considered in Section 5.3.7.1. We advanced a similar argument in Section 4.4.7 for the freezing in of the role of the TCA cycle intermediates as precursors of anabolism prior to the rise of oxygen.

### 8.3.4.4 The Darwinian Threshold and the emergence of sequences

A different level of partitioning than the integration of chromosomes occurred with the inhibition of horizontal transfer of genes across species, described in Section 5.3.7.1. To the extent that it is accomplished by preventing transport of genetic material across the cell boundary, it is similar to integration of the individual's genome, but to the extent that it is accomplished by inherited mechanisms to distinguish native from foreign genetic material, it creates an emergent, species-level[26] identity for groups of genes. Thereafter they take on a new level of shared fate under descent, and can be jointly selected to function together. The inhibition of gene transfer, which at an upper level created a group-identity for the gene complement of a species, was driven by a shift in a different unit of identity at a lower level: the shift from the distribution of sequences that Woese called "statistical proteins" and their equally statistical genes, to the particular sequence as the carrier of phenotype.

The shift from distributions to sequences as the carriers of phenotype is likely to have been one of the most classic of phase transitions in early molecular evolution. As in the transitions in the hierarchy of matter, a common concept such as object identity is transferred between two qualitatively different levels and kinds of organization.

### 8.3.4.5 Multilevel selection: a question of covariance

The foregoing examples all involve stages through which subcellular components would have passed, between starting as heavily randomized entities subject only to distributed selective filters, and later becoming embedded in many-level hierarchical systems where the relations among them also become heritable and thus targets of selection.

A counterpart exists for population processes at and above the level of the organism, recognized by William Hamilton in the 1960s for kin groups [332, 333, 334]. If the organism can be a unit of Darwinian selection that is not captured only in the replication of its components, by the same token populations of organisms may become units of selection above and beyond the selection of the members. To the extent that the process of aggregation changes the mechanisms of variation or the conditions of selection for its members, it may not even be sensible to speak of the evolutionary dynamics of the members except within the context of their relations in the aggregate.

The question whether a group of organisms qualify as a unit with respect to function and Darwinian selection is mathematically simply another application of the covariance criteria for individuality.[27] Steven Frank [273] lays out systematically the origin of Hamilton's rule, expressing fitness as an additive function of regressions on one's own genotype and on the genotypes of others in the population,[28] weighted by the relatedness among the individuals who interact. Hamilton's rule, that covariance between the cost to oneself and the benefit to others can be reinforced by selection, follows as a general consequence of the Price equation [650].[29] Relatedness is a measure of covariance among the members of the population; it may be created by island geography, kin recognition, or many other means of assortation. A population subgroup becomes a proper unit of selection if the covariance

---

[26] In this discussion we use "species" for brevity. The relevant identifier will be a clade-level attribute which could be shared at different levels of relatedness for different genes.

[27] More precisely, the existence of a non-zero covariance defining granularity of function and shared fate in heredity is a *minimal* criterion for the emergence of individuality. Tightly integrated groups, like tightly integrated cells, may have very strong and non-linear couplings, so that many higher moments than covariance are also non-zero.

[28] Thus, fitness is approximated with a linear frequency-dependent model.

[29] The Price equation, in turn, is valid as an accounting identity for any population satisfying the axioms of classical population genetics, and is thus a very general relation [273].

that it possesses can be transmitted under turnover of the population, and maintained either by external constraints or by selection as a long-term property of the group. Determining empirically when this is the case may be a difficult problem, but the criteria are similar to those we have seen for the emergence of cellular or genomic individuality from subcellular components.[30]

### 8.3.4.6 Summary: common principles behind chemical and evolutionary order

The particular revisions of order in the post-cellular world are not directly relevant to our problem of understanding the transition from geochemistry to cells. However, the *nature* of the revision of order shows much more continuity with a pre-cellular world than appears in a very superficial contrast between bulk chemistry and organism selection. Three points in particular stand out.

1. The role played by the Darwinian algorithm is important, but only in a context created by a cascade of levels and forms of individuality that determine the character of variation and selection.
2. The essence of emerging individuality is conditional independence, modularity, and buffering of errors to a degree where control systems can meet criteria of requisite variety. These are the same operative concepts as those behind phase transitions in bulk systems, including dynamical transitions in non-equilibrium chemical kinetics.
3. Whatever advantage drives the emergence of individuality, it did not happen only once to form a clear demarcation between non-life and life. It has happened recursively in small ways to form many layers of structure ranging from subcellular to supra-organismal, and it is an ongoing process by which the structure of evolutionary dynamics changes.

In closing we show how this change in framing leads to a different way of thinking about ecological assembly and about the meaning of autotrophy versus heterotrophy.

### 8.3.5 The ecosystem as community and as entity

The understanding that relations can be transmitted and maintained by selection provides a way to understand ecosystems both as entities in their own right and as communities assembled from species.[31] Ecosystems carry the residual forms of constraint that we argue originally flowed in the sequence (geochemistry $\to$ metabolism $\to$ control) at life's origin. Today, many of these constraints are expressed in evolutionary convergences within coevolving groups of organisms, rather than as properties encapsulated within particular species.

---

[30] The formal equivalence of the mathematics between chromosome selection and group selection is shown in [768].

[31] Note that this does *not* require a characterization of ecosystems as "super-organisms": ecosystems persist through co-evolution of their members and neither compete nor are replicated. It is precisely because we have non-Darwinian notions of entity that we can grant ecosystems the status of entities without needing to dilute the physiological/developmental/Darwinian notion of organisms to express this point.

The modern biosphere comprises something on the order of twenty million species[32] but one nearly universal metabolic chart. The role we think evolution plays as an error-correcting mechanism will determine both what problem we believe is solved in creating ecosystems that contain many interacting species, and the role we imagine selection has played in maintaining the chart. The distinction is between framing the problem as enforcing cooperation and framing it as enabling differentiation.

### 8.3.5.1 The ecosystem as community: reining in autonomy

A perception of the living world that has become entrenched in science since Darwin emphasizes the unity of individuals: the distinctness of species as genomic lineages, the flow of control within each individual from genomes to metabolic capabilities, the assembly of ecosystems as communities of behaviorally and evolutionarily autonomous species.

From this perspective, a universal metabolic chart becomes a *precondition* for the existence of a trophic ecology in which all members of a community can participate. Such a framing suggests historically distinct species lineages brought together somehow and faced with the problem of assembling into a community with sufficient stability and predictability that organisms could become adapted through natural selection. The dynamical problems suggested are the management of unrestrained competition and the formation of higher-level organizations that can facilitate or enforce cooperation. The suggested counterfactual case is a world where ecosystems may consist of subcommunities that do not interact trophically because they use mutually exclusive subsets of core metabolites that limit or preclude their capturing or sharing each others' biomass.

Undoubtedly the individual-oriented view captures many real problems of ecological community assembly, especially in the modern world where ecosystems form from species with long, well-resolved, and independent histories. But does this framework provide the necessary, or even the best, interpretation of the universality of core metabolites?

### 8.3.5.2 The ecosystem as entity: enabling differentiation

From a perspective in which metabolism preceded the informatic layer of genes, Central Dogma control flow, and Darwinian evolution, different observations govern the framing of questions.

Chemoautotrophs and photoautotrophs support biosynthetically complete metabolic networks under the control of single genomes. When integrated genomes, carried on vertically transmitted chromosomes, are the units of heredity, is seems clearly inadequate to try to explain the maintenance of integrity of species against error incursion only in terms of selection on genes. At the same time, unrestrained competition among genes also seems

---

[32] Current best estimates from the scaling of samples as a function of taxonomic level yield estimates of 8.7 million eukaryotic species [566]. The question how to extend this to bacteria and archaea is complicated because simple breeding criteria are not applicable. Curtis *et al.* [172], using somewhat arbitrary difference in 16S rRNA to define groups, estimate on the order of ten million bacteria and archaea. See [104] for a review on both technical problems of sampling and estimation, and also current approaches to choosing species concepts for prokaryotic lineages.

unlikely to be a pervasive problem. At some weak level these effects may arise, but the dominating constraint is the joint regulation of the genome in physiology and the linkage of its gene content during transmission. Genes within a cell have no easy escape from whatever regulatory mechanisms the cell possesses. Genes that compromise their hosts' integrity are almost sure to be lost if they cause the cell to die. The problems of optimization are different from those in a classical Darwinian world of coevolving populations because the levels of integration of the components are different.

From this perspective, in which metabolic productivity at the ecosystem level precedes the evolution of control, and in which autotrophic organisms are known empirically to exist, the more natural question for ecological evolution becomes: why is it favorable in many ecosystems to *distribute* the control of the ecosystem's metabolism under many genomic lineages subject independently to Darwinian selection? The cost of this separation is solving the many problems of coordination and regulation at the population level, which autotrophs provide at the organism level. If those problems are not solved before control is fragmented, all competitors may be lost, while the less risky strategy of autotrophy sustains metabolism.

The suggested comparative question becomes whether most stable ecosystems depend on a layer of primary producers which are autotrophs, to provide a fallback for the stability of metabolic output when higher-level coordination mechanisms among syntrophs perform poorly. Were the only early lineages autotrophic, or was an early partitioning into complementary specialist clades stable? If early lineages were autotrophs, how then could they later have differentiated, and when?

In this framing of the problem of the origin of species, the universal core metabolites are not a merely opportune resource that provides a meeting place for individuals brought together to form ecological communities. Rather, the network functioning as a whole is a reference state for ecosystem-level feedbacks, which provided a foundation for the sorting out of coordination and control mechanisms, both within organisms and among incipient species in populations.

### 8.3.6 Why material simplicity precedes complexity in the phase transition paradigm

The first phase transitions in the emergence of the biosphere were departures from the homogeneity of near-equilibrium states. It may be that, for the same reason as most equilibrium states of matter are not complex, the first ordered dynamical chemical phases could only have had simple order with unique forms. It was the long-range order created by the early transitions that established stable mutual interactions among microscopic components, which provided an infrastructure capable of supporting subsequent phases with a complex array of possible ordered forms.

The hierarchy of matter surveyed in Section 7.4.4 serves as precedent for the suggestion that cascades of phase transitions in which successive symmetries are hidden are the normal route to complexity, rather than single transitions that lead to complexity directly. In the hierarchy of matter, simplicity in the form of more (expressed) symmetry precedes complexity in the form of more (hidden) symmetry. The enormous complexity of the space of chemical molecules and reactions was not formed directly from the vacuum at the Grand Unification scale. A sequence of simpler transitions formed successive layers of order first. Although, among the elementary particles only the proton and neutron (plus massless particles) remained at the 1 MeV energy scale, the condensation of nuclei formed from these produced the large jump to the complexity of the periodic table. The further condensation of atoms and then condensed matter at the 0.1–1 eV scale then produced the very large combinatorial space of chemistry. On this route from states of high symmetry to complex order, in each transition new order is mainly made possible as an outgrowth of the framework of existing order, created by previous transitions.

A similar cascade of nested structures is ubiquitous in biology, whether in the progressing complexity of cell forms, the elaboration of development [180, 227], or the evolution of individuality in multicellular organisms [106]. The premise that order at each stage must have been stable suggests to us that a sequence of simpler bulk transitions must have been the first departures from the Gibbs equilibrium. Only once the order from these transitions was established and had selected a preferred inventory of chemical intermediates, did it provide a foundation for succeeding elaborations of metabolic pathways, condensation and exploitation of vesicles, reinforcement of mechanisms of molecular replication, integration of bioenergy systems, and so forth.

Figure 8.1 illustrates diagrammatically the dependence of different stages in the emergence hierarchy from Chapter 6, as a sequence from more bulk-phase symmetry, toward more structures and more required order.

## 8.4 The nature of the living state

The biosphere is both a unified system unto itself, and a system that shares many substances and attributes with non-living parts of the world. The time-worn questions and paradoxes about how to distinguish life from non-life reflect the difficulty of choosing the right essences and points of distinction.

Should water be considered essential? Is being carbon-based essential? Is *any* aspect of the material substrate essential, or is life defined by dynamical abstractions that could be instantiated in man-made machines or computers, or in exotic forms of organization that might be discovered outside Earth? Is the stability of a pattern and its growth to subsume more matter definitive, and if so, are hurricanes or wildfires then alive? Is cellularization essential to the nature of the living state, as E. B. Wilson maintained [883]? If we take as given that free-living cells are alive, are virions alive? If not, is there a definition by which the viral genome in its whole lifecycle is alive, and what status does that grant to the virion?[33] If virions or viruses are not alive because they only parasitically coordinate cellular processes, why should the same argument not be advanced against obligate intracellular

---

[33] Patrick Forterre [263] argues for the use of the term *virus* to refer to the full lifecycle involving the viral DNA or RNA, and for *virion* as the dormant spore phase of that cycle.

Dynamical phases in Emergence

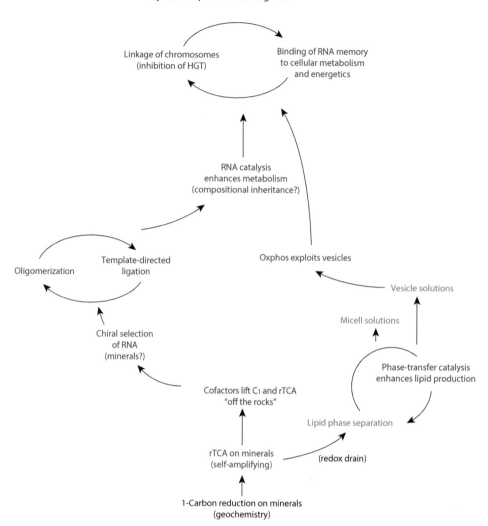

Figure 8.1 The major (phase) transitions in (prebiotic) evolution. Black indicates geochemical processes that are not distinctively biotic, though they may later be displaced by more efficient biotic
counterparts. Green indicates formation of equilibrium ordered phases, understood with classical
thermodynamics. Red entries are those that inherently incorporate dynamics, and in which the order
is principally an order of processes.

parasites which depend on the host for most of metabolism? By *reductio ad absurdum*,
at what point does the dependence of viruses or obligate parasites become qualitatively
different from the dependence of any heterotroph on its ecosystem?

We did not set out to add to the list of formal definitions of life. Our goal was to
bring together the diverse evidence in extant life that suggests specific early stages of

chemical organization, and their role as a foundation for the rise of later complexity and distinctiveness in the biosphere. The interpretation of that evidence has seemed to us to require making certain abstractions explicit: among these, the roles of modularity, requisite variety in hierarchical control systems, and the stability properties conferred by phase transitions have seemed especially important, to avoid becoming lost in a forest of mechanisms and *ad hoc* scenarios for emergence.

As a result of these choices, however, we are inevitably led to a view of what is essential to the nature of the living state, rather different from the views that lead to the questions and paradoxes above. If there is a premise that all the foregoing questions could be said to share, it is that the nature of life can be sought in *properties of components* of the biosphere – whether component entities or component processes. We approach the biosphere, instead, as a *particular system*, somewhat like a superconducting chunk of niobium, or the complex lithosphere, or the brain, are particular systems. Each may be describable by many abstractions, but we wish to avoid mistaking particular abstractions for the actual system, into which different abstractions may provide variously incomplete windows.[34] For us, the question of what is essential to the nature of the living state begins with the emergence and persistence of the realized, material biosphere on Earth, and *all* of the functions that contribute to it. Many of the previous questions can then be reformulated in terms of the roles that different substances or attributes play in maintaining rather than defining life.

### 8.4.1 Replicators: a distinction but not a definition

Since the metabolic core of life consists of quite ordinary organic chemicals, it has been natural for those seeking to define life to focus on attributes that seem best to distinguish living from non-living systems. Therefore many definitions focus on individuals, replication, and selection, either specifically or by invoking Darwinian evolution. We have noted examples such as "Life is that which replicates and evolves" [598], or "Life is a chemical system capable of Darwinian evolution" [403]. Darwinian evolution is unquestionably one of the most *distinctive* mechanisms in the biosphere, and thus is natural to use as a differentiating characteristic for such processes.

However, as we have seen, there is no evidence that individuality or replication were relevant to early chemical organization. Darwinism *per se* predicts no specific relations between individual dynamics and the rich regularities known either in early chemistry or in macroecology and long-term evolution. We are not without scientific understanding of regularities in these areas; they simply have conceptual foundations of their own. In order for Darwinian evolution to *define* life, it would need to entail the other known attributes

---

[34] We believe that our emphasis on the biosphere as a particular system shares a similar perspective with John Searle's objection [719] to attempts to define consciousness in terms of computation. Searle wishes to keep the actual, physical brain before our attention. The assignment of logical symbols to certain brain events may support a mapping of some aspects of brain dynamics to formal models of computation, but these are only particular windows on brain function and are not the brain. Attempts to define consciousness through computation suppose that this abstraction is not only a correct but also a sufficient window, for a concept (consciousness) that is at best only poorly scoped in current conversation.

of the biosphere, at least to the level of explaining why some regularities and not others are present. For a variety of reasons we believe that it not only does not, but *cannot* do this.

### 8.4.1.1 Replication as a means to an end

Selection is more naturally understood as an emergent mechanism to support order for which criteria are defined at the whole-biosphere level. Within the larger context of maintaining chemical order and energy flow, multiple forms and levels of individuality emerge as buffering units for error. Individuality brings into existence replication, competition, and hence selection, and these *contribute to* the mechanics of maintenance. What Darwinian evolution describes is the dynamics of the particular information in *populations of individuals*. The shortcoming of Darwinism as a defining criterion for life reflects the difference between defining the nature of life and identifying attributes of individuality.

### 8.4.1.2 Inverting the order: conflating individuality with life is a category error

Attempting to define life as a consequence – whether dynamically generated or logically entailed – of the reproductive and selective dynamics of individuals puts the relation "creates/entails" in place of the relation "is created/entailed by." We believe the latter is better supported by the evidence in the first six chapters. Life is not created by evolution, through the agency of selection of individuals. Rather, the living state comes into existence by following opportunities for modularization that bring about individual forms of organization. Once these exist, their informational dynamics is the dynamics of replication and selection. The context of life that has led to individuality goes on to define the selective filters through which population dynamics integrates with substrate chemistry and constraints such as ecological stoichiometry [779] to produce long-term stability.

The most basic reason to think that life creates individuality (rather than the other way around) is that no one level or kind of individual is the fundamental unit of life. The more care one takes identifying units of selection, the more complex their diversity and interactions become. The diversity of kinds of individuals is as open ended as the diversity of kinds of covariance that can define relatedness in the Price equation. We must create fitness models far richer and more complex than linear regressions to capture the interdependence among components within a cell or cell lines within a multicellular organism, but in that richer mechanistic context the same selection on relatedness leads from Hamilton's rule for kin selection to Leo Buss' hierarchies of individual forms.

If we associate the nature of the living state with the entire universe of events that contribute to the maintenance of the biosphere, criteria that purport to define life but refer to properties of individuality commit a fundamental **category error**. A pattern of covariance is a different ontological entity than the universe of events whose covariance is measured.

We may understand, then, why classic questions such as "are viruses alive?" rapidly unravel into *reductiones ad absurdum*. Expressions such as "living things," in which "living" is a predicate assumed to segregate entities between the living and the non-living, make life subordinate to entities,[35] where in fact life is super-ordinate to entities and to many other kinds of order as well. The virion grades to the vulture through a sequence of small changes. The paradox of attempting to call one alive, and the other not, does not arise because the boundary between "living things" and "non-living things" has been drawn in the wrong place, but because the syntax "living things" supposes a predicate defined by segregation of things. The problem is that such predicates refer to kinds of individuality rather than the underlying nature of life. The paradox disappears if we ask which forms of individuality contribute to the virion as a spore in a virus lifecycle, the virus as a unit of molecular replication semi-autonomous from cellular metabolism, and the vulture as a complex aggregate form in which the variation and selection of genome, cellular milieu, germline, and soma are mutually constrained through the shared fate of organism selection.

### 8.4.2 Life is defined by participation in the biosphere

Disentangling the concept of life from the concept of individuality frees us from a host of confusions, but it leaves the problem of stating operationally what is the nature of the living state. We believe that ultimately the most productive answer is that being alive is defined by participation in the error-correcting processes that have returned the biosphere reliably to its ordered chemical state for the past four billion years. The fundamental unit of life is not the individual, but the biosphere as a whole.[36] Mechanisms at all levels contribute to its stability. They include constraints on chemical possibility that make transitions into stable ordered phases available. They also include developmental, physiological, and ecological regulatory dynamics, as well as the coevolution of the global inventory of species. Some of these are invariant physical properties of chemistry and physics; others are vertically transmitted and preserved within classic Darwinian units of selection; still others are carried as relations such as trophic flux balance among Darwinian units, which are properties of aggregates such as ecosystems.

Questions such as whether aqueous solvent or a carbon substrate are essential are given meaning by asking whether biological order can be defined from some more basic criteria, to which these contribute in an essential way. The context for essentiality, however, must come from the system level. The incursion of error or deviation in a living system, and the mechanism by which that error is removed – whether that mechanism is physical, chemical, regulatory, or Darwinian – defines an accounting under which the events that make up the biosphere must form a closed system.

### 8.4.3 Covalent chemistry flux and other order parameters

The definition of life may be narrowed somewhat further. Not only is the biosphere stable, it maintains a specific chart of (more or less universal) intermediary metabolism, as well

---

[35] In the different usage that contrasts "living things" with "dead things," this predicate is well defined, because it refers to a stage in the generational cycle of entities that already exist.

[36] To the extent that ecosystems are self-sufficient with only geochemical inputs, they may be taken as equally good fundamental units.

as cells and molecular replication, in varying degrees of coordination. The particularity of these universal order parameters is the other component in a definition of life.

Qualitatively phrased, the criterion that we believe best defines the biosphere as a planetary system is that *it transduces a steady flux of energy through a sector of covalent bond CHNOPS/metal chemistry, which would be inaccessible on a lifeless planet.* On Earth, the biosphere conducts about 100 terawatts of solar and probably less than a few terawatts of geochemical flux,[37] through the order of hundreds of gigatons of compounds formed from about 125 organic substrates.

On first exposure, the claim that the biosphere is *defined* primarily by its function as a redox channel seems an extremely low-level and crude characterization, to entail the diversity of rich structures of life. However, when the problem of actually *making and maintaining* an ordered high-flux energy-flow path through covalent bond chemistry is disaggregated, chemistry provides a much richer and more unique set of supports for such a function than might at first be imagined.

Recall from Section 7.3.3 that a candidate for an order parameter should be a summary statistic, given which other fluctuations are conditionally independent. The **flux through core metabolism** is the most compelling candidate for such a quantity in the biosphere. First, it is a function of the aggregate fluxes of all biota, and is in that sense a summary statistic of the real, physical system. Second, all other downstream structures show some variation in lineages that share the core. Cell membranes and wall architecture have changed or been independently derived among archaea and Gram-positive and Gram-negative bacteria; almost all gene families have experienced non-homologous substitution; species have emerged and gone extinct in profusion, and of course individuals are so fleeting that we defer to poets to find a fit description. Yet all these changes have preserved the core, and we argue below in Section 8.5 that they do so because the filter of natural selection reflects the joint constraints of metabolic completeness and primary productivity at the ecosystem level.

By this criterion, other lesser order parameters may provide summary statistics for subsets of variation. If our interpretation is correct that the microbial tree traces metabolic variations because these were optimal solutions in segregated environments, then the distinction between $CO_2$ and bicarbonate fixation, which is one difference between rTCA and 3-hydroxypropionate organisms, would be a less universal but still strongly determining order parameter for subsequent diversifications in the two sets of clades. A similar partition between oxidant insensitivity and energetic efficiency may define the rTCA loop pathways and the acetyl-CoA (pitchfork) pathways as context-dependent order parameters

---

[37] The heat flux through the mantle is much easier to estimate than the flux of available chemical free energy. In 1978 Jenkins *et al.* [398] estimated $20 \times 10^{19}$ J/yr delivered in global hydrothermal convective heat from helium isotope signatures. In 2001 Butterfield *et al.* [110] estimated $(6 - 12) \times 10^{19}$ J/yr heat flux from on-axis venting at spreading centers, and $(20 - 30) \times 10^{19}$ J/yr from diffuse flows on ridge flanks. These numbers are to be compared to a present-day surface heat flux out of the whole mantle estimated at $38\,\text{TW} \sim 100 \times 10^{19}$ J/yr [453]. Hence a large fraction, 1/3 to 1/2, of mantle heat flux is delivered to seawater at ocean spreading centers. From the moderate to small part of the heat budget that the free energy of hydration contributes to peridotite-hosted hydrothermal systems [18] (which are the more exergonic alteration systems), we estimate that available chemical energy is on the order of a few percent to tens of percent of the heat budget through ridges and ridge flanks.

for subsequent divisions in those clades. Even the degree of conservation of apparently highly conserved deep-branching clades – suggestive that for billions of years these were preserved by a high degree of stabilizing selection apart from a few successful events of species division – may make these clade identities order parameters for both the short-term variations of individuals, and the common properties of the clades.

### 8.4.4 The importance of being chemical

The abstraction of the replicator, which reduces the Darwinian dynamic to its essentials, also de-emphasizes the chemical nature of life. Chemistry can come to seem secondary or even incidental. It is a stage on which the performance of a Darwinian play is the special feature. Having now emphasized the heavily contextual nature of all forms of individuality, we can return to ask what makes chemistry special as the context for the forms in the biosphere.

The assertion that *chemistry matters* to the nature of life has now reversed fortunes twice. A traditional view from molecular biology was captured in the slogan "Chemistry matters," which meant that the particular mechanisms of molecular biology could not readily be guessed or abstracted away from, and needed to be determined empirically. In an argument for the importance of collective effects, Nigel Goldenfeld and Carl Woese [308] challenged the molecular biology view, replying "Or does it?" We agree with their emphasis on collective effects and the pervading systems view that it represents. It is the same view that leads us to emphasize phase transitions as the class of phenomena enabling modularity and buffering. However, we counter that chemistry *does* matter, because it defines a new domain of collective effects. Beyond the details of any particular mechanism, chemistry matters to the nature of the living state conceptually.

Three observations are central.

#### 8.4.4.1 The combinatorial state space of chemistry

The state space of molecules and reactions has unparalleled diversity and structure compared to any other state space in which statistical mechanics has been understood. The current theories of ordered states exist for highly symmetric systems such as gases, liquids, and crystals. A smaller collection of solutions exists for systems with dilute disorder such as impurities, or for local disorder that is pervasive but statistically homogeneous (glasses). The complexity of the chemical state space originates in the ascending complexity of representations of the rotation group to define atomic orbitals in ascending ranks of the periodic table. But the complexity of molecules far exceeds the complexity of orbitals because atomic orbitals interact with spatial locations in molecular orbitals, and with long-range constraints in both molecular forms and concerted reactions.

Reaction-diffusion-front phenomena such as the Belousov–Zhabotinsky or Brusselator (model) systems, as we noted in Section 7.4.4, use little chemical complexity and are driven by concentrations that vary smoothly in space. The chemical order of life is determined, in contrast, by control of kinetics within the reaction network. Therefore it can produce molecules with partial oxidation and partial reduction at adjacent carbon atoms, or can locate oxidant and reductant species within distances of a few atomic diameters of each other in space.

### 8.4.4.2 Separation of spatial and temporal scales

The *theory of rate processes* [304] that derives chemical reaction kinetics from the free energy landscapes of transition states is itself a large-deviations theory. The exponential dependence of both concentrations and reaction rates on temperature and chemical potentials results in a large separation of timescales inherent within the chemical substrate. The resulting very skew distribution of concentrations among molecules in the Gibbs equilibrium determines the "inaccessibility" of much of covalent bond chemistry. This is the CHNOPS/metal chemistry which, having been *made* accessible by life, distinguishes the biosphere from non-living Gibbs states. The even wider distribution of reaction rate constants separates lifetimes of metastable molecules by easily ten or more orders of magnitude. The possibility of changing transition state energies across this scale with catalysts permits the same molecules to serve as either long-lived memory stores or short-lived reactants depending on context. This memory need not be created by collective effects within the reaction network: it is available as a property of the state space directly.

In addition to the range of temporal scales and spatial relations within monomers, chemistry makes available at least two to three orders of magnitude difference in spatial scales. Polymers and physical aggregates such as membranes permit the complex geometries used in catalysis and in isolating microenvironments.

Reaction feedbacks and cooperative effects in networks may then aggregate to produce a *second* large-deviations limit in the dynamics of molecular populations. The combination of state-space complexity and the range of time-sacles and spatial scales opens possibilities for complex order in chemistry that have no parallel in the simpler ordered forms of physical chemistry that depend only on spatial segregation.

### 8.4.4.3 Digital error correction from small-molecule quantum mechanics

The discreteness of chemical states, which for small molecules also limits their diversity, may provide a form of error identification and buffering that is fundamental to the global stabilization of biotic order. This is perhaps the most important conceptual contribution of chemistry to dynamics that we have not seen emphasized.

In the twenty-first century it is difficult to recall that before the Second World War, analogue computers were believed to represent the future of computing. Their lack of flexibility was an initial problem, but after the hybridizing of analogue substrates with digital controllers, special-purpose analogue computers continued to be refined into the 1950s and 1960s. The history of analogue communication is more easily recalled: television broadcast and to a lesser extent telephony remained analogue in most parts of the United States into the last decades of the twentieth century.

The greater flexibility of digital architecture was part of the reason digital computers largely replaced analogue computers after the invention of the transistor, but the main advantage of digital communication came from *universal and economical multi-purpose error correction*. Error correction requires building redundancy into a communication in such a way that, without needing to know the original message, a receiver can recognize and correct error states.

Digital storage and transmission of information permits universal error correction at two stages. First, the integrity of the bits may be monitored, and bit lengths, values, or edges may be sharpened if they are eroded by noise. Second, the patterns of bits may be assigned to messages using an error-correcting code, as we discussed in Section 7.5. When combined with message compression algorithms, both of these stages of error correction may be optimized by general-purpose algorithms, the implementation of which may depend on the nature of the corrupting noise, but not on the statistics of the messages stored or transmitted. All storage or transmission systems can thus share innovations in cost, efficacy, or speed.

Digital substrates in biology are usually recognized when they form combinatorial alphabets for oligomers.[38] The distinction between molecules "in" and "out of" the alphabet provides a first layer of error identification. The standard nucleobases and standard biological amino acids are universal inputs for nucleic acids and polypeptides. Secondary digital systems may then be added: codon assignments with redundancy contribute to the robustness of translation, bases within a gene may become tightly coevolved through constraints of protein folding of the encoded amino acid sequence, or genes may be grouped into regulatory domains such as operons. *All* of these digital alphabets share a feature with the electrical or optical pulses in digital telephony: *they are manufactured at a cost of free energy*. When they are mis-formed, or when they degrade, they must be broken down or excreted and replaced, again at a cost of free energy. The maintenance of the digital alphabet is itself a result of selection and so requires a non-zero *rate* of free energy expenditure. Free energy is a scarce resource in living systems, and must ultimately limit the redundancy that can be provided for error correction.

One digital system in the biosphere does not require energy to produce. This is the limited inventory of small-molecule states – whether used or unused – that exist in the molecular state space. Quantum mechanics ensures that any two molecules of the same kind are *exactly* alike, permitting reliable evolution of catalysts that recognize them or reactions that use them. Among small molecules, error states, whether isomers or altered compositions, are few in number and discretely, qualitatively distinct from the reference molecules. In this feature, small molecules differ almost qualitatively from macromolecules. In most (presumably in all) proteins, mutations at a large number of residues away from the active site or key structural sites may lead to little or no change in function.

The relatively small cardinality and stark difference between different small-molecule states makes distinguishing among these molecules an easier problem to solve robustly than the problem of identifying mutants in a more nearly continuous system such as an

---

[38] The alphabetic character of amino acids and nucleotides has been most emphasized because both polymers are linear chains; the abstraction of an alphabet is less suited to glycobiology because glycans make branched polymers in which topology as well as sequence matter [798]. The inventory of simple sugars can also grow to be very large.

oligomer. In the small-molecule error-identification problem, the discrete state space is the collection of possible molecular states – right or wrong – that must be sensed and acted upon. The selection of which molecules are to be used biotically within this space is a second layer of error correction, analogous to the placement of codewords in sequence space in digital telephony. The second error-correction process requires free energy to maintain, but the specification of the underlying state space does not.

Unlike any error-correction scheme that requires free energy, *error-detection that uses the discreteness of molecular states may be carried out in arbitrarily large volume without constraint from available free energy supply rates.* A chemical system that can identify and detect, and so buffer, many of its biosynthetic errors in the small-molecule network generates a less complex control problem for higher-level processes, from biosynthesis to organism selection.

Suggestive evidence that life makes essential use of the small-molecule network for error buffering comes from our earliest observation: that all of biomass flows from precursors within the citric-acid cycle, or in a few cases from related short pathways such as direct one-carbon reduction. In principle, with the highly evolved complexity of modern genes and enzymes, it should be possible to assemble large molecules from single atoms or functional groups, based only on the selection of reactions by other large molecules. Yet, apart from a small collection of $C_1$ donations (ranging from formyl to methyl groups), aminations, and phosphorylations, biology performs no direct synthesis in this way. Even very refined molecules such as siderophores of only modest complexity are synthesized from precursors such as citric acid moieties.[39]

A question that we think is important, but which we do not know how to answer, is *whether this role of small metabolites as a bottleneck, and putatively an error filter, is essential to the maintenance of a non-equilibrium energy channel such as the biosphere.* Assigning realistic estimates to the number of error states for macrosystems composed of atomic constituents, could one prove that an analogue system, or a system that built macromolecules without passing through a small-substrate stage, could not possess the requisite variety in its controllers to monitor and correct the errors in their own internal configurations?

### 8.4.4.4 Chemistry entails the biosphere best

From the stability perspective, the humble substrate of chemistry appears to provide a more inclusive explanation for diverse properties of the biosphere than higher-level attributes such as evolution. If we can understand why the partitioning of dynamical ordered systems into individuals and generations is a robust solution to the buffering of errors, it follows that Darwinian selection on the individuals captures the remaining variations, providing both stability and adaptation. The universality of intermediary metabolism is putatively a consequence of the coincidence of many features of reaction chemistry and network topology that concentrate fluxes, but also in part an exploitation of the fact that both the molecules and the networks are small. Perhaps this is also why the core has not expanded

---

[39] We thank Alvin Crumbliss for making us aware of this regularity, which is of interest for the reflection it gives of citrate transition-metal chemistry as well as for biosynthesis. See [300] for examples and references.

greatly as a connected network; it has retained the "bowtie" architecture of a densely interconnected core that has not grown, and has elaborated the network with a much less densely crosslinked set of rays emanating from the core.

## 8.5 The error-correcting hierarchy of the biosphere

A biosphere that is organized around the order parameters from a cascade of phase transitions must use an error-correcting hierarchy that combines the buffering within each level with passing of constraint and control "messages" across levels. The cascade of levels in some ways resembles the hierarchy of matter from Section 7.4.4, but because the order at all scales is dynamic, part of the "data" that are preserved in the error-correcting system consists of the error-correction encoding schemes at each level. Two corollaries of the trade-off between complexity, stability, and scale developed in Section 7.5.2 must govern asymptotic reliability in such a system.

1. The progression downward in the hierarchy leads to levels with fewer and less complex possible states of order, which provide the ultimate reference for stability of the whole hierarchy. Progression upward leads to more diverse and more complex possible states which are also less stable individually, though as a set they inherit the stability of the whole. The lower levels tend to be associated with regularities at the ecosystem/whole-biosphere scale, while higher levels tend to be associated with properties of ever more complex forms of individuality.
2. The targets for order at lower levels in the hierarchy define the selection criteria and provide the machinery to generate the more complex solutions at higher levels. Error correction in the biosphere operates concurrently to restore order at all levels. Where more complex but less stable levels appeal to lower levels for error criteria, they do so through the simplified module boundaries and control interfaces discussed in Section 8.2.

These relations suggest that, although the same laws of large numbers are at work in each level of structure, they will lead to different phenomenology across the spectrum from more distributed to more individually localized kinds of order. The most important difference between the hierarchy of matter and the biological cascade will be that in equilibrium matter, the most durable patterns are carried by the most durable entities, while in the biosphere, the most durable patterns are carried by the most ephemeral entities. We close the section and the chapter with a consideration of this important empirical regularity.

### 8.5.1 The main relation: a general trade-off among stability, complexity, and correlation length

We gave a specific form in Section 7.5 to the main trade-offs that the laws of large numbers suggest, in the language of asymptotically optimal error correction. The main relation, from Eq. (7.51), is reproduced here:

$$P_{\text{error}} \sim e^{-D(\mathcal{C}-\mathcal{R})}.$$

For any ordered system whose dynamics tend to preserve the ordered state through time, $P_{error}$ is the probability that a confluence of random events will lose the original ordered form and switch the system into a different state than the historically inherited one.

$D$ is some measure of the system's "correlation length." It stands roughly for the number of independent small-scale components the system can poll to identify and correct errors in the states that a subset of those components may take. $D$ may also be viewed as a measure of the system's "size" relevant to the robustness of the patterns it may carry.

$\mathcal{R}$ determines the "complexity" of ordered forms, measuring (on a logarithmic scale) the growth in the number of possible ordered states with system size $D$: that is, the number of states grows as $n \sim e^{D\mathcal{R}}$. $\mathcal{R}$ is thus also a measure of the information about the state of each of its $D$ components that we could gain if we learn that the system is in a particular one out of its $n$ possible ordered states.

$\mathcal{C}$ is an upper bound to the number of possible ordered states that can be maintained with asymptotically perfect reliability in the limit $D \to \infty$. While more states of order than $n_{max} \sim e^{D\mathcal{C}}$ could be introduced as initial conditions, they would not be preserved for long times by the stochastic dynamics.

These three properties form a basis to classify a number of known ordered thermodynamic phases, as well as their counterparts which we propose for the biosphere.

### 8.5.2  Simple and complex phase transitions

The characteristics of correlation length, complexity, and robustness apply to equilibrium systems as well as to error-correcting dynamical processes. By mentioning a few physical examples, we may illustrate the difference in character among three classes of simple and complex ordered phases.

Here we will speak entirely in terms of fully ordered phases far from the transition point, but in systems with finite size. The correlation length $D$ therefore will be simply a measure of system size, which determines the reliability of its regression toward the average properties of its ordered state.[40]

Making contact with Section 7.5 and more specifically with Box 8.2, ordinary equilibrium thermal systems may be described in the language of decoding as follows. The reductionist description of the microscopic interactions in any physical thermal system determines what are its possible ordered states, and how it relaxes toward them. These small-scale properties of mechanism define both the "code" – in which each ordered state is a "codeword" – and the mechanism by which thermal relaxation "decodes" noisy states of the system by regression toward the average behaviors indicated by the

---

[40] In the modern study of critical phenomena, properties in the immediate neighborhood of the transition point are more often of interest both because these provide the interesting and difficult property of scale-freedom in structure and dynamics, and because they provide the most discriminating characteristics for the theory of the ordered phase. Near the critical point, the notion of a *correlation length* applies when *not all* of the system's degrees of freedom – but rather only a local subset – contribute to mean-regression for any point. Critical theory is usually concerned with the scaling of finite correlation lengths in systems of infinite size. Our discussion here corresponds to the case where we are sufficiently far from the critical point that the whole-system size rather than a dynamically determined correlation length is the main determinant of mean-regression.

order parameter. Just as a given code permits the transmission of any of its allowed codewords and provides support for reliable decoding, the reductionist description permits any of its allowed macroscopic ordered states to emerge as stable solutions supported by mean-regression.[41]

The most important property distinguishing ordered states according to their complexity and stability is the rate function $\mathcal{R}$ that distinguishes simple from complex forms of order. Below are three examples.

### 8.5.2.1 Zero rate, one phase

The simplest ordered materials are those with only one stable ordered state toward which even systems of finite size regress reliably. In equilibrium thermodynamics, these are states such as gases which have unique solutions for expected pressure, internal energy, etc., in response to any given boundary values of volume, temperature, and material content. In coding language, these are the systems with only one codeword, to which the decoder maps all noise states. They have message rate $\mathcal{R} \equiv 0$ and no complexity. Strictly speaking, in communication systems they are trivial and degenerate cases, since the transmitter has no freedom to send any message that the receiver does not already know. Only for these systems, the correlation length $D$ plays no part in the stability/complexity trade-off.

Many of the geological and energetic properties of the Earth have effectively this interpretation as systems with a unique and determined ordered form, with respect to the support of life. The first-order phase transition to lipid vesicles, and perhaps the geometry and topology of lipid-bounded cells, would be other examples of such unique solutions.

If we were seeking a simplified description of the universal metabolic chart which emphasized only its favored member molecules, the chart would be such an ordered phase which serves as an asymptotically reliable reference for living order. In a more mathematically proper treatment using non-equilibrium ensembles, it will be better to regard universal metabolism not as a member of this class, but of the following one.

### 8.5.2.2 Zero rate, many but fixed phases

The next simplest class contains systems which may have more than one possible, degenerate phase, but in which the number or volume of the phases does not depend on the system size. If such phases are discrete, their number $n > 1$ is finite. The Curie binary phase transition of Section 7.4.1 is an example of such a simple transition to ordered states.

If the set of ordered phases is continuous but still simple, as occurs in the theory of symmetry breaking that gives rise to the elementary particles reviewed in Section 7.4.4, we say that the set of ordered states is *finitely generated* but still has measure zero relative to the possible states.

In all of these cases, because the set of ordered phases does not become richer with increasing system size $D$, the number of states $n$ is absorbed in the symbol $\sim$ in Eq. (7.51)

---

[41] In Box 8.2, each of these ordered states is a peak in the "prior" distribution that forms the partition function.

as an overall prefactor. The only difference that results from greater system size is greater stability. The probability to move from one ordered state to another in the case where the states are discrete (the only case appropriately illustrated by the simple code-block example we have been using) decreases as

$$P_{\text{error}} \sim e^{-DC}.$$

In a proper non-equilibrium thermodynamic ensemble treatment, the universal core metabolic network should be an ordered state in this category. The exact amount of flux through the core at any time is a noisy variable not subject to mean-regression, and therefore not predictable as a consequence of symmetry.[42] However, since life does not depend on the exact extent as long as the mean reaction flux is sufficiently stable, for many purposes this simple phase transition provides the same reliable foundation for error correction as the geochemical and geoenergetic boundary conditions in the previous category.

### 8.5.2.3 Finite rate and complex order

The equilibrium systems described by fluctuation formulae with $\mathcal{R} > 0$ have a number of possible ordered states that grows exponentially with system size. For many decades, such systems were not tractable by the standard methods of phase transition theory, but they are coming to be understood as the general category of *glasses*.[43]

Glasses are remarkable ordered phases in which each ordered state may become asymptotically more stable and rigid as $D \to \infty$, because the number of microscopic degrees of freedom to which it may appeal for error correction continues to grow faster than the rate at which the space of available thermal disorder increases. However, the redundancy within each state does not increase as quickly with $D$ as it would for a simple ordered phase such as the Curie ferromagnet or a simple crystal. In this sense, glassy ordered states are "softer" than their crystalline counterparts. In exchange for that softness, the suite of macroscopically distinct states into which a glass may cool continues to grow more complex as the size of the glass increases.

Internal softness combined with increasing degeneracy of states leads glasses to "creep" through their states when viewed at the macroscale [255], in contrast to the discrete "jumps" between states of a simple crystalline ordered phase (as seen in the Curie–Weiss model). Viewed mesoscopically, such creeps take the form of small discrete flips or avalanches, with coherence on spatiotemporal scales larger than the microscale but smaller than the whole system.

It is difficult to use simple models with high symmetry, such as the Shannon sphere-packing model, as descriptions for evolutionary dynamics. Among the physical phase transitions, however, the glasses are the only well-understood class with an "open-ended" number of possible ordered states, and macroscopic dynamics in the form of creep. In

---

[42] This is a case of the time symmetry that is a new effect arising with non-equilibrium phase transitions, mentioned in Section 8.2.6.2 and developed quantitatively in [768] (Chapter 6).

[43] For a detailed discussion of glasses, and the degeneracy and topology of their ordered states, we refer the reader to [255, 556].

this respect at least they resemble the Darwinian dynamics of speciation and ecological rearrangement. It is therefore natural to use the "codewords" in a finite-rate code as limited analogies to the distinguishing phenotypes of different species.

Evolution is complex insofar as the number of states $n \sim e^{D\mathcal{R}}$ at a given correlation length $D$ is large. It is "open ended" to the extent that $D$ may be increased as the system's own dynamics create correlation in aggregates of increasing size. The increase of biotic complexity via aggregation into increasingly hierarchical forms of individuality, as described by Leo Buss [106], is thus an important source of open-endedness for global evolutionary dynamics.

### 8.5.3 Living matter as active data

The Shannon theory of optimal error correction was invented for applications to telephony. It was therefore natural for Shannon – and has been natural for most subsequent expositors – to regard data as passive "signals" created, transmitted, and decoded by a physical system that those signals do not create or maintain.

In living systems, the "signals" are material components that act on one another. For some purposes, this action may be abstracted to a model of signal transmission and error correction, but relevant properties of the system may be carried by other aspects of their material instantiation besides their abstraction as signals.

We will refer to such material bits with symbolic roles as **active data**. The most important role of active data in the biosphere is to create self-decoding "codewords" through redundancy. The mechanics of self-decoding takes advantage of structure that, in equilibrium systems, is given in the microscopic (reductionist) description. In living systems, the creation of the interactions that may be used to implement a self-decoding system is one of the main modes by which low-level, simple order can create the preconditions for the emergence of higher-level, complex order. A few examples below illustrate the enrichment of models of passive signals to models with more materially instantiated active data decoders.

#### 8.5.3.1 In equilibrium: first-order phase transition and thermal relaxation

The Schneider model of molecular recognition reviewed in Section 7.5.5, or more generally systems subject to first-order phase transition, furnish examples of thermal relaxation as a dynamic carried out by active data. A stressed molecule, or a disordered group of molecules with the potential to relax into an ordered state, provide the spatial and geometric boundary conditions to each other that define the low-energy configurations. The laws of chemistry (in the case of condensed matter), perhaps combined with evolutionary selection (for regulatory proteins) dictate the "code" carried by the material components and their interactions. As more molecular components relax into a mutually low-energy configuration, the resulting framework provides a more reliable and less flexible template for the propagation of order. Whether this process results in global long-range order, or local metastable states, depends on the kinetics of the relaxation trajectories.

### 8.5.3.2  *Out of equilibrium: ecosystem-level co-selection*

The concept of active data applies even more intuitively to the process of joint stabilizing selection of the individuals in an ecosystem. Consider, for example, an ecosystem that is in a quasi-steady state of species diversity and species numbers, in which trophic fluxes balance and most species mutations are locally maladaptive. This description is appropriate for ecosystems that are macroscopically stable over time intervals much longer than individual generations. Such ecosystems will undergo fluctuations in the ordinary course of both individual interactions and mutation, but most of these will be removed by physiological regulation by member organisms, changes in interaction opportunities, or stabilizing selection as the ecosystem regresses toward its average composition and function.

In real ecosystems, organisms act on one another as predators, prey, symbionts, ecosystem engineers, or competitors. They physically incorporate – literally, they make up as parts of their bodies – the pools of organic and inorganic resources that must be in stoichiometric balance for the ecosystem-level metabolism to function in a near-steady state. It is the sum of these interactions among (usually stochastically matched) individuals that lead to the events of reproduction or death that change the population state.

In population-dynamic models, all these material events are aggregated and abstracted into a frequency-dependent fitness function [768], perhaps augmented by a model of niche construction [602] and persistence. States of ecological stoichiometric balance are candidates for the long-term steady states toward which individual-level stabilizing selection regresses the population. Interpreted as a coding system, the local responses to excesses or deficits in trophic resource pools, in response to which individuals' actions generate their own and others' frequency-dependent fitness functions, jointly create the direction of regression back toward the target-state for the ecosystem.

In this characterization, the different codewords are a metaphor for the different ecological community states (including species inventories and interspecies relations), which satisfy constraints of metabolic completeness and ecological stoichiometric flux balance. Other criteria may apply as well, such as maximization of primary productivity among local variants.[44]

### 8.5.3.3  *λ-calculus formalization*

Walter Fontana and Leo Buss have created a formal model of macroscopic order generated by stochastic active data [259, 260]. Their system is based on the λ-calculus [336], one of several equivalent systems to implement any algorithmically computable function under the *Church–Turing thesis*.[45]

---

[44] Whether steady states exist, and whether they are determined by simple optimization criteria, depends on *ad hoc* dynamics which must be determined empirically. Howard Odum [603] advocated maximization of power consumed as such a criterion. The constraints on any such maximization principle must be carefully sought before the principle becomes meaningful, and Odum's proposals have been criticized empirically, suggesting that such principles are not simple even if in some version they are true. The criterion of flux balance is much more general if steady states exist, because ecosystems are autotrophic. The only way flux balance can be violated is by finely tuning evolutionary change to compensate for instantaneous imbalances in the input/output characteristics of the ecosystem at each moment.

[45] See Cristopher Moore and Stephan Mertens [565]. Other equivalent models of calculation include the Turing machine, the recursively enumerable functions, combinatory logic, and Markov algorithms.

In Fontana and Buss's system, each element in a well-stirred reactor is a $\lambda$-expression. It is both a data item and a small bit of one-time usable executable computer code, which is partly or completely consumed when it acts on another $\lambda$-expression as its input. Fontana and Buss demonstrated, through elegant use of automated-parser technology, the emergence of stable macroscopic dynamically ordered phases in the production and consumption of $\lambda$-expressions. The population of expressions carries an instantiation of a grammar, which is a property of the relations among the members of self-amplifying reaction sequences.

### 8.5.3.4 Cross-level coupling and the maintenance of species

The canonical example of complex order in the biosphere is the preservation of a species' identity through a large number of generations of individuals.[46] It suggests an interpretation in terms of large-deviations scaling of a complex set of "codewords", because species tenures are much longer than individual generations but not indefinitely long, and in many ecological roles, species are to some degree substitutable.

Take for example a bacterium. The species identity, phenotypic traits, and ecological roles, might be abstractly characterized as the compressed "message" carried forward through time by a clonal population of bacterial cells. Any individual bacterium, however, is not an abstract message, but an actual collection of diverse molecules in quite particular spatial and dynamical, geometric and networked relations. The redundancy of intracellular molecules provides mechanisms of molecular surveillance to maintain and correct the cell state and ensure homeostasis. The cell population is a set of real individuals who may replicate or may die, to be consumed or to decompose. Stabilizing selection in the population expends some fraction of the cells to maintain the population-level collection of genotypes near a stable well-adapted distribution. The events of maintenance, repair, and replication, while perhaps not formally "decoding" the material cell to an abstract message and then re-encoding that message as in the repeater-diagram of Figure 7.11, nonetheless continually replace degraded assemblies of molecules with pristine assemblies. To the extent that bacterial division can concentrate degraded cell components in the "old pole" of the cell [600], the mechanism of reproduction performs some error buffering as part of the population's programmed aging sequence, prior to any case-by-case surveillance by selection.

The basic energy systems, and translation and biosynthetic machinery common to all cells, support the cell's ability to carry a species identity, independently of what phenotypic traits and ecological roles that identity reflects. Some of these universal properties (the core metabolite inventory) derive from elementary chemistry and others (ribosomal translation) descend from the progenote phase of life. Jointly they provide a set of common mechanisms, not particular to any species, which determines the potential capabilities of all cells to carry species identities.

---

[46] For metazoans, species tenures ranging from $10^5$–$10^6$ generations are common. The notion of species in bacteria is more difficult to make unambiguous, but it is likely that comparable numbers of generations furnish a very conservative estimate of the persistence time of any sensibly unchanged bacterial clade.

The separation between a physiologically determined ability to preserve a fixed amount of DNA memory, and an ecologically determined selection for what species identity will be written into that memory, is suggested by an interesting **allometric scaling** of genome length with cell size. Brian Shuter has shown [751] that bacterial genomes (which consist mostly of coding DNA), have lengths that scale as roughly the 1/4-power of cell volume. At least among the species reported in that study (which likely reflect experimental selection bias for similar metabolic strategies and under-representation of extremophiles), 1/4-power scaling with cell volume suggests $(-1)$-power scaling with log-phase metabolic rate or linear scaling with generation time, as these properties also scale allometrically [867, 869]. This result suggests that the number of bases that can be maintained under stabilizing selection is set by metabolic rate or some other property linked to cell volume, independent of the function that the selected bases serve.

On the assumption that not all DNA sequences of 1.5–13 megabases encode viable organisms, purely on grounds of internal coherence and apart from ecologically-determined fitness, the viable bacterial species are again loosely analogous to "codewords" in yet another robust code, not to be confused with the "codewords" that represent stable ecological communities of species. The species codewords must first survive as viable entities using mechanisms of cell physiology. Their interactions then determine the ecological contexts in which they can persist as well-adapted members.

The properties of viability not filtered by cell physiology, as well as ecologically determined aspects of fitness, are passed to natural selection to stabilize, via the bacterial genome. Because the genome is small and stable, compared to the inventory of molecular states of the cell that are dynamically regulated physiologically, the selective filter need only have variety comparable to the genome's variability, and responsiveness comparable to the division rate. Thus the genome naturally forms a module boundary between physiology and ecosystem assembly, which is a simple and stable interface through which selection may transmit ecosystem-derived control signals.

Organisms are sometimes depicted as occupying regions in "morphospace," or filling "niches" as chessmen occupy positions on a chessboard. While such a depiction, if used naively, may fail to reflect the dynamic, constructed nature of biotic niches, some degree of niche availability may be responsible for rapid speciation in the wake of extinctions. If the notion of the single organism's niche is enriched to the notion of jointly stable ecological community-states – which then project onto a subset of viable organism forms – this suggests a way to frame the concept of a niche as a solution within a high-dimensional **constraint satisfaction problem** [401].

It is common that high-dimensional constraint satisfaction problems are *frustrated*, meaning that not all constraints can be simultaneously accommodated, and even the best solutions require trade-offs. Frustrated constraint satisfaction problems tend to lead to a large number of distinct, and about equally good, solutions which violate locally minimal numbers of constraints. Supposing that both physiological viability of a single organism, and joint adaptedness of the members of an ecological community, have properties of high-dimensional, partly random constraint satisfaction problems, we expect that the set of solutions will be sparse among the possible forms, even if the total number of solutions is infinite. This is the basis of our proposal that only a small fraction of 1.5–13 megabase genomes code for viable organisms, and we similarly propose that only a sparse set of possible organism forms are optimally co-adapted within *any* ecosystem.

A wide range of simple constraint satisfaction problems are known to be isomorphic to glassy thermodynamic systems [401, 556], and as we have noted, the glasses share many properties with the ($\mathcal{R} > 0$) error-correcting codes. Therefore, for the same reason as the Shannon sphere-packing codes may be poor models for low-level simple phase transitions such as the emergence of metabolism,[47] they may be appropriate simplified models for the way universal chemistry and cell form generate sparse solutions in viability-space or niche-space.

In the same way as the glassy physical systems creep among particular states, viable species or co-adapted ecological communities would be expected to creep from one to another locally stable form. Constant, ongoing creep is a model for what we have termed "Darwinian churn" in species inventories or ecological assemblies. This is the characteristic dynamic of ($\mathcal{R} > 0$) error-correcting codes which also have finite correlation length $D$. In evolutionary dynamics it has the added complexity that decoding and creep may occur at many levels, so that not only does the preserved codeword (species form or ecosystem structure) drift over time; the coding system that defines which forms are viable or well adapted may also drift. Creep, when viewed on the macroscale, should arise from discrete jumps, or avalanches when viewed at the mesoscale, consistent with the common observation of **punctuated-equilibrium dynamics** emphasized by Stephen Jay Gould [312].

While particular solutions drift, however, *the properties of the coding system as a whole, such as its correlation length $D$ or its capacity $\mathcal{C}$, may remain fixed.* We showed in Section 7.5 that an enormous degeneracy of possible codes, as well as a degeneracy of codewords within each code, was possible in a system for which $D$ and $\mathcal{C}$ were fixed properties determined by the channel noise. The stability perspective would predict that, while any single species or ecological assembly may have finite tenure, the statistical properties of viable organisms, or of stable trophic ecosystems, should remain invariant as a reflection of the lower-level metabolic constraints that stabilize them. This invariance may even hold if the code itself drifts, as long as the levels of coding and their appeal to underlying universal constraints do not change.

We interpret the post-Cambrian stability of trophic ecosystem architectures demonstrated by Dunne *et al.* [205] as an example of this stability of the coding system that supersedes the particular codeword or even the code. At the same time, we recognize that the "rules of the game" can change if fundamental changes are made in the correlation length that defines the code, akin to the rise of new levels of individuality in Buss' characterization. The sensible and relatively rapid change in ecosystem architectures found between the early-Cambrian Chengjiang shales, and the Burgess shales only 20 million

---

[47] They lack simple symmetries that generate continuous, finitely generated manifolds of solutions.

years later, coincided with rapid change in low-level developmental regulatory networks and also oxygen and sulfur profiles in oceans and benthic subsurface [229]. In this brief interval, bioenergetics, development, and ecosystem architecture appear to have undergone a joint change to more complex and intercorrelated forms.

### 8.5.4 Patterns and entities in the biosphere

Both living and non-living matter carry patterns, but the natures of the patterns and the ways they are instantiated are somewhat different. Examples of abiotic patterns include the crystal unit cells of minerals. Examples of biotic patterns include the distinctions between metabolite and non-metabolite small organic molecules, and the roles and fluxes of metabolites in biochemical reaction networks. The patterns themselves can be mathematically abstracted and described whether or not they are instantiated in matter. The goal of a theory of minerals or of biochemistry would be to go beyond description, to predict which patterns, under which circumstances, could be stably instantiated in matter.

The realization of patterns in material objects and processes is consequential in nature because those patterns imprint themselves as boundary conditions on systems. Thus, the pressure dependence of reduced ($FeO$) or oxidized ($Fe_2O_3$) iron oxides, and their solubilities in minerals such as pyroxenes or garnets, determine the redox state of iron and the oxygen fugacity in the mantle. The presence and concentration of universal metabolites determines which anabolic pathways can exist. The role of a pattern in nature depends not only on its existence as a mathematical object, but on the form and stability of the matter in which it is expressed.

Here the abiotic geospheres and the biosphere differ. Durable patterns in the lithosphere, such as the unit cells of zircons or diamonds that furnish evidence about early mantle chemistry, persist for a long time because physical zircons or diamonds are hard, *durable entities*. In contrast, the apparently oldest biotic fossil, the inventory of core metabolites in the universal chart, is instantiated in molecules that turn over many times per second.

The contrast between the durable pattern and the transitory existence of small metabolites extends to a more general *negative covariance between patterns and entities in the biosphere*. Cell form and DNA replication systems appear to have been invented or to have undergone wholesale change at least twice since the last universal ancestor. Ribosomal translation appears to antedate all extant lineages, but ribosomal variants began to be fixed with the emergence of the first major clades from the progenote. These roles are not quite as universal or stable as the roles of TCA cycle intermediates, though they are nearly so. Correspondingly, the cells that instantiate them may divide – and hence, in stationary populations, some fraction must die – as frequently as several times per hour, compared to the more rapid turnover of metabolites.

Species identities, carried in the genomes of clones of cells, are less universal and less persistent than cell form and ribosomes, being changeable through mutation as well as gene transfer. The clonal populations that preserve species identities, in turn, are longer-lasting entities, persisting through many generations of individual cells.

Across these examples, the longer a pattern persists and the more universal it is, the faster the entities that instantiate it turn over. Together with the importance of the emergence of individuality in the maintenance of the biosphere, we regard this negative covariance between the durability of patterns and of entities as a primary regularity of living matter to be understood in terms of non-equilibrium statistical mechanics.

In equilibrium systems, the long persistence of the instantiations of patterns is caused by the durability of entities. This is a cause stemming from the free energy landscapes that create barriers to change. In the biosphere, increasingly fluid turnover of components seems to facilitate more durable instantiation of the patterns they carry. This suggests, for one thing, the inherently kinetic nature of biological pattern, and the secondary role played by material structures in relation to processes of self-renewal. As we saw in Section 8.3, long memory in living states both increases standing variation and inhibits the rapid and strong convergence to average behaviors. Components with long memory in living systems may be likened to the hard mechanical springs, gears, and bearings of mechanical wrist-watches or clocks: they execute the functions of timekeeping up to a point, but they are not stabilized by laws of large numbers and as they degrade, they perform increasingly poorly. Components with short memory in the biosphere may be likened to the cesium atoms or ammonia molecules in atomic clocks: averaging of many small, independent, identical components allows the laws of large numbers to produce a much more stable oscillator than can be produced with a few mechanical components, no matter how refined they are.

# Epilogue

The study of the origin of life is no longer a field in its infancy. The scientific foundation we have to build from today is enormously richer than it was fifty years ago, in facts and also in concepts. The difficulties of origins research are becoming those of a maturing field with a diverse technical knowledge base: a tendency toward fragmentation and the need of more effective ways for researchers to work together as a community.

In attempting to cover this technical material, which provides the needed factual basis for generalizations and constitutes the fascinating detail about life and its planetary context, it is easy to lose sight of the profound changes in point of view that have become possible even within the past 20–30 years. We are struck by how many key points in this argument we did not understand when we were introduced to the field, but which are now fundamental to our view of the origin and nature of life.

- The universality, the historical depth, and the striking economy and conservatism of core metabolism discovered within microbiology; within that, the overarching organizing roles of the TCA cycle and one-carbon reduction.
- The greater continuity that can now be seen between chemical reactions carried out on mineral substrates, those catalyzed by small molecules, and those catalyzed by enzymes.
- The complexity of the concept of individuality and its multifarious realizations, and the complementary importance of ecosystems, culminating in the biosphere as a whole.
- The degree to which a picture can be formed of the interacting dynamics of planetary subsystems during planet formation and early evolution, and within these the role of the aggregate biosphere as a geosphere and its feedback on the other geospheres.
- A better formulation of the nature of thermodynamic limits, integrating in a seamless way the dynamical and inferential interpretations, equilibrium and disequilibrium, and stabilization and error correction.
- An appreciation of the coherence of the concepts of phase and phase transition, the way they underpin and enable reductionist science by dividing continuous scales with hierarchies of ceilings and floors, and the enormous scope of their application, beginning with a comprehensive theory of matter, and we argue, continuing that theory eventually to a theory of life.

- Where the empirical and mathematical tributaries merge, they give us a way to state what the universality of core metabolism signifies and the kind of concept it reflects: that its fluxes are the order parameter of a non-equilibrium phase which serves as a foundation and reference for biological function and innovation at all levels.

The empirical and the mathematical windows on life are equally fundamental in our point of view. Too much in the past, mathematical claims about system properties were viewed as disconnected from practical problems in chemistry. Nowhere has the difference in points of view been more sharply focused than in the polarizing effect that discussions of emergent order have had. That situation, too, may be primed to change, as the need for robustness and selectivity in systems chemistry is driving a search for principles beyond energetic stability analysis and a focus on single pathways.

The view that life must have emerged through a sequence of stages, that the intermediate stages must have been incrementally stable, and that the stability is essentially a process-stability of self-regeneration, is being expressed more widely and more explicitly among origins researchers from many backgrounds [623]. Our effort in Chapter 7 and Chapter 8 has been to show that the conceptual foundations (and a considerable body of technique) for such a theory exist and that these grow continuously out of the thermodynamics that accounts for our hierarchy of matter. The order induced by phase transitions is the epitome of emergent order, and although subtle (it did take nearly a century to understand fully!) it is not malicious.

We hope our introduction to these topics will begin to overturn the perception of life as a paradox of thermodynamics. The biosphere should be the system that will lead us to understand a new domain within thermodynamics – structurally richer than the equilibrium domain and so technically more complex – but conceptually a continuation and not a departure.

A point of view can be important in science. It transforms knowledge into comprehension and directs our exploration. But a point of view is not a substitute for a knowledge of facts, and in this the origin of life is still a very young science. Already, though, the need to grapple with the complex constraints that are known in many disciplines has improved the perspicacity of the questions researchers are asking, and has moved some of the disagreements from the early decades toward a more productive dialogue.

Although researchers remain divided in their priorities, and in their beliefs about which avenues hold out the most promise, it is now possible to lay out a slate of key problems that most will agree are fundamental and define the current frontier of efforts to understand biogenesis. We need to identify sources of order in geochemistry beyond those that are currently recognized. There is still debate over how much memory or selectivity geochemistry must have provided to an emerging biosphere, but the problem of sufficiency and partitioning of selection can be clearly framed. There must be a systems chemistry of RNA, as well as a later population dynamics of RNA (and probably not only of RNA). Microbiology has discovered much about the detailed interconnectedness of geochemistry and metabolism, but the gap from geochemistry to the first integrated cell is still large in many dimensions of structure and function. Every approach to the origin of life will draw, at some stage, on what can be learned about these questions.

At the end, we think our main contribution will have been to offer a reconceptualization of the nature of the living state that has a deeper and wider foundation in current science. Whether the particular sequence of stages in biogenesis that we have proposed in Chapter 6 is roughly correct or turns out to be seriously in error, the scientific contexts laid out in the other chapters should be cornerstones of the eventual correct understanding.

# References

[1] Venenivibrio. Licensed under CC [153] BY-SA 3.0 via Wikimedia Commons. http://commons.wikimedia.org/wiki/File:Venenivibrio.jpg#/media/File:Venenivibrio.jpg.

[2] Anabaena sperica2. Licensed under CC [153] BY-SA 3.0 via Wikimedia Commons. http://commons.wikimedia.org/wiki/File:Anabaena_sperica2.jpg#/media/File:Anabaena_sperica2.jpg.

[3] Published under GNU free documentation license. http://en.wikipedia.org/wiki/GNU_Free_Documentation_License.

[4] Leopard africa, by JanErkamp at the English language Wikipedia. Licensed under CC [153] BY-SA 3.0 via Wikimedia Commons. http://commons.wikimedia.org/wiki/File:Leopard_africa.jpg#/media/File:Leopard_africa.jpg.

[5] 2006-10-25 Amanita muscaria crop, by Amanita_muscaria_3_vliegenzwammen_op_rij.jpg: Onderwijsgekderivative work: Ak ccm. This file was derived from: Amanita muscaria 3 vliegenzwammen op rij.jpg. Licensed under CC [153] BY-SA 3.0 nl via Wikimedia Commons. http://commons.wikimedia.org/wiki/File:2006-10-25_Amanita_muscaria_crop.jpg#/media/File:2006-10-25_Amanita_muscaria_crop.jpg.

[6] Licensed under CC [153].

[7] 20090719 062218 ParameciumBursaria, by Bob Blaylock at en.wikipedia. Licensed under CC [153] BY-SA 3.0 via Wikimedia Commons. http://commons.wikimedia.org/wiki/File:20090719_062218_ParameciumBursaria.jpg#/media/File:200907.

[8] Global volcanism program, 2013. volcanoes of the world, v. 4.3.4. http://dx.doi.org/10.5479/si.GVP.VOTW4-2013, 2013.

[9] Dallas Abbot and William Menke. Length of the global plate boundary at 2.4 Ga. *Geology*, 18:58–61, 1990.

[10] Harold Abelson, Gerald Jay Sussman, and Julie Sussman. *Structure and Interpretation of Computer Programs*. MIT Press, Cambridge, MA, second edition, 1996.

[11] Christoph Adami. Sequence complexity in Darwinian evolution. *Complexity*, 8:49–56, 2002.

[12] Christoph Adami. Information theory in molecular biology. *Phys. Life Rev.*, 1:3–22, 2004.

[13] Hirotugu Akaike. A new look at the statistical model identification. *IEEE Trans. Autom. Control*, 19:716–723, 1974.

[14] A. M. Alayse-Danet, D. Debruyères, and F. Gaill. The possible nutritional or detox-ification role of the epibiotic bacteria of alvinellid polychaetes: review of current data. *Symbiosis*, 4:51–62, 1987.

[15] Bruce Alberts. *Molecular Biology of the Cell*. Garland Science, New York, fourth edition, 2002.

[16] Douglas Allchin. Paul Boyer: bioenergetics and error. *J. Hist. Biol.*, 35:149–172, 2002.

[17] Douglas E. Allen and W. E. Seyfried Jr. Compositional controls on vent fluids from ultramafic-hosted hydrothermal systems at mid-ocean ridges: an experimental study at 400°C, 500 bars. *Geochim. Cosmochim. Acta*, 67:1531–1542, 2004.

[18] Douglas E. Allen and W. E. Seyfried Jr. Serpentinization and heat generation: constraints from Lost City and Rainbow hydrothermal systems. *Geochim. Cos-mochim. Acta*, 68:1347–1354, 2004.

[19] Luís A. Nunes Amaral and Kent Baekgaard Lauritsen. Self-organized criticality in a rice-pile model. *Phys. Rev. E*, 54:R4512–R4515, 1996.

[20] J. P. Amend and E. L. Shock. Energetics of amino acid synthesis in hydrothermal ecosystems. *Science*, 281:1659–1662, 1998.

[21] Jan P. Amend and Everett L. Shock. Energetics of overall metablic reactions of thermophilic and hyperthermophilic archaea and bacteria. *FEMS Microbiol. Rev.*, 25:175–243, 2001.

[22] Jan. P. Amend, Douglas E. LaRowe, Thomas M. McCollom, and Everett L. Shock. The energetics of organic synthesis inside and outside the cell. *Philos. Trans. R. Soc. London, Ser. B*, 368:20120255, 2013.

[23] Jan P. Amend, Karyn L. Rogers, Everett L. Shock, Sergio Gurrieri, and Salvatore Inguaggiato. Energetics of chemolithoautotrophy in the hydrothermal system of Vulcano Island, southern Italy. *Geobiology*, 1:37–58, 2003.

[24] Jakob L. Andersen, Tommy Andersen, Christoph Flamm, Martin M. Hanczyc, Daniel Merkle, and Peter F. Stadler. Navigating the chemical space of HCN poly-merization and hydrolysis: guiding graph grammars by mass spectrometry data. *Entropy*, 15:4066–4083, 2013.

[25] Jakob L. Andersen, Christoph Flamm, Daniel Merkle, and Peter F. Stadler. Max-imizing output and recognizing autocatalysis in chemical reaction networks is NP-complete. *J. Syst. Chem.*, 3:1, 2012.

[26] Jakob L. Andersen, Christoph Flamm, Daniel Merkle, and Peter F. Stadler. Inferring chemical reaction patterns using rule composition in graph grammars. *J. Syst. Chem.*, 4:4:1–14, 2013.

[27] Jakob L. Andersen, Christoph Flamm, Daniel Merkle, and Peter F. Stadler. Generic strategies for chemical space exploration. *Int. J. Comput. Biol. Drug Des.*, 7:225–258, 2014.

[28] Jakob L. Andersen, Christoph Flamm, Daniel Merkle, and Peter F. Stadler. *In silico* support for Eschenmoser's glyoxylate scenario. *Isr. J. Chem.*, in review, 2015.

[29] Don L. Anderson. *New Theory of the Earth*. Cambridge University Press, London, 2007.

[30] P. W. Anderson. More is different. *Science, New Series*, 177:393–396, 1972.

[31] R. B. Anderson. *The Fischer–Tropsch Synthesis*. Academic Press, New York, 1984.

[32] Miho Aoshima and Yasuo Igarashi. A novel oxalosuccinate-forming enzyme involved in the reductive carboxylation of 2-oxoglutarate in *Hydrogenobacter thermophilus* TK-6. *Mol. Microbiol.*, 62:748–759, 2006.

[33] Miho Aoshima and Yasuo Igarashi. Nondecarboxylating and decarboxylating isoc-itrate dehydrogenases: oxalosuccinate reductase as an ancestral form of isocitrate dehydrogenase. *J. Bacteriol.*, 190:2050–2055, 2008.

[34] Miho Aoshima, Masaharu Ishii, and Yasuo Igarashi. A novel biotin protein required for reductive carboxylation of 2-oxoglutarate by isocitrate dehydrogenase in *Hydrogenobacter thermophilus* TK-6. *Mol. Microbiol.*, 51:791–798, 2004.

[35] Miho Aoshima, Masaharu Ishii, and Yasuo Igarashi. A novel enzyme, citryl-CoA lyase, catalysing the second step of the citrate cleavage reaction in *Hydrogenobacter thermophilus* TK-6. *Mol. Microbiol.*, 52:763–770, 2004.

[36] Miho Aoshima, Masaharu Ishii, and Yasuo Igarashi. A novel enzyme, citryl-CoA synthetase, catalysing the first step of the citrate cleavage reaction in *Hydrogenobac-ter thermophilus* TK-6. *Mol. Microbiol.*, 52:751–761, 2004.

[37] Aristotle. *History of Animals*. Translated by D'Arcy Wentworth Thompson. Claren-don Press, Oxford, 1910.

[38] William Ross Ashby. *An Introduction to Cybernetics*. Chapman and Hall, London, 1956.

[39] W. Ross Ashby. Requisite variety and its implications for the control of complex systems. *Cybernetica*, 1:83–99, 1958.

[40] Shreyas S. Athavale, Anton S. Petrov, Chiaolong Hsiao, Derrick Watkins, Caitlin D. Prickett, J. Jared Gossett, Lively Lie, Jessica C. Bowman, Eric O-Neill, Chad R. Bernier, Nicholas V. Hud, Roger M. Wartell, Stephen C. Harvey, and Loren Dean Williams. RNA folding and catalysis mediated by iron(II). *PLoS ONE*, 7:e38024, 2012.

[41] Henri Atlan. Strehler's theory of mortality and the second principle of thermody-namics. *J. Gerontol.*, 23:196–200, 1968.

[42] Thomas R. Ayers. Evolution of the solar ionizing flux. *J. Geophys. Res.*, 102:1641–1651, 1997.

[43] Francis Bacon. *Novum Organum*. 1620.

[44] Francis Bacon. *The New Organon*. Michael Silverthorne and Lisa Jardine, editors. Cambridge University Press, London, 2000.

[45] Scott Bailey, Richard A. Wing, and Thomas A. Steitz. The structure of *T. aquaticus* DNA polymerase III is distinct from eukaryotic replicative DNA polymerases. *Cell*, 126:893–904, 2006.

[46] Per Bak, Chao Tang, and Kurt Wiesenfeld. Self-organized criticality: an explanation of the $1/f$ noise. *Phys. Rev. Lett.*, 59:381–384, 1987.

[47] Per Bak, Chao Tang, and Kurt Wiesenfeld. Self-organized criticality. *Phys. Rev. A*, 38:364–374, 1988.

[48] W. E. Balche and R. S. Wolfe. Specificity and biological distribution of coenzyme M (2-mercaptoethanesulfonic acid). *J. Bacteriol.*, 137:256–263, 1979.

[49] Ruma Banerjee and Stephen W. Ragsdale. The many faces of vitamin B12: catalysis by cobalamin-dependent enzymes. *Annu. Rev. Biochem.*, 72:209–247, 2003.

[50] Arren Bar-Even, Avi Flamholz, Elad Noor, and Ron Milo. Thermodynamic con-straints shape the structure of carbon fixation pathways. *Biochim. Biophys. Acta Bioenergetics*, 1817(9):1646–1659, 2012.

[51] Carlos F. Barbas III. Organocatalysis lost: modern chemistry, ancient chemistry, and an unseen biosynthetic apparatus. *Angew. Chem. Int. Ed.*, 47:42–47, 2008.

[52] Laura M. Barge, Ivria J. Doloboff, Michael J. Russell, David VanderVelde, Lau-ren M. White, Galen D. Stucky, Marc M. Baum, John Zeytounian, Richard Kidd, and Isik Kanik. Pyrophosphate synthesis in iron mineral films and membranes

simulating prebiotic submarine hydrothermal precipitates. *Geochim. Cosmochim. Acta*, 128:1–12, 2014.

[53] Laura M. Barge, Terence P. Kee, Ivria J. Doloboff, Joshua M. P. Hampton, Mohammed Ismail, Mohamed Ourkashanian, John Zeytounian, Marc M. Baum, John A. Moss, Ghung-Kuang Lin, Richard D. Kidd, and Isik Kanik. The fuel cell model of abiogenesis: a new approach to origin-of-life simulations. *Astrobiology*, 14:254–270, 2014.

[54] H. A. Barker and J. V. Beck. The fermentative decomposition of purines by *Clostridium acidi-urici* and *Clostridium cylindrosporum*. *J. Biol. Chem.*, 141(1):3–27, 1941.

[55] J. A. Bassham, A. A. Benson, L. D. Kay, A. Z. Harris, A. T. Wilson, and M. Calvin. The path of carbon in photosynthesis XXI. The cyclic regeneration of carbon dioxide acceptor. *J. Am. Chem. Soc.*, 76:1760–1770, 1954.

[56] Anthony D. Baughn, Scott J. Garforth, Catherine Vilchèze, and William R. Jacobs Jr. An anaerobic-type $\alpha$-ketoglutarate ferredoxin oxidoreductase completes the oxidative tricarboxylic acid cycle of *Mycobacterium tuberculosis*. *PLoS Pathogens*, 5:e1000662, 1–10, 2009.

[57] Monika Beh, Gerhard Strauss, Robert Huber, Karl-Otto Stetter, and Georg Fuchs. Enzymes of the reductive citric acid cycle in the autotrophic eubacterium *Aquifex pyrophilus* and in the archaebacterium *Thermoproteus neutrophilus*. *Arch. Microbiol.*, 160:306–311, 1993.

[58] Henri Bénard. Les tourbillons cellulaires dans une nappe liquide. *Rev. Gén. Sci. Pure Appl.*, 11:1261–1271, 1900.

[59] Gunes Bender, Elizabeth Pierce, Jeffrey A. Hill, Joseph E. Darty, and Stephen W. Ragsdale. Metal centers in the anaerobic microbial metabolism of CO and $CO_2$. *Metallomics*, 3:797–815, 2011.

[60] Steven A. Benner, Andrew D. Ellington, and Andreas Tauer. Modern metabolism as a palimpsest of the RNA world. *Proc. Natl. Acad. Sci. USA*, 18:7054–7058, 1989.

[61] Charles H. Bennett. Logical reversibility of computation. *IBM J. Res. Dev.*, 17:525–532, 1973.

[62] Charles H. Bennett. The thermodynamics of computation – a review. *Int. J. Theor. Phys.*, 21:905–940, 1982.

[63] Ivan A. Berg, Daniel Kockelkorn, W. Hugo Ramos-Vera, Rafael F. Say, Jan Zarzycki, Michael Hügler, Birgit E. Alber, and Georg Fuchs. Autotrophic carbon fixation in archaea. *Nature Rev. Microbiol.*, 8:447–460, 2010.

[64] Claude Berge. *Graphs and Hypergraphs*. North-Holland, Amsterdam, revised edition, 1973.

[65] Frederick Berkovitch, Yvain Nicolet, Jason T. Wan, Joseph T. Jarrett, and Catherine L. Drennan. Crystal structure of biotin synthase, an *S*-adenosylmethionine-dependent radical enzyme. *Science*, 303(5654):76–79, 2004.

[66] J. D. Bernal. *The Physical Basis of Life*. Routledge & Kegan Paul, London, 1951.

[67] J. D. Bernal, editor. *The Origin of Life*, Weidenfeld and Nicolson, London, 1967.

[68] Michael E. Berndt, Douglas E. Allen, and William E. Seyfried Jr. Reduction of $CO_2$ during serpentinization of olivine at 300 °C and 500 bar. *Geology*, 24:351–354, 1996.

[69] R. A. Berner and K. A. Maasch. Chemical weathering and controls on atmospheric $O_2$ and $CO_2$: fundamental principles were enunciated by J. J. Ebelmen in 1845. *Geochim. Cosmochim. Acta*, 60:1633–1637, 1996.

[70] Harold S. Bernhardt. The RNA world hypothesis: the worst theory of the early evolution of life (except for all the others). *Biol. Direct*, 7:23, 2012.

[71] L. Bertini, A. De Sole, D. Gabrielli, G. Jona-Lasinio, and C. Landim. Macroscopic fluctuation theory for stationary non equilibrium states. *J. Stat. Phys.*, 107:635–675, 2002.

[72] L. Bertini, A. De Sole, D. Gabrielli, G. Jona-Lasinio, and C. Landim. On the long-range correlations of thermodynamic systems out of equilibrium. arXiv:0705.2996v1 [cond-mat.stat-mech], 2007.

[73] L. Bertini, A. De Sole, D. Gabrielli, G. Jona-Lasinio, and C. Landim. Towards a nonequilibrium thermodynamics: a self-contained macroscopic description of driven diffusive systems. *J. Stat. Phys.*, 135:857–872, 2009.

[74] Dany J. V. Beste, Bhushan Bonde, Nathaniel Hawkins, Jane L. Ward, Michael H. Beale, Stephan Noack, Katharina Nöh, Nicholas J. Kruger, R. George Ratcliffe, and Johnjoe McFadden. $^{13}$C metabolic flux analysis identifies an unusual route for pyruvate dissimilation in mycobacteria which requires isocitrate lyase and carbon dioxide fixation. *PLoS Pathogens*, 7(7):e1002091, 07, 2011.

[75] Roy A. Black, Matthew C. Blosser, Benjamin L. Stottrup, Ravi Tavakley, David W. Deamer, and Sarah L. Keller. Nucleobases bind to and stabilize aggregates of a prebiotic amphiphile, providing a viable mechanism for the emergence of protocells. *Proc. Natl. Acad. Sci. USA*, early edition:1–5, 2013.

[76] Carrine E. Blank. Phylogenomic dating and the relative ancestry of prokaryotic metabolisms. In J. Seckbach and M. Walsh, editors, *From Fossils to Astrobiology*, pages 275–295. Springer, New York, 2009.

[77] Robert E. Blankenship. *Molecular Mechanisms of Photosynthesis*. Blackwell Science, Malden, MA, 2002.

[78] G. Blatter, M. V. Feigel'man, V. B. Geshkenbein, A. I. Larkin, and V. M. Vinokur. Vortices in high-temperature superconductors. *Rev. Mod. Phys.*, 66:1125–1388, 1994.

[79] Konrad Bloch. *Blondes in Venetian Paintings, the Nine-Banded Armadillo, and Other Essays in Biochemistry*. Yale University Press, New Haven, CT, 1997.

[80] Konstantin Bokov and Sergey V. Steinberg. A hierarchical model for evolution of 23S ribosomal RNA. *Nature*, 457:977–980, 2009.

[81] Ludwig Boltzmann. *Populäre Schriften*. J. A. Barth, Leipzig, 1905. Re-issued F. Vieweg, Braunschweig, 1979.

[82] Ludwig Boltzmann. The second law of thermodynamics. In *Populäre Schriften*, pages 25–50. J. A. Barth, Leipzig, 1905. Re-issued F. Vieweg, Braunschweig, 1979.

[83] Yan Boucher, Christophe J. Douady, R. Thane Papke, David A. Walsh, Mary Ellen R. Boudreau, Camilla L. Nesbø, Rebecca J. Case, and W. Ford Doolittle. Lateral gene transfer and the origins of prokaryotic groups. *Annu. Rev. Genet.*, 37:283–328, 2003.

[84] Bastien Boussau, Laurent Guéguen, and Manolo Gouy. Accounting for horizontal gene transfers explains conflicting hypotheses regarding the position of aquificales in the phylogeny of bacteria. *BMC Evol. Biol.*, 8:272, 2008.

[85] Jessica C. Bowman, Nicholas V. Hud, and Loren Dean Williams. The ribosome challenge to the RNA world. *J. Mol. Evol.*, 80:143–161, 2015.

[86] Samuel A. Bowring and Ian S. Williams. Priscoan (4.00–4.03 Ga) orthogneisses from northwestern Canada. *Contrib. Mineral Petrol.*, 134:3–16, 1999.

[87] Eric S. Boyd and John W. Peters. New insights into the evolutionary history of biological nitrogen fixation. *Frontiers Microbiol.*, 4:201, 2013.

[88]  Paul Boyer. Coupling mechanisms in capture, transmission, and use of energy. *Annu. Rev. Biochem.*, 46:955–1026, 1977.

[89]  Nanette R. Boyle and John A. Morgan. Computation of metabolic fluxes and efficiencies for biological carbon dioxide fixation. *Metab. Eng.*, 13:150–158, 2011.

[90]  Rogier Braakman and Eric Smith. The emergence and early evolution of biological carbon fixation. *PLoS Comp. Biol.*, 8:e1002455, 2012. PMID: 22536150.

[91]  Rogier Braakman and Eric Smith. The compositional and evolutionary logic of metabolism. *Phys. Biol.*, 10:011001, 2013. PMID: 23234798.

[92]  Rogier Braakman and Eric Smith. Metabolic evolution of a deep-branching hyperthermophilic chemoautotrophic bacterium. *PLoS ONE*, 9:e87950, 2014.

[93]  Dan K. Braithwaite and Junetsu Ito. Compilation, alignment, and phylogenetic relationships of DNA polymerases. *Nucleic Acids Res.*, 21:787–802, 1993.

[94]  Jay A. Brandes, Nabil Z. Boctor, George D. Cody, Benjamin A. Cooper, Robert M. Hazen, and Hatten S. Yoder Jr. Abiotic nitrogen reduction on the early Earth. *Nature*, 395:365–367, 1998.

[95]  Ullrich Brandt. Bifurcated ubihydroquinone oxidation in the cytochrome bc1 complex by proton-gated charge transfer. *FEBS Lett.*, 387:1–6, 1996.

[96]  Elbert Branscomb and Michael J. Russell. Turnstiles and bifurcators: the disequilibrium converting engines that put metabolism on the road. *Biochim. Biophys. Acta*, 1827:62–78, 2013.

[97]  William J. Brazelton and John A. Baross. Abundant transposases encoded by the metagenome of a hydrothermal chimney biofilm. *ISME J.*, 1–5, 2009.

[98]  Leon Brillouin. *Science and Information Theory*. Dover Phoenix Editions, Mineola, NY, second edition, 2004.

[99]  James H. Brown. *Macroecology*. University of Chicago Press, Chicago, IL, 1995.

[100]  Michael R. W. Brown and Arthur Kornberg. Inorganic polyphosphate in the origin and survival of species. *Proc. Natl. Acad. Sci. USA*, 101:16085–16087, 2004.

[101]  David E. Bryant, Katie E. R. Marriott, Stuart A. Macgregor, Colin Kilner, Matthew A. Pasek, and Terence P. Kee. On the prebiotic potential of reduced oxidation state phosphorus: the H-phosphinate-pyruvate system. *Chem. Commun.*, 46:3726–3728, 2010.

[102]  Bob B. Buchanan and Daniel I. Arnon. A reverse Krebs cycle in photosynthesis: consensus at last. *Photosynth. Res.*, 24:47–53, 1990.

[103]  Wolfgang Buckel and Rudolf K. Thauer. Energy conservation via electron bifurcating ferredoxin reduction and proton/Na$^+$ translocating ferredoxin oxidation. *Biochim. Biophys. Acta*, 1827:94–113, 2013.

[104]  John Bunge, Amy Willis, and Fiona Walsh. Estimating the number of species in microbial diversity studies. *Annu. Rev. Stat. Appl.*, 1:427–455, 2014.

[105]  Rod Burstall. Christopher Strachey – understanding programming languages. *Higher-Order Symbolic Comput.*, 13:52, 2000.

[106]  Leo W. Buss. *The Evolution of Individuality*. Princeton University Press, Princeton, NJ, 2007.

[107]  Alison Butler. Marine siderophores and microbial iron mobilization. *BioMetals*, 18:369–374, 2005.

[108]  Thomas Butler, Nigel Goldenfeld, Damien Mathew, and Zaida Luthey-Schulten. Extreme genetic code optimality from a molecular dynamics calculation of amino acid polar requirement. *Phys. Rev. E*, 79:00901, 2009.

[109]  A. Butlerow. Formation synthetique d'une substance sucree. *Compt. Rend. Acad. Sci.*, 53:145–147, 1861.

[110] David A. Butterfield, Bruce K. Nelson, Geoffrey Wheat, Michael Mottl, and Kevin K. Roe. Evidence for basaltic Sr in midocean ridge-flank hydrothermal systems and implications for the global oceanic Sr isotope balance. *Geochim. Cosmochim. Acta*, 65:4141–4153, 2001.

[111] Gustavo Caetano-Anollés, Hee Shin Kim, and Jay E. Mittenthal. The origin of modern metabolic networks inferred from phylogenemic analysis of protein architecture. *Proc. Natl. Acad. Sci. USA*, 104:9358–9363, 2007.

[112] Gustavo Caetano-Anollés, Minglei Wang, and Derek Caetano-Anollés. Structural phylogenomics retrodicts the origin of the genetic code and uncovers the evolutionary impact of protein flexibility. *PLoS ONE*, 8:e72225, 2013.

[113] A. G. Cairns-Smith. *Genetic Takeover: and the Mineral Origins of Life*. Cambridge University Press, Cambridge, 1982.

[114] A. G. Cairns-Smith. *Seven Clues to the Origin of Life – A Scientific Detective Story*. Cambridge University Press, Cambridge, 1985.

[115] A. G. Cairns-Smith and H. Hartman, editors. *Clay Minerals and the Origin of Life*. Cambridge University Press, Cambridge, 1986.

[116] Barbara J. Campbell and S. Craig Cary. Evidence of chemolithoautotrophy in the bacterial community associated with *Alvinella pompehana*, a hydrothermal vent polychaete. *Appl. Environ. Microbiol.*, 69:5070–5078, 2003.

[117] Ian H. Campbell. Constraints on continental growth models from Nb/U ratios in the 3.5 Ga Barberton and other Archaean basalt-komatiite suites. *Am. J. Sci.*, 303:319–351, 2003.

[118] I. H. Campbell, R. W. Griffiths, and R. I. Hill. Melting in an Archean mantle plume: heads it's basalts, tails it's komatiites. *Nature*, 339:697–699, 1989.

[119] Sadi Carnot. *Reflections on the Motive Power of Fire*, E. Mendoza, editor. Dover, New York, 1960.

[120] B. J. Carr and M. J. Rees. The anthropic principle and the structure of the physical world. *Nature*, 278:605–612, 1979.

[121] David C. Catling and Kevin J. Zahnle. The planetary air leak. *Sci. Am.*, May:36–43, 2009.

[122] Thomas R. Cech. The RNA worlds in context. *Cold Spring Harb. Perspect. Biol.*, 4:a006742, 2011.

[123] E. Chabrière, M. H. Charon, A. Volbeda, L. Pieulle, E. C. Hatchikian, and J. C. Fontecilla-Camps. Crystal structures of the key anaerobic enzyme pyruvate:ferredoxin oxidoreductase, free and in complex with pyruvate. *Nature Struct. Biol.*, 6:182–190, 1999.

[124] Gregory J. Chaitin. *Algorithmic Information Theory*. Cambridge University Press, New York, 1990.

[125] Patricia P. Chan and Todd M. Lowe. GtRNAdb: a database of transfer RNA genes detected in genomic sequence. *Nucleic Acids Res.*, 37:D93–D97, 2008.

[126] Jean-Pierre Changeux, Philippe Courrége, and Antoine Danchin. A theory of the epigenesis of neuronal networks by selective stabilization of synapses. *Proc. Natl. Acad. Sci. USA*, 70:2974–2978, 1973.

[127] Jean-Luc Charlou and Jean-Pierre Donval. Hydrothermal methane venting between 12°N and 26°N along the Mid-Atlantic ridge. *J. Geophys. Res.*, 98:9625–9642, 1993.

[128] Nyles W. Charon, Russell C. Johnson, and David Peterson. Amino acid biosynthesis in the spirochete leptospira: evidence for a novel pathway of isoleucine biosynthesis. *J. Bacteriol.*, 117(1):203–211, 1974.

[129] Geoffrey Chaucer. *Treatise on the Astrolabe*, Prologue, II 39–40. 1391.

[130] Lubin Chen, Michael L. Johnson, and Rodney L. Biltonen. A macroscopic description of lipid bilayer phase transitions of mixed-chain phosphatidyl-cholines: chain-length and chain-asymmetry dependence. *Biophys. J.*, 80:254–270, 2001.

[131] Peiqiu Chen and Eugene I. Shakhnovich. Lethal mutagenesis in viruses and bacteria. *Genetics*, 183:639–650, 2009.

[132] Xi Chen, Na Li, and Andrew D. Ellington. Ribozyme catalysis of metabolism in the RNA world. *Chem. Biodiv.*, 4:633–655, 2007.

[133] Ludmila Chistoserdova. Modularity of methylotrophy, revisited. *Environ. Microbiol.*, 13(10):2603–2622, 2011.

[134] L. Chistoserdova, M. G. Kalyuzhnaya, and M. E. Lidstrom. The expanding world of methylotrophic metabolism. *Annu. Rev. Microbiol.*, 63:477–499, 2009.

[135] Ludmila Chistoserdova, Julia A. Vorholt, Rudolf K. Thauer, and Mary E. Lidstrom. $C_1$ transfer enzymes and coenzymes linking methylotrophic bacteria and methanogenic archaea. *Science*, 281:99–102, 1998.

[136] Dylan Chivian, Eoin L. Brodie, Eric J. Alm, David E. Culley, Paramvir S. Dehal, Todd Z. DeSantis, Thomas M. Gihring, Alla Lapidus, Li-Hung Lin, Stephen R. Lowry, Duane P. Moser, Paul M. Richardson, Gordon Southam, Greg Wanger, Lisa M. Pratt, Gary L. Andersen, Terry C. Hazen, Fred J. Brockman, Adam P. Arkin, and Tullis C. Onstott. Environmental genomics reveals a single-species ecosystem deep within earth. *Science*, 322:275–278, 2008.

[137] Ahmed S. U. Choughuley and Richard M. Lemmon. Production of cysteic acid, taurine and cystamine under primitive earth conditions. *Nature*, 210:628–629, 1966.

[138] Francesca D. Ciccarelli, Tobias Doerks, Christian von Mering, Christopher J. Creevey, Berend Snel, and Peer Bork. Toward automatic reconstruction of a highly resolved tree of life. *Science*, 311:1283–1287, 2006.

[139] Mark W. Claire, James F. Jasting, Shawn D. Domagal-Goldman, Eva E. Stüeken, Roger Buick, and Victoria S. Meadows. Modeling the signature of sulfur mass-independent fractionation produced in the Archean atmosphere. *Geochim. Cosmochim. Acta*, 141:365–380, 2014.

[140] R. Clausius. On the application of the theorem of the equivalence of transformations to interior work. In T. Archer Hirst, editor, *The Mechanical Theory of Heat*, pages 215–250, Fourth Memoir. John van Voorst, London, 1865.

[141] Jean-Michel Claverie. Viruses take center stage in cellular evolution. *Genome Biol.*, 7:110, 2006.

[142] Donald D. Clayton. *Principles of Stellar Evolution and Nucleosynthesis*. University of Chicago Press, Chicago, IL, 1983.

[143] H. James Cleaves and Stanley L. Miller. The nicotinamide biosynthetic pathway is a by-product of the RNA world. *J. Mol. Evol.*, 52:73–77, 2001.

[144] George D. Cody, Bjorn Mysen, Gotthard Sághi-Szabó, and John A. Tosell. Silicate-phosphate interactions in silicate glasses and melts: I. A multinuclear $\left(^{27}Al, \ ^{29}Si, \ ^{31}P\right)$ MAS NMR and ab initio chemical shielding $\left(^{31}P\right)$ study of phosphorus speciation in silicate glasses. *Geochim. Cosmochim. Acta*, 65:2395–2411, 2001.

[145] G. D. Cody, N. Z. Boctor, J. A. Brandes, T. E. Filley, R. M. Hazen, and H. S. Yoder Jr. Assaying the catalytic potential of transition metal sulfides for abiotic carbon fixation. *Geochim. Cosmochim. Acta*, 68:2185–2196, 2004.

[146] George D. Cody, Nabil Z. Boctor, Timothy R. Filley, Robert M. Hazen, James H. Scott, Anurag Sharma, and Hatten S. Yoder Jr. Primordial carbonylated iron-sulfur compounds and the synthesis of pyruvate. *Science*, 289:1337–1340, 2000.

[147] G. D. Cody, N. Z. Boctor, R. M. Hazen, J. A. Brandes, H. J. Morowitz, and H. S. Yoder Jr. Geochemical roots of autotrophic carbon fixation: hydrothermal experiments in the system citric acid, $H_2O$-($\pm$FeS) ($\pm$NiS). *Geochim. Cosmochim. Acta*, 65:3557–3576, 2001.

[148] Melvin Cohn, N. Av Mitchison, William E. Paul, Arthur M. Silverstein, David W. Talmage, and Martin Weigert. Reflections on the clonal-selection theory. *Nature Rev. Immunol.*, 7:823–830, 2007.

[149] Sidney Coleman. *Aspects of Symmetry*. Cambridge University Press, Cambridge, 1985.

[150] The CMS Collaboration. Evidence for the direct decay of the 125 GeV Higgs boson to fermions. *Nature Phys.*, 10:557–560, 2014.

[151] Matthew D. Collins and Dorothy Jones. Distribution of isoprenoid quinone structural types in bacteria and their taxonomic implication. *Microbiol. Rev.*, 45(2):316–354, 1981.

[152] P. D. B. Collins, A. D. Martin, and E. J. Squires. *Particle Physics and Cosmology*. Wiley, New York, 1989.

[153] Creative Commons. Attribution-noncommercial-sharealike 3.0 unported, April 2015. http://creativecommons.org/licenses/by-nc-sa/3.0/legalcode.

[154] R. C. Conant and W. R. Ashby. Every good regulator of a system must be a model of that system. *Int. J. Syst. Sci.*, 1:89–97, 1970.

[155] Geoffrey M. Cooper. *The Cell: A Molecular Approach*. Sinauer Associates, Sunderland, MA, second edition, 2000.

[156] George Cooper, Chris Reed, Dang Nguyen, Malika Carter, and Yi Wang. Detection and formation scenario of citric acid, pyruvic acid, and other possible metabolism precursors in carbonaceous meteorites. *Proc. Natl. Acad. Sci. USA*, 108:14015–14020, 2011.

[157] Shelley D. Copley. Enzymes with extra talents: moonlighting functions and catalytic promiscuity. *Curr. Opin. Chem. Biol.*, 7:265–272, 2003.

[158] Shelley D. Copley, Eric Smith, and Harold J. Morowitz. A mechanism for the association of amino acids with their codons and the origin of the genetic code. *Proc. Natl. Acad. Sci. USA*, 102:4442–4447, 2005. PMID: 15764708.

[159] Shelley D. Copley, Eric Smith, and Harold J. Morowitz. The origin of the RNA world: co-evolution of genes and metabolism. *Bioorg. Chem.*, 35:430–443, 2007. PMID: 17897696.

[160] Armando Córdova, Magnus Engqvist, Ismail Ibrahem, Jesús Casas, and Henrik Sunden. Plausible origins of homochirality in the amino acid catalyzed neogenesis of carbohydrates. *Chem. Commun.*, 2005:2047–2049, 2005.

[161] John B. Corliss, John A. Baross, and Sarah E. Hoffman. Submarine hydrothermal systems: a probable site for the origin of life. *Oreg. State Univ. Sch. Oceanogr.*, 80-7:1–44, 1980.

[162] J. B. Corliss, J. Dymond, L. I. Gordon, J. M. Edmond, R. P. von Herzen, R. D. Ballard, K. Green, D. Williams, A. Bainbridge, K. Crane, and T. H. van Andel. Submarine thermal springs on the Galapagos rift. *Science*, 203:1073–1083, 1979.

[163] James B. Cotner and Bopaiah A. Biddanda. Small players, large role: microbial influence on biogeochemical processes in pelagic aquatic ecosystems. *Ecosystems*, 5:105–121, 2002.

[164] Thomas M. Cover and Joy A. Thomas. *Elements of Information Theory*. Wiley, New York, 1991.

[165] F. H. C. Crick. Codon-antiodon pairing: the wobble hypothesis. *J. Mol. Biol.*, 19:548–555, 1966.

[166] F. H. C. Crick. The origin of the genetic code. *J. Mol. Biol.*, 38:367–379, 1968.

[167] Francis Crick. Central dogma of molecular biology. *Nature*, 227:561–563, 1970.

[168] Francis Crick. *Life Itself: Its Origin and Nature*. Simon and Schuster, New York, 1981.

[169] F. H. C. Crick and L. E. Orgel. Directed panspermia. *Icarus*, 19:341–346, 1973.

[170] Shane Crotty, Craig E. Cameron, and Raul Andino. RNA virus error catastrophe: direct molecular test by using ribavirin. *Proc. Natl. Acad. Sci. USA*, 98:6895–6900, 2001.

[171] Marie Csete and John Doyle. Bow ties, metabolism and disease. *Trends Biotechnol.*, 22:446–450, 2004.

[172] Thomas P. Curtis, William T. Sloan, and Jack W. Scannell. Estimating prokaryotic diversity and its limits. *Proc. Natl. Acad. Sci. USA*, 99:10494–10499, 2002.

[173] S. Dagley and Donald E. Nicholson. *An Introduction to Metabolic Pathways*. Blackwell Scientific, Oxford, 1970.

[174] Vincent Danos, Jérôme Feret, Walter Fontana, Russell Harmer, and Jean Krivine. Rule-based modelling, symmetries, refinements. *Formal Methods in Systems Biology, Lecture Notes in Computer Science*, volume 5054, pages 103–122. Springer, Berlin, 2008.

[175] M. J. Danson. Central metabolism of the archaea. In M. Kates, D. J. Kushner, and A. T. Matheson, editors, *The Biochemistry of Archaea*, pages 1–24. Elsevier, Amsterdam, 1993.

[176] Claudine Darnault, Anne Volbeda, Eun Jin Kim, Pierre Legrand, Xavier Vernede, Paul A. Lindahl, and Juan C. Fontecilla-Camps. Ni-Zn-[Fe$_4$-S$_4$] and Ni-Ni-[Fe$_4$-S$_4$] clusters in closed and open $\alpha$ subunits of acetyl-CoA synthase/carbon monoxide dehydrogenase. *Nature Struct. Mol. Biol.*, 10(4):271–279, 2003.

[177] Charles Darwin. *On the Origin of Species*. John Murray, London, 1859.

[178] Charles Darwin. *The Life and Letters of Charles Darwin, including an Autobiographical Chapter, Vol. 3*, Francis Darwin, editor. John Murray, London, 1887.

[179] Eric H. Davidson. *The Regulatory Genome: Gene Regulatory Networks in Development and Evolution*. Academic Press, San Diego, CA, 2006.

[180] Eric H. Davidson and Douglas H. Erwin. Gene regulatory networks and the evolution of animal body plans. *Science*, 311:796–800, 2006.

[181] Richard C. Dawkins. *The Selfish Gene*. Oxford University Press, New York, 1976.

[182] M. A de Angelis, M. D. Lilley, E. J. Olson, and J. A. Baross. Methane oxidation in deep-sea hydrothermal plumes of the Endeavour Segment of the Juan de Fuca Ridge. *Deep-Sea Res.*, 40:1169–1186, 1993.

[183] Christian de Duve. *Blueprint for a Cell*. Neil Patterson, Burlington, NC, 1991.

[184] David Deamer. *First Life: Discovering the Connections between Stars, Cells, and How Life Began*. University of California Press, Los Angeles, CA, 2011.

[185] David Deamer and Jack W. Szostak, editors. *The Origins of Life*. Cold Spring Harbor Laboratory Press, Cold Spring Harbor, NY, 2011.

[186] Gerard Deckert, Patrick V. Warren, Terry Gaasterland, William G. Young, Anna L. Lenox, David E. Graham, Ross Overbeek, Marjory A. Snead, Martin Keller, Monette Aujay, Robert Huber, Robert A. Feldman, Jay M. Short, Gary J. Olsen,

and Ronald V. Swanson. The complete genome of the hyperthermophilic bacterium *Aquifex aeolicus*. *Nature*, 392:353–358, 1998.

[187] Veronica DeGuzman, Wenonah Vercoutere, Hossein Shenasa, and David Deamer. Generation of oligonucleotides under hydrothermal conditions by non-enzymatic polymerization. *J. Mol. Evol.*, 78:251–262, 2014.

[188] J. R. Delaney, D. S. Kelley, M. D. Lilley, D. A. Butterfield, J. A. Baross, W. S. D. Wilcock, R. W. Embley, and M. Summit. The quantum event of oceanic crustal accretion: impacts of diking at mid-ocean ridges. *Science*, 281:222–230, 1998.

[189] Michael Denton. The protein folds as platonic forms: new support for the pre-Darwinian conception of evolution by natural law. *J. Theor. Biol.*, 219:325–342, 2002.

[190] Bernard Derrida. Non equilibrium steady states: fluctuations and large deviations of the density and of the current. *J. Stat. Mech.*, page P07023, 2007. arXiv:cond-mat/0703762v1.

[191] Á. S. Dias, R. A. Mills, I. Ribiero da Costa, R. Costa, R. N. Taylor, M. J. Cooper, and F. J. A. S. Barriga. Tracing fluid-rock reaction and hydrothermal circulation at the Saldanha hydrothermal field. *Chem. Geol.*, 273:168–179, 2010.

[192] Jeffrey M. Dick and Everett L. Shock. Calculations of the relative chemical stabilities of proteins as a function of temperature and redox chemistry in a hot spring. *PLoS ONE*, 6:e22782, 2011.

[193] Michael R. Dietric. The problem of the gene. *Compt. Rend. Acad. Sci. Paris*, 323:1139–1146, 2000.

[194] Kang Ding, William E. Seyfried Jr., Zhong Zhang, Margaret K. Tivey, Karen L. Von Damm, and Albert M. Bradley. The in situ pH of hydrothermal fluids at mid-ocean ridges. *Earth Planet. Sci. Lett.*, 237:167–174, 2005.

[195] D. L. Distel, D. J. Lane, G. J. Olsen, S. J. Giovannoni, B. Pace, N. R. Pace, D. A. Stahl, and H. Felbeck. Sulfur-oxidizing bacterial endosymbionts: analysis of phylogeny and specificity by 16S rRNA sequences. *J. Bacteriol.*, 170:2506–2510, 1988.

[196] Mike Dixon-Kennedy. *Encyclopedia of Greco-Roman Mythology*. ABC-Clio, New York, 1998.

[197] Holger Dobbek, Vitali Svetlitchnyi, Lothar Gremer, Robert Huber, and Ortwin Meyer. Crystal structure of a carbon monoxide dehydrogenase reveals a [Ni-4Fe-5S] cluster. *Science*, 293:1281–1285, 2001.

[198] Theodosius Dobzhansky. Nothing in biology makes sense except in the light of evolution. *Am. Biol. Teacher*, 35:125–129, 1973.

[199] M. Doi. Second quantization representation for classical many-particle system. *J. Phys. A*, 9:1465–1478, 1976.

[200] M. Doi. Stochastic theory of diffusion-controlled reaction. *J. Phys. A*, 9:1479–1495, 1976.

[201] Shawn D. Domagal-Goldman, Victoria S. Meadows, Mark W. Claire, and James F. Kasting. Using biogenic sulfur gases as remotely detectable biosignatures on anoxic planets. *Astrobiology*, 11:419–441, 2011.

[202] Matina C. Donaldson-Matasci, Carl T. Bergstrom, and Michael Lachmann. The fitness value of information. *Oikos*, 119:219–230, 2010.

[203] Alexander Donath, Sven Findei, Jana Hertel, Manja Marz, Wolfgang Otto, Christine Schulz, Peter F. Stadler, and Stefan Wirth. Non-coding RNAs. In Gustavo Caetano-Anollés, editor, *Evolutionary Genomics and Systems Biology*, pages 251–293. Wiley-Blackwell, Hoboken, NJ, 2010.

[204] David Duncan. *The Life and Letters of Herbert Spencer*. D. Appleton, New York, 1908. Two volumes.

[205] Jennifer A. Dunne, Richard J. Williams, Neo D. Martinez, Rachel A. Wood, and Douglas H. Erwin. Compilation and network analyses of Cambrian food webs. *PLoS Biology*, 6:e102, 2008.

[206] Rick Durrett. *Essentials of Stochastic Processes*. Springer, New York, 1999.

[207] M. I. Dykman, Eugenia Mori, John Ross, and P. M. Hunt. Large fluctuations and optimal paths in chemical kinetics. *J. Chem. Phys.*, 100:5735–5750, 1994.

[208] F. Y. Edgeworth. An introductory lecture on political economy. *Econ. J.*, 1(4):625–634.

[209] Deeanne B. Edwards and Douglas C. Nelson. DNA-DNA solution hybridization studies of the bacterial symbionts of hydrothermal vent tube worms (*Riftia pachyptila* and *Tevnia jerichonana*). *Appl. Environ. Microbiol.*, 57:1082–1088, 1991.

[210] P. Ehrenfest. Phasenumwandlungen im ueblichen und erweiterten Sinn, classifiziert nach den entsprechenden Singularitaeten des thermodynamischen Potentiales. *Verh. K. Akad. Wet. Amsterdam*, 36:153–157, 1933.

[211] Manfred Eigen. *Steps Toward Life*. Oxford University Press, Oxford, 1992.

[212] Manfred Eigen. *From Strange Simplicity to Complex Familiarity*. Oxford University Press, London, 2013.

[213] Manfred Eigen and Peter Schuster. The hypercycle, Part A. The emergence of the hypercycle. *Naturwissenschaften*, 64:541–565, 1977.

[214] Manfred Eigen and Peter Schuster. The hypercycle, Part C. The realistic hypercycle. *Naturwissenschaften*, 65:341–369, 1978.

[215] Marion Eisenhut, Shira Kahlon, Dirk Hasse, Ralph Ewald, Judy Lieman-Hurwitz, Teruo Ogawa, Wolfgang Ruth, Hermann Bauwe, Aaron Kaplan, and Martin Hagemann. The plant-like C2 glycolate cycle and the bacterial-like glycerate pathway cooperate in phosphoglycolate metabolism in cyanobacteria. *Plant Physiol.*, 142:333–342, 2006.

[216] Marion Eisenhut, Wolfgang Ruth, Maya Haimovich, Hermann Bauwe, Aaron Kaplan, and Martin Hagemann. The photorespiratory glycolate metabolism is essential for cyanobacteria and might have been conveyed endosymbiontically to plants. *Proc. Natl. Acad. Sci. USA*, 105(44):17199–17204, 2008.

[217] Eric H. Ekland and David P. Bartel. RNA-catalyzed RNA polymerization using nucleoside triphosphates. *Nature*, 382:373–376, 1996.

[218] Basma El Yacoubi, Shilah Bonnett, Jessica N. Anderson, A. Swairjo Manal, Dirk Iwata-Reuyl, and Valérie de Crécy-Lagard. Discovery of a new prokaryotic type I GTP cyclohydrolase family. *J. Biol. Chem.*, 281:37586–37593, 2006.

[219] Linda T. Elkins-Tanton. Linked magma ocean solidification and atmospheric growth for Earth and Mars. *Earth Planet. Sci. Lett.*, 271:181–191, 2008.

[220] Linda T. Elkins-Tanton. Formation of early water oceans on rocky planets. *Astrophys. Space Sci.*, 332:359–364, 2011.

[221] Linda T. Elkins-Tanton. Magma oceans in the inner solar system. *Annu. Rev. Earth Planet. Sci.*, 40:113–139, 2012.

[222] Linda T. Elkins-Tanton. Evolutionary dichotomy for rocky planets. *Nature*, 497:570–572, 2013.

[223] Andrew D. Ellington. Experimental testing of theories of an early RNA world. *Methods Enzymol*, 224:646–664, 1993.

[224] Richard S. Ellis. *Entropy, Large Deviations, and Statistical Mechanics*. Springer-Verlag, New York, 1985.

[225] Aaron E. Engelhart, Matthew W. Powner, and Jack W. Szostak. Functional RNAs exhibit tolerance for non-heritable $2'-5'$ vs. $2'-5'$ backbone heterogeneity. *Nature Chem.*, 5:390–394, 2013.

[226] Douglas H. Erwin. *Extinction: How Life on Earth Nearly Ended 250 Million Years Ago*. Princeton University Press, Princeton, NJ, 2006.

[227] Douglas H. Erwin and Eric H. Davidson. The evolution of hierarchical gene regulatory networks. *Nature Rev. Genet.*, 10:141–148, 2009.

[228] Douglas H. Erwin and James W. Valentine. *The Cambrian Explosion: The Construction of Animal Biodiversity*. Roberts and Company, Englewood, CO, 2013.

[229] Douglas H. Erwin, Marc Laflamme, Sarah M. Tweedt, Erik A. Sperling, Davide Pisani, and Kevin J. Peterson. The Cambrian conundrum: early divergence and later ecological success in the early history of animals. *Science*, 334:1091–1097, 2011.

[230] Arthur Conan Doyle. *The Memoirs of Sherlock Holmes*. Simon and Schuster, New York, 2014.

[231] Albert Eschenmoser. On a hypothetical generational relationship between HCN and constituents of the reductive citric acid cycle. *Chem. Biodivers.*, 4:554–573, 2007.

[232] Stewart N. Ethier and Thomas G. Kurtz. *Markov Processes: Characterization and Convergence*. Wiley, New York, 1986.

[233] Katharina F. Ettwig, Margaret K. Butler, Denis Le Paslier, Eric Pelletier, Sophie Mangenot, Marcel M. M. Kuypers, Frank Schreiber, Bas E. Dutilh, Johannes Zedelius, Dirk de Beer, Jolein Gloerich, Hans J. C. T. Wessels, Theo van Alen, Francisca Luesken, Ming L. Wu, Katinka T. van de Pas-Schoonen, Huub J. M. Op den Camp, Eva M. Janssen-Megens, Kees-Jan Francoijs, Henk Stunnenberg, Jean Weissenbach, Mike S. M. Jetten, and Marc Strous. Nitrite-driven anaerobic methane oxidation by oxygenic bacteria. *Nature*, 464:543–548, 2010.

[234] M. C. W. Evans, B. B. Buchanan, and D. I. Arnon. A new ferredoxin dependent carbon reduction cycle in photosynthetic bacterium. *Proc. Natl. Acad. Sci. USA*, 55:928–934, 1966.

[235] Paul G. Falkowski. Evolution of the nitrogen cycle and its influence on the biological sequestration of $CO_2$ in the ocean. *Nature*, 387:272–275, 1997.

[236] Paul G. Falkowski, Tom Fenchel, and Edward F. Delong. The microbial engines that drive Earth's biogeochemical cycles. *Science*, 320:1034–1039, 2008.

[237] James Farquhar, Huiming Bao, and Mark Thiemens. Atmospheric influence of the Earth's earliest sulfur cycle. *Science*, 289:756–758, 2000.

[238] James Farquhar, Marc Peters, David T. Johnston, Harald Strauss, Andrew Masterson, Uwe Wiechert, and Alan J. Kaufman. Isotopic evidence for Mesoarchean anoxia and changing atmospheric sulphur chemistry. *Nature*, 449:706–709, 2007.

[239] James Farquhar, Joel Savarino, Sabine Airieau, and Mark H. Thiemes. Observation of wavelength-sensitive mass-independent sulfur isotope effects during $SO_2$ photolysis: implications for the early atmosphere. *J. Geophys. Res.*, 106:32829–32839, 2001.

[240] Ole Farver, Ernst Greil, Bernd Luwig, Hartmut Michel, and Israel Pecht. Rates and equilibrium of $Cu_A$ to heme $a$ electron transfer in *Paracoccus denitrificans* cytochrome $c$ oxidase. *Biophys. J.*, 90:2131–2137, 2006.

[241] Ole Farver, Peter M. H. Kroneck, Walter G. Zumft, and Israel Pecht. Intramolecular electron transfer in cytochrome cd(1) nitrite reductase from *Pseudomonas stutzeri*; kinetics and thermodynamics. *Biophys. Chem.*, 98:27–34, 2002.

[242] Adam M. Feist, Christopher S. Henry, Jennifer L. Reed, Markus Krummenacker, Andrew R. Joyce, Peter D. Karp, Linda J. Broadmelt, Vassily Hatzimanikatis,

and Berhard O. Palsson. A genome-scale metabolic reconstruction for *Escherichia coli* K-12 MG1655 that accounts for 1260 ORFs and thermodynamic information. *Mol. Syst. Biol.*, 3:121:1–18, 2007.

[243]  E. Fermi. Versuch einer theorie der $\beta$-strahlen. i. *Z. Phys.*, 88:161–177, 1934.

[244]  Enrico Fermi. *Thermodynamics*. Dover, New York, 1956.

[245]  James P. Ferris. Mineral catalysis and prebiotic synthesis: montmorillonite-catalyzed formation of RNA. *Elements*, 1:145–149, 2005.

[246]  J. P. Ferris and L. E. Orgel. An unusual photochemical re-arrangement in the synthesis of adenine from hydrogen cyanide. *J. Am. Chem. Soc.*, 88:1074, 1966.

[247]  J. P. Ferris, P. C. Joshi, K.-J. Wang, S. Miyakawa, and W. Huang. Catalysis in prebiotic chemistry: application to the synthesis of RNA oligomers. *Adv. Space Res.*, 33:100–105, 2004.

[248]  James P. Ferris, Robert A. Sanchez, and Leslie E. Orgel. Studies in prebiotic synthesis: III. Synthesis of pyrimidines from cyanoacetylene and cyanate. *J. Mol. Biol.*, 33:693–704, 1968.

[249]  James G. Ferry and Christopher H. House. The stepwise evolution of early life driven by energy conservation. *Mol. Biol. Evol.*, 23:1286–1292, 2006.

[250]  Martin Ferus, David Nesvorný, Jiří Šponer, Petr Kubelík, Regina Michalčíková, Violetta Shestivská, Judit D. Šponer, and Svatopluk Civiš. High-energy chemistry of formamide: a unified mechanism of nucleobase formation. *Proc. Natl. Acad. Sci. USA*, early edition:1412072111, 2014.

[251]  Georg Feulner. The faint young sun problem. *Rev. Geophys.*, 50:RG2006, 2012.

[252]  Richard P. Feynman. Space-time approach to quantum electrodynamics. *Phys. Rev.*, 76:769–789, 1949.

[253]  Eliane Fischer and Uwe Sauer. A novel metabolic cycle catalyzes glucose oxidation and anaplerosis in hungry *Escherichia coli*. *J. Biol. Chem.*, 278(47):46446–46451, 2003.

[254]  Julia D. Fischer, Gemma L. Holliday, Syed A. Rahman, and Janet M. Thornton. The structures and physicochemical properties of organic cofactors in biocatalysis. *J. Mol. Biol.*, 403:803–824, 2010.

[255]  K. H. Fischer and J. A. Hertz. *Spin Glasses*. Cambridge University Press, New York, 1991.

[256]  R. A. Fisher. *The Genetical Theory of Natural Selection*. Oxford University Press, London, 2000.

[257]  Cyrus H. Fiske and Y. Subbarow. Phosphorus compounds of muscle and liver. *Science*, 70:381–382, 1929.

[258]  W. Fontana. Modeling 'Evo-Devo' with RNA. *Bioessays*, 24:1164–1177, 2002.

[259]  Walter Fontana and Leo W. Buss. The barrier of objects: from dynamical systems to bounded organizations. In John Casti and Anders Karlqvist, editors, *Boundaries and Barriers*, pages 56–116. Addison-Wesley, New York, 1996.

[260]  Walter Fontana, Günter Wagner, and Leo W. Buss. Beyond digital naturalism. *Artificial Life*, 1:211–227, 1994.

[261]  Juan C. Fontecilla-Camps, Patricia Amara, Christine Cavazza, Yvain Nicolet, and Anne Volbeda. Structure-function relationships of anaerobic gas-processing metalloenzymes. *Nature*, 460:814–822, 2009.

[262]  Juan C. Fontecilla-Camps, Anne Volbeda, Christine Cavazza, and Yvain Nicolet. Structure/function relationships of [NiFe]- and [FeFe]-hydrogenases. *Chem. Rev.*, 107:4273–4303, 2007.

[263] Patrick Forterre. Defining life: the virus viewpoint. *Orig. Life Evol. Biosphere*, 40:151–160, 2010.

[264] Yves Fouquet, Pierre Camboun, Joël Etoubleau, Jean Luc Charlou, Hélène Ondréas, Fernando J. A. S. Barriga, Georgy Cherkashov, Tatiana Semkova, Irina Poroshina, M. Bohn, Jean Pierre Donval, Katell Henry, Pamela Murphy, and Olivier Rouxel. Geodiversity of hydrothermal processes along the Mid-Atlantic Ridge and ultramafic-hosted mineralization: a new type of oceanic Cu-Zn-Co-Au volcanogenic massive sulfide deposit. *Geophys. Monogr. Ser.*, 188:321–367, 2010.

[265] G. P. Fournier and E. J. Alm. Ancestral reconstruction of a pre-LUCA aminoacyl-tRNA synthetase ancestor supports the late addition of Trp to the genetic code. *J. Mol. Evol.*, 80:171–185, 2015.

[266] Gregory P. Fournier, Cheryl P. Andam, Eric J. Alm, and J. Peter Gogarten. Molecular evolution of aminoacyl tRNA synthetase proteins in the early history of life. *Orig. Life Evol. Biosphere*, 41:621–632, 2011.

[267] Dionysis I. Foustoukos and William E. Seyfried Jr. Hydrocarbons in hydrothermal vent fluids: the role of chrome-bearing catalysts. *Science*, 304:1002–1005, 2004.

[268] Dionysis I. Foustoukos, Ivan P. Savov, and David R. Janecky. Chemical and isotopic constraints on water/rock interactions at the Lost City hydrothermal field, 30° N Mid-Atlantic Ridge. *Geochim. Cosmochim. Acta*, 72:5457–5474, 2008.

[269] George E. Fox. Origin and evolution of the ribosome. *Cold Spring Harb. Perspect. Biol.*, 2:a003483, 2010.

[270] George E. Fox and Ashwinikumar K. Naik. The evolutionary history of the ribosome. In Lluís Ribas de Pouplana, editor, *The Genetic Code and the Origin of Life*, pages 92–105. Kluwer Academic/Plenum, New York, 2004.

[271] George E. Fox and Ashwinikumar K. Naik. The evolutionary history of the translation machinery. In Lluís Ribas de Pouplana, editor, *The Genetic Code and the Origin of Life*, pages 680–682. Kluwer Academic/Plenum, New York, 2004.

[272] Christine H. Foyer, Arnold J. Bloom, Guillaume Queval, and Graham Noctor. Photorespiratory metabolism: genes, mutants, energetics, and redox signaling. *Annu. Rev. Plant Biol.*, 60:455–484, 2009.

[273] Steven A. Frank. The Price equation, Fisher's fundamental theorem, kin selection, and causal analysis. *Evolution*, 51:1712–1729, 1997.

[274] Steven A. Frank and Montgomery Slatkin. Fisher's fundamental theorem of natural selection. *Trends Ecol. Evol.*, 7:92–95, 1992.

[275] Rosalind E. Franklin and R. G. Gosling. Molecular configuration in sodium thymonucleate. *Nature*, 171:740–741, 1953.

[276] Steven J. Freeland and Laurence D. Hurst. The genetic code is one in a million. *J. Mol. Evol.*, 47:238–248, 1998.

[277] Steven J. Freeland, Robin D. Knight, Laura F. Landweber, and Laurence D. Hurst. Early fixation of an optimal genetic code. *Mol. Biol. Evol.*, 17:511–518, 2000.

[278] M. I. Freidlin and A. D. Wentzell. *Random Perturbations in Dynamical Systems*. Springer, New York, second edition, 1998.

[279] Daniel J. Frost and Catherine A. McCammon. The redox state of Earth's mantle. *Annu. Rev. Earth. Planet. Sci.*, 36:389–420, 2008.

[280] Iris Fry. *The Emergence of Life on Earth: A Historical and Scientific Overview*. Rutgers University Press, New Brunswick, NJ, 2000.

[281] Georg Fuchs. $CO_2$ fixation in acetogenic bacteria: variations on a theme. *FEMS Microbiol. Lett.*, 38:181–213, 1986.

[282] Georg Fuchs. Alternative pathways of carbon dioxide fixation: insights into the early evolution of life? *Annu. Rev. Microbiol.*, 65(1):631–658, 2011.

[283] W. D. Fuller, R. A. Sanchez, and L. E. Orgel. Studies in prebiotic synthesis: VII. Solid-state synthesis of purine nucleosides. *J. Mol. Evol.*, 1:249–257, 1972.

[284] Astrid Gerhardt, Irfan Çinkaya, Dietmar Linder, Gjalt Hulsman, and Wolfgang Buckel. Fermentation of 4-aminobutyrate by *Clostridium aminobytyricum*: cloning of two genes involved in the formation and dehydration of 4-hydroxybutyryl-CoA. *Arch. Microbiol.*, 174:189–199, 2000.

[285] S. Garlick, A. Oren, and E. Padan. Occurrence of facultative anoxygenic photosynthesis among filamentous and unicellular cyanobacteria. *J. Bacteriol.*, 129:623–629, 1977.

[286] G. F. Gause. *The Struggle for Existence*. Williams and Wilkins, Baltimore, MD, 1934.

[287] Gerald L. Geison. *The Private Science of Louis Pasteur*. Princeton University Press, Princeton, NJ, 1995.

[288] Murray Gell-Mann. *The Quark and the Jaguar: Adventures in the Simple and the Complex*. Freeman, New York, 1994.

[289] Murray Gell-Mann and Seth Lloyd. Information measures, effective complexity, and total information. *Complexity*, 2:44–52, 1996.

[290] Murray Gell-Mann and Francis Low. Quantum electrodynamics at small distances. *Phys. Rev.*, 95:1300–1312, 1954.

[291] Andrew Gelman and Cosma Rohilla Shalizi. Philosophy and the practice of Bayesian statistics. *Br. J. Math. Stat. Psychol.*, 66:8–38, 2013. arXiv:1006.3868.

[292] Anja C. Gemperli, Peter Dimroth, and Julia Steuber. Sodium ion cycling mediates energy coupling between complex I and ATP synthase. *Proc. Natl. Acad. Sci. USA*, 100:839–844, 2003.

[293] Howard Georgi. *Lie Algebras in Particle Physics*. Perseus, New York, second edition, 1999.

[294] M. M. Georgiadis, H. Komiya, P. Chakrabarti, D. Woo, J. J. Kornuc, and D. C. Rees. Crystallographic structure of the nitrogenase iron protein from *Azotobacter vinelandii*. *Science*, 257(5077):1653–1659, 1992.

[295] John Gerhart and Marc Kirschner. *Cells, Embryos, and Evolution*. Wiley, New York, 1997.

[296] John Gerhart and Marc Kirschner. The theory of facilitated variation. *Proc. Natl. Acad. Sci. USA*, 104:8582–8589, 2007.

[297] Raymond F. Gesteland, Thomas R. Cech, and John F. Atkins, editors. *The RNA World*. Cold Spring Harbor Laboratory Press, Cold Spring Harbor, NY, 2006.

[298] Wieland Gevers, Horst Kleinkauf, and Fritz Lipmann. Peptidyl transfers in gramicidin S biosynthesis from enzyme-bound thioester intermediates. *Proc. Natl. Acad. Sci. USA*, 63:1335–1342, 1969.

[299] Wieland Gevers, Horst Kleinkauf, and Fritz Lipmann. Erratum: Peptidyl transfers in gramicidin S biosynthesis from enzyme-bound thioester intermediates. *Proc. Natl. Acad. Sci. USA*, 65:249, 1970.

[300] Arun Ghosh and Marvin J. Miller. Synthesis of novel citrate-based siderophores and siderophore-$\beta$-lactam conjugates. Iron transport-mediated drug delivery systems. *J. Org. Chem.*, 58:7652–7659, 1993.

[301] Kingshuk Ghosh, Ken A. Dill, Mandar M. Inamdar, Effrosyni Seitaridou, and Rob Phillips. Teaching the principles of statistical dynamics. *Am. J. Phys.*, 74:123–133, 2006.

[302] Walter Gilbert. The RNA world. *Nature*, 319:618, 1986.

[303] P. Glansdorff and I. Prigogine. *Thermodynamic Theory of Structure, Stability, and Fluctuations*. Wiley, New York, 1971.

[304] S. Glasstone, K. J. Laidler, and H. Eyring. *The Theory of Rate Processes*. McGraw Hill, New York, 1941.

[305] K. Glazyrin, T. Boffa Ballaran, D. J. Frost, C. McCammon, A. Kantor, M. Merlini, M. Hanfland, and L. Dubrovinsky. Magnesium silicate perovskite and effect of iron oxidation state on its bulk sound velocity at the conditions of the lower mantle. *Earth Planet. Sci. Lett.*, 393:182–186, 2014.

[306] J. Peter Gogarten, W. Ford Doolittle, and Jeffrey G. Lawrence. Prokaryotic evolution in light of gene transfer. *Mol. Biol. Evol.*, 19:2226–2238, 2002.

[307] Nigel Goldenfeld. *Lectures on Phase Transitions and the Renormalization Group*. Westview Press, Boulder, CO, 1992.

[308] Nigel Goldenfeld and Carl Woese. Life is physics: evolution as a collective phenomenon far from equilibrium. *Annu. Rev. Condens. Matter Phys.*, 2:375–399, 2011.

[309] Herbert Goldstein, Charles P. Poole, and John L. Safko. *Classical Mechanics*. Addison Wesley, New York, third edition, 2001.

[310] Benjamin Gompertz. On the nature of the function expressive of the law of human mortality, and on a new mode of determining the value of life contingencies. *Philos. Trans. R. Soc. London*, 115:513–585, 1825.

[311] Stephen Jay Gould. *Wonderful Life*. Norton, New York, 1989.

[312] Stephen Jay Gould. *The Structure of Evolutionary Theory*. Harvard University Press, Cambridge, MA, 2002.

[313] Stephen Jay Gould and Elisabeth S. Vrba. Exaptation – a missing term in the science of form. *Paleobiology*, 8:4–15, 1982.

[314] Harry B. Gray. *Chemical Bonds: An Introduction to Atomic and Molecular Structure*. University Science Press, Sausalito, CA, 1994.

[315] D. C. Grenoble, M. M. Estadt, and D. F. Ollis. The chemistry and catalysis of the water gas shift reaction. *J. Catal.*, 67:90–102, 1981.

[316] Laura L. Grochowski, Huimin Xu, Kabo Leung, and Robert H. White. Characterization of an $Fe^{2+}$-dependent archaeal-specific GTP cyclohydrolase, MptA, from *Methanocaldococcus jannaschii*. *Biochemistry*, 46:6658–6667, 2007.

[317] Megan J. Gruer, Peter J. Artymiuk, and John R. Guest. The aconitase family: three structural variations on a common theme. *Trends Biochem. Sci.*, 22:3–6, 1997.

[318] C. Guerrier-Takada, K. Gardiner, T. Marsh, N. Pace, and S. Altman. The RNA moiety of ribonuclease P is the catalytic subunit of the enzyme. *Cell*, 35:849–857, 1983.

[319] Victor Guillemin and Alan Pollack. *Differential Topology*. Prentice Hall, New York, 1974.

[320] Marianne Guiral, Laurence Prunetti, Clément Aussignargues, Alexandre Ciaccafava, Pascale Infossi, Marianne Llbert, Elisabeth Lojou, and Marie-Thérèse Giudici-Orticoni. The hyperthermophilic bacterium *Aquifex aeolicus*: from respiratory pathways to extremely resistant enzymes and biotechnological applications. *Adv. Microb. Physiol.*, 61:125–194, 2012.

[321] Marianne Guiral, Pascale Tron, Corinne Aubert, Alexandre Gloter, Chantal Iobbi-Nivol, and Marie-Thérès Giuici-Orticoni. A membrane-bound multienzyme, hydrogen-oxidizing, and sulfur-reducing complex from the hyperthermophilic bacterium *Aquifex aeolicus*. *J. Biol. Chem.*, 280:42004–42015, 2005.

[322]  Addison Gulick. Phosphorus as a factor in the origin of life. *Am. Sci.*, 43:479–489, 1955.

[323]  Alex Gutteridge and Janet M. Thornton. Understanding nature's catalytic toolkit. *Trends Biochem. Sci.*, 30:622–629, 2005.

[324]  Louis Guttman. The basis for scalogram analysis. In Samuel A. Stouffer, Louis Guttman, Edward A. Suchman, Paul F. Lazarsfeld, Shirley A. Star, and John A. Clausen, editors, *Studies in Social Psychology in World War II Volume IV: Measurement and Prediction*, pages 60–90. Wiley, New York, 1950.

[325]  Marcelo I. Guzman and Scot T. Martin. Photo-production of lactate from glyoxylate: how minerals can facilitate energy storage in a prebiotic world. *Chem. Commun.*, 46:2265–2267, 2010.

[326]  Ernst Haeckel. *Generelle Morphologie der Organismen. Allgemeine Grundzüge der organischen Form-Wissenschaft, Mechanisch begründet durch die von Charles Darwin reformirte Descendenz-Theorie*. G. Reimer, Berlin, 1866.

[327]  David Haig and Laurence D. Hurst. A quantitative measure of error minimization in the genetic code. *J. Mol. Evol.*, 33:412–417, 1991.

[328]  J. B. S. Haldane. The origin of life. *Rationalist Animal*, page 148, 1929. Reprinted as [329].

[329]  J. B. S. Haldane. The origin of life. In J. D. Bernal, editor, *The Origin of Life*, pages 242–249. Weidenfeld and Nicolson, London, 1967.

[330]  Carl H. Hamann, Andrew Hamnett, and Wolf Vielstich. *Electrochemistry*. Wiley, New York, completely revised and updated edition, 2007.

[331]  Keiko Hamano, Yutaka Abe, and Hidenori Genda. Emergence of two types of terrestrial planet on solidification of magma ocean. *Nature*, 497:607–610, 2013.

[332]  William D. Hamilton. The genetical evolution of social behavior. I. *J. Theor. Biol.*, 7:1–16, 1964.

[333]  William D. Hamilton. The genetical evolution of social behavior. II. *J. Theor. Biol.*, 7:17–52, 1964.

[334]  William D. Hamilton. Selfish and spiteful behavior in an evolutionary model. *Nature*, 228:1218–1220, 1970.

[335]  Thomas Handorf, Oliver Ebenhöh, and Reinhart Heinrich. Expanding metabolic networks: scopes of compounds, robustness, and evolution. *J. Mol. Evol.*, 61:498–512, 2005.

[336]  Chris Hankin. *Lambda Calculi: A Guide for Computer Scientists*. Oxford University Press, New York, 1994.

[337]  Ajith Harish and Gustavo Caetano-Anollés. Ribosomal history reveals origins of modern protein synthesis. *PLoS ONE*, 7:e32776, 2012.

[338]  G. F. Hatfull and W. R. Jacobs Jr., editors. *Molecular Genetics of Mycobacteria*, ASM Press, Washington, DC, 2000.

[339]  Shuhei Hattori, Johan A. Schmidt, Matthew S. Johnson, Sebastian O. Danielache, Akinori Yamada, Yuichiro Ueno, and Naohiro Yoshida. $SO_2$ photoexcitation mechanism links mass-independent sulfur isotopic fractionation in cryospheric sulfate to climate impacting volcanism. *Proc. Natl. Acad. Sci. USA*, 110:17656–17661, 2013.

[340]  Mary E. Hawkins, Wolfgang Pfleiderer, Oliver Jungmann, and Frank M. Balis. Synthesis and fluorescence characterization of pteridine adenosine nucleoside analogs for DNA incorporation. *Anal. Biochem.*, 298:231–240, 2001.

[341]  Robert M. Hazen. Chiral crystal faces of common rock-forming minerals. In G. Palyi, C. Zucchi, and L. Cagglioti, editors, *Progress in Biological Chirality*, Chapter 11, pages 137–151. Elsevier, New York, 2004.

[342] Robert M. Hazen. Mineral surfaces and the prebiotic selection and organization of biomolecules (Presidential address to the Mineralogical Society of America). *Am. Mineral.*, 91:1715–1729, 2006.

[343] Robert M. Hazen. Paleomineralogy of the Hadean eon: a preliminary species list. *Am. J. Sci.*, 313:807–843, 2006.

[344] Robert M. Hazen and David W. Deamer. Hydrothermal reactions of pyruvic acid: synthesis, selection, and self-assembly of amphiphilic molecules. *Orig. Life Evol. Biosphere*, 37:143–152, 2007.

[345] Robert M. Hazen and John M. Ferry. Mineral evolution: mineralogy in the fourth dimension. *Elements*, 6:9–12, 2010.

[346] Robert M. Hazen and Dimitri A. Sverjensky. Mineral surfaces, geochemical complexities, and the origins of life. *Cold Spring Harb. Perspect. Biol.*, 2:a002162, 2010.

[347] Robert M. Hazen, Edward S. Grew, Robert T. Downs, Joshua Golden, and Grethe Hystad. Mineral ecology: chance and necessity in the mineral diversity of terrestrial planets. *Can. Mineral.*, in press, 2015.

[348] Robert M. Hazen, Dominic Papineau, Wouter Bleeker, Robert T. Downs, John M. Ferry, Timothy J. McCoy, Dimitri A. Sverjensky, and Heixiong Yang. Mineral evolution. *Am. Mineral.*, 93:1693–1720, 2008.

[349] Steffen Heim, Andreas Künkel, Rudolf K. Thauer, and Reiner Hedderich. Thiol:fumarate reductase (Tfr) from *Methanobacterium thermoautotrophicum* identification of the catalytic sites for fumarate reduction and thiol oxidation. *Eur. J. Biochem.*, 253:292–299, 1998.

[350] Wolfgang Heinen and Anne Marie Lauwers. Organic sulfur compounds resulting from the interaction of iron sulfide, hydrogen sulfide and carbon dioxide in an anaerobic aqueous environment. *Orig. Life Evol. Biosphere*, 26:131–150, 1996.

[351] Bettina Heinz, Walter Ried, and Klaus Dose. Thermal generation of pteridines and flavines from amino acid mixtures. *Angew. Chem. Int. Ed. Engl.*, 8:478–483, 1979.

[352] Eric Herbst. Chemistry of star-forming regions. *J. Phys. Chem.*, 109:4017–4029, 2005.

[353] Gloria Herrmann, Elamparithi Jayamani, Galina Mai, and Wolfgang Buckel. Energy conservation via electron-transferring flavoprotein in anaerobic bacteria. *J. Bacteriol.*, 190:784–791, 2008.

[354] Matthew D. Herron, Armin Rashidi, Deborah E. Shelton, and William W. Driscoll. Cellular differentiation and individuality in the 'minor' multicellular taxa. *Biol. Rev.*, 88:844–861, 2013.

[355] Takeru Higuchi, Lennart Eberson, and Allen K. Herd. The intramolecular facilitated hydrolytic rates of methyl-substituted succinanilic acids. *J. Am. Chem. Soc.*, 88:3805–3808, 1966.

[356] A. Hill and L. E. Orgel. Synthesis of adenine from HCN tetramer and ammonium formate. *Orig. Life Evol. Biosphere*, 32:99–102, 2002.

[357] T. Archer Hirst, editor. *The Mechanical Theory of Heat*. John van Voorst, London, 1865.

[358] Martin F. Hohmann-Marriott and Robert E. Blankenship. Evolution of photosynthesis. *Annu. Rev. Plant Biol.*, 62:515–548, 2011.

[359] James F. Holden and Roy M. Daniel. The upper temperature limit for life based on hyperthermophile culture experiments and field observations. In William S. D. Wilcock, Edward F. DeLong, Deborah S. Kelley, John A. Baross, and S. Craig Cary,

editors, *The Subseafloor Biosphere at Mid-Ocean Ridges*, pages 13–24. American Geophysical Soceity, vol. 144, Washington DC, 2004.

[360] James F. Holden, Melanie Summit, and John A. Baross. Thermophilic and hyper-thermophilic microorganisms in 3–30 °C hydrothermal fluids following a deep-sea volcanic eruption. *FEMS Microbiol. Ecol.*, 25:33–41, 1998.

[361] Nils G. Holm. Glasses as sources of condensed phosphates on the early earth. *Geochem. Trans.*, 15:8, 2014.

[362] W. H. Holmes, I. D. Hamilton, and A. G. Robertson. The rate of turnover of the adenosine triphosphate pool of *Escherichia coli* growing aerobically in simple defined media. *Arch. Microbiol.*, 83:95–109, 1972.

[363] Helge Holo. *Chloroflexus aurantiacus* secretes 3-hydroxypropionate, a possible intermediate in the assimilation of $CO_2$ and acetate. *Arch. Microbiol.*, 151:252–256, 1989.

[364] H. Holo and D. Grace. Polyglucose synthesis in *Chloroflexus aurantiacus* studied by $^{13}$C-NMR. Evidence for acetate metabolism by a new metabolic pathway in autotrophically grown cells. *Arch. Microbiol.*, 148:292–297, 1987.

[365] Michelle D. Hopkins, T. Mark Harrison, and Craig E. Manning. Constraints on Hadean geodynamics from mineral inclusions in >4 Ga zircons. *Earth Planet. Sci. Lett.*, 298:367–376, 2010.

[366] Wim Hordijk and Mike Steel. Detecting autocatalytic, self-sustaining sets in chemical reaction systems. *J. Theor. Biol.*, 227:451–461, 2004.

[367] Wim Hordijk, Stuart A. Kauffman, and Mike Steel. Required levels of catalysis for emergence of autocatalytic sets in models of chemical reaction systems. *Int. J. Mol. Sci.*, 12:3085–3101, 2011.

[368] Wim Hordijk, Mike Steel, and Stuart Kauffman. The structure of autocatalytic sets: evolvability, enablement, and emergence. *Acta Biotheor.*, 60:379–392, 2012. arXiv:1205.0584v2.

[369] Juske Horita and Michael E. Berndt. Abiogenic methane formation and isotopic fractionation under hydrothermal conditions. *Science*, 285:1055–1057, 1999.

[370] N. H. Horowitz. On the evolution of biochemical synthesis. *Proc. Natl. Acad. Sci. USA*, 31:153–157, 1945.

[371] Chiaolong Hsiao, I.-Chun Chou, C. Denise Okafor, Jessica C. Bowman, Eric B. O'Neill, Shreyas S. Athavale, Anton S. Petrov, Nicholas V. Hud, Roger M. Wartell, Stephen C. Harvey, and Loren Dean Williams. Iron(II) plus RNA can catalyze electron transfer. *Nature Chem.*, 5:525–528, 2013.

[372] Chiaolong Hsiao, Srividya Mohan, Benson K. Kalahar, and Loren Dean Williams. Peeling the onion: establishing a chronology of early ribosome evolution. *Mol. Biol. Evol.*, 26:2415–2425, 2009.

[373] Kerson Huang. *Statistical Mechanics*. Wiley, New York, 1987.

[374] Teng Huang and Verne Schirch. Mechanism for the coupling of ATP hydrolysis to the conversion of 5-formyltetrahydrofolate to 5,10-methenyltetrahydrofolate. *J. Biol. Chem.*, 270(38):22296–22300, 1995.

[375] Wenhua Huang and James P. Ferris. Synthesis of 35–40mers of RNA oligomers from unblocked monomers. A simple approach to the RNA world. *Chem. Commun.*, 2003:1458–1459, 2003.

[376] Zhimin Huang, Liang Zhu, Yan Cao, Geng Wu, Xinyi Liu, Yingyi Chen, Qi Wang, Ting Shi, Yaxue Zhao, Yuefei Wang, Weihua Li, Yixue Li, Haifeng Chen, Chen Guoqiang, and Jian Zhang. ASD: a comprehensive database of allosteric proteins and modulators. *Nucleic Acids Res.*, 39:D663–D669, 2011.

[377] Claudia Huber and Günter Wächtershäuser. Activated acetic acid by carbon fixation on (Fe,Ni)S under primordial conditions. *Science*, 276:245–247, 1997.

[378] Claudia Huber and Günter Wächtershäuser. Primordial reductive amination revisited. *Tetrahedron Lett.*, 44:1695–1697, 2003.

[379] Harald Huber, Martin Gallenberger, Ulrike Jahn, Eva Eylert, Ivan A. Berg, Daniel Kockelkorn, Wolfgang Eisenreich, and Georg Fuchs. A dicarboxylate/4-hydroxybutyrate autotrophic carbon assimilation cycle in the hyperthermophilic archaeum *Ignicoccus hospitalis*. *Proc. Natl. Acad. Sci. USA*, 105:7851–7856, 2008.

[380] Philip Hugenholtz, Brett M. Goebel, and Norman R. Pace. Impact of culture-independent studies on the emerging phylogenetic view of bacterial diversity. *J. Bacteriol.*, 180:4765–4774, 1998.

[381] Michael Hügler and S. M. Seivert. Beyond the Calvin cycle: autotrophic carbon fixation in the ocean. *Annu. Rev. Marine Sci.*, 3:261–289, 2011.

[382] Michael Hügler, Harald Huber, Stephen J. Molyneaux, Costantino Vetriani, and Stefan M. Seivert. Autotrophic $CO_2$ fixation via the reductive tricarboxylic acid cycle in different lineages within the phylum *Aquificae*: evidence for two ways of citrate cleavage. *Env. Microbiol.*, 9:81–92, 2007.

[383] J. R. Hulston and H. G. Thode. Variations in the $S^{33}$, $S^{34}$, and $S^{33}$ contents of meteorites and their relation to chemical and nuclear effects. *J. Geophys. Res.*, 70:3475–3484, 1965.

[384] Donald M. Hunten. Thermal and non-thermal escape mechanisms for terrestrial bodies. *Planet. Space Sci.*, 30:773, 1982.

[385] James Hury, Uma Nagaswamy, Maia Larios-Sanz, and George E. Fox. Ribosome origins: the relative age of 23S rRNA domains. *Orig. Life Evol. Biosphere*, 36:421–429, 2006.

[386] Aldous Huxley. *Literature and Science*. Ox Bow Press, Woodbridge, CT, 1991.

[387] Abdul-Aziz Ingar, Richard W. A. Luke, Barry R. Hayter, and John D. Sutherland. Synthesis of cytidine ribonucleotides by stepwise assembly of the heterocycle on a sugar phosphate. *ChemBioChem*, 4:504–507, 2003.

[388] Naoki Irie and Shigeru Kuratani. Comparative transcriptome analysis reveals vertebrate phylotipic period during organogenesis. *Nature Commun.*, 2:248, 2011.

[389] F. Jacob and J. Monod. Genetic regulatory mechanisms in the synthesis of proteins. *J. Mol. Biol.*, 3:318–356, 1961.

[390] Kenneth D. James and Andrew D. Ellington. The fidelity of template-directed oligonucleotide ligation and the inevitability of polymerase function. *Orig. Life Evol. Biosphere*, 29:375–390, 1999.

[391] E. T. Jaynes. Information theory and statistical mechanics. *Phys. Rev.*, 106:620–630, 1957. Reprinted in [680].

[392] E. T. Jaynes. Information theory and statistical mechanics. II. *Phys. Rev.*, 108:171–190, 1957. Reprinted in [680].

[393] E. T. Jaynes. The minimum entropy production principle. *Annu. Rev. Phys. Chem.*, 31:579–601, 1980.

[394] E. T. Jaynes. *Probability Theory: The Logic of Science*. Cambridge University Press, New York, 2003.

[395] James Jeans. *Dynamical Theory of Gases*. Cambridge University Press, London, fourth edition, 2009. Original edition, 1916.

[396] Harold Jeffreys. An invariant form for the prior probability in estimation problems. *Proc. R. Soc. London, Ser. A*, 186:453–461, 1946.

[397] Harold Jeffreys. *Scientific Inference*. Cambridge University Press, London, second edition, 1957.

[398] W. J. Jenkins, J. M. Edmond, and J. B. Corliss. Excess $^3$He and $^4$He in Galapagos submarine hydrothermal waters. *Nature*, 272:156–158, 1978.

[399] Roy A. Jensen. Enzyme recruitment in evolution of new function. *Annu. Rev. Microbiol.*, 30:409–425, 1976.

[400] Ying Ji and Henri-Claude Nataf. Detection of mantle plumes in the lower mantle by diffraction tomography: Hawaii. *Earth Planet. Sci. Lett.*, 159:99–115, 1998.

[401] Haixia Jia, Cris Moore, and Bart Selman. From spin glasses to hard satisfiable formulas. In Holger H. Hoos and David G. Mitchell, editors, *Theory and Applications of Satisfiability Testing, Lecture Notes in Computer Science*, volume 3542, pages 199–210. Springer, New York, 2005.

[402] Fatima D. Jones and Scott A. Strobel. Ionization of a critical adenosine residue in the *Neurospora* Varkud satellite ribozyme active site. *Biochem.*, 42:4265–4276, 2003.

[403] Gerald F. Joyce. Foreword. In David W. Deamer and Gail R. Fleischaker, editors, *Origins of Life: The Central Concepts*, pages xi–xii. Jones and Bartlett, Boston, MA, 1994.

[404] Martin Jung. *Polymerisation in Bilayers*. PhD Thesis, Technische Universiteit Eindhoven, 2000.

[405] Christopher T. Jurgenson, Tadhg P. Begley, and Steven E. Ealick. The structural and biochemical foundations of thiamin biosynthesis. *Annu. Rev. Biochem.*, 78(1):569–603, 2009.

[406] Ville R. I. Kaila, Michael I. Verkhovsky, and Mårten Wikström. Proton-coupled electron transfer in cytochrome oxidase. *Chem. Rev.*, 110:7062–7081, 2010.

[407] Roland G. Kallen and William P. Jencks. The dissociation constants of tetrahydrofolic acid. *J. Biol. Chem.*, 241:5845–5850, 1966.

[408] Alex Kamenev. Keldysh and Doi–Peliti techniques for out-of-equilibrium systems. In I. V. Lerner, B. L. Althsuler, V. I. Fal'ko, and T. Giamarchi, editors, *Strongly Correlated Fermions and Bosons in Low-Dimensional Disordered Systems*, pages 313–340. Springer-Verlag, Heidelberg, 2002.

[409] Masafumi Kameya, Hiroyuki Arai, Masaharu Ishii, and Yasuo Igarashi. Purification of three aminotransferases from *Hydrogenobacter thermophilus* TK-6 – novel types of alanine or glycine aminotransferase: enzymes and catalysis. *FEBS J.*, 277:1876–1885, 2010. PMID: 20214682.

[410] Anastassia Kanavarioti, Pierre-Alain Monnard, and David W. Deamer. Eutectic phases in ice facilitate nonenzymatic nucleic acid synthesis. *Astrobiology*, 1:271–281, 2001.

[411] M. Kanehisa. The KEGG database. *Novartis Found. Symp.*, 247:91–101, 2002. www.genome.ad.jp/kegg/.

[412] Peter D. Karp, Monica Riley, Suzanne M. Paley, and Alida Pellegrini-Toole. The MetaCyc database. *Nucleic Acids Res.*, 30:59–61, 2002. http://ecocyc.org/ecocyc/metacyc.html.

[413] L. Karp-Boss and P. A. Jumars. Nutrient fluxes to planktonic osmotrophs in the presence of fluid motion. *Oceanogr. Marine Biol. Annu. Rev.*, 34:71–107, 1996.

[414] Lee Karp-Boss and Peter A. Jumars. Motion of diatom chains in a steady shear flow. *Limnol. Oceanogr.*, 43:1767–1773, 1998.

[415] Anne-Kristin Kaster, Johanna Moll, Kristian Parey, and Rudolf K. Thauer. Coupling of ferredoxin and heterodisulfide reduction via electron bifurcation in

hydrogenotrophic methanogenic archaea. *Proc. Natl. Acad. Sci. USA*, 108:2981–2986, 2012.

[416] J. F. Kasting and T. P. Ackerman. Climatic consequences of very high $CO_2$ levels in the early Earth's atmosphere. *Science*, 234:1383–1385, 1986.

[417] James F. Kasting and David Catling. Evolution of a habitable planet. *Annu. Rev. Astron. Astrophys.*, 41:429–463, 2003.

[418] James F. Kasting and M. Tazewell Howard. Atmosphere composition and climate on the early Earth. *Philos. Trans. R. Soc. London, Ser. B*, 361:1733–1742, 2006.

[419] James F. Kasting and James B. Pollack. Loss of water from Venus. I Hydrodynamic escape of hydrogen. *Icarus*, 53:479–508, 1983.

[420] J. F. Kasting, K. J. Zahnle, and J. C. G. Walker. Photochemistry of methane in the Earth's early atmosphere. *Precambrian Res.*, 20:121–148, 1983.

[421] Richard F. Katz, Marc Spiegelman, and Charles H. Langmuir. A new parameterization of hydrous mantle melting. *Geochem. Geophys. Geosyst.*, 4:1073, 2003. doi:10.1029/2002GC000433.

[422] Stuart A. Kauffman. Autocatalytic sets of proteins. *J. Theor. Biol.*, 119:1–24, 1986.

[423] Alan J. Kaufman, David T. Johnston, James Farquhar, Andrew L. Masterson, Timothy W. Lyons, Steve Bates, Ariel D. Anbar, Gail L. Arnold, Jessica Garvin, and Roger Buick. Late Archean biospheric oxygenation and atmospheric evolution. *Science*, 317:1900–1903, 2007.

[424] Jonathan Z. Kaye and John A. Baross. High incidence of halotolerant bacteria in Pacific hydrothermal vent and pelagic environments. *FEMS Microbiol. Ecol.*, 32:429–460, 2000.

[425] Patrick J. Keeling and Jeffrey D. Palmer. Horizontal gene transfer in eukaryotic evolution. *Nature Rev. Genet.*, 9:605–628, 2008.

[426] Deborah S. Kelley. From the mantle to microbes: the Lost City hydrothermal field. *Oceanography*, 18:32–45, 2005.

[427] Deborah S. Kelley, John A. Baross, and John R. Delaney. Volcanoes, fluids, and life at mid-ocean ridge spreading centers. *Annu. Rev. Earth Planet. Sci.*, 30:385–491, 2002.

[428] Deborah S. Kelley, Jeffrey A. Karson, Donna K. Blackman, Gretchen L. Früh-Green, David A Butterfield, Marvin D. Lilley, Eric J. Olson, Matthew O. Schrenk, Kevin K. Roe, Geoff T. Lebon, Pete Rivizzigno, and the AT3-60 Shipboard Party. An off-axis hydrothermal vent field near the Mid-Atlantic Ridge at 30° N. *Nature*, 412:145–149, 2001.

[429] Deborah S. Kelley, Jeffrey A. Karson, Gretchen L. Früh-Green, Dana R. Yoerger, Timothy M. Shank, David A. Butterfield, John M. Hayes, Matthew O. Schrenk, Eric J. Olson, Giora Proskurowski, Mike Jakuba, Al Bradley, Ben Larson, Kristin Ludwig, Deborah Glickson, Kate Buckman, Alexander S. Bradley, William J. Brazelton, Kevin Roe, Mitch J. Elend, Ad'elie Delacour, Stefano M. Bernasconi, Marvin D. Lilley, John A. Baross, Roger E. Summons, and Sean P. Sylva. A serpentinite-hosted ecosystem: the Lost City hydrothermal field. *Science*, 307:1428–1434, 2005.

[430] J. M. Keynes. *A Treatise on Probability*. MacMillan, London, 1921.

[431] John Maynard Keynes. *The Collected Writings of John Maynard Keynes:* Volume 4, *A Tract on Monetary Reform*. Cambridge University Press, London, 2012.

[432] P. Kharecha, J. Kasting, and J. Seifert. A coupled atmosphere-ecosystem model of the early Archean Earth. *Geobiology*, 3:53–76, 2005.

[433] Olga Khersonsky and Dan S. Tawfik. Enzyme promiscuity: a mechanistic and evolutionary perspective. *Annu. Rev. Biochem.*, 79:471–505, 2010.

[434] Olga Khersonsky, Sergey Malitsky, Ilana Rogachev, and Dan S. Tawfik. Role of chemistry versus substrate binding in recruiting promiscuous enzyme functions. *Biochemistry*, 50:2683–2690, 2011.

[435] Goro Kikuchi. The glycine cleavage system: composition, reaction mechanism, and physiological significance. *Mol. Cell. Biochem.*, 1:169–187, 1973.

[436] J. Dongun Kim, Augustina Rodriguez-Granillo, David A. Case, Vikas Nanda, and Paul G. Falkowski. Energetic selection of topology in ferredoxins. *PLoS Comp. Biol.*, 8:e1002463, 2012.

[437] Jonsun Kim and D. C. Rees. Structural models for the metal centers in the nitrogenase molybdenum-iron protein. *Science*, 257(5077):1677–1682, 1992.

[438] Juhan Kim, Jamie P. Kershner, Yehor Novikov, Richard K. Shoemaker, and Shelley D. Copley. Three serendipitous pathways in *E. coli* can bypass a block in pyridoxal-5′-phosphate synthesis. *Mol. Syst. Biol.*, 6:436:1–13, 2010.

[439] Mark Kirkpatrick, Toby Johnson, and Nick Barton. General models of multilocus evolution. *Genetics*, 161:1727–1750, 2002.

[440] Marc Kirschner and John Gerhart. Evolvability. *Proc. Natl. Acad. Sci. USA*, 95(15):8420–8427, 1998.

[441] Charles Kittel and Herbert Kroemer. *Thermal Physics*. Freeman, New York, second edition, 1980.

[442] Robin D. Knight, Steven J. Freeland, and Laura F. Landweber. Selection, history and chemistry: the three faces of the genetic code. *Trends Biochem. Sci.*, 24:241–247, 1999.

[443] Robin D. Knight, Steven J. Freeland, and Laura F. Landweber. Rewiring the keyboard: evolvability of the genetic code. *Nature. Rev. Genet.*, 2:49–58, 2001.

[444] Andrew H. Knoll. *Life on a Young Planet*. Princeton University Press, Princeton, NJ, 2003.

[445] Martin Kochmański, Tadeusz Paszkiewicz, and Sławomir Wolski. Curie–Weiss magnet – a simple model of phase transition. *Eur. J. Phys.*, 34:1555–1573, 2013.

[446] John B. Kogut and Mikhail A. Stephanov. *The Phases of Quantum Chromodynamics: From Confinement to Extreme Environments*. Cambridge University Press, Cambridge, 2004.

[447] A. N. Kolmogorov. New metric invariant of transitive dynamical systems and endomorphisms of Lebesgue spaces. *Dokl. Akad. Nauk SSSR*, 119:861–864, 1958.

[448] A. N. Kolmogorov. Entropy per unit time as a metric invariant of automorphism. *Dokl. Akad. Nauk SSSR*, 124:754–755, 1959.

[449] Andrey Kolmogorov. On tables of random numbers. *Sankhyā Ser. A*, 25:369–375, 1963. Reprinted in [450].

[450] Andrey Kolmogorov. On tables of random numbers. *Theor. Comput. Sci.*, 207:387–395, 1998.

[451] Dilip Kondepudi and Ilya Prigogine. *Modern Thermodynamics: From Heat Engines to Dissipative Structures*. Wiley, New York, 1998.

[452] Eugene V. Koonin and William Martin. On the origin of genomes and cells within inorganic compartments. *Trends Genet.*, 21:647–654, 2005.

[453] Jun Korenaga. Initiation and evolution of plate tectonics on Earth: theories and observations. *Annu. Rev. Earth Planet. Sci.*, 41:117–151, 2013.

[454] Arthur Kornberg, Narayana N. Rao, and Dana Ault-Riché. Inorganic polyphosphate: a molecule of many functions. *Annu. Rev. Biochem.*, 68:89–125, 1999.

[455] David C. Krakauer and Joshua B. Plotkin. Redundancy, antiredundancy, and the robustness of genomes. *Proc. Natl. Acad. Sci. USA*, 99:1405–1409, 2002.

[456] H. A. Krebs and W. A. Johnson. The role of citric acid in intermediate metabolism in animal tissues. *Enzymologia*, 4:148–156, 1937.

[457] Supriya Krishnamurthy, Eric Smith, David C. Krakauer, and Walter Fontana. The stochastic behavior of a molecular switching circuit with feedback. *Biol. Direct*, 2:13, 2007. PMID: 17540019.

[458] K. Kruger, P. J. Grabowski, A. J. Zaug, J. Sands, D. E. Gottschling, and T. R. Cech. Self-splicing RNA: autoexcision and autocyclization of the ribosomal RNA intervening sequence of *Tetrahymena*. *Cell*, 31:147–157, 1982.

[459] Thomas S. Kuhn. *The Structure of Scientific Revolutions*. University of Chicago Press, Chicago, IL, 1962.

[460] A. Kuki and P. G. Wolynes. Electron tunneling paths in proteins. *Science*, 236:1647–1652, 2000.

[461] I. S. Kulaev. Biochemistry of inorganic polyphosphates. *Rev. Physiol. Biochem. Pharmacol.*, 73:131–158, 1975.

[462] I. S. Kulaev, V. M. Vagabov, and T. V. Kulakovskaya. *The Biochemistry of Inorganic Polyphosphates*. Wiley, New York, second edition, 2004.

[463] Lee R. Kump and William E. Seyfried Jr. Hydrothermal Fe fluxes during the Precambrian: effect of low oceanic sulfate concentrations and low hydrostatic pressure on the composition of black smokers. *Earth Planet. Sci. Lett.*, 235:654–662, 2005.

[464] Chi-Horng Kuo and Howard Ochman. Inferring clocks when lacking rocks: the variable rates of molecular evolution in bacteria. *Biol. Direct*, 4:35, 2009.

[465] Jorge Kurchan. Six out of equilibrium lectures. In Thierry Dauxois, Stefano Ruffo, and Leticia F. Cugliandolo, editors, *Long-range interacting Systems*, Chapter 2. Oxford University Press, Oxford, 2010.

[466] Noam Lahav. *Biogenesis: Theories of Life's Origin*. Oxford University Press, London, 1999.

[467] Jean-Baptiste Lamarck. *Philosophie Zoologique ou Exposition des Considérations Relatives à l'Histoire Naturelle des Animaux*. Cambridge University Press, London, 2011. Original edition, 1809.

[468] L. D. Landau. Theory of phase transformations. *Zh. Eksp. Teor. Fiz.*, 7:19–32, 1937.

[469] R. Landauer. Irreversibility and heat generation in the computing process. *IBM J. Res. Dev.*, 3:183–191, 1961.

[470] Nick Lane. *Power, Sex, Suicide: Mitochondria and the Meaning of Life*. Oxford University Press, Oxford, 2005.

[471] Nick Lane. Why are cells powered by proton gradients? *Nature Ed.*, 3:18, 2010.

[472] Nick Lane and William Martin. The energetics of genome complexity. *Nature*, 467:929–934, 2010.

[473] Nick Lane and William F. Martin. The origin of membrane bioenergetics. *Cell*, 151:1406–1416, 2012.

[474] Nick Lane, John F. Allen, and William Martin. How did LUCA make a living? Chemiosmosis in the origin of life. *Bioessays*, 32:271–280, 2010.

[475] Nick Lane, William F. Martin, John A. Raven, and John F. Allen. Energy, genes, and evolution: introduction to an evolutionary synthesis. *Philos. Trans. R. Soc. London, Ser. B*, 368:1–5, 2013.

[476] Susan Q. Lang, David A. Butterfield, Mitch Schulte, Deborah S. Kelley, and Marvin D. Lilley. Elevated concentrations of formate, acetate, and dissolved organic

carbon found at the Lost City hydrothermal field. *Geochim. Cosmochim. Acta*, 74:941–952, 2010.

[477] Charles H. Langmuir and Donald W. Forsyth. Mantle melting beneath mid-ocean ridges. *Oceanography*, 20:78–89, 2007.

[478] Pierre Simon Laplace. Mémoire sur la probabilité des causes par les évènements. *Mém. Acad. Sci. Paris*, 6:621–656, 1774.

[479] Pierre Simon Laplace. Mémoire sur les approximations des formules qui sont fonctions de très grands nombres et sur leur application aux probabilités. *Mém. Acad. R. Sci. Paris*, année 1809:353–415, 1810.

[480] Yanm Lei, Shuang Zhang, Peng Chen, Hetao Liu, Huanhuan Yin, and Hongyu Li. Magnetotactic bacteria, magnetosomes and their application. *Microbiol. Res.*, 167:507–519, 2012.

[481] Joseph W. Lengeler, Gerhart Drews, and Hans G. Schlegel. *Biology of the Prokaryotes*. Blackwell Science, New York, 1999.

[482] Richard C. Lewontin. The units of selection. *Annu. Rev. Ecol. System.*, 1:1–18, 1970.

[483] Richard C. Lewontin. *The Genetic Basis of Evolutionary Change*. Columbia University Press, New York, 1974.

[484] Fuli Li, Julia Hinderberger, Henning Seedorf, Jin Zhang, Wolfgang Buckel, and Rudolf K. Thauer. Coupled ferredoxin and crotonyl coenzyme A (CoA) reduction with NADH catalyzed by the butyryl-CoA dehydrogenase/Etf complex from *Clostridium kluyveri*. *J. Bacteriol.*, 190:843–850, 2008.

[485] Ming Li and Paul Vitányi. *An Introduction to Kolmogorov Complexity and its Applications*. Springer, Heidelberg, third edition, 2008.

[486] Zheng-Xue Anser Li and Cin-Ty Aeolus Lee. The constancy of upper mantle $f_{O_2}$ through time inferred from V/Sc ratios in basalts. *Earth Planet. Sci. Lett.*, 228:483–493, 2004.

[487] Li-Hung Lin, Pei-Ling Wang, Douglas Rumble, Johanna Lippmann-Pipke, Erik Boice, Lisa M. Pratt, Barbara Sherwood Lollar, Eoin L. Brodie, Terry C. Hazen, Gary L. Andersen, Todd Z. DeSantis, Duane P. Moser, Dave Kershaw, and T. C. Onstott. Long-term sustainability of a high-energy, low-diversity crustal biome. *Science*, 314:479–482, 2006.

[488] Tracy A. Lincoln and Gerald F. Joyce. Self-sustained replication of an RNA enzyme. *Science*, 323:1229–1232, 2009.

[489] Andrea J. Liu and Sidney R. Nagel. Nonlinear dynamics: jamming is not just cool any more. *Nature*, 396:21–22, 1998.

[490] Yongqing Liu, Jizhong Zhou, Marina V. Omelchenko, Alex S. Beliaev, Amudhan Venkateswaran, Julia Stair, Liyou Wu, Dorothea K. Thompson, Dong Xu, Igor B. Rogozin, Elena K. Gaidamakova, Min Zhai, Kira S. Makarova, Eugene V. Koonin, and Michael J. Daly. Transcriptome dynamics of *Deinococcus radiodurans* recovering from ionizing radiation. *Proc. Natl. Acad. Sci. USA*, 100:4191–4196, 2003.

[491] L. Ljungdahl and H. G. Wood. Incorporation of $C^{14}$ from carbon dioxide into sugar phosphates, carboxylic acids, and amino acids by *Clostridium thermoaceticum*. *J. Bacteriol.*, 89:1055–1064, 1965.

[492] L. Ljungdahl, E. Irion, and H. G. Wood. Total synthesis of acetate from $CO_2$. I. CO-methylcobyric acid and CO-(methyl)-5-methoxybenzimidazolylcobamide as intermediates with *Clostridium thermoaceticum*. *Biochemistry*, 4:2771–2780, 1965.

[493] Elisabeth A. Lloyd. *The Structure and Confirmation of Evolutionary Theory.* Princeton University Press, Princeton, NJ, 1994.

[494] S. J. Lloyd, H. Lauble, G. S. Prasad, and C. D. Stout. The mechanism of aconitase: 1.8 Å resolution crystal structure of the S642A:citrate complex. *Protein Sci.*, 8:2655–2662, 1999.

[495] Jonathan Lombard and David Moreira. Early evolution of the biotin-dependent carboxylase family. *BMC Evol. Biol.*, 11:232:1–22, 2011.

[496] Purificación López-García, David Moreira, and Juli Pereto. Origin and evolution of compartments. In Muriel Gargaud, Phillippe Claeys, Purificación López-García, Hervé Martin, Thierry Montmerle, Robert Pascal, and Jacques Reisse, editors, *From Suns to Life: a Chronological Approach to the History of Life on Earth*, pages 171–174. Springer, Dordrecht, 2006.

[497] James Lovelock. *Gaia: A New Look at Life on Earth.* Oxford University Press, London, 2000.

[498] Donald R. Lowe and Michael M. Tice. Geologic evidence for Archean atmospheric and climatic $CO_2$, $CH_4$, and $O_2$ with an overriding tectonic control. *Geology*, 36:493–496, 2004.

[499] Robert P. Lowell, Peter A. Rona, and Richard P. Von Herzen. Seafloor hydrothermal systems. *J. Geophys. Res.*, 100:327–352, 1995.

[500] Kristin A. Ludwig, Chuan-Chou Shen, Deborah S. Kelley, Hai Cheng, and R. Lawrence Edwards. U–Th systematics and $^{230}$Th ages of carbonate chimneys at the Lost City hydrothermal field. *Geochim. Cosmochim. Acta*, 75:1869–1888, 2011.

[501] Pier Luigi Luisi. *The Emergence of Life: From Chemical Origins to Synthetic Biology.* Cambridge University Press, London, 2006.

[502] Pier Luigi Luisi, Peter Walde, and Thomas Oberholzer. Lipid vesicles as possible intermediates in the origin of life. *Curr. Op. Colloids Interface Sci.*, 4:33–39, 1999.

[503] John E. Lupton, Edward T. Baker, and Gary J. Massoth. Helium, heat, and the generation of hydrothermal event plumes at mid-ocean ridges. *Earth Planet. Sci. Lett.*, 171:343–350, 1999.

[504] Richard A. Lutz, Timothy M. Shank, Daniel J. Fornari, Rachel M. Haymon, Marvin D. Lilley, Karen L. Von Damm, and Daniel Desbruyeres. Rapid growth at deep-sea vents. *Nature*, 371:663–664, 1994.

[505] Vittorio Luzzati and A. Tardieu. Lipid phases: structure and structural transitions. *Annu. Rev. Phys. Chem.*, 25:79–94, 1974.

[506] Shang-Keng Ma. *Modern Theory of Critical Phenomena.* Perseus, New York, 1976.

[507] Robert E. MacKenzie. Biogenesis and interconversion of substituted tetrahydrofolates. In Raymond L. Blakely and Stephen J. Benkovic, editors, *Folates and Pterins*, vol. 1: *Chemistry and Biochemistry of Folates*, pages 255–306. John Wiley & Sons, New York, 1984.

[508] David W. C. MacMillan. The advent and development of organocatalysis. *Nature*, 455:304–308, 2008.

[509] B. Edward H. Maden. Tetrahydrofolate and tetrahydromethanopterin compared: functionally distinct carriers in $C_1$ metabolism. *Biochem. J.*, 350:609–629, 2000.

[510] Robert S. Maier and D. L. Stein. Effect of focusing and caustics on exit phenomena in systems lacking detailed balance. *Phys. Rev. Lett.*, 71:1783–1786, 1993.

[511] Robert S. Maier and D. L. Stein. Oscillatory behavior of the rate of escape through an unstable limit cycle. *Phys. Rev. Lett.*, 77:4860–4863, 1996.

[512] Stephen Maitzen. Stop asking why there's anything. *Erkenntnis*, 77:51–63, 2012.

[513] Thomas Robert Malthus. *An Essay on the Principle of Population*. Cosimo, New York, 2007. Original edition, 1798.

[514] I. Mamajanov and J. Herzfeld. HCN polymers characterized by SSNMR: solid state reaction of crystalline tetramer (diaminomaleonitrile). *J. Chem. Phys.*, 130:134504, 2009.

[515] Michael L. Manapat, Irene A. Chen, and Martin A. Nowak. The basic reproductive ratio of life. *J. Theor. Biol.*, 263:317–327, 2010.

[516] Benoit Mandelbrot. The role of sufficiency and of estimation in thermodynamics. *Ann. Math. Stat.*, 33:1021–1038, 1962.

[517] Craig E. Manning, Stephen J. Mojzsis, and T. Mark Harrison. Geology, age and origin of supracrustal rocks at Akilia, West Greenland. *Am. J. Sci.*, 306:303–366, 2006.

[518] Lynn Margulis and Karlene V. Schwartz. *Five Kingdoms: An Illustrated Guide to the Phyla of Life on Earth*. W. H. Freeman, New York, 1998.

[519] Stephanie Markert, Cordelia Arndt, Horst Felbeck, Dörte Becher, Stefan M. Sievert, Michael Hügler, Dirk Albrecht, Julie Robidart, Shellie Bench, Robert A. Feldman, Michael Hecker, and Thomas Schweder. Physiological proteomics of the uncultured endosymbiont of *Riftia pachyptila*. *Science*, 315:247–250, 2007.

[520] Ana Filipa A. Marques, Fernando J. A. S. Barriga, Valerie Chavagnac, and Yves Fouquet. Mineralogy, geochemistry, and Nd isotope composition of the Rainbow hydrothermal field, Mid-Atlantic Ridge. *Miner. Deposita*, 41:52–67, 2006.

[521] Pablo A. Marquet, Andrew P. Allen, James H. Brown, Jennifer A. Dunne, Brian J. Enquist, James F. Gillooly, Patricia A. Gowaty, Jessica L. Green, John Harte, Steve P. Hubbell, James O'Dwyer, Jordan G. Okie, Annette Ostling, Mark Ritchie, David Storch, and Geoffrey B. West. On theory in ecology. *BioScience*, 64(8):701–710, 2014.

[522] P. C. Martin, E. D. Siggia, and H. A. Rose. Statistical dynamics of classical systems. *Phys. Rev. A*, 8:423–437, 1973.

[523] William F. Martin. Hydrogen, metals, bifurcating electrons, and proton gradients: the early evolution of biological energy conservation. *FEBS Lett.*, 586:485–493, 2012.

[524] William Martin and Michael J. Russell. On the origin of cells: an hypothesis for the evolutionary transitions from abiotic geochemistry to chemoautotrophic prokaryotes, and from prokaryotes to nucleated cells. *Philos. Trans. R. Soc. London, Ser. B*, 358:27–85, 2003.

[525] William Martin and Michael J. Russell. On the origin of biochemistry at an alkaline hydrothermal vent. *Philos. Trans. R. Soc. London, Ser. B*, 362:1887–1926, 2007.

[526] William Martin, John Baross, Deborah Kelley, and Michael J. Russell. Hydrothermal vents and the origin of life. *Nature Rev. Microbiol.*, 6:805–814, 2008.

[527] William F. Martin, Filipa L. Fousa, and Nick Lane. Energy at life's origin. *Science*, 344:1092–1093, 2014.

[528] Berta M. Martins, Holger Dobbek, Irfan Cinkaya, Wolfgang Buckel, and Albrecht Messerschmidt. Crystal structure of 4-hydroxybutyryl-CoA dehydratase: radical catalysis involving a [4Fe-4S] cluster and flavin. *Proc. Natl. Acad. Sci. USA*, 101(44):15645–15649, 2004.

[529] C. N. Matthews and R. D. Minnard. Hydrogen cyanide polymers, comets and the origin of life. *Faraday Discuss.*, 133:393–401, 2006.

[530] John S. Mattick and Michael J. Gagen. The evolution of controlled multitasked gene networks: the role of introns and other noncoding RNAs in the development of complex organisms. *Mol. Biol. Evol.*, 18:1611–1630, 2004.

[531] Daniel C. Mattis and M. Lawrence Glasser. The uses of quantum field theory in diffusion-limited reactions. *Rev. Mod. Phys*, 70:979–1001, 1998.

[532] Fabio Mavelli and Pier L. Luisi. Autopoietic self-reproducing vesicles: a simplified kinetic model. *J. Phys. Chem.*, 100:16600–16607, 1998.

[533] Ernst Mayr. Where are we? *Cold Spring Harbor Symp. Quant. Biol.*, 24:1–14, 1959.

[534] Ernst Mayr. *The Growth of Biological Thought: Diversity, Evolution, and Inheritance*. Harvard University Press, Cambridge, MA, 1985.

[535] Ernst Mayr. A natural system of organisms. *Nature*, 348:491, 1990.

[536] Ernst Mayr. More natural classification. *Nature*, 353:122, 1991.

[537] Ernst Mayr. The objects of selection. *Proc. Natl. Acad. Sci. USA*, 94:2091–2094, 1997.

[538] Ernst Mayr. Two empires or three? *Proc. Natl. Acad. Sci. USA*, 95:9720–9723, 1998.

[539] Catherine A. McCammon. Mantle oxidation state and oxygen fugacity: constraints on mantle chemistry, structure, and dynamics. In R. D. Van Der Hilst, J. D. Bass, J. Mates, and J. Trampert, editors, *Earth's Deep Mantle: Structure, Composition, and Evolution*, pages 219–240. American Geophysical Union, Washington, DC, 2005.

[540] Gordon McCleod, Christopher McKeown, Allan J. Hall, and Michael J. Russell. Hydrothermal and oceanic pH conditions of possible relevance to the origin of life. *Orig. Life Evol. Biosphere*, 24:19–41, 1994.

[541] Thomas M. McCollom. Methanogenesis as a potential source of chemical energy for primary biomass production by autotrophic organisms in hydrothermal systems on Europa. *J. Geophys. Res.*, 104:30729–30742, 1999.

[542] Thomas M. McCollom. Laboratory simulations of abiotic hydrocarbon formation in Earth's deep subsurface. *Rev. Mineral. Geochem.*, 75:467–494, 2013.

[543] Thomas M. McCollom and Jeffrey S. Seewald. A reassessment of the potential for reduction of dissolved $CO_2$ to hydrocarbons during serpentinization of olivine. *Geochim. Cosmochim. Acta*, 65:3769–3778, 2001.

[544] Thomas M. McCollom and Jeffrey S. Seewald. Carbon isotope composition of organic compounds produced by abiotic synthesis under hydrothermal conditions. *Earth Planet. Sci. Lett.*, 243:74–84, 2006.

[545] Thomas M. McCollom and Everett L. Shock. Geochemical constraints on chemolithoautotrophic metabolism by microorganisms in seafloor hydrothermal systems. *Geochim. Cosmochim. Acta*, 61:4375–4391, 1997.

[546] Thomas M. McCollom, Barbara Sherwood Lollar, Georges Lacrampe-Couloume, and Jeffrey S. Seewald. The influence of carbon source on abiotic organic synthesis and carbon isotope fractionation under hydrothermal conditions. *Geochim. Cosmochim. Acta*, 74:2717–2740, 2010.

[547] W. F. McDonough and S.-S. Sun. The composition of the Earth. *Chem. Geol.*, 120:223–253, 1995.

[548] James D. McGhee and Peter H. von Hippel. Formaldehyde as a probe of DNA structure. I. Reaction with exocyclic amino groups of DNA bases. *Biochemistry*, 14:1281–1296, 1975.

[549] Kathleen E. McGinness and Gerald F. Joyce. In search of an RNA replicase ribozyme. *Chem. Biol.*, 10:5–14, 2003.

[550] Christopher P. McKay, Carolyn C. Porco, Travis Altheide, Wanda L. Davis, and Timothy A. Kral. The possible origin and persistence of life on Enceladus and detection of biomarkers in the plume. *Astrobiology*, 8:909–919, 2008.

[551] Donald L. Melchior, Harold J. Morowitz, Julian M. Sturtevant, and Tian Yow Tsong. Characterization of the plasma membrane of *Mycoplasma Laidlawii*. *Biochim. Biophys. Acta*, 219:114–122, 1970.

[552] Herman Melville. *Moby-Dick; or, The Whale*. Modern Library, New York, 1992.

[553] Gregor Mendel. Experiments on plant hybridization. *J. R. Hortic. Soc.*, 26:1–32, 1901. English translation.

[554] César Menor-Salván and Margarita R. Marin-Yaseli. Prebiotic chemistry in eutectic solutions at the water-ice matrix. *Chem. Soc. Rev.*, 41:5404–5415, 2012.

[555] John W. Merck. Volcanism I: sources and composition of magma. GEOL212: planetary geology lecture notes, 2014. www.geol.umd.edu/~jmerck/geol212/lectures/13.html.

[556] Marc Mezard, Giorgio Parisi, and Miguel Angel Virasoro. *Spin Glass Theory and Beyond*. World Scientific, Singapore, 1987.

[557] S. L. Miller. Production of amino acids under possible primitive Earth conditions. *Science*, 117:528–529, 1953.

[558] S. L. Miller and D. Smith-Magowan. The thermodynamics of the Krebs cycle and related compounds. *J. Phys. Chem. Ref. Data*, 19:1049–1073, 1990.

[559] Peter Mitchell. Coupling of phosphorylation to electron and hydrogen transfer by a chemi-osmotic type of mechanism. *Nature*, 191:144–148, 1961.

[560] S. J. Mojzsis, G. Arrhenius, K. D. McKeegan, T. M. Harrison, A. P. Nutman, and C. R. L. Friend. Evidence of life on Earth before 3,800 million years ago. *Nature*, 384:55–59, 1996.

[561] Jacques Monod. *Chance and Necessity*. Knopf, New York, 1971.

[562] Jacques Monod, Jean-Pierre Changeux, and Francois Jacob. Allosteric proteins and cellular control systems. *J. Mol. Biol.*, 6(4):306–329, 1963.

[563] Thierry Montmerle and Sylvia Exström. Hertzsprung–Russell diagram. *Encyclopedia of Astrobiology*, pages 749–754. Springer, Berlin, 2011.

[564] S. Moorbath. Evolution of Precambrian crust from strontium isotopic evidence. *Nature*, 254:395–398, 1975.

[565] Cristopher Moore and Stephan Mertens. *The Nature of Computation*. Oxford University Press, London, 2011.

[566] Camilo Mora, Derek P. Titensor, Sina Adl, Alistair G. B. Simpson, and Boris Worm. How many species are there on earth and in the ocean. *PLoS Biol.*, 9:e1001127, 2011.

[567] Eduardo Moreno and Christa Rhiner. Darwin's multicellularity: from neurotrophic theories and cell competition to fitness fingerprints. *Curr. Opin. Cell Biol.*, 31:16–22, 2014.

[568] W. J. Morgan. Convection plumes in the lower mantle. *Nature*, 230:42–43, 1971.

[569] Harold J. Morowitz. Proton semiconductors and energy transduction in biological systems. *Am. J. Physiol.*, 235:R99–R114, 1978.

[570] Harold J. Morowitz. *Energy Flow in Biology*. Ox Bow Press, Woodbridge, CT, 1979.

[571] Harold J. Morowitz. *Foundations of Bioenergetics*. Academic Press, New York, 1987.

[572] Harold J. Morowitz. *Beginnings of Cellular Life*. Yale University Press, New Haven, CT, 1992.

[573] Harold J. Morowitz. Phenetics, a born-again science. *Complexity*, 8:12–13, 2003.

[574] H. J. Morowitz, J. D. Kostelnik, J. Yang, and G. D. Cody. The origin of intermediary metabolism. *Proc. Natl. Acad. Sci. USA*, 97:7704–7708, 2000.

[575] Michael R. Morrow, John P. Whitehead, and Dalian Lu. Chain-length dependence of lipid bilayer properties near the liquid crystal to gel transition. *Biophys. J.*, 63:18–27, 1992.

[576] Daniel Mueller, Stefan Pitsch, Atsushi Kittaka, Ernst Wagner, Claude E. Wintner, and Albert Eschenmoser. Chemistry of alpha aminonitriles: aldomerization of gly-colaldehyde phosphate to racemic hexose 2,4,6-triphosphates and (in presence of formaldehyde) racemic pentose 2,4-diphosphates: rac-allose 2,4,6-triphosphate and racemic ribose 2,4-diphosphate are the main reaction products. *Helv. Chim. Acta*, 73:1410–1468, 1990.

[577] Ute Müh, Irfan Cinkaya, Simon P. J. Albracht, and Wolfgang Buckel. 4-hydroxybutyryl-CoA dehydratase from *Clostridium aminobutyricum*: characterization of FAD and iron–sulfur clusters involved in an overall non-redox reaction. *Biochemistry*, 35(36):11710–11718, 1996.

[578] Armen Y. Mulkidjanian, Pavel Dibrov, and Michael Y. Galperin. The past and present of the sodium energetics: may the sodium-motive force be with you. *Biochim. Biophys. Acta*, 1777:985–992, 2008.

[579] Armen Y. Mulkidjanian, Michael Y. Galperin, and Eugene V. Koonin. Co-evolution of primordial membranes and membrane proteins. *Trends. Biochem. Sci.*, 34:206–215, 2009.

[580] Armen Y. Mulkidjanian, Michael Y. Galperin, Kira S. Makarova, Yuri I. Wolf, and Eugene V. Koonin. Evolutionary primacy of sodium bioenergetics. *Biol. Direct*, 3:13, 2008.

[581] H. J. Muller. The relation of recombination to mutational advance. *Mutat. Res.*, 1:1–9, 1964.

[582] Ursula Munro, John A. Munro, John B. Phillips, and Wolfgang Wiltschko. Effects of wavelength of light and pulse magnetism on different magnetoreception systems in a migratory bird. *Aust. J. Zool.*, 45:189–198, 1997.

[583] Bjorn O. Mysen. An experimental study of phosphorus and aluminosilicate speciation in and partitioning between aqueous fluids and silicate melts determined in-situ at high temperature and pressure. *Am. Mineral.* 96:1636–1649, 2011.

[584] Bjorn O. Mysen and George D. Cody. Silicate-phosphate interactions in silicate glasses and melts: II. Quantitative, high-temperature structure of P-bearing alkali aluminosilicate melts. *Geochim. Cosmochim. Acta*, 65:2413–2431, 2001.

[585] John F. Nagle. Theory of the main lipid bilayer phase transition. *Annu. Rev. Phys. Chem.*, 31:157–195, 1980.

[586] Shu-ichi Nakano, Durga M. Chadalavada, and Philip C. Bevilacqua. General acid-base catalysis in the mechanism of a hepatitis delta virus ribozyme. *Science*, 287:1493–1497, 2000.

[587] David L. Nelson and Michael M. Cox. *Lehninger Principles of Biochemistry*. W. H. Freeman, New York, fourth edition, 2004.

[588] Anna Neubeck, Nguyen Thanh Duc, David Bastviken, Patrick Crill, and Nils G. Holm. Formation of $H_2$ and $CH_4$ by weathering of olivine at temperatures between 30 and 70°C. *Geochem. Trans.*, 12:6:1–10, 2011.

[589] Marc Neveu, Hyo-Joong Kim, and Steven A. Benner. The "strong" RNA World hypothesis: fifty years old. *Astrobiology*, 13:391–403, 2013.

[590] Friedrich Nietzsche. *The Will to Power*. C. G. Naumann, Leipzig, 1901.

[591] Poul Nissen, Joseph A. Ippolito, Nenad Ban, Peter B. Moore, and Thomas A. Steitz. RNA tertiary interactions in the large ribosomal subunit: the A-minor motif. *Proc. Natl. Acad. Sci. USA*, 98:4899–4903, 2001.

[592] Wolfgang Nitschke and Michael J. Russell. Hydrothermal focusing of chemical and chemiosmotic energy, supported by delivery of catalytic Fe, Ni, Mo/W, Co, S and Se, forced life to emerge. *J. Mol. Evol.*, 69:481–498, 2009.

[593] Wolfgang Nitschke and Michael J. Russell. Beating the acetyl coenzyme A-pathway to the origin of life. *Philos. Trans. R. Soc. London, Ser. B*, 368:20120258, 2013.

[594] W. Nitscke, D. M. Kramer, A. Riedel, and U. Liebl. From naptho- to benzo-quinones – (r)evolutionary reorganizations of electron transfer chains. In P. Mathis, editor, *Photosynthesis: from Light to the Biosphere,* vol. 1, pages 945–950. Kluwer Academic Press, Dordrecht, 1995.

[595] Harry F. Noller. On the origin of the ribosome: co-evolution of sub-domains of tRNA and rRNA. In Raymond F. Gesteland and John F. Atkins, editors, *The RNA World*, pages 137–156. Cold Spring Harbor Laboratory Press, Plainview, New York, 1993.

[596] Harry F. Noller. On the origin of the ribosome: co-evolution of sub-domains of tRNA and rRNA. In Raymond F. Gesteland and John F. Atkins, editors, *The RNA World*, pages 197–219. Cold Spring Harbor Laboratory Press, Plainview, New York, 1999.

[597] Yehor Novikov and Shelley D. Copley. Reactivity landscape of pyruvate under simulated hydrothermal vent conditions. *Proc. Natl. Acad. Sci. USA*, 110:13283–13288, 2013.

[598] Martin A. Nowak and Hisashi Ohtsuki. Prevolutionary dynamics and the origin of evolution. *Proc. Natl. Acad. Sci. USA*, 105:14924–14927, 2008.

[599] Allen P. Nutman, Vickie C. Benett, Clark. R. L. Friend, Frances Jenner, and Yusheng Wan. Eoarchaean crustal growth in West Greenland (Itsaq Gneiss Complex) and in northeastern China (Anshan area): review and synthesis. *Earth Accret. Syst. Space Time*, 318:127–154, 2009.

[600] Thomas Nyström. A bacterial kind of aging. *PLoS Genet.*, 3:2355–2357, 2007.

[601] Patrick J. O'Brien and Daniel Herschlag. Catalytic promiscuity and the evolution of new enzymatic activities. *Chem. Biol.*, 6:R91–R105, 1999.

[602] F. John Odling-Smee, Kevin N. Laland, and Marcus W. Feldman. *Niche Construction: The Neglected Process in Evolution*. Princeton University Press, Princeton, NJ, 2003.

[603] Howard T. Odum and Richard C. Pinkerton. Time's speed regulator: the optimum efficiency for maximum output in physical and biological systems. *Am. Sci.*, 43:331–343, 1955.

[604] Katsuhiko Ogata. *Modern Control Engineering*. Prentice-Hall, New York, fifth edition, 2010.

[605] Gary J. Olsen and Carl R. Woese. Ribosomal RNA: a key to phylogeny. *FASEB J.*, 7:113–123, 1993.

[606] Jonathan O'Neil, Richard W. Carlson, Don Francis, and Ross K. Stevenson. Neodymium-142 evidence for Hadean mafic crust. *Science*, 321:1828–1831, 2008.

[607] Lars Onsager. Reciprocal relations in irreversible processes. I. *Phys. Rev.*, 37:405–426, 1931.

[608] Lars Onsager. Reciprocal relations in irreversible processes. II. *Phys. Rev.*, 38:2265–2279, 1931.

[609] L. Onsager and S. Machlup. Fluctuations and irreversible processes. *Phys. Rev.*, 91:1505, 1953.

[610] A. I. Oparin. *Proiskhozhdenie zhizy*. Moskovski Rabochii, Moscow, 1924. In Russian.

[611] Alexander I. Oparin. The origin of life. In J. D. Bernal, editor, *The Origin of Life*, pages 199–234. Weidenfeld and Nicolson, London, 1967.

[612] Aharon Oren. Microbial life at high salt concentration: phylogenetic and metabolic diversity. *Saline Syst.*, 4:2:1–13, 2008.

[613] Leslie E. Orgel. Prebiotic chemistry and the origin of the RNA world. *Crit. Rev. Biochem. Mol. Biol.*, 39:99–123, 2004.

[614] Leslie E. Orgel. The implausibility of metabolic cycles on the early Earth. *PLoS Biology*, 6:e18, 2008.

[615] J. Oró. Mechanisms of synthesis of adenine from hydrogen cyanide under possible primitive Earth conditions. *Nature*, 191:1193–1194, 1961.

[616] J. Oró and A. Kimball. Synthesis of adenine from ammonium cyanide. *Biochem. Biophys. Res. Commun.*, 2:407–412, 1960.

[617] J. Oró and A. Kimball. Synthesis of purines under possible primitive Earth conditions I: adenine from hydrogen cyanide. *Arch. Biochem. Biophys.*, 94:217–227, 1961.

[618] J. Oró and A. Kimball. Synthesis of purines under possible primitive Earth conditions II: purine intermediates from hydrogen cyanide. *Arch. Biochem. Biophys.*, 96:293–313, 1962.

[619] Sijbren Otto, Jan B. F. N. Engberts, and Jan C. T. Kwak. Million-fold acceleration of a Diels–Alder reaction due to combined Lewis acid and micellar catalysis in water. *J. Am. Chem. Soc.*, 120:9517–9525, 1998.

[620] M. Paecht-Horowitz, J. Berger, and A. Katchalsky. Prebiotic synthesis of polypeptides by heterogeneous polycondensation of amino-acid adenylates. *Nature*, 228:636–639, 1970.

[621] Bernhard O. Palsson. *Systems Biology*. Cambridge University Press, Cambridge, MA, 2006.

[622] Eric T. Parker, H. James Cleaves, Michael P. Callahan, Jason P. Dworkin, Daniel P. Glavin, Antonio Lazcano, and Jeffrey L. Bada. Prebiotic synthesis of methionine and other sulfur-containing organic compounds on the primitive Earth: a contemporary reassessment based on an unpublished 1958 Stanley Miller experiment. *Orig. Life Evol. Biosphere*, 41:201–212, 2011.

[623] Robert Pascal, Addy Pross, and John D. Sutherland. Towards an evolutionary theory of the origin of life based on kinetics and thermodynamics. *Open Biol.*, 3:130156, 2013.

[624] Matthew A. Pasek and Dante S. Lauretta. Aqueous corrosion of phosphide minerals from iron meteorites: a highly reactive source of prebiotic phosphorus on the surface of the early Earth. *Astrobiology*, 5:515–535, 2005.

[625] Matthew A. Paseka, Jelte P. Harnmeijerb, Roger Buick, Maheen Gulla, and Zachary Atlas. Evidence for reactive reduced phosphorus species in the early Archean ocean. *Proc. Natl. Acad. Sci. USA*, 110:10089–11194, 2013.

[626] Matthew A. Pasek, Jacqueline M. Sampson, and Zachary Atlas. Redox chemistry in the phosphorus biogeochemical cycle. *Proc. Natl. Acad. Sci. USA*, 43:15468–15473, 2014.

[627] A. A. Pavlov and J. F. Kasting. Mass-independent fractionation of sulfur isotopes in Archean sediments: strong evidence for an anoxic Archean atmosphere. *Astrobiology*, 2:27–41, 2002.

[628] Alexander A. Pavlov, Lisa L. Brown, and James F. Kasting. UV shielding of $NH_3$ and $O_2$ by organic hazes in the Archean atmosphere. *J. Geophys. Res.*, 106:23267–23287, 2001.

[629]  A. A. Pavlov, M. J. Mills, and O. B. Toon. Mystery of the volcanic mass-independent sulfur isotope fractionation signature in the Antarctic ice core. *J. Geophys. Res.*, 32:L12816, 2005.

[630]  L. Peliti. Path-integral approach to birth-death processes on a lattice. *J. Phys. (Paris)*, 46:1469, 1985.

[631]  L. Peliti. Renormalization of fluctuation effects in $a + a \to a$ reaction. *J. Phys. A*, 19:L365, 1986.

[632]  Juli Peretó. Out of fuzzy chemistry: from prebiotic chemistry to metabolic networks. *Chem. Soc. Rev.*, 41:5394–5403, 2012.

[633]  S. Petersen, K. Kuhn, T. Kuhn, N. Augustin, R. Hékinian, L. Franz, and C. Borowski. The geological setting of the ultramafic-hosted Logatchev hydrothermal field $\left(14°45'\text{N}, \text{Mid} - \text{AtlanticRidge}\right)$ and its influence on massive sulfide formation. *Lithos*, 112:40–56, 2009.

[634]  Anton S. Petrov, Chad R. Bernier, Eli Hershkovitz, Yuzhen Xue, Chris C. Waterbury, Chiaolong Hsiao, Victor G. Stepanov, Eric A. Gaucher, Martha A. Grover, Steven C. Harvey, Nicholas V. Hud, Roger M. Wartell, George E. Fox, and Loren D. Williams. Secondary structure and domain architecture of the 23S and 5S rRNAs. *Nucleic Acids Res.*, 14:7522–7535, 2013.

[635]  Anton S. Petrov, Chad R. Bernier, Chiaolong Hsiao, Ashlyn M. Norris, Nicholas A. Kovacs, Chris C. Waterbury, Victor G. Stepanov, Stephen C. Harvey, George E. Fox, Roger M. Wartell, Nicholas V. Hud, and Loren D. Williams. Evolution of the ribosome at atomic resolution. *Proc. Natl. Acad. Sci. USA*, 2014.

[636]  Susan M. Pfiffner, James M. Cantu, Amanda Smithgall, Aaron D. Peacock, and David C. White. Deep subsurface microbial biomass and community structure in Witwatersrand basin mines. *Geomicrobiol. J.*, 23:431–442, 2006.

[637]  S. Pilgram, A. N. Jordan, E. V. Sukhorukov, and M. Büttiker. Stochastic path integral formulation of full counting statistics. *Phys. Rev. Lett.*, 90:206801, 2003.

[638]  Sandra Pizzarello and Arthur L. Weber. Prebiotic amino acids as asymmetric catalysts. *Science*, 303:1151, 2004.

[639]  Andrey V. Plyasunov and Everett L. Shock. Thermodynamic functions of hydration of hydrocarbons at 298.15 K and 0.1 MPa. *Geochim. Cosmochim. Acta*, 64:439–468, 2000.

[640]  Anja Poehlein, Silke Schmidt, Anne-Kristin Kaster, Meike Goenrich, John Vollmers, Andrea Thürmer, Johannes Bertsch, Kai Schuchmann, Birgit Voigt, Michael Hecker, Rolf Daniel, Rudolf K. Thauer, Gerhard Gottschalk, and Volker Müller. An ancient pathway combining carbon dioxide fixation with the generation and utilization of a sodium iongradient for ATP synthesis. *PLoS ONE*, 7:e33439, 2012.

[641]  Andrew Pohorille, Karl Schweighofer, and Michael A. Wilson. The origin and early evolution of membrane channels. *Astrobiology*, 5:1–17, 2005.

[642]  Andrew Pohorille, Michael A. Wilson, and Christophe Chipot. Membrane peptides and their role in protobiological evolution. *Orig. Life Evol. Biosphere*, 33:173–197, 2003.

[643]  Joseph G. Polchinski. Renormalization group and effective lagrangians. *Nucl. Phys. B*, 231:269–295, 1984.

[644]  Anthony M. Poole, Daniel C. Jeffares, and David Penny. The path from the RNA world. *J. Mol. Evol.*, 46:1–17, 1998.

[645]  Karl R. Popper. *Logik der Forschung*. Julius Springer Verlag, Vienna, 1935.

[646]  Karl R. Popper. *The Logic of Scientific Discovery*. Hutchinson, London, 1959. Translation of [645].

[647] Matthew W. Powner, Béatrice Gerland, and John D. Sutherland. Synthesis of activated pyrimidine ribonucleotides in prebiotically plausible conditions. *Nature*, 459:239–242, 2009.

[648] Matthew W. Powner, John D. Sutherland, and Jack W. Szostak. The origins of nucleotides. *SYNLETT*, 14:1956–1964, 2011.

[649] Steve Pressé, Kingshuk Ghosh, Julian Lee, and Ken A. Dill. The principles of maximum entropy and maximum caliber in statistical physics. *Rev. Mod. Phys*, 85:1115–1141, 2013.

[650] G. R. Price. Fisher's 'fundamental theorem' made clear. *Ann. Hum. Genet.*, 36:129–140, 1972.

[651] Giora Proskurowski, Marvin D. Lilley, Jeffery S. Seewald, Gretchen L. Früh-Green, Eric J. Olson, John E. Lupton, Sean P. Sylva, and Deborah S. Kelley. Abiogenic hydrocarbon production at Lost City hydrothermal field. *Science*, 319:604–607, 2008.

[652] William B. Provine. *The Origins of Theoretical Population Genetics*. University of Chicago Press, Chicago, IL, 2001.

[653] E. M. Purcell. Life at low Reynolds number. *Am. J. Phys.*, 45:3–11, 1973.

[654] Henry Quastler. *The Emergence of Biological Organization*. Yale University Press, New Haven, CT, 1964.

[655] John M. Quick. *Statistical Analysis with R*. Packt Publishing, Birmingham, 2010.

[656] Efraim Racker and Walther Stoeckenius. Reconstitution of purple membrane vesicles catalyzing light-driven proton uptake and adenosine triphosphate formation. *J. Biol. Chem.*, 249:662–663, 1974.

[657] Petronella C. Raemakers-Franken, Roy Bongaerts, Roel Fokkens, Chris van der Drift, and Godfried D. Vogels. Characterization of two pterin derivatives isolated from *Methanoculleus thermophilicum*. *Eur. J. Biochem.*, 200:783–787, 1991.

[658] Petronella C. Raemakers-Franken, Frank G. Voncken, Jaap Korteland, Jan T. Keltjens, Chris van der Drift, and Godfried D. Vogels. Structural characterization of tatiopterin, a novel pterin isolated from *Methanogenium tationis*. *Biofactors*, 2:117–122, 1989.

[659] Stephen W. Ragsdale. Enzymology of the Wood–Ljungdahl pathway of acetogenesis. *Ann. N Y Acad. Sci.*, 1125:129–136, 2008.

[660] Stephen W. Ragsdale and Manoj Kumar. Nickel-containing carbon monoxide dehydrogenase/acetyl-CoA synthase. *Chem. Rev.*, 96:2515–2540, 1996.

[661] Stephen W. Ragsdale and Harland G. Wood. Enzymology of the acetyl-CoA pathway of $CO_2$ fixation. *Crit. Rev. Biochem. Mol. Biol.*, 26:261–300, 1991.

[662] S. W. Ragsdale, J. E. Clark, L. G. Ljungdahl, L. L. Lundie, and H. L. Drake. Properties of purified carbon monoxide dehydrogenase from *Clostridium thermoaceticum*, a nickel, iron-sulfur protein. *J. Biol. Chem.*, 258(4):2364–2369, 1983.

[663] Burki Rajendar, Arivazhagan Rajendran, Zhiqiang Ye, Erko Kanai, Yusuke Sato, Seiichi Nishizawa, Marek Sikorski, and Norio Teramae. Effect of substituents of alloxazine derivatives on the selectivity and affinity for adenine in AP-site-containing DNA duplexes. *Org. Biomol. Chem.*, 8:4949–4959, 2010.

[664] Kalervo Rankama and Thure Georg Sahama. *Geochemistry*. University of Chicago Press, Chicago, IL, 1950.

[665] Jason Raymond, Janet L. Seifert, Christopher R. Staples, and Robert E. Blankenship. The natural history of nitrogen fixation. *Mol. Biol. Evol.*, 21:541–554, 2004.

[666] Eoghan P. Reeves, Jill M. McDermott, and Jeffrey S. Seewald. The origin of methanethiol in midocean ridge hydrothermal fluids. *Proc. Natl. Acad. Sci. USA*, 111:5474–5479, 2014.

[667] Howard M. Reid. *Introduction to Statistics: Fundamental Concepts and Procedures of Data Analysis*. SAGE publications, Washington DC, 2014.

[668] Anna-Louise Reysenbach and Everett Shock. Merging genomes with geochemistry in hydrothermal ecosystems. *Science*, 296:1077–1082, 2002.

[669] J. M. Rhodes. Mantle melting and origin of basaltic magma. Notes: GEO-321. Igneous & Metamorphic Petrology, March 2005. www.geo.umass.edu/courses/geo321/Lecture.

[670] Ignasi Ribas. The sun and stars as the primary energy input in planetary atmospheres. In A. G. Kosovichev, A. H. Andrei, and J.-P. Rozelot, editors, *Proc. International Astronomical Union Symposium No. 264, 2009*, pages 3–18. Cambridge University Press, Cambridge, 2010.

[671] A. Ricardo, M. A. Carrigan, A. N. Olcott, and S. A. Benner. Borate minerals stabilize ribose. *Science*, 303:196, 2004.

[672] William J. Riehl, Paul L. Krapivsky, Sidney Redner, and Daniel Segrè. Signatures of arithmetic simplicity in metabolic network architecture. *PLoS Comput. Biol.*, 6:e1000725, 2010.

[673] G. Rieley, C. L. Van Dover, D. B. Hedrick, and G. Eglinton. Trophic ecology of *Rimicaris exoculata*: a combined lipid abundance stable isotope approach. *Mar. Biol.*, 133:495–499, 1999.

[674] Jorma Rissanen. *Stochastic Complexity in Statistical Inquiry*. World Scientific, Teaneck, NJ, 1989.

[675] Michael P. Robertson and Gerald F. Joyce. The origins of the RNA world. *Cold Spring Harb. Perspect. Biol.*, 4:a003608, 2010.

[676] Michael P. Robertson and William G. Scott. The structural basis of ribozyme-catalyzed RNA assembly. *Science*, 315:1549–1553, 2007.

[677] João F. Matias Rodrigues and Andreas Wagner. Evolutionary plasticity and innovations in complex metabolic reaction networks. *PLoS Comput. Biol.*, 5:e1000613:1–11, 2009.

[678] Rajat Rohatgi, David P. Bartel, and Jack W. Szostak. Kinetic and mechanistic analysis of nonenzymatic, template-directed oligoribonucleotide ligation. *J. Am. Chem. Soc.*, 118:3332–3339, 1996.

[679] Rajat Rohatgi, David P. Bartel, and Jack W. Szostak. Nonenzymatic, template-directed ligation of oligoribonucleotides is highly regioselective for the formation of $3'$–$5'$ phosphodiester bonds. *J. Am. Chem. Soc.*, 118:3340–3344, 1996.

[680] R. D. Rosenkrantz, editor. *Jaynes, E. T.: Papers on Probability, Statistics and Statistical Physics*. D. Reidel, Dordrecht, 1983.

[681] Paul J. Rothwell and Gabriel Waksman. Structure and mechanism of DNA polymerases. *Adv. Protein Chem.*, 71:401–440, 2005.

[682] Bertand Russell. *A History of Western Philosophy*. Simon & Schuster, New York, 1967.

[683] Michael J. Russell. Downward-excavating hydrothermal cells and Irish-type ore deposits: importance of an underlying thick Caledonian prism. *Trans. Inst. Min. Metall.*, B87:168–171, 1978.

[684] Michael J. Russell. Mining, metallurgy and the origin of life. *Miner. Indust. Int.*, 1009:4–8, 1993.

[685] Michael J. Russell. First life. *Am. Sci.*, 94:32–39, 2006.

[686] Michael J. Russell. The alkaline solution to the emergence of life: energy, entropy, and early evolution. *Acta Biotheor.*, 55:133–179, 2007.

[687] Michael J. Russell and A. J. Hall. The emergence of life from iron monosulphide bubbles at a submarine hydrothermal redox and pH front. *J. Geol. Soc. London*, 154:377–402, 1997.

[688] Michael J. Russell and Allan J. Hall. The onset and early evolution of life. *Geol. Soc. Am. Memoir*, 198:1–32, 2006.

[689] Michael J. Russell and William Martin. The rocky roots of the acetyl-CoA pathway. *Trends Biochem. Sci.*, 29:358–363, 2004.

[690] Michael J. Russell, Laura M. Barge, Rohit Bhartia, Dylan Bocanegra, Paul J. Bracher, Elbert Branscomb, Richard Kidd, Shawn McGlynn, David H. Meier, Wolfgang Nitschke, Takazo Shibuya, Steve Vance, Lauren White, and Isik Kanik. The drive to life on wet and icy worlds. *Astrobiology*, 14:308–343, 2014.

[691] Michael J. Russell, Roy M. Daniel, and Allan J. Hall. On the emergence of life via catalytic iron-sulphide membranes. *Terra Nova*, 5:343–347, 1993.

[692] Michael J. Russell, Allan J. Hall, Adrian J. Boyce, and Anthony E. Fallick. On hydrothermal convection systems and the emergence of life. *Econ. Geol.*, 100:419–438, 2005.

[693] M. J. Russell, A. J. Hall, A. G. Cairns-Smith, and P. S. Braterman. Submarine hot springs and the origin of life. *Nature*, 336:117, 1988.

[694] M. J. Russell, A. J. Hall, and W. Martin. Serpentinization as a source of energy at the origin of life. *Geobiology*, 8:355–371, 2010.

[695] W. J. Rutter. Evolution of aldolase. *Fed. Proc.*, 23:1248–1257, 1964.

[696] Carl Sagan and Christopher Chyba. The early faint young sun paradox: organic shielding of ultraviolet-labile greenhouse gases. *Science*, 276:1217–1221, 1997.

[697] Carl Sagan and George Mullen. Earth and Mars: evolution of atmospheres and surface temperatures. *Science, New Series*, 177:52–56, 1972.

[698] Raffaele Saladino, Giorgia Botta, Samanta Pino, Giovanna Costanzo, and Ernesto Di Mauro. From the one-carbon amide formamide to RNA all the steps are prebiotically possible. *Biochimie*, 94:1451–1456, 2012.

[699] Raffaele Saladino, Giorgia Botta, Samanta Pino, Giovanna Costanzo, and Ernesto Di Mauro. Materials for the onset: a story of necessity and chance. *Frontiers Biosci.*, 18:1275–1289, 2013.

[700] Maria do Céu Santos and Manuel A. S. Santos. Structural and molecular features of non-standard genetic codes. In Gina M. Cannarozzi and Adrian Schneider, editors, *Codon Evolution: Mechanisms and Models*, pages 258–270, Oxford University Press, New York, 2012.

[701] Takaaki Sato, Hiroyuki Imanaka, Naeem Rashid, Toshiaki Fukui, Haruyuki Atomi, and Tadayuki Imanaka. Genetic evidence identifying the true gluconeogenic fructose-1,6-bisphosphatase in *Thermococcus kodakaraensis* and other hyperthermophiles. *J. Bacteriol.*, 186:5799–5807, 2004.

[702] Brandon Schmandt, Kenneth Dueker, Eugene Humphreys, and Steven Hansen. Hot mantle upwelling across the 660 beneath Yellowstone. *Earth Planet. Sci. Lett.*, 331:224–236, 2012.

[703] Johan A. Schmidt, Matthew S. Johnson, and Reinhard Schinke. Carbon dioxide photolysis from 150 to 210 nm: singlet and triplet channel dynamics, UV-spectrum, and isotope effects. *Proc. Natl. Acad. Sci. USA*, 110:17691–17696, 2013.

[704] Philippe Schmitt-Kopplin, Zelimir Gabelica, Régis D. Gougeon, Agnes Fekete, Basem Kanawati, Mourad Harir, Istvan Gebefuegi, Gerhard Eckel, and Norbert Hertkorn. High molecular diversity of extraterrestrial organic matter in Murchison

meteorite revealed 40 years after its fall. *Proc. Natl. Acad. Sci. USA*, 107:2763–2768, 2010.

[705]   Thomas D. Schneider. Theory of molecular machines I: channel capacity of molecular machines. *J. Theor. Biol.*, 148:83–123, 1991.

[706]   Thomas D. Schneider. Theory of molecular machines II: energy dissipation from molecular machines. *J. Theor. Biol.*, 148:125–137, 1991.

[707]   Barbara Schoepp-Cothenet, Clément Lieutaud, Frauke Baymann, André Verméglio, Thorsten Friedrich, David M. Kramer, and Wolfgang Nitschke. Menaquinone as a pool quinone in a purple bacterium. *Proc. Natl. Acad. Sci. USA*, 106:8549–8554, 2005.

[708]   J. William Schopf. Microfossils of the early Archaean apex chert: new evidence of the antiquity of life. *Science*, 260:640–646, 1993.

[709]   Laurier L. Schramm. *Emulsions, Foams, and Suspensions: Fundamentals and Applications*. Wiley, New York, 2005.

[710]   Matthew O. Schrenk, John R. Kelley, Deborah S. anf Delaney, and John A. Baross. Incidence and diversity of microorganisms within the walls of an active deep-sea sulfide chimney. *Appl. Environ. Microbiol.*, 69:3580–3592, 2003.

[711]   E. Schrödinger. *What is Life? The Physical Aspect of the Living Cell*. Cambridge University Press, New York, 1992.

[712]   Mitch Schulte, David Blake, Tori Hoehler, and Thomas McCollom. Serpentinization and its implications for life on the early Earth and Mars. *Astrobiology*, 6:364–376, 2006.

[713]   Gerrit J. Schut and Michael W. W. Adams. The iron-hydrogenase of *Thermotoga maritima* utilizes ferredoxin and NADH synergistically: a new perspective on anaerobic hydrogen production. *J. Bacteriol.*, 191:4451–4457, 2009.

[714]   Michael Schutz, Barbara Schoepp-Cothenet, Elisabeth Lojou, Mireille Woodstra, Doris Lexa, Pascale Tron, Alain Dolla, Marie-Claire Durand, Karl Otto Stetter, and Frauke Baymann. The naphthoquinol oxidizing cytochrome bc1 complex of the hyperthermophilic knallgasbacterium *Aquifex aeolicus*: properties and phylogenetic relationships. *Biochemistry*, 42:10800–10808, 2003.

[715]   Alan W. Schwartz. Phosphorus in prebiotic chemistry. *Philos. Trans. R. Soc. London, Ser. B*, 361:1743–1749, 2006.

[716]   Gideon E. Schwarz. Estimating the dimension of a model. *Ann. Stat.*, 6:461–464, 1978.

[717]   Esther M. Schwarzenbach, Gretchen L. Früh-Green, Stefano M. Bernasconi, Jeffrey C. Alt, and Alessio Plas. Serpentinization and carbon sequestration: a study of two ancient peridotite-hosted hydrothermal systems. *Chem. Geol.*, 351:115–133, 2013.

[718]   Anja Schwögler and Thomas Carell. Toward catalytically active oligonucleotides: synthesis of a flavin nucleotide and its incorporation into DNA. *Org. Lett.*, 2:1415–1418, 2000.

[719]   John R. Searle. *The Mystery of Consciousness*. New York Review of Books, New York, 1997.

[720]   Henning Seedorf, W. Florian Fricke, Birgit Veith, Holger Brüggemann, Heiko Liesegang, Axel Strittmatter, Marcus Miethke, Wolfgang Buckel, Julia Hinderberger, Fuli Li, Christoph Hagemeier, Rudolf K. Thauer, and Gerhard Gottschalk. The genome of *Clostridium kluyveri*, a strict anaerobe with unique metabolic features. *Proc. Natl. Acad. Sci. USA*, 105:2128–2133, 2008.

[721] Daniel Segré, Dafna Ben-Ali, and Doron Lancet. Compositional genomes: prebiotic information transfer in mutually catalytic noncovalent assemblies. *Proc. Natl. Acad. Sci. USA*, 97:4112–4117, 2000.

[722] Daniel Segré, Doron Lancet, Ora Kedem, and Yitzhak Pilpel. Graded autocatalysis replication domain (GARD): kinetic analysis of self-replication in mutually catalytic sets. *Orig. Life Evol. Biosphere*, 28:501–514, 1998.

[723] Daniel Segré, Barak Shenhav, Ron Kafri, and Doron Lancet. The molecular roots of compositional inheritance. *J. Theor. Biol.*, 213:481–491, 2001.

[724] Teddy Seidenfeld. Why I am not an objective Bayesian: some reflections prompted by Rosenkrantz. *Theory Decision*, 11:413–440, 1979.

[725] Teddy Seidenfeld. Entropy and uncertainty. In I. B. MacNeill and G. J. Umphrey, editors, *Foundations of Statistical Inference*, pages 259–287. Reidel, Boston, MA, 1987.

[726] Javier Seravalli, Yuming Xiao, Weiwei Gu, Stephen P. Cramer, William E. Antholine, Vladimir Krymov, Gary J. Gerfen, and Stephen W. Ragsdale. Evidence that NiNi acetyl-CoA synthase is active and that the CuNi enzyme is not. *Biochemistry*, 43(13):3944–3955, 2004.

[727] W. E. Seyfried Jr. and Kang Ding. Phase equilibria in subseafloor hydrothermal systems: a review of the role of redox, temperature, pH and dissolved Cl on the chemistry of hot spring fluids at mid-ocean ridges. In Susan E. Humphris, Robert A. Zierenberg, Lauren S. Mullineaux, and Richard E. Thomson, editors, *Seafloor Hydrothermal Systems: Physical, Chemical, Biological, and Geological Interactions*, pages 248–272. Geophysical Monograph. American Geophysical Union, Washington, DC, 1995.

[728] W. E. Seyfried Jr., Kang Ding, and M. E. Berndt. Phase equilibria constraints on the chemistry of hot spring fluids at mid-ocean ridges. *Geochim. Cosmochim. Acta*, 55:3559–3580, 1991.

[729] Cosma Rohilla Shalizi. Dynamics of Bayesian updating with dependent data and misspecified models. *Electron. J. Stat.*, 3:1039–1074, 2009.

[730] Timothy M. Shank, Daniel J. Fornari, Karen L. Von Damm, Marvin D. Lilley, Rachel M. Haymon, and Richard A. Lutz. Temporal and spatial patterns of biological community development at nascent deep-sea hydrothermal vents (9°50′N, East Pacific Rise). *Deep-Sea Res. II*, 45:465–515, 1998.

[731] Claude E. Shannon. Communication in the presence of noise. *Proc. IEEE*, 86:447–457, 1949.

[732] Claude Elwood Shannon and Warren Weaver. *The Mathematical Theory of Communication*. University of Illinois Press, Urbana, IL, 1949.

[733] Robert Shapiro. Small molecule interactions were central to the origin of life. *Q. Rev. Biol.*, 81:105–125, 2006.

[734] Anurag Sharma, George D. Cody, and Russell J. Hemley. *In situ* diamond-anvil cell observations of methanogenesis at high pressures and temperatures. *Energy Fuels*, 23:5572–5579, 2009.

[735] Jia Sheng, Li Li, Aaron E. Engelhart, Jianhua Gan, Jiawei Wang, and Jack W. Szostak. Structural insights into the effects of 2′–5′ linkages on the RNA duplex. *Proc. Natl. Acad. Sci. USA*, 111:3050–3055, 2014.

[736] Peter P. Sheridan, Katherine H. Freeman, and Jean E. Brenchley. Estimated minimal divergence times of the major bacterial and archaeal phyla. *Geomicrobiol. J.*, 20:1–14, 2003.

[737] B. Sherwood Lollar, G. Lacrampe-Couloume, G. F. Slater, J. Ward, D. P. Moser, T. M. Gihring, L.-H. Lin, and T. C. Onstott. Unravelling abiogenic and biogenic sources of methane in the Earth's deep subsurface. *Chem. Geol.*, 226:328–339, 2006.

[738] B. Sherwood Lollar, T. D. Westgate, J. A. Ward, G. F. Slater, and G. Lacrampe-Couloume. Abiogenic formation of gaseous alkanes in the Earth's crust as a minor source of global hydrocarbon reservoirs. *Nature*, 416:522–524, 2002.

[739] Takazo Shibuya, Tsuyoshi Komiya, Kentaro Nakamura, Ken Takai, and Shigenori Maruyama. Highly alkaline, high-temperature hydrothermal fluids in the early Archean ocean. *Precambrian Res.*, 182:230–238, 2010.

[740] Takazo Shibuya, Miyuki Tahata, Kouki Kitajima, Yuichiro Ueno, Tsuyoshi Komiya, Shinji Yamamoto, Motoko Igisu, Masaru Terabayashi, Yusuke Sawaki, Ken Takai, Naohiro Yoshida, and Shigenori Maruyama. Depth variation of carbon and oxygen isotopes of calcites in Archean altered upper oceanic crust: implications for the $CO_2$ flux from ocean to oceanic crust in the Archean. *Earth Planet. Sci. Lett.*, 321:64–73, 2012.

[741] Takazo Shibuya, Miyuki Tahata, Yuichiro Ueno, Tsuyoshi Komiya, Ken Takai, Naohiro Yoshida, Shigenori Maruyama, and Micheal J. Russell. Decrease of seawater $CO_2$ concentration in the Late Archean: an implication from 2.6 Ga seafloor hydrothermal alteration. *Precambrian Res.*, 236:59–64, 2013.

[742] Takazo Shibuya, Motoko Yoshizaki, Yuka Masaki, Katsuhiko Suzuki, Ken Takai, and Michael J. Russell. Reactions between basalt and $CO_2$-rich seawater at 250 and 350°C, 500 bars: implications for $CO_2$ sequestration into the modern oceanic crust and the composition of hydrothermal vent fluid in the $CO_2$-rich early ocean. *Chem. Geol.*, 359:1–9, 2013.

[743] Graham Shields-Zhou and Lawrence Och. The case for a neoproterozoic oxygenation event: chemical evidence and biological consequences. *GSA Today*, 21:4–11, 2011.

[744] Cristal Shih, Anna Katrine Museth, Malin Abrahamsson, Ana Maria Blanco-Rodriguez, Angel J. Di Bilio, Jawahar Sudhamsu, Brian R. Crane, Kate L. Ronayne, Mike Towrie, Antonn Vlček Jr., John H. Richards, Jay R. Winkler, and Harry B. Gray. Tryptophan-accelerated electron flow through proteins. *Science*, 320:1760–1762, 2008.

[745] Everett L. Shock and Harold C. Helgeson. Calculation of the thermodynamic and transport properties of aqueous species at high pressures and temperatures: standard partial molal properties of organic species. *Geochim. Cosmochim. Acta*, 54:915–945, 1990.

[746] Everett L. Shock and Mitchell D. Schulte. Organic synthesis during fluid mixing in hydrothermal systems. *J. Geophys. Res.*, 103:28513–28527, 1998.

[747] Everett L. Shock, Harold C. Helgeson, and Dimitry A. Sverjensky. Calculation of the thermodynamic and transport properties of aqueous species at high pressures and temperatures: Standard partial molal properties of inorganic neutral species. *Geochim. Cosmochim. Acta*, 53:2157–2183, 1989.

[748] Everett L. Shock, Melanie Holland, D Arcy Meyer-Dombard, Jan P. Amend, G. R. Osburn, and Tobias P. Fischer. Quantifying inorganic sources of geochemical energy in hydrothermal ecosystems, Yellowstone National Park, USA. *Geochim. Cosmochim. Acta*, 74:4005–4043, 2010.

[749] Everett L. Shock, Thomas McCollom, and Mitchell D. Schulte. The emergence of metabolism from within hydrothermal systems. In Juergen Wiegel and Michael

W. W. Adams, editors, *Thermophiles: The Keys to Molecular Evolution and the Origin of Life*, pages 59–76. Taylor and Francis, London, 1998.

[750] J. William Shopf, editor. *Life's Origin: the Beginnings of Biological Evolution*. University of California Press, Berkeley, CA, 2002.

[751] Brian J. Shuter, J. E. Thomas, William D. Taylor, and A. M. Zimmerman. Phenotypic correlates of genomic DNA content in unicellular eukaryotes. *Am. Nat.*, 122:26–44, 1983.

[752] V. Shuvalov. Atmospheric erosion induced by oblique impacts. *Meteoritics Planet. Sci.*, 44:1095–1105, 2009.

[753] Bettina Siebers, Henner Brinkmann, Christine Dörr, Britta Tjaden, Hauke Lilie, John van der Oost, and Corné H. Verhees. Archaeal fructose-1,6-bisphosphate aldolases constitute a new family of archaeal type class I aldolase. *J. Biol. Chem.*, 276:28710–28718, 2001.

[754] Herbert A. Simon. The architecture of complexity. *Proc. Am. Philos. Soc.*, 106:467–482, 1962.

[755] Herbert A. Simon. The organization of complex systems. In Howard H. Pattee, editor, *Hierarchy Theory: The Challenge of Complex Systems*, pages 3–27. George Braziller, New York, 1973.

[756] Herbert A. Simon. *The Sciences of the Artificial*. MIT Press, Cambridge, MA, third edition, 1996.

[757] Kai Simons and Julio L. Sampaio. Membrane organization and lipid rafts. *Cold Spring Harb. Perspect. Biol.*, 3:a004697, 2011.

[758] Y. Sinai. On the notion of entropy of a dynamical system. *Dokl. Akad. Nauk SSSR*, 124:768–771, 1959.

[759] N. A. Sinitsyn and Ilya Nemenman. Universal geometric theory of mesoscopic stochastic pumps and reversible ratchets. *Phys. Rev. Lett.*, 99:220408, 2007.

[760] N. H. Sleep, A. Meibom, Th. Fridriksson, R. G. Coleman, and D. K. Bird. $H_2$-rich fluids from serpentinization: geochemical and biotic implications. *Proc. Natl. Acad. Sci. USA*, 101:12818–12823, 2004.

[761] Alexander Smirnov, Douglas Hausner, Richard Laffers, Daniel R. Strongin, and Martin A. A. Schoonen. Abiotic ammonium formation in the presence of Ni-Fe metals and alloys and its implications for the Hadean nitrogen cycle. *Geochem. Trans.*, 9:5, 2008.

[762] Eric Smith. Self-organization from structural refrigeration. *Phys. Rev. E*, 68:046114, 2003.

[763] Eric Smith. Thermodynamic dual structure of linearly dissipative driven systems. *Phys. Rev. E*, 72:36130, 2005.

[764] Eric Smith. Thermodynamics of natural selection I: energy and entropy flows through non-equilibrium ensembles. *J. Theor. Biol.*, 252:185–197, 2008.

[765] Eric Smith. Thermodynamics of natural selection II: chemical Carnot cycles. *J. Theor. Biol.*, 252:198–212, 2008.

[766] Eric Smith. Thermodynamics of natural selection III: Landauer's principle in chemistry and computation. *J. Theor. Biol.*, 252:213–220, 2008.

[767] Eric Smith. Large-deviation principles, stochastic effective actions, path entropies, and the structure and meaning of thermodynamic descriptions. *Rep. Prog. Phys.*, 74:046601, 2011.

[768] Eric Smith and Supriya Krishnamurthy. *Symmetry and Collective Fluctuations in Evolutionary Games*. IOP Press, Bristol, 2015.

[769]  Eric Smith and Harold J. Morowitz. Universality in intermediary metabolism. *Proc. Natl. Acad. Sci. USA*, 101:13168–13173, 2004.

[770]  Eric Smith, Supriya Krishnamurthy, Walter Fontana, and David C. Krakauer. Non-equilibrium phase transitions in biomolecular signal transduction. *Phys. Rev. E*, 84:051917, 2011.

[771]  Lee Smolin. *The Life of the Cosmos*. Phoenix, London, 1997.

[772]  Theodore P. Snow and Benjamin J. McCall. Diffuse atomic and molecular clouds. *Annu. Rev. Astron. Astrophys.*, 44:367–414, 2006.

[773]  Filipa L. Sousa and William F. Martin. Biochemical fossils of the ancient transition from geoenergetics to bioenergetics in prokaryotic one carbon compound metabolism. *Biochem. Biophys. Acta*, 1837:964–981, 2014.

[774]  Roger W. Sperry. Mind, brain and humanist values. In John R. Platt, editor, *New Views on the Nature of Man*, pages 71–92. University of Chicago Press, Chicago, IL, 1965.

[775]  Alexander S. Spirin. Energetics and dynamics of the protein synthesizing machinery. In Horst Kleinkauf, Hans von Dören, and Lothar Jaenicke, editors, *The Roots of Modern Biochemistry: Fritz Lippmann's Squiggle and its Consequences*, pages 511–533. Walter de Gruyter, Berlin, 1988.

[776]  Vijayasarathy Srinivasan and Harold J. Morowitz. Analysis of the intermediary metabolism of a reductive chemoautotroph. *Biol. Bull.*, 217:222–232, 2009.

[777]  Vijayasarathy Srinivasan and Harold J. Morowitz. The canonical network of autotrophic intermediary metabolism: minimal metabolome of a reductive chemoautotroph. *Biol. Bull.*, 216:126–130, 2009.

[778]  P. F. Stadler, S. J. Prohaska, C. V. Forst, and D. C. Krakauer. Defining genes: a computational framework. *Theory Biosci.*, 128:165–170, 2009.

[779]  Robert W. Sterner and James J. Elser. *Ecological Stoichiometry: The Biology of Elements from Molecules to the Biosphere*. Princeton University Press, Princeton, NJ, 2002.

[780]  Patrick Stover and Verne Schirch. The metabolic role of leucovorin. *Trends Biochem. Sci.*, 18(3):102–106, 1993.

[781]  William E. Strawderman. Sufficient statistics: theoretical background. *Wiley StatsRef*, 2014.

[782]  Bernard L. Strehler and Albert S. Mildvan. General theory of mortality and aging. *Science*, 132:14–21, 1960.

[783]  Lubert Stryer. *Biochemistry*. Freeman, San Francisco, CA, second edition, 1981.

[784]  David P. Summers and Sherwood Chang. Prebiotic ammonia from reduction of nitrite by iron(II) on the early Earth. *Nature*, 365:630–633, 1993.

[785]  Melanie Summit and John A. Baross. Thermophilic subseafloor microorganisms from the 1996 North Gorda Ridge eruption. *Deep-Sea Res. II*, 45:2751–2766, 1998.

[786]  Melanie Summit and John A. Baross. A novel microbial habitat in the mid-ocean ridge subsurface. *Proc. Natl. Acad. Sci. USA*, 98:2158–1263, 2001.

[787]  Roger E. Summons, Linda L. Jahnke, Janet M. Hope, and Graham A. Logan. 2-methylhopanoids as biomarkers for cyanobacterial oxygenic photosynthesis. *Nature*, 400:554–557, 1999.

[788]  Shunryu Suzuki. *Zen Mind, Beginner's Mind*. Weatherhill, New York, 1973.

[789]  Vitali Svetlitchnyi, Holger Dobbek, Wolfram Meyer-Klaucke, Thomas Meins, Bärbel Thiele, Piero Römer, Robert Huber, and Ortwin Meyer. A functional

Ni-Ni-[4Fe-4S] cluster in the monomeric acetyl-CoA synthase from *Carboxydothermus hydrogenoformans*. *Proc. Natl. Acad. Sci. USA*, 101(2):446–451, 2004.

[790]  Eörs Szathmáry and John Maynard Smith. *The Major Transitions in Evolution*. Oxford University Press, London, 1995.

[791]  Jack W. Szostak. The eightfold path to non-enzymatic RNA replication. *J. Syst. Chem.*, 3:2, 2012.

[792]  F. A. Tabita. Research on carbon dioxide fixation in photosynthetic microorganisms (1971–present). *Photosynth. Res.*, 80:315–332, 2004.

[793]  F. A. Tabita, T. E. Hanson, H. Li, S. Satagopan, J. Singh, and S. Chan. Function, structure, and evolution of the RubisCO-like proteins and their RubisCO homologs. *Microbiol. Mol. Biol. Rev.*, 71:576–599, 2007.

[794]  Ken Takai, Tetsushi Komatsu, Fumio Inagaki, and Koki Horikoshi. Distribution of archaea in a black smoker chimney structure. *Appl. Env. Microbiol.*, 67:3618, 2001.

[795]  Ken Takai, Duane P. Moser, Tullis C. Onstott, Nico Spoelstra, Susan M. Pfiffner, Alice Dohnalkova, and Jim K. Fredrickson. *Alkaliphilus transvaalensis* gen. nov., sp. nov., an extremely alkaliphilic bacterium isolated from a deep South African gold mine. *Int. J. Syst. Evol. Microbiol.*, 51:1245–1256, 2001.

[796]  Dan S. Tawfik. Messy biology and the origins of evolutionary innovation. *Nature Chem. Biol.*, 6:692–696, 2010.

[797]  F. J. R. Taylor and D. Coates. The code within the codons. *Biosystems*, 22:177–187, 1989.

[798]  Maureen E. Taylor and Kurt Drickamer. *Introduction to Glycobiology*. Oxford University Press, London, third edition, 2011.

[799]  Robin Teufel, Johannes W. Jung, Daniel Kockelkorn, Birgit E. Alber, and Georg Fuchs. 3-Hydroxypropionyl-coenzyme A dehydratase and acroloyl-coenzyme A reductase, enzymes of the autotrophic 3-hydroxypropionate/4-hydroxybutyrate cycle in the *Sulfolobales*. *J. Bacteriol.*, 191:4572–4581, 2009.

[800]  Rudolf K. Thauer, Kurt Jungermann, and Karl Decker. Energy conservation in chemotrophic anaerobic bacteria. *Bacteriol. Rev.*, 41:100–180, 1977.

[801]  Rudolf K. Thauer, Anne-Kristin Kaster, Henning Seedorf, Wolfgang Buckel, and Reiner Hedderich. Methanogenic archaea: ecologically relevant differences in energy conservation. *Nature Rev. Microbiol.*, 6:579–591, 2008.

[802]  D'Arcy Wentworth Thompson. *On Growth and Form*. Dover, New York, complete revised edition, 1992.

[803]  G. Thompson, M. K. Tivey, and S. E. Humphris. Deducing patterns of fluid flow and mixing within the TAG active hydrothermal mound using mineralogical and geochemical data. *J. Geophys. Res.*, 100:12527–12555, 1995.

[804]  Feng Tian, Owen B. Toon, Alexander A. Pavlov, and H. De Sterck. Transonic hydrodynamic escape of hydrogen from extrasolar planetary atmospheres. *Astrophys. J.*, 621:1049–1060, 2005.

[805]  Jing Tian, Ruslana Bryk, Manabu Itoh, Makoto Suematsu, and Carl Nathan. Variant tricarboxylic acid cycle in *Mycobacterium tuberculosis*: identification of $\alpha$-ketoglutarate decarboxylase. *Proc. Natl. Acad. Sci. USA*, 102:10670–10675, 2005.

[806]  Michael Tinkham. *Introduction to Superconductivity*. Dover, New York, second edition, 2004.

[807]  Margaret Tivey. How to build a black smoker chimney. *Oceanus*, 41, 1998.

[808]  Margaret Kingston Tivey. Generation of seafloor hydrothermal vent fluid and associated mineral deposits. *Oceanography*, 20:50–65, 2007.

[809] Margaret Kingston Tivey, Debra S. Stakes, Terri L. Cook, Mark D. Hannington, and Sven Petersen. A model for growth of steep-sided vent structures on the Endeavour Segment of the Juan de Fuca Ridge: results of a petrologic and geochemical study. *J. Geophys. Res.*, 104:22859–22883, 1999.

[810] Count Lyof N. Tolstoy. *Anna Karénina*. Thomas Y. Crowell, New York, 1914. Translated by Nathan Haskell Dole.

[811] Hugo Touchette. The large deviation approach to statistical mechanics. *Phys. Rep.*, 478:1–69, 2009.

[812] Edward N. Trifonov. The triplet code from first principles. *J. Biomol. Struct. Dyn.*, 22:1–11, 2004.

[813] Rebecca M. Turk, Nataliya V. Chumachenko, and Michael Yarus. Multiple traslational products from a five-nucleotide ribozyme. *Proc. Natl. Acad. Sci. USA*, 107:4585–4589, 2010.

[814] Yuichiro Ueno, Matthew S. Johnson, Sebastian O. Danielache, Carsten Eskebjerg, Antra Pandey, and Naohiro Yoshida. Geological sulfur isotopes indicate elevated OCS in the Archean atmosphere, solving faint young sun paradox. *Proc. Natl. Acad. Sci. USA*, 106:14784–14789, 2009.

[815] Yuichiro Ueno, Shuhei Ono, Douglas Rumble, and Shigenori Maruyama. Quadruple sulfur isotope analysis of ca. 3.5 Ga Dresser Formation: new evidence for microbial sulfate reduction in the early Archean. *Geochim. Cosmochim. Acta*, 72:5675–5691, 2008.

[816] Jos Uffink. Can the maximum entropy principle be explained as a consistency requirement? *Studies Hist. Philos. Mod. Phys.*, 26B:223–261, 1995.

[817] Jos Uffink. The constraint rule of the maximum entropy principle. *Studies Hist. Philos. Mod. Phys.*, 27:47–79, 1996.

[818] H. Edwin Umbarger and Barbara Brown. Threonine deamination in *Escherichia coli* II: evidence for two L-threonine deaminases. *J. Bacteriol.*, 73(1):105–112, 1957.

[819] The UniProt Consortium. Ongoing and future developments at the universal protein resource. *Nucleic Acids Res.*, 39(suppl 1):D214–D219, 2011.

[820] D. A. Usher and A. H. McHale. Hydrolytic stability of helical RNA: a selective advantage for the natural $3',5'$- bond. *Proc. Natl. Acad. Sci. USA*, 73:1149–1153, 1976.

[821] M. F. Utter and H. G. Wood. Mechanisms of fixation of carbon dioxide by heterotrophs and autotrophs. *Adv. Enzymol. Relat. Areas Mol. Biol.*, 12:41–151, 1951.

[822] Kaimars Vafiya. *Duality, Bosonic Particle Systems and Some Exactly Solvable Models of Non-Equilibrium*. PhD Thesis, Universiteit Leiden, 2011.

[823] Nilesh Vaidya, Michael L. Manapar, Irene A. Chen, Ramon Xulvi-Brunet, Eric J. Hayden, and Niles Lehman. Spontaneous network formation among coperative RNA replicators. *Nature*, 491:72–77, 2012.

[824] Patrick van Beelen, Joannes F. A. Labro, Jan T. Keltjens, Wim J. Geertz, Godfried D. Vogels, Wim H. Laarhoven, Wim Guijt, and A. G. Haasnoot. Derivatives of methanopterin, a coenzyme involved in methanogenesis. *Eur. J. Biochem.*, 139:359–365, 1984.

[825] Cindy Lee Van Dover and Brian Fry. Microorganisms as food resources at deep-sea hydrothermal vents. *Limnol. Oceanogr.*, 39:51–57, 1994.

[826] David A. Vanko and Debra S. Stakes. Fluids in oceanic layer 3: evidence from veined rocks, hole 735B, Southwest Indian Ridge. In Richard P. Von Herzen, Jeff Fox, Amanda Palmer-Julson, and Paul T. Robinson, editors, *Proceedings of the Oceanic Drilling Program, Scientific Results*, Vol. 118, pages 181–215. Texas A & M University, Houston, TX, 1991.

[827] Gabriele Varani and William H. McClain. The G-U wobble base pair: a fundamental building block of RNA structure crucial to RNA function in diverse biological systems. *EMBO Rep.*, 1:18–23, 2000.

[828] Vera Vasas, Eörs Szathmary, and Mauros Santos. Lack of evolvability in self-sustaining autocatalytic networks: a constraint on metabolism-first path to the origin of life. *Proc. Natl. Acad. Sci. USA*, 107:1470–1475, 2010.

[829] Krassimir Vassilev, Marina Dimitrova, and Sevdalina Turmanova. Catalytic activity of histidine-metal complexes in oxidation reactions. *Synth. React. Inorg. Met.-org. Nano-met. Chem.*, 43:243–249, 2013.

[830] J. Craig Venter, Karin Remington, John F. Heidelberg, Aaron L. Halpern, Doug Rusch, Jonathan A. Eisen, Dongying Wu, Ian Paulsen, Karen E. Nelson, William Nelson, Derrick E. Fouts, Samuel Levy, Anthony H. Knap, Michael W. Lomas, Ken Nealson, Owen White, Jeremy Peterson, Jeff Hoffman, Rachel Parsons, Holly Baden-Tillson, Cynthia Pfannkoch, Yu-Hui Rogers, and Hamilton O. Smith. Environmental genome shotgun sequencing of the Sargasso Sea. *Science*, 304:66–74, 2004.

[831] Sergio Verdú. Fifty years of Shannon theory. *IEEE Trans. Inf. Theory*, 44:2057–2078, 1998.

[832] Vladimir I. Vernadsky. *Geochemistry and the Biosphere*. Synergetic Press, Santa Fe, NM, 2007.

[833] Kalin Vetsigian, Carl Woese, and Nigel Goldenfeld. Collective evolution and the genetic code. *Proc. Natl. Acad. Sci. USA*, 103:10696–10701, 2006.

[834] Ann M. Vickery and H. Jay Melosh. Atmospheric erosion and impactor retention in large impacts with application to mass extinctions. In V. L. Sharpton and P. D. Ward, editors, *Global Catastrophes in Earth History*, pages 289–300. Geological Society of America, Boulder, CO, 1990.

[835] Alexander V. Vlassov, Sergei A. Kazakov, Brian H. Johnston, and Laura F. Landweber. The RNA world on ice: a new scenario for the emergence of RNA information. *J. Mol. Evol.*, 61:264–273, 2005.

[836] Alexey N. Volkov, Robert E. Johnson, Orenthal J. Tucker, and Justin T. Erwin. Thermally-driven atmospheric escape: transition from hydrodynamic to Jeans escape. *Astrophys. J. Lett.*, 729:L24, 2011.

[837] K. L. von Damm. Seafloor hydrothermal activity: black smoker chemistry and chimneys. *Annu. Rev. Earth Planet. Sci.*, 18:173–204, 1990.

[838] J. von Neumann. Probabilistic logics and the synthesis of reliable organisms from unreliable components. *Automata Stud.*, 34:43–98, 1956.

[839] Diter von Wettstein, Simon Gough, and C. Gamini Kannangara. Chlorophyll biosynthesis. *Plant Cell Online*, 7(7):1039–1057, 1995.

[840] Julia A. Vorholt, Ludmila Chistoserdova, Sergei M. Stolyar, Rudolf K. Thauer, and Mary E. Lidstrom. Distribution of tetrahydromethanopterin-dependent enzymes in methylotrophic bacteria and phylogeny of methenyl tetrahydromethanopterin cyclohydrolases. *J. Bacteriol.*, 181:5750–5757, 1999.

[841] Julia M. Vraspir and Alison Butler. Chemistry of marine ligands and siderophores. *Annu. Rev. Marine Sci.*, 1:43–63, 2009.

[842] L. J. Waber and H. G. Wood. Mechanism of acetate synthesis from $CO_2$ by *Clostridium acidi-urici. J. Bacteriol.*, 140(2):468–478, 1979.

[843] Günter Wächtershäuser. Before enzymes and templates: a theory of surface metabolism. *Microbiol. Rev.*, 52:452–484, 1988.

[844] Günter Wächtershäuser. Pyrite formation, the first energy source for life: a hypothesis. *Syst. Appl. Microbiol.*, 10:207–210, 1988.

[845] Günter Wächtershäuser. Evolution of the first metabolic cycles. *Proc. Natl. Acad. Sci. USA*, 87:200–204, 1990.

[846] C. H. Waddington. Canalization of development and the inheritance of acquired characters. *Nature*, 150:563–565, 1942.

[847] H. Wakamatsu, Y. Yamada, T. Saito, I. Kumashiro, and T. Takenishi. Synthesis of adenine by oligomerization of hydrogen cyanide. *J. Org. Chem.*, 31:2035–2036, 1966.

[848] George Wald. Life in the second and third periods: or why phosphorus and sulfur for high-energy bonds. In M. Kasha and B. Pullman, editors, *Horizons in Biochemistry*, pages 127–142. Academic Press, New York, 1962.

[849] P. Walde, R. Wick, M. Fresta, A. Mangone, and P. L. Luisi. Autopoietic self-reproduction of fatty acid vesicles. *J. Am. Chem. Soc.*, 116:11649–11654, 1994.

[850] Sara Imari Walker, Martha A. Grover, and Nicholas V. Hud. Universal sequence replication, reversible polymerization and early functional biopolymers: a model for the initiation of prebiotic sequence evolution. *PLoS ONE*, 7:e34166, 2012.

[851] Shuning Wang, Haiyan Huang, Johanna Moll, and Rudolf K. Thauer. NADP+ reduction with reduced ferredoxin and NADP+ reduction with NADH are coupled via an electron-bifurcating enzyme complex in *Clostridium kluyveri*. *J. Bacteriol.*, 192:5115–5123, 2010.

[852] Shinya Watanabe, Michael Zimmermann, Michael B. Goodwin, Uwe Sauer, Clifton E. Barry 3rd, and Helena I. Boshoff. Fumarate reductase activity maintains an energized membrane in anaerobic *Mycobacterium tuberculosis*. *PLoS Pathogens*, 7:e1002287, 2011.

[853] J. D. Watson and F. H. C. Crick. A structure for deoxyribose nucleic acid. *Nature*, 171:737–738, 1953.

[854] James D. Watson, Tania A. Baker, Stephen P. Bell, Alexander Gann, Michael Levine, and Richard Losick. *Molecular Biology of the Gene*. Pearson, New York, seventh edition, 2014.

[855] Duncan J. Watts. *Everything is Obvious: How Common Sense Fails Us*. Crown Business, New York, 2012.

[856] Arthur L. Weber. Energy from redox disproportionation of sugar carbon drives biotic and abiotic synthesis. *J. Mol. Evol.*, 44:354–360, 1997.

[857] Arthur L. Weber. Sugars as the optimal biosynthetic carbon substrate of aqueous life throughout the universe. *Orig. Life Evol. Biosphere*, 30:33–43, 2000.

[858] Arthur L. Weber. Chemical constraints governing the origin of metabolism: the thermodynamic landscape of carbon group transformations under mild aqueous conditions. *Orig. Life Evol. Biosphere*, 32:333–357, 2002.

[859] Arthur L. Weber. Kinetics of organic transformations under mild aqueous conditions: implications for the origin of life and its metabolism. *Orig. Life Evol. Biosphere*, 34:473–495, 2004.

[860] Arthur L. Weber and Sandra Pizzarello. The peptide-catalyzed stereospecific synthesis of tetroses: a possible model for prebiotic molecular evolution. *Proc. Natl. Acad. Sci. USA*, 103:12713–12717, 2006.

[861] Steven Weinberg. Phenomenological Lagrangians. *Physica A*, 96:327–340, 1979.

[862] Steven Weinberg. *The First Three Minutes*. Basic Books, New York, second edition, 1993.

[863] Steven Weinberg. *The Quantum Theory of Fields, Vol. I: Foundations*. Cambridge University Press, Cambridge, 1995.

[864] P. P. Weiner, editor. *Leibniz Selections*. Charles Scribner, New York, 1951.

[865] Joshua S. Weinger, K. Mark Parnell, Silke Dorner, Rachel Green, and Scott A. Strobel. Substrate-assisted catalysis of peptide bond formation by the ribosome. *Nature Struct. Mol. Biol.*, 11:1101–1106, 2004.

[866] Pierre Weiss. L'hypothèse du champ moléculaire et la propriété ferromagnétique. *J. Phys. Theor. Appl.*, 6:661–690, 1907.

[867] Geoffrey B. West and James H. Brown. The origin of allometric scaling laws in biology from genomes to ecosystems: towards a quantitative unifying theory of biological structure and organization. *J. Exp. Biol.*, 208:1575–1592, 2005.

[868] Geoffrey B. West, James H. Brown, and Brian J. Enquist. A general model for ontogenetic growth. *Nature*, 413:628–631, 2001.

[869] Geoffrey B. West, William H. Woodruff, and James H. Brown. Allometric scaling of metabolic rate from molecules and mitochondria to cells and mammals. *Proc. Natl. Acad. Sci. USA*, 99:2473–2478, 2002.

[870] Frank H. Westheimer. Why nature chose phosphates. *Science*, 235:1173–1178, 1987.

[871] Laura Reiser Wetzel and Everett L. Shock. Distinguishing ultramafic- from basalt-hosted submarine hydrothermal systems by comparing vent fluid compositions. *J. Geophys. Res.*, 105:8319–8340, 2000.

[872] Harold B. White. Coenzymes as fossils of an earlier metabolic state. *J. Mol. Evol.*, 7:101–104, 1976.

[873] Robert H. White. Structures of the modified folates in the extremely thermophilic archaebacterium *Thermococcus litoralis*. *J. Bacteriol.*, 175:3661–3663, 1993.

[874] Robert H. White. Structures of the modified folates in the thermophilic archaebacteria *Pyrococcus furiosus*. *Biochemistry*, 32:745–753, 1993.

[875] William White. *Geochemistry*. Wiley, New York, 2013.

[876] Andrew R. Whitehill, Changjian Xie, Xixi Hu, Daiqian Xie, Hua Guo, and Shuhei Ono. Vibronic origin of sulfur mass-independent isotope effect in photoexcitation of $SO_2$ and the implications to the early Earth's atmosphere. *Proc. Natl. Acad. Sci. USA*, 110:17697–17702, 2013.

[877] Herbert S. Wilf. *Generating Functionology*. A. K. Peters, Wellesley, MA, third edition, 2006.

[878] David A. Williams. Gas and dust in the interstellar medium. *J. Phys., Conf. Ser.*, 6:1–17, 2005.

[879] George C. Williams. Pleiotropy, natural selection, and the evolution of senescence. *Evolution*, 11:398–411, 1957.

[880] George C. Williams. *Adaptation and Natural Selection*. Princeton University Press, Princeton, NJ, 1966.

[881] R. J. P. Williams. The fundamental nature of life as a chemical system: the part played by inorganic elements. *J. Inorg. Biochem.*, 88:241–250, 2002.

[882] R. J. P. Williams and J. J. R. Fraústo Da Silva. Evolution was chemically constrained. *J. Theor. Biol.*, 220:323–343, 2003.

[883] Edmund Beecher Wilson. *The Cell in Development and Inheritance*. Macmillan, New York, third edition, 1925.

[884] J. Tuzo Wilson. A possible origin of the Hawaiian islands. *J. Phys.*, 41:863–868, 1963.

[885] K. G. Wilson and J. Kogut. The renormalization group and the $\varepsilon$ expansion. *Phys. Rep., Phys. Lett.*, 12C:75–200, 1974.

[886] Ludwig Wittgenstein. *Tractatus Logico-Philosophicus*. Routledge & Kegan Paul, London, 1922. Translated by C. K. Ogden.

[887] Carl R. Woese. *The Genetic Code: The Molecular Basis for Genetic Expression*. Harper and Row, New York, 1967.

[888] Carl R. Woese. There must be a prokaryote somewhere: microbiology's search for itself. *Microbiol. Rev.*, 58:1–9, 1994.

[889] Carl R. Woese. Default taxonomy: Ernst Mayr's view of the microbial world. *Proc. Natl. Acad. Sci. USA*, 95:11043–11046, 1998.

[890] Carl R. Woese. The universal ancestor. *Proc. Natl. Acad. Sci. USA*, 95:6854–6859, 1998.

[891] C. R. Woese. Interpreting the universal phylogenetic tree. *Proc. Natl. Acad. Sci. USA*, 97:8392–8396, 2000.

[892] C. R. Woese. On the evolution of cells. *Proc. Natl. Acad. Sci. USA*, 99:8742–8747, 2002.

[893] Carl R. Woese and George E. Fox. The concept of cellular evolution. *J. Mol. Evol.*, 10:1–6, 1977.

[894] Carl R. Woese and George E. Fox. Phylogenetic structure of the prokaryotic domain: the three primary kingdoms. *Proc. Natl. Acad. Sci. USA*, 74:5088–5090, 1977.

[895] C. R. Woese, D. H. Dugre, W. C. Saxinger, and S. A. Dugre. The molecular basis for the genetic code. *Proc. Natl. Acad. Sci. USA*, 55:966–974, 1966.

[896] Carl R. Woese, Otto Kandler, and Mark L. Wheelis. Towards a natural system of organisms: proposal for the domains Archaea, Bacteria, and Eucarya. *Proc. Natl. Acad. Sci. USA*, 87:4576–4579, 1990.

[897] Carl R. Woese, Otto Kandler, and Mark L. Wheelis. A natural classification. *Nature*, 351:528–529, 1991.

[898] Carl R. Woese, Gary J. Olsen, Michael Ibba, and Dieter Söll. Aminoacyl-tRNA synthetases, the genetic code, and the evolutionary process. *Microbiol. Mol. Biol. Rev.*, 64:202–236, 2000.

[899] Friedrich Wöhler. Über künstkliche Bildung des Harnstoffs. *Ann. Phys. Chem.*, 88:253–256, 1828.

[900] Yuri I. Wolf, L. Aravind, Nick V. Grishin, and Eugene V. Koonin. Evolution of aminoacyl-tRNA synthetases – analysis of unique domain architectures and phylogenetic trees reveals a complex history of horizontal gene transfer events. *Genome Res.*, 9:689–710, 1999.

[901] J. Tze-Fei Wong. A co-evolution theory of the genetic code. *Proc. Natl. Acad. Sci. USA*, 72:1909–1912, 1975.

[902] David Wu, Kingshuk Ghosh, Mandar Inamdar, Heun Jin Lee, Scott Fraser, Ken Dill, and Rob Phillips. Trajectory approach to two-state kinetics of single particles on sculpted energy landscapes. *Phys. Rev. Lett.*, 103:050603, 2009.

[903] Hai Xu, Yuzhen Zhang, Xiaokui Guo, Shuangxi Ren, Andreas A. Staempfli, Juishen Chiao, Weihong Jiang, and Guoping Zhao. Isoleucine biosynthesis in *Leptospira interrogans* serotype lai strain 56601 proceeds via a threonine-independent pathway. *J. Bacteriol.*, 186(16):5400–5409, 2004.

[904] Michael Yarus. Getting past the RNA world: the initial Darwinian ancestor. In John Atkins, Ray Gesteland, and Tom Cech, editors, *Cold Spring Harbor Perspect. Biol.*, pages 1–8. Cold Spring Harbor Laboratory Press, Cold Spring Harbor, NY, 2011.

[905] Yuk Yung, M. Allen, and J. P. Pinto. Photochemistry of the atmosphere of Titan: comparison between model and observations. *Astrophys. J. Suppl. Ser.*, 55:465–506, 1984.

[906] Kevin J. Zahnle. Photochemistry of methane and the formation of hydrocyanic acid (HCN) in the Earth's early atmosphere. *J. Geophys. Res.*, 91:2819–2834, 1986.

[907] Kevin Zahnle and James F. Kasting. Mass fractionation during transonic escape and implications for loss of water from Mars and Venus. *Icarus*, 68:462–480, 1986.

[908] Jan Zarzycki, Volker Brecht, Michael Müller, and Georg Fuchs. Identifying the missing steps of the autotrophic 3-hydroxypropionate $CO_2$ fixation cycle in *Chloroflexus aurantiacus*. *Proc. Natl. Acad. Sci. USA*, 106:21317–21322, 2009.

[909] Jing Zhao, Lin Tao, Hong Yu, JianHua Luo, ZhiWei Cao, and YiXue Li. Bow-tie topological features of metabolic networks and the functional significance. *Chin. Sci. Bull.*, 52:1036–1045, 2007.

[910] D. Zhou and Robert H. White. 5-(p-aminophenyl)-1,2,3,4-tetrahydroxypentane, a structural componentof the modified folate in *Sulfolobus solfataricus*. *J. Bacteriol.*, 174:4576–4582, 1992.

[911] Weibiao Zou, Ismail Ibrahim, Pawel Dziedzic, Hendrik Sundén, and Armando Córdova. Small peptides as modular catalysts for the direct asymmetric aldol reaction: ancient peptides with aldolase enzyme activity. *Chem. Commun.*, 2005:4946–4948, 2005.

[912] G. Zubai and T. Mui. Prebiotic synthesis of nucleotides. *Orig. Life Evol. Biosphere*, 31:87–102, 2001.

# Index